FOUNDATIONS OF BEHAVIOR GENETICS

FOUNDATIONS OF
Behavior genetics

JOHN L. FULLER, Ph.D.

Professor of Psychology,
State University of New York at Binghamton,
Binghamton, New York

WILLIAM ROBERT THOMPSON, Ph.D.

Professor of Psychology,
Queens University, Kingston,
Ontario, Canada

with 172 *illustrations*

THE C. V. MOSBY COMPANY

Saint Louis 1978

Printed in the United States of America

The C. V. Mosby Company
11830 Westline Industrial Drive, St. Louis, Missouri 63141

Library of Congress Cataloging in Publication Data

Fuller, John L
    Foundations of behavior genetics.

    Includes bibliographical references and index.
    1. Behavior genetics. I. Thompson, William
Robert, 1924- joint author. II. Title.
[DNLM: 1. Genetics, Behavioral. QH457 F965f]
QH457.F84   591.5   78-4199
ISBN 0-8016-1712-X

GW/CB/B 9 8 7 6 5 4 3 2 1

Preface

Foundations of Behavior Genetics follows our first book, *Behavior Genetics*, by nearly twenty years. In 1960 when the original volume appeared, we hoped it might have a catalytic effect on an area of investigation that had aroused considerable interest but which lacked focus. Since that time, a Behavior Genetics Association has been organized, which, together with the journal, *Behavior Genetics*, founded in 1970, provides an opportunity for the presentation of research and for discussion among individuals from many disciplines. Many colleges and universities now offer courses in the genetics of behavior. The literature has burgeoned to the point where complete coverage in a single volume is impossible.

The result of our effort to survey the field in the late 1970s is a book larger than the first but at the same time more selective. In our view the major goal of behavior genetics is to increase our understanding of the etiology of individual and group differences. We have devoted more space to the behavioral effects of normal genetic variation than to the consequences of inherited neurological defects and chromosomal aberrations. Similarly we have written more about insects and vertebrates with their complex behavior than about such invertebrates as *Paramecium* and *Caenorhabitis*, whose behavioral repertoires are rather simple. In the insects and vertebrates we have concentrated more on be-

havioral variation of evolutionary significance than on the exotic neuromotor mutants of fruit flies and mice. Work in these areas is important, but it is of greater interest to the neurobiologist and pathologist than to the ethologist, evolutionary biologist, and psychologist for whom we are chiefly writing.

This book has four sections. Part I, Selected Genetic Principles, is an introduction to genetic topics most relevant to the analysis of behavior. The core of the book, a description of and comment on research findings, is divided into Part II, Experimental Behavior Genetics, and Part III, Human Behavior Genetics. The methods used by workers in these fields are based on common principles, but their details differ enough to require separate chapters on techniques. Furthermore, the behavioral phenotypes that are studied in animals and in humans are seldom directly comparable. There are exceptions; genes affecting taste aversion for phenylthiourea are found in mice and people. We do not, however, foresee family and twin studies of verbal fluency and schizophrenia in any species other than our own. We do not hold that there are no unifying principles that apply to animals and humans alike or that the comparative method is valueless. Thus Parts II and III are largely arranged in parallel so that an instructor using this book as a text can, for example, take up Chapter 9 on animal learning and

v

Chapter 14 on human intelligence in sequence. The arrangement also facilitates the use of this text in courses that stress either the experimental or the human studies. In Part IV, Psychology, Biology, and Behavior Genetics, we provide an overview and attempt to project future trends. In writing this chapter we found that much of the newer research has strengthened the evidence for concepts that were expressed by such pioneers as Dobzhansky in the 1940s and 1950s and were summarized by us in 1960.

Though a specialty, behavior genetics is also an integrative science. Its domain potentially encompasses all aspects of animal and human behavior relevant to our knowledge of (1) differences among individuals and groups and (2) the DNA-guided processes that convert a zygote to a mature adult with species-specific behavior patterns. Research in both of these areas identifies coactions and interactions between genes and environments. The reader will find in this book almost as much discussion of environmental as of genetic influences on behavioral individuality. This is consistent with our belief that behavior genetics and developmental psychology are simply two ways of looking at the same phenomena. Workers in either field must be aware of the concepts and findings of the other.

We are grateful to the authors and publishers who granted permission for the reproduction of copyrighted material. All such sources are cited in the text or in the Bibliography. W. R. T. was aided in the preparation of this book by a Senior Fellowship from the Canada Council for 1972 to 1973 while on sabbatical leave from Queens University. J. L. F. wrote two chapters while on sabbatical leave from the State University of New York at Binghamton in 1976.

Joseph M. Horn and James R. Wilson critically reviewed our original draft, and many of their recommendations have been adopted. Helpful suggestions were received also from other individuals, particularly Bruce C. Dudek and Peter J. Donovick. However, we are fully responsible for all conclusions and any errors that may remain.

We are grateful to Vicky Malcolm and Patricia Doloway, who typed the major portion of the manuscript. Elizabeth Bouchard assisted in compiling references and Sarah Bottger in the development of Chapter 9. Mary Thompson provided moral support and assisted in proofreading and checking references. Ruth Fuller contributed editorial advice and aided in bibliographical research, proofreading, and indexing.

John L. Fuller
William Robert Thompson

Contents

IV Summary and conclusions

I
Selected genetic principles

1

Scope of behavior genetics

Behavior genetics is a science that has aroused more than its share of controversy. Some disputation is inevitable in a developing field of knowledge because scholars in all honesty interpret the same facts in different ways, but in the field of behavior genetics a more important source of conflict arises from the social and political aspects of the subject. Pastore (1949) has persuasively argued that the attitudes of scientists on the issues are affected by their liberal or conservative social views. It is not a coincidence that genetics has been the biological science most prostituted in both Fascist and Communist states. Men are different, but man has not always been open-minded in seeking for the source of these differences.

In our first book, *Behavior Genetics* (1960), we wrote:

The vigor of the nature-nurture controversy has declined in America since the 1920's: thus fewer scientists can be classified as "hereditarians" or "environmentalists."

We further stated that the extreme views of that period were now of historical significance only. We were mistaken in our view. In particular the issue of possible racial differences in intelligence is argued with heavy political overtones. To deal adequately with this topic would extend this book beyond reasonable limits. Some of the flavor of the debate can be found in Cancro (1971), Ehrman, Omenn, and Caspari (1972), and Jensen (1972). Further comment will be deferred to Chapters 14 and 18.

NATURE-NURTURE PROBLEM

Perhaps at this time there is no one who classifies behavior into two categories, innate and learned. The dichotomy, carried to its logical conclusion, would define innate behavior as that which appears in the absence of environment, and learned behavior as that which requires no organism. Verplanck (1955) has exposed the absurdities of slightly less extreme positions. The dichotomy is not in the kind of behavior studied (the dependent variable), but in the independent variables that are manipulated or observed. Here a clear distinction can be made between genetic factors that are transmitted from parents to offspring in the gametes and nongenetic factors that are not. This distinction, of course, limits the genetic contribution to extremely small packets of molecules in the nuclei of sperm and ovum.

Three kinds of questions may be raised with respect to the nature-nurture relationship (Anastasi, 1958a). What are the effects of heredity on behavior? How large are these effects? What mechanisms are involved? Answers to all these questions span the fields of genetics, physiology, and psychology. The purpose of this book is to present current thought on all of them.

The development of any trait always involves genetic and environmental determinants, but the variation between individuals is sometimes almost entirely due to one or the other type of factors. In common speech and in many genetic investigations a threefold classification of the characteristics of an organism has been used (Dahlberg, 1953).

1. A trait is called hereditary if most of the variation within a population is associated with differences in genetic endowment. As an example, the agglutinogens of red blood cells are directly controlled by genes with which they have a one-to-one relationship. Even here, cattle twins have the same blood type more frequently than predicted from

3

genetic theory. The proffered explanation is the transfer of blood-forming elements through a common circulation in the placenta (Owen, 1945).

2. A nonhereditary or acquired trait has little or no genetically determined variance. Customs and language are conventional examples.

3. Variation in a third group of traits is significantly affected by both genetic and environmental factors. Skin color, body size, and most characteristics that vary quantitatively over a wide range belong in this coaction category.

The use of the convenient terms *hereditary traits* and *acquired traits* should not lead to the erroneous conclusion that they are fundamentally different from *coaction traits*. An organism develops from a deceptively simple-appearing cell containing complex molecules in specific patterns. To produce a blood cell antigen, a gene requires a supply of nutrients, oxygen, and other essentials. The point is that genetic determination of antigens is bound up so intimately with development that it cannot be modified without abolishing the organism. Thus the observable variation of cellular antigens is almost wholly genetic. On the other hand, if one were to compare the acquisition of language in a population of normal children and children with trisomy 21 (Down syndrome), one would find most of the variance attributable to a genetic defect. Although exposed to usually adequate stimulation, the Down child fails to acquire the behavior that fits him for ordinary social living. The greater the demands of his culture for communication and abstraction, the greater the extent of his incapacity.

It is important to note that allocating variation to heredity and environment is not the same as evaluating the relative role of the two kinds of factors in a particular individual. What the geneticist calls the heritability of a trait refers to a specific population in a particular situation. It is not a fixed value for a trait and can fluctuate widely. This leads to paradoxes such as these: a trait of low heritability can be drastically affected by the mutation of a single gene, and changes in environment can greatly alter a trait of high heritability.

GENES, SOMATOPHENES, AND PSYCHOPHENES

The visible and physical traits of an organism comprise its phenotype. Classically the phenotypes chosen by geneticists for study have been characteristics such as pigmentation, body size, or shape and more recently molecules of protein or other organic compounds. We shall refer to these as *somatophenes* (from *soma*, body, and *phene*, appearance). Somatophenes may be divided further into *chemophenes* (e.g., type of hemoglobin) and *morphenes* (e.g., number of ridges in a fingerprint). Since genes themselves are chemical structures, it has proved easier to relate gene action to chemophenes than to morphenes, even when the latter are inherited in simple Mendelian fashion. We shall often refer to behavioral phenotypes as *psychophenes*. There is no implication in such a separation of phenotypes that psychophenes are somehow independent of a physical substrate. The term merely signifies that psychophenes are measured directly or indirectly from behavior. Again we recognize two classes: *ostensible* psychophenes are based on the occurrence, frequency, and intensity of an objectively defined bit of behavior. An example is the number of fecal boli deposited in an open field during 5 minutes by a rat. An investigator may choose instead to measure an *inferred* psychophene, rat emotionality, which is measured by a defecation score either singly or in combination with other measures. The distinction is an important one in the study of behavior generally.

The genetic portion of behavior genetics is wholly conventional. Mendel's laws of segregation and independent assortment, chromosomal mechanics, DNA and RNA chemistry, and population and quantitative genetics are the basic supports for this specialty, as they are for other branches of genetics. But behavior is not a coat color, hemoglobin molecule, or length of a bone. Once an organism matures, these durable somatophenes remain relatively constant over long periods. They can often be determined as precisely in a dead as in a live animal. Many are ascertainable from a sample of tissue or body fluid.

Psychophenes are very different. Behavior is a sequence of events, not a physical structure. Behavior, of course, is organized, but explaining it in terms of psychological structures is a recourse to analogy that is not particularly helpful. An organism's structure places limitations on the kind of events in which it can participate, but it does not completely determine them. Behavioral events are joint functions of an organism and its surroundings at a point in time. One cannot measure the social dominance of a solitary male mouse or estimate the IQ of a sleeping or drunken man.

Given the ephemeral nature of psychophenes, one may ask if a behavior genetics really exists. Obviously we believe that it does, or this book would not have been written, but we cannot minimize the problems of using events rather than structures as phenotypes. Even a complex somatophene can be considered as an orderly aggregation of molecules, which in turn are made up of atoms and ultimately of subatomic particles. But one cannot decompose a response to an item in an IQ test or a bout of fighting between two male mice into molecules.

How can we deal with this matter? One way is to take the outcome of a behavioral test as an ostensible psychophene. Thus we can carry out a genetic analysis of the number of fights in paired encounters or the number of correct items on an IQ test. Frequently, of course, such measures of performance are regarded as representative of an inferred trait. The IQ score measures intelligence; success in fighting denotes aggressiveness. Though convenient, this approach has obvious weaknesses. Transported to a radically new environment, the "intelligent" person is confused; the "aggressive" mouse runs from a cat. But if its limitations are kept in mind, the trait concept facilitates communication, particularly when validation of the trait in a variety of situations is successful. It makes good sense to study behavior that will be predic-

tive outside the limited area of the testing room, and, if we can find genetic effects on traits of wide generality, the results are likely to be important.

An obvious pitfall in the trait approach is to use it in an explanatory capacity. A mouse does not attack because it is aggressive; it is aggressive because it attacks. If now we can find some somatophene such as body weight or amount of testosterone in the blood that is highly correlated with fighting success, we may undertake a study of the genetics of this correlated character. In doing so we will have deserted behavior genetics proper, but the results may be relevant to our understanding of the behavior. The obverse of this approach is the study of the behavioral correlates of a well-defined heritable variation. There are many studies of the behavior of phenylketonuric children (an inherited metabolic disorder), albino mice, and yellow-bodied fruit flies. The behavioral dimension usually adds little to the knowledge of the transmission and biochemistry of these conditions. Nevertheless, understanding the physiological and biochemical substrate of such deviant individuals may lead to correction of behavioral problems. Behavior-genetic analysis (Hirsch, 1967) can be a powerful tool in psychophysiological research, supplementing such techniques as brain lesions, drug administration, and electrical stimulation.

Behavior genetics seems to offer more to psychology than to genetics, since psychophenes are inferior to somatophenes as markers for genes. The neglect of genetic factors by psychologists can have serious consequences. At the simplest level the genetic specification of experimental animals can eliminate sources of variability that may cause discrepancies between experimenters. Genetic variants of psychological interest may not be duplicable by any other means. Perhaps most important, acquaintance with genetic diversity of man and other species is essential for an understanding of individual differences. Individuality is not "error"; neither is it entirely a matter of differential reinforcement. We are a long way from a complete formulation of the laws of psychological development that will explain it, but certainly behavior genetics will play an important role in the achievement of this objective.

SOME CRITICISMS OF BEHAVIOR GENETICS

Psychologists sometimes object to such phrases as "the inheritance of aggressiveness" or "the genetics of intelligence." It is not a trait that is inherited, say the critics, but some structure which in turn affects behavior through transactions with the environment (Kuo, 1929; Anastasi and Foley, 1948). The point is well made, but there is an inconsistency in the critics' reference to the inheritance of body size or organization of the brain. These characteristics are not transmitted in the genes but arise from gene-environment transactions. They are epigenetic. "Inheritance of intelligence" implies a little more than "inheritance of body size," but not much more. The really sticky problem is still the boundary between physical structures and psychic events, and bringing genetics into the picture does not complicate this matter.

The criticism is sometimes made that the heritability of a behavioral variant cannot be considered as proved until a gene has been located and a specific physiological mechanism discovered. Such demonstrations are certainly desirable, but it is doubtful that so complete an explanation has been attained for any complex morphological trait. The triumphs of developmental genetics are yet to come; molecules are easier to work with. Progress in this direction poses another set of questions at a more sophisticated level. The delay in discovering a chemical basis for gene action did not prevent genetics from moving forward on the basis of statistical rather than mechanistic associations between genes and traits. In a parallel fashion psychology has made great progress in relating behavior to previous experience without success in explaining learning in physiological terms (Lashley, 1950). Agranoff (1972), for example, states that the detailed mechanisms of behavioral plasticity are "the most obscure frontiers and, accordingly, hypotheses run wild." Behavioral techniques may well prove to be the most sensitive (perhaps the only) method for detecting certain genetic differences, just as they are now the only way of determining whether a rat has learned a maze.

It is an interesting speculation that gene action and learning are fundamentally similar. The gene-controlled pattern of body form tends to remain constant throughout life in spite of the rapid overturn of the constituent atoms (Schoenheimer, 1942). In this constant resynthesis of protoplasm, the modifications that have been impressed by learning are retained along with those determined by genes. We remember our childhood with molecules that were not in our bodies when we experienced it. Learning may be something like mutation (Davis, 1954), and it can be viewed as the process of completing the differentiation of the nervous system in greater detail and more adaptively than can be accomplished through gene encoding alone (Katz and Halstead, 1950; Hydén, 1970).

BEHAVIOR GENETICS AND THE DOCTRINE OF INSTINCTIVE BEHAVIOR

Behavior that seems to appear in relatively perfected form without practice is popularly called instinctive. Psychologists tend to avoid the word "instinct," partly because it has been misused and partly because the kinds of behavior that most psychologists study are greatly influenced by learning. Among insects and the lower vertebrates, however, many complex patterns of behavior are executed without much evidence for learning, and even human babies are born with a repertoire of coordinative patterns. The neural programs for these behaviors must be dependent on a genetically encoded program. One might expect that the ways in which the genetic coding becomes manifested as a behavioral pattern and the dependence of the process on environmental stimulation would be a major enterprise of behavior geneticists. Actually, it has not been so, but a start is being made. Studies such as Marler's (1970) on the development of song in the white-crowned sparrow (Chapter 11) are not genetic in a narrow sense, but they do illustrate that heredity may place major restrictions on the capacity of an animal to acquire a specific form of behavior.

PROBLEMS IN THE CHOICE OF BEHAVIOR TRAITS TO STUDY

An infinite number of measurements may be made on the body or the behavior of an organism. In a sense each of them may be considered as a character whose inheritance may be studied. In practice the geneticist selects characters that are convenient and will provide maximum information concerning other characters. Such a correlated set of characters defines a trait. The choice is often simpler among physical characters than it is among behavioral characters. No theoretical issues are raised when one studies the inheritance of body length. The dimensions of temperament and personality, however, have not been standardized. Many psychologists have dealt with this question from a variety of viewpoints, and a book larger than this would be required to deal adequately with the subject. (For sample discussions see Murphy, 1947; Thurstone, 1947; Anastasi, 1948, 1958a; Cattell, 1955; and Allport, 1966.) Traits that have been used in behavior genetics range from specific motor components of fish courtship to susceptibility to perceptual illusions and scores on Stanford-Binet tests. Surprisingly, genetic effects have been shown at both extremes of complexity. We shall return to this subject in Chapter 18, where we consider whether genetics can assist in defining behavioral traits that correspond to biological units.

SUBJECTS FOR BEHAVIOR GENETICS

Success in biological research often depends on proper selection of material. Genetic studies require a variable species, one which is prolific, easily maintained, and with a small number of large-sized chromosomes so that hereditary factors can be manipulated and directly observed. The fruit flies, drosophilae, fit these specifications. Man fails on all counts except variability. Yet because of the particular interest in the study of man, human subjects have been much more commonly used in behavior genetics than drosophilae.

An advantage of *Drosophila*, other insects, fish, and, to a lesser degree, birds is that the gene-behavior-trait relationship is more precise than is typically true of mammals. The behavioral characters selected for study in the lower phyla and orders are usually specific movements in response to specific stimuli. The advantage from the biological side is countered by the difficulty in generalizing to the kinds of individual differences that are characteristic of man. Man is a mammal, and there are many parallels between the development of behavior in subhuman mammals and in human infants before the beginning of speech. This probably explains the predilection of psychologists for mammals as subjects for behavior genetics. Biologists, less concerned with generalization to man and more interested in the evolutionary aspects of behavior, have done most of the experiments with the lower species.

Among nonhuman mammals, the house mouse, *Mus musculus*, is now the favorite subject for genetics. Over 200 named mutations are known, many distinctive inbred strains are available, and the chromosomes are individually identifiable. The behavior of mice has been fairly well studied, though not nearly as thoroughly as that of rats. Compared with their larger cousins, mice are less convenient for some psychological and physiological procedures. The formal genetics of rats is less well known than that of mice, but it is questionable whether knowledge of the mode of inheritance of coat color and developmental anomalies is as valuable for behavior genetics as information on the physiological correlates of behavior. On the whole, rats and mice have equal advantages from a scientific point of view. For experiments in which either species would be satisfactory, mice may be favored for economic reasons.

Cats and dogs are the oldest domestic animals. The worldwide distribution of these species and the existence of many specialized breeds provide a ready-made source of material for behavior genetics. These carnivores give an impression of greater individuality than rodents, but this impression may reflect our greater intimacy with them. Dogs have considerable use (Scott and Fuller, 1965) because of their highly developed social behavior. Cats have not been used in behavior genetics to our knowledge, although the extensive knowledge of feline neurophysiology should make them useful for certain problems. Scattered references will be found to behavior genetics research on other species of mammals, but these reports are incidental to other studies. Subhuman primates would seem to have advantages for research on the inheritance of intelligence, but the difficulties of laboratory rearing and the relatively low fecundity have discouraged attempts in this direction.

To what extent is it possible to formulate general principles from genetic experiments performed on diverse species? The problem is similar to that faced by Tolman (1932) in writing *Purposive Behavior in Men and Animals* and Beach (1947) in his cross-species survey of sexual behavior. When a sufficiently large spectrum of species is observed, principles emerge that would not be evident in more limited studies. Fortunately the mechanisms of gene transmission are practically identical in all the organisms we shall consider. The primary physiological action of genes is also believed to be broadly similar in all species, though more complicated structures involve more steps between primary gene action and the completed character. There is no reason to expect that the specific gene systems controlling courtship behavior will be the same in rats and dogs, but the manner in which control is exerted is probably as similar as the effects of hormones on the behavior in the two species. It is safe to conclude that the problems of generalizing from comparative genetic studies are of the same order as those encountered in synthesizing results of experiments in different species on the effects of early experience or brain lesions.

SOME METHODOLOGICAL PROBLEMS

In this section we shall be concerned with some of the broader methodological problems of behavior genetics. In general, heredity as an independent variable can be incorporated into the design of a psychological experiment just as one introduces physiological or experimental factors. The dependent variable can be any form of behavior that interests the investigator. The simplest experiment is to take two groups of different heredity, treat them alike in all other respects, and administer a behavior test. The results are compared against the prediction from the null hypothesis that the groups differ no more than two independent samples drawn from the same population. If the null hypothesis is not supported, evidence for heritability of the behavior variation has been obtained.

But although logically identical with other experimental procedures in psychology, behavior genetics has certain peculiarities. The differential treatments (distribution of genes) precede the existence of the subjects of the experiment. In fact, genetic control may extend back many generations before the birth of the actual subjects of an experiment. The need for long periods of treatment (selective breeding and/or inbreeding) is inherent in this area. Another feature is the impossibility of manipulating genes directly. The distribution of genes to subjects is essentially random and is controlled by the experimenter only in a statistical sense. Since genes are not observed directly, their presence is deduced from their effects. At first thought, the argument for their existence may seem circular. Traits are ascribed to genes whose presence is proved by the existence of the trait. Fortunately, the gene theory rests on a more ample foundation, which is described briefly in Chapter 2. The worker in behavior genetics must understand chromosomal behavior as well as organismic behavior in order to design his experiments.

Two major strategies have been employed in behavior genetics (Scott and Fuller, 1963). The genotypic approach starts with a known difference in heredity and evaluates its influence on behavior. A gene substitution or a chromosomal variant is analogous to other kinds of treatment used by experimental psychologists. In the phenotypic approach an attempt is made to discover the genetic factors (if any) responsible for observed variation in behavior. The two strategies tend to involve different methodologies: the genotypic approach leads to developmental and physiological investigations of the pathways between genes and behavior. The phenotypic approach tends to emphasize the quantitative interactions of genetic and environmental influences in populations. Claims have been made that one or the other strategy is the more productive (Wilcock, 1969, 1971; Thiessen, 1971). Actually both strategies have limitations, and neither alone will solve all the problems in this area.

The heredity of an organism is fixed at the moment of fertilization. This imposes a limitation on experimental design. One can present stimulus A before or after stimulus B and can train subjects before or after a cortical ablation, but genes cannot be changed in the middle of the life span. Thus there is no way of teasing apart the effects produced by the genic control of contemporary metabolism and the effects due to genic determination of growth and differentiation. The latter effects are inevitably confounded with conditions during development.

A special concern of behavior genetics is the avoidance of nonrandom association between environmental and hereditary factors. The fact that human families share experiences as well as genes clearly leads to difficulties of data interpretation. The problem also occurs in experimental behavior genetics, at least in birds and mammals that give parental care. Uteri, compared with the external world, may provide protection against many stimuli, although recent experiments (Thompson, 1957a; Joffe, 1969) have reopened the question of effects of prenatal experience on later behavior. Differences in postnatal family environment are of even greater potential significance. Cross-fostering of the young of one strain to the dam of

another permits evaluation of effects of different nutrition and type of maternal care. Experimental regulation of litter size can be used to control the nature of early experience, degree of competition, and the like. Statistical corrections can be applied to allow for the fact that members of a litter share experiences unique to that litter. These techniques have their parallels in human studies, but they cannot be applied with the same rigor.

Concern for the environment in behavior genetics research goes beyond avoidance of heredity-environment correlations. Simply providing the same defined, controlled environment for each genetic group is not enough for good design. Conditions must not only be uniform for all groups, but also favorable to the development of the behavior of interest. An unsuitable rearing system may modify or even completely suppress the manifestations of a genetic difference (Howells, 1946; Freedman, 1958; Henderson, 1970; Fuller and Herman, 1974).

Perhaps the most important conclusion to be drawn is that research in behavior genetics cannot be isolated from research in the development of behavior, the area traditionally known as genetic psychology. Hereditary-environment interactions are more than a statistical abstraction. They can be observed and analyzed in experiments in which genetic and experiential factors are varied simultaneously in controlled fashion.

APPLICATIONS OF BEHAVIOR GENETICS

Behavior genetics has relationships with both parent sciences. Behavioral characters, because they are so environment sensitive, are unsuitable for most research of interest to formal genetics. Nevertheless, some application has been made of behavioral tests to the detection of genetic differences not discernible from morphology (Reed, Williams, and Chadwick, 1942; Kaplan and Trout, 1969). Considerable effort has been expended on mating behavior, particularly in *Drosophila*, because of the importance of sexual selection in evolutionary theory. Human geneticists have been concerned with the inheritance of mental deficiency and psychiatric disorders in order to provide genetic counseling. To a minor degree behavior genetics has found applications in applied animal breeding.

Undoubtedly, more research in this area has been motivated by interest in behavior than by interest in genetics. Even investigators who consider individual differences a nuisance use littermate controls, co-twin controls, and purebred stocks to reduce genetic sources of variability in their material. More significant for our purposes are attempts to employ genetics as a research tool (Scott, 1949). Such uses go beyond the demonstration of heritability of a particular kind of behavior.

The repetition of a procedure with different strains is a means of extending or limiting generalizations based on a single type of experimental animal. Comparisons between domesticated and wild rats, for example, have demonstrated important psychophysiological differences within the same species (Richter, 1952, 1954).

The argument has been made that emphasis on the genetic variability within organisms (both interspecific and intraspecific) leads to a "fatal overparticularization" (Thiessen and Rodgers, 1967). Yet it can be asserted that only generalizations robust enough to be manifested in spite of individual variation deserve to be called general laws (Vale and Vale, 1969).

The use of mutant stocks or of strains selected for special behavioral characteristics provides material for physiological psychology that cannot be duplicated by surgery, electrical stimulation, drugs, or other techniques. Inherited factors are perhaps more likely to contribute to our understanding of individual differences in intact organisms.

Finally, behavior genetics has a potential contribution to education, psychiatry, clinical psychology, and other professions that deal firsthand with a variety of human problems. Heritability of a deleterious deviation does not mean that it cannot be ameliorated. If heredity does play a role, recognition of the fact and understanding of the intermediate physiological mechanisms may be the most direct way to a satisfactory treatment. A behavioral disorder associated with a correctable metabolic defect would call for a rational rather than a symptomatic therapy.

2

Transmission of genetic information

Although the application of hereditary principles to the development of various varieties of domesticated plants and animals goes back into prehistory, a comprehensive theory of inheritance is the product of the twentieth century (Dunn, 1951). Present-day genetics represents the fusion of two lines of investigation, one dealing with the processes of cell division and fertilization, the other concerned with crossing variant types and analyzing the characteristics of the offspring by statistical methods.

Chapters 2 through 5 contain an elementary account of selected topics in genetics that are particularly important in behavioral studies. Among these topics, some areas such as selection theory and the genetics of quantitative characters are underrepresented in elementary biology courses.

Major areas of genetics that have only a peripheral relationship to behavior have been omitted in this brief summary. Obviously our choice is arbitrary, and those who are stimulated to seek a more complete account are advised to find it in one of several excellent general or specialized textbooks listed in the Bibliography (Falconer, 1960; Watson, 1970; Cavalli-Sforza and Bodmer, 1971; Stern, 1973; Levine, 1978).

ORGANIC PATTERNS AND LIFE CYCLES

The fertilized ova of different mammals appear much alike under the microscope and can be identified only by a trained microscopist. If it were possible to provide adequate nourishment and protection so that development would proceed on the microscope stage, these cells would be observed to divide, to increase in mass until a microscope could no longer be used, and to di-

verge in form until their identities as human, seal, horse, or rat would be apparent. The result of the developmental process identifies the source of each ovum because it is a biological axiom that each species reproduces its own kind.

The regularity of the development tempts the observer to compare it with the unfolding of a predetermined form, as exemplified in the Chinese paper flowers that expand into intricate patterns when placed in water. But the analogy is incorrect. The process is not an unfolding but the carrying out of a series of reactions that are encoded in the genes and perhaps in other cellular elements. The result is a structure that adheres to the characteristic pattern of its species but varies in detail from other members of the species. Both the constancy and the variability of the overall organization have their basis in the functions of genes.

Patterns are observable in living organisms at many levels from size factors expressed over the whole body down to the configuration of protein molecules. It is convenient to begin a consideration of genetics at the intermediate level of the cell. Every higher organism is an aggregate of cells, some of which are highly specialized in structure and function. The central, denser-appearing *nucleus* is more uniform in different tissues than the outer portion, which is known as *cytoplasm* (Fig. 2-1). This cytoplasm may be stretched into a nerve fiber several feet long specialized for conducting electrical pulses or compressed into a cube in the thyroid gland, where it is the site of hormone synthesis. Experiments on separation of nucleus and cytoplasm have shown that the nucleus is essential for the continued existence of the cell as an organized system. Apparently it controls

11

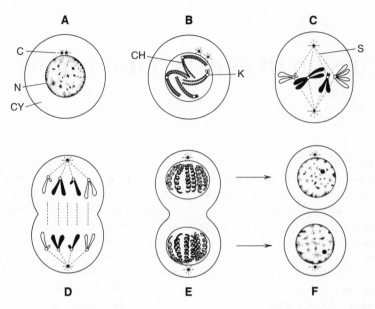

Fig. 2-1. Diagram of cell division and mitosis. *C*, Centriole; *N*, nucleus; *CY*, cytoplasm; *CH*, chromosome; *K*, kinetochore; *S*, spindle. (From *Principles of human genetics*, 3rd ed., by Curt Stern. W. H. Freeman & Co., San Francisco. Copyright © 1973.)

the synthesis of molecules that are necessary for cytoplasm. The nucleus also plays a unique part in the process of cell division known as *mitosis*.

Chromosomes and mitosis

At the time of cell division the nucleus undergoes complete reorganization, and in this period it is possible to observe elongated bodies known as *chromosomes* (colored bodies), which have the capacity to absorb certain dyes. Each species has a characteristic number of chromosomes: the mouse, 40; rat, 42; dog, 78; corn, 20; *Drosophila melanogaster*, 8; potatoes, 48; and man, 46.

In sexually reproducing species the chromosomes occur in pairs, with certain exceptions related to sex determination. For example, in the human female 23 pairs match in size, shape, and banding when examined by fluorescence microscopy. In the human male we find 22 of the same pairs (collectively called *autosomes*) plus 2 unmatched chromosomes, the larger designated X, the smaller, Y. The X chromosome is found to correspond to the unique pair in the female. Fig. 2-2 shows a photograph of the chromosomes

of a male arranged to demonstrate these facts. In humans the complements of each sex may be summarized as:

Female 44A + 2X = 46 total
Male   44A + X + Y = 46 total

In standard nomenclature the number of chromosomes in one complete set is the *haploid* number (23 in humans, 4 in *D. melanogaster*); the total of two sets is the *diploid* number (46 in humans, 8 in *D. melanogaster*). Later we shall see that deviations from normal diploidy often have important behavioral consequences. In the majority of animals males have the XY genotype as in humans, but in birds and butterflies females are found to have the mismatched chromosomes.

In ordinary cell division, as in the growth of an embryo or the replacement of worn-out skin, each chromosome duplicates itself as the cell divides so that each daughter cell comes to possess a complete set of 46 chromosomes. The process known as *mitosis* is illustrated in Fig. 2-1, which shows the successive events of chromosome division and the formation of a new nucleus, starting with

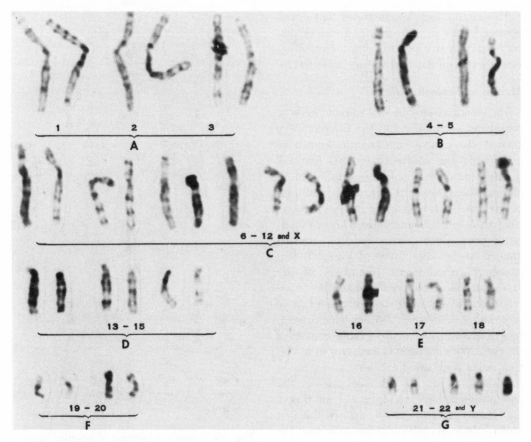

Fig. 2-2. Chromosomes of a human male: separated, photographed, and arranged in pairs as a karyotype. (From McKusick, V. A. 1972. Study guide in human genetics. Prentice-Hall, Inc., Englewood Cliffs, N.J.)

the resting stage (not truly an inactive nucleus, but only one that is not dividing). Not all the features of the diagram are seen in every cell, but the onset of mitosis is usually indicated by division of the *centriole* (Fig. 2-1, *A*). The chromosomes first appear as elongated bodies with a specialized region, the *kinetochore*, which serves as an attachment point for *spindle fibers* (Fig. 2-1, *B*). Later the chromosomes become more condensed, the centrioles move to opposite poles of the cell, and spindle fibers running from centrioles to kinetochores are seen. The nuclear membrane breaks down in this stage, which is called *metaphase* (Fig. 2-1, *C*). The remaining diagrams (*D* to *F*) illustrate the separation of each chromosome from its newly replicated partner, the reestablishment of

the nuclear boundaries, and the eventual separation of the daughter cells.

The significant result of mitosis is the duplication of chromosomes to produce a series of pairs, followed by the separation of each pair. It is probably significant that the nuclear membrane breaks down during mitosis so that cytoplasmic constituents are available for the synthesis of new chromosome material. Also important is the fact that, although each daughter cell receives the same chromosomes* and hence the same genetic factors, the two cells may eventually be markedly different.

*Exceptions to equality of chromosome numbers in somatic cells occur, but their significance is difficult to evaluate. Certainly these differences have not been found to be related to cell function (Srb and Owen, 1953).

This simple fact demonstrates that development is not an unfolding of an inner pattern but an active process of interaction between extracellular and intracellular forces.

Meiosis and crossing-over

The cell divisions of the somatic cells of organisms are mitotic, but the production of germ cells involves a variation known as *meiosis*. In the course of meiosis two cell divisions occur with only one duplication of chromosomes; hence the chromosome number is exactly halved in sperm and ova as compared with somatic cells. The essentials of the process are simple, and they are diagramed in the upper part of Fig. 2-3. *Homologous chromosomes* (members of the same pair, one of maternal and another of paternal origin) approach each other and come to lie side by side (Fig. 2-3, *A*). This pairing, known as *synapsis*, involves close contact of the corresponding parts of each chromosome. Centrioles and spindle fibers are seen in meiosis as in mitosis. Before pairing, the chromosomes have reduplicated so that a four-strand structure is formed. In the first meiotic division each set of four is reduced to a group of two; in the second meiotic division the pairs divide again so that the original four chromosomes are distributed one each to four germ cells.

The most important feature of meiosis is the reshuffling of chromosomes and the consequent appearance of combinations in the offspring which are unlike those in the parents. The redistribution is possible because of different arrangements of the chromosome pairs as they line up in the metaphase stage of the first meiotic division. Consider Fig. 2-3 again, this time comparing the upper and lower portions. In a species with two pairs of chromosomes there are two alternative arrangements, *A* and *A'*. In this figure the chromosomes of maternal origin are shown in outline, those from the father as solid areas. Four types of daughter cells, *B*, *B'*, *C*, and *C'*, are produced in equal numbers, and four corresponding types of gametes are found (*D* and *D'*). It can readily be shown that the number of possible types of gametes is 2^n,

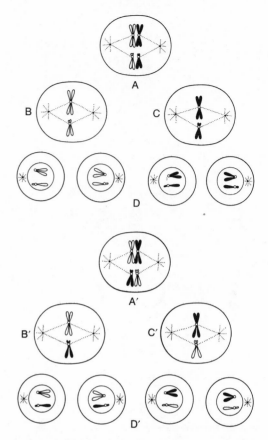

Fig. 2-3. Simplified diagram of meiosis, including the two meiotic divisions. *A* and *A'*, The two alternative arrangements of the chromosome pairs on the first meiotic spindle. *B* to *D* and *B'* to *D'*, Second meiotic divisions and the different types of reduced chromosome constitutions of the gametes. (From Principles of human genetics, 3rd ed., by Curt Stern. W. H. Freeman & Co., San Francisco. Copyright © 1973.)

where *n* is the number of pairs of chromosomes. When *n* is 23, the number of gametic types is 8,388,608.

Expressed in another fashion, each human parent has the potentiality of producing over 8 million distinct types of germ cells. The probability of any particular combination occurring in a mating between two specified individuals is the product of 8,388,608 by itself. It is probable that no human beings except identical twins have ever been genetic duplicates. If complicating effects such as crossing-over are considered, the possibilities of recombination become much greater.

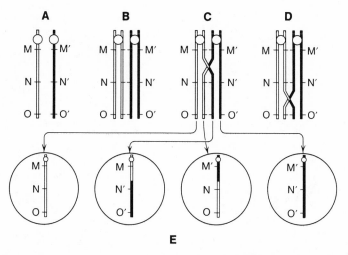

Fig. 2-4. Crossing-over. **A,** A pair of homologous chromosomes heterozygous for three pairs of loci, *M, M'; N, N';* and *O, O'.* **B,** Four-strand stage. **C,** Crossing-over between two of the four strands in the region between *M, M'* and *N, N'.* **D,** Same crossing-over in the region between *N, N'* and *O, O'.* **E,** The four types of reduced chromosome constitutions of the gametes resulting from crossing-over in **C.** (From Principles of human genetics, 3rd ed., by Curt Stern. W. H. Freeman & Co., San Francisco. Copyright © 1973.)

Stern (1973) has estimated that a single human pair have the potentiality of producing 20^{24} different types of children, a number far greater than the total number of human beings who have ever existed. Of course, much of this genetic variability may have little importance for behavior, but on purely logical grounds, uniqueness of heredity is as much a fact as uniqueness of experience.

Homologous chromosomes are not merely similar externally but are comparable part by part. Thus, during the intimate contact of synapsis, each section, or *locus,* within a chromosome of maternal origin is associated with the corresponding locus in a chromosome of paternal origin. In Fig. 2-4 a single pair of chromosomes is depicted, the maternal by outline and the paternal by a solid bar. The chromosomes are different (heterozygous) at three loci: *M, N,* and *O* on one chromosome and *M', N',* and *O'* on the other. Synapsis is shown in *A* and duplication in *B.* Crossing-over between *M* and *N* is diagramed in *C* and between N and O in *D.* When the intertwining is followed by breakage and recombination of parts, the resultant chromosomes are a composite of maternal and paternal contributions (middle chromo-

somes of *E*). This process obviously increases the possibilities for recombinations of hereditary factors in meiosis. Were it not for crossing-over, the chromosome rather than the gene would be the unit of heredity. Crossing-over may be double, triple, or even more complex, and this variation complicates calculations of crossover frequencies.

GAMETOGENESIS

The process of gamete formation, *gametogenesis,* is similar in male and female insofar as the nuclear processes in meiosis are concerned, but the cytoplasmic events are modified in relation to the different functions of sperm and ovum. The two processes are diagramed side by side in Fig. 2-5. In spermatogenesis the germinal cells that line the walls of tubules in the testis are known as *spermatogonia.* The cells in which synapsis occurs (*A* to *E*) are *primary spermatocytes;* the cells containing dyads (*F* and *G*) are *secondary spermatocytes;* and the final products of meiosis are four *spermatids,* which metamorphose into *spermatozoa.* The head of a spermatozoon is composed almost entirely of chromosomes. Thus a male's contribution to the substance of his offspring is compressed

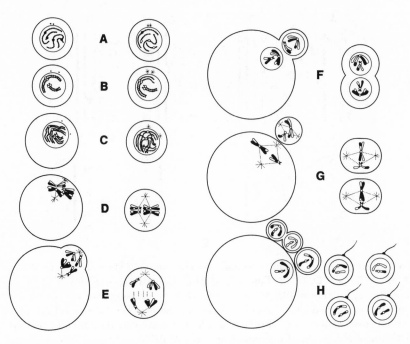

Fig. 2-5. Diagram of meiosis and formation of gametes. *Left*, oogenesis; *right*, spermatogenesis. *A* to *C*, Chromosome pairing and crossing-over. *D* and *E*, First meiotic division. *F*, Products of first division. *G*, Second meiotic division. *H*, Egg with polar bodies *(left)*, sperm cells *(right)*. (From Principles of human genetics, 3rd ed., by Curt Stern. W. H. Freeman & Co., San Francisco. Copyright © 1973.)

into a few cubic micra (1 micron = 1/1000 millimeter).

In *oogenesis* the division of the cytoplasm at the two meiotic divisions is unequal (Fig. 2-5, *E* to *G*). Practically all the cytoplasm is retained within one of the cells, the secondary oocyte. The smaller cell, containing a full set of chromosomes (*F* and *G*) is known as a polar body. It may divide again. A second polar body is produced at the second meiotic division (*H*). Both polar bodies normally degenerate, since their one function is to serve as a repository for excess nuclear material. Thus when the ovum is fertilized by a sperm, the somatic number of chromosomes is reconstituted.

The ovum contains a considerable amount of cytoplasm and sometimes much stored food in addition. Thus the possibility exists that the maternal gamete contributes more to the determination of biological characteristics than does the sperm. Possibly the broad patterns of development are encoded in the cytoplasm, and variations on the main theme in the genes. It is impossible to prove the point one way or the other until cells are synthesized with nuclei of one species and cytoplasm of another. The weight of the evidence is that each parent contributes equally to genetic variation. In mammals maternal influences operate through the health of the mother during pregnancy, the quality of her milk, and the adequacy of her care for her helpless young. These may be important sources of variation, but they are classified as environmental rather than genetic factors.

Sex determination

The previous section on gametogenesis demonstrates that the formulae for the germ cells of most sexually reproducing species may be written as:

> **Ovum** A + X
> **Sperm** A + X or A + Y

A represents the full set of autosomes:

22 pairs for man, 19 pairs for mouse, 3 pairs for *Drosophilia melanogaster*, and so on.

If one assumes that the two kinds of sperm are equally likely to fertilize an ovum, the two possible zygotes would be formed in equal numbers: females, 2A + 2X; males, 2A + X + Y. Approximate equality is attained in general, but significant exceptions do occur. In man, for example, there is a reliable excess of male over female births, although the reasons are unknown (Stern, 1973).

The basic method of achieving sexual differentiation by a dimorphism of one type of germ cell is generally similar throughout the animal kingdom. However, the developmental processes that intervene between chromosomal allocation and the mature male or female are variable. Some of these variations of interest to contemporary behavior genetics are described in Chapter 17.

GENES: UNIT FACTORS IN INHERITANCE

The greatest difference between modern genetic concepts and traditional folk ideas of heredity is that today we conceive of the transmission of individual particles of hereditary material from parents to offspring, whereas older ideas postulated a blending of the hereditary elements of the two sexes. The basis for the modern view derives from experiments on garden peas performed by an Austrian monk, Gregor Mendel (1865). Mendel deduced the segregation of maternal and paternal factors (Mendel's first law) from an analysis of the ratios of different types of offspring from crosses between pure lines differing in unitary physical characteristics. The regularity of these ratios could only be explained by postulating that the probability of a given gamete carrying the factor of maternal origin was one half, and the probability that it carried the paternal factor was the same. The discovery of these lawful relationships was made without knowledge of the chromosomal basis of meiosis. In fact, the relationship between genetic segregation and meiosis was not clearly recognized until the

early twentieth century (Sutton, 1903). Once recognized, it provided a physical base for the phenomena studied by geneticists. We shall now see how this established law can be used to test whether variability in a new characteristic arising in a species is dependent on segregation at a single locus.

Varitint-waddler mouse

This rather quaintly named mouse variety first appeared in a cross between a solid-colored black and a solid-colored brown strain, both of which had bred true for many generations (Cloudman and Bunker, 1945). These more lightly pigmented and spotted mice with locomotor defects produced progeny with the same characteristics. The change, therefore, was a *mutation*, a persistent alteration of one of the hereditary units. Cloudman and Bunker assigned the symbol *Va* to the mutant gene and the symbol *va* to its unmutated normal counterpart. *Va* and *va* are *alleles*, genes that differ in their physiological properties, although they occupy the same locus on a chromosome. Since every mouse has two complete sets of chromosomes, its genetic formula with respect to the *Va* locus can be *Va/Va*, *Va/va*, or *va/va*. Each of these formulae symbolizes a different *genotype*. *Genotype* in its most general sense is the complete assemblage of genes that an individual possesses. Genotypic formulae are never complete, since it is impossible to identify all genes present.

Individuals with identical genes at corresponding loci are *homozygous* at this locus (for example, *va/va* and *Va/Va*). *Heterozygous* individuals have unlike genes at corresponding loci (for example, *Va/va*).

Each genotype in this series is identifiable from the physical characteristics or phenotype of its possessor. A *va/va* mouse may be of many colors and forms, depending on the remainder of its genotype, but it is certain that it will not be a varitint-waddler. The typical waddler *(Va/va)* shows piebald spotting, erratic ducklike locomotion, circling, and head shaking. Mice homozygous for *Va* are practically all white and are so hyperexcitable at 14 days of age that jarring the

---

Prediction model based on single-gene hypothesis

The ratio below is obtained if it is assumed that the varitint-waddler character is produced by a single gene. Cross is between normal and varitint-waddler mice.

Genotype of V-W mouse, $Va/+$, will produce equal numbers of gametes of the following types:

	Va	$+$
Genotype of normal mouse, $+/+$, can produce only one type of gamete: $+$	$Va/+$	$+/+$

Types of zygotes formed

Predicted ratio of offspring (zygotes) is ½ $Va/+$: ½ $+/+$.

---

cages in which they live may induce violent convulsions. Most Va/Va animals are sterile, and the heterozygotes show reduced fertility.

An alternative way of writing genotypic formulae is to use the symbol $+$ for the "wild-type" or "normal" allele present in the general population. Ordinary mice are $+/+$, varitint-waddlers, $Va/+$, and homozygous defectives, Va/Va.

Testing genetic hypotheses

Genetic theory can be used to predict the outcome of breeding heterozygous waddlers to normal mice. If we assume, as is in fact true, that the gene is located on an autosome, it will make no difference whether the waddler parent is male or female. This assumption must be checked by comparing the results of reciprocal crosses in which the waddler gene is introduced through the paternal and maternal sides. The model for prediction is shown above.

If the normal mice resulting from this cross are mated inter se, all the offspring will be normal, even though each has a waddler grandparent. If waddlers are mated with waddlers, the gametes and resultant zygotes will be as shown on the opposite page.

These two procedures, (1) crossing heterozygotes to homozygotes and obtaining a 1:1 ratio in the offspring and (2) crossing heterozygotes with heterozygotes and obtaining a 1:2:1 ratio, are the basic devices of experi-

mental genetics. More complicated designs are extensions of the same principles.

Cloudman and Bunker carried out these procedures with the results shown in Table 2-1.

In two of the crosses agreement with prediction is good. The significance of the difference between the 325 $Va/+$ obtained and the 332.5 predicted can be evaluated by the chi-square test. The formula used in computing chi-square is

$$\chi^2 = \Sigma \frac{(O - P)^2}{P}$$

where

O = observed number in each class of offspring
P = predicted number in each class

For the $Va/+$ by $+/+$ cross, the calculation is

$$\chi^2 = \frac{(325 - 332.5)^2}{332.5} + \frac{(340 - 332.5)^2}{332.5} = 0.338$$

with a p value of more than 0.5. Obviously the data agree satisfactorily with the hypothesis.

The $Va/+ \times Va/+$ cross produces far too few homozygous white mice. Chi-square is calculated as follows:

$$\chi^2 = \frac{(18 - 96)^2}{96} + \frac{(236 - 192)^2}{192} + $$
$$\frac{(132 - 96)^2}{96} = 86.96 \ p < 0.001$$

Prediction model based on single-gene hypothesis

The ratio below is obtained if it is assumed that the varitint-waddler character is produced by a single gene. Cross is between two varitint-waddler mice. Each parent will produce equal numbers of Va and $+$ gametes.

		Gametes of first parent	
		Va	$+$
Gametes of second parent	Va	Va/Va	$Va/+$
	$+$	$+/Va$	$+/+$

Types of zygotes formed

Predicted ratio of offspring (zygotes) is ¼ Va/Va : ½ $Va/+$: ¼ $+/+$.

Table 2-1. Results of breeding experiments on the varitint-waddler mouse*

Type of mating	Litters	Individuals	Phenotypic classes		
			White defective	Varitint waddler	Wild type
Waddler × Wild type	93	665	Pred. 0 Obs. 0	332.5 325	332.5 340
Wild type† × Wild type†	7	52	Pred. 0 Obs. 0	0 0	52 52
Waddler × Waddler	62	386	Pred. 96 Obs. 18	192 236	96 132

*From Cloudman, A. M., and L. E. Bunker, Jr. 1945. The varitint-waddler mouse. J. Hered. **36**:259-263.
†These animals each had a waddler parent.

Since the discrepancy is too large to be attributed to random sampling, an explanation must be given, or the hypothesis of single gene inheritance must be rejected. It is known that the homozygous white mice are biologically inferior. Thus it is reasonable to assume that many Va/Va mice succumb before birth and never enter the statistics. From the number of $Va/+$ and $+/+$ individuals produced, we would expect 123 homozygous whites [$^1/_3 \times (236 + 132)$]. Presumably all but 18 of these failed to attain a stage of development permitting their classification. Numerous examples are known of genes with lethal or sporadically lethal effects that disrupt ratios calculated from simple assumptions. Statistical predictions must take biological realities into account.

The contrast between the particulate or Mendelian theory of heredity and a blending theory is illustrated by the data in the second row of Table 2-1. None of the 52 offspring of normal parents, but with defective grandparents, were themselves defective. Thus the progeny of a wild-type-by-waddler mating are either defective and capable of transmitting the waddler gene, or they are normal with no such gene to pass on. None are intermediate-strength waddlers either in behavior or in genetic potentiality, as would be predicted if genetic determiners were blended. We will see later that the Mendelian mechanism operating on numerous independent genes can result in intermediate individuals, but the principles illustrated by the varitint-waddler trait still hold for the inheritance of each pair of genes.

Dominant and recessive genes

In the varitint-waddler mouse there is a simple relationship between the number of *Va* genes and their phenotypic expression. The disruption of normal function is more extreme in *Va/Va* animals than in *Va/+* mice. This seems very reasonable and represents additive effects of the mutant gene on the phenotype. Such additivity is not found universally. Consider another example from mouse genetics, this time a coat-color character. When mice from a purebreeding black stock are bred to those from a purebreeding brown stock, the offspring are all black. Breeding the first filial generation hybrids (F_1) inter se yields a second filial generation (F_2) comprised of approximately three-fourths black and one-fourth brown animals. In situations of this type, black *(B)* is said to be *dominant* to brown *(b)*, which is called *recessive*. The genetic situation is depicted in Table 2-2.

The 3:1 ratio is merely a variant of the 1:2:1 ratio previously described. It is characteristic of the F_2 generation whenever dominant inheritance is found. The distinction between dominance and nondominance is not always clear-cut or fundamental. It depends in part on the state of knowledge of the effects of the gene, since if the heterozygote can be distinguished phenotypically, dominance is not complete. The degree of dominance is very important wherever there is natural or artificial selection. Suppose that it was desired to eliminate the *b* gene from the F_2 population produced by crossing blacks with browns. Removing the animals with a brown phenotype would eliminate only one half of the brown genes. Those in the heterozygotes would be detected only when brown offspring turned up in the next generation.

In general the zygosity of an animal with a dominant phenotype can be determined only from its ancestry or its progeny. A black mouse coming from a long line of inbred ancestors who have bred true for black is probably homozygous; one whose father or mother was brown is certainly heterozygous. In any other circumstance the diagnosis must be uncertain. The progeny test is most efficiently made by breeding the animal whose genotype is in question to a homozygous recessive *(b/b)*. The prediction for a mating of the type $B/b \times b/b$ is that one half of the offspring will show the dominant character, and the probability that all of a series of *n* progeny will be of this type is $\frac{1}{2}^n$. When $n = 7$,

Table 2-2. Predicted results of crossing pure lines of black and brown mice: black is dominant

Generation	Parent (sex)	Phenotype	Genotype	Gametes
Original pure lines	Male or female	Black	*B/B*	*B*
	Male or female	Brown	*b/b*	*b*
F_1 generation	Male or female	Black	*B/b*	½ *B*, ½ *b*
	Male or female	Black	*B/b*	½ *B*, ½ *b*
F_2 generation		¾ black	¼ *B/B*, ½ *B/b*	
		¼ brown	¼ *b/b*	

$p = 0.0078$, so that a series of 7 offspring of the dominant type is a strong indication that the tested individual is in fact B/B, not B/b.

Because test matings are not possible with human beings, human geneticists are on the lookout for small effects of genes in a heterozygous state. In some cases it is possible to identify individuals carrying "recessive" genes that produce anomalies when homozygous (Neel, 1949; Hsia, 1967) and to provide information to prospective parents who come from families with serious hereditary defects.

Multiple alleles

At many loci there are more than two known alleles, and potentially every locus might be polyallelic. An instructive example is the *agouti (A)* locus in the house mouse (Green, 1966). The alleles at this locus have been detected by their effects on the distribution of pigment in the body hair. Table 2-3 lists the phenotypic effects of some of the alleles most frequently encountered in laboratory strains.

The table indicates that two members of the series have effects on nonpigmentary characteristics. Both $A^y/-$ and $A^{vy}/-$ mice are inactive, hyperphagic, and obese as compared with nonyellow littermates. Such mul-

tiplicity of effects is called *pleiotropism*. Actually the effects of the A locus genes on pigment distribution, though conspicuous and convenient tags for identification, probably give a very incomplete picture of the physiological significance of this locus. There is evidence that it controls properties of cell membranes and thus participates in life processes at a very basic level (Woolf, 1972).

The existence of multiple alleles greatly increases the possibility of genetic variation. With two alleles there are three possible genotypes; with three alleles, six genotypes; and with the six listed alleles of Table 2-3, twenty-one genotypes. The number of phenotypes for the six A alleles (excluding the lethal A^y/A^y) is only six, however, because of the dominance effects characteristic of this locus.

Two or more independent loci

Transmission of genes that lie in separate chromosomes is independently determined during gametogenesis. Again, we owe the discovery of this principle to Mendel (1865), and the rule by which the characters determined by such genes are transmitted is known as Mendel's law of independent assortment. This is well illustrated by a diagram such as Fig. 2-6, the lower portion of

Table 2-3. Phenotypic effects of some alleles* at agouti locus in house mouse

Symbol name		Pigment effects	Other effects
A^y	Yellow	All hair pigment yellow	Homozygotes die prior to implantation; heterozygotes obese, inactive, hyperphagic
A^{vy}	Viable yellow	Varies from all yellow to agouti; usually mottled yellow	Homozygotes viable; obese, inactive, hyperphagic
A^w	White-bellied agouti	Hairs of back are black with subapical yellow band (agouti pattern); underside white	
A	Agouti	Like A^w with darker belly	Usually regarded as "wild type"
a^t	Black and tan	Black back and cream belly	Recessive to A on back, but dominant for belly color
a	Nonagouti	Hairs are unbanded; solid black or brown, depending on other genes present	No well-established behavioral or physiological differences from wild type

*Alleles are listed in order of dominance.

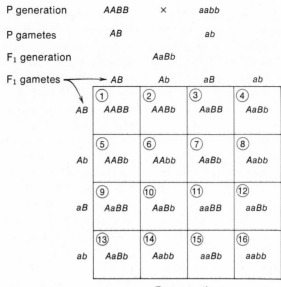

P generation   AABB   ×   aabb

P gametes   AB   ab

F₁ generation   AaBb

F₁ gametes

F₂ generation

Fig. 2-6. Combinations of two independently segregating genes in the F₂ of a hybrid between pure strains. Both male and female in the F₁ can produce four kinds of gametes; hence there are sixteen combinations, although some of these are identical. Actually, there are nine different genotypes in the ratio $1:1:1:1:2:2:2:2:4$. The phenotypic ratios observed in the F₂ of a dihybrid cross vary according to the type of physiological interaction between the genes.

which is known as a *Punnett square*. This example shows a dihybrid (two-locus) system. The diagram represents the F₂ from a cross between two genotypes, *AABB* and *aabb*. Along the edges are the expected proportions of male and female gametes from the F₁. Each cell in the diagram corresponds to an expected one sixteenth of the F₂ population. It is also possible to compute the expected genotypic proportions algebraically, as is generally more convenient when more than two or three independent loci are considered simultaneously.

The phenotypic results of a dihybrid cross depend on physiological interactions between nonallelic genes. The simplest situation is complete independence of physiological effects, yielding the $9:3:3:1$ ratio of phenotypes that led Mendel to his discovery of the law of independent assortment. For example, two of the loci affecting coat pigmentation in mice, *B* (*B*, black; *b*, brown) and *D* (*D*, dense pigmentation; *d*, dilute pigmentation) are on separate chromosomes. Crossing a purebreeding dense black strain (*BBDD*)

by a dilute brown strain (*bbdd*), one obtains an F₁ generation (*BbDd*) that can be intercrossed to obtain an F₂. The predicted proportions of the two phenotypic characters are simply the combined probabilities of the independently determined color and density characteristics. The expected ratios in each case are 0.75 dominant and 0.25 recessive phenotypes. Thus the probability of simultaneous expression of the two dominants (black dense) is $0.75 \times 0.75 = 0.05625$ ($9/16$); of one dominant and one recessive (black dilute or brown dense) is $0.75 \times 0.25 = 0.1875$ ($3/16$); and of a double recessive (dilute brown), $0.25 \times 0.25 = 0.0625$ ($1/16$).

The dihybrid phenotypic ratios may be modified if the genes show complementary action. Consider, for example, a cross between a pigmented black strain of mice (*BBCC*) and an albino strain carrying the brown allele at the *b* locus (*bbcc*). In animals homozygous for the *c* allele, all pigment production is suppressed; only animals with a *C* gene express other pigment genes. Computing expected ratios of phenotypes as before, we obtain $9/16$ black mice, $3/16$ brown mice, and $4/16$ albinos. The phenotype of an albino mouse gives no clue to its status at the b locus. Such interactions between nonallelic genes are known as *epistatic* effects.

LINKAGE

The mechanics of meiotic cell division impose certain limitations on independent assortment of genes. Homologous chromosomes synapse and separate as units, so that all genes of a single chromosome tend to segregate together. This is a tendency rather than an absolute law because of the crossing-over phenomenon described earlier.

Sex-linkage

It is convenient to introduce linkage in general with a discussion of sex-linkage. The transmission of genes included in the X or Y chromosome is inextricably bound up with the sex-determining properties of these chromosomes. The consequences of sex-linkage differ depending on whether the gene in question is on the X or Y chromosome, whether it is dominant or recessive, and whether crossing-over from X to Y is ever possible. Sex-linkage is easy to detect, and examples have been described from many species.

Y-linkage inheritance has been reported for the "porcupine men" who lived in England during the eighteenth and nineteenth centuries. There is, however, some question of the accuracy of the records on which this famous pedigree is based (Stern, 1973). At any rate, Y-linked chromosomal effects on behavior have been reported rarely (Selmanoff, Maxson, and Ginsburg, 1976).

A gene located on the X chromosome of a mammal may be present in a single or double dose in females, but only in a single dose in males. (The reverse is true in butterflies and birds, where the female is heterogametic.) Such a gene may behave as a recessive in females and as a dominant in males, where the normal allele cannot be simultaneously present. One of the most thoroughly studied cases of this type is "red-green" color blind-

ness. Actually this is not a single defect but a composite of several distinct but related anomalies of color vision, whose individual attributes can be neglected in an elementary account.

Color blindness is transmitted in a crisscross fashion from fathers through daughters to grandsons. Nonaffected males never pass the gene on to children of either sex, but nonaffected females who are carriers of the gene show their heterozygosity by bearing on the average 50% of color-blind sons. Likewise, one half of the daughters of such women prove to be carriers like their mothers. Color-blind women have color-blind fathers and often color-blind brothers. If a color-blind woman marries a normal-visioned man, her sons are all color-blind, while her daughters have normal vision like their father. All these facts fall into place if we assume that the condition is brought about through the intermediacy of a recessive gene on the X chromosome. The diagrams of Fig. 2-7 represent the probabilities of different types of offspring from individuals bearing the color-blindness gene.

Sex-linkage is probably not often significantly involved in behavior genetics. Most behavioral traits depend on the interaction of the many genes, which are probably distributed over a number of chromosomes. The effect of genes located on the X chromosome might be difficult to separate from the effects of a much larger number of genes located on autosomes. Nevertheless, appropriate tests for sex-linkage should be introduced into experimental designs whenever feasible.

This is an appropriate place to contrast sex-linked genes with *sex-limited* characters. All that sex-linkage implies is that the genes involved are located on the X or the Y chromosome. Sex-limited characters are restricted in manifestation to only one sex, although genes influencing the character may be carried in both sexes. It is well known that bulls transmit genetic factors that affect the milk production of their daughters, even though the bull has no functional mammary glands. This need not imply that the genes concerned have no function in males. Genes

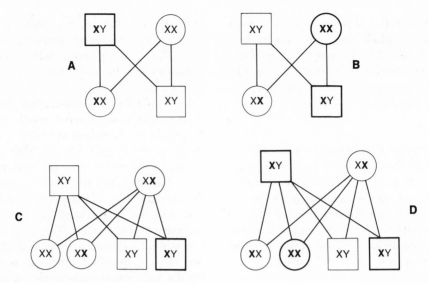

Fig. 2-7. Transmission of red-green color blindness in man. The gene is carried on the X chromosome and is expressed in females only when homozygous. X chromosomes containing the gene are shown in heavy letters; afflicted individuals are shown in heavy outline. **A,** Color-blind man × normal woman. **B,** Normal man × color-blind woman. **C,** Normal man × carrier. **D,** Color-blind man × carrier.

work at the cellular level; their effect on milk production is indirect.

Autosomal linkage

Genes that lie close together on a chromosome will tend to go together in gametogenesis (Fig. 2-3) unless there is a crossover between them (Fig. 2-4). Such associations can be detected by discrepancies between the observed outcome of certain crosses and expectations based on the law of independent assortment. Consider, for example, two dominant genes *A* and *B* that are found together in a strain breeding true for their phenotypic manifestations, (A) and (B). We desire to know whether the two loci are linked or on separate chromosomes.

Our task is simplified if we have available another stock breeding true for the recessive phenotypes (a) and (b). We proceed as follows:

Parental cross

Genotypes	*AABB*	× *aabb*
Phenotypes	(AB)	(ab)
Gametes	*AB*	*ab*

F₁

Genotype	*AaBb*			
Phenotype	(AB)			
Gametes	*AB*	*Ab*	*aB*	*ab*

Potentially the F₁s can produce four types of gametes, each in equal numbers if the loci are unlinked. To discover the actual proportions of the four gametic types, we backcross the F₁ with the double-recessive parent (test cross). Four kinds of zygotes are possible:

Genotypes	*AaBb*	*Aabb*	*aaBb*	*aabb*
Phenotypes	(AB)	(Ab)	(aB)	(ab)

Note that the phenotypes correspond precisely to the gametes produced by the F₁; a count of the backcross offspring gives the information needed to confirm or reject linkage between the loci. Independent assortment will yield equal numbers of each phenotype; linkage will produce a higher proportion of the combinations present in the original lines. The outcome of such an experiment might be this:

Phenotypes	(AB)	(Ab)	(aB)	(ab)
Number	38	9*	11*	42

TOTAL = 100

The individuals in the groups marked with the asterisks are known as *recombinants*. Since 20% of the test cross offspring belong in these classes, the recombination rate is 20%.

In the preceding example, genes *A* and *B* are said to be *coupled*, since they are in the same chromosome. Hence the phenotypic traits associated with each will tend to be correlated positively. A different association between the phenotypes results if *A* and *B* are in *repulsion* (in opposite homologous chromosomes); thus *Ab/Ab* and *aB/aB*. Carrying out the same operations as before, we obtain from a test cross of the F_1 40% *Ab/ab*, 40% *aB/ab*, 10% *AB/ab*, and 10% *ab/ab*. Now there is a negative correlation between the phenotypes associated with *A* and *B*. The association between characters related through a common chromosome is inconsistent and usually close to zero in the population at large; within families the association may be significant.

Chromosome maps

Linkage measurements have contributed greatly to the science of genetics. If it is assumed that close linkage indicates proximity on a chromosome, recombination frequencies can be used to construct chromosome maps. In *Drosophila* it has long been possible to confirm microscopically the validity of maps based on this assumption and thus strengthen faith in results with other species where such visual checking is technically more difficult.

In mammals the correspondence between linkage groups and individual chromosomes as viewed under the microscope has been more difficult to establish, but advances in recognition of individual chromosomes through fluorescent staining methods ensure that such correlations will be established for the species of greatest interest to geneticists.

In the mouse a large number of genes with neurological and metabolic effects that modify behavior have been assigned to linkage groups (Green, 1966). Genes detectable only by specific behavioral tests are also amenable to mapping, although relatively few have been studied. One example is the location of the *asp* locus, *audiogenic seizure prone*, on linkage group VIII (chromosome 4) of the mouse (Collins, 1970a). This locus, which controls initial-trial seizure susceptibility (see p. 102 for further information), is loosely linked with the *b* locus, which determines coat color. A recombination value of 0.40, though significant, is too high to make the visible coat color a useful marker for the presence, or absence, of the cryptic behavioral gene, *asp*. Marker genes closely linked to deleterious genes would be particularly valuable in man for purposes of genetic counseling and guidance. An example of the search for such linkages is a study by Elston, Kringlen, and Namboodiri (1973).

These investigators looked for linkages between a number of blood group phenotypes, whose inheritance is well established, and manifestations of psychosis, schizophrenia in particular. Positive evidence for linkage of psychiatric symptoms with three of the blood group systems was found, but the association was not clear enough to be used clinically. There was no evidence that any blood group alleles were directly associated with psychosis; they seemed only to be possible markers for chromosomes that predisposed their bearers to mental disorders. The classical method of gene mapping by means of computing linkage from crosses is difficult to apply in humans except for those genes located on the X chromosome. New methods based on somatic hybrids between human and mouse cells in tissue culture have enabled geneticists to determine the location of at least one gene on each of the 23 pairs of human autosomes (McKusick and Ruddle, 1977).

3

Physiological and developmental genetics

The mechanisms for the transmission of genetic information from one generation to the next are well understood. The manner in which this information is applied to the specification of a somatophene, particularly in complex multicellular organisms, is still largely unexplained. The problem is particularly acute with purely behavioral phenotypes (psychophenes) whose physical basis is unknown. But the situation is far from hopeless; much of the basic chemistry of gene action, which has been determined largely from experiments with microorganisms, appears applicable, with modifications, to flies, mice, or men. In this chapter we shall sketch the broad outlines of the processes that intervene between genes and phenotypes of varying degrees of complexity. Additional details may be found in textbooks of genetics and in Watson (1970).

CHEMICALS OF HEREDITY

In 1944 Avery, McLeod, and McCarty demonstrated that a nonvirulent strain of pneumococcus bacteria (type R) could be transformed to a virulent strain (type S) by growing it in a medium containing deoxyribonucleic acid (DNA) from the S strain. The transformed bacteria continued to breed true as type S. The experiment demonstrated that the physiological and structural differences between the two strains were dependent on DNA and suggested that this substance had a major role to play in heredity. DNA is a polymer made up of aggregates of similar subunits as a chain is made up of links. A small number of different subunits can be combined in an infinite number of ways, providing the diversity needed to explain biological variability. Supporting evidence for its genetic role gained over the next ten years included the following: (1) most DNA is found in the chromosomes; (2) the amount of DNA per diploid set of chromosomes is constant for a given species; (3) the amount of DNA per haploid set of chromosomes in the gametes is half the diploid amount; and (4) DNA is metabolically stable as required for accurate transmission of information, and its synthesis is inhibited by x rays and other mutagenic agents.

In 1953 Watson and Crick demonstrated by physical methods that the DNA molecule was a double helix whose properties were perfectly suited to the biological functions required for the genetic material. A schematic representation of the Watson-Crick model is shown in Fig. 3-1. Each strand in the helix consists of a linear sequence of *nucleotides*, each of which in turn contains a nitrogenous base, a five-carbon sugar (deoxyribose), and phosphoric acid. In DNA there are four different nucleotides corresponding to two pyrimidine bases, cytosine (*C*) or thymine (*T*), and two purine bases, guanine (*G*) and adenine (*A*). The structural characteristics of these bases are such that the complementary strands are held together by hydrogen bonds of two types, $A = T$ or $G \equiv C$. It follows that the number of molecules of adenine equals that of thymine, and, similarly, the amounts of guanine and cytosine are the same. It is also evident that the order of nucleotides in one strand is related to the order in its complementary strand by a simple rule. If the order in strand 1 is CGAATA, that of strand 2 must be GCTTAT.

The *replication* of DNA involves the breaking of the hydrogen bonds, the uncoiling of the helix, and the reassembly of each strand of a new complementary helix from free nucleotides in the nucleoplasm. The

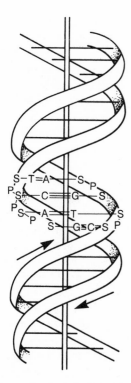

Fig. 3-1. Watson-Crick model of DNA. The molecule (double helix) is approximately 20 A in diameter; the axial distance required for each complete turn is 34 A (10 nucleotides per turn). *A*, Adenine; *C*, cytosine; *G*, guanine; *P*, phosphoric acid; *S*, deoxyribose; *T*, thymine. (Modified from Watson, J. D., and F. H. C. Crick. 1953. Cold Spring Harbor Symp. Quant. Biol. **18:**123-131.)

process is catalyzed by the enzyme *DNA polymerase.*

The second important chemical for the transmission of genetic information is ribonucleic acid (RNA). The RNAs differ from DNA in the sugar component (ribose instead of deoxyribose); in the substitution of the pyrimidine, uracil, for thymine; and in form (single- rather than double-stranded). RNA is synthesized in the nucleus on DNA templates in the presence of the enzyme *RNA polymerase*, but its functions are performed in the cytoplasm where the synthesis of proteins from amino acids takes place. Three varieties of RNA differing in size and function are recognized.

Messenger RNA (mRNA) comprises only 1% to 3% of the total, but it is of critical importance, since the arrangement of its nu-

cleotide bases provides a template specifying the nature of the proteins that play such a dominant role in cell function. Ribosomal RNA (rRNA) comprises 75% to 85% of total RNA. It is found in combination with proteins in structures called *ribosomes*, which may be free in the cytoplasm or attached to a network of membranes called the *endoplasmic reticulum.* Aggregations of ribosomes (polyribosomes) provide a structural base for mRNA during the active synthesis of proteins. Transfer RNA (tRNA), or soluble RNA (sRNA), is a small form of RNA whose function is to convey amino acids from solution to the ribosomes for incorporation into proteins. Since there are 20 amino acids making up the primary cell proteins, there are at least 20 species of tRNA, one for each amino acid. Additional nucleotides have been found in the tRNAs, but their significance has not been clearly established.

PROTEIN SYNTHESIS

Proteins are complex molecules of bewildering diversity that have many roles in the living cell. Like the nucleic acids they are polymers, but their subunits are amino acids that are joined by peptide linkages. They are constituents of cell membranes, enzymes, contractile fibers, and other structures found in living organisms. With the possible exception of identical twins, probably no two humans have exactly the same set of protein molecules, a fact that interferes with successful transplantation of tissues between individuals. The reason, of course, is the great diversity of DNA molecules that regulate the synthesis of the proteins. Two steps are recognized in this process: *transcription*, which involves transferring information encoded in DNA to mRNA, and *translation*, which involves the actual synthesis of polypeptides (linear aggregates of amino acids that may combine further to form larger proteins) by the combined action of mRNA, rRNA, and tRNA.

Genetic code

A correspondence between the linear arrangement of nucleotides in DNA and that of amino acids in proteins implies a code that can be translated. The experiments by which this code was deciphered are of great interest, but the reader is referred to other sources for a more complete account (Watson, 1970; Levine, 1978). It is now generally agreed that amino acids are coded by triplets of nucleotides. The language of genetics is thus based on an alphabet of 4 letters, A, T, G, and C (with U substituting for T in RNA); words, or *codons*, of 3 letters each, specifying amino acids; and sentences long enough to define a sequence of amino acids that can function adequately in a living organism. Table 3-1 shows translation of the 64 possible triplets from 4 letters taken 3 at a time. It will be noted that the genetic language has "punctuation marks": three of the triplets are terminators of messages, and one doubles as the codon for methionine and as an initiator. Also there are multiple codons specifying some of the amino acids (leucine appears six times). The significance of this variability is unknown.

We are now in a position to see that an error in replication of a DNA strand could change the codon in an mRNA and lead to the substitution of a new amino acid in the protein for which that mRNA is the template. The mutation that produced the varitint-waddler mouse referred to earlier must have been of this nature, although its molecular basis is not known. The effects of such a substitution of amino acids vary greatly. If the change involves a physiologically active site on the protein, it might even be lethal; other changes may be undetectable except by highly refined methods of protein chemistry.

An understanding of the genetic code also allows us to determine the enormous capacity of the system to convey messages. A mRNA molecule in the colon bacillus contains 900 to 1500 nucleotides corresponding to peptide chains of 300 to 500 amino acids. It is possible to compute the number of possible different chains of these lengths, given the availability of 20 amino acids for each and every link. Dr. Howard Dintzis of the Johns

Table 3-1. RNA genetic code: the 64 triplets and their corresponding amino acids*

Letter 1	Letter 2	Letter 3†			
		U	C	A	G
U	U	Phe	Phe	Leu	Leu
U	C	Ser	Ser	Ser	Ser
U	A	Tyr	Tyr	Term	Term
U	G	Cys	Cys	Term	Tryp
C	U	Leu	Leu	Leu	Leu
C	C	Pro	Pro	Pro	Pro
C	A	His	His	GluN	GluN
C	G	Arg	Arg	Arg	Arg
A	U	Ileu	Ileu	Ileu	Met
A	C	Thr	Thr	Thr	Thr
A	A	AspN	AspN	Lys	Lys
A	G	Ser	Ser	Arg	Arg
G	U	Val	Val	Val	Val
G	C	Ala	Ala	Ala	Ala
G	A	Asp	Asp	Glu	Glu
G	G	Gly	Gly	Gly	Gly

*Note that the DNA code would be complementary, with thymine replacing uracil.

†*Ala*, alanine; *Arg*, arginine; *AspN*, asparagine; *Asp*, aspartic acid; *Cys*, cysteine; *Glu*, glutamic acid; *GluN*, glutamine; *Gly*, glycine; *His*, histidine; *Ileu*, isoleucine; *Leu*, leucine; *Lys*, lysine; *Met*, methionine; *Phe*, phenylalanine; *Pro*, proline; *Ser*, serine; *Thr*, threonine; *Tryp*, tryptophane; *Tyr*, tyrosine; *Val*, valine; *Term*, terminator or "nonsense" codon.

Hopkins University Medical School estimates the number of potential combinations as larger than the number of electrons in the visible universe. But such theoretical calculations are less important to the behavioral geneticist than estimates of the actual variation in gene frequency among different groups of individuals. We shall return to this matter in the next chapter.

The overall scheme of transcription and

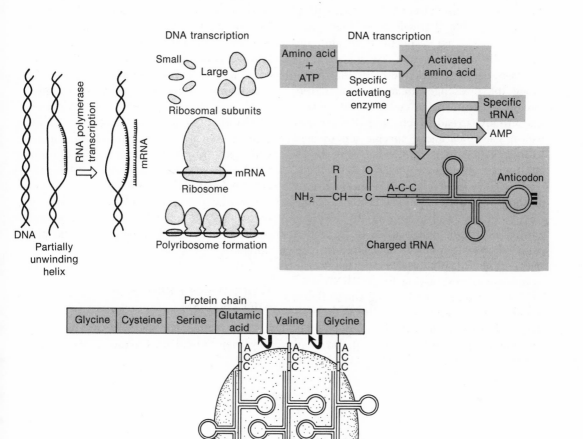

Fig. 3-2. A scheme for protein synthesis. Messenger, ribosomal, and transfer RNAs are transcribed from DNA, and all participate in protein synthesis. The actual site of protein synthesis is the polyribosome. Messenger RNA and transfer RNA pair through their codon and anticodon regions and ensure a faithful reading of the message. One cistron is shown. (From Biology, fourth edition by Willis H. Johnson, Louis E. Delanney, Thomas A. Cole, and Austin E. Brooks. Copyright © 1966 and 1972 by Holt, Rinehart & Winston. Copyright © 1956 and 1961 by Holt, Rinehart & Winston. Reprinted by permission of Holt, Rinehart & Winston.)

translation is shown diagramatically in Fig. 3-2. The partially unwound DNA helix transcribes its information to each of the three types of RNA following the rules of complementarity. Thus the DNA nucleotide sequence CTT CAG CCT directs the synthesis of an mRNA segment GAA GUC GGA. Reference to Table 3-1 will show that these codons correspond respectively to glutamic acid, valine, and glycine. Messenger RNA forms complexes with ribosomes that move along the messenger in a direction determined by an initiation codon. As it moves, the mRNA is read by charged tRNA molecules, each bearing a specific amino acid corresponding to its anticodon. The tRNA anticodons in our example are CUU, CAG, and CCU, respectively. As a ribosome moves in the direction of the arrow in Fig. 3-2, the tRNAs attach in the appropriate positions and release their amino acids to be linked with others by peptide bonds. The resultant polypeptide terminates when a nonsense codon not corresponding to any tRNA anticodon is encountered. Fig. 3-2 shows that a number of ribosomes may be attached simultaneously to a single mRNA molecule, each reading a portion of the code in its turn and directing the synthesis of a separate polypeptide of the same kind.

Our knowledge of these processes has come largely from microorganisms whose phenotypes are enzymes and other protein molecules. It is believed that the same processes take place in mammalian cells, but the control functions are more complicated. All cells of a human or a mouse probably contain the same types of DNA, but their products may be very different in neurons, muscle fibers, and gland cells. This implies that every gene is not active in every cell. The study of gene control systems in multicellular animals is an active field of research with some implications for psychology that will be discussed a little later. For the present, however, we shall turn to certain genetic disturbances of metabolism (metabolic errors) that have helped in the understanding of the relationship between genes and phenotypes.

GARRODIAN ERRORS OF METABOLISM

Early in the twentieth century a British physician, Archibald Garrod, proposed that

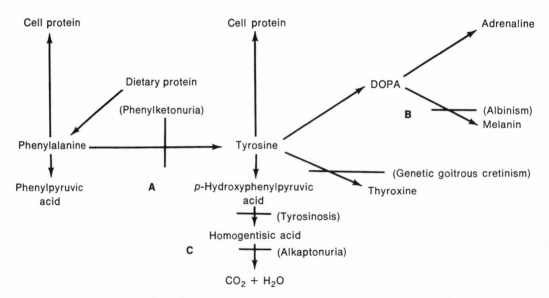

Fig. 3-3. Phenylalanine and tyrosine metabolism in man. (From Levine, L. 1973. Biology of the gene, 2nd ed. The C. V. Mosby Co., St. Louis.)

a number of familial disorders were caused by inherited "errors of metabolism" (Garrod, 1923). Today we recognize that these errors and many others discovered later are the result of incorrect specification of proteins (often enzymes) which lead to physiological malfunction. In general, such errors are inherited as autosomal recessives; heterozygotes are normal. One of the better-known examples concerns defects in aromatic amino acid metabolism, as shown in Fig. 3-3. Each of the three separate blocks in this set of related reactions produces an impairment.

A block at *A* prevents the conversion of phenylalanine to tyrosine by hydroxylation. As a result the metabolism of phenylalanine proceeds by a normally secondary pathway, producing large amounts of phenylpyruvic and other keto acids. Because these products are excreted in the urine, this disease is called *phenylketonuria* (PKU). The developing brain is damaged by the keto acids, and the victims of the disease are usually severely retarded mentally. In Chapter 17 we shall see how the physiological analysis of PKU has led to a rational therapy for the condition.

A block at *B* prevents the conversion of tyrosine to the pigment melanin and results in *albinism*, a defect common in all vertebrate classes. Recently it has been found that the visual pathways in albinos of several species have abnormal connections in the central nervous system. Further discussion of the behavioral differences associated with albinism will be found in Chapter 6.

The third block, at *C*, results in the disease called *alkaptonuria* (black urine), one of the original conditions studied by Garrod. The metabolism of tyrosine is halted at an intermediate step, homogentisic acid. Exposed to the air, this substance turns black; in addition it accumulates in the joints, causing arthritis. In contrast with PKU and albinism, there are no known neurological effects.

These three conditions thus illustrate three different ways by which a metabolic error produces a pathological effect. In PKU the damage is due to a by-product produced in excessive amounts when metabolism is shunted to an abnormal route; in albinism the difficulties are traceable to a lack of the normal end product (why the visual tracts run askew is not known); and in alkaptonuria the problem is associated with the accumulation of an intermediary compound normally present just proximally to the block.

MOLECULAR DISEASE

Deviant DNAs may produce anomalies in molecules that are not enzymes. Sickle cell anemia, a genetic disease most common in individuals of African descent, is an example. The hemoglobin of the persons homozygous for this disease differs from normal in electrical charge and led Pauling et al. (1949) to characterize the condition as a molecular disease. Heterozygotes produce a mixture of normal and sickle hemoglobin. The cause of the abnormal electric charge was later traced to substitution of valine for glutamic acid at a specific location in the molecule (Ingram, 1957).

The concept of metabolic errors and molecular disease has notable explanatory powers, and many persons have sought to extend it to mental disorders and to variations within the so-called normal range. Williams (1956) has argued that much maladaptive behavior, such as alcoholism, is a behavioral expression of partial genetic blocks leading to "genetotrophic disease." Williams suggests that it might well be possible to compensate for these blocks by adjusting the diet to contain more or less of the components involved in the disturbed system. We shall return to a consideration of possible inherited metabolic deviations in the psychoses in Chapter 16. For the present it is fair to say that no biochemical error has been clearly linked to any of the major or minor psychiatric entities. Nevertheless, the genetotrophic hypothesis is plausible, and there are many reasons for failure to detect such conditions, even if they exist (Kety, 1959; Robins and Hartman, 1972).

CONTROL OF GENE ACTION

The processes of replication, transcription, and translation as described previously carry us only a short way along the path from gene to psychophene. An organism, particularly the complex ones that behavior geneticists usually study, is much more than a mixture of specific proteins. Development is somehow guided into a highly specific pattern of differentiated cells, and within this pattern the genes that express themselves in brain, liver, and skin cells, for example, are very different.

Much of the research on mechanisms of control of gene action has been performed with microorganisms, where problems of cell differentiation do not complicate matters. An outstanding example is the control of lactose utilization by the colon bacillus, discovered by Jacob and Monod (1961a).

Their work and that of others has led to a model of control of RNA transcription shown in Fig. 3-4. The model distinguishes two major kinds of control systems, inducible and repressible.

Inducible systems (upper part of Fig. 3-4) generally regulate the production of enzymes that split compounds with a release of energy. In bacteria it is commonly found that the genes for the various enzymes in such metabolic systems lie close to each other. Since these genes code for the structure of the enzymes, they are known as *structural genes*

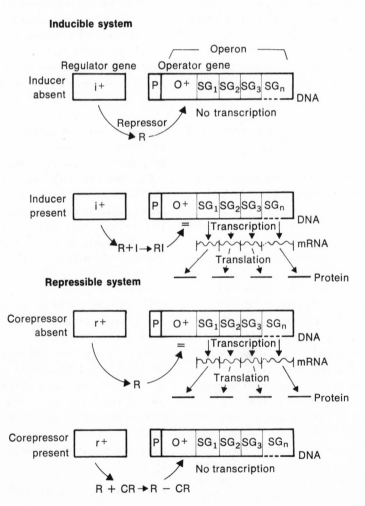

Fig. 3-4. Operon model for induction and repression. (From Levine, L. 1973. Biology of the gene, 2nd ed. The C. V. Mosby Co., St. Louis.)

(SG_{1-3} in Fig. 3-4). Adjacent to each group of structural genes is an *operator gene*, O^+, which, when activated, causes its related structural genes to transcribe to mRNA. The entire unit is called an *operon*. In an inducible system the operator is generally repressed by a nuclear constituent, R, the product of a regulator gene, i^+ or P. When an inducer, I, is introduced (I is generally the substance to be split), it combines with the repressor, the operator is "derepressed," and transcription and translation follow.

Repressible systems are generally found associated with synthetic enzymes. Here the regulator gene, r^+, produces an incomplete repressor that does not block the operon from functioning. With the addition of a corepressor, C, typically the final synthetic product, the operon is inactivated. This will be recognized as an example of negative feedback; accumulation of product slows down synthesis, and exhaustion of product stimulates it.

GENETIC REGULATION IN MULTICELLULAR ANIMALS

Intracellular processes similar to those described for bacteria are almost certainly also found in multicellular animals. However, the range of regulatory influences is much greater, since physically remote cells can communicate with one another by chemical messengers such as hormones, neurotransmitters, and antibodies. The relatively simple operon models involving repression and induction do not suffice to explain the complex coordinated syntheses that go on in the specialized cells of multicellular organisms. The model of Britten and Davidson (1969) integrates information from molecular genetics and developmental biology and attempts to account for gene regulation in these more specialized cells.

The Britten-Davidson model is based on the following six postulates: (1) the differentiation of cells is often mediated by simple, external chemical signals; (2) a given state of differentiation requires the integrated action of a very large number of nonlinked genes; (3) a significant number of genomic sequences are transcribed in nuclei but are absent from the cytoplasmic RNA; hence an intranuclear function is suggested; (4) the genome of higher cell types is much larger than that of bacteria, so that additional DNA is available for control functions; (5) these genomes contain large fractions of repetitive nucleotide sequences that are not found in bacteria and could participate in intranuclear information processing; and (6) these repetitive sequences are transcribed according to cell type–specific patterns.

A schematic version of the model is shown in Fig. 3-5. Sensor genes (S) bind with and respond to agents such as hormones that induce specific patterns of activity in the cell's genome. Integrator genes (I) are associated with sensors and produce activator RNAs (*dashed lines* in Fig. 3-5) that form complexes with receptor genes (R). These in turn are linked with producer genes (P), which correspond to structural genes that transcribe to the messenger RNAs that are involved in the synthesis of enzymes and structural proteins. A battery of producer genes essential for some complex biological process can be activated simultaneously when an appropriate inducing agent is present. In Fig. 3-5 two different possible arrangements for achieving this result are depicted. Although the details of the model have not been completely verified and, in fact, have been modified in later publications (Britten and Davidson, 1971; Davidson and Britten, 1973), something like it must exist in order to explain cell differentiation and cyclical activation.

The Britten-Davidson model seems particularly relevant to instances in which stimulation of young developing organisms produces long-lasting behavioral effects. Genetic variation in the intragenomic communication system could cause variation in the cellular response to external stimuli. If these differences in cellular response persist in some permanent form, one might find long-lasting differences between genotypes.

There are few attempts to apply this type of theoretical analysis to behavior. One interesting example is a study of scent marking by

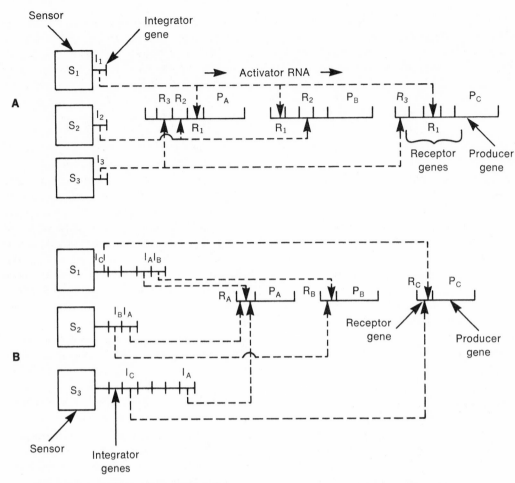

Fig. 3-5. Two proposed models for the control of gene expression in higher cells. The diagrams schematize the events that occur after the three sensor genes, which are sensitive to specific stimuli, have initiated transcription of their integrator genes. Activator RNAs diffuse (*dashed lines*) from their sites of synthesis, the integrator genes, to receptor genes. In diagram **A** a redundancy in receptor genes is postulated so that combinations of producer genes P_A, P_B, and P_C are activated according to the overlapping specificities of their sets of receptor genes. In diagram **B** there is redundancy in the integrator genes and a specific receptor gene for each producer. Under either hypothesis stimulation of S_1 leads to transcription of P_A, P_B, and P_C; stimulation of S_2 to transcription of P_A and P_B; and stimulation of S_3 to transcription of P_A and P_C. (From Britten, R. J., and E. H. Davidson. 1969. Science **165:**344-357. Copyright 1969 by the American Association for the Advancement of Science.)

Mongolian gerbils (Thiessen and Yahr, 1970). Males of this species mark their territory with sebum from a ventral scent gland. The response is androgen dependent and can be elicited in castrated males by very small implants of testosterone propionate into the hypothalamus. The behavioral inductive effect of the hormone is lost, however, when actinomycin D is injected along with it. This antibiotic prevents DNA-directed transcription onto mRNA. Thiessen and Yahr suggest that the normal hormonal effect is due to the stimulation of DNA transcription in central neurons. The resultant RNA somehow triggers the combination of responses that comprise marking behavior. Obviously, as the authors state, the links between transcription and marking must be determined before the theory can be accepted. Developmental genetics and developmental psychology are beginning to influence each other.

TIMING OF GENE ACTION

Although an organism's genotype is determined at fertilization, its phenotype changes dramatically during development. Such changes at the molecular, cellular, and organ levels imply that different portions of the genotype become quiescent or active as some synthetic processes cease and others begin. A well-known example is hemoglobin synthesis in human beings. In the early stage of uterine life, a fetal form of hemoglobin (HbF) is produced by the combination of two alpha and two gamma chains, each type of chain being controlled by its own locus. Late in the fetal stage and continuing into infancy, the gamma locus becomes inactive, and two other loci, beta and delta, take over the synthesis of peptide chains that combine with alpha chains to form adult hemoglobin. By 1 year of age, 98% of hemoglobin is HbA (2 alpha + 2 beta), and 2% is HbA_2 (2 alpha and 2 delta).

Developmental changes of this kind are not unique to blood. In the nervous system many biochemical shifts occur during development (Benjamins and McKhann, 1972). In the newborn rat brain, enzymes of the glycolytic pathway permit anaerobic use of glucose, but this capability is lost within the first 2 weeks. The situation is adaptive for the young rat when it is exposed to anoxia. Here again the data are interpretable in terms of scheduling of genic activity. We conclude that genes may influence behavior in two ways. Genes active early in life during the differentiation and growth of neurons, glia, and other elements may affect the permanent patterning of the nervous system. Their behavioral effects may long outlast their biochemical activity. Genes involved in the synthesis of metabolically active molecules may affect behavior by modifying the cellular milieu throughout the period of life in which they are active.

The mechanisms through which temporal changes in gene activity might operate are several. At the level of DNA to RNA transcription, segments of chromosomes can be stimulated by various factors. In the extremely large chromosomes in the salivary glands of some fly larvae, it is possible to correlate puffing in various regions with specific stages of protein synthesis (Berendes, 1965). Puffing has been demonstrated to be characteristic of regions of DNA transcription; hence the locus of control here seems to be at the gene itself. It is probable that similar processes occur in vertebrates during differentiation. But there may be other mechanisms. Tyler (1969) has evidence for "masked mRNA" that can remain silent for a long period and then be activated by specific stimulation. Here the locus of control would be shifted from DNA transcription to mRNA translation into polypeptide chains.

Among the stimuli that conceivably could activate the DNA or masked RNA of neurons are chemical neurotransmitters, ionic changes in cells, and electric currents. It is almost self-evident that molecular processes of learning and those of cell differentiation and gene activation must have elements in common. But behavior genetics in the narrow sense has not yet made substantial contributions to the pulling together of these areas. The situation may be changing as the emphasis of behavior genetic research shifts from determining heritabilities to the investigation of the development of "species-specific" behavior (Thiessen, 1972).

PLEIOTROPY

Although each gene may have a single biochemical function, its effects are not limited to a unit function at the structural, physiological, or behavioral level (Russell, 1963). Descriptions of the effects of identified genes in mammals are at present better phrased in terms of specific tissues affected and specific stages of development than in terms of chemical reactions. When a tissue (cartilage, for example) enters into the development of many structures, all will be affected by a gene substitution affecting the tissue. The multiple effects are an example of pleiotropy, in this case, secondary pleiotropy, since all stem from one basic process.

Not all cases of pleiotropy have been traced back to a single common origin. The dominant white spotting series of genes in the house mouse (W and W^v) produce, in addition to their effects on pigment, anemia and impaired gonadal function (Russell, 1963). Careful embryological studies have traced the gene action backward in development without convergence on a common process. Russell, however, is not convinced that the genes at this locus show primary pleiotropism and suggests that they may act on the three biological systems through a single chemical reaction. Perhaps the contrast between primary and secondary pleiotropism is a little forced, since one can never prove that divergent but correlated phenomena are not related by some undiscovered unitary process. At any rate, the distinction is not important in behavior genetics. The significant fact is that in complex organisms the consequences of a gene substitution may be manifest in a number of apparently independent functions. Yet these deviations will be correlated because of dependence on a common genetic mechanism.

DUPLICATE PHENOTYPES, DIFFERENT GENOTYPES

Phenotypic similarity need not imply genetic identity. Shaker mice are a well-known example. These animals have choreic head movements and some tendency to run in circles. Gates (1934) proved that there are two distinct genetic forms, both inherited in a recessive fashion, that are phenotypically alike. Shaker 1 has normal alleles at the shaker 2 locus and vice versa. The mating between the two is represented as follows:

> Parents sh-$1/sh$-1, $+/+$ × $+/+$, sh-$2/sh$-2
> Both are shakers.
>
> F_1 sh-$1/+$, $+/sh$-2
> All are "nonshakers."

In the F_1 each mutant gene is counteracted by its normal allele obtained from the other side of the cross. The fact that the F_1 is normal while the parents are not indicates that the two shaker genes do not have the same function. It may be inferred that each locus is concerned with a different step in morphogenesis of the nervous system and that blocking the process in either place produces the same result.

In human genetics it is not feasible to make test crosses to determine whether two similar phenotypes are also genotypically alike. Sometimes, however, genetic analysis of a series of pedigrees of clinically similar cases leads to the discovery that inheritance of a particular defect in one family follows a pattern of dominance, in another recessivity, and in still another, recessive sex-linkage. This appears to be true of retinitis pigmentosa, a chronic progressive degenerative disease of the retina (Sorsby, 1953). This type of evidence proves genetic heterogeneity but does not indicate whether the several genes have the same or different physiological effects.

PHENOCOPIES

Treatments by physical agents can produce effects on phenotype comparable to those of a known mutant gene. Landauer (1945) injected insulin into developing hen's eggs and produced a large number of rumpless chicks phenotypically similar to the rumpless birds produced by a pair of autosomal recessive genes. Goldschmidt (1938) produced aberrant types of *Drosophila* by exposing developing larvae of wild-type stock to elevated temperatures for various periods of time at particular stages of development. The resultant anomalies were often faithful replicas of recognized inherited mutant phenotypes. Rubella (German measles) in pregnant women leads to a high proportion of malformed children who may be indistinguishable from those bearing genetically induced deformities.

The production of phenocopies experimentally seems to have some promise as a means of elucidating the mode of action of mutant genes. There must be something in common between two agents that produce the same result, and, if the copy is very precise, the point of action may be the same.

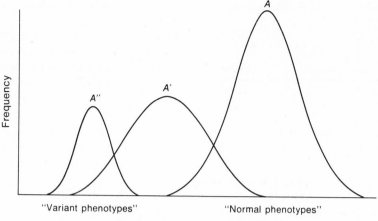

Fig. 3-6. The statistical nature of genotypic control of phenotype is depicted in these three curves. Each represents the distribution of phenotypes in a related genotype. The range of variation in *A'* is much greater than in the "normal" *A* or "variant" *A"* genotype and overlaps both. Hence phenotype is not an absolute guide to genotype.

The fact that similar phenotypes (e.g., shaker 1 and shaker 2 mice) can be produced by mutations at different loci, however, indicates the need for caution in this approach.

STATISTICAL NATURE OF GENOTYPIC CONTROL

A living organism is a dynamic system whose range of adaptation is a function of its genotype as a whole. This concept is diagramed in Fig. 3-6. Genotype *A* is the modal, or "normal," form that has essentially the same phenotype under a variety of environmental conditions. Genotype *A"* is a variant that is always clearly distinguishable from the mode. Genotype *A'* is extremely variable and cannot be perfectly separated from either *A* or *A"* on the basis of phenotype alone. The *A'* individuals falling far to the left are variants; those on the right are "normal overlaps." Genotypes control differentiation and development in a statistical sense. Sometimes the choice between alternative phenotypes is regulated with practically all-or-none precision. In other situations nongenetic factors are more critical.

The terms *penetrance* and *expressivity* are often used with respect to the action of genes. Penetrance is a measure of the prob-

ability that a genotype can be identified by its phenotypic effects. Penetrance values are provisional, since new effects may be discovered and individuals reclassified on the basis of obscure characters. The more general term, expressivity, refers to the intensity of the phenotypic manifestation. For example, one form of jaundice is inherited as a dominant autosomal character. All afflicted individuals have fragile red blood corpuscles, but the amount of red-cell breakdown in vivo varies. Clinically, some individuals appear healthy, whereas others have a severe, often fatal, disorder (Snyder and David, 1953).

The environment of a gene includes the effects of the other genes present, although none of them may be identifiable as specific contributors to the trait concerned. This is well exemplified in an experiment reported by Runner (1954). The fused (*Fu*) gene, which produces structural anomalies in the tail and other parts of the skeleton of mice, was placed on two different genetic backgrounds by repeatedly crossing fused individuals into two inbred lines. The results are shown in Table 3-2. It is evident that the gene was more effective in modifying the BALB/c phenotype than the C57BR/a phenotype. Furthermore, a remarkable differ-

Table 3-2. Effect of genetic background on penetrance of *Fu* in heterozygotes*

Strain	Source of *Fu* gene	
	Mother	Father
BALB/c	72%	88%
C57BR/a	34%	65%

*From Runner, M. N. 1954. Inheritance of susceptibility to congenital deformation-embryonic instability. J. Natl. Cancer Inst. **15**:637-649.

ence between the reciprocal crosses was found, particularly in the C57BR/a strain. The reciprocal cross difference could be explained as due to cytoplasmic factors present in the ova or by modifying genes in the Y chromosome (Gruneberg, 1952). A more likely explanation is that the uterine environment of *Fu*/+ mothers is less favorable for the expression of the gene than that of +/+ mothers.

Variations in expression of a gene are presumably caused by small forces acting at random. By manipulating the environment systematically, it is possible to shift the modal phenotype. The incidence of skeletal anomalies in the mouse can be shifted by such procedures as (1) producing temporary anoxia in the mother at the seventh day of gestation, (2) transferring ova of one strain to the uterus of another, or (3) simply allowing the mothers to grow older (Runner, 1954). Effects such as these are functions of genotype, but the anomalies themselves are not inherited in the usual sense of the word. What has been inherited is a reactive system that responds in a characteristic way to stress.

We shall refer again and again to the relationship between gene and character in our survey of behavior genetics. There is a great gap between enzymes controlling phosphorylation reactions and a brain capable of solving differential equations. Yet genes are concerned with both, and both are problems for behavior genetics in the broadest sense.

GENIC BALANCE AND X CHROMOSOME INACTIVATION

In the previous sections we have concentrated on normal complements of chromosomes with each autosomal locus being represented twice in the genome. We have seen that when a mutant gene is present on one chromosome, its homologue on the other may suppress its phenotypic effects completely or partially. We call such a mutant gene recessive and interpret recessivity as indicating that the wild-type gene produces enough product to compensate for the deficient mutant. Thus development and function in the heterozygote are essentially normal.

What happens when there is an excess of genetic material instead of a deficit? Such a situation can arise when homologous chromosomes do not separate during meiosis (nondisjunction). Half of the resultant gametes will have a missing chromosome; the other half will have a double dose. If the gamete that is diploid for one chromosome fertilizes a normal gamete of the opposite sex, the zygote will have three rather than two sets of some of its genes. One might think that more is better, but this is not true. An extra chromosome 21 in man (trisomy 21) is responsible for Down syndrome (formerly called mongolism), which is characterized by specific physical and behavioral signs, including severe retardation. This syndrome and others resulting from disturbances of chromosomal balance are discussed in Chapter 17. Trisomy of the larger chromosomes in man induces more severe retardation and deformity and is frequently lethal. Apparently the extent of the dislocation of development is a function of the amount of excess genetic material, and chromosome 21 is one of the smallest.

We shall now consider the sex chromosomes, X and Y. The Y chromosome in mammals appears to contain little or no genetic information except for that concerned with masculinizing development. The X chromosome is much more diversified. In humans over a hundred pathologies, major and minor, have been found to be X-linked. Clearly the one X chromosome of a male, provided it does not carry one of these deleterious mutations, is capable of guiding normal development. When such a mutation is present it is expressed in a male, even though the gene is recessive in females. This situation poses a problem for the hypothesis of genic balance. If one X chromosome is enough to ensure normal development in a male, why does the double dose in females not produce problems? The answer is that an inactivation process occurs in the early embryo that leaves only one active X chromosome in each female somatic cell. X inactivation was first demonstrated in mice and has some implications for behavior genetics. It has been employed as a procedure for investigating brain-behavior relationships in fruit flies and mice (Chapter 7).

The clue that led to the concept of X inactivation was provided by certain female cats and mice. These animals, heterozygous for recessive coat color genes, are always mottled or blotched. On the average the recessive allele is expressed over one half the body area. Lyon (1962) proposed that the phenomenon could be explained by a random inactivation of one or the other X chromosome at an early embryonic stage. If the two chromosomes carried different alleles, the result would be a mosaic of two functionally different kinds of cells. When the functional difference involves pigmentation, the mosaicism is apparent externally. However, it is not restricted to pigmentation. In human females heterozygous for X-linked muscular dystrophy, one half of the muscle fibers are normal, one half dystrophic.

In human cells the inactivation of an X chromosome leaves a deeply staining Barr body (named for its discoverer) attached to the nuclear membrane. The Barr body is a consistent constituent of female cells, and it is also found in male cells of aberrant chromosome count such as XXY. Teleologically, X inactivation has the function of making normal males and females, and even individuals with unusual complements of X chromosomes, equivalent with respect to genetic information.

Since the inactivation process appears to be random, two individuals with the same initial genotype, such as female monozygotic twins, could have different active genes in parts of the body involved with the control of behavior and thus be more divergent than male monozygotics. This possibility assumes, of course, that there are genes on the X chromosome with important behavioral effects and that alternative alleles are relatively common. For most of behavior genetics, the significance of X inactivation is that it calls to our attention the necessity for genic balance to ensure normal development.

4

Population genetics

Up to this point we have considered the transmission patterns of genes from parents to their offspring and the relationship between the genotype of an individual to its phenotype. This chapter is concerned with genotypes and phenotypes in a larger context, that of a population. And we shall be concerned with changes in genotype in successive generations, changes that are the essential basis of organic evolution.

The word population has a very special meaning in genetics, one which is somewhat difficult to define briefly and precisely, largely because populations are fluid aggregations of individuals often in a state of flux. We propose the following as a working basis: a genetic population is a subgroup of a species characterized by a gene pool significantly different from other divisions of the species and maintained in its identity by geographical, ecological, or cultural isolation. This definition is broad enough to cover major races of mankind, the diverse breeds of domesticated animals, and the numerous varieties and subspecies that are characteristic of most widely distributed species of plants and animals. Boundaries between populations are typically zones of mixing in which individuals of different groups mate. The classification of populations may be hierarchical; we may consider all Caucasoid peoples as one population or subdivide down to a point of comparing the genetic composition of Britons with English and Welsh surnames (Roberts, 1942, cited by Stern, 1973). We shall return to a consideration of the concept of race later in this chapter. For the present we shall develop a model system that involves interactions between a number of dynamic processes. Although population genetics may involve laboratory experiments, its main

thrust is toward the analysis of changes in the gene pool of natural groups. To this end a considerable body of theory has been built up to deal with the complexities of the real world. We shall present a much simplified account of the subject; one which we hope will enable students to apply their knowledge to behavioral problems. More detailed accounts can be found in Li, 1955; Falconer, 1960; Crow and Kimura, 1970; and Cavalli-Sforza and Bodmer, 1971.

GENE POOL AND GENETIC EQUILIBRIUM

Imagine that all ova and sperm of a given population were thrown together, thoroughly mixed, and allowed to form zygotes which then developed into the next generation. We are assuming here that the species must reproduce sexually. If we consider any particular autosomal locus A, with two alleles, A' and A'', a gamete may contain either of these but not both. We now symbolize the proportion of A' as p and that of A'' as q, with $p + q = 1$. The model can deal with more than two alleles at a locus, but their introduction would make the equations more complex without illustrating any new principles. If there are no barriers to the fertilization of either type of ovum by either type of sperm, the probabilities for the formation of each possible type of zygote are given by the products of the frequencies of the combining gametes. Such a random-mating system is called *panmixia*. The outcome is shown in Table 4-1 as a Punnett square. Since the two ways of forming heterozygotes yield identical zygotes, the proportions of the three genotypes are $p^2(A'A') + 2pq(A'A'') + q^2(A'A'')$. This is the familiar binomial expansion of $(p + q)^2$. When $p = q = 0.5$, as in the F_2 of

Table 4-1. Panmixia in a one-locus two-allele system*

		Male gamete	
		A'(p)	**A''(q)**
		Zygotes	
Female	A'(p)	$A'A'$(p²)	$A'A''$(pq)
gamete	A''(q)	$A''A'$(pq)	$A''A''$(q²)

*The frequencies of genes and genotypes are in parentheses. $p + q = 1$.

two purebreeding lines (one homozygous for A', the other for A''), these proportions are $0.25(A'A') + 0.50(A'A'') + 0.25(A''A'')$. If $p = 0.9$, the proportions are $0.81(A'A') + 0.18(A'A'') + 0.01(A''A'')$.

It is important to note that the attainment of these frequencies of combinations bears no relationship to the source of the genes if they combine at random. The gametes could come from two pure stocks, one $A'A'$ and the other $A''A''$, or from a population that has practiced panmixia for many generations. The result in either case, provided the assumptions are met and the population is large, is a predictable genetic equilibrium. In small populations, of course, sampling variation can lead to substantial deviations. We shall stay with large populations and state that genetic equilibrium, a condition of transgenerational stability of the frequency of both genes and genotypes, is attainable with a single generation of random mating regardless of possible previous assortment of the genes. This fact, discovered independently by two scientists (Hardy, 1908, and Weinberg, 1908), is known as the Hardy-Weinberg law and is the starting point of further extensions of population genetics.

Perhaps the fertilization of the ova of starfishes occurs in the random manner assumed in our model because the gametes of these animals are shed into seawater, where they combine. In most of the animals studied by behavior geneticists, mating occurs between individuals, and in some species monogamous sexual relations are the rule. In human beings and many other species, generations overlap and thus complicate matters.

Despite these deviations from the idealized model, many genetic systems fit the Hardy-Weinberg equations well. In Chapter 12 we shall illustrate its use in testing a genetic hypothesis in humans, thus circumventing our inability to perform genetic experiments in our own species.

DEVIATIONS FROM EQUILIBRIUM

There are other, more serious weaknesses in the panmictic model as descriptive of actual populations. Gametes, and particularly zygotes, vary in viability; errors in replication of DNA result in new alleles; the mixing process in our model lake is imperfect; genes may be transferred between pools and thus alter the value of p and q. Six ways in which deviations from genetic equilibrium may occur are formally recognized. We shall discuss each separately and then consider how they interact.

Mutation

As we have learned, DNA replication is subject to error at a very low rate; the novel allele is called a *mutation*. We shall designate the mutation rate of A' to A'' by u. This means that the proportion of alleles changing from A' to A'' per generation is pu. Mutations are potentially reversible, and we symbolize the backward rate, A'' to A', by v. Hence qv is the proportion of genes reverting from A'' to A'. If p and u are relatively large with respect to q and v, the mutational process will move the gene pool toward a higher proportion of A''. As A'' increases, the backward mutations tend to balance the forward ones. At equilibrium, the rates will be equal: $pu = qv$; $q = u/(u + v)$; $p = v/(u + v)$. Note that the equilibrium values for p and q are independent of the absolute mutation rates and depend only on their relative sizes. The so-called spontaneous mutation rate in higher animals is low and is generally measured indirectly. It varies from locus to locus, but values between 10^{-5} and 10^{-6} (between 10 and 1 per million) are often reported. In general u and v differ markedly; thus mutation by itself would produce equilibria in which the more mutable allele is much rarer than the stable one. In the ordinary breeding experiments of behavior genetics, mutations are considered rare enough so that they do not affect the result. However, these rates are per locus. If an organism has 10^4 genes with an average mutation rate per locus of 10^{-5}, 1 out of 10 gametes will carry a new mutation. Thus, over longer periods of time, as in sublines of common origin maintained apart for a long period, behavioral differences can arise that may be due to mutation (Denenberg, 1959; Ciranello et al., 1974).

Selection

All zygotes formed by the random combination of gametes must contribute equally to the next generation if genetic equilibrium is to be maintained across generations. This assumption is, of course, not true in general and may possibly never be true. The consequence of differential viability (or fertility) of some genotypic combinations is selection against alleles that produce these genotypes. If A'' is recessive to A' and $A''A''$ individuals reproduce less often than $A'A'$ or $A'A''$, successive gene pools will contain fewer and fewer A'' alleles; their only chance for survival will be in heterozygous combination with A'. The varitint-waddler gene *(Va)* of Chapter 2 survives only because geneticists counteract natural selection against the gene by counterselection of mice known to be carriers.

Selection acts on the phenotype, but its genetic consequences depend on the resultant changes in genes and genotypes. Hence the relationship between genotype and phenotype is of fundamental importance. A dominant gene of high penetrance is exposed to selection in practically every individual who carries it, but a recessive gene is so exposed only when it is homozygous. Environment also plays a role. Advances in medical techniques permit the survival of individuals with genetic defects that might have been fatal earlier in our history. If such individuals are capable of normal reproduction, selection is essentially eliminated. Sensory defects such as myopia and deafness would have been a severe handicap to humans in the hunting and gathering stage; today eyeglasses and hearing aids compensate for the inadequacies of genes, and these conditions have increased.

Selection in the population-genetics sense should not be confused with individual survival, which is a necessary but not a sufficient condition for gene survival. The genes of an individual who leaves no offspring die just as completely as those of the fetus aborted *in utero*, regardless of the physical vigor and behavioral competence of their possessor. And the choice is not simply between fertility and sterility. If different genotypes also differ in average number of descendants, selection will favor the genes of the more prolific group. The degree of disadvantage of a particular allelic combination is expressed in terms of the relationship of its average fertility to that of the fertility of possible alternative genotypes. This *index of selection* is symbolized by s. Consider, for example, the situation in which A is fully dominant to a. Table 4-2 shows the original proportions and the results of one generation of selection. The adaptive value, or Mendelian fitness, of each genotype is simply $1 - s$. Note that if $s = 1$, none of the aa genotypes leaves descendants, and all a genes are in the heterozygotes. If s is small, say 10^{-3} (1/1000), the change in gene frequency per generation will be slight. But given time, even low values of s will lead to elimination of a gene unless there are compensating factors. Also noteworthy is that selection against a complete recessive is fairly efficient when q is large, since q^2 individuals, each with two a genes, are exposed. When q is small, however, q^2 is very small, and most of the a genes are sheltered from selection within heterozygotes. For example, when $q = 10^{-3}$, only one individual per million is exposed to selection. The number of heterozygotes per million is $10^6 \times 2 \times 0.999 \times 0.001$, or 1998. Thus only 2 of the total of 2000 a genes are accessible to selection. The great bulk of genetic variability is hidden from detection by phenotypic inspection alone.

Fig. 4-1 shows the effect over generations of various intensities of selection and type of

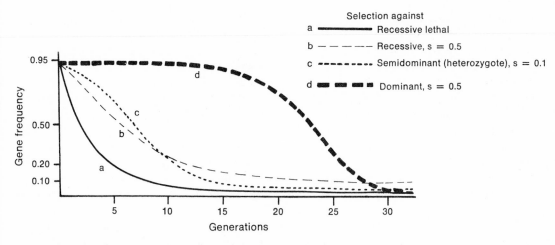

Fig. 4-1. Selection rate for various genetic situations. (From Levine, L. 1973. Biology of the gene, 2nd ed. The C. V. Mosby Co., St. Louis.)

Table 4-2. Selection against recessive allele at a single locus

Genotype frequency	AA	Aa	aa	Frequency of a*
Base population Generation 0	p_0^2	$2p_0 q_0$	q_0^2	q_0
Fitness	1	1	$1-s$	
After one generation	p_0^2	$2p_0 q_0$	$q_0^2(1-s)$	
When $s=1$				$q_1 = \dfrac{q_0}{1+q_0}$
When $s<1$				$q_1 = \dfrac{q_0(1-sq_0)}{1-sq^2}$

*The formula for q_1 is derived by substituting $1-q_0$ for p_0 in the equation giving the composition of the first selected generation in terms of the parental population. The sum of these terms reduces to $1-sq_0^2$, indicating that the selected generation is deficient in offspring of the recessive class. Since each heterozygote contributes one recessive gene, and each homozygote, two, the total number of recessive alleles in the selected generation is $p_0q_0 + q_0^2 - sq_0^2$. This simplifies to $q_0(1-sq_0)$.

dominance, starting each curve with an initial frequency of 0.95 for the allele selected against. Can you explain why curve *a* is so flat from the tenth to the thirtieth generation? And why the decline in the frequency of a dominant allele that is exposed to selection in both homozygous and heterozygous combinations is so slow when the allele is common?

There are important implications of the inaccessibility of fully recessive genes to selection. By extending the analysis shown in Table 4-2 to additional generations, it is easy to demonstrate that q_n, the frequency of *a*

after *n* generations, is:

$$q_n = \frac{q_0}{1+nq_0} \tag{1}$$

Suppose, for example, that one wished to reduce the frequency of a deleterious recessive gene from 10^{-2} to 10^{-3} by preventing the breeding of all individuals showing the trait. Such a change would result in the lowering of the incidence of the trait from 1:10,000 to 1:1,000,000. Rearranging equation 1, one obtains:

$$n = \frac{q_0 - q_n}{q_0 q_n} \tag{2}$$

Solving for n yields 900 generations. If there were a method for detecting heterozygous carriers of the unwanted allele, the efficiency of selection could be increased. An important area of human behavior genetics is directed toward this objective, which can be very useful in genetic counseling.

The equations relating changes in gene frequency to the number of generations of selection become more complex when fitness of the homozygous recessive is greater than zero. Solutions may be found in specialized texts such as Li (1955), Crow and Kimura (1970), and Wilson and Bossert (1971). Given data for values for changes in gene frequencies over one or more generations, it is possible to compute s. Such computations may have behavioral significance if s can be related to sensory and motor capacities or to other traits that may affect fitness.

Kinds of selection. All organisms are subject to natural selection, which operates through all factors influencing the survival of genes. Darwin (1859) made this principle the cornerstone of his theory of organic evolution, although his ideas of genetic transmission were completely erroneous. Natural selection has no preordained direction; the genotypes with the best survival potential in one environment may be inferior in another. And, of course, differential survival of phenotypes has no genetic consequences unless the phenotypes are based on genetic variability within a population.

Artificial selection was practiced by man long before genetics emerged as a science, and it resulted in the production of improved (from the human point of view) varieties of domestic animals and crop plants. Psychologists have used selection to produce strains of rats differing in maze learning (Tryon, 1929, 1940), in emotionality (Hall, 1938; Broadhurst, 1960b), and in other characteristics. Artificial selection is ordinarily *directional*, since it is guided by previously assigned criteria of what is desired.

Natural selection may also be directional if an environmental change renders an extreme phenotype more fit than those nearer the population mean. More frequently, natural selection works against all extremes, and it is then called *stabilizing* selection. An animal breeder culling undersized and oversized individuals from his stocks is practicing the same thing.

Intensity of selection may be defined as the proportion of individuals in one generation who transmit their genes to the next. Clearly, selection can be much more intense if one parental pair produces many offspring. Very intensive selection might seem to be the way to achieve the most rapid results in a directional project, since only the most extreme phenotypes would contribute. But, as we shall explain more fully later, funneling all propagation through a few individuals narrows the amount of genetic variation on which selection can operate. For the long pull, moderate intensity of selection is usually more productive.

Migration

If there is mixing between two populations with different gene frequencies, the new composite population will have a new set of frequencies determined by the initial values in the two original groups and the proportional contribution of each. Let q_x and q_y symbolize the frequency of an allele of interest in populations x and y, respectively. Let M equal the proportion of x individuals in a new population, z, formed by a mixing of x and y. Then:

$$q_z = Mq_x + (1 - M)q_y \qquad (3)$$

If the values of q_x, q_y, and q_z are determined empirically, it is possible to compute M and thus estimate the degree of admixture of two originally separate populations. The equation is simplified if either q_x or q_y is zero. For example, the contribution of persons of European descent to produce the American black population has been estimated by a number of investigators with variable results. Perhaps the best estimate is that about 22% of genes in present-day blacks are of non-African origin. Reed (1969) has dealt critically with the difficulties of applying the simple model to real populations. More important than the exact values is the principle that population genetics can provide information on mating patterns of natural populations, human and animal, that are not accessible to direct observation.

Genetic drift

Perfect mixing of the gametes in our hypothetical gene pool is unlikely in practice. Ova and sperm from one source will tend to stay together so that samples from one region will not represent the whole population. In a very large population, deviations in opposite directions will over the long run be equal, so that the gene and genotype frequencies remain stable in the population as a whole. However, if a small, nonrepresentative sample were used to seed a new pool (a previously uninhabited area), the gene frequencies of old and new populations would be different and would continue so. Successive serial sampling and colonization of several new areas could result in a number of discrete groups of the same origin, although differing significantly from their source and from each other. Such sampling variations will be more important when very small numbers of individuals found a new population. Until recently mankind was characterized by numerous small religious or geographical isolates that were reproductively isolated. Many of the features that differentiate human racial groups must have been established in such isolates through drift, although others could be the result of different selective pressures associated with climate, food resources, and the like.

Assortative mating

The mating of individuals with others of similar phenotype more frequently than would be predicted by random association is called positive *assortative mating*. Negative assortative mating is, of course, a disproportionate incidence of matings between individuals of differing phenotypes. Positive assortative mating for physical and psychological characteristics of man is well documented (Conrad and Jones, 1940; Vandenberg, 1972). Positive assortative mating for morphological traits in *Drosophila melano-*

gaster has been demonstrated by Parsons (1965), and in mice both positive and negative assortment have been shown in laboratory studies (Mainardi, Scudo, and Barbieri, 1965; Yanai and McClearn, 1972, 1973).

Assortative mating is based on phenotypes, but it will have genetic consequences if genes contribute substantially to the variability of the characteristics on which it is based. As we shall demonstrate later, genes do contribute to many kinds of behavioral variation, and assortative mating must be considered in applying mathematical models to real populations. By itself, however, assortative mating does not alter gene frequencies, only genotypic ones. In the extreme it could result in a bimodal distribution of the phenotype on which assortment is based and the establishment of two relatively independent populations.

Inbreeding

Inbreeding may be defined as the mating of individuals more closely related than the average of the population. In humans such matings are usually called *consanguineous*. The definition implies that the measure of inbreeding must be relative to some base population in which all individuals are nonrelated. It also implies that some inbreeding will occur in panmixia, since a perfectly random system will not proscribe matings between close relatives such as parent and offspring or brother and sister. Inbreeding has been an important tool for geneticists, since it is used to establish the pure lines on which much of experimental genetics depends.

To understand the consequences of inbreeding it is convenient to introduce two new terms. Genes are considered to be *identical by function* when they are alike at the molecular level and perform identically in the cell. Genes are *identical by descent* when they can be traced back to one specific ancestral gene that has replicated to form identical descendants. Obviously, *identity by descent* also connotes *identity by function*, but the opposite is not true. Note also that there must be some arbitrary base from which identity by descent is determined. All human beings alive today are descended from a small band of *Homo sapiens* from whom we are separated by perhaps 10,000 generations. Many mutations have occurred during this time; some have survived, and others have been lost, but in a broad sense most of our genes are alike by descent from the remote past. For our purposes, however, it is more useful to consider closer relationships. Thus, before considering the quantitative aspects of inbreeding, we shall consider the measurement of relationships.

Coefficient of relationship. An objective measure of genetic relationship between two individuals is the probability that their genes at a given locus are identical by descent. This will also be the average proportion of all their genes that are alike by descent. We shall calculate this coefficient for a few common situations: parent-offspring, sib-sib, and uncle-nephew. In the following diagram we deviate from the standard method of designating alleles and use a series of capital letters, unique for each of four homologous genes in a pair of unrelated individuals.

	Father		Mother	
Parental genotypes	AB		WX	
Possible offspring genotypes	AW	AX	BW	BX

It can be seen readily that each offspring must have one gene in common with each parent and no more than one gene. Hence the coefficient of relationship between a child and either parent is ½.

The computation of the relationship between sibs is slightly more complicated but still straightforward. Sibling 1 must receive either A or B from the father; the probability that sibling 2 will receive the same gene is ½; the probability that they will be different is also ½. The same reasoning applies to the maternal contribution. Combining the probabilities of identity and nonidentity of the two parental contributions we have:

Sibs 1 and 2		Degree of relationship
Alike in maternal and paternal contribution	½ × ½ = ¼	1
Alike in maternal, different in paternal	½ × ½ = ¼	½
Different in maternal, alike in paternal	½ × ½ = ¼	
Different in both maternal and paternal contribution	½ × ½ = ¼	0

The average relationship is the same as parent with offspring, but siblings may be more or less closely related at particular loci, depending on chance.

A simple method of determining relationship is illustrated in Fig. 4-2. The letters designate individuals; the connecting lines, transmission of a gene preceded by a segregation. To find the relationship (probability of having genes identical by descent) between any pair of individuals, I_1 and I_2, we count the number of sectors, k, in each separate pathway connecting them, and record the lengths, $n_1, n_2 \ldots n_k$. Their relationship is simply:

$$R_{I_1,I_2} = \sum_1^k (\tfrac{1}{2})^{n_i} \qquad (4)$$

For the siblings, D and E, there are two paths *(1,2)* and *(3,4)*, each of length 2. Their relationship is $(\tfrac{1}{2})^2 + (\tfrac{1}{2})^2 = \tfrac{1}{2}$. Similarly, the relationship of H to E is ¼. As an exercise determine the relationship between H and I; I and J. These computations assume that none of the individuals in the pedigrees are inbred. If some are, the probability that their descendants will have genes identical by descent is increased.

Coefficient of inbreeding. The quantitative measurement of inbreeding follows the same line of reasoning we have used to calculate relationships between individuals. The inbreeding coefficient, commonly sym-

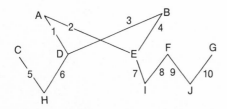

Fig. 4-2. Pedigree for calculation of degree of relationship. (See text for explanation.)

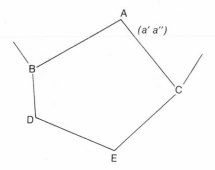

Fig. 4-3. *A* is a common ancestor of *C* and *D*, parents of *E*. The inbreeding coefficient of *E*, $F_E = (\frac{1}{2})^4 (1 + F_A)$. (See text for explanation.)

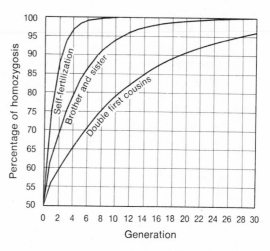

Fig. 4-4. Percentage of homozygosis in successive generations under three different systems of inbreeding. (From Russell, W. L. 1941. Inbred and hybrid animals and their value in research. In Snell, G. B., ed. 1960. Biology of the laboratory mouse. Dover Publications, Inc., New York.)

bolized as *F*, is the probability that the two homologous genes of an individual are identical by descent. When they are, the individual may be called *autozygous* at the locus. One can think of *F* as a measure of the relatedness of the maternal and paternal contributions to the genome. The more closely related the parents, the greater is the probability of autozygosity in their offspring.

Consider, for example, the pedigree of Fig. 4-3. *A* is an ancestor of both *C* and *D*, whose offspring *E* is, therefore, inbred. Five segments in the lines of descent form a closed path, including *A* and *E*. The probability is $(\frac{1}{2})^5$ that *A*'s gene *a'* was transmitted in each of the five segregations, so that *E*'s genotype is *a'a'*. There is an equal probability for autozygosity of *a''*, and the probability of one or the other outcome is the sum of the two, $(\frac{1}{2})^4 = \frac{1}{16}$. If *A* were already inbred, the inbreeding coefficient of *E* would be higher, since both *B* and *C* would receive genes identical by descent for all of *A*'s autozygous loci. Thus, the general formula for the inbreeding coefficient of *i* is

$$F_i = \sum_1^k (\tfrac{1}{2})^{n-1} (1 + F_a) \qquad (5)$$

where

    n = number of segregations in each ring
    k = number of closed paths
    F_a = inbreeding coefficient of common ancestor

Inbreeding does not directly change gene frequency, but it may have large indirect effects. As the probability of autozygosity in-

creases, recessive alleles will be more frequently exposed to selection than in a panmictic population. Homozygosity and consequent phenotypic expression of rare recessive alleles is substantially more frequent in the offspring of consanguineous matings. And regardless of selection, continued inbreeding within lines results in loss of genetic variability as more and more loci become autozygous. Geneticists have made use of this fact to develop inbred lines whose members are essentially genetically identical. Fig. 4-4 shows the rate of attaining homozygosity with several regular systems of inbreeding. The particular assortment of alleles fixed in a group of inbred lines derived from the same base population is essentially random, though subject to constraints imposed by the necessity of survival and propagation.

Inbred lines of rats and mice are in a sense a sampling of populations of commensal rodents that have been domesticated by man quite recently. If they were a true random sample, the genomes of an array of such strains could be considered to be representative of the original species. Actually, inbred strains come in an assortment of coat colors almost never seen in their feral and commensal relatives, and have been subjected to selection for docility and adaptability to confinement or for special phenotypic characteristics such as susceptibility to cancer. Inbred strains are usually smaller, less vigorous, and less fertile than noninbred lines or than their F_1 hybrids. They do, however, have the great advantage for geneticists of genetic uniformity within lines that usually leads to a high degree of phenotypic resemblance. Exceptions to this principle will be noted later. Despite the somewhat artificial nature of inbred strains and their generally reduced vigor as compared with heterogeneous stocks, they are extremely valuable for experimental behavior genetics.

Interactions between factors

In real populations, selection, mutation, mixing, assortative mating, and inbreeding may be acting simultaneously. Drift may have important influences on gene frequencies if the population is small. The quantitative aspects of such interactions are discussed in the specialized texts previously cited. In general, theory is ahead of experimentation and even farther ahead of our knowledge of actual populations. Nevertheless, there is growing interest in what Bruell (1970) has called "population behavior genetics." Here we shall merely touch on a few of the ways in which the concepts of population genetics have been important to behavior genetics.

Mutation-selection equilibria are of particular relevance in the case of genes that produce mental defect with resultant reductions in fitness. Such genes are regularly removed from the population as their carriers fail to leave offspring, but they continue to be added through new mutations. In a population at equilibrium, the rate of removal must equal the rate of mutation. If one makes some simplifying assumptions, it can be shown that, for a dominant gene with complete penetrance, the mutation rate, u, is approximately equal to one half the product of the selection index by the gene frequency, p.

$$u = \frac{ps}{2} \qquad (6)$$

For a completely recessive gene with frequency, q, the comparable expression is:

$$u = q^2 s \qquad (7)$$

And for an X-linked recessive (q_m = frequency of the gene in males):

$$u = \frac{q_m s}{3} \qquad (8)$$

In each of these equations the right-hand term represents the product of the selection index by the proportion of individuals showing the trait. Dominant genes and X-linked genes in males are exposed to selection in all individuals carrying them; recessive genes are so exposed primarily in the homozygous state. One can use these expressions to measure mutation rates indirectly in a population if data are available on the incidence of the genetic anomaly and the intensity of selection against it. The accuracy of the estimate depends on the validity of the assumptions (1) of genetic equilibrium and (2) of the lack of deleterious effects of "recessive" genes in heterozygotes.

When one homozygote, say AA, is superior to another, say $A'A'$, and at least equally as fit as the heterozygote, selection leads steadily to the elimination of A'. What happens when selection favors the heterozygote over both homozygotes? A classic example of such a situation is the gene for sickle hemoglobin, *HbS*. In portions of Africa where falciparum malaria is or has been common, heterozygotes (*HbA/HbS*) resist the disease better than do individuals with normal hemoglobin (*HbA/HbA*). *HbS/HbS* individuals suffer from severe anemia and die in infancy or, at least, before reproducing. The geographical distribution of *HbS* and some other genes with similar effects on resistance to malaria has been found to correspond closely with high-incidence areas of the disease. Equilibrium in such cases is attained when the proportional rate of removal of each allele is the same for both. Consider a situation in which the fitness of AA is $(1 - s)$, and that of $A'A'$ is $(1 - s')$. Gene frequencies are p and q, respectively. The proportion of allele A accessible to selection is $2 p^2/p = 2 p$. The number of homozygotes

(AA) is p^2, and each individual removed from the population carries two genes. A similar calculation can be made for A'. Then, at equilibrium:

$$2\,ps = 2\,qs' \tag{9}$$

$$p = \frac{s'}{s + s'}, \text{ and } q = \frac{s}{s + s'} \tag{10}$$

Such systems are said to show *balanced polymorphism* because they ensure the continued survival of more than one allele at a locus. The store of genetic variability so maintained can be an important resource in adapting to environmental change. Under such conditions the relative values of s and s' will shift, and a new equilibrium of gene frequencies will be established.

Natural populations appear to be very heterogeneous genetically (Lewontin and Hubby, 1966), and this heterogeneity is likely to be based in part on the balance system just described. However, other mechanisms could also lead to the maintenance of genetic diversity. In a nonhomogeneous world the relative fitness of genotypes fluctuates; AA may sometimes outpropagate $A'A'$ and at other times or in other places do more poorly. Whatever its source, one expects that a highly variable population would be more successful in exploiting new environments and in coping with unusual stresses.

The population genetics models described in this chapter have necessarily been simplified. Assortative mating and inbreeding, particularly in small populations, introduce complexities that must be considered in research, particularly with humans who do not mate and propagate to suit an ideal experimental design. For our species, in particular, the concept of a gene pool in a somewhat unstable equilibrium is central to much behavior genetic research.

5

Genetics of quantitative attributes

Thus far we have been considering the relationship between genotype and phenotype as deduced from experiments in which the substitution of one allele for another leads to a change of the phenotype from one class to another. Thus a wild-type house mouse of agouti coat color has an AA genotype; changing one of these genes to A^{vy} produces a yellow coat because of the suppression of black pigment formation. If we know the dominance relationships of each allele for a single locus, we can determine the frequency of each by the Hardy-Weinberg formula, provided, of course, that the population is in genetic equilibrium.

Genes with clearly defined phenotypic effects are often called *major genes*, and many are known by their influence on structural and biochemical traits. Often such genes have pleiotropic effects on behavior; thus albino mice are less active than pigmented mice in a brightly illuminated open field. The merits of looking for behavioral effects produced by well-defined mutations are great. The progress of genetics as a science is based on its adoption of a particulate theory of heredity to replace the older ideas of blending of the germinal material from the two parents in their offspring. Mendel demolished this concept when he recovered both parental phenotypes in typical form from the F_2 of a cross between purebreeding lines.

Yet there are characteristics that superficially appear to be transmitted as the blending model would predict. Children resemble their parents in height, and, when the parental difference is great, the children are usually intermediate between them. The offspring of two parents of intermediate height will, on the average, also be intermediate, although they may vary considerably in both

directions. There are no sharp discontinuities in the frequency distribution of height that enable one to classify humans as "tall", "intermediate," or "short" on a rational basis. Neither can the outcomes of matings between individuals of different heights be predicted from the original form of the Mendelian laws, which require specification of a limited number of phenotypic categories.

In the early part of the twentieth century there was a period of controversy between Mendelians and biometricians regarding the proper method for genetic analysis of continuously distributed traits, such as height, body proportions, and behavior. Some biometricians believed that mechanisms other than the nuclear factors that transmitted Mendelizing traits must be involved in the inheritance of quantitative traits with an essentially continuous distribution. Fisher (1918) demonstrated, however, that the observed resemblance between relatives in such traits was explicable by a model which assumed that the degree of their development was dependent on effects of segregation at a number of loci with cumulative effects on the phenotype. Mather (1949) introduced the name *polygenes* for genes that cannot be identified individually but whose existence is inferred on the basis of the transmission patterns of quantitative phenotypes. This definition essentially holds today, although in a few instances loci affecting quantitative characteristics have been mapped through their linkage with major genes (Thoday, 1961).

Sometimes the term *polygenic trait* is used, but one must not infer that major genes and polygenes operate in separate realms. Current hypotheses for genetic regulation of intelligence uniformly stress the involvement of many loci. Yet there are a great number of biochemical and neurological mutations that interfere with nervous system function and depress IQ test scores moderately or severely. For the most part (there are exceptions; see Chapters 14 and 17), the influence of such mutations on intelligence is nonspecific; a gene is identified by its biochemical and morphological effects rather than by a unique psychological effect. Since such genes are individually rare, they account for only a small portion of the heritable component of intelligence, but they are of great concern to persons involved with problems of mental deficiency. Polygenes affecting intelligence are not detectable by abnormal products or gross alterations in structure but can be demonstrated by statistical means, just as Mendel demonstrated the existence of genetic factors with absolutely no knowledge of their physical nature.

We have suggested that traits with discontinuous distributions are likely places to look for major gene control, whereas those with continuous distributions usually involve polygenes. As generalizations go, this is reasonably accurate, but there are exceptions. A basically continuously distributed trait, such as the activity of an enzyme involved in neural transmission, may have minor effects on behavior until it falls below a critical threshold. The population will then be divided on a behavioral scale into two phenotypic classes, afflicted and normal but on a biochemical scale its variation could fit the normal curve.

Conversely, a single locus might produce classical Mendelian ratios for a biochemical characteristic that has a small input into brain function, an input which is insufficient to split the population into discontinuous phenotypic groups. Fig. 5-1 represents the two situations. We shall see later that after several decades of research, monogenic and polygenic models of inheritance are still competing as explanations for such traits as audiogenic seizures in mice (Chapter 6) and schizophrenia in man (Chapter 16).

For a behavior geneticist with physiological interests, polygenic systems are less satisfying for study than monogenic ones. The specific locus approach offers the potentiality of relating primary gene products to behavior through the intermediacy of development, differentiation, and cell function. The advantages of this strategy have been stressed by several writers (e.g., Caspari, 1964; Merrell, 1965; Hawkins, 1970), and it has also been strongly criticized (Wilcock, 1969). We shall leave the mediation of this clash of views to students as they evaluate the results of each method. Actually the two strategies, specific locus and biometrical, are complementary, since each is uniquely suitable for certain problems. Behavior is an attribute of an organism as a whole, and hun-

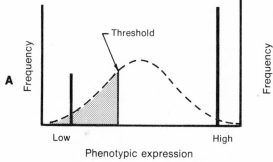 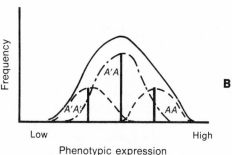

Fig. 5-1. **A,** Continuous frequency distribution (*dashed line*) of a somatophene and discontinuous distribution (*heavy bars*) of a psychophene whose two states are determined by the value of the somatophene relative to a threshold. **B,** Discontinuous distribution (*heavy bars*) of a somatophene with continuous distribution (*broken lines*) of a related psychophene. *Broken lines* represent distribution for each of three genotypes; *solid line* is the sum of the three curves.

dreds of genes are active in the differentiation and functioning of sense organs, effectors, and the nervous and endocrine systems. In a sense all behavior is under polygenic influence, although at times one locus can produce an enormous difference between individuals. As long as psychologists are interested in individual differences in complex phenotypes, such as rate of learning, levels of activity and emotional arousal, and tendencies to explore or to seek mates, they will be involved in polygenic systems, and their synergism with environmental stimuli. The sharper analytical properties of the specific locus approach will find usefulness in the study of extreme behavioral deviations, and such investigations can contribute to understanding of normal functions.

In this chapter we shall introduce only as much of the methodology of quantitative genetics as is essential to comprehension of later chapters. A minimum of proofs will be given, and these for simple systems. Other methods will be introduced later in connection with discussion of specific studies. Useful guides to this area of genetics include Reeve and Waddington (1952), Falconer (1960), Burdette (1962), and Mather and Jinks (1971).

MEASURING QUANTITATIVE ATTRIBUTES

Since individuals cannot be arbitrarily assigned to classes on the basis of a measurement (X_i) of a continuously distributed trait, their positions are defined in relation to the population mean $(\overline{X})$. Immediately it is clear that the genetic analysis of quantitative traits must involve the specification of a reference population. It will often be convenient in our discussions to express measurements as deviations from the mean $(x_i = X_i - \overline{X})$. The dispersion of measurements within the population is commonly expressed in terms of variance $(V$ or $s^2)$. The standard deviation (s) is the square root of the variance and is expressed in the same units as X. Variance is the average of the squared deviations of a set of observations from their mean. The relationships for a sample of N individuals in terms of both X and x are summarized here:

	Raw scores (**X**)	Deviation scores (x)
Mean	$\overline{X} = \dfrac{\Sigma X_i}{N}$	$\overline{x} = \dfrac{\Sigma x_i}{N} = 0$
Variance	$s^2 = \dfrac{\Sigma (X_i - \overline{X})^2}{N}$	$s^2 = \dfrac{\Sigma x_i^2}{N}$

Textbooks of statistics should be consulted for additional information. For example, the best estimate of the variance uses $N - 1$ as the denominator rather than N, as given here.

If a population is distributed symmetrically about its mean, approximately 67% of the observations will fall within one standard deviation from $\overline{X}$, and about 96% within two standard deviations. We shall be concerned primarily with variances because of certain mathematical properties that are useful for genetic analysis, in particular their property of additivity.

Basically, all forces determining the phenotype (P) of an individual can be subsumed under a genetic (G) or an environmental (E) category. Each category is extremely complex and must be subdivided for analytical purposes. We may express the simple model as follows:

$$P = G + E \qquad (1)$$

All three terms are best expressed in relation to the population mean. It follows that the phenotypic variance can likewise be divided into components:

$$V_p = V_g + V_e + V_{ge} \qquad (2)$$

The V_{ge} term is added to take care of situations in which the effects of E are quantitatively different for different genotypes. The fundamental problem of quantitative behavior genetics is to partition V_p into its components so as to estimate the proportional contributions of genes and life histories to population variability. All methods make use of the phenotypic resemblance between relatives as a means of estimating V_g. In animal experiments inbred strains and selected lines can be compared and crossed in orderly fashion. In human behavior genetics the investigator compares the degree of similarity of close relatives with that of individuals taken at random. In both cases attention must be given to possible genotype-environment correlations. It is often possible in animal experiments to reduce such correlations by strictly controlling rearing and test procedures. With humans an attempt must be made to evaluate the degree of correlation and consider it in any partitioning of variance into genetic and environmental components. Genotype-environment correlations may even have interest in their own right.

Before further formal development of the model, it is of interest to consider special situations that illustrate some fundamental points. If one were to measure behavior in a highly inbred strain of fruit flies or mice, all or nearly all of the phenotypic variation would be attributable to environmental sources, since genetic homogeneity abolishes V_g, leaving only V_e. Similarly, differences between monozygotic (one-egg or identical) co-twins are a pure measure of V_e. Conversely, if one could rear genetically different individuals exactly alike, all variance between them would be V_g. In practice this is impossible to achieve, although a close approximation can be reached in the laboratory.

These extreme examples demonstrate that V_p, V_g, and V_e are properties of defined populations in specified situations. Ratios between them, such as *heritability*, which will be defined later in this chapter, can neither be constants for a given trait nor indicate the relative contribution of G and E to the phenotype of a particular individual. Quantitative genetics is inextricably bound up with the concept of populations.

GENOTYPIC VALUES: ADDITIVE AND DOMINANCE EFFECTS

Our model for the genetics of quantitative attributes is a one-locus system with two alleles, *A* and *A'* (Falconer, 1960). Fortunately, it can be extended with reasonable accuracy to polygenic systems with multiple alleles at each locus. In Fig. 5-2 we depict a cross between two purebreeding lines, P_1, which is homozygous *A'A'*, and P_2, which is homozygous *AA*. The average phenotype of P_1 is $-a$; that of P_2 is $+a$. Throughout this chapter lowercase symbols will designate phenotypes rather than recessive genes. Plain capital letters represent alleles associated with positive deviations from the mean; primed capitals represent alleles producing negative deviations. In Fig. 5-2 the parental strains are shown as equidistant from a fictitious midparent (*M*) with a phenotype of zero. The mean phenotype of the F_1 hybrid is shown on the P_2 side of *M* at point *d*, in this diagram at $0.4a$. If *A* and *A'* combined additively (no dominance), the F_1 mean would coincide with *M*. If the F_1 were on the left of *M*, *d* would be negative, and if it were on the right of P_2, as shown in the lower part of the figure, we would call the phenomenon *overdominance*.

An extension of the model to a two-locus system is shown in Fig. 5-3, and further expansion to *n* loci is straightforward. We show the simplest situation, the F_2 of a cross between two purebreeding strains, one *AABB* and the other *A'A'B'B'*. The phenotypic effects of the two loci are equal and cumulative. Thus either *AA* or *BB* contributes $a/2$

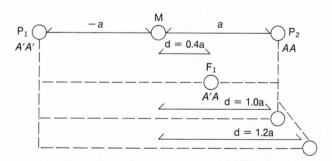

Fig. 5-2. Genotypes and phenotypes: a model of quantitative inheritance.

Female gametes \ Male gametes	AB	AB′	A′B	A′B′
AB	*AABB* a	*AABB′* $\frac{a + d}{2}$	*AA′BB* $\frac{a + d}{2}$	*AA′BB′* d
AB′	*AABB′* $\frac{a + d}{2}$	*AAB′B′* 0	*AA′BB′* d	AA′B′B′ $\frac{d - a}{2}$
A′B	*AA′BB* $\frac{a + d}{2}$	*AA′BB′* d	*A′A′BB* 0	*A′A′BB′* $\frac{d - a}{2}$
A′B′	*AA′BB′* d	*AA′B′B′* $\frac{d - a}{2}$	*A′A′BB* $\frac{d - a}{2}$	*A′A′B′B′* $-a$

Fig. 5-3. Extension of model to a two-locus system and further expansion to n loci.

to the phenotype; AA' or BB' contributes $d/2$; $A'A'$ or $B'B'$ contributes $-a/2$. The achieved average phenotypes are simply the sums of the effects of the two loci and are shown in the lower portion of each of the 16 cells in Fig. 5-3. Adding these values and dividing by 16 yields $d/2$ as the mean phenotype of the whole array.

In more detail the 16 cells constitute six phenotypic classes that are listed here with the frequency of each: $1(a)$, $4\left(\frac{a + d}{2}\right)$, $4\left(\frac{d}{2}\right)$, $2(\text{zero})$, $4\left(\frac{d - a}{2}\right)$, $1(-a)$. Note that if $d = 0$, the mean phenotype also becomes 0, and the distribution is symmetrical: $1(a)$, $4\left(\frac{a}{2}\right)$, $6(\text{zero})$, $4\left(\frac{-a}{2}\right)$, $1(-a)$. A valuable exercise is to extend this model to the two backcrosses, ($F_1 \times P_1$ or B_1) and ($F_1 \times P_2$ or B_2), and the F_2. You should obtain means of $\frac{(d - 2a)}{4}$ for B_1 and $\frac{(d + 2a)}{4}$ for B_2. What do you find for F_2? The graphic representation of this calculation is shown in Fig. 5-4.

The thoughtful reader will recognize that nature may be much more complicated than this model. A pure line may carry a mix of increasing alleles and decreasing alleles (e.g., $AAB'B'$); there may be strong dominance at the A locus and none at the B locus; interlocus interactions may result in making the value of BB partially dependent on conditions at the A locus. Unraveling such prob-

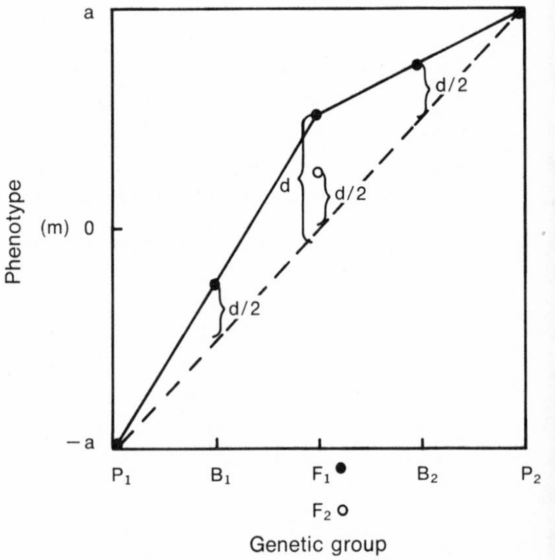

Fig. 5-4. Graphic representation of phenotypes of crosses shown in Fig. 5-3.

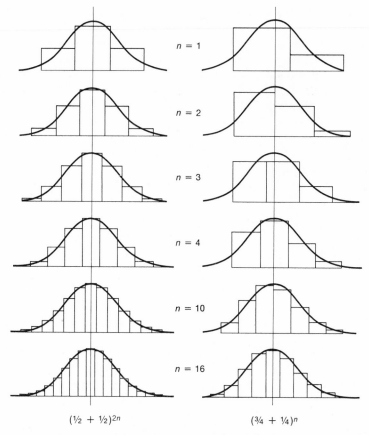

$(\frac{1}{2} + \frac{1}{2})^{2n}$ $(\frac{3}{4} + \frac{1}{4})^{n}$

Fig. 5-5. Binomial distributions for n pairs of genes with equal effects and superimposed normal curves with equal means, equal area, and equal variance. *Left,* No dominance. *Right,* Complete dominance in the same direction in all pairs of genes. (From Lush, J. L. 1945. Animal breeding plans, 3rd ed. Iowa State University Press, Ames, Iowa. Reprinted by permission.)

lems is the concern of biometrical geneticists. We shall stay with the simple model, which has proved adequate in a number of studies.

As the number of effective loci influencing a quantitative characteristic increases, systematic changes occur in the dispersion of the phenotypic measurements. One can generalize from the dihybrid case of Fig. 5-3 to a system of n loci with an increasing and a decreasing allele at each. There are $2n + 1$ genotypes in terms of the numbers of increasing genes present. Actually there are 3^n different genotypes, but many may be assumed to be equivalent functionally. The proportions of each of these "equivalent genotypes" are given by the successive terms of the expansion $(p + q)^{2n}$ when all plus

alleles have the same frequency, p, and all negative ones the frequency q $(q = 1 - p)$. The distribution converges on the normal as n becomes larger (Fig. 5-5). The effect of dominance is to produce marked skewness at low values of n, but the distortion from symmetry is low when n is 8 or more (right-hand side of Fig. 5-5). If gene frequencies are unequal, variability is somewhat reduced; if the various loci have unequal effect, variability is increased.

TRANSMISSION OF QUANTITATIVE TRAITS

In the following four sections the model of quantitative gene action we have just developed is combined with the simple population genetics of genetic equilibrium and random mating. The algebra is straightforward, but the answers may not be obvious without working through the calculations. We have supplemented the algebraic treatment with an arithmetic example to facilitate understanding of the material. However, it must be recognized that adopting other values for p, q, and d or introducing assortative mating and selection would yield different numerical solutions.

The important points to be understood are (1) populations are genotypically and phenotypically stable in the absence of selection, in spite of the fact that the offspring of phenotypically extreme individuals tend to be less extreme than their parents (regression toward the mean), and (2) the phenotype of an individual is an imperfect guide to its breeding value when allelic frequencies are unequal and dominant gene action is present. The first of these points is demonstrated in Fig. 5-6, which depicts gene transmission and its phenotypic consequences in a two-allele system from a parent generation, *G-1*, to its offspring, *G-2*.

The two alleles are designated *A* (frequency = p) and *A'* (frequency = q). The average phenotypic scores of the three possible genotypes are *a* (*AA*), *d* (*AA'*), and $-a$ (*A'A'*). For the numerical computations that accompany the algebra we have set $a = 1$, $d = 0.4$, $p = 0.75$, and $q = 0.25$. The three left-hand boxes contain descriptions of G-1 in terms of genotypes, phenotypes,

G-1	From specified genotypes Female gametes	From random mating population Male gametes		G-2 By parentage	G-2 By genotype
		A p(0.75)	*A'* q(0.25)		
Genotype phenotype frequency Ph × F		*A* p(0.75)	*A'* q(0.25)	Average phenotype frequency Ph × F	Genotype phenotype frequency Ph × F
A A a(1.0000) p²(0.5625) p²a(0.5625)	*A* p(1.0)	*AA* a(1.0000) (0.75)	*A A'* d(0.4000) (0.25)	pa + qd(0.8500) p²(0.5625) (0.4781)	*A A* a(1.0000) p²(0.5625) (0.5625)
A A' d(0.4000) 2pq(0.3750) 2pqd(0.1500)	½ p(0.50)	a(1.0000) (0.3750)	d(0.4000) (0.1250)	½[(p − q)a + d] (0.4500)	*A A'* d(0.4000) 2pq(0.3750) (0.1500)
	½ q(0.50)	d(0.4000) (0.3750)	− a(−1.000) (0.1250)	2pq(0.3750) (0.1688)	
A'A' −a(−1.0000) q²(0.0625) − q²a(−0.0625)	*A'* q(1.0)	*A A'* d(0.4000) (0.75)	*A'A'* − a(−1.000) (0.25)	pd − qa(0.0500) q²(0.0625) (0.0031)	*A'A'* −a(−1.0000) q²(0.0625) (−0.0625)
Sum (p − q)a + 2pqd (0.6500)				Sum (p − q)a + 2pqd (0.6500)	Sum (p − q)d + 2pqd (0.6500)

Fig. 5-6. The transmission of genes and phenotypes in a random mating population. Genes of a two-allele system are represented by uppercase letters. The average phenotypes of the genotypes are represented by lowercase letters. Arbitrarily assigned numerical values for the phenotypes follow the algebraic symbols in parentheses. Allele frequencies (p for *A*, q for *A'*) have also been assigned numerical values. The left-hand boxes show the three possible genotypes and their frequency in the population. The individuals represented by these boxes are designated as females and are assumed to mate at random with males of the same population. The position of the sexes could be reversed without changing any other feature of the figure. See text for further explanation.

frequencies, and the product of the last two. Adding these products yields the mean phenotypic value of the entire population; $(p - q)a + 2pqd$, or 0.65 with our numerical constants.

Note that the sum of the genotypic frequencies is always 1. Therefore, multiplying the mean phenotype of a genotype by its frequency and adding the products for all genotypes yields the population mean directly. Also note that the term $(p^2 - q^2) a = (p + q) (p - q) a = (p - q)a$.

G-1 individuals can produce two kinds of gametes, A-bearing (0.75) and A'-bearing (0.25). The possible combinations between male and female gametes are shown in the central part of Fig. 5-6 within heavy lines, along with the corresponding phenotypes and their frequency of occurrence. To the right of the central section is a representation of the G-2 offspring from each genotypic category in G-1. Consider, for example, the expected progeny from AA parents. They can produce only A gametes, which have probability p of combining with another A, producing an AA offspring with phenotype a, and a probability q of combining with A' to produce a heterozygote with phenotype d. The sum of these two outcomes is $pa + qd$ (0.85), which is a reduction from the parental value. Since we are assuming equal fertility of all genotypes, the relative contribution of AA parents to G-2 is the same as their original frequency (0.5625), and the product $(0.85 \times 0.5625 = 0.4781)$ represents their weighted contribution to the total G-2 phenotypic score.

The same type of analysis can be applied to the offspring of G-1 genotypes AA' and $A'A'$. The dashed lines in Fig. 5-6 separate gametes and G-2 zygotes according to their origin from G-1. Note that heterozygotes produce equal numbers of two types of gamete and that all three genotypes are included in their progeny. The offspring of the heterozygotes are variable, but their mean deviates only slightly from that of their parents (0.4500 vs. 0.4000 on our numerical scale). The greatest change is found in the offspring from $A'A'$ parents (0.05 as com-

pared with −1.00). The reason should be clear. Since the dominant gene (A) is the more common, a majority of A' gametes will combine to form heterozygotes that benefit from dominant gene action. But although each of the three genotypes fails on the average to produce offspring as deviant as the parents were, the population mean is unchanged; the gains and losses of the three groups exactly balance for a net change of zero. Thus, when G-2 (in the last column) is arranged by genotype rather than by ancestry by combining appropriately the probabilities within the heavy lines and multiplying by the frequency of the parental classes, the array is identical in detail with that of G-1. The population parameters remain stable, even though the genes and their resultant phenotypes are continuously recombining in a manner which ensures that the offspring of deviates (homozygotes) are less extreme on the average than their parents.

To see how the magnitude of this *regression to the mean* depends on p, q, and d, one can calculate the average transgenerational change in phenotype as follows:

G-1	G-2	(G-2) − (G-1)	
a	$pa + qd$	$q(d - a)$	(3)
d	$\dfrac{d + (p - q)a}{2}$	$\dfrac{(p - q)a - d}{2}$	(4)
$-a$	$pd - qa$	$p(d + a)$	(5)

It is instructive to insert several different values for the constants in these equations and see how they affect the magnitude of the regression effect.

Breeding value: transmittable effect of a genotype

As stated earlier, there are advantages in expressing the relationships between phenotypes as shown in Fig. 5-6 in terms of deviations from the population mean. By following this procedure, we define the *breeding value* of an individual, which can be measured by the phenotypes of its offspring. Genotypic values, on the other hand, cannot be directly measured except in the special case of highly inbred lines. We begin by rewriting the G-1 genotypes and the G-1 and

G-2 phenotypes as deviations from the mean, $(p - q)a + 2\,pqd$, with the same algebraic and numerical values used in Fig. 5-6:

G-1 genotype	G-1 phenotype	G-2 phenotype	
AA	$2q(a - pd)$	$q[a + (q - p)d]$	(6)
	(0.35)	(0.20)	
AA'	$(q - p)a + (p^2 + q^2)d$	$\dfrac{(q - p)a + (1 - 4pq)d}{2}$	(7)
	(−0.25)	(−0.20)	
$A'A'$	$-2p(a + qd)$	$-p[a + (q - p)d]$	(8)
	(−1.65)	(−0.60)	

Let us now look at the average effects of genes A and A' in a random population and see how they lead to a simplification of these rather complex relationships. *Average effect* is defined as the expected phenotype of the offspring from a gamete with a specified gene such as A or A'. An A-bearing gamete, for example, will unite with another A with probability p to produce an a phenotype and with an A' to produce a d phenotype with probability q. The average effect of A' can be calculated similarly, and both are summarized here with our arbitrary equivalents:

Gamete	Expected progeny	−	Population mean	=
A	$pa + qd$	−	$(p - q)a + 2pqd$	=
	(0.85)		(0.65)	
A'	$pd - qa$	−	$(p - q)a + 2pqd$	=
	(0.05)		(0.65)	

Average effect	
$q[a + (q - p)d] = q\alpha$	(9)
(0.20)	
$-p[a + (q - p)d] = -p\alpha$	(10)
(−0.60)	

The bracketed terms in the full algebraic expressions are identical and are commonly symbolized as α, which has the numerical value of 0.80 in our example. The breeding value of a genotype, sometimes called its *additive value*, is logically the sum of the average effects of its two alleles. Thus:

Genotype	Breeding value	Mean of offspring	
AA	$2q\alpha$ (0.40)	$q\alpha$ (0.20)	(11)
AA'	$(q - p)\alpha$ (−0.40)	$\dfrac{(q - p)}{2}\alpha$ (−0.20)	(12)
$A'A'$	$-2p\alpha$ (−1.20)	$-p\alpha$ (−0.60)	(13)

The reason that the expected mean of the offspring is half the breeding value is that a given parent contributes only one-half of the genotype of each offspring, and under random mating the average breeding value of its mates will be zero. We do not need to know the actual genotype of an individual to estimate its breeding value from the mean of its offspring's phenotypes. Of course, large families are required for precision, and the basic assumption of random mating and zero correlation of genotype and environment must be met.

Dominance deviations

The preceding section demonstrated that the breeding or additive (A) value of a genotype differs from its average phenotype or genotypic value (G). This difference is called the *dominance deviation* (D) and can be calculated simply from the difference between the values of G (equations 6 to 8) and A (equations 11 to 13) for each genotype. For the AA genotype, for example:

$$D_{AA} = 2q(a - pd) - 2q\alpha = -2q^2d \quad (14)$$
(Numerical value in our example = −0.05.)

Similarly:

$$D_{AA'} = 2pqd(+0.15) \quad (15)$$
$$D_{A'A'} = -2p^2d(-0.45) \quad (16)$$

It can be demonstrated that A and D are not correlated and that the mean value of each is zero. The importance of differentiating between the two kinds of genetic effect is that they contribute unequally to the resemblance between different types of relatives and hence affect the interpretation of phenotypic correlations. Furthermore, the relative importance of D and A has been interpreted from an evolutionary point of view. The association of a large dominance deviation with a trait is taken as evidence that the trait has been subjected to positive selection (Bruell, 1964).

Note that the dominance deviation (D) and the value we have called d, deviation of the F_1 from the midparent, are not the same. The formula for the additive component includes a term in d, so that part of the effect of a dominant gene is transmitted. If $d = 0$, the dominance deviation disappears completely, and the breeding values are $A_{AA} = 2qa$, $A_{AA'} = (q - p)a$, and $A_{A'A'} = -2pa$. Quantitative genetics is simpler when dominance is eliminated. One way of doing this is by scaling, which is considered a little further on.

It is appropriate here to note that there are two major terminologies for quantitative genetics. In this chapter we have employed with slight modifications the system of Falconer (1960). Roberts (1967) and McClearn and DeFries (1973) have also followed this system. Many authors, however, employ a somewhat different terminology that may be confusing, since some letters (e.g., D) have unlike meanings. We do not in this book deal extensively with this alternate system and recommend that the student who encounters it in the literature refer to such sources as Broadhurst (1960b) and Mather and Jinks (1971).

Epistatic deviations

The model we have developed based on a single locus can be generalized to a polygenic system as

$$G = A + D + I \qquad (17)$$

where G, A, and D are summed over all loci. The new term, I, is inserted to provide for nonadditive interactions between two or more loci. Such interactions define the term *epistasis*, a phenomenon that is probably common for behavioral traits which involve synergistic actions of many genes. In practice, however, it is often impracticable to distinguish I, and its variance may be pooled with the error term. The concepts of breeding value and dominance deviations, combined with the quantitative expression of degree of genetic relationship, provide a basis for the estimation of heritability from the phenotypic resemblance of various kinds of relatives. We shall return to this topic after a digression on the problem of scaling.

On scales and transformations

We have stated in the preceding sections that the equations of quantitative genetics are

simpler in the absence of dominance. The assertion of dominance is not as simple as one might think. Suppose, for example, that we have a strain of rat (P_1) that averages four section entries during an open field test. We breed this strain to another (P_2) which averages 64 entries in the same test and obtain an F_1 that scores 25. Since this is 9 points below the midparent, we conclude that low activity is dominant. There may be good reasons, however, to express scores in terms of the square root of the number of entries. Now we have average scores: P_1, 2; F_1, 5; and P_2, 8. Dominance has disappeared. Were we to transform the scores into logarithms to base 10, we would obtain P_1, 0.60; F_1, 1.40; and P_2, 1.81, with dominance of high activity.

A common reason for scale transformation is the requirement of many statistical techniques, such as analysis of variance, that the groups to be compared have reasonably homogeneous variances. This requirement is often in conflict with the expression of data in natural form such as the number of sectors entered in the open field. Commonly large quantities fluctuate more than small ones with a corresponding inflation of the variance in an active as compared with a lethargic strain. This is well illustrated by exploratory activity scores in a Hebb-Williams (1946) maze of two inbred strains of mice and their hybrids (Thompson and Fuller, 1957) (Table 5-1). Note particularly that the variances of

Table 5-1. Exploratory activity in C57BR (P_2) and A (P_1) mice and their hybrids

Genetic group	Raw scores (X_i)		Transformed scores (X_i)$^{1/2}$	
	Mean	Variance	Mean	Variance
P_2	532	22,274	22.9	9.74
B_2	396	22,259	18.9	30.15
F_1	303	10,420	17.0	12.23
F_2	288	23,613	16.1	29.60
B_1	148	11,816	11.0	27.58
P_1	11	80	1.0*	16.48*

*There were many zero scores in this group. An adjustment was made by considering the distribution of $X_i^{1/2}$ to be of truncated normal form.

the raw scores for the genetically homogeneous groups, P_1 and P_2, vary more than 700-fold. When the data are transformed to square roots, the ratio falls to 1.69. Both these variances, as well as that of the F_1, are attributable wholly to environment. The transformation not only helps to meet the requirement for homogeneity of variances, but it allows one to estimate V_e by pooling the information from P_1, P_2, and F_1. We shall return to these data later and will only point out here that on the transformed scores the variances are considerably higher in the three groups in which segregation of P_1 and P_2 alleles was possible than in the three genetically homogeneous groups.

Judgment must be used in normalizing a distribution by means of a scale transformation, since nonnormality is one way of recognizing genetic effects. But if skewness is found in the distribution of test scores from a highly inbred, genetically homogeneous line, it is probably due to some feature of the test, and a formula for normalization derived from such data can be applied with caution to data from other populations.

Wright (1952) described four criteria for scaling and gave examples of their application. The best scale is one on which the effects of both genetic and environmental factors are as additive as possible. This desideratum may be sought by various methods.

1. A scale may be derived on which the variances of the purebreeding (homozygous, if possible) strains and their F_1 are as similar as possible. The assumption is that a scale which works well with environmental variance (the only one operative if the strains are truly homozygous) will also serve well for genetically produced variance.

2. A second type of transformation uses Laplace's principle that a variable compounded from the effects of many small factors acting independently should be normally distributed irrespective of the frequency distribution of each individual component. This involves transforming a scale in such a way that the relative rank of each individual is maintained while the new distribution follows the normal curve. Scott and Fuller (1965) em-

ployed stanine scores based on this principle in their study of behavioral differences among breeds of dogs.

3. Scales may be developed that allow two major factors to operate additively.

4. Scales may be based on the relationships of the means of the parental ($\overline{P}_1$, $\overline{P}_2$) and F_1 ($\overline{F}_1$) groups and so calculated that the means of the backcrosses ($\overline{B}_1$ and $\overline{B}_2$) and the $F_2 (\overline{F}_2)$ fit the following identities:

$$\overline{B}_1 = \frac{\overline{P}_1 - \overline{F}_1}{2}$$

$$\overline{B}_2 = \frac{\overline{P}_2 + \overline{F}_1}{2}$$

$$\overline{F}_2 = \frac{\overline{P}_1 + \overline{P}_2 + 2\,\overline{F}_1}{4}$$

There is no simple rule, and the various criteria may conflict with each other. Each situation must be evaluated separately and a solution found that is genetically and psychologically defensible. The important point is that experimenters be aware of the problem because faulty scaling can lead to erroneous conclusions.

HERITABILITY

The heritability of a phenotype is briefly defined as the proportion of its variance in a specified population attributable to additive genetic factors. The concept was developed by animal and plant breeders who were concerned with predicting the results of selection. It has been taken over frequently in behavior genetics, sometimes with a less limited sense than is implied by the preceding definition. For example, *heritability in the broad sense* is the ratio of the total genotypic variance (A + D + I) to the phenotypic variance. Sometimes the result of a cross between two pure lines is used to compute a heritability. Although such an estimate may be perfectly valid, it is extremely limited in generality, since the population on which it is based is atypical.

Analysis of variance model

The basic theorem of variance analysis is that a total variance can be expressed as the sum of the variances of a set of independent determinants. If the determinants are correlated (not independent) or if their effects are not additive, the relationships become more complex. The fundamental equation for a genetic analysis is:

$$P_i = G_i + E_i + f(G_i, E_i) \quad (18)$$

The expected or average phenotype of individual i is P_i; G_i is the mean phenotypic score of all individuals of the specified genotype; and E_i the mean score of all individuals exposed to i's particular environment. The interaction term is inserted to take care of the possibility that a particular combination of G_i and E_i will affect P_i in a way not predictable from their average effects. All terms are expressed as deviations from their means and are regarded as independent. We shall assume that the interactions are relatively small within the range of G and E, which permit a meaningful estimate of heritability, and that including $f(G_i, E_i)$ with E_i creates no serious error. When E is manipulated over a wide range of conditions in an experiment, this assumption is not justified. In fact, some of the more interesting and important research in behavior genetics is concerned with the interaction term and will be discussed later.

The corresponding variances are expressed as

$$V_p = V_g + V_e + r_{ge}V_g^{\frac{1}{2}} V_e^{\frac{1}{2}} \quad (19)$$

in which r_{ge} is the correlation between G and E. In an experiment with animals, r_{ge} can be reduced to a low and probably insignificant value by such techniques as cross-fostering, but in natural populations it may be substantial. Neglecting the correlation term, therefore, heritability in the broad sense, h_b^2 is:

$$h_b^2 = V_g/V_p \text{ or } V_g/(V_g + V_e) \quad (20)$$

Heritability, h^2, sometimes called *heritability in the narrow sense*, is defined similarly. Breaking down G into its major components, A and D, we write:

$$P_i = A_i + D_i + E_i \quad (21)$$
$$h^2 = V_a/V_p = V_a/(V_a + V_d + V_e) \quad (22)$$

The difference between the two heritabilities will be large if dominance deviations are substantial. Broad heritability, perhaps better called *degree of genetic determination*, is an indicator of the relative contribution of all genetic differences to phenotypic variation in a population. Narrow heritability, generally lower, is an indicator of the predictability of an individual's phenotype, given knowledge of the phenotypes of his close relatives.

It is obvious that changing V_a, V_d, or V_e will alter the value of h^2. In relation to this point it may be recalled that the dominance deviation is very sensitive to the nature of the measurement scale. Thus heritability is not an attribute of a trait with a fixed value that more and more refined methods will define with greater and greater precision. It is a characteristic of a particular population with respect to a specific phenotype, and we shall encounter many divergent estimates of h^2 for seemingly similar psychophenes. Because of this lack of constancy and because the term is often misinterpreted, it has been suggested that behavior geneticists avoid the concept, but there is as yet no satisfactory alternative for quantifying the extent of genetic influence on continuously distributed characters. Those who understand the assumptions inherent in the measurement process will not be misled.

It is instructive to enumerate some of the factors that can decrease heritability. Consider an inbred line of mice in which all loci are essentially homozygous and V_g is close to zero. Heritability likewise approaches zero, although all the genes are still functioning and regulating the development of somatophenes and psychophenes. Directional selection may reduce the frequency of some alleles so that the selected population becomes uniform with respect to loci affecting the selected trait. Again, heritability of that trait will be reduced. An increase in V_e occasioned by greater diversity of environmental conditions will likewise decrease h^2. The heritability of a trait may also be lowered by shifting a population from an "expressive" environment to a "suppressive" one (Henderson, 1970). An expressive environment is one that provides essential stimulation for the development of a particular trait.

The reverse of these circumstances will increase heritability, as will the increase of genetic variance by mutation or immigration. A parameter subjected to so many influences is certain to vary widely. Lerner (1954) has theorized, however, that natural selection favors a relative constancy of h^2, a phenomenon called *genetic homeostasis*. The argument is that genetic variability within a population allows some of its members to cope with changes in the environment and to pass on the genic basis for superior coping to their descendants.

Measuring heritability

The determination of the heritability of a psychophene may be compared with a psychological experiment in which genotypes are the treatments. Our task is to determine the relative contribution of this treatment to the total variance. In behavior genetics the technique of analysis of variance is employed widely. Most readers will be familiar with the use of this procedure as a test of the null hypothesis, which states that the differences between experimental groups (genotypes) do not exceed those to be expected from a set of random samples from the tested population. We shall go a little beyond this application and estimate variance components that have genetic meaning.

Since heritability is defined as a ratio of variances, V_a/V_p, it is clear that it can be estimated if these two quantities can be measured. V_p is simply the observed phenotypic variance of our sample; we have considered measurement of V_a in previous sections and shown that A is the portion of the phenotype predictable from the phenotype of an ancestor (or ancestors). Another way of expressing this is $V_a = COV_{ap}$, where the covariance is between the breeding value and the phenotypic value of individuals. Since $COV_{ap}/V_p = b_{ap}$ (b_{ap} is the regression of breeding value on phenotypic value), one can write:

$$h^2 = b_{ap} \qquad (23)$$

This form is appropriate for determining heritability from measurements on parents and their offspring.

The variance and covariance approaches both involve determining the phenotypic similarity of relatives and comparing it with the proportion of genes that they have in common.

Regression analysis

It may be helpful here to remind the student of the derivation and meaning of covariance and regression. Consider two variables, X and Y, which occur in pairs. The covariance of the two, COV_{xy}, is

$$COV_{xy} = \frac{\Sigma(X_i - \overline{X})(Y_i - \overline{Y})}{N - 2} \qquad (24)$$

where N is the number of pairs. The regression coefficient, b_{yx}, is:

$$b_{yx} = \frac{COV_{xy}}{V_x} \qquad (25)$$

In concrete terms, b_{xy} is the average deviation of Y_i from $\overline{Y}$ per unit deviation of X_i from $\overline{X}$. The expected value of Y in individual i is:

$$\hat{Y}_i = \overline{Y} + b_{yx}(X_i - \overline{X}) \qquad (26)$$

Suppose that Y is a measure of IQ made on sixth-grade children, and X is a measure of IQ on each of their fathers. We might obtain data of this nature: $\overline{X} = 100$, $\overline{Y} = 110$, and $b_{yx} = 0.25$. The expected IQ of a child whose father scored 120 is:

$$\hat{Y} = 110 + 0.25(120 - 100) = 115 \qquad (27)$$

Since the father contributes only one half the child's genes, the regression coefficient should be multiplied by 2 to estimate heritability of IQ ($2 \times 0.25 = 0.50$). The general principle is to multiply the phenotypic regression between relatives by the reciprocal of their coefficient of relationship. If scores of both parents are available, the regression of offspring on midparent, b_{om}, is a direct estimate of h^2.

In many species parents provide more than genes to their offspring, and the assumptions of random mating implicit in the basic model are often violated. Hence, caution must be used in the interpretation of heritability estimates made by this method. In mammals it is frequently found that mothers have a greater effect on their off-spring than do fathers. Scott and Fuller (1965), for example, found this to be true of a majority of the behavioral measures obtained from hybrids between breeds of dogs. The difficulties are not insurmountable, however, and we shall encounter various procedures such as cross-fostering in animals and retro-spective studies of adoptive children in which an attempt has been made to separate genetic and experiential transmission from parent to offspring.

Intraclass correlation

A schematic analysis of variance is shown in Table 5-2. The table is based on data from k treatments with n subjects in each. M_b and M_w represent the mean-square deviations between and within groups. In the familiar F test, the ratio, M_b/M_w, is used to test the null hypothesis. Inspection of the table will show that the higher the value of V_b, the vari-ance component attributable to group differ-ences, the larger the F ratio. Another way of expressing the relative importance of V_b is the ratio, $V_b/(V_b + V_w)$, the intraclass corre-lation coefficient. We shall designate this as r_i; it is often symbolized by t. A very similar statistic, ω^2 is more appropriately used with some experimental designs (Hays, 1963; Wahlsten, 1972). Both statistics denote the proportion of variance in the experimental data attributable to differences between groups. Convenient methods of computing r_i from an analysis of variance are included in

Table 5-2. If the group means are more un-like than random samples from a homoge-neous population, r_i is positive with an upper limit of 1.0 when members of each group are identical and all variation is found between groups. A nonsignificant F ratio in an analysis of variance means that r_i is not reliably dif-ferent from zero and that group differences may be accounted for by sampling variation. When the group means are more similar than predicted from random sampling theory, r_i is negative, although it can reach its limit of -1.0 only when $n = 2$. Such negative values suggest either sampling biases or interactions between group members, which accentuate divergence of individuals within a group. For example, pairs of male mice from an inbred strain often develop dominance-subordina-tion relationships when housed together. If pair-raised males are tested later for aggres-sion against a standard opponent, one would expect wide differences between the domi-nant and subordinate members of a pair but little variation among the pair averages.

In a genetic experiment genes are the treatments, and individuals who receive the same genes, that is, families or pure lines and their hybrids, are the treatment groups. Genes cannot be manipulated directly, but their distribution is controlled in a statistical sense by mating patterns. Full sibs, for ex-ample, have half their genes in common, just as parent and offspring. With full sibs there is no basis for designating one as X and the other as Y, and we use the intraclass correla-

Table 5-2. Simple analysis of variance for k groups with n subjects each

	Source of variation		
	DF	MS	Components of mean square
Between treatments	$k - 1$	M_b	$V_w + nV_b$
Within treatments	$k(n - 1)$	M_w	V_w

$$V_b = (M_b - M_w)/n$$
$$F = M_b/M_w$$
$$r_i = (M_b - M_w)/[M_b + (n - 1)M_w]$$

or

$$r_i = (F - 1)/[F + (n - 1)]$$

tion rather than a regression coefficient to estimate heritability. It might be thought that r_i for full sibs would be multiplied by two to obtain h^2, as was done for the parent offspring relationship, since the degree of relationship, 0.5, is the same. Actually, the cases are dissimilar. A family is not a random mating system, and sibs share one half the dominance deviation and also a common family environment. The intraclass correlation for full sibs is $(\frac{1}{2}V_a + \frac{1}{4}V_d + V_{ec})/V_p$, where V_{ec} represents the effects of a common environment. Doubling r_i for full sibs, therefore, overestimates h^2 to an unknown degree. In animal studies half-sib correlation, usually one sire mated to two females, is often used to estimate heritability, since it is free of this bias (Table 5-3). The special case of monozygotic twins will be considered in Chapter 12.

Table 5-3. Covariances, regression (b) or correlation (r_i), and heritability for relatives*

Relationship	Covariance	Heritability (h^2)
Offspring-one parent	$\frac{1}{2}V_a$	$2b$
Offspring-midparent	$\frac{1}{2}V_a$	b
Half sibs	$\frac{1}{4}V_a$	$4r_i$
Full sibs	$\frac{1}{2}V_a + \frac{1}{4}V_d + V_{ec}$	$<2r_i$
Monozygotic twins	$V_a + V_d + V_{ec}$	$<r_i$

*Modified from Falconer, D. S. 1960. Introduction to quantitative genetics. The Ronald Press Co., New York.

Classical cross

The classical Mendelian procedure of crossing purebreeding lines has been employed for quantitative as well as qualitative characters. By means of suitable analytical techniques, such experiments provide estimates of additive, dominance, and epistatic effects in polygenic systems. A concise summary of the procedures and their underlying assumptions has been made by Wright (1952).

Table 5-4. Biometric analysis of cross between two purebred lines*

| Population | Mean | | Variance† | |
	Symbol	Theoretical value	Symbol	Theoretical value
P_1	$\overline{P}_1$		V_{P1}	V_e
P_2	$\overline{P}_2$		V_{P2}	V_e
Midparent	$\overline{P}_M$	$\frac{1}{2}(\overline{P}_1 + \overline{P}_2)$	V_{PM}	$\frac{1}{2}V_e$
$F_1(P_1 \times P_2)$	$\overline{F}_1$		V_{F1}	V_e
$F_2(F_1 \times F_1)$	$\overline{F}_2$	$\frac{1}{2}(\overline{P}_M + \overline{F}_1)$	V_{F2}	$V_a + V_d + V_e$
$B_1(P_1 \times F_1)$	$\overline{B}_1$	$\frac{1}{2}(\overline{P}_1 + \overline{F}_1)$	V_{B1}	$\frac{1}{2}V_a + V_d + V_e + \Sigma\,AD$
$B_2(P_2 \times F_1)$	$\overline{B}_2$	$\frac{1}{2}(\overline{P}_2 + \overline{F}_1)$	V_{B2}	$\frac{1}{2}V_a + V_d + V_e - \Sigma\,AD$

*From Wright, S. 1952. The genetics of quantitative variability. In E. C. R. Reeve and C. H. Waddington, eds. Quantitative inheritance, Her Majesty's Stationery Office, London.

†*Notes:* Nonadditive interactions removed by means of an appropriate scale.

Variances V_a and V_d refer to F_2 even when used for backcrosses.

$\Sigma\,AD$ = nonlinear additive-genetic interaction.

V_e estimated as the mean of V_{P1}, V_{P2} and V_{F1}.

$V_a = 2V_{F2} - (V_{B1} + V_{B2})$ or $V_{F2} - V_{F1}$ (if V_d is relatively small).

$V_d = V_{F2} - V_a - V_e$.

Minimal number of segregating units = $\dfrac{R^2}{8V_a}$, where R = distance between $\overline{P}_1$ and $\overline{P}_2$.

$h^2 = \dfrac{V_a}{V_a + V_d + V_e}$.

Table 5-4 summarizes the theoretical means and variances for such a cross. We shall apply the model to Thompson and Fuller's (1957) data on exploratory activity in C57BR and A mice, which have already been presented in Table 5-1. The raw scores were unsuitable for the analysis because of the high positive correlation of the means and variances. A square-root transformation produced reasonable homogeneity of the variances, though there were still problems because of a high number of zero scores in strain A. Data from this strain have therefore been omitted from the following computations. These computations follow the notes for Table 5-4:

$$V_e = \tfrac{1}{2}(9.74 + 12.23) = 10.99$$
$$V_a = (59.20 - 57.73) = 1.47$$
$$V_d = 17.14$$
$$h^2 = 0.05$$

Using the differences between the F_2 amd F_1 variances to estimate V_g, one has:

$$V_g = 29.6 - 12.23 = 17.37$$
$$V_g/V_p = 0.59 \text{ (coefficient of genetic determination)}$$

These data are neither the best nor the worst for the biometric analysis of a psychophene. The reality of a genetic effect on exploratory activity is shown clearly by the orderly sequence of means; the higher the proportion of C57BR genes within a group, the more sections of the test maze were entered. It is also evident from the transformed scores that the variances of the segregating groups, F_2, B_1, and B_2 were significantly higher than those of the genetically homogeneous groups, P_1, P_2, and F_1. The two backcrosses were, however, as variable as the F_2, thus creating uncertainty as to the applicability of Wright's equations.

Often the variances of the segregating groups are no greater than that of the F_1. Tryon (1940a) after finding that the learning scores of the F_2 hybrids of a cross between maze-bright and maze-dull rats were distributed almost identically to those of the F_1, elected to discontinue further biometrical studies. The "Tryon effect" was also found for many behavioral characteristics in crosses between two dog breeds, basenjis and cocker spaniels (Scott and Fuller, 1965). One must conclude that the precision of this form of biometrical analysis is less than one would hope. Discrepancies could arise from several sources. In a polygenic system the hybrids (F_2 and backcrosses) will include a large number of qualitatively different genotypes, each of which may interact in a specific and unpredictable way with the rearing environment. Then, too, organisms are not simply molded passively by an environment. They can modify it, or they can modify their behavior. In general, heterozygotes seem more capable of adaptive reactions than do homozygotes. If such adaptation involves suppression of the more extreme psychophenes, behavioral uniformity could be favored by genetic heterogeneity.

Diallel cross

The diallel method of biometric analysis was adapted to behavior genetics by Broadhurst (1960b) and has been employed frequently since that time. A complete diallel design consists of a set of all possible intercrosses between n purebreeding lines and yields n^2 genotypes. Of these, n will be continuations of the original parental strains; there will also be $(n^2 - n)/2$ different hybrids, each represented by its two reciprocal forms. Partial diallels can also be carried out, though they yield less genetic information. The theory of this form of biometric analysis is highly developed, and original sources may be consulted for procedures (e.g., Hayman, 1954, 1958; Griffing, 1956; Broadhurst, 1967; Mather and Jinks, 1971).

An example of the method based on Fulker's (1966) observations on male mating speed in six lines of the fruit fly, *Drosophila melanogaster*, is shown in Table 5-5. The table shows the average number of each type of female inseminated during a standard testing period. The experiment was replicated to obtain an estimate of environmental variance. In such a table the rows and columns corresponding to a given strain can be averaged to obtain a measure of its general combining action, which corresponds in general

Table 5-5. Replicated diallel cross of mating speed in male *Drosophila melanogaster**†

Paternal line	Maternal lines						Statistics	
	6CL	Ed	Or	W	S	F	V_r‡	COV_r§
6CL	*1.4*	3.6	2.2	3.2	2.6	3.0	0.759	0.547
	1.2	2.6	2.6	3.8	3.4	3.2		
Ed	4.0	*3.0*	3.7	3.4	3.2	3.2	0.172	0.001
	3.2	*3.8*	4.6	4.0	2.8	4.2		
Or	2.3	3.4	*1.8*	3.4	2.4	2.8	1.120	0.903
	1.6	4.6	*0.8*	4.0	1.6	3.8		
W	3.2	4.4	3.8	*3.0*	2.4	3.6	0.218	0.020
	3.4	3.0	3.2	*2.2*	3.6	4.2		
S	2.4	3.6	2.0	2.4	*1.2*	2.4	0.707	0.585
	3.2	4.0	2.2	4.6	*1.2*	3.8		
F	3.3	4.0	3.2	4.6	2.0	*2.8*	0.524	0.382
	3.8	4.2	2.8	3.4	3.6	*1.8*		

*From Fulker, D. W. 1966. Mating speed in male *Drosophila melanogaster:* a psychogenetic analysis. Science **153:** 203-205. Copyright 1966 by the American Association for the Advancement of Science.
† Mating speed was measured by the number of females fertilized during 12 hours. Maximum possible score = 6. Scores of inbreds on the main diagonal are in italics.
‡ V_r, Variance of the row (and its corresponding column) containing data from all groups with a common parent strain.
§ COV_r, Covariance of the row (and its corresponding column) with scores of the variable parents.

to the additive genetic effect. The deviations of particular cells from the expectations based on the average performance of the hybrids of the two parents is a measure of specific combining ability attributable to dominance and genic interaction. The differences between the reciprocal hybrids detect maternal and other effects.

The diallel analyses of Hayman (1954) and Jinks (1954) yield a larger number of parameters: D, the additive genetic variation; H_1 and H_2, which are indicators of dominance; F, an indicator of the relative proportions of dominant and recessive alleles; and E, the environmental variation. Fulker's analysis of his data showed, in addition to E, only highly significant additive effects (D) with strong directional dominance (H) for higher mating speed. In this situation a variance-covariance diagram based on the results of the diallel is instructive. The variance of each of the six arrays of scores from all crosses with a recurrent parent is calculated, averaging over reciprocals and replicates. The covariances of each array with the nonrecurrent parent scores (the pure lines lying on the main di-

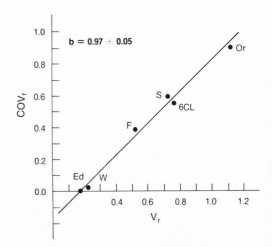

Fig. 5-7. Variance-covariance diagram for mating speed in male *Drosophila melanogaster*, showing regression of COV_r and V_r for number of females fertilized for a replicated diallel cross of six inbred lines. (From Fulker, D. W. 1966. Science **153:** 203-205. Copyright 1966 by the American Association for the Advancement of Science.)

agonal in Table 5-5) are also calculated. If the recurrent parent contributes a large number of dominant genes, its hybrids will all tend to be like that parent; hence the variance of the array will be small. The covariance with the nonrecurrent parents will also be small. The opposite holds for the offspring of the lines that contribute predominately recessive alleles. When the data are plotted as in Fig. 5-7, the strains fall on a straight line of unit slope. Strains *Ed* and *W* carry mostly dominant alleles; *Or* contributes mainly recessives. The others are intermediate.

As with other methods of biometric analysis, a diallel cross permits computation of broad and narrow heritabilities that can be used to predict the results of selection and, with certain assumptions, to deduce the probable evolutionary history of a behavioral trait.

Realized heritability

One of the reasons for estimating heritability is the prediction of the efficacy of selection. Unless V_a is substantial relative to V_d and V_e, progress under directional selection is expected to be low. Realized heritability can be determined from three values, the mean score for the original population, $\overline{X}_0$; the mean score of the individuals selected for breeding, $\overline{X}_s$; and the mean score of the offspring in the following generation, $\overline{X}_1$ (Fig. 5-8). The selection differential, $\overline{X}_s - \overline{X}_0$, is the gain that would be achieved if the phenotype were determined only by the genotype.

The realized gain is $\overline{X}_1 - \overline{X}_0$. Realized heritability is:

$$h_r^2 = \frac{(\overline{X}_1 - \overline{X}_0)}{(\overline{X}_s - \overline{X}_0)} \tag{28}$$

In practice there are several procedures to be followed. An unselected control population must be maintained under similar conditions to distinguish between the effects of the selection and fortuitous environmental variation. Frequently another line selected in the opposite direction is established to show whether the upward and downward changes are symmetrical. If one of the objectives is to look for other characters correlated with the criterion trait, it is important to have multiple selected lines in order to test the reliability of the correlations. Finally, the selective process should be carried out over several generations so that chance fluctuations can be averaged out. One would not expect h_r^2 to remain stable over long periods, since selection, if it is successful, will change the genotypes of the population. As the population becomes more homogeneous, heritability must decline.

An example of successful selection for a psychophene is Siegel's (1965) experiment on mating ability in male domestic fowl. The subjects of this study were tested with a panel of random-bred females and scored for the number of completed matings. The progress of the selection over six generations is depicted in Fig. 5-9. Although progress was irregular, the separation between the high line and the low line was well established by the fourth generation. The realized heritabilities for each generation are shown at the bottom of the figure. Although the progress of selection appears to be symmetrical with respect to direction, the realized heritability was 0.18 ± 0.05 for high success matings and 0.31 ± 0.11 for low success. Equality of rate of progress in both directions is explained by the reduced intensity of selection in the low line. The rate of progress of selection is related to the product of heritability by the size of the selection differential; in this example, the two varied inversely so as to produce near equality of results.

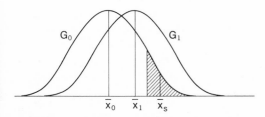

Fig. 5-8. Realized heritability. $\overline{X}_0$, mean of original population (G_0); $\overline{X}_S$, mean of selected group (shaded portion of G_0); $\overline{X}_1$, mean of the offspring of the selected group (G_1).

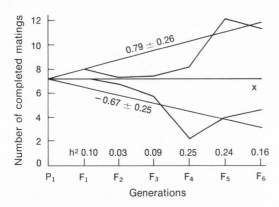

Fig. 5-9. Bidirectional selection for number of completed matings. The *x* denotes the mean of reciprocal crosses between lines in the F_6 generation. Linear regression lines were fitted to the generation means in both directions. (From Siegel, P. B. 1965. Genetics **52**:1269-1277.)

CORRELATED CHARACTERS

Selecting for one phenotypic trait often brings about changes in other characters not entering into the selection criterion. Castle (1941), MacArthur (1949), and Lewis and Warwick (1953) have described behavioral changes associated with selection for physical characters. Many psychologists have selected for one behavioral character and found accompanying changes of quite a different sort. In interpreting these results, consideration must be given to the genetic significance of phenotypic correlations (Thompson, 1957b). Four sources of correlation may be recognized.

1. The correlated traits may derive from a common functional dependence on a particular gene *(gene communality)*. In other words, the gene has pleiotropic effects. The relationship may be lineal:

Gene → Selected trait → Correlated trait

or collateral:

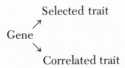

In the lineal relationship any environmental factor that suppresses the selected trait will also affect the correlated trait. This is not necessarily true when the relationship is collateral.

2. Genetic linkage *(chromosomal communality)* results in phenotypic correlations within families. Indeed, phenotypic correlations are the means by which linkages are detected. In large random-breeding populations, chromosomes of types *AB, Ab, aB,* and *ab* occur in numbers proportional to the products of the gene frequencies, and no correlation is found. In more limited populations, as in most behavior genetics experiments, the possibility of linkage must always be considered.

3. Phenotypic correlations may be generated whenever the mating system is nonrandom. Such correlated characters may be said to have *gametic communality*. The mechanisms that bring this about are as varied as the processes of assortative mating and selection. Many examples are found in domestic animals in which a particular color is selected as a kind of identification tag for the breed, along with a variety of functionally unrelated traits. In human populations, a correlation will be built up between two heritable traits if individuals high in one tend to choose mates high in the other without regard to standing on the first trait (Price, 1936; Bartlett, 1937). This phenomenon, cross-homogamy, probably characterizes some human societies.

Under special conditions phenotypic correlations arise from certain physiological properties of genetic systems. Most populations are genetically variable; hence attempts to change the population mean by selection are usually successful. But it is a mistake to consider the original population as really unselected or to believe that selection can actually deal with one criterion only. More vigorous and adaptable individuals leave on the average more progeny, and their genes increase in proportion. Over the generations, combinations of genes have been built up that produce the maximum average fitness of the species. This does not lead to genetic uniformity because natural conditions are so variable that the fitness value of a particular genotype fluctuates from generation to generation. The great bulk of the population is intermediate genotypically and phenotypically with respect to the possible range of variation. The most common genotypes are balanced for maximum average fitness, while the extreme genotypes are a safety factor in the event of environmental changes. If selection for an extreme phenotype upsets the genic balance, a correlation may be generated between the selected character and such attributes as low fertility (Lerner, 1958). Examples cited to illustrate disturbed genic balance usually refer to selection for traits of economic or esthetic significance to the breeder. Reduced fertility has sometimes been observed in stocks selected for behavioral characters, and this may represent a similar phenomenon.

4. A common response of two traits to environmental variation is the final source of phenotypic intercorrelations. A vitamin deficiency would affect both sensory functions and general activity; good home environments favor both physical growth and intelligence. If environmental factors were always recognized, these correlations would not be confused with genetic correlations. Actual experiments must be carefully scrutinized to ensure against misinterpretations.

The stability of the three types of genetic correlations varies. Gene communalities are generally very stable in a constant environment under any mating system. Chromosomal communalities are unstable to an extent dependent on crossover frequency. In a small-scale experiment lasting only a few generations, it may be impossible to distinguish these two forms of communality. Gametic communalities are disrupted by a single generation of random mating and are thus readily identifiable in laboratory situations. Since human beings do not mate randomly with respect to behavioral traits, gametic communalities in this interesting species are less easily recognized.

The study of environmental communalities is, of course, the main task of all behavioral science except behavior genetics. For this one area it is desirable to minimize environ-

mental variability while maximizing genetic variability. This procedure reduces the significance of environmental communalities. Phenotypic correlations that persist in a constant environment are probably genetic. Conversely, correlations in a highly inbred strain are safely inferred to be of environmental origin.

Phenotypic correlations are often used to analyze the organization of behavior. Animals selected for timidity, maze learning, activity, and the like are given various physiological and psychological tests. Positive correlations with the selected trait have been assumed to indicate a functional relationship dependent on gene communality or pleiotropy. It should be obvious from the preceding discussion that correlations within small selected populations may be caused by any one type of communality or by a combination of them. To use correlations to prove a functional relationship based on common genes, one must study randomly bred populations raised under uniform conditions or demonstrate the particular association of characters in a number of independently selected lines.

Genetic correlations

As used in quantitative genetics, the term *genetic correlation* (r_a) applies only to phenotypic associations based on gene communality. Consider, for example, that selection for criterion trait X is accompanied by a change in trait Y. Is this an indication that the genotypic change responsible for the change in X is also responsible for the change in Y? In other words, is there a genetic correlation, and how large is it?

The derivations of the formulae for genetic correlations are well summarized by Falconer (1960). Any phenotypic correlation (r_p) can be considered as the sum of two terms

$$r_p = h_x h_y r_a + e_x e_y r_e \qquad (29)$$

where

h_x and h_y = square roots of heritabilities of X and Y

e_x and e_y = square roots of environmentalities of X and Y, $e^2 = 1 - h^2$

r_e = environmental correlation

From this fundamental theorem it is possible to devise methods for calculation of r_a that are basically similar to those used for heritabilities but make use of the covariance between trait X in one individual and trait Y in another (crossed covariance). Thus for parent (p) and offspring (o):

$$r_a = \frac{COV_{x_p y_o}}{\sqrt{COV_{x_p x_o} COV_{y_p y_o}}} \quad (30)$$

Genetic correlations can vary widely from phenotypic correlations, as is clearly shown in Table 5-6, based on Siegel's (1965) selection for mating ability in cockerels. There is no consistent ordering of the two sets of values. The table contains some expected results: one would predict that genes increasing mating success would operate through strengthening components of breeding behavior such as mounts and treads. But the meaning of the high negative correlation with sperm concentration is less obvious.

THRESHOLDS AND POLYGENIC SYSTEMS

Sometimes the nature of measurement in science produces dichotomous classifications when the underlying phenomenon of genetic interest is a continuous variable. For ex-

Table 5-6. Phenotypic and genetic correlations of unselected characteristics with mating ability, the selected trait, over six generations*

Unselected characteristic	Phenotypic correlation	Genotypic correlation
4-week weight	0.07	0.31
8-week weight	−0.01	0.07
20-week weight	−0.01	0.17
Breast angle	−0.01	−0.12
Aggressiveness	0.09	0.08
Semen volume	−0.04	−0.38
Sperm motility	0.01	0.23
Sperm concentration	0.01	−0.67
Number of courts	0.60	0.36
Number of mounts	0.94	0.71
Number of treads	0.97	0.82

*From Siegel, P. B. 1965. Genetics of behavior: selection for mating ability in chickens. Genetics **52**:1269-1277.

ample, the differential resistance of two strains of mice to a toxic drug could be measured by comparing the percentage of mice killed at several dose levels. Actually, there is little difference in physiological efficiency between an animal that just survives and one which dies after a long period of illness. Nevertheless, for quantitative measurement we prefer the objectivity of the life-death dichotomy to subjective appraisal of the degree of illness produced in the survivors or estimation of the rate of dying in those who succumb.

Dichotomous classifications also occur in behavior genetics. An investigator may be interested in hereditary differences in an ability that can be tested only in trained subjects. A proportion of individuals may fail to meet the criterion of training that is necessary before a quantitative test can be given. Experimental psychologists in similar situations discard the rats that do not run the maze, the dogs that do not learn delayed response. The behavior geneticist must use all the subjects in every genetic subgroup, provided they are physically normal, if he is to secure a true estimate of the attributes of the population. Thus a pass-fail classification may be employed, even when it is obvious that the attribute of the animal being measured is distributed continuously.

An excellent explanation of the threshold hypothesis is found in Wright's (1934a) study of polydactyly in guinea pigs. Fuller, Easler, and Smith (1950) applied Wright's concepts to the differences in audiogenic seizure susceptibility found in various inbred strains of mice. Although the genotype in each strain is fixed so that it can be considered the same in all individuals, susceptibility to convulsions is predictable only in a statistical sense. The various strains do not "breed true" for susceptibility. When tested under standard conditions, about 99% of DBA/2, 80% of DBA/1, 35% of A, and 0.5% of C57BL were classified as convulsers. There is nothing absolute about these percentages. Changing the age of testing, the nature of the sound stimulus, or the method of handling can shift the values up or down. Under controlled conditions, however, each genotype is characterized by a constant percentage of individuals who surpass the threshold for convulsions.

The threshold hypothesis can be expressed in a somewhat more formal fashion. We shall denote the unknown physiological basis for susceptibility by the symbol X and assume that X varies continuously over a wide range of values. When $X < X_t$, a nonconvulser phenotype is produced; when $X > X_t$, a convulser phenotype results. The value of X for a particular individual i of genotype g is

$$X_i = X_g + X_{e_i} \qquad (31)$$

where

X_g = average value of X for all individuals of genotype g

X_{e_i} = sum of environmental influences on character X in i

The influences may be positive or negative, and the mean of X_{e_i} within strain g is zero.

The requirement for a resistant phenotype is expressed as

$$X_g + X_{e_i} < X_t$$

and for a susceptible phenotype as

$$X_g + X_{e_i} > X_t$$

When the value of X_g is close to a threshold, fluctuations in X_{e_i} are extremely important in determining whether X_i will be above or below the threshold. When the genotypic mean is far from a threshold, environmental factors have less influence on the observed phenotype.

Quantitative estimates of X_g are possible if certain assumptions are made. We assume that X_{e_i} is normally distributed within a genotype with a mean of zero and standard deviation of 1.0. Furthermore, we shall consider that genetic and environmental effects on X are additive over the limited range of X that is near a threshold. Referring back to equation 31, we can predict that when X_{e_i} is normally distributed, 68.2% of individuals will lie between $X_g + 1$ and $X_g - 1$. Similarly, 95.45% will be in the range of $X_g \pm 2$, and 4.6% will be beyond this range. If X_g is lo-

cated 2.0 units below the threshold, X_t, it is apparent that one half of the 4.6% will be over the threshold and show a distinct phenotype. In practice this reasoning is reversed, and values of X_g are computed from the observed percentages of the two phenotypes by means of the inverse probability transformation (Wright, 1920, 1952).

Some of the features of polygenic-threshold systems are illustrated in Fig. 5-10 (Fuller, Easler, and Smith, 1950). Animals resistant to audiogenic convulsions in a standard test are considered to be above the threshold. The three curves in each section of the figure represent the distribution of **X** in a highly susceptible strain, DBA, a resistant strain, C57BL, and their F_1 hybrid. These

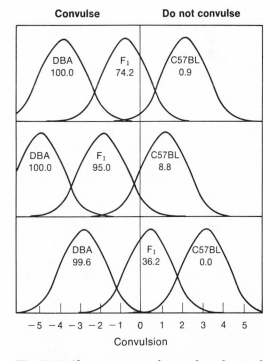

Fig. 5-10. Changes in convulsive risk on first trial (numerals under curves indicate percent) associated with conditions shifting physiological susceptibility by one standard deviation. The abscissa is a scale of physiological susceptibility. Each genotype is assumed to vary normally about some point on this scale. The convulsive risk is dependent on the proportion of the curve to the left of an arbitrary threshold. (From Fuller, J. L., C. Easler, and M. E. Smith. 1950. *Genetics* **35**:622-632.)

curves have been drawn assuming additive genetic effects so that the hybrid is exactly intermediate to the parent strains. The top set of curves represents the results in the main experiment. The effects of changing the testing conditions so that the threshold is raised by one unit are shown in the middle section. Reducing the threshold by one unit produces the results shown in the bottom section of the figure. A significant feature of the system is that a given amount of change on the X scale has quite different effects on the phenotypic ratios of genotypes lying near or far from X_t. Changes in the proportion of convulsions in the DBAs are insignificant, but they are large in the hybrids. In general, threshold characters are most sensitive both to environmental and genetic effects when they are near threshold, that is, their frequency is near 50%. This produces a complication in selecting for or against such characters, since their heritability decreases as the selected populations move away from a midpoint (Dempster and Lerner, 1950).

One further point should be emphasized. Crosses between strains that are high and low with respect to a behavioral trait may yield ratios in the F_1 and F_2 and backcrosses that closely approximate Mendelian ratios for a single factor showing dominance (Witt and Hall, 1949). This similarity may lead to the adoption of a hypothesis of single-factor determination. However, these results are much like those which will be obtained in a polygenic threshold system if one strain is more distant from the threshold than the other. The predictions are not exactly the same, but the precision of measurement is usually not sufficient to decide between the two hypotheses in a small-scale experiment. One property of the polygenic threshold system is that the backcrosses tend to be somewhat closer to the parental types than to the F_1, so that an appearance of dominance in opposite directions may be found. Repeated backcrossing to the strain that appears to carry the recessive factors is one method of arriving at a decision between the polygenic and single-gene hypotheses and should always be employed before one or the other is

adopted. When such procedures are impossible, Falconer's (1965) method for estimating liability to a disease based on its incidence in relatives may be useful.

CONCLUDING REMARKS

Although we have reviewed a considerable number of biometric procedures that have been employed in behavior genetics, much has been excluded. Some topics, twin methods, for example, will be considered in more detail later. Nothing has been said concerning the errors of estimate for heritabilities and genetic correlations. Individuals contemplating such measurements can find this information in the cited references and in texts on quantitative genetics. For the more complex analyses, the computational burden is heavy and can be greatly reduced by access to a computer with the appropriate programs.

Despite the brevity of this treatment, some readers may wonder about the purpose of biometrical analysis. Certainly there are advantages in working with phenotypes associated with segregation at a single locus; here it is possible, at least theoretically, to follow the path from a DNA molecule through various complexities of somatophenes to a related psychophene. Estimates of V_a, V_d, and V_e do not directly lead to this kind of reductionist approach. Instead they are chiefly useful for guiding artificial selection of animals and forecasting or reconstructing the genetic history of natural populations including human ones.

Within the field of genetics, quantitative genetics, although based on Mendelian principles, has developed independently of other branches to a considerable degree (Caspari, 1967, 1972). Similarly, in behavior genetics there are workers who apply biometric techniques and those who prefer to study the effects of individual genes on behavior. Actually the two methods address themselves to different questions, and each is superior in its own domain. The two approaches have been called the phenotypic and the genotypic (Scott and Fuller, 1963). When one starts with a psychophene, it is almost axiomatic that a variety of genes will influence it and that a biometric analysis will be appropriate. Caspari (1972) refers to Hirsch and Boudreau's (1958) investigation of phototaxis in *Drosophila melanogaster*, which demonstrated that each of the three major chromosomes of this species carries genes modifying the behavior. The genotypic approach is illustrated by Benzer's (1967, 1973) isolation of a large number of genes affecting phototactic behavior, and his use of these as a "genetic dissection" of the nervous system. And finally, it seems clear that the phenotypes employed in these two investigations, though both labeled as phototaxis, were really very different from each other (Rockwell and Seiger, 1973). We shall return to these experiments in Chapter 7 as case studies in experimental behavior genetics.

II

Experimental behavior genetics

Experimental behavior genetics involves observations on animals whose ancestry and rearing have been regulated by an investigator. Potentially, any species and any behavior of that species could be the subject of research, but in practice the majority of investigators have restricted their attention to the traditional material of animal geneticists, various species of *Drosophila*, laboratory strains of mice and rats, and the smaller domestic animals. In part this is a matter of convenience and in part a recognition of the fact that prior knowledge of the genetics and behavioral repertoire of a species is important for the design of fruitful experiments.

In Chapter 6 we consider the general objectives and materials of experimental behavior genetics. Chapters 7 and 8 deal with the genetic analysis of a number of simple behavior patterns, primarily in drosophilae or in mice. Selected case histories illustrate the strategy and general nature of the results of such studies. In Chapter 9 we review studies in learning and memory as related to genetics. Here we are concerned in particular with the generality of psychological principles and the possibility that behavior-genetic analysis can be a means of dissecting the learning process into components that have biological significance. Chapter 10 summarizes research on interactions between genotypes and environment, particularly environment during the period of most active psychological development. Here again we encounter the problem of generalization of principles deduced from observations on a limited genetic sampling to all members of a species.

In Chapter 11 attention is given to social behavior that involves interactions between two or more individuals. A pair or a group is, of course, the basic unit for phenotypic description, but genotypes are individual characteristics. Behavior-genetic analysis of social behavior, therefore, encounters complexities not found when the individual is both the phenotypic and the genotypic unit. Social behavior is of particular importance in evolution, since it has implications for assortative mating and, in birds and mammals, for the differential survival of offspring. To the extent that sexual and care-giving behaviors are genotype dependent, behavior will play a role in natural selection.

When appropriate we call attention to related human studies that are treated in more detail in Part III. In some instances the parallelisms are clear, but, despite the essential identity of genetic mechanisms in human beings and other sexually reproducing species, there are important psychological differences between humans, mice, and fruit flies. Nevertheless, experimental behavior genetics has developed to the point that it can illuminate problems that cannot be studied experimentally in human beings. And it can be a source of hypotheses of wide generality applicable to humans as well as to other species.

6

Experimental behavior genetics: goals and methods

In Chapter 1 we contrasted the genotypic and phenotypic approaches to behavior genetics research. Now with a background in genetic principles we can consider these strategies in more detail. The subject matter for both is the same, but starting from the right or from the left side of the following diagram has a number of consequences.

Genotype $\longleftrightarrow$ Somatophene $\longleftrightarrow$ Psychophene

If we enter from the left, the natural unit of behavior is the variation associated with a gene substitution (Ginsburg, 1954). For example, the phenylketonuria gene primarily affects amino acid metabolism and produces a variety of behavioral symptoms, mental retardation, and, frequently, convulsions, motor disabilities, and temper tantrums. Since these all depend on the PKU gene, they form a natural unit or syndrome in the genetic sense. But no psychologist would attempt to classify the forms of human behavior by making the basic units correspond to the inherited forms of mental retardation.

Admittedly the concept of "natural units" is easier to apply when one starts with genes than when one starts with psychophenes. There is no standard system of classifying behavior, although Nissen's (1958) "axes of behavioral comparison" are useful. His *functional* axis compares behaviors with respect to their biological utility (e.g., food getting and mating). The *descriptive* axis is concerned with the mechanics of behavior, such as locomotor functions and sensory capacities. The *explanatory* axis includes several levels of analysis, phylogenetic history, physiological and biochemical correlations, and explanations in terms of purely behavioral

observations and concepts. In this book we have classified behavior in functional terms and attempted to explain them by the application of genetic techniques.

Even though genes have major effects on behavior, it should be apparent that natural units derived from genes and those derived from psychophenes will not be related in any one-to-one fashion. We call this the *principle of noncongruence*.

THREE QUESTIONS OF EXPERIMENTAL BEHAVIOR GENETICS

Once an appropriate psychophene has been defined (this is crucially important), there are three main kinds of genetic questions to be asked: Is the psychophene transmitted genetically? If so, how do the genes produce their behavioral effects? Finally, how are these genes distributed in time and space?

Sometimes the transmission problem has already been solved. Numerous identified mutations in all well-studied species have obvious or suspected pleiotropic effects on behavior. In such instances, an experimenter can go directly to the second two questions. More frequently, presumed behavioral differences between strains or natural populations are identified, but their mode of transmission is not clear. In such instances there is generally no somatophenic marker for the psychophene, at least none that is obvious at the start. Methods for the identification of such cryptic loci and fitting the data to a Mendelian model are discussed in a later section. Relating psychophenic variation to segregation at a single locus is an advantage,

since it opens the possibility of explaining behavioral variation in terms of defined biochemical reactions. Unfortunately, it is often
not possible to assign psychophenic variation
to segregation at one or two loci. Given the
noncongruence model we espouse, this is the
expected outcome. However, the biometric
methods described in Chapter 5 are available
to partition the observed variance into environmental and genetic components. Although such analyses are less useful in leading back to the physiological bases of gene action on psychophenes, they do permit deductions regarding the adaptive value of behavioral characteristics and predictions concerning the effectiveness of selection on them.
Furthermore, a genetic correlation between
two psychophenes is presumptive evidence
that their variability is controlled by the same
group of genes.

There are two main methods by which experimenters have sought to discover the
pathways between genetic variation and behavior. One method is to search for correlations between somatophenes and psychophenes. The possible genetic interpretations
of such correlations were discussed in Chapter 5 (pp. 68 to 70). Correlations attributable to pleiotropic effects of genes may suggest not only functional relationships between somatophenes and psychophenes but
common developmental processes for correlated behavioral measures. By the same reasoning, similar psychophenes showing different patterns of correlation with genotypes
probably have unlike biological bases.

The method of retrograde analysis may be
useful when behavioral phenotypes are
sharply divergent from the normal range.
The investigator looks at earlier and earlier
stages of development, possibly going back
into the fetal period, searching for the first
and presumably causal somatophenic or psychophenic differentiation between normal
and aberrant individuals.

Problems of gene survival and distribution
in space and time are also the province of experimental population behavior genetics.
Laboratory studies of this type have been
performed primarily with insects, where

large populations can be maintained at reasonable cost. Field studies of natural populations of insects and rodents, though not as
strictly controlled as is possible in the laboratory, have demonstrated that the behavioral
effects of genes are often critical factors in
evolution. The distribution of genes in a population can also be a means of deducing social structure in natural populations that is in-
accessible to direct observation (Chapter 11).

The differences in objectives between
physiological and population behavior genetics imply differences in the kinds of genes
that are of interest. For the physiological approach, a rare mutant that must be maintained by special procedures can be excellent
material for investigating such behavior as
food intake or locomotion. The fact that the
mutant could not survive in nature is imma-
terial, as long as it is a preparation with
unique properties. The population geneticist
is more interested in the fine tuning of behavior that is mediated by systems less amenable to classic Mendelian analysis.

The three questions of experimental behavior genetics we have discussed are primarily genetical in content. On the psychological side there is a fourth question: How
universal or, conversely, how genotype
specific are psychological principles? A tremendous amount of psychological theory is
based on experiments with white rats and
college students. Sometimes the white rats
are of different strains or the students from
different backgrounds; then the question
arises as to the extent that principles derived from experiments on one group are
generalizable to another.

MATERIALS OF EXPERIMENTAL
BEHAVIOR GENETICS

The practice of experimental behavior
genetics, including the choice of experimental design, is shaped by the availability of suitable genetic stocks. The most
used types are inbred strains and their hybrids, heterogeneous stocks, selected lines,
and mutations on various backgrounds. In
Drosophila, stocks with chromosome inversions or transpositions have proved useful

(Hirsch, 1967). In this chapter we discuss the advantages and limitations of these stocks in general terms with illustrations from some classical experiments.

Inbred strains

The genetic homogeneity achieved by continued close inbreeding provides one of the most useful techniques for behavior genetics. Possibly natural selection for viability permits the maintenance of some heterozygosity, but it is certainly small and can usually be neglected. The investigator who employs these strains, however, must not assume that removing genetic variance necessarily reduces phenotypic variance. Evidence exists that homozygous individuals are less well buffered against minor environmental agents, and inbred animals are sometimes no more uniform in response than random-bred subjects (McLaren and Michie, 1956). The use of F_1 hybrids between inbred strains retains the advantages of genetic uniformity while adding the advantages of superior developmental and physiological homeostasis. Most of the evidence in support of the hypothesis of greater heterozygote stability is derived from physiological and morphological studies (Lerner, 1954; Yoon, 1955).

There are complications in applying the concept of developmental homeostasis to behavioral characteristics (Mordkoff and Fuller, 1959; Hyde, 1973). Increased behavioral variability could actually facilitate physiological homeostasis, which is really the important consideration. Another complication arises in threshold systems. If the hybrid between two inbred strains happens to fall in a critical range, small environmental differences can produce large phenotypic effects. For example, almost all C57BL/6 mice are resistant to audiogenic seizures on first trial at 30 days of age, and almost all DBA/2 mice are susceptible. Their F_1 hybrids are less predictable; separate groups had risks of 74%, 95%, and 36% (Fuller, Easler, and Smith, 1950). Here the heterozygotes are more variable both within and between samples. In general, F_1 hybrids are more

stable than inbreds in experiments in which environmental variation can be considered as chance fluctuations or "error" (Hyde, 1973). There is evidence, however, that F_1 hybrids are more *responsive* to major modification of their environment (Henderson, 1972) and that heterosis is shown most clearly in homeostatic behaviors (Barnett and Scott, 1964). For experiments designed to detect genetic-environment interactions, F_1 hybrid mice may be particularly suitable.

Once made homozygous, inbred lines retain their genetic characteristics for long periods of time. Over many generations mutations will occur, and the phenotype may change. Experience suggests that the drift will be slow during one investigator's research career. Thompson (1953b) compared the open field activity of a number of inbred strains from the Jackson Laboratory. Very large differences were found among the scores of the fourteen inbred strains and one mutant strain. Details of the results are summarized in Chapter 10 (p. 161 and Tables 10-4 and 10-5). Here it is interesting to compare the ordering of the strains with Heston's (1949) chart showing their origin. The three C57 strains descended from Lathrop's colony and occupied the three top ranks. The two DBA strains, although separated for many years, did not differ significantly. BALB/c and A have common ancestry, and both scored near the bottom. C3H, derived from a cross between the ancestral DBA and BALB/c lines, was intermediate between them in activity. Strain TC3H is descended from C3H through an ovum transferred to another strain to free the line from the mammary-tumor milk virus. The difference in exploratory behavior between these strains, though large, did not reach the 0.05 level of significance. Thompson's data indicate that relative behavioral performance in mice is correlated with genetic events occurring over a hundred generations before his subjects were born.

Other comparisons might show less constancy over generations. Certainly the prudent investigator using inbred lines as standards should take steps to avoid subline differentiation, which is likely when stocks are separated over a period of generations. Comparisons with the results of other workers will be facilitated if breeding stocks are regularly replaced from a mammalian genetics center. Between replenishments a controlled mating system as shown in Fig. 6-1 can be used to prevent diversification within a single colony.

Inbred lines are valuable for experiments on the inheritance of quantitative psychophenes. Their value in reducing phenotypic variance in other types of study differs from case to case, and their advantages, if any, may not be worth the cost in reduced vigor. Employing a number of such strains for comparative studies can be readily justified. Each strain may be considered as an individual that can be duplicated, and the replicates exposed to a number of experimental treatments without concern for interactions between one procedure and another. Comparing a number of strains in a variety of environments often provides a striking demonstration of heredity-environment interaction (Chapter 11) and tests the generality of principles derived from a single strain.

Inbreeding is not restricted to *inbred strains*. It cannot be avoided in a small closed colony, since the choice of mates is limited, and eventually all animals become related to one another. The rate of inbreeding increase per generation in a closed population mated at random is approximately

$$\Delta F = \frac{1}{8N_m} + \frac{1}{8N_f}$$

where N_m and N_f are the numbers of breeding males and females respectively. With a systematic mating system, taking equal numbers of offspring from each mating and pairing them with the least related available mate, the rate of inbreeding can be reduced to approximately one half the value given by the preceding formula (Falconer, 1960).

Heterogenic stocks

In contrast with inbred lines, heterogenic stocks are bred deliberately for high genetic variability. They are particularly useful as the

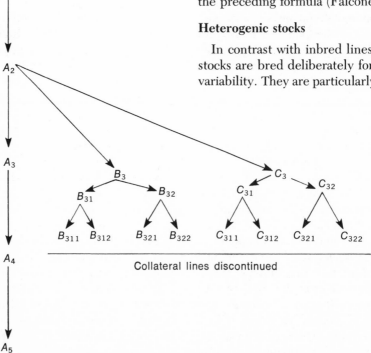

Fig. 6-1. A mating system for routine maintenance of an inbred line. A_1, A_2, B_2, etc. are brother-sister pairs. The strain is maintained by a single direct line, $A_1 \rightarrow A_4$. At A_2 collateral pairs are taken, bred for two generations to produce experimental subjects, and then discontinued. The process is repeated as needed.

basis for planned selection, for observations on the range of phenotypic variation within a species, for measuring genetic correlations between psychophenes and between psychophenes and somatophenes, and as backgrounds for ascertaining the generality of the behavioral effects of mutant genes.

Some of these purposes can be achieved by employing F_2 hybrids between inbred lines, but phenotypic correlations obtained in such crosses may be attributable to linkage rather than to a true genetic correlation. Furthermore, there cannot be more than two alleles at any locus in an F_2 population, and the frequency of any possible allele must be 0, ½, or 1. Noninbred stocks such as those supplied by commercial producers, though genetically variable, have an unknown amount of inbreeding and may have been selected for characteristics making them adaptable to mass production at low cost. Thus there has been a move in behavior genetics toward synthesized heterogenic lines based on crosses between a number of independently derived inbred lines (McClearn, Wilson, and Meredith, 1970; Fuller, 1975). Each of the two stocks described by these authors is descended from an eight-way cross between inbred lines and is maintained by quasi-random mating, in which all fertile matings are represented equally as parents, and no mated pairs share common grandparents.

Such stocks are more vigorous than typical inbreds, as reflected in more rapid development, higher fertility, and larger litters (McClearn, Wilson, and Meredith, 1970). These authors found their HS stock (heterogenic stock) to be generally more active than its inbred constituents in a wide variety of situations, although on some tests one or two of the inbred strains scored higher. Since the HS stock contained many segregating loci, one might predict that its phenotypic variance would be larger than that of an inbred strain. Again the data confirmed this expectation, but not uniformly. This inconsistency in outcome is not surprising, considering the dual sources of variance, heredity and environment. McClearn, Wilson, and Meredith point out that the phenotypic correlation, r_p,

between two traits, x and y, depends on the heritability of each, their genetic correlation, and their environmental correlation.

$$r_p = h_x h_y r_a + e_x e_y r_e \qquad (1)$$

When heritabilities are low, widely disparate, or the genetic correlation is low, moderate values for r_E will lead to higher rather than lower phenotypic correlations in inbreds. Empirically the correlations between various measures of activity were frequently higher in inbred than in HS mice (McClearn, Wilson, and Meredith, 1970). Such a situation implies strong common environmental influences on the correlated characteristics.

Heterogenic stocks will probably see more use in behavior genetics research in the future. They complement rather than compete with inbred strains.

Selected lines

The objective of most selection programs in behavior genetics is the production of a psychophene that is adapted to certain experimental procedures. The goal is an emotionally reactive or nonreactive rat (Broadhurst, 1960b), an active or inactive mouse (DeFries and Hegmann, 1970), a maze-bright or maze-dull rat (Tryon, 1940a), or a rapid- or slow-avoidance learner (Bignami, 1965). Occasionally the selection criterion is a somatophene believed to have correlations with behavior, brain cholinesterase (Roderick, 1960), or brain weight (Fuller and Herman, 1974). Domestic animals selected for behavioral and other characteristics also find some use (e.g., Willham, Cox, and Karas, 1963; Scott and Fuller, 1965). Success in changing a behavioral phenotype by means of genetic selection is a priori evidence for heritability of the criterion character, provided precautions have been taken to rule out effects due to environmental factors and genetic drift.

In principle the expected phenotypic response, R, is the product of the heritability of the criterion character and the selection differential, S, the difference between the mean of the base population and that of the individuals selected as parents.

$$R = h^2 S \qquad (2)$$

It is clear that from this equation that high heritability which is dependent on additive genetic variation and an "expressive" environment will facilitate progress. With other factors equal, a high selection differential will also speed progress. In practice, however, there may be conflicts between simultaneously maintaining high h^2 and high S.

In the typical laboratory selection study, the base population is small, and the number of selected individuals for a line is still smaller. Under such conditions the amount of inbreeding in a selected line can increase rapidly, so that further progress is blocked. The problem of adequate size of the selected lines is accentuated by the desirability of carrying out two complete replications, each consisting of a control, or unselected line, with corresponding high and low scoring lines. Such duplication is insurance against accidents and also important in evaluating the significance of correlated responses. In small, inbred populations, fixation of noncriterion characteristics may occur by chance, giving rise to a phenotypic correlation with the criterion. Such a correlation has no functional significance, and it can be misleading. When the same correlations between traits appear repeatedly in independently selected lines, however, the case for genic communality becomes stronger. Good practice in a selection program involves maintaining as much genetic variability in the selected lines as is compatible with the objectives and limitations of resources.

Course of selection. The psychologist selecting for a quantitative psychophene is dealing with the simultaneous effects of genes at many loci. Under the simplest set of assumptions, the outcome can be predicted from principles already described. Consider a system of n pairs of genes, each plus allele producing one unit of positive effect, each minus allele one unit of negative effect, and no dominance. Fig. 6-2 represents the course of such an idealized experiment under intense selection, starting from a population in which the frequency of each plus allele is 0.5. It is interesting to note that this figure approximates the results of actual experiments such as those of Tryon (1940a). As the number of involved loci increases, the effect of selection on each individual gene is less because the selector cannot know whether he is choosing $+^a$, $+^b$, $+^c$, or any other allele. However, the rate of change of the population mean is not dependent on the number of loci (N), but is proportional to N times the average effect of all loci. Genotypes will become fixed more slowly, and progress will continue over more generations when N is large. Although there may have been no individuals in the base population who possessed the maximum number of plus alleles, such

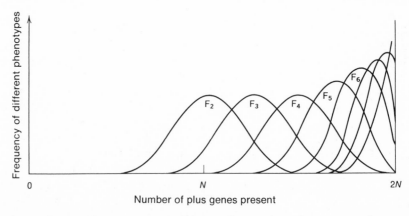

Fig. 6-2. Distribution of successive generations under intense selection toward an extreme, with few mistakes from dominance or from environmental causes and with no epistasis. (From Lush, J. L. 1945. Animal breeding plans, 3rd ed. Iowa State University Press, Ames, Iowa. Reprinted by permission.)

individuals may appear as plus alleles gradually displace their negative counterparts. In this statistical sense selection can create new phenotypes by increasing the probability of gene combinations that would occur infrequently under a system of random mating.

Selection procedures. Falconer (1963) provides an excellent introduction to the practice of selection in mammals, and much of his material is applicable also to other groups. Essentially the behavior geneticist who employs selection has two variables under his control, the selection differential and the mating system. The need for keeping an adequate number of breeders to maintain a line places a limit on intensity. If it is desired to make selections from a pool of 50 animals (25 males and 25 females) per generation, and a female can be counted on for 10 young, then at a minimum, the top 5 and the bottom 5 females must be set aside. For insurance against losses, the criterion should be broadened. If the criterion character is distributed normally and the top and/or bottom 25% are selected, the differential will be about 1.3 standard deviations; with a 50% cutting point the differential will be about 0.8 standard deviation. High fecundity facilitates selection, since it permits more intensive culling. By using polygamous matings, the intensity of selection on the male side can be made more intense, and the entire process accelerated at the cost of increased inbreeding.

Selection can be based on individual performance, on average family performance, or on some combination, such as taking the highest or lowest individuals from each available family. Selection on the basis of the performance of progeny is often practiced in farm animals, and it has been used in behavior genetics to select for phenotypes such as brain weight whose determination requires sacrifice of the animal (Fuller and Herman, 1974). When h^2 is high, the phenotype of an individual is an accurate guide to its genotype, and individual selection is effective. When h^2 is low, selection of whole families on the basis of sibling averages will give faster results. A consequence of family selection is increased inbreeding, since more and more of the parents will eventually be closely related even though brother-sister matings are avoided. As a population becomes more inbred, family selection becomes relatively more effective than individual selection, and in a set of isogenic lines, selection within lines is totally useless. Lerner (1958) has published diagrams comparing the efficiency of individual and family selection for various degrees of inbreeding, family size, and heritability. Falconer (1963) estimates that a laboratory selection program with rodents can be expected on the average to change a character by between 0.2 and 0.5 standard deviation per generation and that this rate of response may continue for five to ten generations. Eventually progress ceases, although genetic variability may still be present, as indicated by the success of reverse selection.

Heritability determines the optimum intensity of selection. When it is high, progress is more rapid with intense culling. If it is low, intense culling may impede progress because many of the selected individuals will owe their position to environmental rather than genetic factors. If there are nonlinear genotype-environment interactions, the most extreme phenotypes might come from a "sensitive" genotype whose average is near the mean but which is extremely variable.

A simple system for retarding the inbreeding process in small selected populations is shown in Fig. 6-3. The foundation stock should be composed of unrelated individuals. Inbreeding increases at the rate of 0.0625 per generation under this system with two pairs (Fig. 6-3, *A*) and at half the rate with four pairs (*B*).

An ingenious method for mass screening and selection was developed by Hirsch and Tryon (1956). Essentially the technique consists of a procedure for permitting observations of a large number of subjects at one time and simultaneously separating subgroup classes based on their cumulative score on a series of tests. The scheme can be illustrated by an apparatus designed by Hirsch to sepa-

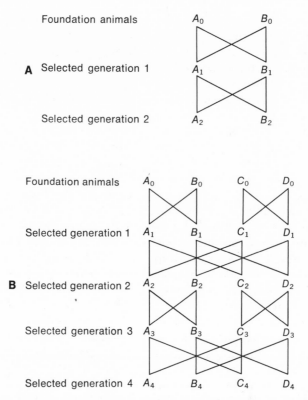

Fig. 6-3. Mating systems to retard inbreeding in selection experiments. A_0, B_0, C_0, and D_0 represent pairs of unrelated animals. A_1 is a pair whose father came from A_0 and mother from B_0. Other symbols have similar meaning. **A,** Two-pair system. Two pairs are retained in each generation, and males from one are mated to females from the other. **B,** Four-pair system. Four pairs are retained in each generation, and alternate generations are mated in a similar pattern.

rate drosophilae on the basis of their geotactic response (Fig. 6-4).

The number of steps of selection can be extended without limit, although in the diagram it is restricted to three. In this particular device, separation is achieved automatically by the design of the apparatus, but the same plan of screening can be used for any selection based on a 2-point scoring system. Methods for computing the reliability of the measurements are given in the original paper. Mass screening procedures are particularly important in working with *Drosophila*, since the small size of individuals makes them difficult to handle. The advantages of their large production of offspring are negated unless procedures are available for measuring individual differences reliably and rapidly.

Natural populations

A natural population can be considered as a heterogeneous group that has been subjected to natural selection. Comparisons between the emotional behavior of wild and domesticated rodents were certainly among the earliest in the field of behavior genetics. In 1913 Yerkes published a study of the inheritance of wildness, savageness, and timidity. Four 5-point rating scales were used. The subjects were captured wild rats and tame hooded animals from the Harvard colony. The contrast between the two parent lines, Yerkes pointed out, was extremely marked and could definitely be attributed to genetic factors. In the tame strain, females were less gentle and tame than males. Coburn (1922) reported work that was essen-

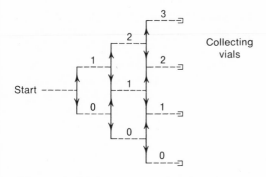

Fig. 6-4. Representation of a set of pathways in a vertically oriented sheet of plastic. Arrowheads represent devices for the discouragement of retracing. Drosophilae start at the left and move to the right. For each upward turning at a choice point, the individual receives a score of 1; for each downward turn, a score of 0. The pathways lead together in such a manner that individuals having the same cumulative scores are grouped together at the end of each trial. (Modified from Hirsch, J., and R. C. Tryon. 1956. Psychol. Bull. **53**:402-410.)

tially a continuation of Yerkes' study but used mice instead of rats. Wild gray and tame strains were rated for wildness and savageness, timidity being omitted. There was a striking difference between the wild and tame strains in these traits, and, as before, females were found to be wilder than males.

Dawson (1932) also studied the inheritance of wildness and tameness in mice, but, instead of a rating scale, he used an objective measure, the average time taken to traverse a 22-foot runway in three trials. A barrier was moved along behind the subjects to prevent retracing. Reliability of the measure over the whole population was 0.92. The subjects were laboratory-reared offspring of wild mice (*Mus musculus*) and three strains of tame mice. Again, marked behavioral differences between the genotypes were found, indicating the influence of hereditary factors. Stone (1932) reached the same conclusion using rats.

A good deal of work was also done on the endocrine differences between wild and domestic rats. Rogers and Richter (1948) found striking differences in the size of adrenal

glands in two wild strains of rats (Norways and Alexandrines) and a tame strain (domesticated Norways). Most of the difference was in the cortical portion of the gland, rather than the medulla. F_1 hybrids resembled the tame parent in adrenal size. Evidently wild animals depend heavily on adrenal function for survival, the cortex of this gland playing an important adaptive role in response to stress. It is significant that domestic rats survived adrenalectomy, but wild ones tended not to survive it.

Further endocrine differences were also found. Wild rats have a great deal more adrenal cholesterol than tame animals. Capture causes temporary loss of cholesterol with a return to normal in 24 hours (Nichols, 1950). Richter, Rogers, and Hall (1950) showed that salt replacement failed to compensate for adrenalectomy in wild but did so in tame rats, and Woods (1954) found that fighting caused a decrease in adrenal ascorbic acid content in domesticated but not in wild rats. Finally, Richter and Uhlenhuth (1954) demonstrated that gonadectomy had little effect on the activity of wild Norways but a strong reducing effect on that of domesticated Norway animals. The authors suggested that running activity is controlled in wild rats largely by the adrenals and in tame animals by the gonads and that during the process of domestication, a gradual shift occurs from adrenal to gonadal control of many bodily functions.

Clearly, domestication has had striking effects on the physiological and behavioral characteristics of animals, even though the basic behavioral repertoires are very similar in wild and domesticated individuals.

The behavioral genetics of natural populations in the field has been less investigated, although this situation might seem to provide the best model for the situation in human genetics. The reason for the neglect is not lack of interest but technical difficulties in making the necessary observations. Bruell (1970) has urged that "population behavior genetics" be expanded into the field. In Chapter 11 we shall discuss the beginnings of this aspect of our subject.

Mutant genes and mutant stocks

Conceivably any gene substitution could have pleiotropic effects on behavior. The nature of these effects can be tested by comparing the psychophenes of individuals carrying different alleles that are identifiable by their somatophenic effects. In such situations the formal genetics are already available, and attention can be focused on the manner in which the alternative alleles produce their behavioral effects. The nature of the genetic background on which the effect of the allelic substitution is studied is often important. This background may be an inbred strain genome into which the mutant gene is introduced by means to be described later. Strains of this type in which both mutant and nonmutant individuals, differing only at a single locus, occur are known as *congenic strains*. Phenotypic differences between the two forms can be ascribed with confidence to events set in motion at the locus of interest.

There is, however, some cost in attaining this precision, since the effects of a gene substitution may be specific to certain backgrounds. For example, Caviness, So, and Sidman (1972) studied correlations between behavior and neurological characteristics in reeler mice (*rl/rl*) in a congenic line and in an F_1 cross between two inbred strains. On the inbred background the reelers survived for too short a time to permit satisfactory behavioral studies. On the F_1 background the neurological aberrations remained, but the mutants were vigorous enough to permit extensive observations. It is also possible to observe the effects of a mutant gene on a heterogeneous background, where its behavioral effects will be randomly associated with those of other segregating loci. The loss of precision because of background variability may be compensated by the greater ability to generalize any findings.

It is fairly simple to introduce a dominant gene into an inbred stock that carries its recessive allele by repeatedly backcrossing trait bearers into the inbred line. With each backcrossing the proportion of "foreign" chromosomes decreases by one half. The introduced chromosome bearing the mutant allele will gradually exchange material with its homologue from the inbred line until only the locus of interest and adjacent chromosomal segments differentiate the two phenotypic classes.

If the gene is an autosomal recessive, progress is slower. After each cross into the inbred line, brother-sister matings must again be made to produce trait bearers (phenotype R) that can be crossed into the inbred line (phenotype D). The plan is as follows:

Generation 1 Trait bearers (R)   × Inbred line (D)
            a/a   ×   +/+

Generation 2   All D
            +/a   ×   +/a
                    (brother-sister)

Generation 3 R offspring from 2   Inbred line (D)
            a/a   ×   +/+

Subsequent generations repeat the alternation of 2 and 3.

Sometimes the bearers of a homozygous recessive gene are sterile or almost so. In such cases the gene of interest has to be transmitted through heterozygotes. It is still possible to put such a gene onto a standard background, although the process is laborious.

Generation 1 Known heterozygote × Inbred line
            +/a   ×   +/+

Generation 2
      a. Offspring will be ½ +/a and ½ +/+.
         Mate to known +/a in order to identify
         the +/a individuals.
      b. Tested heterozygote × Inbred line
            +/a   ×   +/+

Subsequent matings are repetitions of 2a followed by 2b. Matings between tested heterozygotes yield 25% mutant offspring.

The number of generations of crossing into an inbred line that is necessary for a particular experimental use is a matter of judgment. E. L. Green (1966) provides tables of the degree of background homozygosity attained by various mating systems. The possibility of confusing pleiotropic effects of a gene with effects due to a closely linked gene introduced along with it is rather large. After

50 crosses, genes with a recombination frequency of 0.02 would fail to be separated in 36.4% of all lines developed by this method.

For genes with robust effects on the phenotype, the matter of background genotype is less important. The obese gene *(ob)* in the mouse produces a striking effect on the regulation of eating, regardless of the backgrounds in which it is observed. For many studies the advantages of placing such a gene on an inbred background may not be worth the trouble. However, in Chapter 8 we describe experiments in which the physiological effects of *ob* were strikingly different in two inbred strains.

Mutant gene studies in *Drosophila* and mice

Because of the great variety of identified mutants in *Drosophila melanogaster*, it has been a favored species for the unit-gene approach. Among the mammals, house mice offer the greatest possibilities in terms of described mutations and specialized stocks. We

have chosen a study of Merrell (1949) to illustrate the application of the mutant gene technique to a behavioral problem in *D. melanogaster*. The influence of four X-linked genes on mating success was ascertained in competitive situations. X-linked genes are particularly convenient because the mutant phenotypes of recessive genes are directly observable in the hemizygous males, thus obviating the necessity for test matings to identify heterozygotes. The following is Merrell's mating system:

Generation 1 Mutant male × wild-type inbred female
Generation 2 Female heterozygote × wild-type male
(both from 1)
Generation 3 Male mutant from 2 × wild-type female
(inbred)
Generation 4 Repeat of 2
Generation 5 Repeat of 3

In Merrell's experiment the procedure was complicated by a requirement of developing stocks with all possible combinations of mutant and wild-type genes at four loci. Eventually sixteen sublines were produced, each

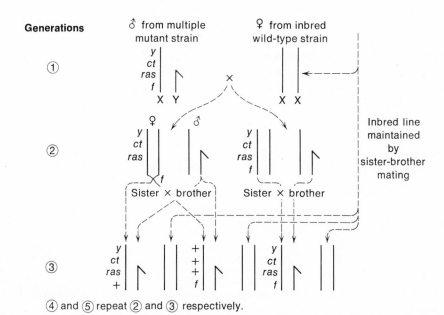

Fig. 6-5. A method for obtaining various combinations of the sex-linked genes, *y, ct, ras,* and *f,* on a constant genetic background by repeated backcrossing of the multiple mutant stock into an inbred strain and isolation of the various crossover classes as they occur. On the left, for example, is shown a crossover that produces a *y-ct-ras* line and an *f* line. (Modified from Merrell, D. J. 1949. Genetics **34**:370-389.)

with a specific set of wild-type and mutant alleles at the raspberry (eye color), cut wings, yellow (body color), and forked (bristles) loci. Fig. 6-5 illustrates the procedure for producing two of these lines by taking advantage of a crossover between *ras* and *f*.

Both male-choice and female-choice mating experiments were conducted, and the results were evaluated by the types of offspring produced. Male choice involves placing mutant males with a mixture of mutant and wild-type females and noting which are inseminated. Female choice can be identified by the use of genetic markers as illustrated in the following example. Wild-type females heterozygous for *rs* were placed with wild-type and raspberry males. The kinds of offspring resulting from successful matings of the two kinds of males are shown in Table 6-1. Note that the appearance of mutant females indicates success of the mutant male.

The results of the female-preference experiments showed that wild-type males were much more successful than yellow males and moderately more successful than cut and raspberry males. Forked males were equal to wild-type in mating capability. In general, the effects of multiple mutant genes were additive, although *ct ras* males were superior to both *ct* or *ras* males, implying some compensatory effect of this particular combination. The behavior of the females was primarily responsible for nonrandom mating. So-called male choice is better interpreted as caused by rejection of the less vigorous males by the less receptive type of female. In line

with this hypothesis, male choice was most evident in the least vigorous males; furthermore, it was consistently in the same direction for all males tested with the same combinations of females. Merrell concluded that within *D. melanogaster* there is selective, but not assortative, mating.

In a later experiment, Merrell (1953) showed that the rate of elimination of mutant genes from a population followed time curves predicted from the effects of each gene on mating success in the earlier "choice" experiments.

Other studies of the effect of single-gene substitutions on the behavior of drosophilae have been reviewed by Parsons (1967a), Wilcock (1969), and Rockwell and Seiger (1973b). Some of them will be considered in Chapters 7 and 11.

Behavioral pleiotropy in mice

Systematic observations of the effects of a number of mutant genes on the behavior of mice have been made (Denenberg, Ross, and Blumenfield, 1963; van Abeelen, 1963a, b,c; Thiessen, Owen, and Whitsett, 1970). Van Abeelen constructed an ethogram consisting of distinguishable forms of behavior shown by either solitary mice, pairs of males, or male-female pairs. Typical elements of these ethograms were grooming, digging, wrestling, and copulation. Four mutants, yellow *(A^y)*, pink-eyed dilution *(p)* *(p/p)*, brown *(b)* *(b/b)*, and jerker *(je)* *(je/je)*, were compared with nonmutants. Typical of the findings were reduced "exploratory" and

Table 6-1. Evaluation of effects of *ras* on success in mating*

		Gametes			
		Mutant male		Wild-type male	
		ras	Y	+	Y
Gametes (heterozygous wild-type female)	*ras*	†*ras/ras*	*ras*/Y	*ras*/+	*ras*/Y
	+	*ras*/+	+/Y	+/+	+/Y

*Based on data from Merrell, D. J. 1949. Selective mating in *Drosophila melanogaster*. Genetics 34:370-389.
†Raspberry females produced only when mutant male is successful.

"comfort" behaviors in the mutants. Feeding, fighting, and mating were little affected. Thiessen, Owen, and Whitsett observed 14 mutants, all on an inbred background, using a battery of tests designed to measure variation in activity, sensory sensitivity, and sensory preferences. They, too, reported effects of certain genes on such behaviors as open field activity, geotactic response, latency to escape from water, and amount of wheel running. Denenberg, Ross, and Blumenfield found hairless mice *(hr/hr)* less active in an open field and unable to perform on water escape. Pintails *(Pt/+)* extinguished more rapidly in shock escape. No behavioral differences were found associated with the short-ear *(se/se)* or pale-ear *(ep/ep)* genotypes.

Studies of this type have been criticized by Wilcock (1969) on the grounds that most behavioral differences between mutant and nonmutant animals are "direct peripheral effects of the mutant gene and are thus devoid of psychological significance." It must be admitted that simply correlating a mutant gene with a quantitative or qualitative variation in activity, sensory capacity, or even learning ability does not in itself have much explanatory value. But it is not necessary to stop with the correlation. If something is known about the total somatophenic effects of a mutant gene, there is a possibility of relating these to the physical substrate of the observed psychophene. We may conclude that hairless mice failed on a water-escape test because they lacked buoyancy, but other cases are not as obvious. Genes named for their effect on coat color may have very important effects on the nervous system. Albinism is a case in point and will be considered in detail in Chapter 7.

A number of interpretations are possible when a battery of behavioral tests are applied to a set of genic substitutions. A case may be made that the tests responding similarly to a particular substitution share a dependence on a common biological substrate. One might also reason that all genes affecting a particular behavior are interacting to produce the biological substrate on which behavior depends. In such a procedure the mutant is essentially a "black box" (though it comes in various colors and forms) with its inner workings concealed. When both somatophenic and psychophenic measurements are included in the test battery, there is a possibility that items of each type will covary with a particular gene substitution. Such clusters could provide clues to the functional relationships within the black box.

The researcher adopting the mutant survey strategy is faced with choosing among a potentially infinite number of phenotypes of both classes, and he can measure only a few. A guiding principle is that mutant genes are useful in behavioral research when they produce biochemical and structural effects related to important behavior systems (Fuller, 1967). Some mutant genes produce physiological changes that could be attained in no other way, and, in such cases, they can be valuable as adjuncts to other experimental procedures such as lesioning or drug administration.

Another approach to the study of behavioral pleiotropy of recognized genes is *retrograde analysis*. The technique, which is widely used in physiological genetics, involves a search for the earliest deviation of the mutant from the normal phenotype. In behavior genetics retrograde analysis has been applied to the wabbler-lethal mouse by Thiessen (1965). Although it is particularly suitable for detecting the embryological basis of neurological anomalies that lead to behavioral deficits, it could be employed at the purely behavioral level.

Identifying cryptic loci

Loci are typically identified and named for their somatophenic effects. The great value of linking a behavioral variation to a named locus is the possibility of tracing its cause backward to a specific biochemical process. The advantages and limitations of the genotypic approach have been discussed earlier. Here we may add the observation that the information obtained from studies of the usual mutants may contribute little to our understanding of the genetic basis of behavioral differences among phenotypically normal individuals. Fruit flies with bar eyes or vestigial wings and hairless mice do not survive in nature.

The attractiveness of the single-locus approach for developmental-physiological analysis is real, however, and has motivated the search for *cryptic* genes that are recognizable, at least initially, only by their behavioral effects. Although all behavior is fundamentally under polygenic control, it is not illogical to suppose that differences in the psychophenes of individuals or inbred strains could be caused by gene substitution at one or very few loci.

The simplest procedure for detecting such loci is to classify one's subjects into discrete classes, carry out a standard set of crosses, and note whether the proportions in each class conform with Mendelian expectations. Typically, two inbred strains clearly distinct in behavior are chosen as the initial subjects. An example of this approach is the inheritance of saccharin preference in crosses between C57BL/6J and DBA/2J mice (Fuller, 1974). The data, as shown in Table 6-2, are compared with the expectations for control of preference by a single locus with the high-preference allele dominant.

Two features of the table are noteworthy. First, there is a small amount of overlap in preference behavior between the original inbred strains. The expectations for high- and low-preference individuals in the backcrosses and F_2 were adjusted to allow for overlap; thus the ratios of high : low individuals are not precisely 1 : 1 in the backcross to DBA/2 or 3 : 1 in the F_2. Second, the offspring from the first backcross to DBA/2 were separated into high preferrers (presumably heterozygous) and low preferrers (presumably homozygous recessive) and crossed again with DBA/2. This was done because it is possible to simulate Mendelian ratios in a polygenic system operating on a developmental threshold (Wright, 1934). In this instance the second backcrosses yielded high and low preferrers in proportions consonant with the single-locus hypothesis and a new locus, *Sac*, was proposed with at least two alleles, Sac^b from C57BL/6 and Sac^d from DBA/2.

A later study (Ramirez and Fuller, 1976) indicates that the Sac^b allele is probably rare in laboratory mice. In a heterogeneous population of laboratory mice from an eight-way cross (one ancestor was C57BL/6), amount of saccharin intake was found to be highly heritable, but biometric analysis suggested polygenic control of variation.

A somewhat different approach was used by Whitney (1973) to test for single-locus influences on handling-induced vocalization (squeaking) in another heterogeneous stock of mice from an eight-way cross. One of its progenitor strains, I^s/Bi, was a high vocalizer; the remainder were low. The frequency of vocalization and its distribution among families matched that predicted by the hypothesis that squeaking depended on a single dominant gene with a frequency of 0.125, the expected contribution from strain I^s. Two locus models were also tested but did not fit the observed data.

Table 6-2. Observed and expected frequency of high (75% and over) and low (under 75%) preference for 0.1% saccharin in crosses between C57BL/6J (*B*) and DBA/2 (*D*)*

	Observed		Predicted[†]	
Group	**High**	**Low**	**High**	**Low**
B	32	1	—	—
D	5	25	—	—
F_1	25	3	—	—
B × F_1	14	1	14.0	1.0
D × F_1	37	25	34.0	28.0
F_2	50	12	45.9	16.1
‡D × L	7	37	7.1	36.9
‡D × H	18	19	20.3	16.7

*Based on data from Fuller, J. L. 1974. Single-locus control of saccharin preference in mice. J. Hered. **65:** 33-36.

†Predictions based on single-locus model with high-preference dominance. Adjustments have been made for partial phenotypic overlap between presumed homozygous parental strains. None of observed values differ significantly from those predicted.

‡High preferrers (*H*) and low preferrers (*L*) from F_1 × D were backcrossed a second time to D to strengthen case for single locus control of preference.

Any method which depends on classification of individuals into discrete categories requires that the boundaries of the groups be prescribed in advance of measuring the segregating generations. This procedure was followed in Fuller's (1974) study of saccharin preference, but even here an allowance had to be made for misclassification. Most psychophenes are so remote from the primary action of genes that one-to-one correspondence is unlikely. The more closely it is approached, however, the stronger the case for a simple mode of transmission. In the following chapters we shall see that the different techniques for measuring psychophenes can lead to different genetic hypotheses.

Analysis of multimodal distributions. Categorization of individuals becomes difficult when behavior is highly variable within pure lines, and the scores of segregating groups are distributed with indistinct evidence of bimodality or trimodality. Such overlapping distributions do not negate cryptic major gene influences, but they do require special analytical methods.

Assume that we have a set of Mendelian crosses between two purebreeding strains. The avenue of approach is to determine first the empirical total frequency distributions, f_x, of the nonsegregating groups, P_1, P_2, and F_1. On a single-locus hypothesis the frequency distributions of the backcrosses and F_2 are:

$$f_{B_1} = \frac{1}{2}(f_{P_1} + f_{F_1})$$
$$f_{B_2} = \frac{1}{2}(f_{P_2} + f_{F_1})$$
$$f_{F_2} = \frac{1}{4}(f_{P_1} + 2f_{F_1} + f_{P_2})$$

Collins (1967, 1968a) has proposed a general, distribution-free, nonparametric method for cases in which assumptions of normality and equal environmental variances in the groups are not met. Obtained distributions can be compared with those predicted by one- and two-factor models. An application to ethanol consumption in hybrids between C57BL/6J and DBA/2J mice indicated that a two-factor model fit the data very well; a one-factor model did not (Fig. 6-6).

If the assumption of normal distribution of environmental variance is considered appro-

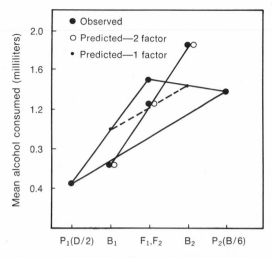

Fig. 6-6. Distribution-free analysis of consumption of ethanol by C57BL/6J and DBA/2 mice and their hybrids. Note that a one-factor model does not fit the data; a two-factor model does. (From Fuller, J. L., and R. L. Collins. 1972. Ann. N.Y. Acad. Sci. **197**:42-48.)

priate, a variety of genetic hypotheses can be tested on data from Mendelian crosses between inbred strains (Elston and Stewart, 1973). The model giving the best fit to the data can be provisionally accepted. The methods are based on graphic and numerical analytic procedures developed by Stewart (1969a,b) and applied to a number of anatomical characteristics of mice by Stewart and Elston (1973).

The tests are based on the principle that multimodal frequency distributions in a set of measurements may indicate an underlying discontinuity attributable to gene segregation. The position of the modes is ascertained by arraying individual scores in rank order and plotting the cumulative frequency against the phenotypic values. A smooth S-shaped curve indicates probable polygenic control; a curve with one or more discontinuities may indicate segregation. By differentiating the cumulative curve and graphing the results, the regions of maximum slope can be visually identified. In Fig. 6-7 this procedure is applied to a portion of Fuller's (1974) data on saccharin preferences in mice. The complete data are shown in Table 6-2. Fig. 6-7 shows clearly that the

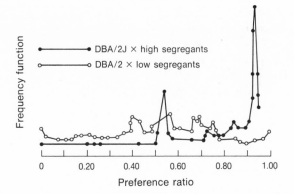

Fig. 6-7. Frequency distribution of preference ratios for 0.1% saccharin in crosses between DBA/2 and high- or low-preference segregants from a prior backcross. Each individual is represented as a point in the curve with abscissa, X_0, corresponding to its preference score. The ordinate shows frequency calculated from the midpoint formula,

$$\frac{28}{3X_3 + 2X_2 + X_1 - X_{-1} - 2X_{-2} - 3X_{-3}},$$ where X_{-3}, X_{-2}, . . . X_3 are scores for consecutive individuals arranged in rank order, low to high. The more closely the scores are packed, the higher the ordinate. Thus in this figure about one half of the high-segregant crosses are found on the high peak at the right, as predicted from a one-factor theory. The low-segregant crosses showed no definite peaks, as was also predicted. Both genotypes showed a secondary rise between 0.40 and 0.60, scores that reflect neither preference nor aversion.

backcross to the low-preference segregants yielded a flat frequency function with a small hump in the 0.4-to-0.6 range that indicates equal acceptance of saccharin and water. In contrast, the backcross to the high-preference segregants showed a significant peak in the 0.9-to-0.98 range, containing approximately 50% of the subjects. Further details may be found in the legend for Fig. 6-7.

The technique of Stewart and Elston (1974) confirms by an objective procedure the earlier empirical division of the mice in the high-preference segregant backcross into two categories and strengthens the case for the hypothesis that a single locus affects the response of mice to saccharin solutions. The method should have many applications in behavior genetics.

Recombinant inbred strains

Recombinant inbred (RI) strains of mice were developed by Bailey (1971) for the purpose of studying the genetics of tissue transplantation and were introduced to behavior genetics by Oliverio, Eleftheriou, and Bailey (1973b). An RI line is produced by a series of

brother-sister matings, starting with the F_2 of a cross between two highly inbred and unrelated progenitors, P_1 and P_2. In the course of time each line becomes fixed for one or the other allele at any locus at which the two ancestral strains differed. Thus, on the average, an RI strain will have one half of its genes from P_1 and one half from P_2. If a set of RI lines are clearly divided into two classes based on a behavioral test, one resembling P_1 and the other P_2, one can argue that a single locus is responsible for the original strain difference. Further genetic analysis is made possible by the fact that each of Bailey's RI lines has been characterized by means of skin grafting tests as carrying a P_1 or P_2 allele at a number of histocompatibility (H) loci. If the classification of RI strains based on behavior matches that of a specific H locus, it is likely that the behavioral effect may be mediated by that H allele or by a closely linked gene. Fig. 6-8 shows that the avoidance performances of seven RI lines were clearly separated into two groups, one like their BALB/c progenitor, the other like C57BL/6. The distribution pattern of good and poor

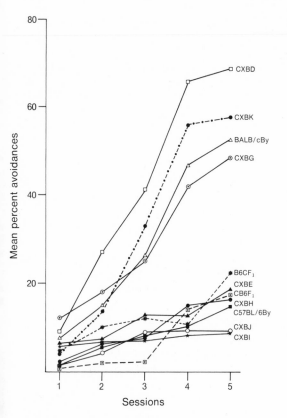

Fig. 6-8. Mean percentage of avoidances per session for 5 days of seven RI lines, their two ancestral strains, and two reciprocal F_1 hybrids. (From Oliverio, A., B. E. Eleftheriou, and D. W. Bailey. 1973. Physiol. Behav. **11:**497-501.)

avoidance was found to correspond exactly with that of alternative X alleles at the H-25 locus on chromosome 9. On this basis Oliverio, Eleftheriou, and Bailey proposed a new locus, *Aal* (active avoidance learning), which is either identical or closely linked with H-25.

SUMMARY

This review of materials and methods of experimental behavior genetics is not exhaustive, but it does demonstrate that in a number of instances evidence has been found for one- or two-locus systems with important behavioral effects. In addition to statistical tests of compatibility with a simple model, it is always desirable to link a cryptic gene, identified through a psychophene, with a somatophene and to locate it on a chromosome. When this is accomplished, of course, one no longer has a cryptic gene.

In the following chapters we shall constantly return to the methodological problems discussed in this chapter. With all the advantages of inbred lines, defined heterogeneous stocks, RI lines, and the like, it has not been easy to prove unequivocally the superiority of either a Mendelian or polygenic interpretation of variation in psychophenes. Once a test procedure is changed or a new set of strains observed, the genetics of transmission may also change. We shall find examples in the case histories of Chapter 7.

7

Individual behavior: four case studies

To introduce the experimental genetics of behavior, we have chosen to illustrate a variety of research strategies by four rather diverse case studies: phototaxis in *Drosophila*, audiogenic seizures and paw preference in mice, and the behavioral correlates of albinism in several species. In the first three, analysis starts with a behavioral phenotype of interest and employs genetic techniques for acquiring a deeper understanding of the phenomenon. The fourth example represents the genotypic approach; the mode of inheritance is already known, and the nature and magnitude of the behavioral effects of the albino gene are the goals.

Phototaxis is movement toward (positive) or away from (negative) a source of light. More complex than a reflex, since it involves a response of the organism as a whole, a taxis seems to be little affected by learning, although its strength may be affected by physiological state and environmental factors. *Drosophila melanogaster* is typically positively phototaxic, undoubtedly because such a response contributes to survival under natural conditions. In spite of natural selection, however, there is still great diversity in the strength of the response among laboratory stocks, making the phenotype well suited for genetic analysis. It also turns out that the type of procedure used for measuring phototaxis has a great influence on the results obtained.

One advantage of *Drosophila* is its adaptability to mass separation techniques that makes it unnecessary to handle and record the behavior of individuals one by one. Thus selection experiments with fruit flies are possible on a scale that is impracticable with even the smallest, most prolific, and least expensive mammals or birds. This fact plus the advantages of working with a species with four

pairs of chromosomes and a great variety of specialized stocks provide behavior geneticists with unique opportunities for research.

Audiogenic seizures and paw preference in mice have no obvious adaptive significance; their interest for behavior geneticists is based on the distinctiveness of their phenotypic expression. Most mice are either susceptible or resistant to a sound-induced convulsion at a given time, and they are either right or left pawed. Such distinctive psychophenes are good candidates for the classical techniques of Mendelian crosses with the objective of explaining the behavioral variation by a simple one-locus model. Actually, matters are not that simple.

The albino gene is probably the mutation most widely studied by psychologists, though few experiments include nonmutant controls. Because so much of the literature and theory of experimental animal psychology is based on the white rat, it is particularly important to know what kinds of biases may have been introduced by this choice of a standard animal. Albino individuals have been described from a great number of species, but their rarity in nature indicates that the mutants are severely disadvantaged. Obviously, in the sheltered laboratory they survive.

PHOTOTAXIS IN *DROSOPHILA*

A taxis is an oriented movement in relation to a gradient or a directional field of force. Inherent in the concept is a degree of automaticity similar to a reflex but involving coordinated locomotor activity. Actually, the direction of a taxis whether toward or from a light source is not invariant. The same individual may be phototactically negative or positive, depending on its physiological state. Furthermore, as we shall see in this section, both natural and laboratory populations are heterogeneous with respect to the direction and strength of tactic responses.

Among the parameters affecting phototactic responses are age, diurnal rhythms, desiccation, wavelength and intensity of the stimulus, temperature, and level of excitation as produced by mechanical stimulation prior to the test (Rockwell and Seiger, 1973a). Various combinations of these factors may change the sign or intensity of the phototaxis. Regarding taxic behavior as a simple automatism is naive.

Rockwell and Seiger recognize three major designs, each with variations, that have been employed in research on phototaxis of *Drosophila*. Design 1 involves movement toward a light source whose gradient is parallel to the plane of movement of the flies. Strength of phototaxis can be measured by (1) rate of approach, (2) position after a fixed time period, or (3) fractionation of the tested population in a series of trials (usually five) so that the flies responding positively zero, one, two, three, four, or five times are contained in separate bottles. Design 2 involves movement of flies in a field illuminated by a light perpendicular to the plane of movement. Design 3 requires the flies to make a series of choices in a classification maze (Hirsch and Boudreau, 1958; Hadler, 1964a). At the end of the maze the subjects enter different chambers, depending on the number of photonegative and photopositive responses they have made.

It is obvious from the preceding discussion that comparisons between results obtained with different techniques must be made with care. Furthermore, conclusions with respect to the adaptive advantages of phototaxis must consider the relationship between laboratory test procedures and the ecological conditions under which flies exist in a natural state.

Mutant gene effects

Some of the earliest behavior-genetic analyses were directed toward the effects of mutant genes on phototaxis in *Drosophila*. The methodology was a simple version of design 1. Flies were placed in a tube with a light directed along it from one end, and their rate of progression toward the light was measured. This procedure does not separate the photokinetic effect of light (stimulating locomotion) from its phototactic effect (directing the locomotion). Some of the earlier workers, including McEwen (1918, 1925) and Cole (1922), compared stocks containing various mutations but did not distinguish the effects of conspicuous genes characterizing the stocks from other genetic differences that accumulate when stocks are separated for many generations.

A marked improvement in genetic technique was achieved by Scott (1943), who prepared congenic inbred stocks segregating for either white (w) or brown (b) eye color. The red-eyed wild-type and the mutants in each stock could be assumed to be genetically identical except for the named gene and its closely linked neighbors. Phototaxis was scored by the median rate of crawling over a measured course. White-eyed flies were less phototactic than their congenic siblings, but brown flies did not differ from the wild type of the same stock. Particularly notable was the large difference in running time between the wild-type flies from the congenic brown stock (11.5 seconds) and the wild type from the white stock (32 seconds). Effects of the two mutant genes were being observed on very different genetic backgrounds.

Scott's experiment did not separate the effects of photokinesis and phototaxis. This was accomplished by Brown and Hall (1936), who used a Y tube, one arm of which was illuminated. Flies inserted into the stem of the Y were observed at the choice point against an opal red light to which they did not respond. The threshold was defined as the level of illumination at which entries into the light and the dark tube were equal. Full and bar-eyed (eyes reduced in size) flies with red or white pigmentation were tested. The results are summarized in Fig. 7-1, which shows a linear relationship between the logarithm of threshold light intensity and the surface area of the eye. Apparently, *white* achieves its effect on phototaxis through reducing the size of the eye just as *bar* does. In the Brown and

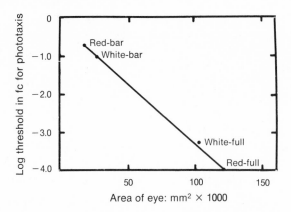

Fig. 7-1. Mean light-intensity threshold for phototaxis in *Drosophila melanogaster*, mutants and wild-type, as a function of the surface area of the eye. The threshold is a simple function of area irrespective of eye color. (Modified from Brown, F. A., Jr., and B. U. Hall. 1936. J. Exp. Zool. **74**:205-221.)

Hall experiment, genes act like surgical instruments to reduce the size of a receptor and thereby to lessen the neural input to the central nervous system. The importance of the study lies in localization of the significant gene action in the periphery rather than in the central nervous system and the demonstration that two genes with different external features affect behavior through the same mechanism.

This line of investigation has been continued by Pak, Grossfield, and White (1969) with induced mutations as the genetic tools for physiological investigations. Their object was to generate a series of nonphototactic mutations, each of which could produce a defect in one step of the visual pathway, and to use them as a set of probes to analyze the visual system. Mutations were induced in males by ethyl methanesulfonate. These treated flies were bred to virgin females of an attached-X stock, yielding F_1 males with a paternal X chromosome. Dark-preferring hybrid males were backcrossed repeatedly to the attached-X stock, producing male nonphototactic flies with the mutated X chromosome. Seven sublines were produced by these elegant manipulations, and electroretinograms (ERGs) were obtained from each. Six sublines had normal ERGs, indicating that their nonphototactic behavior was caused by factors outside the visual system.

One had an abnormal ERG pointing to a defect in the visual pathway. Thus similar psychophenes may result from unlike somatophenes.

Genetic dissection of behavior

By adding a technique for producing genetically mosaic flies to procedures for inducing mutations and for mass screening of behavioral variants, a genetic dissection of behavior is possible (Benzer, 1967, 1973; Hotta and Benzer, 1972). Mosaic flies have different genotypes in various parts of the same individual and are produced by breeding to a stock with an unstable ring chromosome, X_R. During the first division of the zygote of a female, XX_R, the X_R may be lost; the descendants of such a cell are genetically and phenotypically male. If the X_R stock is crossed with a stock carrying a recessive mutant gene affecting body color on its X chromosome, parts of each offspring will be female, $X_R X_M$, and parts male, X_M. The mutant gene is phenotypically expressed only in the male parts, and it will thus serve as a marker for the location of other X-linked mutant genes known only by their behavioral effects.

The use of these mosaics in dissection of behavior is based on the fact that the fate of a cell, whether it forms eye, bristle, or gut, is determined by its location in the embryonic blastoderm. The separation of the $X_R X_M$ and

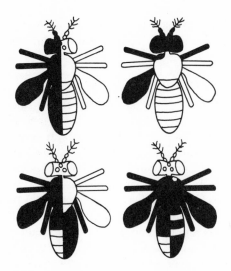

Fig. 7-2. Typical *Drosophila melanogaster* mosaics. Body parts derived from the XX cell line are shaded. Parts derived from the XO line are unshaded and express recessive mutant characters uncovered by the loss of the other X. (From Hotta, Y., and S. Benzer. 1972. Nature **240**:527-535.)

X_M regions in the blastoderm may have many orientations: hence the varieties of mosaicism shown in Fig. 7-2.

The probability that the $X_R X_M - X_M$ boundary in the adult will fall between any two structures is proportional to the distance apart of their cells of origin in the blastoderm. Parts originating from the same cells will always be concordant in color, parts from neighboring cells will usually be concordant, and parts from remote cells will be concordant only by chance. Hotta and Benzer produced mosaics in which the genotype of the eye was tagged with color genes; the flies also carried radiation-induced genes for nonphototaxis and absence of the ERG. The color gene expression and absent ERG mapped to the same region of the blastoderm, indicating that the absence of phototaxis was attributable to conditions intrinsic to the eye.

The method described is applicable to any behavioral difference produced by an X-linked mutation. For a description of its use on such exotically named genetic defects as "drop-dead" and "ether-a-go-go," see Benzer (1973). The parts of the fruit fly are certainly more autonomous with respect to their behavior than are the regions of a mammal with a more centralized and hierarchical nervous system and a means of spreading chemical signals swiftly to all cells by secreting hormones into the blood. Thus one must be cautious in taking the results of research of this kind as a general model for the relationship between genes and behavior. Nevertheless, it is one of the more impressive accomplishments of behavior genetics.

Selection for phototaxis

Traditionally, fruit flies have been described as positively phototactic. As noted earlier, such a statement is not appropriate unless a considerable number of experiential and environmental conditions are specified. It also turns out that populations of flies are not behaviorally homogeneous and that a portion of their variability is of genetic origin. Most of the selection studies have employed a classification maze based on the design of Hirsch and Tryon (1956; see Fig. 6-3). Hirsch and Boudreau (1958), Hadler (1964a), and Walton (1970) selected positively and negatively phototactic strains of *Drosophila melanogaster*. Selection has also been successful in *D. pseudoobscura* (Dobzhansky and Spassky, 1967; Dobzhansky, Spassky, and Sved, 1969) and *D. subobscura* (Kekić and Marinković, 1974). In *D. melanogaster*, heritability was high: estimated as 0.566 by Hirsch and Boudreau (1958) and as from 0.412 to 0.668 by Hadler (1964b). The power of selection to modify the average phenotype is shown by Hadler's (1964a) report that after fifteen generations, his positive phototactic strain averaged 3.90; his negative strain, 12.60; and his unselected stock, 7.88. In his apparatus, a fly making all turns toward light would score zero; all toward dark, 17.

Realized heritabilities for phototaxis in Dobzhansky and Spassky's (1967) experiment with *D. pseudoobscura* were considerably less (0.08 to 0.10), but excellent separation was nevertheless attained. Selection can be very effective even with low heritabilities if the selection differential is high and inbreeding is avoided.

Successful selection, however, does not result in strains that are isogenic for the genes relative to phototaxis. In *D. melanogaster*, reversing the direction of selection at G-27 was successful. In *D. pseudoobscura*, simply relaxing artificial selection after twenty generations and allowing the strains to breed promiscuously led to convergence in phototactic scores that was almost as rapid as the earlier divergence (Dobzhansky, Spassky, and Sved, 1969). These authors consider that phototactic neutrality, with an underlying genetic capacity for rapid change in either direction, is an example of adaptive genetic homeostasis.

It is an oversimplification to consider that selection can only operate to move populations toward extreme reactions. Lines of *D. subobscura*, a species with light-dependent mating, have been selected for preference for low, medium, or high light intensity (Kekić and Marinković, 1974). Heritabilities were low and similar in magnitude to those reported by Dobzhansky and Spassky (1967).

A Hayman analysis of gene action in a 7×7 diallel design, including selected photopositive and negative strains and wild type and mutant stocks, demonstrated additive and dominance effects (photonegativity more dominant) with evidence for reciprocal cross differences (Walton, 1970).

The experiment by Dobzhansky and Spassky (1967) can be considered as a model for social mobility. After populations were selected for positive and negative phototaxis and similarly for geotaxis, exchanges were made between them, and the resultant offspring evaluated. For example, the culls (themselves geopositive) from a geonegative population were mixed in 20:80 proportion with a geoneutral population. The ingress of this proportion of phenotypically geopositive immigrants caused the recipient population to become more geonegative. Why the paradoxical effect? Dobzhansky and Spassky attribute it to the fact that the donor population as a whole was high in "geonegative genes." Even its geopositive members shared this characteristic, and their deviant status was apparently due to nongenetic influences. Analogies with migration between social classes in humans have been made (Dobzhansky, 1967).

Phototaxis and mating behavior

Could selection for differences in phototactic behavior affect the choice of mates? If so, the possibility exists that strains selected for opposite direction of phototaxis would be sexually isolated and could thus be incipient races. In a test for sexual isolation with strains of *D. pseudoobscura* selected for photopositive and photonegative behavior, Del Solar (1966) did find a highly significant majority of homogametic (positive $\times$ positive or negative $\times$ negative) matings as compared with heterogametic (positive $\times$ negative) ones. These tests, carried out in observation chambers, may overstate the degree of isolation in more natural surroundings. We have already noted that continued artificial selection (forced sexual isolation) was necessary to maintain photopositive and photonegative lines in population cages (Dobzhansky, Spassky, and Sved, 1969). However, phototactic behavior may influence mating success differently depending on light conditions. Although *D. pseudoobscura* does mate in either light or darkness, and photopositive and photonegative lines were equally successful in light, photopositive strains were clearly impaired in darkness (Marinković, 1974).

The role of phototaxis in nature

Pittendrigh (1958) conducted both ecological and laboratory studies of two species, *D. pseudoobscura* and *D. persimilis*. *Persimilis* occupies the colder and wetter sections of California. Both species occur in the Sierra Nevada, but *persimilis* is more characteristic of woodlands, *pseudoobscura* of the drier meadows. *Persimilis* has more difficulty in maintaining water balance, but it does not in laboratory tests select moist environments reliably. Also, *persimilis* is more photopositive than *pseudoobscura*, a difference that at first glance seems to be in the wrong direction, since it inhabits shady areas. Pittendrigh explains the paradox by hypothesizing that *pseudoobscura*, living in exposed conditions, must continuously locate the moister parts of its habitat in order to avoid dehydration. Darker areas are generally damper; hence the negative response to light. A conspicuous stimulus detectable at a distance has become an agency for water conservation.

Partial confirmation of the Pittendrigh hypothesis has been obtained from a comparison of nine strains of *pseudoobscura* and six of *persimilis* (Spassky and Dobzhansky, 1967; Rockwell and Seiger, 1973). However, large intrapopulation and interpopulation variability was found in both studies.

These two species occupy different regions in the range of phototactic behavior, probably imposed by natural selection. And just as in the selected laboratory populations, genetic heterogeneity remains in a degree that permits rapid change in the median level of phototaxis when conditions change. In these flies one sees adaptation at two levels. Individuals adapt by changing their location according to light intensity. Populations adapt as environmental conditions favoring photopositive or photonegative behavior become critical factors for successful propagation and thus produce changes in the proportions of plus and minus genes. The process, fortunately, is inefficient. A residue of genetic variability remains as insurance against a reversal of selective pressures requiring a change in the direction of response.

Geotactic behavior

Although we have concentrated our attention on phototactic behavior in this section, it is important to note that similar results have been obtained when geotaxis has been the behavior of interest. Particularly significant, but not yet duplicated for phototaxis, is a series of studies with *D. melanogaster* that permit the assignment of genes modifying geotaxis positively or negatively to each of the three major chromosomes of this species (Hirsch, 1959; Erlenmeyer-Kimling, Hirsch, and Weiss, 1962; Hirsch and Erlenmeyer-Kimling, 1962). Many readers should find these papers interesting for the descriptions

of technical procedures and the demonstration that polygenes, even though not identifiable separately, can to an extent be localized physically.

Summary

The behavior-genetic analysis of phototaxis in *Drosophila* has ranged over characteristics as diverse as the electroretinogram of individual flies to the heterogeneity of habitats in the mountains of California. Many of the issues discussed in this section occur over and over again. From the point of view of geneticists, *Drosophila* has many advantages, but its small size and limited behavioral repertoire make it less attractive to psychologists.

Following are the most important accomplishments of the research reviewed here:

1. Mutant genes may permit the assignment of a behavioral variant to a particular part of the organism: receptor, central nervous system, or effectors. This has been termed *genetic dissection of behavior.*

2. Both laboratory and natural populations are heterogeneous for loci with marked effects on phototaxis. Although heritability varies, even when it is low, directional selection leads to great diversity within ten to fifteen generations.

3. Photopositivity and photonegativity are not fixed traits but are affected by many environmental variables as well as by genotype.

AUDIOGENIC SEIZURES IN MICE

If a 3-week-old mouse of a so-called audiogenic seizure–susceptible strain is placed in an enclosure and exposed to the sound of an electric doorbell (intensity, 90+ decibels) 12 to 18 inches distant, the mouse will probably startle and then remain quiet briefly. After a variable period (5 to 25 seconds), it will begin to run about the periphery of the enclosure and accelerate to maximum speed. Sometimes this wild running seizure terminates with the mouse becoming quiet again; more frequently in a susceptible strain, the mouse falls on its side and continues a running type of limb movement for several seconds in a clonic seizure. Again the seizure may terminate with the intensity of limb movements decreasing. After a variable quiet period the mouse stands and walks a bit unsteadily but appears to be little affected after 2 or 3 minutes. Sometimes a running seizure stops for 20 to 30 seconds and begins again. These delayed running seizures may or may not lead to a convulsion. If no seizure occurs within 60 seconds, continued bell ringing is almost never effective.

In some strains the clonic phase culminates in a rigid extension of limbs and trunk: the tonic seizure. Frequently mice fail to recover from these intense convulsions, probably because of anoxia due to cessation of breathing. If one observes large numbers of animals, a few seizures of intermediate intensity will be recognized, but the large majority of responses can be classified as no seizure, wild running, and clonic or tonic seizure.

Function of audiogenic seizures

Unlike the phototactic responses of *Drosophila*, which we selected for special attention, audiogenic seizures have no obvious function. Perhaps they are a manifestation of neuropathology brought about by relaxation of selective pressures in the course of domestication. At any rate, they are not restricted to house mice; they have been described in rats, rabbits, and deer mice. We shall, however, restrict this account to work with mice, since it has been more extensive, more varied in approach, and involves more genetics. For scientists, the attraction of this phenotype lies in the ease of measurement with simple equipment, the precision of classification of responses, the existence of reliable differences in frequency of seizures among existing inbred strains, and the readiness with which seizure susceptibility can be modified by appropriate treatments.

Whether these seizures are an appropriate model for any specific form of human pathology (e.g., epilepsy) may be argued. What is clear is that they illustrate many of the general issues involved in ascertaining the mode of inheritance of a psychophene, the physiological links between gene and psychophene,

age changes in the manifestation of genetic differences, and the joint influence of environmental and genetic factors on the phenotype.

Strain differences

Hall (1947) reported that most 30-day-old DBA/2J mice convulsed within 60 seconds when exposed to the sound of a doorbell and that most C57BL/6J were resistant. These two strains have continued to be favorite subjects for research. More extensive sampling of inbred strains and F_1 hybrids has shown that some have intermediate seizure risks and should not be classified as either resistant or susceptible.

For example, Fuller and Sjursen (1967) sampled 40 mice from each of eleven inbred strains. Every animal was exposed to bell ringing at weekly intervals from 21 to 42 days of age, unless meanwhile it died of a seizure. At 3 weeks of age the seizure risk in three strains ranged from 75% to 100%. In three other strains it was from 10% to 45%, and in five it was zero. In these five "resistant strains," the risk at 4 weeks ranged from 10% to 90%, and it also increased for two of the three "intermediate strains." Some strains with high seizure risks had few deaths; in others, only about 10% survived the series of tests.

In four strains, 50% or more of the subjects died as a result of convulsion. In the remaining seven strains, there were sufficient survivors to indicate that at six weeks of age susceptibility could still be very high (four strains) or very low (three strains).

Data as a whole indicated that the changes in seizure susceptibility and severity over the 4-week period followed a characteristic course for each strain. Any genetic analysis must be concerned with these age-related changes, as well as with the possible effects of repeated exposure. The latter has proved to be extremely important.

Acoustic priming

At the time of the Fuller-Sjursen study, it was recognized that there were persistent effects of auditory stimulation on seizure susceptibility, but these were generally considered to be transitory. In fact, repeated challenges by sound were routinely used by most investigators who were trying to determine the mode of inheritance of seizure susceptibility. Furthermore, a convention arose of testing subjects at 30 days of age, which happened to be optimal for demonstrating a difference between the DBA/2J and the C57BL/6J strains.

In 1967, however, Henry found that resistant C57BL/6J mice could be made highly susceptible if they were exposed to bell ringing between 15 and 24 days of age and tested 3 days later. At almost the same time, Iturrian and Fink (1968) and Fuller and Collins (1968b) reported the same priming effect (so named by Henry) in other strains.

The importance of priming to research on seizures is obvious, and it has been studied in a variety of strains. The following account is a composite of several studies. Priming can be demonstrated as early as 14 days of age at about the time that auditory reflexes are apparent. The maximum age at which priming can induce susceptibility seems to vary considerably; in the SJL strain it extends to at least 4 months. In the same strain the onset of seizure susceptibility after a 1-minute exposure to sound requires 30 to 48 hours (Fuller and Collins, 1968b). The priming effects can be restricted to one side by plugging an ear with glycerin during the first exposure to the sound stimulus (Fuller and Collins, 1968a). Such unilaterally primed mice will convulse on a test trial only if stimulated through the previously open ear; they are resistant if that ear is blocked with glycerin. The data suggest that priming depends on changes in parts of the auditory system associated with input from one ear, and not on changes in overall level of emotionality.

At this time the most widely accepted explanation for acoustic priming involves hypersensitization of portions of the auditory system (Gates, Chen, and Bock, 1973; Henry and Saleh, 1973; Willott and Henry, 1974). The hypersensitization is believed to be due to reduction in input to the system because of damage to the organ of Corti. Supporting this hypothesis is the fact that puncturing the tympanic membrane, which produces a similar reduction in input, also has a priming effect (Chen, Gates, and Bock, 1973).

The recognition of acoustic priming has forced a reevaluation of genetic studies in which susceptibility to seizures was determined by multiple exposures to the sound stimulus. It also has led to comparisons between the genetic bases of spontaneous and priming-induced seizures. At the psychophenic level they appear identical, but this does not mean that priming produces the same state in "resistant" strains that genes do in "susceptible" strains. For that matter, both kinds of susceptibility may themselves be polymorphic.

Mode of inheritance

Armed with information on the audiogenic seizure phenotype, its developmental characteristics, and the priming phenomenon, we can consider research on the mode of inheritance of susceptibility. It might seem a simple matter, given inbred strains differing reliably in seizure risk at certain ages, to distinguish between competing hypotheses. Basically there are three major candidates. Susceptibility in house mice may be inherited in a simple Mendelian pattern, all susceptible strains being fixed for one allele, and all resistant strains for others. Instead, it may be polygenically determined by many factors with small individual effects, none of which is necessary or sufficient in itself. A third possibility is genetic polymorphism in which several independent genes produce seizures that are indistinguishable by behavioral observation, though they may be caused by different physiological factors. Combinations of these hypotheses are also possible. We shall see later that a similar choice of genetic models is available for human disorders such as schizophrenia. One expects that such issues would be difficult to resolve with human data because of the impossibility of performing true genetic experiments. It may be disappointing to learn that even in the mouse agreement on the mode of inheritance has not been easy to reach.

Part of the problem stems from the fact that some investigators have defined susceptibility on the basis of a single trial; others have given a series of trials and classified an animal as susceptible if it convulsed on any one of them. Even if there were no such thing as acoustic priming that induces susceptibility, the multitrial procedure yields results very different from the single-trial method. For example, an individual with a constant risk of 50% for seizing on one trial will be expected to have at least one seizure in 94.75% of a series of four tests. The risk of a seizure changes dramatically over time, and most investigators have standardized on a particular age to ensure consistency in their data. The cost of such standardization is loss of information concerning genetic influences on the developmental processes underlying the observed orderly changes in susceptibility.

Table 7-1 summarizes ten studies of the mode of inheritance of audiogenic seizures. The diversity of results is in part explicable by the differences in techniques just discussed. It may also reflect the fact that similar behavioral phenotypes may have different genetic and physiological correlates (Ginsburg, 1954; Fuller and Collins, 1970). A reasonable conclusion from this discordant collection of results seems to be that in some instances segregation at a single locus can account for the difference in susceptibility between specified strains under prescribed test conditions (Collins and Fuller, 1968). That this locus is not the only determinant is

Table 7-1. Summary of ten studies on mode of inheritance (MOI) of audiogenic seizure susceptibility*

Reference	Genetic method	Procedure	MOI
Witt and Hall (1949)	Cross of two inbred strains	Repeated trials	Single locus, susceptibility dominant
Fuller et al. (1950)	Cross of two inbred strains	Single and repeated trials	Polygenic, susceptibility a quantitative trait
Frings et al. (1956)	Cross of selected lines	Repeated trials	Polygenic
Ginsburg and Miller (1963)	Crosses of four inbred strains	Repeated trials	Two locus, susceptibility mainly dominant, modified by background genotype
Lehmann and Boesiger (1964)	Cross of selected lines	Single trial	Susceptibility recessive, single locus
Collins and Fuller (1968)	Cross of two inbred strains	First and second trial analyzed separately	Trial 1, single locus, susceptibility recessive; trial 2, probably polygenic
Schlesinger et al. (1966)	Cross of two inbred lines	Single trial	Polygenic
Henry and Bowman (1970)	Cross of two inbred lines	Postpriming trial only; subjects etherized during priming	Polygenic, direction of dominance varied with age of subjects
Schlesinger and Griek (1970)	Cross of two inbred lines at four ages	Single trial	Intermediate, polygenic; resistance partially dominant
Chen (1973)*	Cross of two inbred lines	Postpriming trial only	Polygenic? intermediate dominance

*With the exception of Chen (1973), crosses were made between a high seizure–risk (on first trial) and a low-risk strain. Chen crossed two strains with low risk on first trial and differing risks on second trial.

shown by the fact that inbred strains have been fixed at intermediate levels of first-trial risk (Fuller and Sjursen, 1968). The two-locus theory of Ginsburg and Miller (1963) includes a proviso that the *A* and *B* genes have epistatic relationships and that their effect on seizure susceptibility is affected by the background genotype. The most compelling evidence for the polygenic basis of seizure susceptibility comes from an experiment in which susceptible progeny from a cross between "susceptible" strain DBA/2J and "resistant" strain C57BL/6J were repeatedly backcrossed to the resistant line (Fuller, unpublished). The seizure risk over four trials was about 70% in F_1 hybrids and less than 1% in the C57BL strain. By the fifth backcross (only susceptible hybrids were bred) the frequency of susceptibles had fallen to about 1%, a clear contradiction to a simple single-locus hypothesis, which predicts a constant risk of 70%.

The issue of monogenic versus polygenic inheritance of a psychophene is not fundamental, and the progress of behavior-genetic analysis of that psychophene is not impeded by the failure to settle on one or the other hypothesis. Apparently both modes of inheritance can be demonstrated for audiogenic seizure susceptibility of the mouse. When it is possible to detect on a polygenic background the effects of segregation at a single locus, there are, of course, clear advantages for researchers interested in the mediating processes between gene and psychophene (Ginsburg, 1967).

Genetic techniques have been employed to determine the degree of independence of first-trial seizure susceptibility (seizure proneness) and susceptibility after priming. In a heterogeneous stock of mice there was no reliable relationship between the family incidence of seizures on first trial and the increment in this family due to priming on a second test (Fuller, 1975). From the same stock, lines have been selected that are spontaneously seizure-prone, resistant to spontaneous seizures but priming prone, and both seizure and priming resistant (Chen and Fuller, 1976). There are problems with interpreting such data, since there is no way of detecting a seizure-prone priming-resistant animal.

Physiological evidence supports the belief that priming does not simply serve as a substitute for the genes which make DBA/2 and LP mice spontaneously susceptible. When one inferior colliculus (a part of the brain receiving a crossed input from the auditory tracts) was lesioned in primed C57BL susceptible mice, and the lesioned animals were later tested with only one ear open, susceptibility was much greater on the ipsilateral than the contralateral side. In contrast, unilateral lesions in spontaneously susceptible DBA/2 mice gave no protection on either side (Henry, Wallick, and Davis, 1972). Priming sets in motion processes which produce a psychophene apparently identical to that associated with seizure-prone genotypes. Are the two psychophenes really the same? The answer depends on the level of analysis that we choose.

Physiological basis of seizure susceptibility

Demonstrations of genetic variation among mice in their response to intense high-frequency sound are numerous and convincing, but they do not by themselves explain why some animals convulse and others simply startle and pursue their normal activities. Inbred strains and selected lines are valuable aids in the search for the physiological bases of the dramatic seizure syndrome. Each inbred strain has a predictable risk, and one can for practical purposes eliminate genetic variation as a source of individual differences. The differences among strains are comparable to those which might be found among individuals in a heterogeneous population. Thus the physiologically minded investigator can impose the effects of age, brain lesions, drugs, and the like on an array of strains and hybrids and look for differences attributable to genotype.

There is some disagreement as to whether audiogenic susceptible mice are also more sensitive to other convulsive agents. Audiogenic seizure–prone strains have been reported to have lower thresholds for chemically induced (pentylenetetrazol [Metrazol]) and electrically induced seizures (Schlesinger and Griek, 1970). In a different set of strains, however, Castellion, Swinyard, and Goodman (1965) found no association between electroshock threshold and audiogenic susceptibility.

That the brain amines are related to susceptibility is more certain. Amounts of both norepinephrine (NE) and serotonin (5-HT) are higher in the brains of resistant C57BL than in the brains of susceptible DBA/2 (Schlesinger and Griek, 1970). Moderate doses, below the incapacitating level, of reserpine and tetrabenzene that deplete brain amines also increase seizure risk and intensity in both strains. Their effect on amine levels was greater in the DBA/2 mice than in the resistant C57BL/6 and the F_1 hybrid between these strains. Parachlorophenylalanine, which interferes with 5-HT synthesis, and alphamethyltyrosine, which similarly reduces NE production, prolonged the period of seizure risk in the DBA/2 strain. As might be predicted, increasing the brain amines by administering monamine oxidase inhibitors protected against seizures. Other pharmacological studies support the importance of amines in the regulation of seizure resistance (Lehmann, 1967).

That 5-HT rather than NE is the critical substance is supported by the finding that the increased susceptibility produced by reserpine can be counteracted by administration of 5-hydroxy-DL-tryptophan, a precursor of 5-HT (Boggan and Seiden, 1973). A serotonergic system is also implicated in the genesis of audiogenic seizures in ordinarily resistant mice made susceptible by prolonged exposure to ethanol early in life (Yanai, Sze, and Ginsburg, 1975). Here, too, 5-hydroxy-DL-tryptophan protected against reserpine enhancement, and L-dopa, a precursor of NE, was ineffective.

The two loci that differentiate the C57BL and DBA groups of strains according to Ginsburg and Miller (1967) have been inferentially associated with two biochemical systems (Ginsburg et al., 1969). Their A locus appears to modulate nucleoside triphosphatase activity in the brain, particularly in the hippocampus. Genetic studies suggest that their B locus influences glutamic acid decarboxylase (GAD) activity. Both systems are important in the balance of excitation and inhibition in the brain.

A somewhat different genetic approach is the exposure of resistant strains to x-irradiation and screening the offspring of the irradiated animals for audiogenic seizure susceptibility. If such individuals breed true, there is presumptive evidence for a mutation affecting neural excitability. Several of these lines have been produced (Ginsburg, 1967). In one of them the level of GAD activity is like the original C57BL/6 ancestor until 25 days of age, when it rises to the characteristic DBA/2 level. Coincident with this chemical shift is an increased seizure risk characteristic of the DBA.

Results of this kind do not exactly fit the conventional notion of metabolic errors that set their possessors definitely apart from "normal" individuals. Instead they point to mutant genes whose function is the fine tuning of development. Phenotypic differences between mutant and parent strain may be limited to brief periods in their life history, and their importance to the welfare of their possessor may be dependent on just what environmental stresses are encountered during that time. At present the detection of such mutations, as those described by Ginsburg et al. (1969), depends on chemical analysis or exposure to an intense sound at a suitable age. But it would be unwise to assume that their sole function is the regulation of audiogenic seizure susceptibility.

Audiogenic seizures as a model system

In many ways the investigation of audiogenic seizures in mice by psychologists, geneticists, physiologists, and pharmacologists resembles the multipronged attack by other scientists on the etiological factors in psychiatric disorders of man. Obviously, audiogenic seizures are not like psychoses. But we believe that the careful study of this model system is instructive for students whose major interest lies either in the genetic or the environmental etiology of aberrant human behavior. In both humans and mice we find genes and experience inextricably interwoven in the production of behavioral phenotypes. Neither nature nor nurture in isolation can adequately explain why individuals differ in susceptibility to stress over their life span.

PAW PREFERENCE IN MICE

The third case study of this chapter demonstrates that behavior-genetic analysis can be useful in ascertaining that a form of behavioral variation is not heritable. Logically, attribution of behavioral variation to an environmental factor cannot be regarded as proven unless genetic sources of variation have been excluded by (1) the use of genetically homogeneous strains as subjects or (2) assurance of random sampling from a population of specified genetic characteristics.

The paw preference of hungry mice can be readily determined by requiring them to reach into a narrow tube to withdraw pieces of food (Collins, 1968b, 1970b). Most mice show definite right or left paw preference. In a wide variety of inbred strains and hybrids the probability of either outcome is approximately 0.50. Inbred strains such as DBA/2J

and C57BL/6J, which differ markedly on many behavioral characteristics, are practically identical with respect to paw preference (Fig. 7-3). Since inbred strains are genetically homogeneous, there can be no doubt that the variance in paw preference is environmentally determined.

Further evidence for nonheritability of the direction of laterality has been obtained from selection experiments (Collins, 1969). Conceivably, even within an inbred strain there might be residual heterozygosity that is manifested in paw preference. The supposition is unlikely, given the uniformity of the data from several strains; nevertheless, it has been tested. Breeding exclusively from right-pawed or left-pawed C57BL/6J mice resulted in offspring with the same 50%-50% split in their laterality. This result not only rules out a genetic explanation, but excludes any simple hypothesis of maternal influence or parental tutoring.

The environment does have an influence. When C57BL/6J mice were reared in biased worlds (cages constructed asymmetrically so that it was much easier for a resident to use one paw than the other for food gathering), they continued to use that paw when tested in an unbiased choice apparatus (Collins, 1975). This result might seem to complete

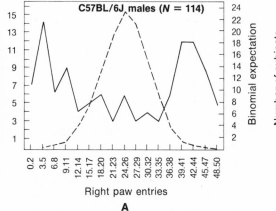

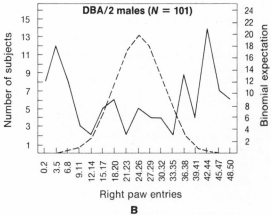

Fig. 7-3. Frequency distribution of lateral preference scores for two inbred strains of mice. Preference was measured by the number of times the right paw was used to withdraw food from a narrow tube. *Dashed line* in both **A** and **B** represents the expected frequencies if individual preferences did not exist. (From Collins, R. L. 1968. J. Hered. **59**:9-12.)

the demonstration of the mode of environmental determination. However, there is more to tell. Among control subjects reared in unbiased worlds, females were more strongly lateralized than males. Having an XX or XY genotype does not influence the direction of preference but does modulate its strength. More interesting is the finding that mice reared in biased worlds, tested, retrained with an opposite bias, and retested split into two groups. On the second test one half shifted preference readily to the reverse condition; the other half tended to retain the laterality consistent with their original training. When the data from the easy shifters and the difficult shifters were separated, a reanalysis of the first test results showed that the easy shifters had been less strongly lateralized by their original training. Collins believes that originally these mice were trained contrary to their preexisting tendencies for right or left lateral preference. A biased world can force all its inhabitants to a uniform preference, but it is more effective in individuals with a preset concordant bias. Also, a biased world is not essential for the development of paw preference; remember that strong lateral preference developed in mice whose experience was limited to standard unbiased cages.

Since genetics has been excluded as the source of directional preferences, and environmental factors have been shown to be effective but not essential, what can we conclude? One way to regard the situation is to postulate a mechanism (presumably genetically programed in the brain) that is unstable and that has an equal probability of tilting toward right or left dominance. From this point of view, paw preference is not inherited, but the capacity—perhaps even the necessity—to develop a paw preference is. In the mouse, the probabilities of right or left preference are equal, but unequal probabilities are consistent with the model and could lead in other species to a preponderance of dextrality or sinistrality. Birnbaum (1972) has discussed this concept of a random phenotype, applied it to morphological data,

and suggested its relevance to behavioral phenotypes.

Collins (1970b) has extended his analysis to human data and argues that in human beings, as in mice, the direction of laterality is determined by nongenetic factors. This extension has been vigorously contested by Nagylaki and Levy (1973) and will be considered in Chapter 13. For our present purpose this controversy is not important, since the mouse data are extremely consistent and convincing. Their significance is the demonstration of phenotypic differences not attributable to genetic variance, differences that are responsive to directed training but which are not dependent on any obvious external push. The result is a seeming paradox: maximal phenotypic variation with minimal genotypic variation within strains and minimal phenotypic variance between strains that are very dissimilar genetically. Such relationships are probably unusual, but they can occur, and their possibility must be kept in mind in the interpretation of data from animals or humans.

ALBINISM AND BEHAVIOR

Our fourth case history is concerned with attempts to specify and explain behavioral variation associated with albinism. As noted earlier, the mode of inheritance of albinism, autosomal recessive, is clearly established, and research can be concentrated on the psychophenic correlates of this distinctive somatophene. In addition to their lack of melanin pigment in skin and eyes, albinos are characterized by a diminution of tyrosinase and dopa-oxidase.

Albinism occurs in many vertebrates, but its rarity in natural populations testifies that it impairs fitness. An albino mouse would be a conspicuous prey for a hungry cat; an albino cat might be easily avoided by potential victims. In laboratories albinos are very common, and the volume of research on white rats and mice almost certainly exceeds that performed with pigmented members of these species. Occasionally concern has been expressed over the dangers of generalizing from a few species and particularly from a mutant (Beach, 1950), but with little effect on actual practice.

Escape behavior

A mouse dropped into a tub of water swims vigorously until it is either fatigued or locates a means of escape. In the water-escape task, a ladder is hung in the tub opposite the point of immersion, and the rate at which escape time decreases on repeated trials is taken as a measure of learning. Winston and Lindzey (1964) measured escape times in a number of pigmented and albino inbred strains and in the F_1, F_2, and backcross progeny of some of them. They found albinos to be consistently poorer in performance. Before concluding, however, that the behavioral effects were direct consequences of being homozygous for the albino gene, it was necessary to exclude other interpretations. Conceivably both of the albino strains used by Winston and Lindzey carried other recessive genes, closely linked with the c locus, which interfered with escape learning. The white coat and pink eyes could be merely external markers for the genes that really were affecting escape behavior. It was pointed out, for example, by Meier and Foshee (1965), that some albino strains (noninbred) perform as well as pigmented mice on the water-escape task. Although suggestive, this fact neither disproves a pleiotropic effect of albinism on water escape in the Winston and Lindzey study nor proves the existence of linked genes, which are the true effectors.

The use of congenic lines made it possible to decide between the two interpretations. An albino mutant appeared in the well-established black C57BL/6J strain. By back-crossing these mutants (c/c) to the stem line $(+/+)$, heterozygotes were obtained and interbred to yield c/c, $c/+$, and $+/+$ offspring in the expected $1:2:1$ ratio. This process can be repeated indefinitely, and it provides the experimenter with pigmented and albino animals differing at only the c locus. These congenic albinos were also inferior to their pigmented siblings in water

escape, and it was concluded that the albino locus actually is an effective agent in producing a behavioral change (Fuller, 1967).

Open-field behavior

Albino mice and rats have also been compared with pigmented individuals in the open field. This apparatus has long been popular for the measurement of "emotionality" and "reactivity" (Chapter 10). It consists of an enclosed arena, scaled to the size of the animal to be tested, which is subdivided into squares. Two scores are generally obtained: the number of squares entered during a standard time period (ambulation) and the number of fecal pellets deposited. Albino mice are less active and defecate more than pigmented ones whether the gene is on a heterogeneous background (DeFries, Hegmann, and Weir, 1966; DeFries, 1969) or a congenic one (Fuller, 1967; Henry and Schlesinger, 1967). But this is true only if the open field is brightly illuminated. When the two types of animals are tested under red illumination, the difference disappears, as shown in Table 7-2. Apparently, the unpigmented eye of the albino allows so much light to reach the retina that the brightly illuminated open field is aversive, and for most mice the prepotent response to moderately aversive stimuli is freezing.

Table 7-2. Effect of level of illumination on open-field behavior of albino (c/c) and pigmented $(+/-)$ mice*

	Number tested		Activity†		Defeca-tion‡	
Illumination	c/c	$+/-$	c/c	$+/-$	c/c	$+/-$
White light	39	37	8.8	12.9	2.10	1.95
Red light	38	38	13.3	14.1	1.73	1.76

*From DeFries, J. C., J. P. Hegmann, and M. W. Weir. 1966. Open-field behavior in mice: evidence for a major gene effect mediated by the visual system. Science **154:** 1577-1579. Copyright 1966 by the American Association for the Advancement of Science.

†, square root of total number of squares entered.

‡, square root of (total boluses + ½).

How important is the effect of the c locus on open-field activity relative to the effect of genes without conspicuous external manifestations? DeFries and Hegmann (1970) made such calculations as a part of their long-term selection program for high- and low-activity mice. About 12% of the additive genetic variance for ambulation and 26% of that for defecation was attributable to the albino locus. Of course, these values apply only to their particular stock and test procedures. In any natural population albinism is so rare that the contribution of the c locus to any kind of behavioral variation is close to zero.

Gene dosage and behavior

Up to the present point we have considered only two alleles at the c locus: albinism (c) and full pigmentation $(C$ or $+)$. Actually, there are others with intermediate effects on tyrosinase activity and coat color. Thiessen, Lindzey, and Owen, (1970) compared the effects of five such alleles on eight sensorimotor tasks. Some of their mutants had reduced coat color but pigmented eyes. Their objective was to learn whether the behavioral consequences were quantitatively related to the degree of inactivation of the pigment-forming system. Significant effects of genotype were found in six of the eight tests, but they were not clearly related to tyrosinase activity. Some of the behavioral differences were explicable by photophobia of the pink-eyed mutants, but others were not. Consequently, the authors suggested that the c locus when homozygous may "disrupt sensitivity to any environmental cue of a negative nature."

In all experiments reviewed thus far, comparisons have been made between mutants (c/c) and pigmented mice that may be either $(+/c)$ or $(+/+)$. The latter two genotypes are distinguishable by breeding tests but appear to be phenotypically identical. That this identity does not extend to behavior has been demonstrated by Henry and Haythorn (1975). These investigators produced animals of the three genotypes on an inbred background by appropriate breeding procedures and compared them on audiogenic seizure susceptibility, ease of acoustic priming, threshold for an auditory evoked potential (AEP), and rate of growth. Their results, shown in Table 7-3, clearly demonstrate that at 16 days of age albinos, compared with $+/+$, were retarded in growth and had high AEP thresholds. Heterozygotes were less affected but were still significantly lighter and less sensitive to sound than the $+/+$ pigmented mice. At 21 days of age the differences in weight and AEP threshold were greatly attenuated; the mutants and heterozygotes seemed to have caught up. The audiogenic seizure test, however, detected delayed effects attributable to the albino gene. At 16 days albino mice had a measur-

Table 7-3. Effects of albino gene on developmental indices in congenic C57BL/6J mice*

Measure	Age	$+/+$	$+/c$	c/c	Probability
Audiogenic seizures, $\dfrac{N \text{ convulsed}}{N \text{ tested}}$	16 days	1/47	0/47	11/36	0.001†
Audiogenic seizures, mean intensity§	21 days	0	20	45	0.005‡
Body weight, grams	16 days	9.97	8.18	6.89	0.0001‡
Body weight, grams	21 days	8.96	7.96	7.5	0.05‡
AEP|| threshold, decibels	16 days	22	28	37	0.0001‡

*From Henry, K. R., and M. M. Haythorn. 1975. Albinism and auditory function in the laboratory mouse. I. Effects of single gene substitutions on auditory physiology, audiogenic seizures and developmental processes. Behav. Genet. 5:137-149.
†, chi-square test.
‡, linear trend analysis of variance.
§, intensity score based on wild run, 25; clonic seizure, 50; tonic seizure, 75; fatal seizure, 100.
||, auditory evoked potential.

able seizure risk; both homozygous and heterozygous pigmented mice were resistant. However, when the pigmented mice were retested at 21 days, the heterozygotes had more severe seizures than the homozygotes. A single albino gene, not detectable by external criteria, can modify the timetable of development in such a way as to make a mouse more susceptible to acoustic priming. One wonders whether further research will reveal additional cases of cryptic genes that heighten vulnerability to stress during sensitive periods of development. The behavioral effects of such genes would be evident only if a stressful event occurred at a particular point in development.

Albinism and the nervous system

The somatophenic effects of albinism are not solely peripheral. In mammals, mutations at this locus have striking effects on the routing of nerve fibers in the visual pathways of the central nervous system (Guillery, 1974). As shown in Fig. 7-4, an illuminated point in the central portion of the visual field produces on the retinas of both eyes an image that generates a train of impulses in a neuron of the optic nerve. Some of these neurons, those originating in the medial part of the eyeball, cross to the opposite (contralateral) side at the optic chiasma. Others from the lateral portion of the retina bend at the chiasma and remain on the same side (ipsilateral). The routing of the neurons ensures that information from both eyes about a given point in the visual field arrives in adjacent areas of the lateral geniculate nucleus, a relay station in the visual pathway. Here the information is processed and passed along to the visual cortex. Note that in pigmented animals the left half of the cortex receives the neural representation of the right visual field, and vice versa.

Guillery and colleagues (Guillery, Amorn, and Eighmy, 1971; Guillery, 1974) and Lund (1965) among others have demonstrated that in albinos of several species (and in related *c*-locus mutants such as Siamese cats) there is a deficit in the ipsilateral projection of the retina on the lateral geniculate. Axons that normally form this projection cross to the opposite side. The result is that the superposi-

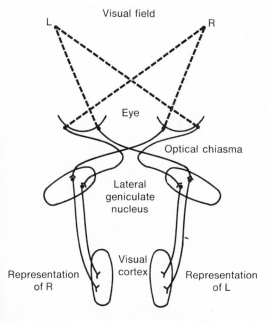

Fig. 7-4. Diagramatic representation of the visual pathways in pigmented mammals.

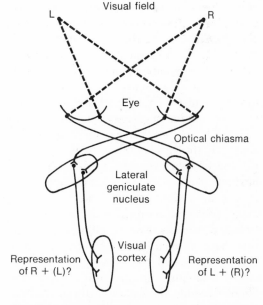

Fig. 7-5. Diagramatic representation of the visual pathways in an albino mammal.

tion of information from the two eyes in the lateral geniculate cannot occur (Fig. 7-5). It would appear that the visual cortex must receive a muddled message regarding the visual environment.

Actually, this miswired visual system works better than one might expect. Siamese cats and albino rats are not blind and seem to have vision adequate for their ordinary needs. Apparently the nervous system copes with the problem of the misdirected nerve fibers either by suppressing one of the conflicting sources or by rerouting some of the geniculate-cortical fibers (Guillery, 1974).

Tasks that are more demanding than simple form discrimination reveal deficiencies in albinos. Hooded rats (which have pigmented eyes) show much better interocular transfer of a visual discrimination learned with one eye occluded than do albinos (Sheridan, 1965). Albino and hooded rats learned a black-white discrimination equally well, but albinos were much more seriously impaired in relearning the skill after bilateral ablations of the visual cortex (Meyer, Yutzey, and Meyer, 1966).

A particularly interesting use of the *c* locus for investigating the nature of the processes involved in guiding the direction of growth of nerve fibers has been reported by Guillery et al. (1973). In flecked mice a portion of the chromosome bearing the *c* locus has become attached to the X chromosome. In females of the $X.c/X.+$ genotype, one or the other X chromosome is randomly inactivated early in development (p. 38). This leads to patches of white spotting on the coat and in the retina, which on the average involve half the cells. If the ipsilateral-projecting retinal neurons take their course depending on their intrinsic genotype, one would expect that flecked mice would have about one half as many ipsilateral neurons as do normal controls and twice as many as albinos. Actually they have as many or possibly more than normal pigmented mice. It appears that the effect of albinism on the routing of nerve fibers depends on factors external to the neurons. From the neuron's point of view

these factors are in the environment; from the organism's point of view they are internal and clearly heritable. What we classify as genetic or environmental control will then depend on the level of organization with which we are dealing.

Evaluation of studies on albinism

The preceding case history illustrates the usefulness and the limitations of the genotypic approach to behavior genetics. Demonstrations of behavioral differences between albino and pigmented animals are chiefly valuable as warnings that experimental results with rare (in nature) mutants must be generalized with caution.

Behavioral studies of mutants do not seem likely to lead to a better understanding of the processes whereby adaptive behavior patterns have evolved or how genetic variation in behavior is maintained. The well-documented interaction between lighting conditions and albinism with respect to open-field activity is also of minor psychological interest, although it does serve as a reminder that genes with less conspicuous somatophenic effects may also contribute to variation in sensitivity to environmental change.

Finally, although neuropsychological studies of visual processing with albinos must be regarded as suspect for characterizing the capacities of nonmutants, such studies do have a unique value. The *c*-locus mutations have proved to be precise instruments for changing the internal connections of the central nervous system in ways that no drug or scalpel could duplicate. For the developmental neuropsychologist the fact that albinos are rare in nature and rather poor representatives of their species is no handicap. By introducing a perturbation into the visual system, he can learn more about how it operates normally. In the same fashion the demonstration that heterozygotes for the albino gene are retarded in some aspects of development should lead to their use in the analysis of the ontogeny of behavior. Thus the albino gene has proven value for developmental and physiological studies.

SUMMARY

What general conclusions may be drawn from the four case histories that deal with diverse behavior and species? Perhaps the most important are these.

1. Genotypic effects on behavior are easily demonstrated in both laboratory and natural populations. The only important exception is within inbred strains that are genetically homogeneous.

2. It has proved relatively easy to change the average behavior of a genetically heterogeneous population through selective breeding. Although rare and extreme phenotypes may become common as a result of artificial selection, in nature intermediate phenotypes seem to have greater survival value. Genetic heterogeneity enables a population to adjust its average phenotype upward or downward as environmental conditions change.

3. Behavior-genetic analysis can be useful in demonstrating that a form of behavioral variation, paw preference in mice, for example, is not genotype dependent.

4. Behavioral development always involves the coaction of a genotype and an environment. Thus the effects of a gene substitution on behavior depend on the external milieu to which the organism is exposed. In an experiment the dividing line between genotype and environment depends on the nature of the questions being asked.

5. Genetic variants, particularly mutants, are useful in physiological studies of behavior. Genes can produce neurological or biochemical conditions that are unachievable by other means.

6. Although gene substitutions frequently produce striking modifications of behavior, there is no point-to-point equivalence of genes and behavior patterns.

8

Ingestive behavior

Animals need water and nutrients to survive. In general, intake of these substances is related to metabolic needs. The adult organism, if food and water are available, maintains a relatively constant body size. An animal deprived of food or water can usually be depended on to learn a task that will make the missing substance available. A great deal of learning and motivation theory in psychology is based on results obtained with severely deprived animals. The severe deprivation technique has been criticized as too limiting by Moran (1975), who suggests more use of "free behavior situations" by experimental psychologists. Clearly, a very small proportion of animal or human learning outside the laboratory occurs in individuals maintained at 80% of their normal body weight.

The reason for the practice is, of course, that a food- or water-deprived animal is more responsive to stimuli emanating from or associated with that substance and does not waste the experimenter's time by reacting to stimuli irrelevant to the solution of the problem to be learned. Indeed, an animal that does not react selectively to food when depleted of nutrients would be unlikely to survive. Genetic variation in the efficacy of severe deprivation as a motivator could scarcely be tolerated. Under less stringent conditions, however, there might be room for variation in preference for flavors, the temporal pattern of ingestion, and the set point for regulation of the amount of food or water ingested. In fact, there might be an advantage to a species in retaining genes that predispose to such diversity, whose very existence would be undetected in experiments using severe deprivation. In free behavior situations, variation in ingestive behavior attributable to

genes has been reported consistently, and examples of such findings are the subject matter for the first part of this chapter.

In the second portion of the chapter we consider the ingestion of a substance that is not usually a significant part of the diet of any animal. Ethyl alcohol (ethanol) supplies calories, but people consume it for its behavioral effects rather than as a source of energy. Alcoholism is a serious social problem in many human groups. Although cultural factors are certainly responsible for much of the variation between individuals in their use of ethanol, there has been considerable speculation regarding a possible genetic contribution to such usage. (See Chapter 17 for a discussion of evidence on this issue from human studies.) The roles of genes and experience, both on spontaneous intake of ethanol and on its subsequent behavioral effects, have been extensively studied in laboratory animals. Some of the findings may have relevance for the problem of drug abuse in general and could lead eventually to a better understanding of the nature of addiction.

Throughout this chapter emphasis is placed on the use of genetic variation to investigate the physiological and psychological bases of ingestion, a very important kind of behavior. Mutants, inbred strains, selected lines, and heterogeneous stocks have all been employed in the search. To a considerable extent, research on genetic factors in ingestive behavior has been motivated by the desire to learn more about the etiology of human disorders such as obesity and alcoholism. It should be stressed that animal models are always imperfect replicas of human disorders and that, sometimes, their resemblances are superficial. Still, the use of such models is the only way in which many problems, particularly those of early development and prevention, can be studied experimentally. At the very least, animal model research is a fruitful source of hypotheses to be tested by observations on humans.

INTRASPECIFIC DIFFERENCES IN TASTE PERCEPTION

Before proceeding to the main theme of this chapter, the regulation of ingestive behavior, it is of interest to review intraspecific variation in taste preference and sensitivity. Taste and odor are important factors in the acceptability of food and fluids and are probably an important determinant of differences in food habits. It will become apparent as we consider genetic factors in obesity and alcohol consumption that some of the effects observed could be mediated through the sense of taste.

Data on taste preference is commonly expressed as a preference ratio, $R = \dfrac{T}{T + S}$, where T is the intake of a test solution or substance and S is the intake of water or of a standard food. The term *preference* is somewhat misleading in this context. A value of 0.5 can indicate either inability to detect the test substance or indifference to the consequences of ingesting it. Also, values of the ratio significantly below 0.5, interpretable as aversion or "negative preference," are designated by a positive number. To reduce the confusion inherent in this measure, Klein and DeFries (1970b) proposed a preference index, $I = 2(R - 0.5)$. Using this index, complete preference for the test substance is scored as 1.0, indifference as 0.0, and complete avoidance as -1.0. Although this index seems to have advantages, it has not yet been widely adopted. Klein and DeFries also proposed a taste sensitivity index, the absolute value of the preference index irrespective of sign. This index also has potential advantages for taste research, since it depends on the intensity and not the direction of the response to the test substance.

Considerable effort has been devoted to the search for strain differences in the acceptance of propylthiocarbamide (PTC) and related compounds. All of these have antithyroid properties and are fairly toxic. These substances have attracted interest in human genetics because the ability to taste PTC in low concentrations is inherited as an autosomal dominant, and the frequency of the alleles has often been determined in population surveys (see Chapter 13 for additional information). Interestingly, a similar polymorphism exists in mice, where, as in humans, high sensitivity is inherited as an autosomal dominant (Klein and DeFries, 1970a). A fivefold variation in the threshold for discrimination of PTC was found in a set of six inbred strains (Hoshishima, Yokoyama, and Seto, 1962). A similar strain difference in the mean aversion threshold to another toxic, bitter substance, cycloheximide, has been reported in rats (Tobach, Bellin, and Das, 1974). Taste blindness was so serious in one strain that some individuals ingested a lethal dose.

Saccharin is a nonnutritive synthetic substance that many humans use as a sugar substitute. Highly acceptable to laboratory rats and mice, it is widely used in psychology laboratories as a reinforcer to test theories of motivation. Individual variation in saccharin preference is prevalent in rats, and Nachman (1959) was able in two generations of bidirectional selection to shift the preference ratio from 0.68 to 0.85 in his high line and down to 0.39 in his low line. Among mice, strain differences in saccharin preference have been reported frequently (Fuller and Cooper, 1967; Capretta, 1970; Pelz, Whitney, and Smith, 1973). The latter authors estimated a coefficient of genetic determination between 0.81 and 0.94 and heritability between 0.55 and 0.68. Fuller (1974) proposed that the difference between high saccharin–preferring C57BL/6J and low-preferring DBA/2J mice was attributable to a single locus *(Sac)* with the high-preference allele, Sac^b, dominant over Sac^d. A more extensive sampling of inbred strains and of a heterogeneous stock of mice indicates that the variation cannot entirely be explained by segregation at a single locus (Ramirez and Fuller, 1976). In this study the coefficient of genetic determination was 0.78, which is close to the value found by Pelz, Whitney, and Smith, (1973). Ramirez and Fuller computed h^2 from parent-offspring regression in their heterogeneous stock for consumption of 0.1% saccharin (0.52) and 3% sucrose (0.32). The concentrations were chosen to be as equally acceptable as possible to the subjects. The genetic correlation between consumption measures for the two sweet fluids was 0.93 ± 0.04, indicating that common genes must be contributing to the intake variability of both. Stockton and Whitney (1974) also noted that the ranking of their inbred strains on glucose and sucrose preferences was similar to the ranking of the same strains on saccharin (Pelz, Whitney, and Smith, 1973). Since saccharin and the sugars are very different chemically, the similarity is not readily explained.

The other taste modalities, sour and salty, have been explored less by behavior geneticists. However, Hoshishima, Yokoyama, and Seto (1962) reported large differences in threshold preference for sodium chloride and acetic acid in mouse strains. Probably a directed search would discover many taste polymorphisms among animals.

The existence of variations in sensitivity to and preference for a small number of substances has been clearly demonstrated. How important this variability is in the life of laboratory rodents or their wild ancestors is not as clear. In fact, the high heritability of saccharin preference probably indicates that neither preference, aversion, nor indifference to this substance has much effect on fitness. It is true that directional dominance for high-saccharin preference is shown in crosses with C57BL mice and that such dominance is considered to indicate a positive relationship to fitness. But in this respect the C57BL strains may be extreme deviants (Ramirez and Fuller, 1976). We shall see later in this chapter that they also deviate markedly from other strains in their high consumption of alcohol. To experimenters interested in analyzing mechanisms, genetic deviants are attractive, but it is risky to generalize from them to populations.

DRINKING BEHAVIOR

Conventionally, animals are expected to drink when they are thirsty, a condition brought about by a shortage of body water. The physiological bases of drinking are complex, and there are still debates concerning the relative importance of various control mechanisms. Texts in physiological psychology and motivation should be consulted for discussions of these issues. Many experiments on the physiological basis of drinking behavior involve rather drastic procedures, such as severe water deprivation, injection of hypertonic saline solutions, or withdrawal of blood. The information gained from such studies may be more relevant to backup emergency mechanisms that function during stress than to patterns of drinking when water is continuously available and there is no significant dehydration of the body as a whole. The dedicated beer or coffee drinker takes in more water than he needs and disposes of the excess through the kidneys. Peripheral stimuli such as dryness of the pharynx, attributes of a liquid such as temperature and taste, social facilitation, and habitual temporal patterns of activity are certainly involved in the day-to-day patterning of drinking. Genetic variation in drinking might well involve responses to some of these factors rather than responses to changes in osmotic pressure or volume of body fluids.

There are great differences in the amount and frequency of drinking among species. Camels have pouches along their digestive tracts that can store water, enabling them to go for long periods without drinking, and they can tolerate considerable loss in blood plasma volume. A few desert rodents can subsist when necessary on the water obtained from the metabolism of foodstuffs. The laboratory rat and mouse require water, most of which they ingest at intervals during feeding periods, the so-called prandial drinking. However, all members of a species do not drink identical amounts. Strain differences are commonly found; mutant genes are known to induce polydipsia (excessive drinking), selection can produce lines with high or low water intake, and genetic differences are implicated in the degree to which fluid intake can be modified by adding flavors to water. Thus, even in responding to the basic need for water, one finds variation to a surprising degree.

Normal variation

The daily intake of freely accessible water among seven inbred strains of mice varies more than twofold (Kutscher, 1974). Females of the highest strain (SWR/J) had an average daily intake of 11 ml; those of the lowest strain (A/J), 4.7 ml. Only a part of the variation was accounted for by differences in body weight. Under partial food deprivation one strain (C3H) became polydipsic, three (SWR, CBA, DBA/2) kept water intake constant, and three (BALB/c, A, C57BL/6) decreased their drinking. These strains are all considered physiologically normal, though SWR mice become polydipsic at advanced ages. What, then, is the "normal" response to food deprivation in "the mouse"? It appears that there are at least three possibilities, depending on one's choice of subjects.

Genetic effects on normal drinking behavior of mice have also been detected by parent-offspring and sibling correlations in a heterogeneous stock (Ramirez and Fuller, 1976). The heritability of free water intake (from parent-offspring regression) was 0.44. Differences in body weight accounted for only 10% of the variance. The addition of saccharin (0.1%) or sucrose (3%) increased fluid intake, but the size of the increase was related to prior water intake. The high genetic correlation (0.84 ± 0.07) between water and saccharin solution intake suggests that the same gene-influenced mechanisms are regulating the ingestion of both.

Selection was highly effective in raising the water intake of rats and moderately effective in lowering intake (Roubicek and Ray, 1969). Heritabilities varied from 0.31 to 0.07, depending on the temperature at which the animals were maintained and the direction of selection. It should not be surprising that selection for high water intake was more effective at a maintenance temperature of 35° C

and for a low intake at 22° C. The heritability of a behavior is determined in part by the adaptive value of a directional change.

In summary, genetic variation in the regulation of drinking has been clearly demonstrated in rats and mice living under ordinary laboratory conditions and free of any obvious pathology. We cannot at this time explain why such variability exists, though we may hypothesize that it might have adaptive value for a species living in the wild under somewhat inconstant ecological conditions. At any rate, physiologists and psychologists studying the nature of behavioral regulation of water intake or using water deprivation as a motivating technique should be aware of the fact that "normal" rodents of the same species differ in their drinking patterns and in their mode of adjustment to environmental changes.

Polydipsia

The line between polydipsia and the higher range of normal intake of water is somewhat arbitrary. Polydipsia is one part of the syndrome of diabetes insipidus; the other component is a large volume of dilute urine (polyuria). At least four strains of mice, DE (Chai and Dickie, 1966), MA (Hummel, 1960), DI (Naik and Valtin, 1969), and SWR (Kutscher and Miller, 1974) show polydipsia that is generally more severe in older animals. The Brattleboro rat also develops diabetes insipidus (Saul et al., 1968). The physiological bases of the disorder vary: cystic degeneration of the pituitary (MA mice); faulty salt resorption in the kidneys (SWR mice); a hypothalamic disorder (Brattleboro rat). Diabetes insipidus is also a pleiotropic effect of the semidominant mutation, oligosyndactyly (*Os*), which is named for its effect on fusion of the middle digits of the feet (Falconer, Latyszewski, and Isaacson, 1964). In this mutant the kidneys are not only small, with 20% of the normal number of glomeruli, but they are also refractory to antidiuretic hormone.

The polydipsias associated with diabetes insipidus are of greater interest to physiologists than to psychologists. Other polydipsias, such as those produced by adding saccharin to water and the schedule-induced polydipsia of Falk (1961), are not as clearly related to physiological defects. Schedule-induced polydipsia is the excessive intake of water by food-deprived animals when food pellets are made available from a device operated by the animal that pays off at intervals of 45 to 180 seconds. Between pellet deliveries the subjects drink excessive volumes of water. Explanations of the phenomenon range from physiological factors such as dehydration produced by the dry food to the hypothesis that drinking serves as a timing cue for operating the feeder. Whatever the explanation, it cannot be applied universally. C57BL/6 mice do not become polydipsic under the presumed optimal conditions; DBA/2 mice do so reliably (Symons and Sprott, 1976). The occurrence of schedule-induced polydipsia in hybrids between these strains indicates that the difference is due to segregation at a single locus (Table 8-1). It is likely that similar genetic effects will eventually be found in other species.

It is currently unclear as to whether the differences found among "normal" strains or among families of heterogeneous stocks represent minor physiological deviations that are similar in kind to those reported for the polydipsic strains and mutants. Certainly the variation must have a physical basis, but it might be only indirectly related to homeostatic mechanisms of water balance. The experimenter who reinforces behavior with water should be aware of the kinds of variation in drinking behavior that exist and consider them in the interpretation of data.

FOOD CONSUMPTION
Variation in selected and inbred lines

The literature on food consumption and the regulation of eating is voluminous. Food deprivation is a common procedure for equalizing motivation when preparing animals for experiments in learning. Although species differences in patterns of eating are well recognized—herbivores forage regularly for a large part of the day; carnivores gorge when they make a kill and may desist from eating for long periods—it seems generally to be assumed that all members of a species, a few mutants excepted, are essentially the same with respect to eating behavior. Of course, big animals eat more than small ones, and young growing animals need more calories per gram than conspecifics of equal weight. Thus most genetic investigations of eating behavior have dealt with extreme obesity associated with rare (in nature) recessive mutations. This approach has been valuable in analyzing the mechanisms involved in the regulation of body weight. Such regulation must involve at least two components, an initiator that activates eating and a satiety indicator that signals the termination of a meal.

There is strong evidence, however, for considerable genetic variability in the spontaneous intake of food among domestic and laboratory animals given free access to it. Selection for rapid rate of growth has been successful in mice (Bielschowsky and Bielschowsky, 1953, 1956; Falconer and Latyszewski, 1952; Timon and Eisen, 1970; Timon, Eisen and Leatherwood, 1970), in swine (Fowler and Ensminger, 1960), and in chickens (Siegel and Wisman, 1966). In all these lines, rapid growth seems to be more

Table 8-1. Occurrence of schedule-induced polydipsia in C57BL/6J and DBA/2J mice and their hybrids*

Group	Observed polydipsia +	Observed polydipsia −	Predicted† polydipsia +	Predicted† polydipsia −
C57BL/6J (B)	0	4	—	—
DBA/2J (D)	3	0	—	—
B × D (F_1)	3	0	—	—
F_1 × B	6	5	5.50	5.50
F_1 × D	10	1	11.00	0
F_1 × F_1 (F_2)	19	9	20.25	6.75

*From Symons, J. P., and R. L. Sprott. 1976. Genetic analysis of schedule-induced polydipsia. Physiol. Behav. **17:**837-839.
†Predictions made for segregating generations on the basis of polydipsia requiring a dominant gene (Sip^d) derived from DBA/2J strain.

dependent on heightened food intake than on more efficient conversion of nutrients (Fuller, 1972). The physiological basis of the enhancement of eating is unknown, but the evidence points more strongly to the brain than to the liver. Increased eating among rapid growth lines also appears to be, in part, specific to conditions, such as diet quality, under which genetic selection is carried on.

Inbred lines that have not been subjected to selection for level of food intake also differ in their response to dietary manipulations. Increasing the fat content of the diet caused young C3H and A strain mice to become obese; in contrast young C57BL and I strain animals were unresponsive. Some of the latter two strains even suffered severe weight loss at the higher fat concentrations (Fenton and Dowling, 1953). Les (1968) observed the growth of several inbred strains at different densities: 1, 2, 3, 4, 6, or 8 per cage. Density-dependent effects on growth were observed in strains A/J and C57BL/6J, but not in AKR/J, C3H/HeJ, or DBA/2J. These differences may represent indirect effects of inherited differences in social interaction, which are expressed in many other facets of behavior.

The observed behavioral variation among the lines selected for weight gain and the "normal" inbred strains are not dramatic, but they are large enough to warrant attention in any research that involves food deprivation and food reinforcement. Perhaps under the severe conditions frequently employed in psychological experiments any genetic differences may be overridden, but this is a risky assumption.

Fat mutants

A number of rodents develop extreme obesity, usually as a result of hyperphagia, often accompanied by diabetes mellitus, sterility, and a reduced life expectancy. Bray and York (1971), reviewing much of the research on these animals, separated the obesities according to their genetic basis as (1) effects of mutation at a single locus, (2) characteristics of certain inbred strains, (3) characteristics of certain hybrids, and (4) charac-

teristics of some species of desert rodents when fed on standard laboratory feed. The mutant obesities, since they are based on a single, specific genetic change rather than on complex polygenic relationships, have received the most attention from investigators. In this section we shall concentrate on the fatty rat *(fa/fa)* (also known as the Zucker rat), the obese mouse *(ob/ob)*, the diabetes mouse *(db/db)*, and the yellow (A^y) and viable yellow $(A/^{vy})$ mice. All but the latter pair are inherited as autosomal recessives. Interest in these animals has been stimulated by their phenotypic resemblance to severe human obesities and diabetes mellitus. Although it is probable that only a small proportion of human obesity is attributable to the kinds of metabolic errors found in these mutants, they have proved useful as probes for the investigation of basic mechanisms of regulation. Since these mutants phenotypically resemble the obese rats produced by lesions in the ventromedial hypothalamus and the obese mice produced by gold-thioglucose treatment, there has been particular interest in determining if these diverse conditions involve fundamentally the same functional disturbance.

Fatty rats. Fatty rats appeared spontaneously in a stock being selected for a high rate of growth (Zucker and Zucker, 1961). Like fat mice they are hyperphagic; males are rarely fertile, and females are sterile. Fatty rats have difficulty in regulating caloric intake when their diet is diluted by cellulose or made more concentrated by the addition of fat (Bray and York, 1972). Amphetamine is less effective in reducing the food intake of fatties than of normals. Although some of these regulatory difficulties are shared with ventromedial hypothalamic–lesioned (VMH) rats, there are differences that point to basically different mechanisms for the two forms of obesity.

Other evidence also points to striking differences between VMH-lesioned and *fa/fa* rats. Vagotomy abolishes the surgically induced, but not the genetically based, obesity (Opsahl and Powley, 1974). VMH-lesioned rats reject quinine to a greater extent than either fatty or normal rats that do not differ from each other (Cruce et al., 1974). When food pellets were made available from a bar-pressing device, the response rates of both VMH-lesioned and *fa/fa* subjects were higher than those of lean rats. As fixed ratios were introduced requiring from 4 to 256 responses per pellet, genetically obese animals increased their response rate most rapidly, followed closely by normals. VMH-lesioned rats increased their response rates little, if at all (Greenwood et al., 1974). These authors believe that the genetically obese rat is better than a lesioned rat as a laboratory model for human obesity.

Fat yellow mice. The obesity of the yellow mouse $(A^y/+)$—the homozygote dies in utero—has been known for many years. Weitze (1940) showed that a yellow mouse in parabiosis (surgical union which allows some mingling of body fluids) with a normal one did not become fat. He ascribed the inactivity and obesity to a chemical deficiency, probably hormonal, that was alleviated through access to the blood constituents of the normal partner. Yellow mice eat more and are less active than nonmutant controls (Dickerson and Gowen, 1947). The viable yellow mutant (A^{vy}/A^{vy}) or $(A^{vy}/+)$ varies in coat color from yellow through mottled yellow to almost pure agouti (Woolf, 1965a, 1971). Like other yellows, they are less active than nonmutant littermates and eat more (Fuller, 1972). Interestingly, their rate of gain during the period of rapid growth is positively correlated with the amount of yellow hairs in the coat (Woolf, 1971). The striking color difference is evidently a surface symptom of a much more fundamental metabolic disturbance. Although viable yellow mice are moderately hyperphagic, they do not manifest the degree of regulatory disturbance that characterizes the obese and diabetes mutants in the mouse (Fuller, 1972).

When the A^y gene is combined with homozygous dwarfism (dw/dw) a fat, dwarf mouse results (Woolf, 1965b). Since dwarfs are deficient in thyrotropin and somatotropin, this result demonstrates that A^y obesity is not attributable to an imbalance of these hormones. Yellow and nonyellow dwarfs respond similarly to injected somatotropin, indicating that responses to this hormone are not impaired by the A^y gene. At present there is no adequate physiological explanation of the hyperphagia and inactivity of yellow mutant mice. The difficulties of tracing the gross phenotypic effects back to an enzyme (or enzymes) are well expressed in an article by Woolf and Pitot (1973). They found that both the A^y and A^{vy} genes produced quantitative effects on four hepatic enzymes, but the amount and even the direction of the change varied with the background genotype. It is apparent that even mutant genes with marked phenotypic effects interact with the total genome at a very basic level of expression.

Obese and diabetes mice. The obese mutant mouse *(ob/ob)* was described in 1950 (Ingalls, Dickie, and Snell) and has been studied intensively ever since. Reared under standard conditions, obese mice are less active and eat and drink more than normal littermates. These sterile, relatively short-lived animals usually attain a body weight two to three times that of normal mice. Because they sometimes have very high blood sugar levels, they have been referred to as obese-diabetic mice. The obesity and characteristic infertility of males may be partially prevented by restricted food intake (Lane and Dickie, 1954). Food restriction also increases life span (Lane and Dickie, 1958). In an ingenious experiment Mayer (1953) bred animals homozygous for both waltzing *(v/v)* and obesity and found that the combination produced a very active mouse which did not become grossly fat.

All these facts support the idea that much of the characteristic syndrome is a secondary effect of overeating and inactivity. In a sense, most of the bad consequences of the gene are attributable to its effects on the regulation of eating in relation to activity. In turn, of course, these effects can be considered to be sequelae of gene-determined metabolic disorders within cells.

The search for a primary biochemical deficit that might explain the behavioral deficit has not been entirely successful. In an early article Guggenheim and Mayer (1952) considered the defect to be failure to oxidize pyruvate and acetate, but this was disputed by Parson and Crispell (1955). Yen, Lowry, and Steinmetz (1968) reported slower oxidation of glucose by *ob/ob* tissues in vitro and favored an explanation in terms of cell permeability or membrane transport. Hypothalamic norepinephrine is elevated in obese mice of both sexes, and hypothalamic dopamine appears to be elevated in males, but the functional significance of this finding is unclear (Lorden, Oltmans, and Margules, 1975). Even with this clearly Mendelizing trait with definite physiological and behavioral indicators, the attempt to define a path from gene to behavior runs into difficulties at the first step of primary gene action, still remote from the neural substrate of the behavior.

Several deficiencies of intake regulation were demonstrated by Fuller and Jacoby (1955). When food was diluted with an inert filler, normal mice adjusted their intake upward in order to maintain a constant supply of calories. Obese mice did this less well. They were also less efficient in adjusting intake to need when a bitter substance was added to the diet or fat was substituted to make it both calorie rich and highly palatable. Both obese and control mice reacted similarly with decreased or increased intake on the first day of such changes. However, after a few days the normal littermate controls readjusted their eating to approximate their usual caloric requirements. Obese mice also readjusted in an appropriate direction, but less effectively, as though the regulatory mechanism were sticking. In other words, obese animals seemed to respond predominantly to the sensory characteristics of their food; normals were also regulating on the basis of caloric need (Table 8-2).

Continuous recording of the eating behavior of genetically obese mice revealed an absence of the 24-hour cycle of eating and quiescence characteristic of normal mice (Anliker and Mayer, 1956). Mature obese mice ate in short, irregularly spaced bursts, similar but not identical to the pattern of rats and mice with VMH lesions. Another peculiarity of obese mice is the absence of food deprivation–induced enhancement of saccharin intake (Fuller, 1972). Both obese mice and their normal littermates prefer 0.1% saccharin over water to about the same degree. When food is removed, the intake of the saccharin solution by normals increases threefold or more. Obese mice continue to drink saccharin at the same rate as when food is present.

Since obese mice initially respond normally to changes in the physical quality of their food, sensory deficits seem to be ruled out. Thus these data point toward a failure to respond efficiently to internal signals of caloric need. We shall return to a direct test of this hypothesis after considering another fat mutant mouse, the diabetes *(db/db)*. This mutant resembles the obese mouse in adiposity, hyperphagia, polydipsia, inactivity, and lack of the deprivation enhancement of saccharin intake (Fuller, 1972). The gene was

Table 8-2. Nutritive intake (in percentage of stable control level) of obese and normal mice given special diets*

Diet	Group	First day	Stabilized	Return to standard
Bitter	Obese	56	72	109
Bitter	Normal	58	93	148
Added fat	Obese	177	126	79
Added fat	Normal	168	95	84

*Based on data from Fuller, J. L., and G. A. Jacoby, Jr. 1955. Central and sensory control of food intake in genetically obese mice. Am. J. Physiol. **183:**279-283.

named *diabetes* because on certain genetic backgrounds it leads to degeneration of the islets of Langerhans in a fashion similar to the human disease.

Parabiosis between fat mutants. In parabiosis two animals are grafted together so that they share to some degree a common circulation. Thus humoral factors present in one member of a pair can pass into the other. The degree of communal circulation in the experiments to be described was, however, insufficient for the transfer of nutrients in large amounts; each parabiont had to feed itself in order to survive. The durability of a parabiotic graft is dependent on tissue compatibility between the subjects. Unless the parabionts are nearly identical genetically, as within an inbred strain, there is mutual rejection and necrosis of tissues at the site of joining. Herein lies the importance of placing mutations such as *diabetes* and *obese* on the same inbred background, so that parabiosis in all possible combinations can be achieved.

Coleman and Hummel (1969) joined adult

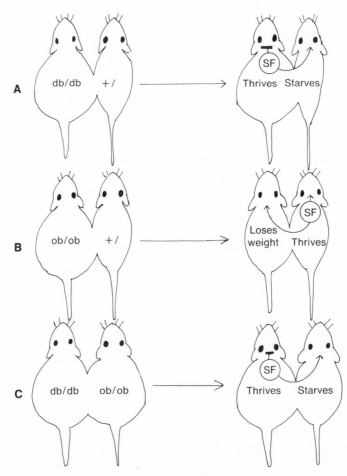

Fig. 8-1. Results of parabiosis experiments between combinations of normal, diabetes, and obese mice. (See text for explanations.) Parabiont pairs of like genotypes all survive and maintain their usual physical and behavioral characteristics. *SF*, postulated satiety factor of unknown origin. Its point of action is here tentatively designated as the brain. (Based on data from Coleman, D. L., and K. P. Hummel. 1969. Am. J. Physiol. **217**:1298-1304; Chlouverakis, C. 1972. Horm. Metab. Res. **4**:143-148; and Coleman, D. L. 1973. Diabetologia **9**:294-298.)

db/db mice on a C57BL/KsJ background with nondiabetes mice of the same strain. They probably anticipated one of three results: (1) the diabetes mouse would remain diabetic and the normal one, normal; (2) the diabetes mouse would improve because it received some missing substance from its normal partner; or (3) the normal animal would become diabetic because of a toxic material emanating from the mutant. Actually, none of these happened. The diabetes mice remained hyperphagic and obese, but their normal partners lost weight steadily as a result of not eating. They literally starved to death in the midst of plenty (Fig. 8-1, *A*). Normal mice paired with normals tolerated parabiosis with no problems.

This result was surprising in the light of reports of parabiosis between *ob/ob* and normal mice (Chlouverakis, 1972; Hausburger, 1958). In these pairings the direction of parabiotic influence was from normal to mutant. The obese member usually survived, but it ate less and gained weight more slowly (or even lost weight) (Fig. 8-1, *B*). That the effect was not due to intolerance of obese mice to parabiosis was evidenced by the survival of obese-obese pairs with continued hyperphagia and growth of both members. On the basis of overt behavior and many physiological and pathological indicators, the diabetes and obese genes produce identical phenotypes; in parabiotic experiments they are diametrically opposite.

Parabiosis between *ob/ob* and *db/db* had to await the availability of congenic stocks of both genes on the same inbred background. When the experiment was finally conducted (Coleman, 1973), the obese members of the pairs stopped eating and eventually died of starvation, while the diabetes ones thrived (Fig. 8-1, *C*). Coleman hypothesized that the results could be explained by assuming that diabetes and normal mice produce a satiety factor, SF, that in normals suppresses neural mechanisms concerned with eating but in diabetics is ineffective. Obese mice cannot produce SF but do react appropriately when it is supplied. Normal mice, of course, both produce and react to the substance, and thus match their food intake to caloric demands. The postulated sources and transfer of SF are shown on the right-hand side of Fig. 8-1. The horizontal bars represent inactivation or insensitivity.

This elegant series of experiments demonstrates the value of mutant genes in the physiological analysis of behavior. It involves genes with reliable physiological effects on inbred backgrounds. The measures of growth and food intake are highly reliable, and there is a tremendous body of information from other kinds of experiments concerning the physiological and biochemical basis of the behavior of interest.

Another general point illustrated by research on fat mutants is the influence of background genotype on the phenotypic consequences of a mutant gene or genes. We have previously commented on the behavioral similarities of diabetes and obese mice, emphasized that many of the damaging physiological and structural correlates can be ameliorated if access to food is restricted, and argued that it is the failure to regulate behavior that leads to pathology. Both the diabetes and the obese genes, when homozygous, produce much more severe pathology in C57BL/KsJ mice than in the closely related and similar C57BL/6J mice (Coleman and Hummel, 1973). Thus far, however, there is no hint from observations of the normal members of these strains as to why they differ in response to the mutant genes. Clearly the total genotype influences the phenotypic response to both external factors (e.g., acoustic priming discussed in Chapter 7) and internal factors such as mutant genes.

ALCOHOL CONSUMPTION AND SENSITIVITY

Human beings are unique as a species whose members (some of them) actively seek ethyl alcohol (ethanol) to consume. For some humans alcohol may actually provide a substantial proportion of their calories, but the explanation for the popularity of the substance is based on its pharmacological properties rather than its energy content. The toxic effects of overdoses and the behavioral effects of even moderate doses are well known and will not be discussed in detail here.

Although humans stand alone as devotees of alcoholic beverages, some animals, such as the drosophilid flies that feed on fermenting vegetable matter, ingest considerable quantities of ethanol. Robert Frost commemorated the incidental behavioral effects of such consumption in a mammalian species with his poem, "The Cow in Apple Time." As a result of the ubiquity of ethanol and other related alcohols, many animals are equipped with an enzyme, alcohol dehydrogenase (ADH), that catalyzes the oxidation of ethanol to acetaldehyde. This highly toxic substance is, in turn, oxidized to acetate or carboxylated to pyruvate. These products are in turn involved with other parts of the organism's metabolic system, particularly the tricarboxylic acid cycle. Later in this section we shall review attempts to relate genetic differences in consumption of and preference for ethanol in terms of enzymatic activities.

Interest in the free-choice consumption of ethanol by laboratory animals is undoubtedly based on the possibility that they can provide a model system for experimentation on the factors which regulate intake and produce individual differences in pharmacological and psychological reactions. (Lester, 1966; Lester and Freed, 1972). This area of research is subsumed under the general heading of psychopharmacogenetics, a triple scientific hybrid of relatively recent origin. Essentially, psychopharmacogenetics deals with inherited differences in responses to drugs that are particularly effective in modifying behavior. This offshoot of behavior genetics has obvious implications for clinical medicine (Eleftheriou, 1975; Mendlewicz, 1975). Our concern will be concentrated on the contribution of genetics to an understanding of individual and group differences in what McClearn (1968a, 1972) has called the *ethanol intake control system* (EICS). Clearly this approach is closely related to problems in the control of water and food intake that were reviewed earlier in this chapter.

Alcohol-related behavioral phenotypes

Before discussing experimental results in detail, it is instructive to review alternative procedures for measuring alcohol-related behavioral phenotypes. These have been reviewed critically by Eriksson (1969b). Perhaps the most common is the preference ratio, $\frac{A}{A + W}$, where A is the intake volume or weight of an alcohol solution, and W is the intake of water during a specified period of time. We discussed characteristics of this measure in the section on taste preference, pp. 113 and 114. The ratio fluctuates with the concentration of the alcohol solution (10% by volume is most commonly used) and is also affected by the organism's demands for water. Another measure is the absolute amount of ethanol ingested per unit of body weight; for example, grams per 100 grams of body weight. Sometimes data are given in the form of the volume or weight of the alcohol solution per unit of body weight; these can, of course, be converted to absolute units. Another measure, the proportion of ingested calories derived from ethanol, is less used. It takes account of the fact that alcohol is a source of energy as well as pharmacological effects. In all these procedures two bottles are generally provided, one with water. It is customary to reverse the positions of the alcohol and water bottles systematically. Eriksson (1969b) has shown that this practice increases intrasubject variability without modifying group differences. He advocates that position reversal be scheduled at intervals long enough to permit subjects time to learn the positions of the two stimuli. Multiple-choice procedures have been designed for special purposes (Rodgers and McClearn, 1962; Fuller, 1964; Satinder, 1972). Finally, differences in consumption may be measured when the only source of water is "contaminated" with alcohol, and no choice is possible.

The variety of methods reflects the complexity of measuring a behavioral phenotype in order to account for confounding factors such as position preference, differences in the need for water and calories, and variations in taste sensitivity and preference.

Sometimes, the alcohol solutions may be sweetened in an attempt to overcome the apparent aversive taste properties of the pure substance. The preference ratio, like other ratio measurements, is not normally distributed and should be transformed prior to using the data for quantitative genetic analysis. Fortunately, in practice the various methods of reporting data are correlated positively to a high degree. The reliability of the genetic effects is so great that major conclusions based on one system of measurement are generally confirmed by others. Still, it is possible for a low preference–ratio subject who drinks a large volume of fluid to ingest more ethanol calories than a high preference–ratio subject with a low intake. In evaluating the literature, it is important that measurement techniques be considered critically.

Individual and strain differences

It is instructive to begin our consideration of individual and strain differences in alcohol-related behavior with Williams' genetotrophic theory of disease (Williams, 1956; Williams, Berry, and Beerstecher, 1950). Briefly, this theory maintained that individuals could inherit metabolic blocks that predisposed them to alcoholism, which was defined rather loosely as the consumption of an amount producing injury. Such a genetic block, or partial block, could lead to diminished enzyme production, impairment of the ability to use one or more essential nutrients, and thence to an augmented appetite for alcohol. How the nutritional deficit produced its behavioral effects was never specified, but there are numerous possibilities.

Several experiments were reported in support of this hypothesis. Williams, Berry, and Beerstecher (1949) found that rats on a vitamin-deficient diet drank more alcohol in a free-choice situation than they did on an adequate basic diet. Two strains, O and H, were used. Feeding O rats with a vitamin supplement reduced alcohol consumption, but the same regimen had little effect on H rats. Alcohol solutions provide calories: increased alcohol "preference" might simply be a response to a calorie deficiency induced by the fact that vitamin-free diets are less palatable than stock diets. To combat this criticism, Williams, Pelton, and Rogers (1955) gave rats on a vitamin-deficient diet a choice between water and a 10% sucrose solution and followed this with a choice between water and 10% ethanol. The malnourished rats drank more sucrose solution and more alcohol solution than controls, but consumption of the two substances was not correlated. In its original form, particularly its emphasis on nutritional deficiency as a prime factor in alcoholism, the genetotrophic theory has not fared well (Lester, 1966). However, the idea of a biochemical basis for variation in free-choice alcohol consumption still guides much of the research described next.

The genetic basis of individual differences in voluntary alcohol consumption has been demonstrated by successful selection (Mardones, 1952; Mardones, Segovia and Hederra, 1953; Eriksson, 1968). Eriksson, starting from an outbred population of rats, produced a heavy drinking line (alcohol addicted, or AA) and a light drinking line (alcohol nonaddicted, ANA). These lines have been widely used in research by workers at the Finnish Foundation for Alcohol Studies. The heritability of free-choice alcohol consumption was 0.263 for males and 0.371 for females. Line differences accounted for 65.5% of total variance in ethanol intake.

It is not necessary to breed selectively in order to find strains of laboratory rodents that differ in alcohol preference. Given a choice over a wide range of concentrations (1.25% to 20%) versus water, G-4 rats had consistently higher preference ratios than Wistars (Myers, 1962). Inbred strains of mice differ strikingly in their preference ratios for ethanol (McClearn and Rodgers, 1959, 1961; Fuller, 1964). In both of these studies mice of the same four basic strains were compared. However, different sublines separated for many generations were used. In one experiment (McClearn and Rodgers, 1961) the measure was the simple preference ratio for 10% ethanol versus water. In the other (Fuller, 1964) six bottles were available with ethanol concentrations of 0.5%, 1%, 2%, 4%, 8%,

Table 8-3. Alcohol preference of four inbred mouse strains by two methods*

Strain[†]	McClearn and Rodgers (1961) Preference ratio[‡]		Fuller (1964) Preference score[§]	
	Mean	SD	Mean	SD
C57BL	0.76	0.10	1.78 (6.0)	0.20
C3H	0.19	0.13	1.61 (4.1)	0.21
A	0.14	0.05	1.39 (2.4)	0.25
DBA	0.09	0.03	1.08 (0.8)	0.21

*From McClearn, G. E., and D. A. Rodgers, 1961. Genetic factors in alcohol preference of laboratory mice. J. Comp. Physiol. Psychol. **54**:116-119. Copyright 1961 by the American Psychological Association. Reprinted by permission. Fuller, J. L. 1964. Measurement of alcohol preference in genetic experiments. J. Comp. Physiol. Psychol. **57**:85-88. Copyright 1964 by the American Psychological Association. Reprinted by permission.

†Crgl sublines were used in the McClearn and Rodgers study, J sublines for the Fuller study.

‡Preference ratio = $\frac{A}{A + W}$ (A, volume of 10% alcohol consumed; W, volume of water consumed).

§Preference score = $1 + \log_{10}$ of median concentration (in percent) of alcohol drunk in a six bottle–choice test. Conversion to percent is shown in (   ).

and 16%. The score was the logarithm of ten times the alcohol concentration, in percent, below which one half of a subject's fluid was derived. As Table 8-3 shows, the ranking of the strains was identical in the two studies despite the opportunity for genetic drift during separation of the lines and the widely disparate nature of the phenotypic measure.

Mode of inheritance

The pattern of inheritance of alcohol consumption in crosses between inbred and selected lines fits the model of a polygenic system, since no clear segregation of phenotypes is found in the backcross and F_2 generations. It should be noted, however, that even within an inbred strain, individual variability in preference and consumption is conspicuous, and the detection of segregation would be difficult. Such variation indicates the importance of individual differences in

developmental history and in conditions surrounding the test situation.

Table 8-4 presents the results of two classical crosses between a high- and a low-preference inbred mouse strain. In one study (Whitney, McClearn, and DeFries, 1970) the F_1 has the same mean as the low parent; in the other (Fuller and Collins, 1972) it is intermediate but closer to the high parent. Variances of the segregating generations are not consistently higher than those of the inbreds and F_1s, so that computation of heritability based on variances (Chapter 5) is not applicable. Thus the values for h^2 in this table were obtained from an approximate equation presented by Whitney, McClearn, and DeFries (1970). The higher value for the Fuller and Collins data is attributable to their use of two parental lines deliberately selected to be extreme. Fuller (1964) also computed h^2 from a half diallel of four strains (no reciprocal crosses) as 0.39, which is one half the intraclass correlation. The results of this experiment are summarized in Table 8-5.

It is of interest that Brewster (1968) computed an h^2 of 0.82 from the McClearn and Rodgers data (1961) and 0.86 for Fuller's (1964). The arguments for a lower value as summarized by Whitney, McClearn, and DeFries (1970) illustrate the problems of choosing a best value when one is dealing with real data instead of idealized models. Brewster (1968) also reanalyzed Fuller's (1964) half-diallel data and constructed a variance-covariance diagram (Fig. 8-2). The rationale of this diagram has been explained by Broadhurst (1967) and is discussed in Chapter 5. The good fit to a straight line with a slope of one indicates no significant deviation from additivity on the scale of measurement. Incomplete dominance is indicated by the intercept of the line on the ordinate above the origin. The greatest proportion of dominant genes is found in C57BL, the most recessives in A. Thus crosses with C57BL tend to behave like C57BL; offspring of crosses of other strains with A tend to behave like their non-A parent.

Brewster (1968) compared alcohol intake and preference ratios under free choice from

Table 8-4. Alcohol preference in two classical crosses*†

Generation	Whitney, McClearn, and DeFries (1970) Arcsin transform of preference		Fuller and Collins (1972) Ethanol intake: ml/day	
	Mean	Variance	Mean	Variance
P_1‡	1.27	0.35	1.38	0.39
B_1	0.87	0.14	1.87	0.35
F_1	0.64	0.04	1.52	0.53
F_2	0.70	0.04	1.25	0.62
B_2	0.64	0.04	0.63	0.13
P_2§	0.64	0.07	0.42	0.05

*From Whitney, G., G. E. McClearn, and J. C. DeFries. 1970. Heritability of alcohol preference in laboratory mice and rats. J. Hered. **61**:165-169; Fuller, J. L., and R. L. Collins. 1972. Ethanol consumption and preference in mice: a genetic analysis. Ann. N.Y. Acad Sci. **197**:42-48.

†h^2 calculated by formula of Whitney et al. $h^2 = \dfrac{\frac{1}{2}(a)^2}{\frac{1}{2}(a)^2 + V_e}$; $a = \text{mean}_{B_1} - \text{mean}_{B_2}$; V_e = average variance of P_1, F_1, and P_2. For Whitney et al., $h^2 = 0.15$; for Fuller and Collins, $h^2 = 0.70$.

‡P_1 = C57BL/1Bi (Whitney et al.) = C57BL/6J (Fuller and Collins).

§P_2 = JK/Bi (Whitney et al.) = DBA/2J (Fuller and Collins).

Table 8-5. Half-diallel table of alcohol scores for four inbred strains of mice and their hybrids*†

Strain	DBA/2J	A/J	C57BL/6J	C3HeB/J
DBA/2J	1.0800†	1.0700	1.5200	1.1700
	0.0437‡	0.0389	0.0434	0.0393
A/J		1.3900	1.7100	1.3500
		0.0665	0.0888	0.0452
C57BL/6J			1.7800	1.5900
			0.0415	0.0498
C3HeB/J				1.6100
				0.0442

*From Fuller, J. L. 1964. Measurement of alcohol preference in genetic experiments. J. Comp. Physiol. Psychol. **57**:85-88. Copyright 1964 by the American Psychological Association. Reprinted by permission.

†Alcohol score = 1 + $\log_{10}$ of median concentration (in percent) of alcohol drunk in a six bottle–choice test.

‡Variance of score.

the Maudsley reactive (MR) and nonreactive (MNR) rats, which had been selected for differences in emotionality. Further discussion of these animals will be found in Chapter 10. The MNR rats drank more heavily, and the F_1 was similar to them. Erikkson (1969a) computed parent-offspring correlations from a set of apparently random matings in a population drawn from his AA and ANA lines. The overall correlation was 0.40; for somewhat mysterious reasons offspring tended to resemble their like-sexed more than their opposite-sexed parent. In contrast to these relatively high estimates of heritability, parent-offspring correlation in a heterogeneous population of mice did not differ significantly from zero (Whitney, McClearn, and DeFries, 1970). On the basis of all available data, Whitney et al. concluded that the heritability of alcohol preference in laboratory rodents is of the order of 0.10 to 0.15, though higher values are obtained when experiments are conducted with highly divergent strains. Such values are usually considered

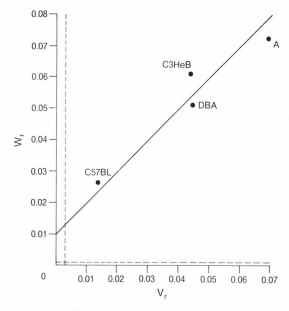

Fig. 8-2. Variance: covariance diagram for ethanol preference in mice (see Table 8-5). The abscissa is formed from the variances of the horizontal arrays of the half-diallel table, the means for the hybrids being entered below the leading diagonal and the ordinate from the co-variances of the horizontal arrays with the values in the leading diagonal. *Dotted lines* are the axes corrected for variance due to environmental causes. (From Brewster, D. J. 1968. J. Hered. **59**:283-286; based on data from Fuller, J. L. 1964. J. Comp. Physiol. Psychol. **57**: 85-88.)

"low," but they are adequate for successful selection. As Erikkson has demonstrated, only time and consistent bidirectional selection pressures are needed to redistribute the genes from a heterogeneous population and produce two populations with practically nonoverlapping distribution.

Neural and behavioral sensitivity to ethanol

Thus far we have concentrated on the genetics of alcohol intake. Also of interest is the possibility of inherited differences in sensitivity to the pharmacological effects of this substance. It is plausible that such variations in sensitivity are the basis for the observed behavior differences. Indeed, striking differences have been found between the alcohol-preferring C57BL mouse strain and the alcohol-avoiding BALB/c strain (Kakihana et al., 1966). Sleep times following a standard intraperitoneal dose of ethanol were 38 and 138 minutes, respectively. The differences

cannot be explained by more rapid clearance of alcohol from the system of C57BL, but are rather due to the threshold concentration for neural depression. The brain concentration of alcohol at waking was 430 ± 29 mg/100 grams brain in C57BL and 287 ± 17 in BALB/c.

Selection for long and short sleep times has been successful (McClearn and Kakihana, 1973). After fourteen generations the mean duration of sleep after intraperitoneal injection of ethanol (3.4 grams/kg) was 97 minutes for a long sleep (LS) line and just under 12 minutes for a short sleep (SS) line. The difference in neural sensitivity seems to be specific to alcohols and does not extend to hypnotics in general (Heston et al., 1973). The activities of the primary metabolizing enzymes, ADH and ALDH, are virtually identical in the LS and SS lines (Heston et al., 1974). Intracerebral injections of salsolinol, a substance that may be formed in brain by a reaction between acetaldehyde

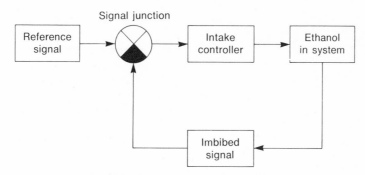

Fig. 8-3. McClearn's schematic automatic closed-loop negative-feedback control system for alcohol preference. (Modified from McClearn, G. E. 1972. Int. Symp. Biol. Aspt. Alc. Consumpt. **20:**113-119.)

and dopamine, caused longer sleep in LS than in SS mice (Church, Fuller, and Dudek, 1976). Offered a choice between a highly acceptable glucose-saccharin solution and the same solution with 4% of added ethanol and water, SS mice consistently drank more of the alcohol mixture than did LS (Fuller, Church, and Dann, 1976). Both lines, however, drank more of the glucose-saccharin than of the alcohol mixture; taste preference does not seem adequate to explain the alcohol intake of the SS mice or alcohol avoidance of the LS.

Another aspect of specific neural sensitivity is the occurrence and intensity of a withdrawal reaction after cessation of prolonged high-level intake of ethanol. The occurrence of withdrawal symptoms is frequently considered to be an intrinsic characteristic of addiction. Since laboratory rodents, even alcohol preferrers, are not easily induced to ingest on their own enough alcohol to produce reliable withdrawal effects, other methods such as the continuous inhalation of alcohol vapors have been devised. With this procedure, mice were selected for high and low vulnerability to convulsions after removal from an inhalation chamber (Goldstein, 1973). Likewise, a comparison of the intensity of withdrawal convulsions among alcohol-preferring (C57BL/6J) mice and alcohol-avoiding (BALB/cJ, DBA/2J) mice demonstrated less severe reactions in the preferring strain (Goldstein and Kaki-

hana, 1974). Parallel strain differences were found in sensitivity to reserpine, which depletes brain catecholamine stores.

Although many details remain to be worked out, it now appears that the well-documented genetic differences in alcohol-related behavior are associated with differences in central nervous system sensitivity to ethanol. The data also suggest more specifically that reactions of ethanol, or its products, with catecholamines may be the physiological basis for the behavioral differences. It is certain that this area of research will continue to be active for some time.

It is also clear that preference and aversion are not absolute; even among alcohol preferrers that are less sensitive to its effects, there appears to be a mechanism for limiting the intake of that substance. The following section considers some of the possibilities for such regulation.

Ethanol intake control system

Both the genetic data on alcohol intake and the results of experiments on its environmental manipulation led McClearn (1968a, 1972) to postulate an ethanol intake control system. The schematics of the hypothesized system are shown in Fig. 8-3. The concept is the familiar one of negative feedback. A signal produced by the imbibing of alcohol is matched with a reference signal that is set at a point determined in part by genotype. If the imbibed signal is less than the reference signal, alcohol, if available, will be ingested until balance is attained. When matching occurs, ingestion ceases.

Such a conceptual system is a help to behavior-genetic analysis of alcohol-related behavior. Are strain and individual differences attributable to variations in the reference set point, perhaps zero for a DBA/2 mouse and between 2 to 4 mg/gram brain for a C57BL/6 mouse? Or does the difference lie in the imbibed signal, possibly in the production of a chemical from ethanol that serves as a controller of intake through its action on certain neurons? Answers to these questions are not now available. However, considerable effort has been applied to the search for physiological and psychological correlates of alcohol preference and consumption.

Biochemical correlates of alcohol-related behavior

Explaining the function of a complex system in terms of the functions of its components is the familiar *reductionist approach* to science. Following this tradition, attempts have been made to correlate differences in the preference for and consumption of alcohol with differences in biochemistry. Since ADH is the rate-limiting enzyme for the metabolism of ethanol, it seemed plausible that its activity could be the key to the behavioral phenomena. In fact a sample of six inbred mouse strains showed a high correlation between liver ADH activity and alcohol preference (Rodgers et al., 1963). However, the correlation broke down in the F_2 offspring of a cross between high-preference C57BL and low-preference DBA (McClearn, 1968a). Although ADH activity appears to account for 10% of the variance in ethanol preference in a heterogeneous mouse stock (McClearn, 1972), studies of the rate of alcohol metabolism in vivo show such slight differences among strains that it is unlikely that they are the source of the behavioral differences (Sheppard, Albersheim, and McClearn, 1966).

The evidence for behavioral significance of differences in acetaldehyde metabolism is somewhat stronger. The concentration of acetaldehyde in the blood during ethanol metabolism is higher in DBA/2 than in C57BL mice (Schlesinger, Kakihana, and Bennett, 1966; Sheppard, Albersheim, and McClearn, 1970; Schneider et al., 1973). Treating C57BL mice with disulfiram (Antabuse), a drug that inhibits ALDH, reduced their ethanol consumption (Schlesinger, Kakihana, and Bennett, 1966). Also, in rats, the blood concentration of acetaldehyde after ingestion of a standard amount of ethanol is higher in a nondrinking strain (ANA) than in a drinking strain (AA) (Eriksson, 1973). Eriksson suggests that acetaldehyde has a strong inhibitory effect on brain metabolism because of interference with enzymes related to catecholamines.

Behavioral and biochemical genetics have proved to be a potent combination for the investigation of alcohol-related behavior in laboratory animals. Since the biochemistry of humans is very similar, some of the findings may be applicable directly to our own species. Of course, any explanation of behavioral variation in terms of the activity of an enzyme is incomplete without consideration of the total system in which it operates. The possible link between acetaldehyde and brain catecholamines is a step in this direction, since the amines are related to reinforcement processes. It is also possible to work from the other direction, seeking behavioral correlates of alcohol-related activity and looking for common physiological bases of the set of associated behavior patterns.

Psychological correlates of alcohol-related behavior

Explaining genetic differences in alcohol preference and consumption in terms of psychological traits may be designated as the *intrapsychic approach*, in contrast to the reductionist approach just considered. Actually, the two approaches are complementary rather than contradictory. A little reflection leads to the conclusion that no single biochemical factor can really explain behavior, although the efficiency of a key biochemical process may be critical in modifying a complex phenotype.

A popular hypothesis is that alcohol is ingested because of its tension-reducing properties; thus the amount ingested is an indirect measure of an individual's tension level. Adamson and Black (1959) predicted that rats with intermediate consumption of ethanol would learn an avoidance task more effectively than either high- or low-intake animals. This prediction was based on a general finding that very low or very high states of arousal interfere with learning. After determining individual ethanol preferences in a free-choice situation, avoidance-learning tests were conducted during forced abstinence from alcohol. The results supported the idea that differences in emotionality underlie individual variation in volitional drinking of ethanol. When alcohol consumption was forced during the period of learning, the inverted U–shaped relationship between consumption and learning broke down. Although this study does not distinguish between environmental and genetic factors underlying individual differences, it does support the validity of the tension-reducing hypothesis for alcohol consumption.

In a more formal genetic investigation, factor analysis of a battery of emotionality tests given to four inbred strains of mice indicated that two factors, disorganization and audiogenic reactivity, were most consistently related to the consumption of ethanol (Poley, Yeudall, and Royce, 1970). A comparison of alcohol preference in lines of rats selectively bred for emotionality (Maudsley reactive and nonreactive) and for avoidance learning (Roman high avoidance and low avoidance) found that the MR and RHA, presumably the most susceptible to stress, drank most heavily (Satinder, 1972). These selected lines are discussed more thoroughly in Chapters 9 and 10.

Sensory and energetic factors. Other explanations for genetic variation in alcohol-related behavior invoke physiological factors more complex than enzymes but less global than emotionality. For example, could the preference differences found in inbred strains be due simply to the taste or odor of ethanol? The fact that odor plays some role is demonstrated by the elimination after olfactory bulbectomy of the usual aversion to alcohol found in BALB/c mice (Nachman, Larue, and LeMagnen, 1971). The operation had no effect on the alcohol preference of C57BL mice. Since intact BALB/c mice avoided ethanol on initial contact, learning was apparently not involved. When injection of toxic lithium chloride was paired with ingestion of sucrose or saccharin solutions, both strains acquired an aversion to the sweet substances on subsequent tests. When an ethanol solution was paired with lithium chloride, C57BL mice were deficient in learning the aversion. Nachman, Larue, and LeMagnen concluded that BALB/c mice avoid alcohol because of its odor; the preference shown by C57BL must be attributable to postingestional factors.

Is alcohol ingestion simply a more efficient way of obtaining calories for some strains of mice? This possibility was advocated by Lester and Greenberg (1952), but experimentation showed that when their food supply is restricted, strains differing widely in alcohol

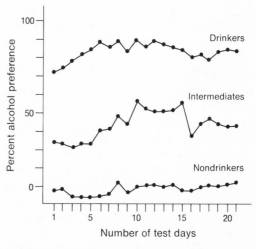

Fig. 8-4. Mean daily alcohol preference ratios of rats from three groups during 3 weeks of free choice between water and a 5% v/v ethanol solution. Prior to the experiment all subjects received only the alcohol solution for 10 days. Forced intake did not modify the genetically determined preferences. (From Eriksson, K. 1969. Ann. Zool. Fenn. **6:**227-265.)

preference do not differ in their capacity to obtain energy from ethanol (Rodgers et al., 1963). Adding sucrose to alcohol solutions increases intake in most strains, but strain rankings for amount ingested are unchanged (Rodgers and McClearn, 1964). The enhancing effects of sucrose on alcohol consumption seem to be related to its sweet taste rather than to its caloric content.

Phenotypic stability. Finally, let us consider the stability of the alcohol-related psychophenes. Even nondrinking strains such as ANA rats or DBA/2 mice consume alcohol solutions when they are the only source of fluid. Under such a regimen ANA rats gradually increase their intake of ethanol but resume their nondrinking status when again offered free choice (Erikkson, 1969b). The data shown in Fig. 8-4 indicate that forced ingestion of alcohol has no sustained influence on the drinking pattern of these strains. Similarly, forced consumption of sweetened alcohol does not increase later ethanol preference in either alcohol-accepting SS or alcohol-rejecting LS mice (Fuller, Church, and Dann, 1976).

Such stability of phenotype is not universal. Exposure of juvenile BALB/c mice to ethanol does result in adults with alcohol preference higher than that of unexposed controls (Kakihana, 1965). Enhancement was not found in other low-preference strains subjected to the same treatment. Randall and Lester (1975) reported that DBA/2 mice reared as weanlings with C57BL/6 adults drank more at maturity than DBA/2 reared with their own strain. Conversely, a reduction in intake was found in C57BL housed for a 7-week period with DBA. Changing the social environment reduced, but did not reverse, the typical strain difference in level of intake found consistently in these two strains (Table 8-6). The authors favor an explanation in terms of learning, a kind of "peer pressure," but other possibilities cannot be excluded. In general, genotype seems to prevail over experience in setting the hypothetical reference signal of McClearn's EICS.

The stability of an individual's preference is perhaps of more general interest than the

Table 8-6. Effect of postweaning social environment on alcohol preference*

Parental strain	Adult companions—4 to 10 weeks			
	DBA/2J		C57BL/6J	
	N	Intake†	N	Intake
C57BL/6J	12	6.5 ± 0.9	7	12.8 ± 2.3
DBA/2J	11	1.1 ± 0.2	13	2.6 ± 0.7

*From Randall, C. L., and D. Lester, 1975. Social modification of alcohol consumption in inbred mice. Science **189:**149-151. Copyright 1975 by the American Association for the Advancement of Science.
†Grams/kg/day.

constancy of strain averages. Over long periods of observation, marked fluctuations in intake were observed in some mice by McClearn (1972), who called the high-consumption periods "periodic drinking." Since the genotype is constant, the fluctuations must be ascribed to something in the environment or to individual developmental history. A similar phenomenon was noted in LS mice offered a 4% ethanol solution sweetened with glucose and saccharin (Fuller, Church, and Dann, 1976). It is likely that the probability of developing the episodic pattern of drinking is affected by genotype, but too few animals have been observed continuously for long periods to prove the point.

SUMMARY

The input of genetics to the study of consumatory behavior has been substantial. Mutants, the polydipsic and obese, have been the favored subjects for the study of eating and drinking. Although strain differences in consumption of food and water are not difficult to find, they have not been as useful for the analytic attack on physiological mechanisms of regulation. From studies with mutants, it is clear that somatophenes and psychophenes can appear very similar externally yet prove to be caused by independent mutations which operate through different intermediary steps. Also, it is evident that the details of the phenotypic effect of a mutant gene depend on the background genotype in which it operates. The obese and diabetes mice are an instructive illustration of these principles. One hopes that more of this kind of useful mutant will be found and studied.

Mutants have played a negligible role in the behavior-genetic analysis of alcohol preference. Instead, selected lines of rats and inbred strains of mice have proved the most useful subjects. There is little reason to doubt that a similar rich lode of genetic variation can be found by exploring reactions of animals to other drugs of psychological interest. As an example, genetic differences in susceptibility to morphine have been reported both in rats (Nichols and Hsiao, 1967) and mice (Erikkson and Kiianmaa, 1971; Shuster, 1975). Further examples of differences in the behavioral response to drugs will be found in the chapters immediately following. The past few years have demonstrated that it is easy to find genetic differences in the effects of drugs on experimental animals; the task of future investigators is to learn how genes produce these differences.

9

Learning ability in animals: genetic aspects

Learning may be defined as the process by which the probability of a response to a stimulus is changed as a direct result of experience with that stimulus or with a similar one, provided the effects of fatigue, trauma, or nonspecific modifications of metabolism can be excluded. Experimental psychologists have traditionally devoted much effort to the study of learning in animals, predominantly in the albino rat. However, other species from the octopi to the great apes have provided important data. We can recognize two major objectives for such studies: (1) a search for general principles that will apply to many species, including man, and (2) a search for physiological explanations through comparisons of the learning process in species that differ in neural structures or by comparisons among animals of the same species whose nervous systems have been modified experimentally. The concentration of experimental psychologists on animal learning is explicable by the obvious importance of learning in humans. Ideas generated by laboratory studies of conditioning of salivary secretion in dogs and the rate of bar pressing by rats to obtain food pellets have found wide human application in education, psychotherapy, and other applied areas of psychology.

Despite much effort and significant accomplishment, there are signs of disillusionment with the idea that a universal set of principles can be found which will explain all learning and which will, when properly applied, enable an experimenter to teach an organism to do anything if it has the sensory capacity to detect the stimulus and the motor capacity to execute the required act. A considerable body of research supports the view that genetically, organisms are prepared for cer-

tain associations, unprepared for others, and contraprepared for still others (Seligman and Hager, 1972).

Comparative studies of learning have often been influenced by the assumption that learning ability increases progressively in vertebrate phylogeny. Thus, by correlating the complexity of brain structures with behaviorally defined specific learning abilities, one could possibly ascertain a morphological feature associated with each ability. Difficulties are encountered in putting this scheme into practice because of the great variation in sensory and motor characteristics among animals; this prohibits testing all species on common tasks with common equipment. Furthermore, some of the learning tasks that are most convenient for the experimental psychologist do not sharply separate the larger taxonomic groups. Classical conditioning, active and passive avoidance behavior, and elimination of errors in simple mazes are often acquired as rapidly by animals "low" in the phylogenetic scale as by primates, including humans (Warren, 1973). More complex tasks, such as those involving conditional discrimination, delay of response, or acquisition of learning sets, are more promising as indicators of overall species differences in learning ability, but up to now they have not yielded unequivocal evidence for the ranking of species on a single scale of intelligence (Warren, 1973; Riopelle and Hill, 1973). One problem in making such comparisons has been the neglect of individual differences among members of a species; one or two individuals should never be used as representative of an entire taxonomic group. Warren and Baron (1956) found very large variations in the acquisition of a learning set among four cats. Among cocker spaniels the ability to delay response varied from 0 seconds to 2 minutes or more (Scott and Fuller, 1965). It is impossible to determine whether genes or experience played the major role in producing these striking differences in performance.

In the behavior-genetic analysis of learning, both genes and experience must be controlled. Furthermore, testing for individual and strain variation in learning within a species, provided one avoids animals with sensory or neurological defects, circumvents the problem of trying to compare animals of extremely different physical characteristics. The same apparatus and training procedures can be used for all subjects. In planning experiments, behavior geneticists have asked themselves three kinds of questions:

1. How large a role do genes play in the production of individual differences in learning ability?

2. To what extent is the ability to learn one task correlated with the ability to learn others?

3. What are the mechanisms through which gene substitutions affect learning? Do they involve associative processes in the central nervous system or peripheral functions related to sensory capacities, strength of motivation, and the like?

The first question is primarily one for quantitative genetics; the second and third involve a functional analysis of the learning process in biological as well as psychological terms.

HISTORICAL BEGINNINGS

Learning ability was perhaps the first animal psychophene to be investigated formally by geneticists. Comparisons between strains (in the early studies not well defined genetically) and selection techniques were employed. Bagg (1916, 1920) compared the performance of a number of coat-color strains of mice on a maze and on a multiple-choice apparatus. Performances were extremely variable, and his conclusions on inheritance were correspondingly cautious. Nevertheless, he concluded that males were superior

to females and that yellow mice were duller than other breeds. Bagg reported a sibling correlation of 0.50, which corresponds to their degree of relationship. Vicari (1929) compared the reaction time of mice in a simple three-unit maze with two inbred strains, DBA and BALB, and two mutant stocks, Japanese Waltzer and Myencephalic Blebs. Although the results were too complex to prove any simple genetic hypothesis, they suggested multifactorial inheritance of running time. Surprisingly, little research on the learning ability of neurological mutants has been done since.

The pioneer in selecting animals on the basis of learning ability was Tolman (1924). Starting with a rather heterogeneous stock of rats, he selectively bred "bright" and "dull" animals on the basis of a composite performance score (based on time, errors, and number of perfect runs) in a maze. DeFries (1967) calculated the heritabilities of these measures in the first generation as errors, 0.93; time, 0.57; and number of perfect runs, 0.61. Tolman did not continue his project, but the attempt was of historical importance in influencing Tryon to undertake his classical experiments on selection for maze running ability in rats.

SELECTION AND MAZE LEARNING

Three major selection experiments have used rats as subjects and the number of errors in a complex maze as a criterion (Tryon, 1929, 1940a; Heron, 1935; Thompson, 1954). None of these programs included replicated selected lines, unselected controls, or breeding systems designed explicitly to reduce inbreeding and thus conserve genetic variance. One should not, however, be too critical. Any selection program requires a long-term commitment of time and resources, which are always in short supply. And, although the design of these studies makes their biometric analysis impossible, each produced important results. Perhaps their most significant effect was the demonstration that maze learning ability is heritable, though the physical basis of the genetic effects is still not clear.

Tryon program

Tryon initiated his selection program in 1926 with three stated objectives: to produce a maze-bright and a maze-dull line of rats, to investigate the mode of inheritance, and to identify the major biological and psychological correlates of the selection criterion. He tallied the total errors made during trials 2 through 19 in a 17-unit multiple T-maze, and mated low scorers with low, high with high. Progress toward establishing maze-bright and maze-dull rats over eight generations is shown in Fig. 9-1. At the end of this period there was little overlap between the two lines, and the importance of genetic factors on performance was clearly established.

Crosses between the two strains yielded an F_1 whose average error score was intermediate to those of the parental strains. The F_2 was similar to the F_1 and did not show the

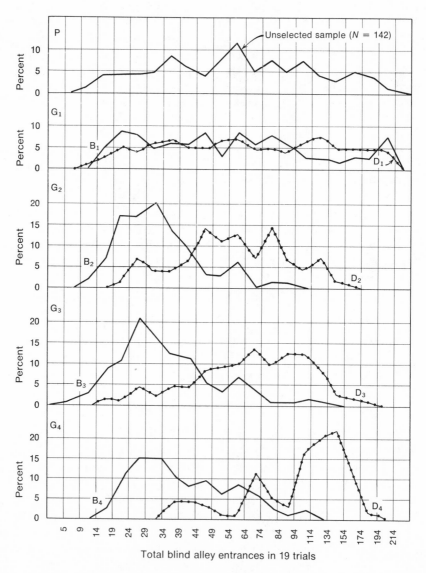

Fig. 9-1. Selection for maze brightness *(B)* and maze dullness *(D)* in rats over eight generations, G_1 to G_8. (From Tryon, R. C. 1942. Individual differences. In F. A. Moss, ed. Comparative psychology. Prentice-Hall, Inc., Englewood Cliffs, N.J.)

increased variance that Tryon expected as a result of genetic segregation (Fig. 9-2). He concluded that a polygenic system was involved, but he did not continue with a quantitative analysis of the genetic parameters. If a very large number of loci were involved in producing the differences between the bright and dull strains, Tryon's sample sizes may not have been large enough to detect differences in genetic variance (Bruell, 1962).

Generality of results

Did Tryon's selection result in all-around more intelligent and less intelligent rats or in two strains that performed well and poorly respectively only in a 17-unit automatic T-maze? It turned out that their abilities, or lack of the same, were rather specific. Searle (1949) obtained thirty measures of learning, emotionality, and activity on ten rats of each strain. In three out of five maze measures,

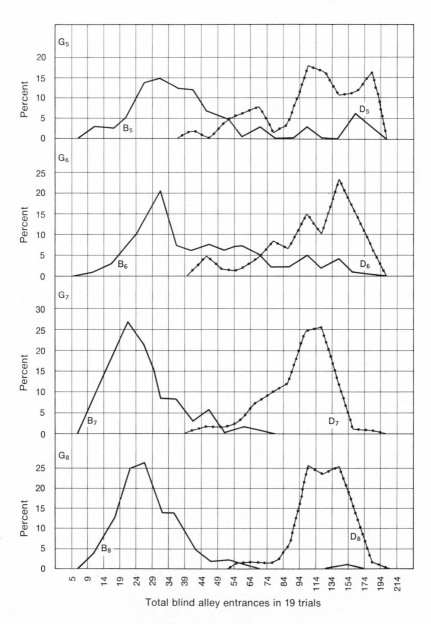

Total blind alley entrances in 19 trials

Fig. 9-1, cont'd. For legend see opposite page.

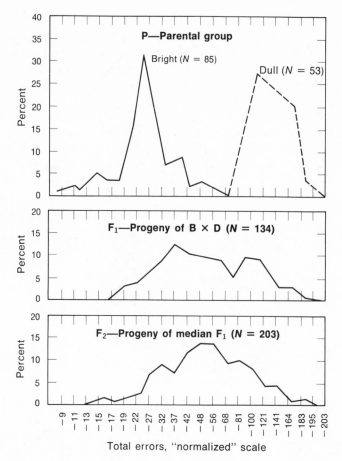

Fig. 9-2. Results of crossing selectively bred "maze-bright" and "maze-dull" rats. (From Tryon, R. C. 1940. Genetic differences in maze learning in rats. In National Society for the Study of Education, 39th yearbook. Public School Publishing Co., Bloomington, Ill.)

the Tryon maze dulls (TMD) were either equal or superior to the brights (TMB). TMB rats were reported to be more food driven, lower in motivation to escape from water, more timid in an open field, and less distractable. TMDs appeared to be fearful of the noisy mechanical features of the maze used in the original selection, and Searle suggested that an emotional reaction might be the underlying cause of the differences in learning. Searle's findings may explain the greater variance of the F_1s relative to the parental strains that is evident in Fig. 9-2. If TMBs are food motivated and relaxed, they will uniformly learn well in an automatic maze. TMD rats, being less motivated and more fearful, will all learn poorly. Thus, one

would expect their F_1 hybrids, intermediate in these dimensions, to perform correspondingly. In this middle range, chance might play a large role in the success or failure of individuals and result in an increased variance.

Searle analyzed his data in terms of the intercorrelations for performance on thirty variables within strains and across strains. As shown in Table 9-1, the intercorrelations within strains were high; those between strains were insignificant and negative. Tryon's selection had resulted in an aggregation of behavioral differences rather than separation along a single dimension. If positive correlations between performance variables indicate dependence on common traits, one

Table 9-1. Intercorrelations of thirty performance variables within and between TMB and TMD rat strains*

Group	Number of rats	Number of intercorrelations	Average intercorrelation
Within TMB	10	45	0.587
Within TMD	10	45	0.533
TMB with TMD	20	100	−0.188

*From Searle, L. V. 1949. The organization of hereditary maze-brightness and maze-dullness. Genet. Psychol. Monogr. **39:**279-325.

could use such data to characterize the TMB and TMD strains. The two trait structures, however, would not correspond.

Further investigations of the importance to maze learning of food motivation and fear of apparatus have been carried out with the S_1 (descendants of TMB) and S_3 (descendants of TMD) rat strains. Differences in maze learning were still apparent after the strains had been maintained for about twenty years without selection (Wolfer, 1963). The differences in error scores were not abolished by varying the degree of food deprivation. Deprived TMDs began eating more quickly than deprived TMBs, but total food consumption was similar; TMDs drank more water and had the higher basal metabolism (Wolfer et al., 1964). In contrast, Rowland and Woods (1961) actually obtained lower error scores in TMD rats, which they tested in a replica of the original Tryon maze.

Cognitive explanations of strain differences

There is as yet no simple explanation of the differences in learning ability between the Tryon strains. Because the selected lines were not replicated, we cannot exclude the possibility that many of the observed correlations resulted from the random fixation of different sets of alleles in the TMB and TMD lines, which were not functionally related to the maze-learning criterion but which did lead to other behavioral differences. Tryon himself (1930 to 1941) spent many years attempting to account for the differences between brights and dulls by his method of component analysis, which was based on the

pattern of errors and speed of running in various portions of the maze. He distinguished ten components: direction set, food pointing, shortcut tendencies, counter tendency, centrifugal swing, adaptation, lassitude, exit gradient, initial-inertia gradient, and conflict. The first five, which relate to general orienting ability, were considered to be major components; the second five involve motivational factors and were considered minor. The relative importance of specific components varied both for TMBs and TMDs at various stages of learning, and the change from early to late component patterns occurred sooner in TMBs. This strain also employed food pointing and exit gradient to a greater degree during the transition period. Tryon interpreted this to mean that "brights elicit the more abstract cognitive types earlier than do the dulls."

Some confirmation of this view was obtained by Krechevsky (1933), who devised an apparatus in which the choice of a path leading to food could be made by using either spatial or visual cues. Actually the cues were irrelevant, and consistently following either type would be correct only half the time. Rats tested in the apparatus tended to develop either a visual or a spatial "hypothesis," and such preferences became sharper over a 14-day testing period. Random-bred rats divided about equally as spatial or visual hypothesizers, but TMDs used more visual and TMBs more spatial cues. Spatial cues were considered to be more internalized than visual ones and were thus related to a greater capacity for cognitive abstraction.

Wherry (1941) also undertook the investi-

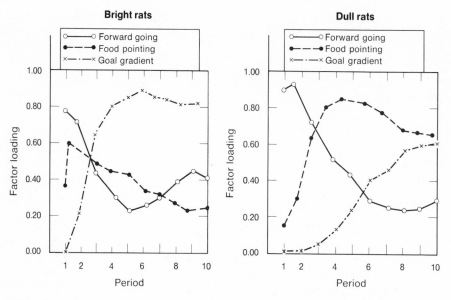

Fig. 9-3. Changes in components of maze learning ability during training shown by bright and dull rats. (From Wherry, R. J. 1941. J. Comp. Psychol. **32**:237-252.)

gation of the abilities involved in maze learning by rats with the TMB and TMD lines as subjects. From his data, he extracted three factors: forward-going tendency, food pointing, and goal gradient. The relative use of these three components by the two strains of animals changed over the course of training. On early trials they were similar, but goal gradient became dominant for TMBs relatively early; it was never as strong in TMDs, who relied more on food pointing (Fig. 9-3). Wherry related his findings to those of Krechevsky, arguing that forward going and goal gradient became dominant for TMBs rather than food pointing, which is more exteroceptive and stimulus-bound.

Minnesota and McGill maze ability selections

Two other major selection experiments involving performance on a complex maze are available for comparison with the Tryon results. Heron (1935, 1941) selected rats for low and high error scores on an automatic maze very much like Tryon's. There was little separation between the lines after generation six, and fluctuations in performance of the two lines from generation to genera-

tion were correlated, suggesting variations in the environment during the course of the study. The Heron maze dulls (HMD) ran more slowly in the maze, although this did not altogether account for their poorer performance in terms of errors (Heron, 1941).

Graves (1936) compared the fifth generation of HMB and HMD in a Stone multiple T-maze, in activity in a rotating cage, in reaction time to escape electric shock, and in strength of drive on the Columbia obstruction apparatus. Correlations between the test scores were consistently low, and the HMB line was consistently superior in performance only on the maze task, which was most similar to the criterion used in selection.

When the strains were compared in rate of bar pressing for food with intermittent reinforcement, HMB was consistently higher and was more resistant to extinction during the first 3 days of that procedure (Heron and Skinner, 1940; Skinner, 1940). Incidentally, this experiment was one of the first to employ the apparatus which has become well known as the Skinner box.

Still other behavioral and physiological differences between the strains have been sought. HMB rats ate more rapidly when food was made available for limited periods only, but they were more readily disturbed by stimulation while eating (Kruse, 1941). Attempts to detect differences in basal metabolism and brain weight between the lines were inconclusive (Heron and Yugend, 1936; Silverman, Shapiro, and Heron, 1940).

A third selection experiment at McGill University attempted to meet two requirements that had been neglected in the earlier studies (Thompson, 1954). In particular, selection was to be made on operationally defined "intelligence" unconfounded with other traits, and homozygosity was to be achieved by inbreeding. To meet the first requirement, the Hebb-Williams maze was used. It consists of a square enclosure with a removable wire-mesh top, a starting box in one corner, and a goal box diagonally opposite. Barriers of various lengths interposed in a number of ways between the starting and goal boxes constituted the problems. Prior to testing on a standard series of 24 problems, rats were given a lengthy period of habituation with simple problems up to a predetermined criterion. The individual problems in the Hebb-Williams maze are somewhat analagous to the items in a human intelligence test, and the habituation procedure ensures that all subjects are familiar with the testing situation.

Results for six generations are shown in Fig. 9-4. The mean score of the McGill bright (MMB) line was significantly different from MMD by generation three, and there was very little overlap in error scores thereafter. Unfortunately, inbreeding was commenced too early, and problems with infertility cropped up in later generations.

Again it is of interest to consider correlated traits in the selected lines. No differences were found in emotionality of the two lines as measured by (1) urination and defecation in an open field and (2) latency of leaving a home cage to reach food at the end of an elevated open runway 3 feet away (Thompson and Bindra, 1952). MMB rats were slightly and unreliably more motivated by food. In an enclosed maze, dulls explored more (Thompson and Kahn, 1955), but there

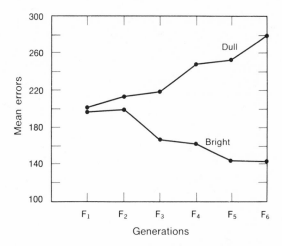

Fig. 9-4. Mean error scores of bright and dull rats selectively bred for performance on the Hebb-Williams maze. (From Thompson, W. R. 1954. Proc. Assoc. Res. Nerv. Ment. Dis. **33:**209-231.)

were no differences in exploratory activity on a simple elevated maze (Thompson, 1953a). It was concluded that differences in maze performance between these strains could not be attributed to emotional or motivational variables.

Unsuccessful selection experiments

In order to present a balanced picture, two unsuccessful attempts to select for learning ability are recorded here. McDougall (1927, 1938) had claimed that the offspring of trained female rats learned better than controls. He attributed his results to a direct influence of training on the germ cells, a Lamarckian type of explanation, but his critics claimed that accidental selection might have occurred. An unsuccessful attempt to change performance on a light discrimination task by selective breeding was presented as a rebuttal to the critics (Rhine and McDougall, 1933), but the small scale of the study and the extreme fluctuations between generations negate any definite conclusions.

Another negative case has been reported by Kuppusawny (1947). Using a simple water maze and breeding over ten generations, this investigator found no directional changes due to selection. Since neither the experimental procedure nor the results are adequately described, it is impossible to determine the reasons for the failure.

Genetics and maze learning: an overview

On the whole, the evidence for the importance of genetic factors in the maze performance of rats is extremely strong. Quantitative estimates of heritability are not possible with currently available data, but the rapid divergence between the high and low error lines, particularly in the Tryon and Thompson experiments, indicates that it must be substantial.

Correlations between error scores on mazes, the criterion for selection, and other performances are consistently low and often nonreliable. The animals have not been selected for general problem solving ability and should not be designated as bright or dull except in a limited sense. Motivational and physiological differences between selected lines have been described, but they do not seem to be the complete explanation for the variations in performance. There probably are inherited differences in associative capacity between specific maze cues and the responses that result in poor or good learning of complex maze patterns.

Maze running is a suitable test of learning ability for rodents that characteristically burrow and run along tunnels or trails. It certainly is not adaptable to comparing learning abilities across species. But even for rats, maze running is too complex to be readily analyzed in terms of physiology and structure, and therefore it is unsuitable for discovery of mediating processes between genes and learning. Hence it is not surprising that behavior geneticists have turned to other tasks, such as escape from or avoidance of nocuous stimuli, for the behavior-genetic analysis of learning ability.

ESCAPE AND AVOIDANCE LEARNING: SELECTION AND STRAIN DIFFERENCES

Training animals to escape or to avoid stimuli such as foot shock or immersion in cold water is a common technique in psychology laboratories. That animals are reinforced by the reduction of stimulation which we humans subjectively call painful may seem so obvious that it requires little demonstration. Escape behavior often seems to be reflexive, and, in fact, aversive stimuli are defined by their elicitation of an escape response. Avoidance behavior is not clearly reflexive but is generally learned, and its explanation is more complex.

In the laboratory a common procedure is to signal the imminent application of foot shock (the unconditioned stimulus, or US) by an originally neutral stimulus (the conditioned stimulus, or CS). After a number of trials when the CS occurs, the subject makes an avoidance response by crossing from one side of a cage to the other (shuttle-box technique), turning a wheel, climbing a pole, or any other act arbitrarily prescribed by the experimenter. Sometimes a discrimination is superimposed on the avoidance or escape response. The subject has the choice of two compartments to move into at either the CS or the US, but in only one, perhaps the lighted one, will there be no shock.

Just how does escape behavior transform into avoidance? A common view is that by pairing a CS with an aversive US, the CS acquires a new property, the elicitation of a state of fear. The reduction of fear by the avoidance response (the CS is generally terminated at the same time) reinforces that response and increases the probability that it will occur again when the CS is presented.

It is also possible to arrange an experiment so that a response must be inhibited in order to avoid aversive stimulation. For example, a mouse may be placed on a small platform over an electrically charged grid floor. Most mice step down within a few seconds and are unpleasantly surprised. Replaced on the platform after a single trial, they typically avoid passively by remaining on the platform longer than they did the first time. An advantage of the passive avoidance procedure is that the interval between the learning and the test trial can be varied; thus the duration of the inhibition of the step-down response can be used as an indicator of memory.

Selection for avoidance learning

Bignami (1965) reported successful selection of rats for rapid and slow acquisition of shock avoidance in an automated shuttle box. These strains are now known as RHA (Roman high avoidance) and RLA (Roman low avoidance). At the beginning of the program the average number of errors to criterion was 89.2 for males; after five generations the averages were 173.5 for RLA and 49.8 for RHA. No effect on learning was found as a result of cross-fostering between the strains. Thus it is clear that selection can influence avoidance as well as maze learning.

Other behavioral characteristics of the Roman strains have been reported. For example, RHA rats were more active than RLAs in a brightly illuminated open field, but differences in defecation, considered by some to indicate emotionality, were small and unreliable (Holland and Gupta, 1966). Daily exposure to shock from 1 to 15 days increased activity in a shuttle box in both strains but facilitated avoidance only in RHAs (Satinder and Hill, 1974). Thus the difference between the strains was enhanced by the stressful early experience. The shock threshold for a detectable motor reaction was lower in the RHAs, suggesting that they learn better because of higher shock sensitivity.

A more direct approach to the possible role of variation in shock sensitivity to strain differences in avoidance learning is to compare strains, not with equal shock strength, but with shock adjusted to each strain so that an equal unconditioned response is elicited. When this was done by adjusting shock level to twice the strength, which elicited a flinch reaction, the RHAs still outperformed the RLAs (Satinder and Petryshyn, 1974). However, the strength of effect (proportion of variance attributable to genotypic differences) was reduced from 0.38 under equal shock to 0.09 under individual specific shock. Even when shock intensity was high enough to evoke a vigorous running response, the

RHAs continued their superiority. These results indicate that superior avoidance is partially but not completely attributable to greater shock sensitivity.

Strain differences: rats

Each experimental psychologist generally adopts a particular strain of rat for his experimental work. Although this should result in consistency between experiments performed in the same laboratory, it may also result in consistent discrepancies with the work of investigators using different strains. Researchers who have looked for strain differences have often found them. Myers (1959) varied CS (tone or buzzer), manipulandum (bar press or wheel turn), time of training (day or night), and strain (Wistar or Sprague-Dawley). Each variable had significant interactions with at least one of the others, and the direction of strain differences could be reversed by changing the nature of the warning signal. Long-Evans hooded rats outperformed Sprague-Dawleys in active avoidance in a variety of training schedules (Nakamura and Anderson, 1962), and even strains of the same origin (Long-Evans) that had been separated and bred by different vendors were reliably different in their tests. Shaefer (1959) reported strain differences in avoidance without an explicit warning stimulus (Sidman procedure). The relative standing of Wistar and Long-Evans rats was not changed by cross-fostering, though that procedure did impair performance of both strains (Oliverio, Satta, and Bovet, 1968).

Strain differences in avoidance learning can be further analyzed by varying the conditions of the task systematically. Barrett, Leith, and Ray (1973) used a Y-maze for a discriminated avoidance task that was administered to F_{344} and ZM rats. Switching on a light served as a CS, and subjects had 10 seconds to enter the safe arm, which was also illuminated. Over a wide range of shock intensities, F_{344} rats learned to avoid consistently in 30 or fewer trials, but ZMs usually waited for shock and only then escaped. But although ZM rats failed to learn to avoid, they did learn to select the lighted branch of the Y almost as well as F_{344} rats. Both strains learned where to run; only one learned when to run. Injecting subjects with amphetamine or scopolamine improved the avoidance performance of ZMs markedly, although these drugs had no effect on the correctness of the spatial discrimination. The authors interpret their data as indicating contrapreparedness (Seligman, 1970) of ZMs for an avoidance response. But, although they are preprogramed to freeze at the CS, giving them low avoidance scores regardless of the intensity of shock that will follow, ZM rats are not inferior in discrimination learning. A later study extended the generality of these findings by testing additional strains (Caul and Barrett, 1973).

Does superior avoidance in a shuttle box indicate the more rapid formation of an association between the CS and the avoidance response? Apparently not. Katzev and Mills (1974) compared the performance of three strains of rats under three conditions: inescapable shock, escapable shock, and avoidance. In the first procedure the shock duration was so short that it could not be escaped by shuttling; in the second, shock could not be avoided but once administered

could be escaped by running to the opposite side of the chamber. Fischer rats in both situations crossed frequently when the CS was turned on, although this response was completely ineffective in modifying the shock. Long-Evans and Lewis rats seldom ran until shock was actually administered. Finally, shifting to the avoidance procedure increased the frequency of crossing when the CS was presented in all strains, but their relative position remained unchanged (Fig. 9-5). It should be noted that the increment in number of running responses produced by adding the avoidance contingency was somewhat less than the rate established by classical conditioning with inescapable shock. Katzev and Mills believe that strain differences in avoidance learning are based on differences in the unconditioned response to a warning signal rather than to differences in associative learning capacity.

Still another possible explanation for strain differences in avoidance learning is that emotionality (fearfulness) varies with genotype and that the more fearful subjects might be more motivated to learn an avoidance response. This hypothesis is not supported by experiments with the Maudsley nonreactive

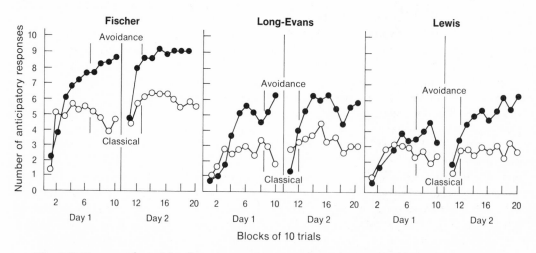

Fig. 9-5. Mean number of shuttle responses during CS for each strain during blocks of 10 trials under two conditions. *Classical:* avoidance impossible during CS; escape possible during shock. *Avoidance:* no shock received if shuttle response made during CS. (From Katzev, R. D., and S. K. Mills. 1974. J. Comp. Physiol. Psychol. **87:**661-671. Copyright 1974 by the American Psychological Association. Reprinted by permission.)

(MNR) and reactive (MR) strains. MNR rats made 87% avoidances as compared with 66% for MRs, which are presumably more emotional (Levine and Broadhurst, 1963). Joffe (1964) also found the MNR strain superior in learning to avoid in a shuttle box. It is possible that differences in avoidance performance may have been affected by physical factors. The MRs were heavier, and heavy rats have been shown to have higher shock thresholds (Pare, 1969).

Strain differences: mice

The house mouse has been a favorite species for the study of the mode of inheritance of avoidance performance, and this topic will be considered in a later section. Here we are concerned with research bearing on the nature of strain differences and their possible psychological and physiological explanations.

Fuller (1970) compared four inbred strains in a shuttle box in which shocks were presented at fixed intervals without a warning stimulus. For actively trained groups, crossing between the two sides delayed shock onset for 20 seconds. For passively trained groups, crossing resulted in a delayed shock; remaining on one side avoided shocks completely. Each subject had a yoked control; both received shocks at the same time. As predicted, actively trained mice shuttled more frequently than their passively trained strainmates. However, the ordering of the four strains in terms of activity was the same regardless of whether they were experimentals or controls and whether they were actively or passively trained. Training simply shifted a genetically determined level of activity upward or downward, and the strains that learned the active avoidance best were the ones that learned passive avoidance most poorly. Although there was no CS in this study, it can be interpreted very similarly to the Katzev and Mills (1974) experiments described previously.

The rate of avoidance learning is affected by the strength of the US, and strain differences in learning might be attributable to differing sensitivities to electric shock. Supporting this hypothesis, Carran, Yeudall, and

Royce (1964) found marked variation in avoidance learning among three strains of mice at a low shock level but none when the voltage was increased to the practicable maximum. Shock sensitivity does not, however, appear to be the sole factor acting to produce strain differences in avoidance performance. Wahlsten (1972) compared four strains on a jump-up shock avoidance task with two kinds of equalization of conditions. Some groups were compared at a shock level of 180 μamp, which was constant for all strains. Significant differences were found both in latency of the first escape and in total errors prior to reaching criterion. Other groups were tested at shock levels that had been found previously to produce a standard jumping response in other members of their strain. With this procedure the strain differences in latency disappeared, but differences in error scores persisted, though to a reduced degree. The proportion of variance in error scores attributable to genotype was 0.176 under equal shock and fell to 0.117 under equal jump.

Strain differences in appetitive learning

High correlations between response rates in both reinforced and nonreinforced conditions have been found in tasks motivated by access to food rather than by escape from shock. When mice of four strains were trained to press a bar for food pellets (each bar press yielded one pellet), they varied significantly in rate of pressing (Padeh, Wahlsten, and DeFries, 1974). After the behavior was well established, a discriminative signal, light plus tone (TL), was added to denote 55-second periods of reinforcement. During alternate 55-second periods, the light and tone were off ($\overline{TL}$), and the food delivery mechanism was inoperative. All mice learned to discriminate between the two conditions, but the response rates in TL and $\overline{TL}$ were highly correlated. Strain A was lowest in both conditions; strains BALB and HS (heterogeneous stock) were highest. If the efficiency of learning is measured by the proportion of all responses during the TL period, $\dfrac{n_{TL}}{n_{TL} + n_{\overline{TL}}}$, the A females were the best learners. However, from another point of view these superior discriminators were the most inefficient, since they received less food per unit of time than the more active and less selective strains.

The authors emphasize the point that superior performance on simple tasks where speed or vigor of response is important does not imply enhanced associative learning ability, and they ask whether associative learning should be considered as the apex of adaptation. Biological fitness may depend more on quick reflexes and vigorous exploration than on exquisite discriminatory ability.

Have genetic differences in associative learning been demonstrated?

The studies just reviewed demonstrate that genetic differences in avoidance learning are slightly, if at all, related to differences in the associative process between CS and US. Rat strains that learn quickly when to respond are no better than slow responders in discriminating between alternative routes to safety. Similarly, strains that show the widest variation in response rate between reinforced and nonreinforced periods in an appetitive task are simply those strains whose activity is greatest under either condition. Whether strain differences in performance are found depends frequently on the test procedure that is chosen, and it is not safe to make a sweeping generalization on the basis of strain comparisons made under a single standard procedure.

The general principle which emerges most clearly from the avoidance-escape learning studies is that genes produce differences in the probability of alternative initial responses to a CS or a US. If avoidance is made contingent on the response that is most probable for a particular strain, learning is fast. If avoidance is contingent on a response incompatible with the preprogramed reaction, the animal may be called dull. But such a designation may be the consequence of preprograming of the experimenter to explain behavioral variation in terms of learning differences.

In our opinion the animal data do not permit an unequivocal answer to the question posed in the heading of this section. Probably there are genetic differences in associative learning, but a great deal of the variation in performance that has been ascribed in the literature to differences in learning ability can be explained in other ways. Perhaps to some extent the answer depends on the definition of learning that is adopted and whether it is considered in a narrow sense as a unitary general process or in a broad sense as including any combination of processes that enhances the acquisition of a specified response.

IS A GOOD MEMORY HERITABLE?

All tests for learning require the demonstration that experience in a specified situation changes the probability that a particular form of behavior will occur at a later time under similar conditions. In some way or other, information on the outcome of behavior in the earlier situation must be retained over a period of time that may be measured in seconds, minutes, hours, or even years. Often it is not easy to judge whether failures to remember such outcomes are due to failure to store the information, failure to retrieve the information, or to the deterioration of the information in storage. It is clear, however, that memory capacity could be a factor affecting individual and strain differences in learning. A common idea is that different mechanisms may be involved in short-term memory (STM) and long-term memory (LTM) and that even these categories can be further subdivided. If so, it is reasonable that different genetic substrates might underlie the two mechanisms and that inherited variation in ability to learn might be a function of the degree to which STM and LTM were involved in the task used to evaluate learning.

One way to test this idea is to vary the intertrial interval (ITI) between training trials. If trials are very close together (*massed practice*), STM will be relatively more important than LTM; with *distributed practice* (long ITI) the opposite is true. Following the same line of reasoning and giving training in a number of separated sessions of massed trials, improvement within sessions can be interpreted as a function of STM; decrements between sessions to a failure to transfer information to LTM.

This memory model is oversimplified; nevertheless, it has provided the framework for some well-known studies of the genetics of learning. On the basis of an extensive series of experiments on shuttle-box avoidance in mice, Bovet, Bovet-Nitti, and Oliverio (1969) found that the optimal spacing of trials for learning differed among strains. Particular emphasis was placed on strain C3H/HeJ, which appeared to forget between sessions of 50 massed trials. In contrast, strain DBA/2J mice showed no such decrement between sessions and, in fact, appeared to learn faster when the intersession interval was 24 hours than when it was 5 to 120 minutes. Similarly, the retention of passive avoidance following a single shock was reported to be maximal for C3H/He for 1 to 30 minutes and absent at 24 hours. In DBA/2 response inhibition was weak up to 30 minutes after shock and improved sharply thereafter. The authors characterized C3H/He as depending on STM for learning and DBA/2 as relying on LTM.

Other investigators have found that the spacing of training trials affects the direction of strain differences on learning tasks. In a one-way active escape-avoidance task and a passive-avoidance task, DBA/2J mice performed better with massed trials; C57BL/6J were superior with distributed trials (Wimer et al., 1968). It should be noted that the characterization of the DBA/2 strain with respect to STM and LTM is just opposite to that reported by Bovet, Bovet-Nitti, and Oliverio (1969).

Learning differences between the Tryon rat strains have also been related to the scheduling of training trials and by inference to differences in memory characteristics (McGaugh and Cole, 1965). TMB (bright) and TMD (dull) rats were trained in a Lashley III maze at two ages (30 days or over 150 days), and two ITIs (30 seconds and 30 minutes). All groups except the young TMDs made fewer errors with distributed trials. Their

errors were attributed to less efficient "consolidation" and storage of the memory trace. The implication is that LTM develops more slowly in the TMD strain. Later studies with these strains showed that the differences in learning the Lashley maze could be abolished by injecting a strychninelike drug that was hypothesized to increase the rate of memory storage in the TMDs (McGaugh, Westbrook, and Burt, 1961).

The fact that modification of training schedules affects strains differently has been amply demonstrated, but explanations in terms of memory mechanisms have not stood up well. Other investigators have either been unable to confirm strain differences in STM and LTM with the shuttle-box technique (Rucker, 1973) or have provided alternate explanations of strain differences in terms of strain-specific responses to the CS and, in the case of strain C3H, to the unsuitability of a visual CS for a mouse with pattern blindness due to inherited retinal degeneration (Duncan, Grosen, and Hunt, 1971).

Comparison of the effectiveness of massed and distributed training trials for learning is not the most direct procedure for detecting differences in memory processes between strains. The method of successive reversals in a T-maze was employed by Alpern and Marriot (1972a,b) to compare STM in three strains of mice. Subjects were trained to a criterion of five out of six correct turns to escape shock. Then a "sign trial" was given in which the subject was shocked on the previously correct side and thus informed that it should henceforth turn in the opposite direction. Mice could be trained to reverse reliably when the interval between sign trial and test trial was 5 seconds. Once the reversal was mastered, the interval was varied over the range of 5 seconds to 120 minutes. Strains A and DBA mice were maximally proficient with delays up to 1 minute; CBAs performed best with a 10-minute delay; no mice performed above chance with delays of 20 or more minutes. Although the reversal-learning technique is a more straightforward approach to the measurement of memory than the spacing of training trials, the results can be interpreted in several ways. It is not clear whether the findings of Alpern and Marriott should be explained by strain differences in memory storage or retrieval or possibly by some complex nonspecific aftereffect of the sign-trial shock. However, enough has been done to demonstrate the value of the behavior-genetic analysis of memory capacities. Certainly the validity of hypotheses based on a single strain must be suspect until they are confirmed with a wide selection of available genotypes.

MODE OF INHERITANCE

That genes have something to do with differences in the learning ability of animals has been demonstrated repeatedly by selection and by the prevalence of strain differences. In this section we shall consider genetic experiments that have been directed toward quantitative specification of the mode of inheritance of these differences. As emphasized previously, the extent of such differences, and even their direction, is frequently a function of the particular test used to evaluate learning ability. Thus the experiments discussed in this section should not be regarded as pertaining to a general, unitary factor called intelligence. Rather, their results must be considered task specific, at least for the present.

For this reason the material in this section has been grouped according to the task used to evaluate learning. The genetic techniques have been predominantly biometric. Diallel crosses have been most popular; parent-offspring regression and rate of change under selection have also been used. Inbred strains of rats and mice are the most common subjects because of their genetic uniformity and the availability of numerous diverse types. In evaluating experiments employing these strains, it is important to recall that the proportion of genetic variance observed is inflated compared with estimates from more natural populations. The contribution of specific loci such as albinism (Tyler and McClearn, 1970) and retinal degeneration (Duncan, Grosen, and Hunt, 1971; Fuller, Brady-Wood, and Elias, 1973) to

learning task performance has been evaluated. Thus far such gene effects seem to be explained by peripheral influences on sensory systems, and there seems to be little need to invoke central factors (Wilcock, 1969). A single-locus model has been proposed to account for the difference between two inbred strains of mice in learning an avoidance task (Oliverio, Eleftheriou, and Bailey, 1973b). This gene has not been associated with any somatophene, so its mode of action is presently unknown.

Runway and water-escape learning

One of the simplest measures of learning is the reduction over repeated trials of the time to traverse a runway in order to avoid shock or to obtain food or water. Instead of using shock in a runway, the experimenter may use water in a swimway. Water escape requires a minimum of equipment and represents a hazard that small rodents might encounter in their natural habitat.

Tyler and McClearn (1970) measured starting latencies and running times of food-deprived HS mice (a defined heterogeneous stock) in a 3-foot alley. Subjects were reinforced by food for 5 days and extinguished for 3 days. Heritabilities of running time during acquisition were calculated from the regression of litter means on parental means and ranged from 0.21 to 0.41 over the 5 days. Similar results were obtained for starting latency during acquisition. During extinction, heritabilities of these measures fell rapidly to a nonsignificant value. Each individual's record of acquisition of the running response was analyzed by fitting its scores to the polynomial equation $\hat{S}_t = a + bt + ct^2$, where $\hat{S}_t$ is the expected score for trial t. The coefficients were interpreted as follows: a, initial level; b, amount learned; and c, rate of change or approach to asymptote. Heritabilities for running time acquisition were a, 0.41 ± 0.12; b, 0.30 ± 0.10; and c, 0.26 ± 0.14. A high genetic correlation indicated that the same genes were contributing to running and starting time differences.

The stability of heritability estimates over successive days is notable. Although continued training tended to make an individual's performance more predictable from day to day and although almost all subjects ran faster on later trials, the contribution of genes to variability remained relatively constant. Similar stability was observed in dogs trained to inhibit movement while unrestrained (Scott and Fuller, 1965). There are contrary findings. Kerbusch (1974) observed in mice a gradual, but irregular, decline in the heritability of running time to escape shock over 20 trials, and Willham, Karas, and Henderson (1964) found in pigs that the genetic contribution to variance in an avoidance task rose and then fell over five blocks of trials.

The heritability of water-escape time in mice has been estimated by four methods in a set of related experiments (Festing, 1973a, b, 1974). Strain comparison, rate of change under selection, parent-offspring correlation, and a 5×5 diallel cross all yielded low values (less than 0.1) for h^2. However, the coefficient of genetic determination in the diallel study was high (0.74), indicating substantial dominance effects and heterosis and suggesting that rapid escape times were related to fitness during the evolution of the species.

Similar results were obtained by Hyde (1974) with a 4×4 diallel in mice. The genetic contribution to variance of escape times ranged from 0.19 to 0.47; about one third was attributable to additive effects. Hyde's subjects were also tested on a T-maze, but there was no genetic correlation between performances on the two tasks.

Heterosis and directional dominance for rapid escape were also found in a 3×3 diallel (no reciprocal crosses) with rats (Heinze, 1974). The latency of stepping from a brightly illuminated chamber into a dark compartment was measured at three ages. Such latencies are often used to evaluate learning and retention after a single trial in which stepping into the dark chamber is punished by electric shock. Hybrids resembled their faster parent, often showing overdominance; inbreds, but not hybrids, reacted more slowly as they aged. This experiment is not a

learning study, since each subject was tested only once, but it demonstrates important genetic effects on the unlearned behavior, which is the baseline for evaluating learning.

Overall the data on inheritance of escape learning consistently indicate heterosis for rapid initial escape and rapid improvement over trials. Such findings are evidence of directional selection for a character related to fitness.

Active avoidance

Pigs are not commonly used as subjects in behavior genetics, but representatives of two breeds readily learned to avoid shock in a shuttle box with a buzzer as the CS (Willham, Cox, and Karas, 1963; Willham, Karas, and Henderson, 1964). Half-sib and full-sib correlations based on day 3 avoidance performance yielded within-breed heritability of 0.45 ± 0.12. In the second study emphasis was placed on the relative importance of between- and within-breed genetic effects on several parameters of acquisition and extinction of the avoidance response. Breed differences increased to a maximum on day 3 of acquisition, then decreased and disappeared during extinction. Litter effects were one and one-half to three times as large as breed effects, indicating considerable heterogeneity of learning speed within each breed.

The full-sib, half-sib design was applied to avoidance learning in genetically heterogeneous CD1 mice by Oliverio (1971). He also mated heterogeneous males with inbred females. Since all genetic variance in the offspring was attributable to the sires, h^2 could be estimated from the paternal regression. Heritability estimates of 0.515 and 0.490 from the two methods are very close to those obtained from the pig studies (Willham, Cox, and Karas, 1963; Willham, Karas, and Henderson, 1964).

Using inbred strains of mice, Collins (1964) reported on a 5×5 diallel analysis of shuttle avoidance. His major finding was strong heterosis of high avoidance scores. Out of twenty reciprocal F_1 hybrids, fourteen scored higher than the better parent, and only one scored below the midparental value. A summary of

Table 9-2. Diallel studies of shuttle avoidance in mice and rats

Authors	Species	Size of diallel	Genetic information*
Collins (1964)	Mice	5 × 5	Highly significant GCA and SCA; heterosis favoring high scores
Messeri, Oliverio, and Bovet (1972)	Mice	5 × 5	Variable MOI, but frequently high scores dominant without heterosis; genetic correlation with wheel running: -0.42 ± 0.09
Oliverio, Castellano, and Messeri (1972)	Mice	3 × 3 plus F_2 and F_3	Variable MOI among crosses, but complete dominance most frequent; significant genetic correlation with maze learning and wheel running; h^2 for avoidance from sib correlation: 0.48 ± 0.08
Royce, Yeudall, and Poley (1973)	Mice	6 × 6	Sexes similar; additive variance, 0.503; dominance, 0.430; environmental, 0.066; h^2 about 0.50 on days 1 and 2, about 0.40 on day 3
Wilcock and Fulker (1973)	Rats	8 × 8	For trial blocks 1, 2, and 6, respectively, additive variance was 0.57, 0.69, 0.79; dominance, 0.23, 0.22, 0.12; environmental, 0.20, 0.09, 0.02; directional dominance for low scores in early trials; shifted toward high scores later

*GCA, General combining ability; SCA, specific combining ability; MOI, mode of inheritance.

this and other diallel studies is shown in Table 9-2.

Experimental and analytic procedures of these studies differ in details. Nevertheless, the consistent finding of both additive and dominant gene effects is striking. Directional dominance and often overdominance of high avoidance scores are also reported. The measurement procedures, unfortunately, do not allow a separation of genetic effects attributable to initial levels of activity, initial response to shock, and association of the CS with the avoidance response. The importance of test procedures is emphasized by the inconsistency of strain rankings reported by various investigators. In the diallel experiments on escape and avoidance learning reviewed in this chapter, there were thirteen strain pairings that were replicated in from two to five independent studies. In only two of these pairings was there consistency in the relative ranking by the investigators. Some of the divergence may be attributable to subline differentiation, but procedural details are probably the major cause of the diversity.

The paradox of consistency among studies in the basic mode of inheritance and inconsistency in the relative quality of performance of individual strains is perplexing. It will be resolved only by large-scale experiments in which variations in learning tasks are superimposed on sets of identical diallels.

Passive avoidance

Passive avoidance can be measured by delay of a step-through response on a test trial following a single training trial in which the response was punished by shock. Kerbusch (1974) studied passive avoidance learning in a 5 × 5 diallel of mice. Length of delay on the training trial was interpreted as a measure of "dark preference" and was transmitted in a primarily additive fashion with some dominance without heterosis. Test latency and increase of latency (a measure of learning) showed only maternal and dominance effects. Contrary to expectations, directional dominance was not found. Thus the mode of inheritance appears to be different in active and passive avoidance learning.

Appetitive tasks

Responding for food has been little used in behavior genetic research, in contrast with its wide use in other areas of experimental psychology. An exception is Smart's (1970) investigation of lever-pressing rates by food-deprived C57BL/Tb and A2G/Tb mice and their F_1 hybrids. Though A2G had the highest operant rate of bar pressing, C57BL and F_1 mice increased their rates more rapidly with continuous reinforcement. With fixed-ratio (FR) schedules, hybrids had the highest rate, with C57BLs second. However, at very high ratios (FR 100), A2Gs continued to respond, and C57BLs extinguished. When delayed responses were selectively reinforced, A2Gs were most efficient. As with other studies, no genetic differences were found on extinction.

There is no obvious simple explanation for these results, and it is unlikely that clarification will come soon. Operant procedures undoubtedly have potential for the investigation of individual and strain differences, but there are serious logistic problems in employing them for this purpose. The large investment in equipment and time per subject discourages their use in behavior genetics.

The classic multiple-unit maze was combined with a 6×6 diallel mating plan by Hyde (1974). For running time, she found an additive genetic component of 0.50 to 0.59 at several stages with a smaller dominant component (0.04 to 0.09) appearing later. Genetic effects for error scores were smaller. None was found for days 1 and 2 of testing; for days 3 and 4, respectively, additive effects were 0.19 and 0.08, dominant effects, 0.08 and 0.20. Once again the mode of inheritance changed during the learning period, suggesting that at least two distinct processes contributed to the strain differences.

Discrimination learning and reversal

Shock escape in a T-maze was studied in a 6×6 diallel of mice by Stasik (1970). Animals were trained to a criterion of nine consecutive correct responses. There was no heterosis, and inbreds on the average made no more errors than hybrids. Since specific combining ability was significant and general combining ability was not, bidirectional dominance was indicated. Running time was uncorrelated with errors and showed the heterotic effect that others have

reported for similar measures. Inbreds averaged 9 seconds on criterion trials versus 7.1 seconds for hybrids. Thus the genetic evidence supports a distinction between the biological bases of rapid responding and accurate place learning.

The ability to readily reverse a discriminated response is often considered to be an indicator of adaptability to new situations. In a 4×4 diallel with mice, Carran (1972) demonstrated highly significant general combining effects for the number of successful reversals in 50 sessions in a T-maze. Water reinforcement was employed. Specific combining effects were not found, indicating that dominance and epistasis were not important. Later a Mendelian cross between the best performing strain (129/J, with an average of 11.4 reversals) and the poorest (DBA/1J, with 3.8) confirmed the importance of additive gene action and also detected a maternal effect opposite in direction to the influence of transmitted genes (Carran, 1975). Heritability in the narrow sense was estimated as 0.647, which is certainly inflated by the selection of two extremely divergent strains as parents.

Complex learning

More complex learning tasks, such as matching to sample, delayed response, and rate of formation of learning sets, are candidates for behavior genetic analysis. They have not been widely used because the abilities they test are not well developed in the common laboratory rodents that are most suitable for experimental genetics. Furthermore, as with operant conditioning, the expenditure of effort per subject is high. We have already noted that individual variation in complex tasks is common, but the relative contribution of genetic and environmental factors to such differences is unknown.

GENOTYPES, EARLY EXPERIENCE, AND LEARNING

Early experience is widely considered to have a major influence on later learning capacity (Hunt, 1961), though some experiments have shown the impairments produced by poor early environments to be largely transitory (Fuller, 1966). What happens when genotype and early experience are varied systematically? In this kind of experiment the environmental variance is not simply what is left over when genetic variance is extracted; it can be related to environmental features that are controlled experimentally. In such experiments one might seek to determine whether genotype affects vulnerability to experiential deprivation. Or the object may be to discover how robust strain differences in learning are when optimal environments are provided. These questions have their parallels in human behavior genetics where controlled experiments are not possible.

Perhaps the earliest experiment of this kind compared the McGill bright and dull rat strains in the Hebb-Williams maze after rearing in a restricted, standard, or enriched environment (Cooper and Zubek, 1958). Animals raised under the standard condition, similar to that used during the original selection, differed in error scores as predicted. However, after restriction in infancy and adolescence, both strains made many errors; after exposure to a more complex environment during these periods, both performed well. The experiment has flaws. The facts that the MMD rats did not increase errors after rearing in a restricted environment and that the MMBs did not benefit from enrichment suggest that the test may not have been sensitive to variations in problem solving ability at both extremes. Nevertheless, the data indicate that the MGBs require less stimulation to organize their adaptive behavioral responses.

Other studies confirm that the effects of prior experience on learning are genotype dependent. On a 6-unit T-maze, C57BL mice, but not C3H mice, benefited from gentling before testing (Lindzey and Winston, 1962). Mice of three strains exposed to traumatic auditory stimulation at 4 to 7 days made more errors in a maze than did controls (Winston, 1963). Hybrids were found resistant to the deleterious effects of such stimulation (Winston, 1964). The genetic component of variance on a learning task administered to several strains was much less in mice reared in standard laboratory conditions than in those reared in large cages provided with ladders and other paraphernalia (Henderson, 1970). It is apparent that the standard, rather sterile rearing conditions characteristic of most laboratories are designed to conceal genetic differences and possibly the effects of other treatments.

Henderson (1972) looked for genetic variation in the effects of rearing in an enriched-versus-standard environment in a 6 × 6 diallel of mice. Shock escape and water escape were used as learning tasks in order to avoid confounding effects of early experience on learning ability with its possible effects on fear and exploration. These factors play a greater role in avoidance or appetitive learning than in escape learning. Shock escape involved spatial discrimination in a T-maze; water escape employed a visual discrimination. Large additive genetic effects were found on all learning measures, and directional dominance for rapid learning on most. In this respect the Henderson experiment is consistent with other findings. Most significant for our present purpose is that the effects of enriched environment and of maternal environment (measured from reciprocal cross differences) were very small relative to genetic effects and that genotype-environment interactions were negligible.

The results of this study run contrary to a considerable body of data supporting the idea that the effects of early experience on later learning ability are both large and strain specific. Henderson proposes two explanations for the discrepancies. First, many studies reporting benefits have exposed young subjects to visual stimuli and found improved learning when similar stimuli are encountered in a discrimination test. The results are better interpreted in terms of neurological changes underlying perception than in terms of different rates of learning. Second, the experimenters who have found beneficial effects of generalized environmental enrichment have evaluated learning in appetitive or avoidance tasks. Differences in drive level and interference from excessive exploratory activity usually found in postisolates can influence performance negatively. With an escape task, drive level is uniformly high, and interference is less significant.

Henderson's experiment does not negate the influence of early experience on many learned performances. However, it casts doubt on the assumption that these effects are mediated by changes in the capacity for associative learning. The resolution of this doubt will require more large-scale experiments in which the most advanced techniques of genetic analysis are combined with equally advanced procedures for behavioral analysis. Further discussion of this issue can be found in Chapter 18.

SUMMARY

The experiments reviewed in this chapter demonstrate that the influence of genotype on learned performances is significant and complex. The evidence is based on successful selection, reliable differences among genetically defined groups, the outcome of crosses between such groups, and family correlations in heterogeneous populations.

Superiority on one learning task does not imply superiority on others. There is no strong evidence in animals for a general intelligence, or G, factor. Much of the genetic variation in performance on learning tasks is attributable to differences in peripheral sensitivity, in drive level, or in preprogramed tendencies for responding to an alerting stimulus.

An additive-dominant model of inheritance fits the bulk of the experimental data on learning performance. When changes in starting latencies or running speeds are used as learning measures, heterosis and directional dominance for rapid responding are typically found and are interpretable as evidence that these measures are related to genetic fitness. When the number of discrimination errors is used as a criterion of learning, inheritance is generally additive, suggesting selection for intermediate phenotypic expression.

The nature of interaction between genotype and early experience is rather ambiguous as it relates to learning. It seems likely that interactive effects, when they are detected, are mediated through processes other than associative learning.

Despite a considerable literature on the experimental genetics of animal learning, there are many gaps to be filled by innovative research. Such research requires a combination of sophisticated techniques from genetics and experimental psychology. In one direction genetic treatments, particularly of the single-locus type, could be helpful in working back to the physical nature of the engram. Perhaps more feasible at the present time is the use of genetic techniques to test the independence or dependence of such concepts as long- and short-term memory. And it would be interesting to know more about the mode of inheritance of learning set acquisition. Such findings might help to build a bridge between the genetics of animal learning and the genetics of human intelligence.

10

Temperament: activity, reactivity, and emotionality

Animal temperament includes a variety of traits, many of which are similar to those which make up human personality. Existence of these similarities is recognized when a person is described as mousy, foxy, chicken-hearted, hawkish, or dogged. We designate both our acquaintances and our pets as timid, aggressive, emotionally stable, or excitable. This chapter considers the role played by genes in the development of these dispositions and in the production of differences between individuals. You may wish to compare it with Chapter 15, which deals with the complex issues of genetics and human personality.

Ideally the psychophenes selected for the genetic analysis of temperament should be measurable by objective procedures, predictive of behavior in a variety of situations, and stable over time. The facts are that animals habituate to novelty, they learn new responses, and their sensory and physiological processes are modified by experience, especially a stressful experience. Thus a dilemma is created for the behavior geneticist: to obtain data on the large number of subjects required for genetic analysis requires tests that can be administered quickly and yield objective scores. This requirement conflicts with the desire to obtain measures representative of a subject's full repertoire of emotional, exploratory, and social behavior. As a result, literature on the genetics of animal temperament is rather amorphous. Such structure as it possesses has been determined by the decisions of individual researchers to select certain subjects, pretreat them in a particular way, and observe their behavior in one or more standardized procedures. Standardiza-

tion of subjects and procedures is beneficial insofar as it facilitates comparisons between laboratories and validation of findings. But too much reliance on a single test as a measure of "emotionality" or relying on a few well-known inbred or selected lines of mice or rats is too narrowing. Happily, recent research tends to consider data from a variety of genotypes reared under several different conditions and tested in a variety of ways.

We have divided our treatment of the broad area of animal temperament between two chapters. This chapter deals with genetic contributions to variation in the responses of individuals to the physical features of their environment. Chapter 11 is concerned with genetic influences on social behavior, a much more complex phenomenon, since genetic variation in all the interacting participants must be considered. In a social behavior experiment, control of the testing situation passes in part from experimenter to subjects.

SOME MAJOR ISSUES

It does not require scientific training to recognize that wharf and Wistar rats differ strikingly in behavior. Administering a battery of tests to both kinds of rat might yield reliable differences between them on most measures. One of the major questions is whether such intraspecific variation is unidimensional, individuals and groups being distributable along a single scale of emotionality or fearfulness, or whether there are a number of dimensions that are essentially independent. The concept of a single scale of temperament need not imply single-locus or polygenic control of one critical endocrine or neural variable with potent and wide-ranging behavioral effects. Although the sizes and forms of skull, body, and limb of an animal are controlled by independent genes, under natural selection these independent genetic systems become coordinated to produce an animal adapted for a particular habitat and life-style. It is reasonable to suppose that potentially independent behavioral systems are also coordinated through natural selection, so that they form a stable metasystem with a secondary type of unity.

The same issue of dimensionality was encountered during our consideration of learning, and we concluded that heritable differences in the rate of learning are task specific. A considerable amount of experimental evidence also favors a multidimensional structure of animal temperament, though consensus has not been obtained on the number and nature of its components. This problem will be dealt with in more detail later. For the present we shall organize the topic of temperament along empirical lines; that is, we shall try to define the domain by the kinds of test devices and behavioral measures that have been reported in the literature. We start with brief descriptions of some of the procedures that ostensibly measure animal temperament in an objective fashion.

APPARATUS AND PROCEDURES

Considerable ingenuity has been devoted to the design of simple apparatuses that will aid in the objective measurement of activity, reactivity, and emotionality. Fig. 10-1 depicts some of the more commonly used devices. Variations in the dimensions of apparatus are prevalent, sometimes in order to adjust for changes in the size of animals in developmental studies. Measures of latencies and frequencies of behavior patterns expressed within these apparatuses define operationally concepts such as emotionality, timidity, and exploration.

A common measure of activity is the running wheel. The distances run in these devices are often several kilometers per day, greatly exceeding those traversed by animals in laboratory cages and probably much greater than distances covered in nature. Although the recording of activity in home cages over long periods of time is technically feasible, it has been used only rarely in genetic studies, probably because of the high cost of the equipment. Generally in genetic investigations, activity has been recorded for short periods, 2 to 10 minutes, in an open field with a floor marked off into sections. Sometimes it is measured in a field with a pattern of barriers simulating a network of passages, and the resulting datum is designated as an "exploration" score. Locomotion is also measured in straight, Y-shaped, or circular alleys or by the number of crossings between chambers separated by a partial barrier. Measures such as counts of squares entered, sections traversed, and light beams interrupted yield quantitative scores. A human observer can record simultaneously other kinds of behavior, such as rearing, sniffing at objects, and grooming. The size of the apparatus and control of such factors as light intensity, ambient sound level, mode of introduction to the apparatus, and duration of the test have varied considerably. The possible effects of these variations have been reviewed critically by Walsh and Cummins (1976). Within a single laboratory, procedures are generally standardized and permit comparisons between experiments done by different people at different times. Comparisons between different laboratories are more risky.

Other indices of emotionality are based on the assumption that a fearful animal does not investigate novel objects, enter and explore unfamiliar areas, or move away from the walls of an enclosed arena. Long delays in emerging from a starting chamber or descending from a high platform have also been interpreted as timidity. Some measures of temperament do not require special apparatus. The reaction of an animal to capture by a human was an early criterion of "wildness," and it is still employed occasionally.

Psychological arousal, which in ourselves we subjectively denote as anger, fear, or love, is accompanied by physiological changes mediated through the autonomic nervous system. Similarly, in research on animal temperament, defecation, urination,

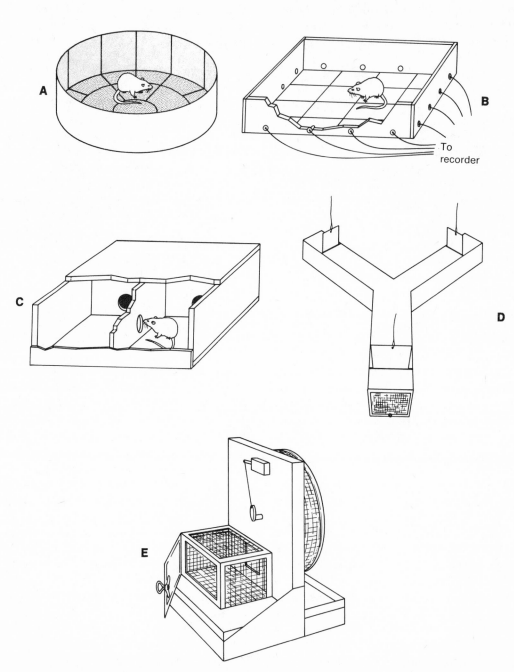

Fig. 10-1. Some devices used in evaluating animal temperament. **A,** Circular open field. **B,** Square open field with photoelectric recording of the number of times a light beam is broken. **C,** Hole-in-wall apparatus for measuring exploration. **D,** Y-maze for exploration or preference tests. Areas at the ends of the arms can be used as starting chambers or to restrain another animal in social preference experiments. **E,** A running wheel with attached living cage for long-term observations on the amount and periodicity of activity. The dimensions of apparatuses of these types vary according to the size of the animals being tested.

and changes in heart and respiration rate and in muscle tension have been used as indicators of emotionality. The last two items are actually mediated by the somatic rather than the autonomic system, but their relation to arousal has been clearly demonstrated. Other physiological processes, including endocrine gland responses and reactions to neurotropic drugs, have been employed in the behavior-genetic analysis of emotionality. Although this is a new field of endeavor, it appears to have great promise.

Attention has also been given to a potential relationship between emotionality and avoidance performance. In the preceding chapter we noted that heritable variation in the nature of the unconditioned response to a painful stimulus seems to be more potent than differences in general ability to associate a CS with a US in accounting for strain differences in avoidance learning. One might predict, therefore, that selection for either emotionality or for avoidance performance would affect the other trait.

The behavior-genetic analysis of animal temperament is made even more complex when temporal changes are considered. Emotional and social behavior is sensitive to the cumulative influences of environment beginning at the moment of zygote formation, long before any behavior is observable. Although there seems to be a general belief that the behavior patterns of adults are less modifiable than those of the young, the difference is certainly one of degree only. Thus the extent to which genotype affects the capacity to modify behavior adaptively at any point in the life cycle is an important area of investigation.

The multiplicity of behavioral phenomena that are subsumed under the heading of temperament creates problems with respect to a systematic presentation. Ideally, each of the main issues would be considered with pertinent data from a variety of species. Actually, it proves easier to consider bodies of data according to species (or by grouping closely related species). The kinds of experiments performed with laboratory rats and mice differ enough from those done with dogs or fruit

flies to warrant separate sections. In the chapter summary we shall see what principles emerge.

HISTORICAL INTRODUCTION

Selection of domestic animals for temperamental qualities certainly precedes the emergence of genetics as a science. The beginnings of a behavior-genetic analysis of temperament in laboratory animals followed shortly on the rediscovery of Mendel's work in the early twentieth century. References to these pioneers and descriptions of their research can be found in *Behavior Genetics* (Fuller and Thompson, 1960). Hall's (1941) review is also valuable for its discussion of this earlier work, much of which is still pertinent today.

The two decades from about 1930 to the early 1950s were characterized by important selection experiments and by the introduction of modes of genetic analysis more complex than simple Mendelian models. Selection for high and low activity in a running wheel was successfully accomplished by Rundquist (1933), and a genetic analysis of the results performed by Brody (1942, 1950). Brody postulated that the difference between the active and inactive strains was attributable mostly to a single locus, with low activity dominant in males and high activity dominant in females. A polygenic model with superimposed sex-hormone effects is more plausible. One of the problems with wheel-running data is their high intrinsic variability unless long periods of adaptation are provided. Gross effects of heredity can be demonstrated easily, but estimation of genetic parameters is difficult. Also, the relationship of such activity to the everyday functioning of an animal is unclear. Running wheels are still used for genetic studies of temperament but usually as part of a battery of tests.

A selection experiment by Hall (1938, 1951) for high and low emotionality of rats in an open field has been extremely influential. Hall evaluated emotionality from the frequency of urination and defecation during 12 daily 2-minute trials in a fairly large, brightly lit open field. Scores could vary from 0 to 12, that is, from no defecation or urination on any day to elimination on all days. This method of scoring gives heavy weight to the animal's adaptation to a novel environment over a period of days. Hall's scoring procedure has not been followed generally by later researchers, though open-field apparatus is found in hundreds of laboratories. Most present-day investigators take a count of fecal boluses as the score for a trial and give only 1, 2, or possibly 4 trials per subject. Such juggling with procedures is not trivial and accounts for some of the contradications in the literature (Walsh and Cummins, 1976).

Activity and emotional differences among nonselected laboratory strains of rats were noted at an early date (Yerkes, 1916; Utsurikawa, 1917). Emotional differences were also reported in mice (Yerkes, 1913; Coburn, 1922; Dawson, 1932; Stone, 1932). A negative relationship between open-field emotionality and aggressive behavior was reported in rats by Hall and Klein (1942). A positive association of these traits was found by Farris and Yeakel (1945), who observed that their gray Norway rats were both more emotional and more aggressive than Wistar albinos. In evaluating such discordant findings, it is important to remember that no generality whatsoever can be inferred from an association based on a sample of two strains.

In mice, Lindzey (1951) found significant differences in urination and defecation among five inbred strains, and Fredericson (1953) found variation in the degree to which strains maintained contact with the wall (thigmotactic behavior) in an arena. High levels of both elimination and thigmotaxis were considered to be diagnostic of timidity and emotionality.

The results of these early experiments were clear: in every case heritable differences in the levels of activity and emotionality were found. Selection for either type of behavior was effective. By the middle 1950s the foundation was laid for more definitive experiments on the mode of inheritance of temperamental traits and on the coactions of genotype and experience in the development of individual and strain characteristics. In-

creasing sophistication in quantitative genetic analysis and improvements in the precision of environmental controls in laboratories facilitated research on the genetics and variability of temperament in animals.

EMOTIONALITY IN RATS AND MICE: GENETIC ASPECTS

The word *emotionality* as used by animal behaviorists has a variety of connotations (Archer, 1973). It may be an inclusive term describing a variety of behaviors elicited by strong stimulation; a drive-energizing state; a stable, genetically influenced property of an individual; or a psychological correlate of certain autonomic responses. Combining the last two definitions, a rat that defecates three times within 2 minutes in an open field is more emotional than one that defecates once. When Broadhurst (1958a,b, 1975), for example, refers to reactive (emotional) and nonreactive (nonemotional) strains of rat, he implies that in the reactive animals the threshold for arousal by stress is lower and that their deviation from a baseline of physiological and psychological function is greater.

The volume of research on genetic factors and temperament in rats and mice is so great that complete coverage is not possible in a general account. Useful reviews from differing points of view are available in Archer (1973), Broadhurst (1960b, 1975), and Walsh and Cummins (1976). We have selected examples of research representing major trends and methods starting with three selection programs: two with rats and one with mice.

Hall selection program

We have previously alluded to Hall's selection program for emotionality because of its historical importance. Hall (1934) was explicit in regarding emotionality as a general excited condition manifested by organic, experiential, and expressive reactions. It was not for him a trait or a faculty; nevertheless, his research led him to conclude that levels of emotional response within individuals were consistent over a period of time. He also reported a negative association between

Table 10-1. Bidirectional selection for emotionality in rats in an open field*

Genera-tion	Emotional strain			Nonemotional strain		
	N	Mean[†]	SD	N	Mean[†]	SD
S_0	145	3.86	3.54	145	3.86	3.54
S_1	40	3.07	3.36	35	0.46	0.77
S_2	18	4.72	4.12	18	1.94	2.28
S_3	65	3.92	3.63	50	1.02	1.30
S_4	84	4.69	3.89	52	1.40	1.43
S_5	75	4.96	3.85	59	0.41	1.18
S_6	48	6.87	3.28	51	0.51	1.13
S_7	72	7.82	3.18	53	0.17	0.47
S_8	77	8.37	2.94	40	1.07	2.46
S_9	85	10.31	2.09	32	1.68	3.25
S_{10}	66	10.41	2.08	22	1.45	3.13
S_{11}	57	10.11	2.39	42	1.05	2.01
S_{12}	47	10.40	2.18	31	1.65	2.53

*From Hall, C. S. 1951. The genetics of behavior. In S. S. Stevens, ed. Handbook of experimental psychology. John Wiley & Sons, Inc., New York.
[†]Number of days out of 12 with defecation or urination.

Table 10-2. Crosses between emotional and nonemotional strains of rats*

Generation (emotional × nonemotional)	Number of days out of 12 with elimination		
	N	Mean	SD
$S_{10} \times S_{10}$	32	4.53	3.84
$S_{11} \times S_{11}$	22	2.81	2.15
$S_{12} \times S_{12}$	27	3.00	2.55

*From Hall, C. S. 1951. The genetics of behavior. In S. S. Stevens, ed. Handbook of experimental psychology. John Wiley & Sons, Inc., New York.

defecation and locomotor activity (now frequently called *ambulation*) and suggested that the role of emotionality was to reduce activity in situations where it would be disadvantageous. Here Hall applied classic evolutionary theory to behavior; species-specific modes of response, like species-typical structures, attained their present form because they increased fitness in the ancestors of present-day individuals.

The heritability of defecation scores was demonstrated by success in selection. The parental generation, consisting of 145 rats obtained from several colonies at the universities of California and Oregon, was divided on the basis of individual scores into two groups: emotional and nonemotional. Means and standard deviations for the successive generations are given in Table 10-1. It is clear that the frequency of defecation increased in the emotional line up to S_9, at which point it appeared to stabilize. Variability in this line decreased steadily, even though the mean increased threefold. Selection for nonemotionality was not successful;

the mean score of the first three selected generations (1.14) is not significantly different from that of the last three (1.33). Variability in the nonemotional line was high relative to the mean. Such asymmetry in response to selection is not uncommon. It might have been due to the particular pool of genes in the base population, to insensitivity of the test at the low end of the scale, or to a physiological limit such that natural selection balanced Hall's artificial selection.

Crosses between the emotional and nonemotional rats yielded the results shown in Table 10-2 (Hall, 1951). The F_1 means are closer to the nonemotional than to the emotional parents. Hall did not report on F_2 or backcross hybrids. He wrote, "Had the two strains been pure the variability of the hybrids would have been smaller than that found . . . it was deemed unfruitful to make further crosses with impure strains" (Hall, 1951, p. 324). In fact, the F_1 hybrids were not significantly more variable than the parental strains, and genetic theory does not predict that they would necessarily be less variable. Hall's work is important as the first well-documented account of selection for emotionality and for his popularization of the open-field test that has been so widely adopted.

Maudsley reactive and nonreactive rats

The selection program inaugurated by Broadhurst (1958b, 1960b) at the Maudsley Hospital in London and continued at the

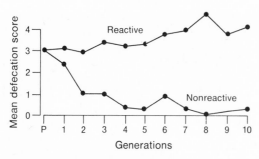

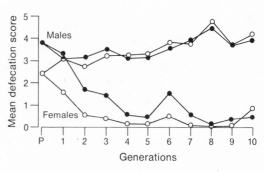

Fig. 10-2. Results of selective breeding for high- and low-defecation scores. Graph shows the mean scores of reactive and nonreactive groups in successive generations of selective breeding. Generations are shown on the abscissa; the ordinate shows the mean number of fecal boluses deposited per trial in the test. (From Broadhurst, P. L. 1960. Experiments in psychogenetics: applications of biometrical genetics to the inheritance of behaviour. In H. J. Eysenck, ed. Experiments in personality, vol. 1. Psychogenetics and psychopharmacology. Routledge & Kegan Paul Ltd., London.)

Fig. 10-3. Results of selective breeding for high- and low-defecation scores: sex difference. Graph shows the mean scores of reactive and nonreactive groups in successive generations, divided according to sex. Generations are shown on the abscissa; the ordinate shows the mean number of fecal boluses deposited per trial in the test. The higher curves represent the reactive strain, the lower, the nonreactive, males being designated by solid points and females by open ones. (From Broadhurst, P. L. 1960. Experiments in psychogenetics: applications of biometrical genetics to the inheritance of behaviour. In H. J. Eysenck, ed. Experiments in personality, vol. 1. Psychogenetics and psychopharmacology. Routledge & Kegan Paul Ltd., London.)

University of Birmingham in many ways replicates and confirms Hall's experiment. There are, however, important differences in testing procedure, breeding plan, and mode of genetic analysis. The two lines, now commonly designated as MR (Maudsley reactive) and MNR (Maudsley nonreactive), have been employed widely in experiments on the behavioral and physiological correlates of emotionality. Broadhurst (1975) cited many papers dealing with these strains and interprets the results as supportive of the thesis that the strains are "characterized by relatively stable differences in a generalized trait of emotional reactivity which expresses itself in many and various ways."

The criterion for the selection program was based on parametric studies of the defecation response as a function of the test situation (Broadhurst, 1957) and the subject's prior experience (Broadhurst, 1958a). As finally standardized, the open field was smaller than Hall's, with a floor marked into approximately equal areas so that a count of areas entered could be converted to ambulation in meters. The field was brightly illuminated (165 foot-

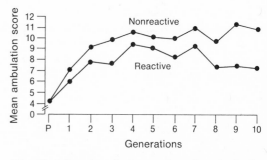

Fig. 10-4. Results of selective breeding for high- and low-defecation scores: effect on ambulation. Graph shows the effect on the mean ambulation scores of reactive groups in successive generations of selective breeding. Generations are shown on the abscissa; the ordinate shows the mean number of meters run per trial in the test. (From Broadhurst, P. L. 1960. Experiments in psychogenetics: applications of biometrical genetics to the inheritance of behaviour. In H. J. Eysenck, ed. Experiments in personality, vol. 1. Psychogenetics and psychopharmacology. Routledge & Kegan Paul Ltd., London.)

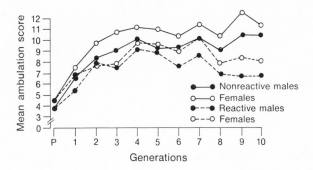

Fig. 10-5. Results of selective breeding for high- and low-defecation scores. Graph shows the effect on the mean ambulation scores of reactive and nonreactive groups in successive generations, divided according to sex. Generations are shown on the abscissa; the ordinate shows the mean number of meters per trial run in the test. (From Broadhurst, P. L. 1960. Experiments in psychogenetics: applications of biometrical genetics to the inheritance of behaviour. In H. J. Eysenck, ed. Experiments in personality, vol. 1. Psychogenetics and psychopharmacology. Routledge & Kegan Paul Ltd., London.)

Table 10-3. Defecation and ambulation in crosses of MR and MNR rat strains*

Generation	Defecation (boluses)				Ambulation (meters)			
	Males		Females		Males		Females	
	Mean	SE	Mean	SE	Mean	SE	Mean	SE
P_1 (MR)	4.6	0.2	4.7	0.2	6.8	0.3	7.7	0.3
B_1	3.7	0.15	2.8	0.2	6.6	0.3	7.7	0.3
F_1	2.5	0.2	1.1	0.2	8.6	0.3	10.1	0.15
F_2	2.8	0.15	1.1	0.1	7.6	0.2	9.3	0.2
B_2	1.4	0.2	0.5	0.1	7.3	0.3	8.9	0.4
P_2 (MNR)	0.04	0.02	0.01	0.01	9.0	0.4	10.1	0.4

*From Broadhurst, P. L. 1960. Experiments in psychogenetics: applications of biometrical genetics to the inheritance of behaviour. In H. J. Eysenck, ed. Experiments in personality, vol. 1. Psychogenetics and psychopharmacology. Routledge & Kegan Paul Ltd., London.

candles), and white noise (78 db) was fed in constantly (Broadhurst, 1960b). Subjects were observed for a 2-minute period on 4 successive days. Two scores were recorded daily: the number of boluses and the distance traversed. In contrast to Hall's the Maudsley system weights the initial responses more heavily and gives less emphasis to the capacity to adapt to an unfamiliar environment. The Maudsley base stock was "of remote Wistar origin" purchased from a dealer. Brother-sister inbreeding was commenced early in the selection program, a choice the experimenter later considered to have been

unwise (Broadhurst, 1960b), since it precluded certain types of genetic analysis.

Despite these differences, the results of Broadhurst's selection were similar to those of Hall. Figs. 10-2 through 10-5 depict changes in mean defecation and ambulation for the first ten generations. In the foundation stock and in the MNR line, males defecated more. In the MR line the sex difference attenuated rapidly. Cross-fostering and reciprocal crosses between lines yielded scant evidence for maternal effects (Broadhurst, 1960a). Results of Mendelian crosses between the MR and MNR strains are shown in

Table 10-3. For defecation the additive genetic component was high; heritability for the combined sexes was estimated as 0.95. For ambulation a significant dominance effect favoring low scores was found, and heritability for the pooled sexes of 0.80. We shall return to the Maudsley lines repeatedly in later sections of this chapter.

Selection for open-field activity in mice

The following selection experiment comes close to meeting the ideal for this type of investigation. The foundation stock was the F_3 of a cross between inbred strains C57BL/6J (pigmented) and BALB/cJ (albino) (DeFries and Hegmann, 1970; DeFries, Wilson, and McClearn, 1970). Earlier observers of these strains had characterized C57BL as a high-activity, low-defecation strain as compared with BALB. A very desirable feature of this experiment was the replication of the high, low, and unselected control lines. On the small scale of practicable laboratory selection experiments, traits independent of the criterion measure may become associated with it through the vagaries of random sampling. Although there is no way of predicting the probability of such functionally meaningless associations, the probability that the same association will occur in two independent lines is very small. An association between traits that occurs in both replicate lines is safely interpretable as based on a functional relationship of both traits to the same genes. Inbreeding was minimized as much as possible to retain genetic variance as a basis for continued progress in selection. The presence of the albino gene from the BALB strain allowed a quantitative estimate of the effect of a single locus on a form of behavior influenced by many genes.

Activity was measured in an illuminated open field for two 3-minute periods during which fecal boluses were counted. After ten generations the two high lines had average activity scores (number of light-beam interruptions) of about 260, the two controls about 200, and the two low lines about 100. Thus the effects of selection were relatively symmetrical, though slightly more effective in the downward direction.

Changes in amount of defecation accompanied the changes in activity, the reciprocal of the effect produced on activity by selection for defecation in the Hall and Broadhurst experiments. But there was evidence that in mice the two traits are not completely interlocked. Both high activity lines had low defecation rates, and both low lines were high; the two control lines split, one moving upward and the other downward, although neither changed much in activity.

Realized heritability for activity was computed from the regression of the divergence in activity scores between the high and low lines on the cumulative selection differential. The obtained value, 0.13 ± 0.02, is low compared with the values obtained by Broadhurst (1960b), but the methods of computation were quite different. The realized genetic correlation between activity and defecation was very high, -0.80 ± 0.13. This means that selection for activity altered defecation scores nearly as effectively as direct selection for defecation rate.

Although selection was based on activity scores without consideration of coat color,

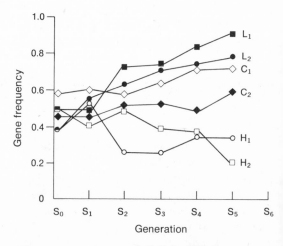

Fig. 10-6. Frequency of allele for albinism in two lines of mice selected for high open-field activity (H_1 and H_2), two selected for low activity (L_1 and L_2), and two unselected controls (C_1 and C_2). (From DeFries, J. C., and J. P. Hegmann. 1970. Genetic analysis of open-field behavior. In G. Lindzey and D. Thiessen, eds. Contributions to behavior-genetic analysis: the mouse as a prototype. Irvington Publishers, Inc., New York.)

the effects of the albino gene, c, were very evident. In the base population the frequency of c was 0.5, and the proportion of albinos was 0.25. At S_{10} the proportion of albinos was near 100% in both low lines, near 30% in controls, and under 10% in the two high lines (Fig. 10-6). We call c a major gene because its phenotypic effects are so clearly apparent. Here it serves as a visible marker for the behavioral selection process operating at a single locus. Only a fraction, probably about 20%, of the response to selection is attributable to changes in gene frequency at the albino locus. We must conclude that similar changes in allelic frequency are proceeding at other loci whose positions and functions are not at present identifiable. Locating such polygenic loci is theoretically possible in mammals but is more easily accomplished in *Drosophila* (see, for example, Hirsch, 1959; Hirsch and Erlenmeyer-Kimling, 1962).

Strain differences

Today the demonstration of strain differences in activity and emotionality, whether measured by defecation, ambulation, or other measures, is by itself of no great scientific interest. Such information does, however, provide basic data for investigations of mode of transmission and research on the physical substrates of behavioral variation. Strain difference studies also have historical interest, since they stimulated research into the development and modifiability of temperament. References to early reports on a number of species may be found in Fuller and Thompson (1960).

Strain comparisons are most valuable when the sample is large enough to be representative of a designated population; results from two or three strains may not be generalizable. One of the first large-scale investigations of this kind compared emotionality (defecation) and food drive in fourteen inbred strains and one mutant stock (Thompson, 1953b). After mild food deprivation, subjects were placed in an open field that contained a dish of wet mash in the center. Tests of 10 minutes were repeated on 6 consecutive days. The results, summarized in Table 10-4,

Table 10-4. Food consumption (hunger drive) and defecation (emotionality) in fifteen mouse strains*

Strain	Food intake† in grams (rank)	Defecation‡ percentage (rank)
TC3H	204 (1)	11 (15)
AKR	199 (2)	34 (13)
C3H	183 (3)	52 (11)
DBA/1	176 (4)	57 (10)
ND	171 (5)	27 (14)
C57BL/6	162 (6)	40 (12)
AK/e	131 (7)	96 (1)
C57BR/a	128 (8)	66 (8)
LP	124 (9)	82 (4.5)
DBA/2	121 (10)	70 (7)
C57BL/10	116 (11)	58 (9)
BALB/c	99 (12)	82 (4.5)
Obese	82 (13.5)	83 (3)
BDP	82 (13.5)	85 (2)
A/J	80 (15)	78 (6)

*From Thompson, W. R. 1953. The inheritance of behavior: behavioral differences in fifteen mouse strains. Can. J. Psychol. 7:145-155. Copyright 1953, Canadian Psychological Association. Reprinted by permission.
† Mean daily consumption of 10 males and 10 females.
‡ Mean daily percentage of defecation by 10 males and 10 females.

showed highly significant variation in the amount of food eaten and percentage of animals defecating. Mice from the more emotional strains ate less; the rank-order correlation between amount of food eaten and percentage of individuals defecating was -0.796 ($p < 0.001$).

The same subjects were subsequently observed in a discontinuous open field, one in which short barriers were inserted to form a network of passages. Exploratory activity of the fifteen strains varied more than twentyfold, as shown in Table 10-5.

Mode of inheritance

The application of quantitative methods of genetic analysis to behavior dates from the late 1950s and early 1960s. The modes of inheritance of two forms of activity, locomotion in the discontinuous open field and in a Y-maze, were investigated in a set of Mendelian crosses between a high-activity strain (C57BR/aJ) and a low-activity strain (A/J)

Table 10-5. Exploratory activity in fifteen mouse strains*

Rank	Strain	Mean number of crossings	Probability of significant difference†	
			<0.05	<0.01
1	C57BR/a	459	>3	>5
2	C57BL/6	361	>7	>10
3	C57BL/10	359	>7	>10
4	DBA/1	334	>8	>12
5	ND	308	>8	>12
6	BDP	286	>10	>13
7	DBA/2	253	>11	>14
8	LP	194	>13	
9	AKR	188	>13	
10	C3H	177	>13	
11	Obese	149	>15	
12	TC3H	117		
13	BALB/c	74		
14	AK/e	60		
15	A/J	20		

*From Thompson, W. R. 1953. The inheritance of behavior: behavioral differences in fifteen mouse strains. Can. J. Psychol. 7:145-155. Copyright 1953, Canadian Psychological Association. Reprinted by permission.
†This column is read as follows: the probability is less than 0.05 that the difference between rank 3 (C57BL/10) and rank 7 (DBA/2) and lower ranks is due to chance. The probability is less than 0.01 that the difference between C57BL/10 and strains of rank 10 and lower is due to chance.

(Thompson and Fuller, 1957; Fuller and Thompson, 1960). The results and their interpretation were reviewed in Chapter 5 as an example of biometric analysis of behavior. Here the coefficient of genetic determination (CGD) was estimated as 0.59, and heritability (h^2) as 0.05. The large difference between these values is attributable to the strong dominance of genes for high activity.

A similar investigation of activity in crosses between C57BL/Crgl and A/Crgl yielded similar results with a CGD of 0.69 (McClearn, 1961). A notable somewhat puzzling feature of the data was an F_1 variance that was higher than that of the parental strains. Heterozygosity has often been considered to induce phenotypic stability by favoring developmental homeostasis. This topic is discussed later in this chapter. We propose an explanation for McClearn's data by invoking the concept of genetic control of thresholds for sensitivity to environmental factors that stimulate activity. In strain C57BL the threshold may be so low that a minimal amount of stimulation produces high activity. In strain A the threshold may be so high that ordinary levels of stimulation are ineffective. If their F_1 hybrid has an intermediate threshold, it would be more responsive to random fluctuations in its milieu; consequently, the variance of the F_1 would increase over the parents.

The mode of inheritance of emotionality in rats as measured by defecation has been studied intensively. Broadhurst (1958b) found that five homozygous strains fell into two groups, high defecators (black-hooded II and albino) and low defecators (brown, brown-hooded, and black-hooded I). In this study the usual negative association between ambulation and defecation broke down. Females were more active, but the size of the sex difference varied significantly among strains. These five strains plus the selected MNR strain, added to increase the phenotypic range, were crossed in a 6 × 6 complete diallel design (Broadhurst, 1960b). A complete Hayman analysis of the ambulation data was conducted; high heritability (0.89) and moderate dominance (average over all loci = 0.23) were found. Although a significant sex × strain interaction was found, analysis of the variance-covariance diagrams for the separate sexes indicated that the genetic regulation of ambulation was essentially the same in males and females. A high proportion of recessive genes was found in the A and MNR strains, not a surprising result, since these are both albino, and all F_1 hybrids with other strains would be pigmented. Albinism, it will be remembered, is associated with low activity in brightly illuminated open fields.

The diallel analysis of defecation scores was less complete because of scaling problems. Heritability was moderately high (0.62), indicating significant additive gene action; differences between strains in their order of dominance were not well defined. Later in the chapter we shall discuss a detailed analysis of the trends in scores over the 4 days of testing. This analysis suggested that different genetic systems are involved at successive phases of the open-field test (Broadhurst and Jinks, 1966).

EXPLORATORY BEHAVIOR

Following the tradition established by Hall and continued by Broadhurst, we have referred to the open-field test as a measure of emotional arousal. This interpretation may be appropriate for the defecation score, but the reduced open-field activity usually found in "emotional" rats and mice suggests inhibition rather than arousal.

Looking at the matter in this way, one can postulate that a novel stimulus simultaneously evokes both an approach (exploratory) and an avoidance (fear) response (Schneirla, 1965). One could, therefore, view observations of behavior in the open field as indicators of genetic variability in the ease of elicitation of either avoidance or approach. Following this reasoning, one would look for heritable variation in the tendency to explore in apparatus less startlingly different from a home cage than is a bright, large open field.

Carr and Williams (1957) observed three inbred strains of rats in a Y-maze. The hooded rats explored more alleys than either Fischer albinos or a black nonagouti strain ($p < 0.01$). The three strains were then crossed in all combinations. Hooded crosses were similar in activity to their hooded parents; the black-albino hybrids had low scores with no indication of heterosis (Williams, Zerof, and Carr, 1962).

In a more elaborate experiment, McClearn (1959) compared six inbred strains of mice on four tests: (1) locomotion in an arena (discontinuous open field), (2) a hole-in-wall test scored by the latency of climbing into an adjacent chamber, (3) an open-field test in which the measure was the amount of time spent in the central area away from the walls, and (4) a barrier test in which subjects had to climb over a low wall in order to move through four connected chambers. Strain differences in all tests were highly significant ($p < 0.001$), and the high concordance of strain rankings on the four tasks indicated that a single temperamental dimension was evaluated by all. McClearn chose to call the dimension "exploratory behavior," but noted that it might have been designated "spontaneous activity" or "the inverse of timidity."

The same tests were then given to reciprocal crosses of the two strains that had ranked high (C57BL/10) and low (A) in the strain comparisons. Both F_1s were like the active parent in the hole-in-wall and barrier tests; they were intermediate to their parents in the arena and open-field tests. In both of the latter tests, behavior was sampled over a relatively long period of time, allowing for dissipation of an initial fear-elicited inhibition of activity. McClearn proposed that this difference in mode of inheritance indicates that exploratory behavior consists of at least two subcharacters whose variability is affected by different genes.

ETHOLOGICAL APPROACHES

Predominantly, experimental psychologists have measured animal temperament by selecting one or two indices such as defecation and ambulation, placing subjects in an apparatus designed to induce these responses, and recording the frequency or latency of their occurrence. The choice of indicator responses is not haphazard; there is always a rationale based on pilot studies or the previous literature. Still, the experimenter makes an arbitrary selection from a subject's total responses and disregards the remainder. The ethological approach is less structured. Subjects are observed for a relatively prolonged period in an environment that provides opportunities for a variety of behavior. The repertoire of activities (the *ethogram*) is cataloged, and each item is named and given an objective description. The frequency and duration of each named act is recorded in an appropriate manner.

Such an ethogram is ostensibly an objective description of what an animal does, without theorizing on possible relationships to constructs such as emotionality, timidity, or exploratory "drive." The atheoretical quality is not absolute, since there are always implied criteria for the separation of the flow of behavior into distinct components. Furthermore, nothing hinders an experimenter from interpreting an ethogram in terms of traits and motivational systems.

The prime example of this approach in behavior-genetic analysis is an illustrated catalog of nearly fifty behavior patterns shown by male mice in an observation chamber furnished with a water bottle, wire rack, and food. Observations were made on solitary subjects, male pairs, and male-female pairs (van Abeelen, 1963a,b,c). Four mutants were compared with nonmutants of the same strain. Differences between mutants and wild type were found for yellow (A^y), brown (bb), and pink-eyed dilute mice (pp). The relationship of the differences to the known physiological effects of these genes was unclear. Not surprisingly, the neurological mutant, jerker $(jeje)$, showed aberrant exploratory and comfort-seeking behavior, but feeding, fighting, and sexual behavior were little, if at all, affected.

In another study involving the behavioral effects of single-locus allelic substitutions in mice, Thiessen, Owen, and Whitsett (1970) used a battery of tests that were intended to measure activity, sensory sensitivity, and

sensory preference. Their measures were mainly in the tradition of experimental psychology but were more varied than is usual for a single experiment. A number of differences between mutant and wild type were found, but aside from those related to the photophobia known to be associated with albinism and a reduced activity level in the mutants that were physically less vigorous, none of the behavioral differences was readily related to the physical expression of the mutant gene.

We interpret both studies as indicating great stability of the genetic mechanisms programing the basic behavioral repertoire of the species. This stability stems from the interactions between genes at many loci and is not greatly disturbed by perturbations at a single locus. Inherited neurological and sensory defects can, of course, impair performance, but the fine tuning of temperament is accomplished by genes with less drastic effects on development. At this point it appears that the study of major gene effects has not led to better understanding of the genetic basis of variability in temperament so commonly found in selection experiments and strain comparisons.

The ethological approach has also been applied to strain and hybrid comparisons (van Abeelen, 1966). The ethograms of two inbred mouse strains (C57BL/6J, DBA/2J) and their F_1 were significantly different in the frequency of many acts such as exploratory sniffing, hair fluffing, food carrying, tail rattling, and lifting one forepaw. These are the kinds of behavior that in human beings we would call acquired mannerisms; their significance in mice is not readily interpretable. In the paired situation the greater frequency of preaggressive behavior in DBA mice may have relevance to fitness.

A more limited set of behavior patterns was the basis for a behavior-genetic analysis of differences in the sequence in which these acts were performed (Guttman, Lieblich, and Naftali, 1969a,b). C57BL/6, DBA/2, and their F_1 hybrids were observed in an arena where a running record was kept of defecation, face washing, fur licking, peeking over barrier, and jumping over barrier. The first twenty-eight acts were analyzed by a computer program for evidence of strain differences in the pattern of appearance of various behaviors. In all three genotypes peeking was found early in the observation period; defecating and washing also started high and tended to fall over time; barrier jumping peaked in midperiod. The results were interpreted in terms of heritable effects on the sequence in which responses occur. The observed sequencing is believed to be adaptive for an animal faced with a novel situation.

BEHAVIORAL AND PHYSIOLOGICAL CORRELATES OF EMOTIONALITY

We have followed common usage in referring to mouse and rat strains with high defecation scores in an open field as "emotional." However, there are important questions to be answered before concluding that the designation is proper. The differences in defecation and ambulation scores among strains are certainly reliable, but are they valid measures of an underlying general trait? Can one predict from an individual's score in the open field its behavior in other circumstances involving stress? Is the defecation index consistent with other autonomic and biochemical changes that occur under strong stimulation? Questions like these occur frequently with respect to the predictive validity of tests of human personality. Within our own species it is impossible to perform fully controlled experiments, and there are ethical limitations on the stressfulness of procedures. With animals we can manipulate genotypes and life histories and possibly find answers to these questions.

Conflicting points of view are well illustrated in four articles concerned with the large body of research on the MR and MNR rat strains (Archer, 1973, 1975; Broadhurst, 1975, 1976). A variety of dependent measures have been studied in these strains; some such as ambulation have been measured routinely along with the criterion for selection: the number of fecal boluses in the open field. Broadhurst (1975) summarized

the data from 104 published studies and personal communications in which the strains were compared. His analysis yielded 131 strain comparisons based on open-field data, 39 based on physiological responses, 23 from endocrine measures, and 83 from pharmacological experiments. On the basis of these data Broadhurst concluded, with mild reservations, that the Maudsley strains represent the genetic extremes of a generalized trait of emotional reactivity that expresses itself in many ways.

This evaluation has been criticized severely by Archer (1975) on the basis that the concept of emotionality is vague (at least it has not been defined operationally by Broadhurst) and that agreement or disagreement of data with the hypothesis is dependent on subjective judgment. We refer the reader to Broadhurst (1976) for a rejoinder. Other points brought up in Archer's critique include the fact that many of the associations reported are based on group means rather than individual scores. Only associations with individuals can provide evidence for common mechanisms. And only when individual genotypes are a random sample of a specified population can the correlations be attributable to gene communality. Archer asks for a sharpening of the definition of emotionality so that the hypothesis of a general emotionality factor is potentially falsifiable and so that it can be compared with other possible explanations. From the tone of his comments it appears that Archer would like to see behavioral variation between strains correlated with physiological or structural differences without recourse to psychological constructs. Following is a sample of empirical research on behavioral and physiological characteristics associated with emotionality. A negative association between open-field defecation and ambulation in rats has been confirmed generally, although its magnitude varies greatly. The relationship seems to be strongest in Wistar and Wistar-derived strains (Archer, 1973). The same tendency for a negative association between high ambulation and high defecation has often been reported for mice, though Bruell (1967) found a small but significant positive association in males ($r = 0.124$, $p < 0.01$) in contrast with females ($r = -0.159$, $p < 0.01$). He suggested that fecal boluses might be a territorial marker and that the sex difference reflected the roles of males and females in maintaining territories.

Change in heart rate is another autonomic function that can be used as an indicator of emotionality. The evidence that defecation and heart rate changes are parts of a unified pattern is weak. Candland, Pack, and Matthews (1967) found that both increased in rats placed in a novel environment, but the amounts of increase were not correlated either for individuals or for treatment groups. Elevated heart rates have been reported for *less* emotional rats (Snowdon, Bell, and Henderson, 1964). In support of the general emotionality hypothesis, Blizard (1971) found the heart rate of the MR rats to be significantly higher than that of the MNR strain. In contrast Harrington and Hanlon (1966) reported no correlation between defecation and heart rate–change scores in the same lines. Since Blizard had larger samples, it is likely that the correlation is real but that each of these two autonomic measures has its own set of genetic determinants.

In mice both basal heart rate and its change over a 3-day testing period varied significantly among inbred strains (Blizard and Welty, 1971). The coefficients of genetic determination were 0.42 for basal rate and 0.21 for adaptation rate. In three strains heart rate measured after shock correlated highly with preshock readings. In six strains the two measures were uncorrelated. Likewise the heart rate responses to handling and to shock were correlated in some strains but not in others. These results suggest caution in considering lability of heart rate as a consistent property of a strain, independent of the stimulating situation. Nevertheless, genotype has a strong influence on the direction and degree of autonomic reactions to disturbance.

Emotionality and response to stress

Extreme stress may lead to actual tissue damage because of severe and prolonged autonomic reactions. Thus there is interest in whether inherited differences in level of emotionality influence vulnerability to stress. The number and size of stomach ulcers after immobilization was essentially the same in MR and MNR rats (Mikhail and Broadhurst, 1965). In contrast with this negative finding, a genetic influence on susceptibility to immobilization-produced ulcers is suggested by the results of a selective breeding experiment (Sines, 1959). Unfortunately this experiment lacked controls for environmental and maternal influences.

Endocrine glands and temperament

The pervasive effects of hormones on physiology and behavior are well known. Could heritable differences in emotionality, exploratory activity, and the like be mediated through the endocrine glands? The thyroid gland is a likely candidate for such a role. Hyposecretion is associated with a lower metabolism and in early life with a profound retardation of development known as *cretinism.* Hypersecretion from the gland or excessive dosage of the hormone increases metabolic rate and may cause disorganization of behavior. The MR rats have less thyroid hormone than do MNR rats, both in the gland and in the blood (Feuer and Broadhurst, 1962a). Uptake of radioactive iodine and the rate of secretion of thyroxin were also slower in the high-defecating strain (Feuer and Broadhurst, 1962b). The MR thyroids presented a histological picture of moderate chronic hypothyroidism. Treatment with an antithyroid agent and with thyrotropic hormone produced complex effects on open-field behavior (Feuer and Broadhurst, 1962c). Selection for low- and high-defecation scores did effect a change in thyroid function, but the change is only one of a number of changes, any of which might plausibly be considered to be involved in producing the total behavioral pattern of the two strains.

The relationship of thyroid activity to emotionality has been approached from the other direction by using mice selected for high- or low-thyroid activity (Blizard and Chai, 1972). Consistent with the data from the

Maudsley rats, mice of the hypothyroid strain defecated more in the open field. However, ambulation scores were also higher in this strain, a fact that was unexpected and unexplained. These experiments demonstrate that genes can mediate behavior through modulation of thyroid activity. It is unlikely, however, that this route is the major way in which emotional behavior is modified by selection.

Other candidates for a mediatory role are the adrenal glands. MR rats have heavier adrenals with a higher content of corticosteroids (Feuer, 1969). The direction of the difference is compatible with the general emotionality hypothesis. On the other hand, no behavioral correlations with adrenal weight, plasma corticosterone, or glandular corticosterone were found in emotional hooded rats and more placid albinos (Ader, Friedman, and Grota, 1967). Large differences in adrenal structure have been reported among inbred strains of mice, but their behavioral significance has not been ascertained (Shire and Spickett, 1968).

The most obvious effect of genetic factors on the endocrine system is the male-female dimorphism. In the Maudsley selection experiment, males of the base population defecated significantly more than females. This difference disappeared in the selected lines, more rapidly in the MR than in the MNR strain (Broadhurst, 1960). Again, one can invoke a threshold hypothesis. The autosomal genotype of the foundation stock produced on the average animals with middle-range scores. On such a genotypic background, male and female hormonal influences were apparent. When the autosomal genotypes were altered by selection, the influence of sex hormones became undetectable because the scale of measurement was insensitive at the phenotypic extremes. For ambulation, a trait not subjected to direct selection, the sex differences found in the base stock (females more active) persisted in the two lines. Gonadal size and hormonal output have been compared in the MR and MNR rats, but differences are either absent or appear unrelated to behavior (Broadhurst, 1975).

In summary, despite the plausibility of endocrine mediation of heritable differences in temperamental traits, evidence that hormones are the major, or even an important, component of the gene-emotionality pathway is not available. However, the situation is more a matter of lack of relevant information than of strong negative evidence. It is our opinion that heritable differences in temperament are more likely to be mediated through variations in neural sensitivity to hormones than through variations in endocrine gland activity. Experiments bearing directly on this issue are discussed in Chapter 11.

PHARMACOGENETICS OF TEMPERAMENTAL TRAITS

We have previously referred to pharmacogenetic investigations of audiogenic seizure susceptibility, learning differences, and reactions to alcohol. It is not surprising, then, that heritable differences have been sought in the effects of drugs on activity, emotionality, and other temperamental attributes. The literature in this field is expanding rapidly and will not be reviewed here extensively. Instead we shall concentrate on a few examples of the combining of genetic and pharmacological techniques that illustrate the potential of the method.

A number of pharmacogenetic experiments with the Maudsley MR and MNR rat strains have been evaluated by Broadhurst (1975) as to whether they are neutral, consistent, or inconsistent with the general emotionality hypothesis. There are problems in making such judgments. If MR rats are below the top of a scale of emotionality, they might be activated by an excitatory drug more easily than MNRs. On the other hand, if MR rats are already at a physiological limit and MNR rats are just below an activation threshold, the MNRs should respond more. Despite such problems, Broadhurst (1975, 1976) concluded that the majority of pharmacological findings favor the hypothesis of a generalized trait of emotionality expressed more strongly in the MR strain.

In mice, strain differences in response

thresholds and slope of dose-response curves are commonplace (Eleftheriou, 1975). By itself the demonstration of a strain difference has minor significance for behavior genetics. However, under some conditions deductions concerning the nature of heritable differences in behavior can be based on pharmacological data. For example, amphetamine reversed the ranking of strains SJL and SWR on a "freezing" factor without altering their relative status on a "disorganization" factor (Satinder, Royce, and Yeudall, 1970). Chlorpromazine had no effect on either factor. The divergence of the factors with respect to modification by amphetamine may indicate that adrenergic systems are of greater significance for freezing than for disorganization.

Striking differences in the behavioral effects of scopolamine, an anticholinergic drug, have been noted in the C57BL/6 and the DBA/2 mouse strains (van Abeelen, 1974). As measured by the amount of rearing in an open field, C57BL/6 mice are consistently more exploratory. Scopolamine injected either intravenously or intraperitoneally depressed rearing in C57BL mice and increased it in DBA, thus bringing the two strains closer together. Physostigmine, an anticholinesterase that enhances cholinergic activity, depressed rearing in both strains (van Abeelen and Strijbosch, 1969). The effect seems to depend on access of the drugs to the central nervous system, since drugs with a similar action that do not pass the blood-brain barrier are ineffective when injected peripherally but are active when injected directly into the hippocampus (van Abeelen, Smits, and Raaijmakers, 1971; van Abeelen et al., 1972). Van Abeelen has proposed that exploratory activity is mediated through a cholinergic system whose function depends on a specific acetylcholine/acetylcholinesterase balance. In C57BL mice the balance is optimal for exploratory activity, and changing either component alters the ratio and disrupts the behavior. In DBA mice acetylcholine is present in excess; scopolamine at the proper dose reduces the surplus and brings the system closer to the optimal

balance. Physostigmine disrupts transmitter balance of C57BL mice (in the opposite direction from scopolamine); it exacerbates the existing imbalance in DBA. Similar pharmacological effects with the two drugs were obtained from two lines of mice selected for differences in rearing behavior (van Abeelen, 1970, 1974). This ingenious theory must be validated by additional experiments, and it may prove to be oversimplified. Nevertheless, this series of studies illustrates how genetic variation can be used as a part of research design in psychopharmacology. Explaining strain differences in behavior in terms of biochemical parameters is one of the goals of behavior genetics. The approach through pharmacology may be the best route to this objective.

The final experiment we shall discuss in this section makes use of the recombinant inbred (RI) strain technique described in Chapter 6. When mice of two parental strains, C57BL/6By and BALB/cBy, their F_1 hybrids, and seven RI strains derived from a previous cross were injected with scopolamine in various doses and placed in a tilting floor apparatus to measure activity, the genetic groups formed two clusters (Oliverio, Eleftheriou, and Bailey, 1973a). One cluster of five groups had low activity following saline injections; four of these showed increased activity after scopolamine. All six groups with high activity after saline showed a decrease after scopolamine. The results are similar to those of the van Abeelen group, which used different strains. Oliverio, Eleftheriou, and Bailey postulate that in their material a single locus on chromosome 4, with two alleles, determines the response to scopolamine. On less firm grounds they propose that a second locus controls the shape of the dose-response curves, which are notoriously variable under the best of conditions. Experiments of this kind have a potential for determination of the physiological basis of heritable differences in response to drugs. A clearly segregating gene with large effects gives an investigator a tool for manipulating phenotypes that is sharper than the polygenic differences which distinguish selected and inbred strains. However, success in correlating a specific biochemical difference with an allelic difference defined by a rather indirect behavioral effect will require a major effort and perhaps a bit of luck.

GENE-ENVIRONMENT INTERACTION

Animal temperament is always the outcome of developmental processes guided by both genes and environment. In this section the topics of gene-environment coaction and interaction are considered in more detail. Coaction is omnipresent and refers to functional and developmental processes that could not occur without both genetic instructions and environmental resources for their execution. Interaction occurs when different genotypes react unequally to identical treatments. Its presence or absence can be detected by an analysis of variance. Interactions can be classified according to whether they affect the *stability* of relatively permanent developmental processes or are manifested in short-term *changes* in behavior that might be observed over successive days in an open field or in learning to avoid shock (Broadhurst and Jinks, 1966). The distinction between stability and change is convenient, but it is not absolute because there is only an arbitrary separation between short- and long-term effects.

Maternal effects

In any species in which parents care for their young, the possibility of nongenetic parental influences is ever present. Such effects are most likely in mammals and in birds. In mammals a standard procedure for detecting maternal influences (both prenatal and postnatal) is the comparison of reciprocal F_1 hybrids. Except for an inevitable difference in the source of the X and Y chromosomes in male F_1 offspring, the reciprocals are genetically identical, and behavioral differences between them can be attributed to one or both of the two classes of maternal influence. Postnatal influences, including contacts with both parents and with other young, are detectable through cross-

fostering of young. Both the reciprocal-crossing and the cross-fostering procedures were applied to the Maudsley selected rat lines after they were well separated. No evidence was found for either prenatal or postnatal maternal influences on the differentiation of the two strains (Broadhurst, 1960a).

Experiments with mice have yielded positive evidence for postnatal maternal effects on defecation scores. When BALB/c and C57BL/6 pups were fostered, either reciprocally or within their own strain, alien-fostered individuals were altered in the direction of their foster mothers. Pups of both strains grew faster when nursing from BALB/c dams (Reading, 1966). Similarly, cross-fostered mice pressed a bar that turned on a cage light at rates more similar to those of their foster than of their biological mothers (Ressler, 1963). Growth and survival were also a function of the foster mother's strain, and the behavioral difference may have been a nonspecific manifestation of differences in vigor. More varied measurements were obtained by Poley and Royce (1970) from same strain– and alien strain–fostered mice. Their battery of tests included avoidance conditioning, running wheel, open field, straightaway alley, and descent from a pole. The study yielded eighteen measures that in turn were reduced to seven factors (Royce, Carran, and Howarth, 1970). Although a number of differences attributable to the foster mother's strain were found, they fit no simple pattern and are not readily interpretable.

It is perhaps surprising that in a few studies of the behavioral and endocrine responses of pups that have been indirectly exposed to stress *in utero*, F_1 hybrids are generally intermediate to the parental strains, but their mean is shifted towards that of the male parent (Joffe, 1965; DeFries, 1969; Treiman, Fulker, and Levine, 1970). Since the fathers in these experiments had no contact with their offspring, a paternal influence seems to be excluded. However, all three of these experiments can be explained by a maternal effect on the offspring that operates in a direction opposite to the mother's genetic contribution (Fulker, 1970). Such a system would act as a buffering device, limiting the extreme phenotypic expression of extreme genotypes; it would be favored by natural selection when intermediate expression increases fitness.

Early experience and genotype

Although maternal care is very important, other aspects of early experience play a significant role in the development of temperamental traits. The effects of handling, litter size, and stressors such as electric shock on later behavior have been studied widely. Another favorite technique is the comparison of behavioral development in enriched and impoverished environments. Naturally, there is interest in the possibility of heritable differences in sensitivity to such manipulations. Examples of strain-treatment interactions have been found in rats (Gauron, 1964) and in mice (Lindzey, Lykken, and Winston, 1960; Lindzey, Winston, and Manosevitz, 1963; Mos, Royce, and Poley, 1973). In general the results of these studies suggest differences in threshold of response in the slope of a dose-response curve rather than diametrically opposite responses to the same treatment.

Nevertheless, genotype-treatment interactions can create problems for scientists trying to arrive at general laws of behavioral development. This is nicely demonstrated in an experiment on the effects of prior experience on open-field behavior in mice (Henderson, 1967). Subjects from four inbred strains and their reciprocal hybrids, sixteen genotypic classes in all, were observed in an open field after being either undisturbed, exposed to an exploratory maze where they were repeatedly stimulated by a loud buzzer, or exposed to the maze and shocked after each presentation of the buzzer. The outcome of the experiment is shown in Table 10-6. The point to be noted is the genotypic specificity of the effects of prior experience on open-field behavior. If we consider each of the sixteen genotypes as a separate experiment on the

Table 10-6. Effects of prior experience on open-field behavior of inbred and hybrid mice*

Strain	General behavior		Relative reactivity of *undisturbed* vs. exposure to *buzzer* or *shock*		Proportion of dominant genes	
	Defecation	Ambulation	Defecation	Ambulation	Defecation†	Ambulation
C57BL/10						
Inbred	Low	High	ns‡	B > S, U	High	High
Hybrids	Low	High	S > U, B	B > S, U		
DBA/1						
Inbred	Medium	High	S > B > U	B > S, U	High	Low
Hybrids	Medium	Medium	S > B > U	B > S, U		
C3H						
Inbred	High	Low	ns	ns	Low	Low
Hybrids	Medium	Low	ns	ns		
BALB						
Inbred	High	Medium	B > S, U	ns	Medium	High
Hybrids	High	Medium	S, B > U	ns		

*From Henderson, N. D. 1967. Prior treatment effects on open field behaviour of mice: a genetic analysis. Anim. Behav. **15**:364-376.
†Results in this column refer to undisturbed and buzzer-exposed groups only.
‡ns signifies that the three treatment groups did not differ significantly.

experiential modification of emotionality by graded amounts of prior stimulation, we find that these experiments contradict one another. Using defecation as our index, we see that in six experiments stimulation had no effect; in two, emotionality increased monotonically with higher stimulation; in three, undisturbed subjects were less emotional than the stimulated groups that do not differ; in two, there was a J-shaped relationship with the shocked pups most emotional; in two, the relationship was U-shaped with undisturbed subjects most emotional; and in one, the relationship was an inverted U (∩). Not one of the genetically homogeneous groups conformed to the average of all hybrids treated as a single population.

The implications of the Henderson experiment are somewhat discouraging with respect to the empirical verification of general principles or laws. Are there such principles, or must we be satisfied with limited rules applicable only to members of a species sharing common genes? The latter prospect has been dubbed "overparticularization" (Rodgers, 1967). The issue is an old one

in psychology that will not disappear quickly. Behavior-genetic analysis is of limited value if it leads only to a vast number of genotype-specific rules. There are other solutions that involve the incorporation of species and genotypic characteristics into the formulations of behavioral principles, but only a beginning has been made in this direction. A necessary point of departure is the realization that some of the discrepancies in the literature of experimental animal behavior have their source in genotypic differences among the subjects.

These matters are explicated in some detail by Henderson (1968). Fig. 10-7 illustrates how an experiment on the effects of early experience (enriched or impoverished) using two strains (A and B) could yield a strong interaction term and lead to a conclusion that the effect of level of stimulation on the behavior of interest was opposite in direction for the two groups. Hypothetical results from the 2 × 2 design are shown by dashed lines. A parametric study employing several levels of treatment would indicate that in both strains the stimulus-response function was U-shaped, as shown by the solid lines. General principles emerge only from experimental protocols designed to sample adequately all relevant independent variables.

Another issue that recurs in behavior-genetic experiments is the validity of the concept of *expressive* and *suppressive* environments during development (Fuller and Thompson, 1960). An example of this distinction is provided by another diallel experiment in which mice were reared either in conventional cages, similar to those commonly found in laboratories, or in large cages furnished with manipulanda, opportunities to climb, and other stimuli (Henderson, 1970). Food-deprived subjects from both environments were placed in a large enclosure where food could be obtained only by an indirect path requiring climbing and balancing on a narrow bridge. In this naturalistic test genetic influences were four times as great in the enrichment groups as in the standard-reared groups (Table 10-7). These results

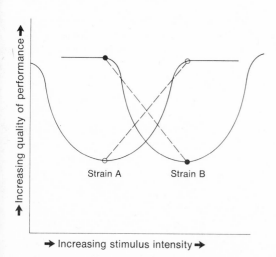

Fig. 10-7. Apparent reverse effect of change in stimulus intensity in two strains (G-E interaction) when only two levels of treatment are observed. Actually strain *A* and strain *B* respond similarly, but in different ranges of stimulus intensity. (Modified from Henderson, N. D. 1968. Dev. Psychobiol. **1:**146-152.)

Table 10-7. Effect of rearing environment on a genetic analysis of a food locating task*†

Source of variation	Percentage of total variance			
	Standard-cage reared		Enriched-cage reared	
Total genotypic		10		40
Additive	4		37	
Specific	6		3	
Total environmental		90		60
Maternal influence	6		4	
Other environmental	84		56	
GENETIC + ENVIRONMENTAL		100		100

*From Henderson, N. D. 1970. Genetic influences on the behavior of mice can be obscured by laboratory rearing. J. Comp. Physiol. Psychol. **72:**505-511. Copyright 1970 by the American Psychological Association. Reprinted by permission.
†The analysis was for general combining ability (additive) applicable to all crosses; specific combining ability refers to genetic influences apparent only in particular crosses.

have relevance not only for genetic experiments but for any investigation of behavioral development. Note that the "other environmental" variance component in Table 10-7 is much higher in the standard-reared group. This number represents noise in the system, variance not attributable to any known source. The higher its value, the less sensitive the testing procedure for the detection of real differences.

We must indicate, however, that genetic effects are not always more apparent in enriched-reared groups compared with standard-reared groups. In fact, in a strain comparison study with mice, Manosevitz and Montemayor (1972) found several exceptions on measures in an open field, an exploration apparatus, and running wheels. Enrichment increased exploration and reduced defecation similarly in all strains. Effects on open-field ambulation and wheel running varied from strain to strain, indicating that a genotype-environment interaction was present.

In MR and MNR rats, daily handling had similar effects on open-field defecation and ambulation in both strains; it had detrimental effects on conditioned-avoidance learning in MR rats and no effect on avoidance in MNR (Levine and Broadhurst, 1963). In another study of these strains, genotype-treatment interactions were not found; handling similar to that used by Levine and Broadhurst had no effect on open-field behavior in either strain, and it improved conditioned-avoidance learning in both (Powell and North-Jones, 1974). Discrepancies of this kind are probably attributable to subtle differences in procedures. Their existence is an admonition to suspend judgment until more data are available.

Genotype and hoarding

Hoarding might have been considered in the previous chapter on ingestive behavior, since it usually involves transport of food and is most easily elicited from food-deprived subjects. For our purposes, however, it is more convenient to treat hoarding as an aspect of rodent temperament, possibly related in some way to fear (Manosevitz, 1965). Genetic effects on hoarding have been demonstrated in rats (Stamm, 1954, 1956) and in mice (Manosevitz and Lindzey, 1967; Manosevitz, 1967). Our interest here lies in possible genotype-treatment interactions involving this species-typical behavior pattern. Stress, produced by immersion in water just prior to testing, doubled the hoarding of JK male mice, and decreased hoarding of C57BL males. Immersion had no effect on hoarding of females. The F_1 hybrid males were also unaffected by prior stress, while hybrid females decreased hoarding (Manosevitz, 1965). When high-hoarding JK mice and low-hoarding C3H mice were reared in enriched environments, the JKs increased their transport of pellets, but no change was seen in C3H mice (Manosevitz, Campenot, and Swecionis, 1968).

Summarizing statement

All the studies in this section support the idea that genotypes react differently to treatments as diverse as handling, rearing in an enriched environment, electric shock, and water immersion. However, variability in procedure, experimental design, and outcome is so great that we have not yet determined the rules of the game that produce or do not produce statistical interactions. Henderson's (1967) experiment involving sixteen genotypes and three levels of stimulation illustrates the need for parametric studies which employ a wide range of genotypes, not simply two strains that happen to differ on a point of interest. Even this experiment raises more questions than it answers. The only safe conclusion is that it is dangerous to generalize to a species from results obtained from one or a few genotypes.

HETEROSIS AND DEVELOPMENTAL HOMEOSTASIS

The concepts of heterosis, or hybrid vigor, and developmental homeostasis (sometimes mistakenly referred to as genetic homeostasis) both involve comparisons between homozygotes and heterozygotes. Both have relevance to all aspects of behavior. Both are particularly well illustrated by research on activity and emotionality. Heterosis is manifested whenever a hybrid outperforms both of its parental strains. Two hypotheses have been advanced for this phenomenon. The compensatory dominance theory starts with the observation that behavioral characteristics which seem to be related to fitness are generally transmitted in dominant fashion. During inbreeding a strain becomes homozygous for a random set of the relevant dominant alleles; thus strain 1 may have the genotype $AAbb$, and strain 2, $aaBB$. The vigor

of each is impaired by the homozygous recessive gene pairs, bb and aa respectively. Their hybrid with the genotype $AaBb$ has a dominant allele at both loci and is, therefore, more vigorous than its parents. Finding consistent dominance in one direction in crosses between inbred lines is interpreted as indicative of a history of selection favoring the dominant phenotype. We can also say that a heterotic psychophene is one that contributes positively to fitness.

The alternative hypothesis is that heterozygosity at a single locus conveys an advantage in fitness as compared with homozygotes. We could also express this idea in terms of the overdominance of fitness. The classic example is the gene for sickle cell hemoglobin in humans (Chapter 4).

Heterosis for behavioral traits, whatever its ultimate genetic explanation, has been found repeatedly. Bruell (1946b, 1967) tested thirteen inbred strains of mice and thirty-one randomly selected F_1 hybrids in running wheels. Although the activity of males was generally higher than that of females, correlations between the sexes were so high that pooled scores were used for the estimation of heterosis. Eighteen of the thirty-one hybrids were crosses between distantly related strains, and in seventeen of these the F_1 mice were more active than the higher parent strain (Fig. 10-8). Among thirteen crosses of closely related strains, heterosis was found in only four. Very similar results were found with activity in an exploration apparatus (Bruell, 1964a). Here the average advantage of F_1 mice from crosses of distantly related strains was 11% over the higher-scoring parent strain; for crosses between closely related strains, only 1%. The strength of heterosis appears to be a function of the number of loci for which the parent strains are discordant.

An experiment by Rose and Parsons (1970) included a smaller range of genotypes (three inbred mouse strains and their hybrids) but covered a range of behavior more closely related to the general emotionality hypothesis: defecation, urination,

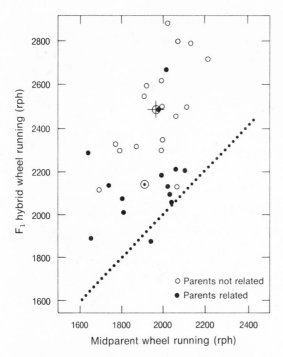

Fig. 10-8. Heterotic inheritance of wheel running in mice. Note that most hybrids scored above the diagonal midparent line and that unrelated hybrids deviated more from the midparent than related hybrids. Means for related and unrelated hybrids are indicated by the *bull's-eye* and *cross*, respectively. (From Bruell, J. H. 1964. Am. Zool. 4:125-138.)

and ambulation in an open field and shock-avoidance learning. There was inconsistent evidence for heterosis in the open field, but the small number of strains may not be representative. However, heterosis would be expected to be less marked for a trait subjected to selection for an intermediate level of expression. Heterosis was found for exploration of the shock apparatus and for the number of conditioned avoidances.

Behavioral heterosis is a common finding. Where it is found in crosses between inbred strains, the implication is that the trait has been subject to natural selection and that the loss of dominant alleles (or alternatively the reduction in heterozygosity) has depressed fitness even in the higher-scoring strains. Heterosis would not be expected for traits not subjected to natural

selection (Bruell, 1967). This line of reasoning is straightforward and has been widely applied in behavior genetics. Caution is desirable, however, in reconstructing the evolutionary history of a species from experiments performed on laboratory strains whose ancestors have lived under specialized conditions for many generations where they have been subjected to vastly different selection pressures.

Developmental homeostasis

The concept of developmental homeostasis is, like heterosis, associated with the effects of heterozygosity and can be explained by either complementary dominance at multiple loci or overdominance at individual loci. Homeostasis, originally defined as the maintenance of constancy in the physiological parameters of an organism, has by extension been applied to both physical and behavioral development and to the genetic composition of populations. Its application to the development of physical characteristics such as body size is usually straightforward; a genetically homogeneous group with small variance is more homeostatic than another homogeneous group with high variance. The same criterion has been applied to behavioral traits, but its appropriateness has been challenged. Higher intragroup variation could reflect greater sensitivity to the environment and result in enhanced stability of the vital functions within an organism (Caspari, 1958; Fuller and Thompson, 1960). Evidence for and against these contrasting views of true homeostasis has been summarized by Fulker, Wilcock, and Broadhurst (1972) and Hyde (1973). There is no need to brand one criterion as true and the other as false. It may be adaptive for development to proceed along a narrow channel regardless of microenvironmental differences such as those encountered by littermates reared together. On the other hand, it could be adaptive to make variable responses to macroenvironmental differences such as exposure to stress or to a strange environment. This dual view is supported by an experiment in which activity and grooming were observed in two strains of mice and their F_1 hybrids. Separate groups were tested at three ambient temperatures (Mordkoff, Schlesinger, and Lavine, 1965). At each temperature the hybrid mice varied less than the inbreds, but their response to differences in temperature was greater.

Comparison of the relative variability of groups with different means creates problems. A common solution is the use of the coefficient of variation (CV), defined as $s/\bar{x}$, where s is the standard deviation, and $\bar{x}$ is the mean. The magnitude of CV is sensitive to changes in scale; transformation of defecation scores from bolus counts, x, to $(x + 0.5)^{\frac{1}{2}}$ or running wheel scores from revolutions (x) to $\log_{10} x$ will alter values of the coefficient drastically. Such transformations are common when it is desired to make variances independent of the mean in preparation for an analysis of variance. If successful, such manipulations of the raw data make it impossible to detect differences in developmental homeostasis between groups. Other problems with the use of CV and possible solutions for them have been discussed by Jones (1966) and Hyde (1973).

Major studies of stability and change in open-field behavior have been performed with diallels of inbred rat strains (Broadhurst and Jinks, 1966; Fulker, Wilcock, and Broadhurst, 1972) and with the MR and MNR selected lines (Broadhurst, 1969). Strain differences in the stability of defecation scores as measured by intralitter variability were found by Broadhurst and Jinks to depend on both dominant and additive gene action. Strains and crosses did not differ in the stability of ambulation scores. Genetic effects on change in open-field behavior over a 4-day testing period were assessed by a regression technique and by conducting separate Hayman analyses of the data for each day. The results are too detailed and complex to present here, but we will summarize the author's conclusions: (1) genes enhancing stability tend to be dominant; (2) variation in the rate of decrease in emotionality is also mediated through dominant genes, indicating a history of selection for rapid habituation to

new environments; and (3) ambidirectional dominance for level of activity in the open field denotes an evolutionary history of selection for an intermediate phenotype. Similar conclusions were reached from the same type of analysis applied to the Maudsley rat strains (Broadhurst, 1969).

A third experiment from the Broadhurst group employed a more complex design with three main objectives: (1) determination of the adaptive and evolutionary aspects of open-field behavior in rats through analysis of gene action (this objective was the major concern of the two studies just reviewed), (2) investigation in an 8×8 diallel of the interaction between genotypes and treatment (handled versus undisturbed), and (3) the use of multivariate analysis to consider simultaneously genetic effects on several phenotypes (Fulker, Wilcock, and Broadhurst, 1972). Three groups of traits were detected by this method: (1) association of good avoidance performance and frequent intertrial crossing in a shuttle box, with high ambulation and low defecation in the open field; (2) a relationship between body build and defecation scores; and (3) a negative relationship between avoidance and intertrial crossings that ran counter to their association in the first group of traits. Heterosis was shown for several behavior patterns, and a maternal effect was apparent for defecation and possibly for intertrial crossings.

Handling produced its anticipated effects: better avoidance, more ambulation, and less defecation. Disappointing to the authors was their failure to detect interactions of treatment with dominant gene action; this they had hoped would provide information on the evolutionary history and adaptive significance of sensitivity to handling.

DIMENSIONS OF TEMPERAMENT

In the discussion of physiological and behavioral correlates of emotionality, we asked how well any one measure such as defecation could predict behavior under conditions other than the open field. We also referred to the emotionality hypothesis which is based on the idea that many behavioral differences among individuals are explicable by their position on a scale of emotionality. The extremes of this scale might be labeled *confidence* and *fearfulness*. The literature on animal temperament, however, refers to types of behavior such as wheel running, exploratory activity, and hoarding, whose variability is not obviously related to a confidence-fearfulness continuum. The correlations between defecation scores and conditioned-avoidance responses, both of which seem intuitively to be related to emotional arousal, are too small to explain the bulk of the variance. It is not surprising, therefore, that efforts have been made to learn whether other dimensions can be delineated that will complement the emotionality hypothesis.

Factor analysis

One approach to this objective is to expose subjects to a wide variety of tests chosen to sample broadly in the temperament-emotionality domain. Intercorrelations between all the measures obtained from the test battery are calculated, arranged in matrix form, and analyzed to learn whether the pattern can be explained by the interplay of a limited number of substantially independent factors. After extracting factors, a descriptive name is commonly given to each, based on the nature of the items that contributed to it. In one sense the number and nature of the factors that emerge from the process are based on objective data and a mathematical analysis of them following prescribed rules. The reliable and robust behavioral correlations that contribute to factors must result from some common influence that affects all correlated measures. These influences are certainly environmental in many instances, but they could also indicate that the correlated behaviors are modulated by the same gene or group of genes. Thus factors could be the psychophenes of choice for genetic analysis (Royce, 1957). Unfortunately, the view that factor analysis is an objective technique that eliminates bias due to preconceived ideas is a bit idealistic. The choice of tests and measurements imposes constraints on the factors that can be found; the naming of factors is sub-

jective and tentative. All in all, the technique is as much an art as a science, but it has an established position in the psychological and social sciences concerned with complex behavior.

Before considering the combination of genetics and factor analysis, we shall review two nongenetically oriented analyses of emotionality. Anderson (1937, 1938a,b) obtained individual measures of "drive strength" and learning in one study and supplemented this by comparing four emotional measures with performance on tests of sexual and exploratory behavior. His battery yielded seven putative measures of exploration, ten of thirst, fourteen of hunger, seven of sex drive, and six of learning ability. Correlations between different supposed criteria of hunger and thirst were not significant; several reliable correlations were found between different measures of exploration and sex drive. Even in the latter two areas the performance of an individual often varied widely on tests purportedly measuring the same drive. Perhaps something is wrong with the concept that an individual rat has a number of characteristic drive states that lead it to act predictably in all tasks with similar incentives. It may also be that our judgment of the similarity of tasks does not correspond with the animal's mode of attacking a problem; hence it is unreasonable to expect performances on these tasks to be correlated.

Willingham (1956) performed a factor analysis on twenty variables representing various aspects of emotionality in mice. He found six factors, which he named elimination, freezing, grooming, reactivity to light, reactivity to experimenter, and emotional maturity.

How does the introduction of a genetic variable affect factor analysis? It is immediately obvious that an analysis based on results from a single inbred strain will yield intercorrelations based solely on environmental covariances. Even moderate inbreeding will tend to accentuate environmental as compared with genetic intercorrelations. Now consider a table of intercorrelations based on a set of inbred strains and their F_1 hybrids.

Again, phenotypic correlations within each genotype arise from environmental variation; those between genotypes may represent genetic communalities. If these intergroup correlations are based on gene communality (Chapter 5), they are of value in analyzing the functional dependence of different forms of behavior on a unitary biological process. Correlations based on gametic communality represent only the different outcomes of a random process that has led to the fixation of a unique set of alleles in each strain. In the standard diallel there is no way of distinguishing between these two possible sources of intercorrelation. If the number of strains employed is large, the gametic associations may be assumed to be random, thus ameliorating their distorting influence.

Nevertheless, the ideal basis for a behavior-genetic factor-analytic study of temperament is a large random sample of a defined population, something that is probably impossible to obtain with available resources. Stocks of known mixed origin that have been bred deliberately to maximize genetic heterogeneity are a practical alternative (McClearn, Wilson, and Meredith, 1970). Examples of such stocks are the HS mice developed by McClearn at the University of Colorado and the Binghamton HET stock (Fuller, 1975). Only one of the factor-analytic studies of rodent temperament used such a stock. McClearn and Meredith (1964) tested 80 mice from a four-way cross of unrelated mouse strains in an open field, hole-in-wall apparatus, barrier apparatus, and Y-maze. Five factors were extracted and designated as (I) ambulation or exploration; (II) tendency to urinate or in males possibly territorial marking; (III) thigmotropism characterized by behavior in the Y-maze and open field; (IV) defecation combined with low activity, the familiar "emotionality" factor; and (V) a weakly defined factor marked chiefly by climbing over barriers. The authors do not claim universality for these factors but believe that they make sense.

The most extensive application of factor analysis to the genetics of emotionality in mice is that of Royce and colleagues. The initial analysis was carried out on thirty-two measures from ten inbred mouse strains (Royce, Carran, and Howarth, 1970). In addition to old standbys such as open field, straight alley, and hole-in-wall, there were tests of pole-descent time, underwater swimming, hoarding of cotton pellets, and tube dominance. Twelve factors were extracted, and six were named. A slightly modified set of tests was used in a massive 6×6 diallel involving 775 subjects (Royce, Poley, and Yeudall, 1973). Each animal received twelve tests over a 1-month period. Eleven interpreted factors were found. Factor I, body weight, was physical; the remainder were behavioral. Eight of these were replicated across three populations, one composed of two inbred strains, another of their F_1 offspring, and a third of the F_2 generation (Poley and Royce, 1973). The names and general characteristics of these eight were (II) motor discharge characterized by rapid emergence and high activity in an open field and alleyway; (III) acrophobia based solely on the pole descent test; (IV) underwater stress based solely on the immersion test; (V) tunneling-1, related to elimination and starting latency in narrow alleys; (VI) audiogenic reactivity; (VIII) autonomic balance based on defecation in four of twelve tests; (IX) territorial marking, primarily urination; and (XI) tunneling-2, characterized by speed in alleyways. The appearance of these factors in the genetically homogenous F_1 sample indicates the environmental origins of the correlation matrices. In a third article a complete Hayman analysis of the results with the diallel was presented (Royce, Holmes, and Poley, 1975). Table 10-8 is a greatly condensed summary of the results for eight factors, some of them separately by sexes. Several of these factors consisted entirely or predominantly of measures from a single test and would not have appeared had that test been omitted. And, although the statistical manipulations did reduce forty-two latencies, bolus counts, distances traversed, and similar measures to eleven factors, the outcome is still complex.

Royce has stated that the objective of fac-

Table 10-8. Summary of genetic effects for eight factors of emotionality*

Factor name	Sex difference in mean score	Sex	Genetic variance†				Type of dominance	Directional dominance
			?	D	H₁	F		
Motor discharge	0	M and F	+	+	0	0	Partial	0
Acrophobia	M > F	M	+	+	+	+	Overdominance	0
		F	0	+	+	+	Overdominance	0
Underwater swimming	0	M	+	0	0	0	Complete	+ for rapid escape
		F	+	b‡	b	b	b	+ for rapid escape
Audiogenic reactivity	0	M and F	+	+	+	+	Complete	+ for rapid escape
Food motivation	F >M	M	0	b	b	b	b	0
		F	0	+	0	+	Partial	0
Territory marking	M > F	M	0	0	c§	0	Overdominance	0
		F	0	b	b	b	b	0
Activity level	F >M	M	+	+	c	0	Partial	0
		F	+	+	c	0	Complete	0
Tunneling-2	0	M	+	+	+	0	Partial	0
		F	+	b	b	b	b	0

*From Royce, J. R., T. M. Holmes, and W. Poley. 1975. Behavior genetic analysis of mouse emotionality. III. The diallel analysis. Behav. Genet. **5**:351-372.

†From a Hayman analysis: ?, presence or absence of significant genetic variation; D, additive effect; H_1; dominance effect; F, covariance effect.

‡b, not calculable because of violation of diallel cross assumptions.

§c, a significant negative correlation computed.

tor analysis is to discover dimensions and to generate testable hypotheses about their relationships. The multiplicity of factors that emerge from his large-scale experiment suggests that only part of the behavioral variation is attributable to differences on a single scale of emotionality. The remainder seems to be situation specific. There is little reason to expect that behavior in an open field would be highly predictive of descent from heights or underwater swimming. Perceptual and motor requirements of these tasks are so different that perhaps they should not be part of the same test battery.

Two factor analyses of rat behavior have been concerned with the relationship between open-field behavior and conditioned-avoidance responding. Holland and Gupta (1966) observed the RHA and RLA rats (selected for good and poor performance in shuttle-box avoidance) in an open field, in an activity cage, and in the conditioned-avoidance test that had been the criterion for their selection. Three correlation matrices were formed: one for each strain separately and

the third for the combined groups. Two apparently similar factors were extracted from each matrix: (I) activity, heavily loaded with ambulation in the open field, rearing in activity cages, and intertrial crossings in the shuttle box and (II) emotionality, with a positive loading of defecation and a negative one for ambulation. The two strains were similar on factor II, but very different on factor I. One might infer that selection for avoidance had modified motor behavior more than emotionality.

Wilcock and Broadhurst (1967) also assayed open field and conditioned avoidance in five inbred strains of rats. Their analysis also yielded two factors: (I) activity with loadings on good avoidance learning and high ambulation and (II) emotionality based on high defecation scores, high latency of escape from shock, and low ambulation.

Summarizing statements

The factor-analytic studies just reviewed support the idea that emotionality is not a single energizing (or inhibiting) property of an animal that predicts its behavior in all circumstances. Separate factors for emotional defecation and activity have been found repeatedly. The kinds of factors that emerge from the larger-scale experiments such as those of McClearn and Meredith (1964) and Royce, Carran, and Howarth (1970) are closely related to the tests employed. The tests were chosen in the first place because they were judged to reflect emotionality. It has yet to be demonstrated that the extraction of factors and the biometric analysis of their modes of inheritance leads us closer to the loci that are responsible for genetic correlations or closer to the physiological mechanisms that mediate between the genes at these loci and behavior. Factors should be anchored to a genetic and physical base if they are to be considered as traits on which natural selection has operated. Selection acts directly on phenotypes and in nature perhaps more frequently on psychophenes than on somatophenes. Ideally, factors should be the psychophenes of choice for genetic analysis, since each is presumed to influence many aspects of overt behavior. We have already given reasons for doubting that factors represent genes or blocks of genes, since factor structures of homogeneous and heterogeneous populations are similar. Perhaps the factors extracted from test batteries do not represent stable emotional properties of an organism but are coactive in the sense that a specific environment has specific effects on emotionality. The responses of an individual to exposure to a novel object, a novel odor, or a novel cagemate may be so different that no general statement about that individual's "emotional reactions to novelty" should be made. The same principle holds for the effects on emotionality of different types of extra stimulation such as handling, exposure to shock, or immersion in water.

Our somewhat pessimistic appraisal of the contribution of factor analysis to behavior genetics is tempered by the realization that the complexity of psychophenes requires complex methods to bring order out of what appears to be a motley assortment of observations. Workers in the field must be aware of the necessity to tie factors to segregating genes. Ideally, factor analysis could define units that include both somatophenes and psychophenes and point the direction for research into the relationships between structure and behavior.

TEMPERAMENTAL DIFFERENCES IN DOGS

In comparison with the much studied laboratory rodents, dogs have been neglected by behavior geneticists. This may seem strange, since human beings have lived communally with dogs for millennia, selected them for behavior as well as for structure, and endowed them, at least anecdotally, with complex, almost human personality attributes. The reasons for the geneticists' neglect are rational, however. In the numbers required for genetic investigations, dogs are expensive to raise and maintain. In comparison to small rodents they are slow breeders. Although many breeds are available, there are no truly inbred strains. Attempts at inbreeding often lead to infertility and increased frequency of malformations.

Nevertheless, there are a number of studies of temperamental characteristics in dogs. Priority might well be given to Pavlov, who, in the late nineteenth century, recognized constitutional differences with respect to ease of conditioning among the dogs in his Leningrad laboratory and speculated that these might be inherited. Breed differences in conditionability were also reported by Stockard, Anderson, and James (1941). Other investigators found highly significant breed and hybrid differences in approach and avoidance behavior (Thorne, 1940, 1944) and in four categories of emotional behavior: avoidance, teasing, approach-avoidance, and wariness (Mahut, 1958). Although environmental controls were rather loose, and rating criteria were somewhat idiosyncratic, there is good reason to believe that much of the observed variation was of genetic origin. Additional details of these experiments are given in Fuller and Thompson (1960).

Scott-Fuller project

The major behavior-genetic experiment with dogs was conducted at The Jackson

Laboratory in Maine from 1946 to 1965 (Scott and Fuller, 1965). During the course of this study, many tests were administered, a high proportion of which were concerned with emotional and social behavior. Here we shall consider results bearing on the heritability of activity and emotionality. The project included two major genetic experiments. In the first, five medium-sized breeds chosen to represent a variety of functional specializations were compared on a battery of tests given in fixed order over the first year of life. The foundation stocks of the five breeds—basenji, beagle, cocker spaniel, Shetland sheepdog, and wirehaired fox terrier—were all purebred. Subjects were reared under conditions as nearly identical as possible. A second experiment comprised Mendelian crosses between basenjis and cocker spaniels, two breeds that differed greatly on the majority of tests. The hybrid data were analyzed to determine the mode of inheritance of the behavioral measures. Reciprocal F_1 hybrids were compared for evidence of maternal effects. The richness and complexity of behavioral responses of dogs is much greater than that of rodents, particularly in the area of subject-experimenter relationships. Extending the period of testing over a full year permitted observation of developmental processes. As examples of the results, we present material from a reactivity test and three tests involving inhibitory training.

Reactivity test

At 17, 43, and 51 weeks of age, each dog was lightly restrained by a harness in an electrically shielded cage and connected to an apparatus that recorded heartbeat, respiration, and muscle action potentials. Undisturbed periods in the apparatus alternated with periods in which an experimenter made a friendly approach, a threatening approach, rang a bell, or administered an electric shock. Throughout the test an observer recorded the occurrence of such activities as movement, vocalization, tail wagging, and drooling.

Thirteen measures that yielded suitable quantitative data were subjected to analysis of variance for breed and litter effects. The latter were relatively small in relation to breed effects. Among the 39 analyses (3 tests × 13 scores), breed effects significant at the 0.05 level were found in 33. In 28 of these the probability of the results occurring by chance was less than 0.001. Intraclass correlations ranged from 0.16 to 0.66. The results for change in heart rate during a handler's friendly approach are typical of the obtained data (Table 10-9). The overall average changes in heart rate were small, except at 34 weeks, but the averages are based on a mix of substantial positive and negative changes that reflect significant genetic effects at each age. From 22% to 41% of the variance is attributable to breed differences. De-

Table 10-9. Change in heart rate (beats/minute) during friendly approach of experimenter

Breed	Age at testing (weeks)			Weighted mean
	17	34	51	
Basenji	1.2	22.4	21.4	15.0
Beagle	0.0	17.4	10.2	9.3
Cocker spaniel	−24.8	−11.5	−10.1	−15.5
Shetland sheepdog	− 9.7	10.2	− 8.4	− 2.6
Fox terrier	26.5	3.2	− 3.8	8.6
Weighted mean	− 1.4	8.4	1.9	3.0
Intraclass correlation	0.413*	0.215*	0.245*	

*$p < 0.01$.

velopmental trends also vary with genotype. If the magnitude of the heart rate change, independent of its sign, is taken as the index of an emotional response, dogs of two breeds tended to become more emotional on repeated tests, two other breeds became less emotional, and one breed fluctuated with no clear trend.

When reactivity was estimated from a pooled score of thirteen variables representing a broad sample of the observations, breed differences accounted for 23%, 27%, and 29% of the variance on successive test administrations. Life experiences seem to accentuate genetic effects rather than attenuate them. The phenomenon could be explained in at least three ways: (1) As animals age, new gene-regulated processes associated with maturation become functional. Early in life these developmental processes are out of step among members of a breed, but in young adults the breed-specific phenotype is well established in most animals. (2) Changes in the reactivity scores over time reflect learning. The kinds of responses that are reinforced are somewhat specific to each breed; hence practice makes animals of a breed more similar to each other. (3) Groups of dogs of the same breed exert a coercive force on each other that reduces intragroup differences and thus accentuates the genetic contribution to variance. Wide, unsystematic fluctuations in successive scores of individual dogs are perhaps most compatible with the first hypothesis.

Inhibitory training: breed comparisons

To adapt the dogs to their role as experimental animals, they were subjected to training procedures involving inhibition of behavior that interfered with easy handling. Three of these procedures were standardized so that they provided quantitative data for genetic analysis. In a sense the tests were models of the inhibitory training that children receive in order to promote their safety and to make them acceptable to society.

From the sixth through the sixteenth week puppies were cajoled into remaining stationary on a platform balance during weighing. Their behavior was rated as active, partly active, or quiet. Initially about 12% of all breeds except cocker spaniels were rated as quiet; in spaniels, 25% were so rated. At 16 weeks 80% of the spaniels were quiet in sharp contrast to only 10% of the terriers. The incidences in the other three breeds were closely grouped at about 25%. Breed differences became accentuated with age, although all puppies were receiving the same training.

At 19 weeks of age the puppies were trained to walk on a leash from their outdoor pens to the laboratory building. Demerits were given for tangling the leash or biting at it, for dragging feet, vocalizing, and so forth. All types of demerits were seen in all breeds, but their relative frequencies were highly breed specific: basenjis bit the leash and jerked at it; beagles howled incessantly; Shetland sheepdogs constantly crowded the trainer, interfering with his movements. All dogs improved their performance, and most animals in all breeds were acceptable after 10 days of training. Nevertheless, the proportion of variance in demerit scores attributable to breed remained constant at 52% from beginning to end.

The Jump Inhibition Test (originally Obedience Test) involved control of a dog by human voice and gesture at distances from about 6 inches up to 12 feet. Subjects were lifted to a stand and given the command "Stay." The experimenter then withdrew to a prescribed distance. After one minute the command "Down" was given, accompanied by a hand gesture. Dogs that failed to jump within 1 minute were lifted down. The amount of time spent on the stand was interpreted as a measure of inhibition. The intraclass correlations of 0.14, 0.19, and 0.18 for days 1 through 3, respectively, indicate a tendency for the breed differences to increase during training.

The evidence from all three types of inhibitory training is clear. Such training is effective in modifying behavior in the intended direction, but it does not make breeds and individuals more alike.

Temperament in hybrids

The reactivity test was given to reciprocal F_1, F_2, and backcrosses of basenjis and cocker spaniels. The outcome for the composite scores based on the means of thirteen separate measures is shown in Fig. 10-9. The most interesting feature of these results is the shift in mode of inheritance with age. At 17 weeks, basenji genes for high reactivity are clearly dominant; by 51 weeks all but one genetic group fall on a line consistent with

strictly additive inheritance. Analysis of the scores from separate items in the reactivity test yielded examples of heterosis, dominance, and additive inheritance.

The three tests of behavioral inhibition gave rather inconsistent results. No genetic effects were apparent in the Jump Inhibition Test. Heritability for the composite leash demerit score was estimated from the F_1 and F_2 variances as 0.44. The validity of this estimate requires the assumption that any demerit, regardless of its nature, is the expression of an underlying emotional state induced by restraint. Variability in the type of demerits is then attributed to response tendencies that are independent of the energizing emotional reaction. If this view is not accepted, the composite score is an artifact with no real existence; therefore separate analyses were carried out for each type of demerit. Fighting the leash showed an additive pattern of inheritance; vocalization showed strong heterosis. Other types of demerits followed no clear pattern of inheritance. These results are compatible with a hypothesis that general emotionality is strongly heritable and that its expression is determined by independent genetic systems which affect the probability of occurrence of specific behavior patterns.

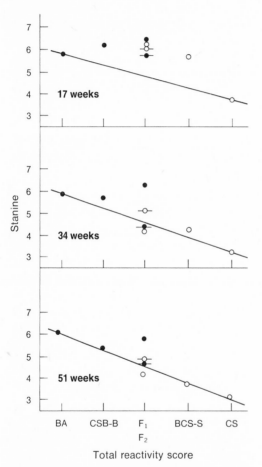

Fig. 10-9. Mean total reactivity scores of hybrids at 17 weeks *(upper)*, 34 weeks *(middle)*, and 51 weeks *(lower)*. *Open circles,* Groups with a cocker spaniel mother or grandmother; *closed circles,* groups with a basenji mother or grandmother. A horizontal line through a point designates an F_2. (From Scott, J. P., and J. L. Fuller. 1965. Genetics and the social behavior of the dog. University of Chicago Press, Chicago.)

Factor analyses of dog behavior

Three factor analyses based on the Scott and Fuller experiments have been reported. Two factors pertaining to emotional reactivity were extracted by Brace (1961). They were interpreted as lability of heart rate change and emotional expression. Earlier, Royce (1955) found six factors relevant to temperament: timidity (including some physiological variables), change in heart rate during contacts with humans, playful aggressiveness, audiogenic reactivity, general activity, and a second timidity factor associated with contacts with people. An analysis of several performance tests yielded three factors of psychological interest (Anastasi et al., 1955). These were described as impulsiveness, docility, and visual exploration.

Brace's factor analysis was the basis of a

search for pattern similarities in behavioral characteristics that were consistent enough to permit allocation of individual dogs into groups that would correspond to acknowledged breeds (Cattell, Bolz, and Korth, 1973). The composition of groups was determined by a computer program, Taxonome, independently of the subjects' genotypes. To an extent the search was successful, indicating that there is some validity to the idea that each of the five breeds in the Scott and Fuller experiment had a distinctive temperament. One can still not judge whether these constellations of traits arise from common dependence on a particular set of genes, on chance associations related to the small number of foundation stock, or on deliberate efforts by breeders to synthesize a desired pattern of behavioral traits. Dog fanciers do not use a single well-defined criterion to choose breeders; instead they select individuals on the basis of conformity to a complex combination of behavioral and physical attributes. Thus a dog breed comes to possess a genotype balanced to guide development into the channels that the breed standard demands. This mode of selection makes it difficult to interpret the genetic significance of correlations based on data from a mixture of purebred animals.

Stable and unstable pointer dogs

The preceding remarks on the consistency of breed temperaments do not imply the absence of substantial variation within dog breeds. Breed fanciers are well aware that some lines are more timid or more aggressive than others. Starting with a few pairs of pointers that had been characterized as either confident or timid, Murphree, Dykman, and Peters (1967a,b) succeeded in establishing two lines designated as stable temperament (line A) and unstable temperament (line E). E-line dogs are less active in an unfamiliar room, tend to freeze during a loud sound, and are much less likely to approach a human. The heart rate of stable dogs averages about 120 beats/minute and declines significantly during petting; in unstable dogs the rate is about 80/minute and

does not vary during petting (Murphree, Peters, and Dykman, 1967). Observations on reciprocal hybrids between the A and E lines failed to detect maternal effects. An extensive program of extra handling of unstable dogs made only a modest reduction of their aversion to humans (Murphree and Newton, 1971). An interesting neurophysiological correlate of the behavioral differences is the near absence in line-E dogs of hippocampal theta activity (trains of relatively high voltage waves at 4 to 8 Hz) during the alert state (Lucas, Powell, and Murphree, 1974). In normal animals orienting and avoiding behavior have been associated with a reduction in hippocampal theta activity; with habituation to a situation, theta returns. This association of a neurophysiological characteristic with a behavioral difference has obvious implications for future research.

Genotype-environment interactions in dogs

A dog's temperament is molded by its experiences, particularly by those which occur during the period from 4 to 15 weeks when manipulative and social competence develop most rapidly. A number of experiments have demonstrated differential effects of experiential restriction between dog breeds (Fuller, 1967). Although variation in activity is a conspicuous part of the postisolation syndrome, change in social behavior is probably of greater significance. Therefore these experiments will be discussed in Chapter 11.

ACTIVITY IN FRUIT FLIES

In the section on phototaxis in drosophilid flies (pp. 94 to 98) we considered the problem of distinguishing among the effects of light on orientation, direction of movement, and rate of locomotion. In the chapter on social behavior we shall encounter a similar difficulty in separating genetic effects on basal, or "spontaneous," activity from effects on specifically sexual responses to other flies. Actually, the concept of spontaneity is hard to define, since one can never exclude the possibility that an animal is responding to some external stimulus. About the best one can do is to reduce stimulation as much as possible and still retain the capacity to observe behavior.

Ewing (1963) applied the technique of genetic selection in a heterogeneous stock of *Drosophila melanogaster* in order to change the rate of migration of flies along a series of linearly arranged chambers equipped with devices to discourage backtracking. Replicate selected lines for high and low activity were established; all separated clearly from unselected controls. H-line flies were faster than L-line flies in dispersing from the center of a circular field, but the lines did not differ when single individuals were tested in a fly-sized version of the open field. Ewing concluded that he had actually selected H-line flies for greater reactivity to neighboring flies. He also found that L-line flies were more reluctant to crawl through the one-way gates between chambers, even in a forward direction.

In an attempt to circumvent these confounding factors, Connolly (1966) tested flies individually in an open field and scored the number of squares entered in 10 minutes. Separation of high- and low-activity lines was rapid, and h^2 was estimated from parent-offspring regression as 0.51 ± 0.10. In three other situations line differences in activity were consistent with the results in the open field.

It does appear that the concept of a basal level of spontaneous, or at least of minimally stimulated, activity has some genetic validity for fruit flies. Such activity may be a complicating factor in evaluating the specificity of genetic differences in many forms of behavior. Actually in all mobile animals locomotor activity is a prerequisite for exploring the environment and often for responding adaptively to it. One suspects that basal activity would generally be subjected to stabilizing selection, since either extremely high or low levels would be maladaptive. At the same time, retention of genetic variability would be advantageous to species like fruit flies which are exposed to a variety of conditions that may require different modes of response for survival. Such variability provides a substrate for rapid artificial selection in the laboratory.

HABITAT PREFERENCE

Habitat preference is included in this chapter because it involves individual reactions to external stimuli that are important in the adaptive process. The existence of such preferences is almost universal. Even the most widely distributed species avoid unsuitable environments; some species are restricted to rare niches that limit their numbers severely. Human beings seem to be an exception, but they have spread widely over the earth by modifying the environment artificially rather than by genetic adaptation to specific conditions. Animals can modify their environment to a degree by constructing shelters, but their capabilities in this regard are limited and rather stereotyped in any one species.

Presumably the behavioral traits that underlie habitat choice are the outcome of a long evolutionary process. For our purposes it is interesting to look at behavioral differences between subspecies that occupy different geographical areas and show distinctive life-styles. An example is the white-footed mouse, *Peromyscus maniculatus*, which has been divided into a variety of subspecies that are mutually interfertile. Some of these, such as *P.m. bairdii*, inhabit open prairie; others, such as *P.m. gracilis* and *P.m. artemesiae*, live in wooded areas and are semi-arboreal. Are these life-styles encoded in the genotypes of these subspecies, or are they culturally transmitted? An answer to this question may be sought by looking at habitat preferences in animals that have been born and reared in laboratories.

Harris (1952) constructed two connecting artificial indoor environments, a "grassland" containing manila folders cut into strips and fastened to the floor, and a "woodland" containing sections of tree trunks with intact bark. Food, water, and activity wheels were available in both, and devices were inserted to measure the time spent in the two habitats. Given this choice, *P.m. bairdii* was predominantly active in grassland, *P.m. gracilis* in the forest. Some form of genetic encoding seems to be operating.

Some years later the effects of prior experience on habitat choice of the prairie species, *P.m. bairdii*, were observed (Wecker, 1963). Animals exposed to simulated grassland early in life chose it over forest in a later test; those with similar early experience in forest failed to show any preference. Apparently the genetic program was still present but could be overridden by early experience. One could also interpret the result in terms of genetic constraints on learning; thus *P.m. bairdii* imprints more readily on stimuli that resemble the habitat of its ancestors.

These results on habitat choice may be compared with earlier experiments on the geotropic orientation of several *Peromyscus* subspecies (Clark, 1936). At the age of 11 to 12 days when their eyes are still closed, mice crawl upward when placed on an inclined plane. The young of semiarboreal subspecies oriented at a lower angle of inclination than did the young of grassland subspecies, such as *P.m. bairdii*. At steeper inclinations semiarboreal subspecies pointed more directly upward. These responses seem to presage adaptation for climbing later in life.

Habitat choice probably involves both perceptual and motor components. It would be well worth while to conduct genetic experiments more complex in design than simple comparisons between subspecies. The results of crosses could, perhaps, throw light on the nature of the selective processes that have shaped the ethograms of these closely related groups which have differing lifestyles.

SUMMARIZING STATEMENTS

Genetic variation has been found in a great variety of behaviors that we have included under the broad heading of temperament. Although we have concentrated our attention on research with laboratory rodents because of its abundance and because its data have been subjected to the most extensive genetic analysis, similar results can certainly be obtained with other species.

The hypothesis of a genetically influenced unidimensional gradient extending from nonreactivity (or low emotionality) to high reactivity (or high emotionality) has been both strongly supported and cogently criticized. It seems reasonable that in a variety of situations an individual who is generally easily aroused will behave differently from an impassive one. Considerable data confirm this belief. However, factor analysis and simple correlational studies indicate that as we increase the kinds of observations we make on a subject, the more dimensions are needed to describe its behavior. There probably is a dimension of threshold behavioral arousal that pervades almost any transaction between an animal and its environment, but in any specific situation so many factors are involved that generalization to another situation is difficult.

Much of the observed data can be explained by postulating that all members of a species possess a common behavioral repertoire, or ethogram, each element of which is elicited by appropriate stimuli. Some elements, such as defecation in the rat, are evoked by almost any stimulus if it is novel. Others such as hoarding require rather special stimuli and are strongly influenced by an animal's physiological state. In every individual there is a priority system that determines the elements of the ethogram which are activated in any situation. These priorities are not the same in all individuals, and they may change as an animal becomes familiar with a situation. The priority system is affected by both genotype and experiential history.

If this concept is correct, the goal of determining a specific number of dimensions of temperament is probably impossible to attain. Dimensions or factors may represent similarities between test situations in their capacity to evoke particular responses, rather than the existence of generalized motivational states labeled emotionality, timidity, or exploratory drive. The organization of the genotype may be more nearly cognate with the elements of the ethogram than with dimensions of temperament. If every genotype is capable of guiding development of the total ethogram, how shall we account for individual differences? Primarily through postulating variation in the thresholds for the separate components and secondarily by noting that at a given instant only one element can be fully expressed. Presumably natural selection has favored establishment of these thresholds at levels that are adaptive in situations most likely to be encountered.

In this formulation generalized traits correspond more to aggregations of responses that occur together because they have been selected as a group than they do to responses that are associated because they have a common physiological basis or depend on the same genes. Of course, the latter possibility is not excluded, but we need more data on genetic correlations from heterogeneous stocks and from properly designed selection experiments before it can be accepted. Thus there is unfinished business in the behavior-genetic analysis of temperament. But regardless of differences of opinion regarding theoretical formulations of the relationship between genes and temperament, one can be sure that heredity has an important role in producing emotional variability among individuals.

11

Social behavior of animals

Social behavior occurs whenever members of a species communicate and react with each other. A minimal amount of social behavior is necessary for all bisexual species; males and females must associate, at least briefly, to ensure fertilization of eggs. Most species have more than this minimal amount of social interaction, but it is still largely concerned with reproduction and care of the young.

Social behavior has been classified in a number of ways, but no one system has been universally adopted. Our classification is similar to that of Scott (1958), but it is more restrictive. Five categories are recognized as reasonably distinct, although a particular act may be considered to belong in more than one.

Affiliative behavior results in the formation of relatively permanent groups that may be as small as a mated pair or as large as a flock of sheep or a hive of bees. The term *social species* is reserved for animals that characteristically live in groups larger than a mated pair with their immature offspring. The cohesion, permanence, and degree of cooperation and coordination among members of such social groups vary widely.

Sexual behavior includes courtship, mating, and postmating activity. In some species it is associated with pair bonding, a special type of affiliative behavior that may be terminated only by the death of a member.

Care-giving, or nurturant, behavior is usually provided by parents, although in the social insects it is the function of sterile females. In many vertebrates nurturance is the special responsibility of females, but there are numerous examples of participation by males. Unfortunately, in the animal species most studied by behavior geneticists, the male's role in caretaking is minimal.

Care-soliciting behavior is particularly characteristic of immature birds and mammals. With nurturant behavior, it forms part of a complementary system.

Agonistic behavior includes fighting, conciliation, retreat, and submission. It is associated with competition for resources such as shelter, mates, and food.

Affiliation may lead to cooperative behavior in which animals act together as though for a common goal, for example, the construction of a beaver dam, the killing of a moose by a wolf pack, or the feeding of nestling robins by both parents. Cooperative behavior increases the fitness of all cooperators and should be favored by natural selection. All forms of social behavior just defined are essentially cooperative, with the exception of agonistic behavior. Usually in the latter the fitness of one individual is increased at the expense of another's. Nevertheless, some agonistic behavior is cooperative, as when the members of a pride of lions act together in excluding a nomadic male from their territory. On the whole, however, the nature of the selective process differs for agonistic behavior as compared with other forms of social behavior. The question as to whether selection for cooperative social behavior acts only on individuals or also on groups as a whole has been hotly debated. A discussion of this issue and the related one of the genetics of altruism can be found in Wilson (1975).

Setting aside the debatable issues of group selection and altruism, it is clear that social behavior is the most critical factor producing differences in success in leaving progeny and thus contributing genes to the next generation. Thus strong selective forces should have led to dominance of the genetic determinants for adaptive social behavior. The consequence would be relatively low heritability for social psychophenes. Still, it is unlikely that genetic variation would be completely lost. And even small variations in sexual behavior that lead to assortative mating or to failure to find a mate have potential for changing the genetic structure of a population in important ways.

As we review the genetics of social behavior, we inevitably encounter the complication of considering the phenotypes of two or more genotypes at the same time. To some extent we can look at each participant separately, relegating the others to the status of environmental stimuli, but this is not a completely satisfactory solution. This issue, with special emphasis on agonistic behavior, has been discussed by Fuller and Hahn (1976).

Two basic approaches have been used in studying the genetics of social behavior. First, there is the traditional methodology of experimental behavior genetics: strain comparisons, intercrossing, mutant comparisons, and selection. The objective of such investigations is to understand the social behavior of present-day species. The second approach is through sociobiology, a fusion of population genetics, ecology, and animal behavior bound together by evolutionary theory. Its objective is the understanding of speciation and the evolution of adaptive social communication and organization. These two lines of endeavor have developed almost independently but seem to be coming closer together. Because the time scales of evolutionary change and of laboratory experimentation are so disparate, it is likely that some separation of these two approaches will continue. Cross-fertilization of ideas between practitioners of each kind of research will be helpful to both.

MATING BEHAVIOR IN *DROSOPHILA*

The genetics of mating behavior has been most intensively studied in the numerous species of *Drosophila*. Much of the research has been motivated by a desire to learn more about the role of mate selection and sexual

vigor in evolution. The availability of a variety of closely related species facilitates this endeavor. More recently attention has been given to the heritable physiological factors underlying mate preference.

Courtship and mating vary somewhat among *Drosophila* species (Spieth, 1952), but the following account based on *Drosophila melanogaster* will suffice for our purposes (Bastock and Manning, 1955; Bastock, 1956). The reactions of a male-female pair are described, although the initial part of the male courtship may be directed toward another male. The whole affair can be characterized as promiscuous activity by the male fly and discriminative passivity on the part of the female. A male about to court approaches a prospective mate and taps her with his forelegs. After a varied interval of seconds or minutes, courtship proper begins. The male orients to the prospective partner, facing her if she is quiet or following if she moves. He vibrates the wing on the side nearest the partner for a few seconds. Several bouts of "singing" may occur. The male then licks the genitalia of his partner and attempts to mount. If these blandishments have been directed at an unreceptive female or at a male, the courtship ends abruptly as the partner moves away or kicks at the suitor. A receptive female stands, spreads her vaginal plates, and copulation follows quickly. Females are receptive from about 2 days after eclosion (emergence from the pupal case) up to a maximum of about 20 days (Manning, 1967a). After copulation, females are generally unreceptive for a period.

From these distinct sex roles two general principles emerge. Successful mating requires adequate courtship to make the female receptive, and differences in the vigor and pattern of the male's performance will determine whether he succeeds in transmitting his genes to the next generation. The female's part is also important. We shall see that she can be discriminating on the basis of her suitor's genotype and even on the basis of whether he belongs to the minority or majority of a group of assorted types of wooers. Since genetic variation in mating behavior has been found in both sexes, analysis of the causes of success or failure in reproduction are often complex.

Speed of mating

A simple quantitative measure of sexual activity is the speed of mating. Manning (1961) placed fifty pairs of an outbred stock of *Drosophila melanogaster* in a bottle and withdrew pairs as they copulated. The first ten pairs mating and the last ten pairs were separated, and each line was selectively bred for twenty-five generations. Fast-mating, slow-mating, and control lines were maintained in duplicate. Selection was successful with a realized heritability of about 30% over the first seven generations. Hybrids were intermediate, indicating a predominantly additive mode of inheritance. When selection was relaxed, mating speeds fluctuated, but the lines retained the characteristics for which they had been selected.

Correlated behavioral changes accompanied selection for mating speed. Somewhat unexpectedly when moved to a new environment, the slow-mating lines were more skittish and much more active than the fast maters. Extreme distractibility rather than reduced interest in sex seemed to explain much of the reduction in mating speed. However, as evidence of increased sexual arousal in the fast-mating males, they courted their own reflections in the apparatus, a behavior not observed in the slow maters. Selection also modified the females: slow maters rejected males more frequently.

In an effort to disentangle the effects of joint selection on the two sexes, procedures were developed for selecting fast- or slow-mating males as tested with standard unselected females. Slow-mating females were also selected on the basis of their receptivity to standard males (Manning, 1963). Selection for fast-mating males and for slow-mating females was ineffective, but a line was produced consisting of slow-mating males with females only slightly changed. The striking difference between these two selection experiments points out the complications of the genetic analysis of social behav-

ior. Since there is a complementarity of responses between the participants, and since the roles of the sexes are different, there are several ways in which mating speed can be varied. Increased general activity may increase mating speed of males; the same shift in females decreases it.

Diallel analyses of mating speed have been reported by Parsons (1964) and Fulker (1966). Parsons found that hybrids mated more quickly than inbreds and that fewer of them remained unmated. Highly significant general combining effects were found (characteristics of a strain associated with all its crosses), plus somewhat weaker specific combining effects (evident in some crosses but not in others). Fulker's data were analyzed by the Hayman-Jinks procedure and were described in Chapter 5 (Fig. 5-7). In both diallels the finding of dominance for short latency to copulation and of heterosis for the same measure suggests a history of selection for rapid mating.

Genes modifying mating speeds are widespread in natural *D. melanogaster* populations. Significant differences were found between flies recently collected from the wild at two Australian stations (Hosgood and Parsons, 1965). Hybrids between the flies from the two sites had intermediate speeds. In every instance the male's behavior appeared to be the factor controlling latency of copulation. Why the populations differ is not known with certainty, but there are marked differences in rainfall at the two collection localities, which could have led to different physiological adaptations.

Duration of copulation may also have a relationship to fitness, since the amount of sperm transferred could affect the number of offspring. Inbred lines of *D. melanogaster* do differ on this measure, and a 4 × 4 diallel cross detected significant general combining ability for this characteristic (MacBean and Parsons, 1967). It is also possible to select for both short and long duration of copulation; the behavior of male flies seems to be the critical factor.

Research on the speed of mating has also been pursued intensively with the western

American species, *D. pseudoobscura*. This species is of particular interest because its natural populations are a mix of varying types of chromosomal arrangements. By inbreeding it is possible to produce lines with only one arrangement present. Flies in these lines are all *homokaryotes;* F[1] hybrids between such lines are *heterokaryotes*. The proportions of the various karyotypes in natural populations are relatively constant, suggesting some homeostatic mechanism. Is it behavioral? The answer appears to be yes. When forty-seven lines of homokaryotic flies of five different arrangements were tested for mating speed, a clear association was found between rapidity of mating of a karyotype and its frequency in nature (Spiess and Langer, 1964). In the most common arrangements (AR and ST) about 60% mated in an hour, in intermediate frequent karyotypes (CH and TL) the corresponding figure was 45%, and in the rare PP variety only 20% had mated within an hour. The case is strong that differences in mating speed are important in maintaining a balance of genetic polymorphisms. The situation is complex, however, since mating propensity is transmitted heterotically (heterokaryotes mate more rapidly than their homokaryotic parents), and such situations lead to equilibria based on the fitnesses of the two homokaryotes (Chapter 4).

Chromosomal arrangements are not the only genetic factors affecting mating speed in *D. pseudoobscura*. Strains of the same karyotype differ significantly in mating speed and in duration of copulation (Parsons, 1967b). A polygenic system appears to be acting in conjunction with the karyotypic controls.

Fast- and slow-mating lines of *D. pseudoobscura* have been selected by Kessler (1968, 1969). The rapid response to selection within five generations indicates considerable additive variance for the trait. When males and females of different lines were placed together, the female's behavior was the chief determinant of mating frequency. As shown in Table 11-1, the combination of fast females and slow males is nearly as efficient as the fast/fast combination. The slow female/fast male pairs were only half as effec-

Table 11-1. Mating efficiency of crosses of selected lines of *Drosophila pseudoobscura**

Mating combination		Number of groups	Mean percent mated	
Females	Males		In 30 min	In 60 min
Control	Control	6	52 ± 6	64 ± 4
Fast	Fast	6	73 ± 3	79 ± 3
Slow	Slow	6	17 ± 3	26 ± 4
Fast	Slow	9	61 ± 3	71 ± 4
Slow	Fast	9	29 ± 3	37 ± 5

*From Kessler, S. 1968. The genetics of *Drosophila* mating behavior. I. Organization of mating speed in *Drosophila pseudoobscura*. Anim. Behav. **16**:485-491.

tive. In this case, being discriminatingly passive is more important than being indiscriminatingly active.

It should be clear by now that research on the mating efficiency of fruit flies is not an exotic form of voyeurism but a way to understanding some of the principles of behavioral evolution. Insofar as mating speed is concerned, Parsons (1974) has summarized the current state of knowledge as follows:

1. There is strong natural selection for rapid courtship and copulation by males.

2. Fast matings seem to be controlled by male behavior; in slower matings the discriminative role of the female becomes more important.

3. Mating speed is positively correlated with fertility and fecundity.

4. In *Drosophila* mating speed is probably the most important component of individual fitness.

Ethological, or sexual, isolation

Important though male vigor and female receptivity may be to individual fitness, their genetic consequences depend on who mates with whom. A species is defined as a closed population whose members share a gene pool and may freely interbreed. Closely related species may interbreed, but their hybrids are generally disadvantaged and do not perpetuate themselves. Therefore the gene pools remain distinct. In the so-called sibling species, morphological differences are so slight that taxonomists may have difficulty in distinguishing them. Nevertheless, hybridization in nature appears to be rare even though it occurs readily in a laboratory setting when males and females of each sibling type are confined together. Given a choice, however, matings are predominantly between conspecifics. In contrast to species, races and subspecies are open populations between which gene flow is potentially free. The distinction is a rather fine one because the very existence of different forms of a species implies a degree of genetic separation. If there are barriers to gene transfer between races, they might be expected in time to become separated enough to become true species.

The restriction, either complete or partial, of interbreeding between races and related species is known as *sexual*, or *ethological*, *isolation*. Individuals do not mate, even though they come in contact with potential mates. A degree of isolation exists within a population when homogamic matings are preferred to heterogamic ones. Sexual isolation is therefore a form of assortative mating, and it can vary quantitatively from complete separation (as between species) to a small statistical bias in favor of like mating with like. An excellent review of this topic is available in Petit and Ehrman (1969).

Evidence for selective mating and isolation is obtained through three types of experiments: male choice, in which one kind of male is placed with two types of females; female choice, in which one type of female is placed with two types of males; and multiple choice, in which males and females of two varieties compete for mates in a free-for-all. Male choice is an inaccurate designation, since it is the female who generally exercises such discrimination as exists in *Drosophila* courtship.

Several quantitative indices have been devised to measure the degree of isolation between genetic groups. A simple and widely used formula devised by Stalker (1942) is

$$I = \frac{C - A}{C + A}$$

where C is the percentage of conspecific females inseminated, and A the percentage of alien females inseminated. An index of +1 indicates complete isolation; one of 0, no isolation; and one of -1, exclusive cross-mating. Given unlimited time, all the females might be inseminated; therefore, experiments of this type should be terminated when 50% of the females have mated. Since the index may vary depending on the type of male used, it is customary to determine separate values for type 1 and type 2 males and calculate a joint isolation index

$$I_j = \frac{(I_1 + I_2)}{2}$$

where the subscripts refer to the type of male tested with a mix of type 1 and type 2 females. Differences in the mating propensities of the females can be measured by:

$$M = \frac{(I_1 - I_2)}{2}$$

For multiple-choice experiments the formula of Malagolowkin-Cohen, Simmons, and Levene (1965) is appropriate. Let the strains be designated 1 and 2; let x_{11}, x_{12}, x_{21}, and x_{22} designate the number of matings of each type, where the first subscript refers to the male and the second to the female partner; and let $N = x_{11} + x_{12} + x_{21} + x_{22}$. Then the isolation index, when there are equal numbers of all types, is $I = (x_{11} + x_{22} - x_{12} - x_{21})/N$. This formula can be restated as shown below.

When the proportions of the two strains are not equal, the formulas must be suitably modified. Levene's solution as presented by Ehrman and Petit (1968) allows the calculation of an isolation index, a male selection index, and a female selection index from the same data. Let there be m_1 males of strain 1,

$$\text{Isolation index} = \frac{\text{number of homogamic matings} - \text{number of heterogamic matings}}{\text{total number of matings}}$$

m_2 males of strain 2, n_1 females of strain 1, and n_2 females of strain 2. Also let the observed number of matings of each type be x_{11}, x_{12}, x_{21}, and x_{22}, where the subscripts are defined as before. For each x_{11} through x_{22}, calculate an x' as follows: $x'_{11} = x_{11}/m_1n_1$, $x'_{12} = x_{12}/m_1n_2$, $x'_{21} = x_{21}/m_2n_1$, and $x'_{22} = x_{22}/m_2n_2$. This step corrects for the unequal numbers of individuals in the four classes. Three indices calculated from these data are:

$$\text{Isolation index } (Z_I) = [(x'_{11} \cdot x'_{22})/(x'_{12} \cdot x'_{21})]^{1/2}$$
$$\text{Male selection } (Z_M) = [(x'_{11} \cdot x'_{12})/(x'_{21} \cdot x'_{22})]^{1/2}$$
$$\text{Female selection } (Z_F) = [(x'_{11} \cdot x'_{21})/(x'_{12} \cdot x'_{22})]^{1/2}$$

Sexual isolation has been demonstrated in many species and subspecies of *Drosophila*, among them *D. prosaltans* (Dobzhanzky and Streisinger, 1944), *D. virilis* (Spieth, 1951), *D. equinoxialis* (Hoenigsberg and Santibanez, 1960b), *D. obscura* (Dobzhansky, Ehrman, and Kastritsis, 1968), and *D. persimilis* (Spiess and Yu, 1975). Mayr (1970) writes, "Ethological barriers to random mating constitute the largest and most important class of isolating mechanisms in animals." As an example of the findings, we shall briefly summarize experiments with *D. paulistorum* (Ehrman, 1964). This species complex ranges from Guatemala to southern Brazil and Trinidad. It is divisible into seven races that are morphologically similar, though chromosomally distinguishable. Even though the ranges of these races overlap to some extent, matings between them are rare. When they do occur, the male offspring are sterile, the females fertile. Although the form of the courtship appears similar in all races, there is a clear preponderance of homogamic matings in both male- and female-choice experiments. This preference for one's own race has the advantage of preventing the wastage of gametes that would result from indiscriminate mating. The degree of isolation is variable and is probably polygenically determined. The degree of genetic difference between groups rather than their geographical separation in nature seems to be the critical factor affecting the strength of isolation. Ethological barriers were least strong between the widely distributed Transitional race and the other races whose ranges surround it.

One might expect that where races of *D. paulistorum* coexist in an area, natural selection might favor sexual isolation, whereas in areas where the races do not come into contact, no such pressure would exist. Ehrman (1965) made an experimental test of this hypothesis by looking at sexual isolation between the races under two conditions. A series of multiple-choice experiments between pairs of races was set up in which each race was represented by either sympatric or allopatric strains. The sympatric strains had originated in areas where the two races overlapped; the allopatric strains came from regions where only one of the races was found. As shown in Table 11-2, the isolation coefficients are consistently higher in the sympatric than in the allopatric combinations. Ehrman concludes that the races are incipient species and that behavioral isolation is an important factor in keeping their gene pools separate.

In a similar study of the *D. obscura* group, ethological isolation was not proportional to morphological or chromosomal differences among the species, to their ability to produce fertile hybrids, or to their geographical separation (Dobzhansky, Ehrman, and Kastritsis, 1968). Sexual isolation may arise from selection acting against heterogamic matings that produce infertile or otherwise disadvantaged offspring; it may also be a residue of past selection or a nonspecific byproduct of genetic divergence through random drift or selection for an unrelated trait.

Selective mating: effect on gene frequencies

Do the results of mate-choice tests enable us to predict the outcome of natural selection operating on a genetically heterogeneous population? It is possible to simulate natural selection by placing a known mix of genetically different flies in a population cage and following the change in gene frequencies over a period of time. If the mating success of the various genotypes in the mix is known, one can see how important a part it plays in

Table 11-2. Isolation coefficients for sympatric and allopatric crosses between races of *Drosophila paulistorum**

Races crossed	Sympatric crosses	Allopatric crosses
Amazonian vs. Andean	0.86 ± 0.05	0.66 ± 0.07
Amazonian vs. Guianian	0.94 ± 0.03	0.76 ± 0.06
Amazonian vs. Orinocan	0.75 ± 0.06	0.61 ± 0.07
Andean vs. Guianian	0.96 ± 0.03	0.74 ± 0.07
Orinocan vs. Andean	0.94 ± 0.03	0.46 ± 0.08
Orinocan vs. Guianian	0.85 ± 0.05	0.72 ± 0.07
Centro-American vs. Amazonian	0.68 ± 0.07	0.71 ± 0.07
Centro-American vs. Orinocan	0.85 ± 0.05	0.73 ± 0.07

*From Ehrman, L. 1965. Direct observation of sexual isolation between allopatric and between sympatric strains of the different *Drosophila paulistorum* races. Evolution **19**:459-464.

the selective process. Merrell (1949) made quantitative estimates in *D. melanogaster* of the effect of four sex-linked recessive mutations on mating success. The mutations were placed on a common background, both singly and in all possible combinations. Male-choice and female-choice experiments were conducted, and the results were evaluated by the types of offspring produced. The use of genetic markers in such experiments is illustrated by the following example (Table 11-3). A wild-type female heterozygous for raspberry *(ras)* was placed with wild-type and raspberry males. The types of offspring that result from the success of each male in achieving insemination are shown in the bottom half of the table. By isolating the females after mating and determining the phenotypes of their offspring, the proportional success of the two kinds of males is readily determined.

Female-preference experiments showed that wild-type males were much more successful in mating than yellow males and moderately more successful than cut or raspberry males. Forked males were equal to wild-type in success. In general the effects of multiple mutant genes were additive, although *ct ras* males were superior to *ct* or *ras* males. The behavior of the females was primarily responsible for nonrandom mating. Where apparent male choice was found, it could be more plausibly interpreted as the rejection of less vigorous males by less receptive types of female. Hence the male was more successful in his courtship of the more receptive fe-

Table 11-3. Determination of mating success in competition between raspberry and wild-type males*

	Males	
Phenotype Genotype	Raspberry *ras*/Y	Wild-type +/Y

		Gametes		
	ras	Y	+	Y
Gametes *ras* from test female +	*ras*/*ras*† *ras*/+	*ras*/Y +/Y	*ras*/+ +/+	*ras*/Y +/Y

*From Merrell, D. J. 1949. Selective mating in *Drosophila melanogaster*. Genetics **34**:370-389.
†The appearance of phenotypically *ras* females among the offspring indicates mating success of the mutant male.

male, though he did not choose her. In line with this interpretation, male choice was most evident in the less vigorous males; furthermore, it was consistently in the same direction for all males tested with the same choice of females. Merrell concluded that within *D. melanogaster* mutants there is selective but not assortative mating. In a later experiment, Merrell (1953) showed that the relative rate of elimination of mutant genes from a population was predictable from the effects of each gene upon success in mating. Generally similar findings have been reported by Hildreth (1962) and Hildreth and Becker (1962).

The possibility that inbreeding could lead to sexual isolation was investigated in *D. melanogaster* by Hoenigsberg and Santibanez (1960a). Male-choice experiments were conducted with three phenotypically wild-type inbred Oregon strains and with two outbred stocks, Oregon and Samarkand. Outbred males showed a very slight preference for homogamic matings. Inbred males oriented and displayed more strongly to females of their own line, and inbred females tended to reject outbred males. It appears that genes capable of producing sexual isolation are present within wild-type populations of *D. melanogaster* and that their effects can be made visible by inbreeding.

Frequency-dependent mating preference

Sexual selection in *Drosophila* is not simply a matter of female preference for one type of male courtship over another. What is now called *frequency-dependent mating*, or the *rare male effect*, was brought to the attention of geneticists by Petit (1958). She found that when wild-type and mutant male *D. melanogaster* competed for females under uniform conditions, one or the other variety had a consistent advantage and obtained a disproportionate share of females. However, the degree and even the direction of the advantage varied systematically with changes in the ratio of mutant to wild-type male. In general, as a variety of male became rare, its relative mating success became greater.

Petit devised a coefficient of mating success to compare competitive ability as the proportions of males changed. Let A = the number of females inseminated by mutants, and a = the number of mutant males in the competition; similarly, let B = the number of females inseminated by wild-type males, and b = the number of wild-type males in the competition. Then the coefficient of mating success for the mutant $K = (A/B) \cdot (b/a)$. When the two kinds of males have equal success, $K = 1$; when the mutants have an advantage, $K > 1$; and when they are disadvantaged, $K < 1$. In general, K was greater than 1 only when the mutant males were relatively rare.

Frequency-dependent mating is not a peculiarity of mutants that are rare in nature. It has been demonstrated in the chromosomal variants of *D. pseudoobscura* (Ehrman, 1966); in three related tropical species, *D. willistoni*, *D. tropicalis*, and *D. equinoxialis* (Ehrman and Petit, 1968); and in wild-type laboratory strains of *D. melanogaster* (Tardif and Murnik, 1975). Minority advantage is shown when flies from distant localities are tested together, even though the mating propensities of the two populations are equal (Spiess and Spiess, 1969).

It is clear that frequency-dependent mating helps to preserve genetic diversity. As a gene becomes rare, the mating success of individuals bearing the gene increases; the result is a dynamic equilibrium with the two types of male coexisting in a relatively fixed ratio. The process is illustrated in Fig. 11-1, where the convergence of frequencies of two eye-color genes that were started in two separate populations is shown. Although the ratios of *or:pr* in the original populations were 80:20 and 20:80, respectively, they came together at an approximately 50:50 ratio within ten generations (Ehrman, 1970b).

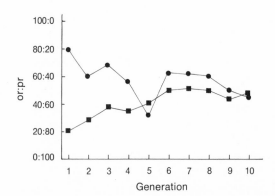

Fig. 11-1. Convergence of gene frequencies over ten generations for orange *(or)* (■) and purple *(pr)* (●) in two populations of *Drosophila pseudoobscura*. Individuals with the rarer phenotype have an advantage in mating competition, so that populations starting with a preponderance of either gene tend to move toward the same equilibrium frequencies. (From Ehrman, L. 1970. Proc. Natl. Acad. Sci. U.S.A. **65:**345-348.)

Frequency-dependent mating implies more than the ability to recognize the difference between genotypes. The female fly must also have some way of knowing which males are rare. She must gain this information rather rapidly because the mating tests last only a few minutes. Although several sensory modalities could convey the necessary information, research has centered on chemical means. Rare males lose their mating advantage when the air of their mating test chamber has been drawn over a large aggregation of males of their own type (Ehrman, 1966, 1969). Introducing acetone or petroleum ether extracts from the rare-type male flies just prior to a mating test also abolishes the rare male advantage. The active substance is probably a lipid or a steroid (Ehrman, 1972; Leonard, Ehrman, and Schorsch, 1974). Previous sexual experience also has a modifying effect on female selectivity. Older females ("widows") tended to mate with males of the same type as their previous consort, thus either enhancing or neutralizing the rare male effect, depending on the female's previous experience (Pruzan and Ehrman, 1974).

Pheromones, neurons, and hormones. The chemicals that inform female *Drosophila* of the relative abundance of rival males are functionally classed as *pheromones*, chemical transmitters of information between members of a species. Their importance extends beyond the rare-male effect to sexual isolation in general. Sexual isolation between the sibling species *D. melanogaster* and *D. simulans* is diminished by surgical excision of cutaneous chemical sensillae (Manning, 1959). Averhoff and Richardson (1974) found that in *D. melanogaster*, individuals were sexually unresponsive to their own pheromones and to those of close relatives. Airborne substances from alien strains were effective in arousing male courtship. They suggest that the diversity of pheromones, coupled with a tendency for negative assortative mating, helps to prevent inbreeding in small populations. Even flies may have incest taboos.

The importance of the genotype of the brain in regulating sexual organization in *Drosophila* has been known for many years. In male-female mosaic flies, behavior follows the sex of the brain rather than that of the genitalia (Morgan and Bridges, 1919). The switch-on of female receptivity that occurs about 2 days after eclosion is associated with stimulation of the brain by hormones from the corpora allata (Manning, 1967a). When *D. simulans* were selected for slow mating, it was found that the effect was due to insensitivity of the brain to the hormone rather than to a deficiency in the corpora allata (Manning, 1968). A polygenic system appears to control the level of sensitivity of the brain to the hormone (Manning and Hirsch, 1971). We shall see in a later section that a similar situation exists in mammals.

The emphasis placed on chemical, tactile, and vibratory stimuli in *Drosophila* courtship varies among the many species of this genus (Spieth, 1952). Thus in *D. subobscura* and *D. auraria*, mating does not take place in the dark. In the *virilis* group, wing vibration is apparently effective at a longer distance than in other species and an odor-dispersing function of wing vibration is postulated for *D. persimilis* and *D. pseudoobscura*. The wing vibrations ("courtship songs") are also patterned differently and probably function in maintaining the separation of these species (Ewing, 1969). The form of courtship songs are also correlated with the mating propensity of crosses among *D. melanogaster*, *D. simulans*, and their hybrids (von Schilcher and Manning, 1975). The light-dependent mating of *D. subobscura* apparently reflects the importance of visual communication between males and females of this species (Maynard Smith, 1956). Clearly, genetic factors affect mating through many different mechanisms.

Summary

This sampling of research on the sexual behavior of *Drosophila* demonstrates the ubiquity of genetic variation among individuals of a local population, among races of the same population, and among the species that comprise this widely distributed genus. The relationship between genes and behavior is reciprocal and dynamic. Different genes lead to different kinds of behavior; different kinds of behavior affect mating success and mate choice; hence behavior determines the kinds of genes that survive. Two kinds of forces seem to operate: the first restricting crossbreeding that would produce disadvantaged hybrids, and the second promoting genetic heterogeneity among the members of natural breeding populations.

CRICKET SONG: GENETIC CONTROL OF COMMUNICATION

The calling songs of crickets, so familiar a part of summer evenings in many parts of the world, are actually a part of male courtship. The chirps and trills of each species have a genetically programed pattern that enables females to identify conspecific males and respond to them. Where the pheromones and wing vibrations of *Drosophila* function for communication over distances measured in millimeters, the strong voices of crickets carry for many meters. The relatively large size of crickets, our detailed knowledge of their nervous system, and the ease with which sound patterns can be recorded, analyzed, and played back makes this group ideal for the study of relationships between genes, neurophysiology, and behavior.

The complex songs of two Australian species of field crickets, *Teleogryllus oceanicus* and *T. commodus*, have been studied most intensively from a genetic point of view (Bentley, 1971; Bentley and Hoy, 1972, 1974). In *T. oceanicus* the song consists of a chirp followed by a long trill of paired pulses; in *T. commodus* the chirp and the initial trill are fused, and the elements of the trill are briefer and closer together than in *T. oceanicus*. Fig. 11-2 depicts the sound-pulse patterns of both species and those of their hybrids. The rigidity of genetic control is shown in three ways: (1) individuals of the same genotype have almost identical calls that vary little under different conditions, (2) different genotypes each have characteristically different songs, and (3) these characteristic songs

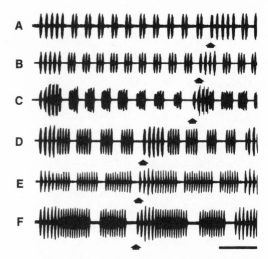

Fig. 11-2. Sound pulse patterns in the male calling songs of *Teleogryllus oceanicus* (**A**), *T. commodus* (**F**), and their hybrids. The songs are composed of phrases that start with a single 4-6 pulse chirp followed by a series of trills containing 2 to 14 pulses, depending on genotype. Each record shows a complete phrase starting at the left and ending at the arrow, where a second phrase begins. The bar at the bottom spans 0.5 second. The hybrid songs shown are *T. oceanicus* ♀ × F₁ ♂ (**B**), *T. oceanicus* ♀ × *T. commodus* ♂ (**C**), *T. commodus* ♀ × *T. oceanicus* ♂ (**D**), and *T. commodus* ♀ × F₁ ♂ (**E**). Hybrid songs are generally intermediate to their parents in such features as number of sound pulses per trill and number of trills per phrase. (From Bentley, D. R. 1971. Science 174:1139-1141. Copyright 1971 by the American Association for the Advancement of Science.)

are produced by individuals who have never heard them. Traits with practically no environmental variance are ideal for genetic analysis; they are rare in the realm of behavior, but cricket songs meet the requirement perfectly.

Bentley and Hoy made precise measurements of the time intervals between various parts of the song and analyzed the inheritance of each parameter. In general the values obtained for hybrids were intermediate to those of their parents, an indication of additive gene action. No parameter appeared to be controlled by segregation of alleles at a single locus. The differences in the calling songs of the reciprocal hybrids (Fig. 11-2, *C*

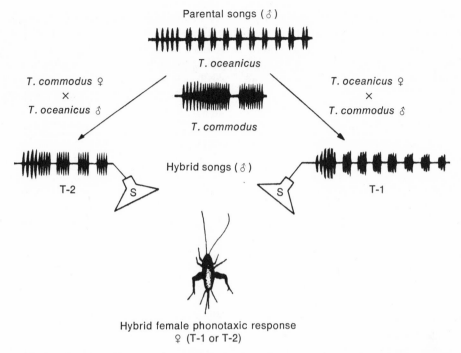

Fig. 11-3. A phonotaxis experiment with hybrid female *Teleogryllus* crickets. Subjects were F_1 hybrids of two types: *T. oceanicus* ♀ × *T. commodus* ♂ *(T-1)* and *T. commodus* ♀ × *T. oceanicus* ♂ *(T-2)*. Male songs of siblings of both types were recorded and played back simultaneously to females. Female choice of song was demonstrated by walking toward one or the other speaker in a single test. A significant majority of positive responses were made to the songs of siblings. The representation of parental songs is for information only; they were not used in the experiment. (From Hoy, R. R., J. Hahn, and R. C. Paul. 1977. Science **195:** 82-83. Copyright 1977 by the American Association for the Advancement of Science.)

and *D*) indicate that some of the genes affecting song pattern are located on the X chromosome (in crickets, males are XO, females XX). But no single feature of the song could be designated as entirely sex-linked; all were multichromosomally and therefore polygenically regulated. Even in this case of extreme genetic determinism there is no isomorphism between the elements of genetics (genes) and the elements of behavior (specific motor patterns).

Males call and females respond. The discriminative response of females is as genetically determined as the song pattern of males. This has been elegantly shown in an experiment in which female F_1 hybrids (*T. oceanicus* female × *T. commodus* male: T_1) and females from the reciprocal cross (*T. commodus* female × *T. oceanicus* male: T_2)

were given a choice of responding to the recorded song of either the T_1 male or the T_2 male (Fig. 11-3) (Hoy, Hahn, and Paul, 1977). Both types of F_1 females showed a clear phonotactic preference (66% to 76%) for the songs of their siblings over those of the reciprocal cross males. Since the differences between the songs of the two F_1s are small compared with those of the parental species, the results are particularly striking and indicate a very precise tuning of the female's receptor-analyzer system. The results also imply that production of song by males and reaction to it by females have a common genetic basis. Complementary selective processes have resulted in producing identical neural rhythmicity in both sexes; how this rhythm is expressed behaviorally is determined by the genes that determine sexuality. Hoy, Hahn, and Paul (1977) speculate that the coupling of female response to male song may be due to either "feature detectors" in the female brain tuned to some specific critical characteristic of the song or to an auditory template against which the pattern as a whole is matched.

The neural circuits underlying the song patterns are not functional at hatching; the elements appear in an ordered, prescribed sequence over the last four molts. Nymphs in the final instar before adulthood can generate nearly perfect patterns of motor discharge for all types of songs, but the neurons whose rhythmic discharge produces the songs are inactive in the nymph unless the mushroom bodies (inhibitory centers) are extirpated.

Hybridization techniques have also been used to investigate the organization of the neural system that generates the rhythms. Most cricket calls consist of a series of evenly spaced chirps; within each chirp is a series of evenly spaced pulses emitted necessarily at a rate faster than the chirp rate. Question: Is there one basic rhythm (the fast sound-pulse rate) that periodically halts briefly because of some intrinsic "fatigue" that produces a refractory period and thus separates the call into chirps? Or are there two interacting centers, one for the pulse rate and the other for the chirp rate? Bentley and Hoy (1972) crossed chirping and nonchirping species of the genus *Gryllus* in an attempt to answer the question. In one of the chirping species they found evidence for two rhythm-generating systems; in another the hypothesis of a single oscillator fits the data better. Two points can be made from these results. First, behaviors that appear to be similar, for example, the generation of chirps, may have a different physiological basis even in closely related species. Second, genetic techniques can be valuable in analyzing complex neurophysiological characteristics.

The cricket research on mating behavior goes beyond the *Drosophila* studies in the elegant analysis of the neurological correlates of the behavior that produces sexual isolation. Differences in sound patterns are easier to quantify than differences in pheromones, although progress in chemistry of the pheromones will surely add to our knowledge of the physiological basis of selective mating and sexual isolation in fruit flies. Both sets of data demonstrate the extent to which sexual isolation has played a role in evolution. For additional examples of the genetics of mating behavior, we turn to the vertebrates. We shall find stereotypy of mating behavior nearly as great as that of the insects just considered, but in some species a somewhat greater influence of early experience on mating preference and, in others, illustrations of the unfortunate consequences of mixing the elements of two different functional patterns in hybrids.

VERTEBRATES
Behavior and sexual isolation

The method of hybridizing closely related species and observing the courtship behavior and mating effectiveness of the hybrids can be applied to vertebrates as well as insects. Stereotypy is characteristic of the mating behavior of most vertebrates just as it is of invertebrates; furthermore, each species has a characteristic courting ritual readily recognizable by an experienced observer. The two examples which follow demonstrate that the species pattern as a whole is critical for the mating success of prospective partners. In hybrids these patterns tend to be disrupted, and the bits and pieces that remain are not very effective.

Interspecific hybrids between the platyfish, *Xiphiphorus maculatus*, and the swordtail, *X. helleri*, can be produced in the laboratory, although they do not interbreed in nature. Fertilization is internal, and the male courtship patterns are highly specialized. In the mating tests swordtails, platyfish, and the F_1 hybrids were tested with females of their own genotype (Clark, Aronson, and Gordon, 1954). Males of other generations were tested with platyfish females who play a rather passive role during courtship. This technique also avoids complications caused by segregation of behavioral traits in both sexes.

Table 11-4. Precourtship patterns of behavior in platyfish, swordtail, and hybrid males*

Genetic group	Number observed	Percentage showing indicated behavior			
		Pecking	Nibbling	Retiring	Backing
Platyfish	33	82	0	73	0
Swordtail	21	0	71	0	95
F_1 (P × S)	5	0	0	20	40
F_2 (F_1 × F_1)	61	39	15	23	67
F_1 × P	10	20	10	10	0
F_1 × S	10	0	40	0	90

*From Clark, E., L. R. Aronson, and M. Gordon. 1954. Mating behavior patterns in two sympatric species of Xiphophorin fishes: their inheritance and significance in sexual isolation. Bull. Am. Mus. Nat. Hist. N.Y. **103**:135-226.

The precourtship behaviors of the two species are clearly differentiated. Platyfish males peck at the sand in the bottom of the test aquarium; swordtail males nibble at the female. In later stages platyfish males approach the female, then suddenly back away with body limp and fins folded, a sequence known as *retiring*. In swordtails the males approach the female by swimming backward and touching her with the tips of their tails. The distribution of these four behavior patterns is shown in Table 11-4. The number of tested F_1s was small because the sex ratio in these hybrids strongly favors females. No simple Mendelian ratios were obtained for the occurrence of any form of behavior in the hybrids, but in the backcrosses and the F_2 there are clear indications of genetic influence on the display patterns.

How effective were the mating displays of the hybrids? These can be evaluated in terms of the percentage of observation periods with each type of male during which the male inseminated a female: platyfish, 80%; swordtail, 39%; F_1, 64%; F_2, 24%; backcross to platyfish, 9%; and backcross to swordtail, 0%. The backcross males were sexually aroused but seldom coordinated the elements of their display into an effective pattern.

Our second example of the inheritance of courtship display in interspecies hybrids comes from the observations of Sharpe and Johnsgard (1967) on two ducks, the mallard *(Anas platyrhyncos)* and the pintail *(Anas acuta)*. These species, though sympatric, sel-

dom hybridize in nature, although the hybrids are fertile. Sharpe and Johnsgard succeeded in rearing 16 pintail-mallard male F_2 hybrids to an age at which they could be intensively studied with respect to plumage characteristics; for 11 of these a comprehensive analysis of courtship display was made from film records.

Rating scales were constructed for both the structural and the behavioral characteristics, with a score of 0 corresponding to the mallard phenotype, and higher scores to the pintail phenotype. The maximum possible scores were 20 for plumage and 15 for courtship behavior. Fig. 11-4 shows the distribution of the two ratings for the 11 males from whom complete data were obtained. It is apparent that plumages range from very good mallard (bird U) to very good pintail (birds D, L, K, and W), with others intermediate. There is a similar wide spread in the behavioral indices. Birds R, V, and F courted like mallards; birds K and W were indistinguishable from pintails in their mating display. The significant correlation ($r = 0.756$) between the two indices indicates either that some genes influence both kinds of traits or that "plumage genes" and "courtship genes" are linked.

All hybrids accurately performed individual display components that are common to the two ancestral species. Display patterns that differ slightly between pintail and mallard were performed by the hybrids in an intermediate form. Display elements characteristic of only one species were exhibited

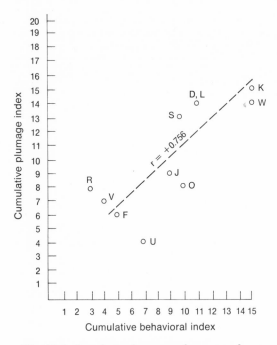

Fig. 11-4. Correlation between plumage and mating-behavior indices obtained for eleven F_2 mallard-pintail hybrids. The lower end of both scales represents the mallard phenotype; the upper end, the pintail. The positive correlation is significant. (From Sharpe, R., and P. Johnsgard. 1967. Behaviour **27**:259-272.)

by some hybrids in perfect form and were completely absent in others. No displays were observed that are not part of the combined mallard-pintail repertoire.

Despite the evidence for the inheritance of complex action patterns, no copulations by the hybrids were observed. As a group they displayed less than did their mallard and pintail male associates in the research area. Hybrids also tended to display to each other rather than to the available mallard and pintail females. The reasons for the lack of sexual competence in these hybrids are not completely known and are undoubtedly complex. It is clear, however, that sexual isolation based on disruption of intersexual communication plays an important role in keeping these two duck species separate. Although the genetic system has not been studied in detail, it seems to be polygenic.

Effects of experience on choice of mate

There was a hint in the *Drosophila* literature that prior experience could modify mate choice (Pruzan and Ehrman, 1974). One might expect that the effects of experience would be greater in vertebrates, where learning plays a more important role than it does in insects. Mating preference in blue-snow geese is a well-studied example; field, laboratory, and theoretical investigations have been coordinated to yield a coherent picture of a complex phenomenon.

Blue and snow geese are color phases of *Anser caerulescens*, a North American species that breeds in Northern Canada. The color dimorphism is so distinct that the two phases were originally classified as separate species, and the birds of mixed parentage and intermediate plumage were considered to be interspecific hybrids. Observations in the field and laboratory have confirmed that there is only one species and that the color dimorphism is due to segregation at a single locus (Cooke and Cooch, 1968). *BB* and *Bb* birds are blue; *bb* are white. However, dominance is incomplete; some *Bb* birds are differentiated from homozygotes by a few white feathers.

Field observers had noted that assortative mating according to color was prevalent on the breeding grounds and hypothesized that exposure to the plumage of its parents might imprint a gosling with a color preference that persisted up to the second year, when pair bonding usually occurs. To test this hypothesis, 300 goose eggs were collected in the wild and hatched in incubators. The young goslings were placed 10 hours after hatching with either blue-, white-, or pink-dyed foster parents for 4 weeks (Cooke, Mirsky, and Seiger, 1972). The young birds were then merged into a large flock and later tested for color preference by observing their approach to white, pink, or blue unfamiliar adults in a choice apparatus. Even though in the mixed flocks the goslings had become acquainted with a range of colors, they showed a clear preference in the test for individuals like their foster parents.

As a follow-up, field and laboratory studies were conducted to learn if these early preferences carried through to adult courtship and mating (Cooke, Finney, and Rockwell, 1976). At a large breeding colony in Manitoba, goslings were marked with color-coded bands that identified the phenotypes of their parents. In later years a number of these banded birds returned to the same breeding area and selected their own mates. From the information on the bands it was possible to determine the relationship between parental color and color of the selected mate. In the eighty-seven pairs of known parentage observed over 2 years there was a strong tendency for geese of white parentage to mate with whites and for those of blue parentage to mate with blues (p < 0.025). Similar results were obtained with incubator-hatched birds reared in confinement with either white, blue, or mixed foster parents. The birds reared by blue-white foster parents mated randomly by color; those reared by matched pairs of foster parents favored the parental color for their mates. When young birds were reared in large mixed flocks where they were exposed to both color phases during development, mating was random with respect to color (Cooke and McNally, 1975).

Selective mating based on imprinting of the parental phenotype has also been reported in pigeons (Warriner, Lemmon, and Ray, 1963). Seiger (1967) used computer simulation to predict the influence of this process on the genetic structure of populations. Given a one-locus, two-allele system with dominance, as in *Anser caerulescens*, there are two possibilities: (1) one allele is eliminated, or (2) the population splits along the lines of the phenotypic difference that is crucial to the imprinting process. Of course, reaching either of these outcomes may take a long time, particularly if imprinting is less than 100% effective and if birds who do not find their ideal mate (one like mother and father) mate with any surplus potential spouses that are available.

Assortative mating in mice. The house mouse provides another example of the influence of early experience on mating preference. Mainardi (1963a) reported that sexually mature female laboratory mice (*Mus musculus domesticus*) that had been reared with both parents approached males of their own strain more readily than males of a distantly related subspecies, *M. m. bactrianus*. Females reared by their mothers alone showed no such discrimination between types of males. Mainardi proposed that infant female mice are imprinted on their fathers and that sexual preference in later life is influenced by this experience. When the same type of female-choice experiment was conducted later with two strains of *domesticus*, an outbred Swiss stock and inbred C57BL, there was a small but significant preference for approach to the alien type of male (Mainardi, 1963b). No such preferences were shown by males given a choice of females. A positive effect of association with parental characteristics was also found when young females were reared with artificially perfumed parents and later allowed to choose between similarly perfumed and ordinary males (Mainardi, Marsan, and Pasquali, 1965). It should be noted that in these studies social attraction rather than actual mating was observed. The results, however, are consistent with those of similar experiments with *Drosophila* in which avoidance of heterogamic mating is characteristic of female choice between own-strain males and those of a different race, whereas heterogamic matings are more frequent when the choice is between very closely related and moderately related males.

Further evidence for assortative mating in mice comes from a series of experiments involving four inbred strains and a heterogeneous stock (Yanai and McClearn, 1972, 1973a,b). Female preference for males of an alien strain was found in two of the four strains tested (Table 11-5). The occurrence or nonoccurrence of choice by a strain of females depends on what the options are. C57BL females mated equally with their own strain or with BALB males but showed clear choice between other combinations. The heterogeneous females mated equally well with all types of males, indicating that sexual

Table 11-5. Female mating preferences in four inbred mouse strains*

Strain of female	Choice of males[†]			
	B vs. C	B vs. D	C vs. D	D vs. C3
BALB *(B)*	0	0	N	N
DBA *(D)*	N	+	+	N
C57BL *(C)*	0	N	+	+
C3H *(C3)*	N	N	N	0

*From Yanai, J., and G. E. McClearn. 1973. Assortative mating in mice. II. Strain differences in female mating preference, male preference, and the question of possible sexual selection. Behav. Genet. 3:65-74.
[†]0, no evidence of female choice; +, female shows preference for alien male; N, combination not tested.

competence was reasonably equal in all strains used.

BALB and DBA females and females from F_1 crosses between these strains were reared with either DBA or BALB foster fathers. Females raised with DBA males had a strong bias against other DBA males in a preferential mating test. No such tendency for negative assortative mating was shown by the females reared with BALB males. The female's own genotype had little influence on choice of mate. Even in these simple mammals, the complexities of mating choice are fascinatingly complex. Choice appears to be lawful, but neither genetic nor environmental determinism provides an adequate explanation.

Variability of sexual competence

Sexual competence is certainly one of the major components of Darwinian fitness, and it must be favored by natural selection in all vertebrates. By now, however, one should not be surprised to learn that considerable variation in the vigor and patterning of sexual behavior can be found within a species.

The genetics of mating behavior in chickens has drawn considerable attention because of its importance to commercial breeding. Wood-Gush (1958, 1960) demonstrated the heritability of mating ability in cockerels and selected high– and low–sexual activity lines. He found that in a standard test the high strain achieved twice as many copula-

tions as the low strain. Raising the cockerels in single-sex or mixed-sex flocks had absolutely no effect on mating scores. Administration of exogenous testosterone did not improve mating success in either the high or the low line. A search for behavioral and physiological correlates of the selection program failed to detect evidence of hormonal insufficiency in the low-mating line; neither did the lines differ in aggressiveness.

In another long-term program of selection for male mating behavior, cockerels were tested eight times with different pullets selected from a standard outbred control stock (Siegel, 1965, 1972). The use of several randomly chosen individuals for evaluating sexual behavior is particularly appropriate for a polygamous species like the domestic fowl. As a criterion of selection, the Cumulative Number of Completed Matings (CNCM) was employed; other behavioral and physiological measures were taken from time to time. Phenotypic changes were slow in the early generations, but a good response was evident by S_4, and it continued through S_{11}, when a major report was published (Siegel, 1972). Over the eleven generations the CNCM increased in the high line from 6.4 to 15 and fell in the low line from 6.4 to 3.2. This appears to reflect a better response for the upward selection, but when the data are standardized by taking account of differences in the variances of high and low scores and differences in the intensity of selection in the two directions, realized heritability upward is 0.16 ± 0.02 and downward, 0.32 ± 0.07. The considerable heritability evident after eleven generations of selection indicates that much additive genetic variance remained. High CNCM scores were correlated positively with volume of semen; as in the study by Wood-Gush (1960), no association was found between mating ability and aggressiveness.

All possible crosses between the two selected lines and the control line were made in order to analyze the genetic system regulating male sexual behavior. (Cook, Siegel, and Hinkelmann, 1972). The results were interpreted as indicating that two different genetic systems were involved, one in the high-line selection and the other in the low-line selection. Progress towards high CNCM scores was dependent on numerous genes, each with small additive effects. Selection for low scores seemed to involve a smaller number of genes, generally with dominant effects and some of them sex-linked.

Genetic analysis of guinea pig sexual behavior. The stereotyped behavior of the male guinea pig is particularly suitable for genetic studies, since reproductive behavior of individuals is qualitatively and quantitatively consistent over long periods of time (Grunt and Young, 1952, 1953). The sexual activity of male guinea pigs varies greatly among individuals. By the use of a rating scale, males were classified as high, medium, and low drive, then castrated and retested. All subjects became sexually inactive within a short time. Injections of testosterone propionate restored sexual behavior in every individual, but only to the level characteristic of that individual preoperatively. Amount of hormone was not the critical factor in producing individual differences; instead, it was the somatic response to the hormone. Variation in the sensitivity of the somatic structures to androgenic hormones could be due to environmental or genetic factors or to a combination of the two.

The importance of genetic factors was shown in a comparative study of male sex behavior in three strains, two inbred and one outbred (Valenstein, Riss, and Young, 1954). On the rating scale both inbred strains scored lower than the heterogeneous stock. A more detailed comparison between the behavior of the two inbred strains was made by comparing the frequency of specific elements of courtship and mating. The results, summarized in Table 11-6, show a number of significant differences between the strains. In general, strain 2 devoted more effort to preliminary courtship, whereas strain 13 spent more time in behavior closely related to copulation.

In other experiments the sexual behavior of females was investigated (Goy and Young, 1956-1957). Subjects from the two inbred lines and the heterogeneous stock were spayed and brought into heat by injections of estradiol benzoate followed by progesterone. This procedure eliminated problems due to asynchrony in the normal estrus cycle. The indicator of estrus was lordosis (arching of the back when the animal was clasped). Table 11-7 summarizes the major findings on five measures. Significant strain differences were found in every category of behavior except the percentage of individuals brought into estrus.

In an experiment in which the dose of estradiol benzoate was varied, the larger doses produced a quicker and longer-lasting response but did not alter the vigor of the be-

Table 11-6. Comparison of inbred strains of guinea pigs on separate components of male sexual behavior*

Group	Frequency of observed behavior per minute						
	Sniffs, nibbles	Nuzzles	Abortive mounts	Mounts	Intromission	Ejaculates	Sum of all scores
Family 2	0.797	1.089	0.465	0.162	0.107	0.007	2.627
Family 13	0.839	0.731	0.287	0.346	0.174	0.019	2.396
t	0.220	2.840	2.24	2.630	1.300	2.500	1.010
p	—	0.010	0.03	0.020	—	0.020	—
Difference favors	—	Family 2	Family 2	Family 13	—	Family 13	—

*From Valenstein, E. S., W. Riss, and W. C. Young. 1954. Sex drive in genetically heterogenous and highly inbred strains of male guinea pigs. J. Comp. Physiol. Psychol. **47**:162-165. Copyright 1954 by the American Psychological Association. Reprinted by permission.

Table 11-7. Strain differences in responses of female guinea pigs to estradiol
benzoate plus progesterone*

Strain	Number	Mean score on tests of sex behavior				
		(1)	(2)	(3)	(4)	(5)
Strain 2	15	3.4	7.2	15.9	1.0	97.8
Heterogenous	21	5.9	4.3	11.7	7.3	96.8
Strain 13	17	5.7	5.0	25.2	37.7	90.2

Description of measures
  (1) Time in hours between progesterone injection and first lordosis
  (2) Number of hours during which lordosis was elicitable
  (3) Duration of maximum lordosis in seconds
  (4) Number of malelike mounting attempts
  (5) Percentage of females brought into heat by hormones

*From Goy, R. W., and W. C. Young, 1956-1957. Strain differences in the behavioral responses of female guinea pigs
to alpha-estradiol benzoate and progesterone. Behaviour **10:**340-354.

havior itself. These results were interpreted
as signifying that, as in males of this species,
individual differences in sexual behavior depend on the somatic response to hormones.
Presumably the structure involved is the
brain.

The differences between the females are
more striking than those found in the males
of the same strains. Interestingly, the measures of vigor (duration of estrus, duration of
lordosis, and mounting by the female) tend to
be higher in the inbred than in the random-
bred females. There may be compensatory
selection for sexually receptive females in
stocks whose males are lacking in sexual
vigor.

A set of classical Mendelian crosses between strains 2 and 13 provided subjects for
an analysis of the mode of inheritance of sexual behavior in males (Jakway, 1959) and females (Goy and Jakway, 1959). F_1 females
were like strain 2 in having a short latency
and long duration of estrus. They were intermediate to the parental strains in the duration of lordosis responses and in the number
of mounts. Hypotheses for the mode of inheritance of each element of mating behavior
were proposed; all are regarded as tentative.
Goy and Jakway concluded that "the vigor of
the mating patterns as we have come to view
it is not inherited as a unitary trait, but rather

as a separate set of genetic factors for each of
the genetically independent behavioral elements."

Among the male groups the F_1 was clearly
superior in mating competence to its inbred
parents and to the other hybrid groups. The
F_1s were much less variable than other
groups. One would expect from classical genetic theory that the F_1 would be more homogeneous than the F_2 and backcrosses because of genetic segregation in the latter
groups. This explanation cannot apply to the
greater phenostability of F_1s relative to their
inbred parents, since all three groups are genetically homogeneous. The F_1 is, however,
the most heterozygous of the tested groups,
with one strain-2 allele and one strain-13
allele for every locus at which the two strains
differ. Heterozygosity seems to lead to increased vigor, and perhaps the sexual success
of vigorous males is relatively independent of
random environmental factors. Jakway's
careful analysis demonstrates the inadequacy
of classical quantitative genetics for measuring the genetic parameters of complex behavior. Since genotype strongly affects developmental homeostasis (that is, the degree to
which environmental perturbations are
modulated so that development proceeds
along a predetermined course), and since
this effect of a genotype on variation cannot

be distinguished from the effect of having multiple genotypes within a group, there is no clear rule for assigning variance to the environmental or genetic categories in experiments with inbred strains. This judgment may be too harsh for behaviors that do not show the strong heterotic and homeostatic characteristics that are found in the sexual performance of male guinea pigs. Nevertheless, caution should be observed in applying simple genetic models to behavior.

Interaction of genetic and experiential factors. Guinea pigs of different genotypes respond differently to sex hormones as measured by the activation of courtship and mating behavior. Do they also react differently to experiential factors? Valenstein, Riss, and Young (1955) reared males of strains 2 and 13 and an outbred stock under two conditions: (1) isolation at the age of 25 days and (2) rearing with a group of females. At the age of 77 days a series of tests with estrous females began. The percentage of males ejaculating during the tests were strain 2: isolates 6%, group reared, 84%; strain 13: isolates 0%, group reared, 57%; and heterogeneous stock: isolates, 71%, group reared, 100%. Conditions of rearing had a much greater effect in the inbred than in the outbred groups. However, when heterogeneous males were isolated at 10 days of age, their performance as adults was significantly inferior to that of socially reared males. The difference between heterogeneous and inbred animals was not in their need for social experience but in the age range during which the experience was effective.

The possibility that the performance of strain 13 males might be improved by providing them with large doses of an androgenic hormone during early development was also tested. No effect of the hormone was observed in an isolated group, but a small improvement was found among the social-reared subjects. The sexual composition of the social group in which males were reared had no effect on their later mating performance.

The experimenters believe that their results accord with the hypothesis that sexual competence involves two semi-independent processes, arousal and organization. The effective organization of sexual responses is dependent on social experience during certain periods of development. These experiences are of a general nature and are not specifically sexual. Strain differences in the organizational process are related to the rate at which animals mature enough to profit from association with other animals. In this sense males of the heterogeneous stock mature earlier than inbred males. The arousal component, as shown overtly by sexual responses in the presence of an estrous female, is not as much affected by experience. Isolated males were obviously aroused in the test situation but did not organize their responses into a coherent, effective pattern. Strain 13 appeared to be genetically deficient in the arousal component, even under optimal social and hormonal conditions.

Mating behavior in the house mouse: genetic aspects. Most genetic research on the sexual behavior of *Mus musculus* has been concentrated on inbred strains and their hybrids. Males have received the most attention, probably because of their more active role in courtship. McGill (1962) observed C57BL, BALB, and DBA males placed with same-strain females in hormonally induced estrus. Sixteen quantitative measures of frequency, latency, and duration of specific behaviors were obtained, and from these a description of the characteristics of each strain was constructed. C57BL males mate rapidly, have a high frequency of intromissions, and, on the average, reach intromission in one third the time required by BALB males. DBA males have difficulty in gaining and continuing intromission, but once it is achieved, they ejaculate quickly. BALB males are slow to mount, require many intromissions before ejaculation, and spend much time rooting (burrowing under the female and lifting her). The same type of analysis of the sexual biogram was made on hybrids between the three strains, and a genetic analysis was carried out on each of the sixteen measures (McGill and Blight, 1963b). The results of the analysis are too complex to summarize briefly. The flavor of the findings is contained in two principles enunciated by McGill (1970): (1) "Conclusions regarding such genetic parameters as mode of inheritance, heritability or degree of genetic determination are specific to the strains studied," and (2) "Genetic conclusions reached in any particular experiment are specific to the total environmental conditions of that experiment."

McGill's second principle was based on the outcome of experiments on the duration of the postejaculatory refractory period for male sexual activity. Male mice of different genotypes who had recently ejaculated were retested with estrous females at regular intervals until a second ejaculation was observed (McGill and Blight, 1963a). The median elapsed time before a second ejaculation occurred was 96 hours for C57BL males and only 1 hour for DBA and C57BL × DBA hybrids. Apparently, there was an extraordinarily large strain difference in an important aspect of sexual competence. The finding was unexpected, since C57BL males reproduce well in breeding colonies. This fact provided a clue for a repetition of the experiment with one modification (McGill, 1970). This time the mating tests were conducted in the male's home cage instead of in a special observation chamber. Now the strain difference in the duration of the refractory period disappeared. The C57BL males in their own familiar territory returned to courtship within an hour after copulation, just as the DBA males did. How important environment is. But environmental determinism is not the answer. There is a real difference between the sexual behavior of DBAs and C57BLs in the observation chamber; the difference must be genetic, and it needs an explanation.

In another series of experiments a search was made for genetic influences on the retention of the ejaculatory reflex after castration (McGill and Tucker, 1964; McGill and Haynes, 1973). In the first study conducted in the glass observation chamber, 0/11 C57BL males, 3/11 DBA males, and 9/12 F_1 males ejaculated at least once after castration. One hybrid male was still showing the

reflex 60 days after surgery. In the second study F_1 and F_2 hybrids maintained an ejaculatory reflex much longer than did the inbred parents. The subjects were tested in their home cages, and the C57BL disadvantage found in the first experiment was no longer seen.

McGill and Haynes raised a question for physiological psychologists. It is generally taught that the sexual reflexes of males are more hormone dependent in rodents than in carnivores and primates. There is even a name for this phenomenon: *cerebralization.* The term implies that the cortex of larger-brained species has taken over the role performed by hormones in smaller-brained species. But from a geneticist's point of view, the data base on which the generalization is built is flawed. Researchers on sexual behavior after castration have used outbred stocks of primates and carnivores and inbred or partially inbred stocks of rodents. Are the differences found attributable to family and species characteristics or to the differences in amount of inbreeding? The difference in hormone dependency between hybrid and inbred male mice in McGill and Haynes's study is as great as that between different orders of mammals. Certainly caution is advisable in promulgating laws based on small, unrepresentative samples of a species.

The same potential risk of overgeneralizing from results with a limited sample of the genotypes of a species is demonstrated by an experiment on the modification of the sexual and agonistic behavior of female mice by neonatal injections of androgen or estrogen (Vale, Ray, and Vale, 1973). Ten of nineteen strain comparisons of behaviors in which any hormonal effect was detected showed a genotype-environment interaction. That is, the direction or intensity of the hormonal influence differed significantly among the strains tested.

Genetics and sexual behavior in animals: summary

The genetics of sexual behavior is almost coextensive with population genetics. Every Mendelian population is maintained in equilibrium or changes its composition depending on the mating choices of its members. Behavioral isolation is one of the most important factors in speciation. The opposing tendency for individuals to select mates that differ slightly from themselves serves to maintain diversity within populations. In vertebrates early experience may be influential in later choice of mates. The genetic control of courtship vigor and female receptivity seems to operate primarily on somatic sensitivity to hormones (and possibly to pheromones) rather than on the supply and composition of the hormones. Probably the most important organ mediating these genetic effects is the brain. Most of these principles apply equally well to insects and to vertebrates of several classes.

CARE GIVING AND CARE SOLICITING

Although care giving and soliciting are crucial to the survival of many vertebrates and invertebrates, they have not been subjected to genetic analysis in the same degree as sexual behavior. Some of the maternal effects on temperament and social behavior that are discussed in this and the previous chapter are probably attributable to differences in the style of maternal care. For the most part, however, experimenters have been content with demonstrating that the individual playing a parent role modifies the behavior of the developing organism; just how the modifications are accomplished is seldom explained. Indeed, the analysis of the nurturant relationship is difficult, since the processes of observation and measurement are likely to disturb the very function that is being observed.

The examples we have chosen for this section deal more with care giving than with care soliciting. The first is concerned with

hygienic behavior in honeybees, the second with broodiness in domestic chickens, and the third with maternal care in several mammalian species.

Hygienic behavior in honeybees

In honeybees, as in many other hymenoptera, the larvae are cared for by sterile females, the workers. The discovery of genetic variation in the nature of care giving in bees came about by chance in the course of investigations on the nature of strain differences in resistance to a contagious disease, American foulbrood, (Rothenbuhler, 1958; 1964a, b; 1967). In the resistant Brown line, worker bees uncovered the cells of diseased larvae and removed them from the hive. Rothenbuhler named this *hygienic behavior*. In a susceptible line, Van Scoy, the workers left inoculated larvae in place, and the disease spread quickly through the hive. When the two lines were crossed, the F_1 workers were nonhygienic. Backcrosses of F_1 queens to Brown males yielded four types of colonies, each descended from a single queen. In approximately one fourth of these colonies, workers showed the complete hygienic pattern; in one fourth uncapping was seen, but the larvae were not removed; and in one half neither component of hygienic behavior was apparent. However, in half of these completely nonhygienic colonies, the workers removed larvae after an experimenter had removed the cap. Thus there were in all four phenotypic classes that appeared in approximately equal frequency in the backcross to the Brown line. The data are compatible with a dihybrid system in which uncapping and removal are disrupted by two independently assorting dominant genes. Rothenbuhler's hypothesis can be illustrated as follows:

	Queens and workers	Males
Van Scoy line, nonhygienic	*UURR*	*UR*
Brown line, hygienic	*uurr*	*ur*
F_1, nonhygienic	*UuRr*	
F_1 female × brown male	¼ *UuRr* (nonhygienic)	
	¼ *Uurr* (remove only)	
	¼ *uuRr* (uncap only)	
	¼ *uurr* (hygienic)	

The most intriguing features of the Rothenbuhler experiments are the apparent single-locus control of specific behavior patterns and the clear separation of genetic control of two elements of a pattern that is functional only when both are present. In the case of cricket songs the discrete characteristics of the calls were also found to vary independently, but each element was under polygenic control; no segregating characteristics were detected in hybrids (Bentley and Hoy, 1972). Another puzzling feature of the bee data is that the double-recessive genotype is the more fit. Almost universally, genes promoting fitness are dominant, or additive if selection favors an intermediate phenotype. For these reasons it is preferable to consider U and R as genes that disrupt a behavior pattern rather than to designate their alleles, u and r, as genes that organize uncapping and removal, respectively. Many instances are known of point mutations that disorganize the development of somatophenes and psychophenes. No gene has been demonstrated to direct a complicated series of motor responses that produce an integrated behavior pattern. The very existence of such a "master gene" seems incompatible with our knowledge of gene action and neurophysiology.

Broodiness in chickens

Economic interest has stimulated much genetic investigation of broodiness in the domestic fowl. Hens who do not take time out from laying to incubate their eggs return more money to their owner. Selection against broodiness has been practiced for years, and certain breeds, White Leghorns, for example, are known as nonbroody. Actually, if proper conditions are established, most fowl of nonbroody strains can be induced to incubate eggs (Burrows and Byerly, 1938). There is no doubt, however, that the threshold of stimulation necessary to produce broody behavior varies tremendously among breeds. Many birds that are nonbroody during their pullet year become broody later. This creates a problem when geneticists try to fit individuals into a di-

chotomous classification. Goodale, Sanborn, and White (1920) noted that the strength of the "broody character" might vary in two ways: (1) in the frequency of broody episodes and (2) in the duration of the episodes. Obviously, broodiness seems to be best considered as a quantitative character, and one would expect it to be controlled by many genes (Lerner, 1950).

The heritability of broodiness is indicated by its response to negative selection over an 18-year period (Hays, 1933, 1940). The average number of broody episodes per broody individual fell from 3.5 to 1.1, and the percentage of broody fowl in the stock was reduced from 86 to 5.

Hays reported no evidence for sex-linked genes affecting broodiness, but several investigators have found the contrary. In fowl the males are homogametic (ZZ), and the females heterogametic (WZ). Females must receive their Z chromosome from their sire. If this chromosome carries factors affecting broodiness, the genotype of the sire will have a detectable influence on the offspring. The data of many investigators as summarized in Table 11-8 support the hypothesis of Z-linkage. Autosomal factors are also involved. If all broodiness-enhancing genes were Z-linked, broodiness in dams and daughters would be uncorrelated, since the female's Z chromosome goes only to her sons. Kaufman (1948) found that 67% of the daughters of broody mothers were themselves broody; only 31% of the daughters of nonbroody mothers were classified as broody.

It has been found repeatedly that no one genetic mechanism accounts exclusively for a particular kind of behavioral variation. Apparently broodiness in some breeds is not sex-linked. It seems certain that selection against this trait has acted on different sets of genes in various breeds.

Maternal behavior in mammals

Scattered observations have been made on breed and strain differences in the care-giving behavior of mammals. Ressler (1962) fostered BALB/c and C57BL/10 mice on male-female pairs of either their own or the alien

Table 11-8. Results of reciprocal crosses between broody and nonbroody breeds of fowl*

Dam	Sire	Percent broody offspring	Reference
Br. Leghorn	*Langshan*	29	Punnett and Bailey, 1920
Langshan	Br. Leghorn	50	Punnett and Bailey, 1920
Wh. Leghorn	*Cornish*	88	Roberts and Card, 1934
Cornish	Wh. Leghorn	37	Roberts and Card, 1934
Rh. Id. Red	*Plymouth Rock*	40, 39	Knox and Olsen, 1938
Plymouth Rock	Rh. Id. Red	12, 46	Knox and Olsen, 1938
Wh. Leghorn	*Plymouth Rock*	42	Knox and Olsen, 1938
Plymouth Rock	Wh. Leghorn	12	Knox and Olsen, 1938
Leghorn	*Greenleg*	78	Kaufman, 1948
Greenleg	Leghorn	0	Kaufman, 1948

*Broody breeds are identified by italics.

strain. On 10 consecutive days the foster parents were briefly removed from the breeding cage, and the pups shifted to the part of the cage farthest from the nest. The parents were then returned, and their time spent in handling and retrieving the pups was recorded. Both strains of pups were more frequently attended by BALB than by C57BL adults. Both strains of adults handled BALB pups more than C57BL pups. These differences may be genetic. They may also, as Ressler suggests, represent transgenerational influence such that the foster parents respond to their adopted pups in the same manner that their parents responded to them.

Differences in the time of building a nest have been observed in races of domestic rabbits (Sawin and Curran, 1952; Sawin and Crary, 1953). Females of race X characteristically prepare a nest before giving birth, whereas other races commonly defer nest building until after parturition. Cannibalism and scattering of the young is more prevalent in race IIIc than in other laboratory stocks. In rabbits, as in other mammalian species, inbreeding decreases the quality of parental care.

Scott and Fuller (1965) observed in dogs that basenji and cocker spaniel mothers spent more time nursing their pups than did their F_1 hybrid daughters. At first thought, this fact might be taken to indicate better care by the purebred dams; more likely the difference is attributable to better milk production of the hybrids. Their pups were more quickly satisfied, ceased sucking (care soliciting) and released the bitch for other activities. In another study, basenji dams were more effective than females of other pure breeds in retrieving puppies taken from the nest. The basenji is more primitive than the other dogs with which it was compared. Its superior performance may be due to less inbreeding or to the absence of mutations that have been perpetuated in most dog breeds to achieve conformity to arbitrary physical standards.

Summary

Although care giving and care soliciting are extremely important in many species and play a major role in the sociobiological approach to the evolution of cooperative behavior and altruism, research on their genetics is scanty compared with that focused on sexual and agonistic behavior. One reason may be that sexual and agonistic behavior are more important in determining which genes are transmitted from one generation to the next. Furthermore, sex and aggression have a peculiar fascination for humans, and science, like literature, is a characteristic human endeavor.

But care giving and care soliciting are more than the means of ensuring that a zygote matures and reproduces in its turn. The communication between giver and receiver provides an extragenetic transmission channel between parent and offspring, a channel whose efficiency is probably influenced by the genotypes of both participants. The extension of nurturant behavior beyond parent and offspring is the basis of complex social organization. It has two peaks, the social insects and human beings, and is worthy of more extensive genetic study.

AGONISTIC BEHAVIOR

Although animals may live together peacefully for long periods of time, they often engage in a considerable amount of competition and even vigorous fighting. Sometimes displays such as the spreading of the peacock's fan, the wren's song, or the wolf's growl and bared teeth suffice to repel an intruder. But display is not always enough. Lethal violence among wild animals is more common than some popular and scientific writers have implied (Wilson, 1975). Wild animals, particularly males, frequently bear scars from past combats. Fighting is not the major occupation of any animal, but most vertebrates and many invertebrates are equipped to battle with conspecifics for mates and with their own or alien species members for physical resources and food.

Fighting is a commonplace occurrence in our domestic animals. Two strange dogs meeting on the street determine their relative status by posture, threats, possibly by pheromones, and, if these are inconclusive, by fighting. Playful fighting during puppyhood can be viewed as a preparation for the challenges of adulthood. In most dog breeds such play stops short of producing serious injury, but in wirehaired fox terriers it can be violent enough to cause death of subordinate members of a litter (Fuller, 1953). In domestic fowl of both sexes a pecking order is established through confrontation and combat. Growth rate and mating success are dependent on a bird's ranking in the hierarchy of its flock. Common house mice, both the laboratory and the wild variety, are not "mousy" in the sense of the dictionary definition: quiet, timid, drab. Gram for gram, male mice fight as fiercely as lions, a trait that

must be considered in the rearing and housing of this important experimental species.

Although there are articles on genetic effects on aggressive behavior in a variety of vertebrate species, only two, the domestic chicken and the house mouse, have been studied intensively enough to permit evaluation of the mode of inheritance, the influence of early experience, and the nature of the stimuli that elicit and inhibit fighting. Our discussion will center on these two very different species.

Domestic fowl

Breed differences in the fighting ability of cocks have been recognized for centuries. In ancient Greece the breeding of gamecocks was an important pastime; the birds from Rhodes and from Tanagra in Boeotia were especially admired. Plato reprimanded the Athenians for their excessive fondness of the cockpit. Scientific verification of these breed differences in both gamecocks and ordinary poultry was obtained in the mid-twentieth century (Fennell, 1945; Potter, 1949; Allee and Foreman, 1955).

In flocks of chickens the male and female dominance orders are separate. The highest individual in either hierarchy is designated as *alpha* (α), the second as *beta* (β), and so on. Birds at the lower end of the scale are often grouped together as *omegas* (ω). Being an alpha has its rewards. In flocks with 3 cockerels and 30 to 36 pullets, high-ranking females laid more eggs; α cockerels mated more frequently (Guhl and Warren, 1946). β males were more active in courting than were α males, but they were less successful in copulation; ω males were sexually inhibited. Significant positive correlations between sexual effectiveness and social aggression ($r = 0.32$, $p < 0.01$) were also found by McDaniel and Craig (1959).

To determine the heritability of social aggressiveness in hens an experiment was carried out over 2 years so that the correlation between mothers and daughters could be determined (Komai, Craig, and Wearden, 1959). Pullets of six strains, three of White Leghorns and one each of White Rock, Black

Australthorp, and Rhode Island Red, were placed in mixed flocks. The status of an individual bird, X, in a flock was determined by the formula

$$R_X = \frac{A + B}{2}$$

where

R_X = indexed rank of bird X
A = percentage of birds dominated by X
B = 100 − percentage of birds that dominate X

Significant mother-daughter correlations in this index were found, and the average heri-

tability for the six strains was 0.34, high enough to predict that selection for social rank should be successful. Evidence for the heritability of rank through sires was also obtained (Tindell and Craig, 1960).

On the basis of these encouraging results, selection for social dominance was initiated in two breeds, White Leghorns and Rhode Island Reds (Guhl, Craig, and Mueller, 1960; Craig, Ortman, and Guhl, 1965). Selection was based on the outcome of contests between a candidate male and a panel of randomly selected opponents, initially from the outbred foundation flock and later from with-

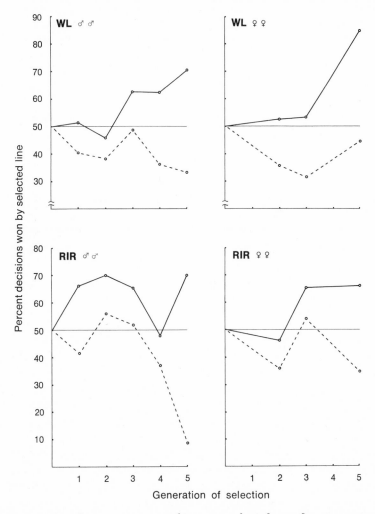

Fig. 11-5. Percentages of pair contests won by strains selected over five generations for high (*solid line*) and low (*dashed line*) social dominance when matched against a series of unselected birds of the same sex. *WL*, White Leghorn; *RIR*, Rhode Island Red. (From Craig, J. V., L. L. Ortman, and A. M. Guhl. 1965. Anim. Behav. 13:114-131.)

in the candidate's own line. The success of the selection procedure is demonstrated by Fig. 11-5. The symmetry of response in both the upward and downward direction and the persistence of high intrastrain variability suggest a polygenic mode of inheritance. The social status of females in the selected lines changed in parallel with the males, although the dominance of the high-line pullets was more apparent in initial encounters than in stabilized flocks.

The expected effect of inbreeding on characteristics related to fitness was found by Craig and Baruth (1965). Five partially inbred lines of White Leghorn fowl derived from an outbred stock were tested against the foundation birds. The mean percentage of inbred-outbred contests won by the inbreds was 28 for a moderately inbred and 18 for a highly inbred group. The disadvantage of the inbred birds was most evident in early encounters; with experience they competed more evenly. As in the selection experiments, a correlated response was found in females, where in mixed flocks the more inbred pullets were disproportionately in the lower ranks.

As with other behavior related to hormones, one can inquire as to whether low-ranking breeds or lines selected for social subordination are deficient in testosterone. Castration of cockerels produces capons that are not only more tender to eat but are less aggressive. Ortman and Craig (1968) used testosterone replacement therapy for capons and testosterone supplementation for pullets to see whether the social status of the low-line birds could be made equal to that of the high line when all individuals had the same amount of male hormone. The results, which by now are expected in such experiments, yield clear evidence that selection has operated on the physiological responsiveness to androgen, rather than on the capacity of the endocrine system to produce the male sex hormone. When heredity, degree of crowding, and stability of flock membership were simultaneously varied, each was found to have an effect on the frequency of social interactions (Craig, Biswas, and Guhl, 1969).

High- and low-dominance strains of White Leghorns differed in the frequency of social reactions as judged by the average distance between neighbors; in Rhode Island Reds similarly selected for dominance, no strain difference in the number of social interactions was found.

Mice

Intermale fighting and competition in the house mouse have been favorite topics for behavior-genetic analysis. The appropriate behavior can be reliably elicited and quantified. Male mice fight vigorously, but they are small enough so that an experimenter can separate combatants in order to prevent serious injury or to delay the establishment of a dominant-submissive relationship within a pair. Female mice rarely fight except in defense of a new litter. However, they can be made aggressive by a combination of perinatal and adult injections of androgens. As with sexual behavior, the agonistic behavior of mice is influenced by prior experience and by the immediate circumstances under which animals meet. Thus there has been marked interest in studying gene-environment interactions with relation to social dominance. Finally it has been possible to deduce something of the social organization of feral mice by combining our knowledge of the agonistic behavior of pairs with surveys of the distribution of genetic polymorphisms within and between localized populations.

Intermale fighting of mice has been described frequently (Scott and Fredericson, 1951; Lagerspetz, 1961). When two strange males are introduced to each other on neutral ground, they typically investigate each other, tentatively at first but more vigorously later. One or both may circle near the opponent with humped back and short rapid steps—*mincing*, in Scott and Fredericson's terminology. Tail rattling is also very common; it indicates a high level of arousal and possibly has a communicative role between the opponents. When two mice attack each other, they wrestle, scratch, and bite for a few seconds, then break away to resume again until one or the other achieves dominance. A defeated animal will escape if possible, but in the small enclosures commonly used for testing, it assumes a submissive posture, standing erect, holding its forepaws towards the aggressor, and remaining motionless. If attacked, the submissive mouse squeaks and jumps away instead of fighting back.

Several methods are used for quantifying agonistic behavior. When individuals of genotype X are matched pairwise with individuals from genotype Y, a score for genotype X can be based on the percentage of encounters won by X mice. In many experiments encounters are arranged among individuals of similar genotypes, and the designation of a winner has no genetic significance. In these cases the intensity of fighting may be graded on a scale with arbitrary divisions ranging from the absence of agonistic behavior to the most intense fighting (Lagerspetz, 1961). Alternatively, latency to attack or to achieve dominance or the proportion of time spent fighting serve as quantitative measures. Indirect evidence of agonistic interaction can be obtained by counting wounds on males that are housed together or the number of offspring sired by each competing male when two or more have access to the same estrous females. Each of these procedures has advantages and disadvantages; the existence of so many methodological variants in procedure and in treatment of data makes it difficult to compare results from different experiments (Fuller and Hahn, 1976).

Strain differences. Lagerspetz and Lagerspetz (1974) tabulated fourteen studies, in each of which from two to fourteen strains of mice were ranked with respect to aggression. Because the techniques of evaluation differ so much among the experiments and because the choice of subjects was so idiosyncratic, little consistency appears in the table. For example, strain C57BL/10J males have been described as more aggressive than BALB/c males by Ginsburg and Allee (1942), Beeman (1947), Fredericson and Birnbaum (1954), and Bauer (1956). King (1957) found the reverse order, and Scott (1942) described C57BL/10 as "pacific" in comparison with other strains, including BALB/c. Southwick

and Clark (1968) placed BALB/c among the high-aggression strains; they did not observe C57BL/10 males, but the related strain, C57BL/6 was definitely less aggressive than BALB/c. Perhaps some of the confusion has been cleared up by Porter (1972), who found that the relative social status of males of these two strains could be altered by early handling that enhanced dominance in BALB/c males but had no effect on C57BL/10 males.

It seems safe to conclude that genotype affects success in competition, but the strength of the effect and even its direction is dependent on the testing procedure and the nature of the criterion chosen as a measure of aggression. One problem in interpreting strain rankings is that some are derived from observations of direct competition between, for example, representatives of strain X and strain Y (X_1 versus Y_1, X_2 versus Y_2, etc.). Other rankings are based on the relative intensity of encounters between X_1 and X_2, Y_1 and Y_2, etc. Can we assume that if strain X males fight more among themselves than strain Y males do, that strain X males will win in competition with strain Ys? Another characteristic of intrastrain competition is that its outcome has no genetic significance. All variance between the members of an X_1/X_2 pair is environmental; whichever one succeeds in establishing dominance or mating with a female makes no difference in the genetic composition of the next generation. The value of inbred strains for aggression research is the reduction of genetic variance, so that experiments using treatments such as hormone administration, variations in early experience, and the like are more efficient. Even here the issue of generalizing from one strain to the species as a whole must be faced.

Competitions between animals from different strains can serve as models for competitions between individual members of a heterogeneous population. An example of the complexity of strain-situation interaction is found in an experiment of Fredericson and Birnbaum (1954). Like-strain pairs of food-deprived mice were observed after being given a single block of laboratory chow. BALB/c mice of both sexes shared the food without fighting; C57BL/10 mice fought vigorously to maintain individual possession. One would predict that with interstrain pairings, the C57BL males would be dominant; actually, the BALB male killed his C57BL opponent in eight of ten pairings.

One advantage of the food competition test is that it can be used with both males and females and that social position may be determined even if fighting does not occur. Manosevitz (1972) compared the success of food-deprived females of strains AKR/J and DBA/2J and their F_1 hybrid in competing for access to wet mash. As shown in Table 11-9, the DBA females were generally dominant over AKR females. More interesting is the strong heterotic effect demonstrated by the overwhelming superiority of the hybrids over both inbred lines. Success in this type of food competition was associated with lower emotionality in an open field and greater activity in running wheels (Manosevitz, Fitzsimmons, and McCanne, 1970).

Strain dominance in one form of competition does not ensure superiority under other conditions. Lindzey, Manosevitz, and Winston (1966) compared two mouse strains on three tests: food competition, spontaneous intermale fighting, and tube dominance. In the latter test two mice from different strains were started forward at opposite ends of a tube too narrow for them to pass each other. The one that retreated was considered subordinate. In thirteen matched pairs the DBA member surpassed the A member 12 times on food dominance, 12.5 times on the number of fights started (one tie), but only once on tube dominance.

Male competition for mates was studied by Levine and associates in the ST (albino) and CBA (agouti) strains. A single female ST mouse was introduced into a cage containing a pair of males, one from each strain. In this mating system the offspring were the same color as their fathers. In 100 litters, 76 were sired by an ST male and 12 by a CBA male; the remaining 12 were of mixed parentage (Levine, 1958). The conclusion seems obvious; ST males have proved their superior fitness by the crucial test: passing on their genes to another generation. But the matter is not that simple. In paired contests between adult ST and CBA males, some with prior social experience and others isolated since weaning, CBAs were superior fighters (Levine, Diakow, and Barsel, 1965). When similar paired encounters were arranged with an estrous female in attendance, CBA males were less aggressive, and the ST partner achieved dominance (Levine, Barsel, and Diakow, 1965). What is the explanation? Sexual distraction? A pheromone? Whatever the answer may be, genetics is involved.

Another possible explanation for the success of ST males in the mating competition is preference for homogamic partners by the ST females. In noncompetitive mating tests, ST males were nearly twice as successful as CBA males in achieving copulation. With CBA females the strain difference in mating success was not found (Levine, Barsel, and Diakow, 1966).

Dominance effects on mating success have been found in other strains (Levine and Lascher, 1965; DeFries and McClearn, 1970). If the results of these laboratory ex-

Table 11-9. Food competition among females of two inbred mouse strains and their F_1 hybrids*

	Competing groups					
	AKR vs. DBA		AKR vs. F_1		DBA vs. F_1	
Contests won	5	15	3	17	3	17
Chi-square	5.0		9.8		9.8	
Probability	<0.05		<0.001		<0.001	

*From Manosevitz, M. 1972. Behavioral heterosis: food competition in mice. J. Comp. Physiol. Psychol. **79**:46-50. Copyright 1972 by the American Psychological Association. Reprinted by permission.

periments can be generalized to natural populations, and if in such populations social status is genetically determined, then position in the social hierarchy is the major component of fitness.

Behind this reasoning are the assumptions that the laboratory strains are equal in fertilizing ability and that the results of staged competitive mating tests are representative of outcomes in more naturalistic surroundings. Horn (1974) tested these premises by conducting competitive mating tests in large enclosures divided into sections by partial barriers. He also overcame the limitations inherent in the offspring–coat color method of determining paternity by using starch-gel electrophoresis to detect alternate types of hemoglobin and esterase-3. RF/J males were found to be more dominant than BALB/cJ, DBA/2J, and C57BL/6J and to sire the majority of offspring. However, part of the RF superiority was due to their higher fertilizing ability, as demonstrated in noncompetitive tests. Among the factors identified by Horn as possibly producing strain differences in fertilizing ability are pregnancy blockage induced by contact with a strange male and a suppressive effect of crowding on sexual behavior detected in BALB males.

We can deduce from these studies that social dominance makes a substantial contribution to the fitness of male mice. Only dominant animals obtain territories through which they move freely and gain access to females (Mackintosh, 1970). But dominance is not the whole story. Female choice might have been a factor in the competitive superiority of ST over CBA males in the experiments of Levine and associates. Gaining access to females is only a preliminary to impregnation, and impregnation does not guarantee offspring. More research like Horn's is needed to disentangle all these factors.

Selection for aggressivity. Differences in aggressivity among inbred strains are the result of chance rather than intentional selection. Selected lines have some advantages for behavior-genetic analysis; response to selection tells something about the mode of inheritance, and the correlated changes in

other characteristics of the selected lines may provide clues to the physiological basis of variation in aggressivity. Lagerspetz (1961, 1964) selected high- and low-aggression lines of mice using a 7-point rating scale for evaluation. The lowest category on this scale is described as "no interest in partner, tries to escape, squeaks if attacked." The highest rating is assigned to an individual that "wrestles fiercely and bites hard enough to draw blood." At the midpoint an animal tail-rattles, noses its partner vigorously, and occasionally attacks briefly. From the base stock, whose males averaged near the midpoint of this scale, two lines were developed, the A (aggressive) and the NA (nonaggressive), with mean ratings of 5.5 and 2.7 for the respective males. Selection was based solely on the behavior of males isolated after weaning and brought together in pairs in a small enclosure. Most of the change in the NA line occurred in the first generation of selection; progress in the A line continued for at least seven generations.

In the open field, A-line animals defecated less and ambulated more. They were more active in revolving drums and were slightly superior in learning a maze. The incentive value of an opportunity to aggress was measured by the latency to cross an electrified grid interposed between the subject and a potential victim, another mouse. A-line males crossed more quickly than NA-line males only when tested immediately after a fight.

In both A- and NA-line males that were reared in isolation, training with submissive opponents increased the mean aggression ratings. Socially reared males did not spontaneously attack their submissive partners (Lagerspetz and Lagerspetz, 1971). A- and NA-line male pups, fostered on dams of the alien line, were compared as adults with controls who were reared by their own mothers (Lagerspetz and Wuorinen, 1965). The results, shown in Table 11-10, show that the line difference persists in the cross-fostered group. However, the mean scores of fostered subjects of both lines were significantly less than those of unfostered subjects. This effect

Table 11-10. Effects of cross-fostering on aggressivity scores of selected mouse strains*

Strain of biological mother	Cross-fostered			True-parent reared			
	N	Mean	SD	N	Mean	SD	Probability
Aggressive line	29	4.87	1.69	16	6.05	1.18	<0.02
Nonaggressive line	19	2.04	1.22	35	2.75	1.24	<0.05
Probability		<0.001			<0.001		

*From Lagerspetz, K. M. J., and K. Wuorinen. 1965. A cross-fostering experiment with mice selectively bred for aggressiveness and non-aggressiveness. Rep. Inst. Psychol. Univ. Turku **17**:1-6.

may have been due to the stress of shifting the pups to a new female. A better control for the effect of cross-fostering would have been fostering within the pup's own line. Nevertheless it is safe to conclude that the differences between the A and NA lines are not caused by differences in the way that the females of the two lines rear their young.

Experiential and environmental effects. A variety of procedures have been used to evaluate the effects of early experience on the aggressive behavior of adults. The variety of methods, experimental design, and strains chosen for study tend to obscure general principles. However, the bulk of evidence supports the proposition that mice raised in groups are less belligerent than those reared in isolation (Scott, 1966). Whether strain differences in aggression are found depends on the nature of their housing prior to testing (Hahn, Haber, and Fuller, 1973).

Maternal influences are also well documented. Southwick (1968) observed both infostered and outfostered A/J (low-aggression) and CFW (high-aggression) mice in same-strain groups. Infostered A mice were slightly, but significantly, more aggressive than nonfostered controls, a result contrary to Lagerspetz and Wuorinen (1965). Fostering A pups on CFW females produced a 52% increment in the chase-attack-fight (CAF) score and an 83% increment over nonfostered controls obtained from a commercial breeder. Fostering had no effect on the CAF scores of the generally more aggressive CWF mice. The reciprocal crosses differed strikingly. Offspring of CWF ♀ × A ♂ were 62% above the CFW mean on the CAF score; offspring of the A ♀ × CWF ♂ crosses were intermediate to the parental lines. What looks like heterosis in the first of these crosses may simply be a byproduct of differences in maternal care.

Increasing the number of individuals who interact is sometimes considered to enhance aggressive behavior. Certainly the number of contacts between individuals, and hence the number of opportunities to aggress, is increased under such circumstances. There could also be more intense competition for resources and a generalized increase in the level of stimulation, which lowers the threshold for fighting. To test these ideas experimentally, Vale, Vale, and Harley (1971) assigned mice of five inbred strains to rearing in isolation or in groups of 2, 4, and 8. The floor area per individual was kept constant in the pens to eliminate crowding as a factor. Fig. 11-6 diagramatically represents selected behavioral and hormonal reactions of the five strains to changes in population size. It is easy to see that the effects were strongest in BALBs and least in As and DBAs. Like other experiments on the effects of early experience on later behavior, this one demonstrates that neglect of genotype-environment interactions can lead to premature generalization.

The effect of crowding on aggression was observed in freely growing populations of three mouse strains by Levin, Vandenbergh, and Cole (1974). In two of the three the level of aggression continued to increase after the population had reached an asymptote. The increase was most definite in BALBs; they also attained a lower population density than the other two strains.

The import of these experiments and of

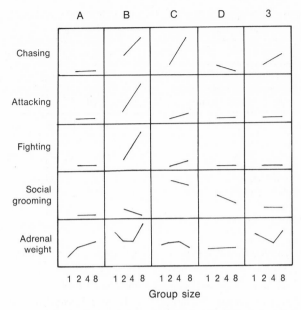

Fig. 11-6. Effect in five mouse strains of group size on four behavioral and one endocrine characteristic associated with aggression. The height and slope of the lines above the base denote the general level of activity and the direction of change as group size increased. It is apparent that some strains are influenced much more than others by an increase in the number of social contacts. *A*, A/J; *B*, BALB/cJ; *C*, C57BL/6J; *D*, DBA/2J; *3*, C3H/He/J. (From Vale, J. R., C. A. Vale, and J. P. Harley. 1971. Commun. Behav. Biol. Part A 6:209-221.)

others reviewed by Scott (1966) and Lagerspetz and Lagerspetz (1974) is clear: differences in aggression among genotypes are functions of the conditions under which observations are made, and the effects of treatments on aggression are functions of the genotypes to which they are applied.

Aggression in female mice. Thus far our discussion of murine aggression has been male oriented; females have appeared on the scene only as prospective mates for whom males compete. Females do fight, however, in defense of a litter. St. John and Corning (1973) found that strain differences in the vigor of such attack against a restrained target male paralleled the aggressiveness of the males in the five strains tested. Attack frequency and intensity peaked at 4 days postpartum and declined rapidly after day 16. Both hormonal factors and stimuli emanating from the pups appear important in maintaining nest defense.

Exogenous androgens given to intact adult females do not induce aggressive behavior as

they do in castrated males (Tollman and King, 1956). However, females that have been injected with androgen during the immediate postnatal period and reinjected with it as adults attack strange males or other androgenized females in a malelike fashion (Bronson and Desjardins, 1968; Edwards, 1968). In mating tests such females may respond to male advances by attack.

One may inquire as to whether the females of "aggressive" and "nonaggressive" strains of mice, who are themselves peaceful, share with their male siblings tendencies for high and low aggression, respectively. Apparently they do. Neonatally androgenized female mice from the Lagerspetz A and NA lines behaved like their male counterparts when injected with testosterone propionate as adults (Lagerspetz and Lagerspetz, 1975). For females of the highly aggressive A strain, the neonatal injections were not essential; injections of androgen to intact adult females induced masculine aggressive behavior.

Vale, Ray, and Vale (1972) injected inbred

Table 11-11. Effects of exogenous neonatal androgen pretreatment on androgen-induced aggression in female mice*

| | Pretreatment | | |
| | Female | | Male |
Strain	Testosterone	Oil	Oil
A/J	0.07† (2/20)‡	0.00 (0/20)	1.75 (5/20)
BALB/cJ	10.07 (13/20)	0.00 (0/20)	20.62 (10/20)
C57BL/6J	0.40 (2/20)	0.00 (0/20)	4.42 (5/20)

*From Vale, J. R., D. Ray, and C. A. Vale. 1972. Interaction of genotype and exogenous neonatal androgen: agonistic behavior in female mice. Behav. Biol. 7:321-334.
†Mean number of attacks per subject.
‡Number responding/number tested.

female mice of three strains with either oil or testosterone propionate in oil at 3 days of age; male siblings were injected with oil as a control. Agonistic behavior was measured later by the dangler technique: presentation of a trained loser that is held by the tail and swung back and forth in front of the animal being tested. The results shown in Table 11-11 are very clear. The number of attacks by the androgenized females is less than that of their congenic males, but the ranking of the strains is the same for both sexes. The genetic substrate for aggressive behavior must be identical in the two sexes, but strain differences between females are apparent only when the substrate is activated by means of hormones at a specific stage of development.

Neonatally administered estrogen has a similar but weaker action in making adult females aggressive. Vale, Ray, and Vale (1973), using the same strains as in the androgenization experiment, found that estradiol benzoate induced attack behavior in BALB/c but not in A or C57BL females. Reference to Table 11-11 shows that BALB males were the most aggressive of the three strains and also that the effect of androgens was strongest in the BALB females.

Ebert and Hyde (1976) successfully selected females from a colony of recent wild origin for high and low intensity of agonistic behavior. The reaction to an inbred female placed in the wild mouse's home cage was rated on a modified Lagerspetz scale with five categories. Duplicate selected and control lines were observed for four generations. At the end of this period the mean ratings were high lines, 3.7 and 3.4; controls, 2.0 and 3.1; and low lines, 1.9 and 2.4. Heritability calculated from parent-offspring regression was 0.17; realized heritability was 0.49. None of the body weight or fertility measures obtained during the course of the experiment correlated with the behavioral measures.

Aggression tests conducted with the males of the lines selected on the basis of female behavior detected differences between the lines, but they were not correlated with the female scores (Hyde and Ebert, 1976). The reason for the discrepancy with the findings of St. John and Corning (1973) and Lagerspetz and Lagerspetz (1975) is obscure. In both of the latter experiments aggression was measured in androgenized females, whereas Hyde and Ebert selected females that had received no hormonal treatments. Their procedures may well have operated on a form of agonistic behavior mediated by physiological mechanisms very different from those involved in spontaneous intermale fighting with its requirement for testosterone. The two types of mechanism would almost certainly be genetically independent.

Hormonal and neurochemical basis of strain differences. For centuries man has used castration to reduce aggressive behavior in males of domestic species that are not needed for breeding. The role that androgens might play in the production of strain differences in male aggression has been of considerable interest. In one of the early experiments of this kind, Bevan et al. (1957) found that doses of testosterone propionate (150 μg/day) too small to stimulate seminal vesicle enlargement were highly effective in restoring aggression of castrated C3H mice; doses of 300 to 600 μg/day given to castrated SWR mice were not only ineffective in restoring aggression but actually depressed it. The results with the C3H mice suggest that in this strain the brain is more sensitive than the reproductive system to the effects of testosterone.

Androgens may not be the only hormones whose effect on aggression depends on genotype. Brain and Poole (1974) reported that ACTH suppresses isolation-induced fighting in TO mice, but not in a number of other strains. They concluded that attempts to influence isolation-induced aggression by manipulation of pituitary-adrenocortical factors must take genetic variation, type of test, and prior experience into consideration.

Comparisons between the A and NA selected lines have been made for a number of neurochemical and endocrine characteristics (Lagerspetz, Tirri, and Lagerspetz, 1967). A-line males had heavier forebrains and testes and higher concentrations of epinephrine in the adrenals. No differences between the lines were found in the weight of the brain stem or in the concentration of serotonin and norepinephrine in the brain. The concept of sympathetic dominance in the A-line mice and parasympathetic dominance in the NA line has been suggested as a general explanatory principle (Lagerspetz, 1964). Orenberg (1975) found a positive association between brain concentration of cyclic adenosine monophosphate (cAMP) and aggressive behavior in four inbred mouse strains and one hybrid. She also reported that the association was present in segregating generations of a cross between high-aggressive, high-cAMP BALB males and low-aggressive, low-cAMP A males.

There are problems in the interpretation of associations based on a relatively small number of inbred strains. Nevertheless, instances of such associations should be followed up, and confirmation sought in a larger selection of strains or in heterogeneous stocks.

Varied observations. In this section we consider three studies dealing from unusual viewpoints with the behavior-genetic analysis of agonistic behavior in mice. Kessler, Harmatz, and Gerling (1975) tested the hypothesis that male-mouse urine contains pheromones, the nature of which determines the probability that another male will attack. Adult males of strains DBA and A who were tested fighters were placed with castrated victims smeared with urine from one of three strains or with water. Target animals treated with DBA urine elicited more fighting than did males wet with C57BR or CBA urine. Mice rely heavily on olfaction; it is not surprising that genetically determined pheromone chemistry may play a role in establishing social rank.

Tellegen and Horn (1972) sought to learn if the opportunity to be aggressive toward another mouse would adequately reinforce the learning of a T-maze and if the effectiveness of such reinforcement for a strain would be related to its aggression ranking on more traditional tests. Comparisons were made between the learning rates of three inbred strains with either food or the opportunity to attack a submissive mouse as reinforcers. Food-rewarded and victim-rewarded subjects were pretrained in an appropriate manner. Both types of reinforcement were effective in all strains, but food was more effective than access to a subordinate mouse. However, in the aggression-rewarded groups a strong warm-up effect was seen over each day's trials; by the third trial of each day mice given an opportunity to aggress were choosing the correct side of the maze almost as often as those trained with food. In RF males, shown to be most aggressive in other tests, aggression came very close to equality with food reinforcement. Under certain conditions fighting seems to be a goal in its own right, not simply the means to an end.

The final article reviewed in this section deals with the somewhat neglected topic of genetic mechanisms. For the most part, the behavior-genetic investigation of agonistic behavior has been based on strain differences. Selection experiments and interstrain crosses support the idea that aggressivity is a quantitative character with polygenic inheritance. However, the kind of aggression usually observed in mice is correlated with maleness, and maleness requires a Y chromosome. The question arises as to whether this chromosome bears genes that influence agonistic behavior. Supporting this idea is the fact that in reciprocal crosses between high-aggression DBA/1Bg and low-aggression C57BL/10Bg the transmission of intense fighting follows the DBA/1 Y chromosome (Selmanoff, Maxson, and Ginsburg, 1976). In crosses of DBA/2Bg with C57BL/10 or of DBA/1 with C57BL/6Bg the association broke down. To explain these discrepancies, the investigators postulate the absence of the aggression-enhancing factor in the DBA/2Bg Y chromosome and a suppressive factor in the C57BL/6Bg. The matter is obviously complex, and the phenomenon may be a peculiarity of one cross without general implications. In the light of interest in the possible effects of an extra Y chromosome on aggression in man, the Y chromosome of mice merits additional study, although possibly maternal effects will prove to be more important than paternal ones. Southwick (1968) found a strong maternal effect in reciprocal crosses between a high- and low-aggression strain. Nevertheless, it is interesting to have the Y chromosome considered as having a function other than that of shifting the development of the reproductive system from a female to a male configuration.

Genetics and social organization

The research reviewed thus far demonstrates that fighting ability in laboratory strains is influenced by genotype. Here we consider the agonistic behavior and social organization of wild mice and the role that genetics has played in gaining this knowledge. To learn what goes on among free-living mice requires a combination of behavioral observation, population genetics, and demography. Descriptions of the social behavior and

organization of *Mus musculus* can be found in Crowcroft and Rowe (1963), Reimer and Petras (1967), and Selander (1970). Populations are divided into tribes composed of a dominant male, several females, and several subordinate males. A tribe holds a territory whose size depends on the terrain and food supply. A territory may be very small even in the absence of physical barriers. In chicken barns with abundant food and shelter, the mean area of territories was about 2m^2 (Selander, 1970). The formation of a territory is primarily due to a male's dominance status, but once it is established, the resident females may contribute to its defense. Tribes are stable over several generations at least; the dominant role seems to pass from older to younger males within the tribe. Intertribe migration is rare and involves only females. The effective breeding size of such a unit has been estimated at five or less if the subordinate males do not breed and two to three times greater if they do (Petras, 1967).

The existence and persistence of these isolated breeding units, or *demes*, is demonstrated by the nonrandom distribution of the alleles for certain polymorphic proteins, esterases, and hemoglobin. Fig. 11-7 is the plan of a large chicken barn in which traps were set out in a grid pattern. (Selander, 1970). The symbols described in the figure legend represent the three possible genotypes for two alleles at the esterase-3 locus, and their location denotes where these genotypes were found. In some areas there are disproportionate numbers of mice homozygous for the slow-migrating form of the enzyme *(solid circles);* in others there are aggregations of mice homozygous for the faster-moving form *(open circles);* and there are areas with a mix of the two forms, including many heterozygous individuals *(dots in circles).*

The mosaic-like distribution of the different genotypes is evidence for deme structure with a high degree of sexual isolation between adjacent demes. Differences in the genetic composition of demes is explained by the fact that each is founded by a few individuals, probably a single pair, who can bring with them only a limited set of the genes

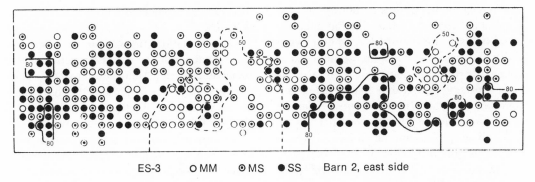

ES-3   ○ MM   ⊙ MS   ● SS   Barn 2, east side

Fig. 11-7. Distribution of esterase-3 *(ES-3)* genotypes in mice collected from traps arranged in a grid within a large barn (7.3 × 58 m). The 50% and 80% isofrequency lines for the S allele (slow migration) are shown. The patchy nature of the distribution of the two alleles is evidence for genetic isolation between adjacent breeding groups or demes. Thus near the middle there is a region where the fast-migrating *(M)* allele is more common. On the extreme right and left are areas where the S allele is more frequent. (From Selander, R. K. 1970. Am. Zool. **10:**53-66.)

present in the total population. As long as mating is restricted within the deme, inbreeding must occur with a consequent reduction in heterozygosity. The coefficient of inbreeding will be higher in the smaller demes. Reasoning from a theoretical model of the relationship between deme size, inbreeding, and its effects on the number of heterozygotes, Petras (1967) used population survey data for biochemical variants to estimate the deme size. His estimations have been criticized by Selander (1970) on a number of grounds, the most important of which is that heterotic effects could influence genotype frequencies enough to invalidate computations of deme size based on the deficiency of heterozygotes found in some surveys. Favoring this view is the fact that the heterozygote deviation from the expected equilibrium value was negative in large barns and positive in small ones. Selander ascribes the deficiency of heterozygotes in the large barns to the division of the population into isolates within which inbreeding occurs. The positive deviation in the small barns attests to greater fitness of heterozygotes under the more severe environmental stresses that are characteristic of these habitats. Since small-barn populations are unstable, they do not survive long enough to become inbred.

With the same objective of studying sexual isolation in mice, Klein and Bailey (1971) ob-

served the survival of skin grafts between the offspring of wild males and C3H/HeJ females. The males had been collected from six Michigan farms, none of them closer than 2 kilometers from any other. Skin grafting is an extremely sensitive and roughly quantitative test for genetic similarity. Grafts between members of an inbred strain are accepted; those between different strains are rejected. The less individuals are related, the faster they reject each other's skin graft. In the mouse this complex histocompatibility system is regulated by genes at 30 or more loci. Klein and Bailey found that grafts between littermates survived an average of 17.1 days; between offspring of sires from the same farm, 13.8 days; and between offspring of sires from different farms, 10.6 days.

These techniques of biochemical and immunological population genetics provide an indirect look into the nature of social organization in the wild house mouse. They confirm the territorial nature of the species and the existence of sexual isolation within the species based on behavioral as well as physical barriers. They also point out the limitations of much aggression research conducted with inbred strains. Wild mice live in tribes, not in isolation, in pairs, or in unisex groups. In the attempt to maximize environmental and genetic controls, relevance may have been sacrificed.

The wild mouse studies do not tell us anything about the genes that may be affecting social status or, for that matter, whether variability in social rank has any genetic component. It probably does, but there are few data. Certainly genetic polymorphisms are common in wild *Mus musculus.* Selander (1970) cites evidence that the variance in protein polymorphisms among all inbred strains of mice is only a fraction of that existing in wild populations. Of course, he is referring to phenotypically competent mice, not to the exotic mutants that are maintained in laboratories for the benefit of genetic and medical research.

Does variation in wild mice extend to genes that influence agonistic behavior and social dominance? Almost certainly yes. They are very likely the same ones that have been distributed erratically to our common inbred strains. Future research might well be directed toward learning how these genes operate on the sensitivity of the nervous system to hormones and to aggression-eliciting stimuli.

III
Human behavior genetics

12

Methods in human behavior genetics

The application of genetics to the study of human behavior differs from its application to animals in at least two major ways. In the first place, as we pointed out earlier (Fuller and Thompson, 1960), experimenters who use animal material to study behavior, whether bacteria, *Drosophila*, or mice, are usually oriented to the solution of basic problems. Consequently, choice of methods, subjects, and behaviors is dictated more by convenience than by considerations of economic or social interest. By contrast, human behavior geneticists have often focused on problems that have actual or potential relevance to the management of human society. Thus they have given much of their attention to such characters as intelligence or mental illness that may add to or detract from the usefulness of individuals in society.

Such an orientation has characterized human behavior geneticists throughout the entire history of the field. Darwin, Galton, and Pearson were all very much aware of the great practical significance latent in the new sciences of evolution and genetics; it was precisely this awareness that led to the formation of the eugenics movement in England and in the United States. Many of the promulgations that emerged from the movement now seem to many people to have been hasty and ill conceived. Some of them we will take up in a later chapter. Suffice it to say at this point that human behavior genetics, of all disciplines in science, is always in danger of edging into the political domain. To the extent this happens, it becomes less of a science. Consequently, there is a real onus on all workers in the area to avoid, as far as possible, any political biases that may erode their impartiality as scientists and to follow closely the dictates of facts and logic.

As we have just indicated, behavior geneticists who use animals have largely remained aloof from practical problems. Their concern has been rather with the construction and testing of basic models. Thus the behaviors they have studied have been simple and easily measurable but usually of little social relevance. Indeed, such traits often have had little interest to other branches of biology and psychology. There is no logical reason, however, why this should be so, and it is encouraging to find, for example, a growing concern with characters that may have a direct bearing on fitness (Chapters 11 and 18). No doubt, as the field of animal behavior genetics progresses, more and more effort will be devoted to the study of traits that have social value. If this is accomplished, the goals of human and animal genetic research will become more closely aligned with each other than they have been in the past. Certainly, there is no reason why work with animals cannot contribute usefully to our understanding of traditional social problems such as education and the etiology and cure of mental disorders.

If work with animals and humans can be considered to be more or less on a continuum with respect to their goals, much the same can be said in respect to the methodological problems that each field encounters. It is true, certainly, that control is more difficult to exert with human beings. This fact led Hogben (1933) to remark, "A human society may be crudely compared to a badly managed laboratory in which there are many cages containing a pair of rats and their offspring." Thus in human populations we are seldom able to achieve more than statistical control over mating systems or environmental conditions. This is more easily accomplished with animals. However, even with such subjects, control is probably imperfect. Many animal species, for example, have mating preferences and habits that may frustrate the design goals of the experimenter. Likewise, we never have any guarantee that animal subjects will not differentially select from some broad environmental condition we impose on them those microaspects of it that are most compatible with their genetic makeup. Some revealing examples of such selection have been ably put forward by Lerner (1968).

However, even granting that the problems of human behavior genetics are not of a different kind from those involved in work with animals, they must be considered to be far more troublesome and to demand, therefore, the application of special methods not commonly used with animal populations. Basically, they may be broken down into two types of approach, depending on whether we

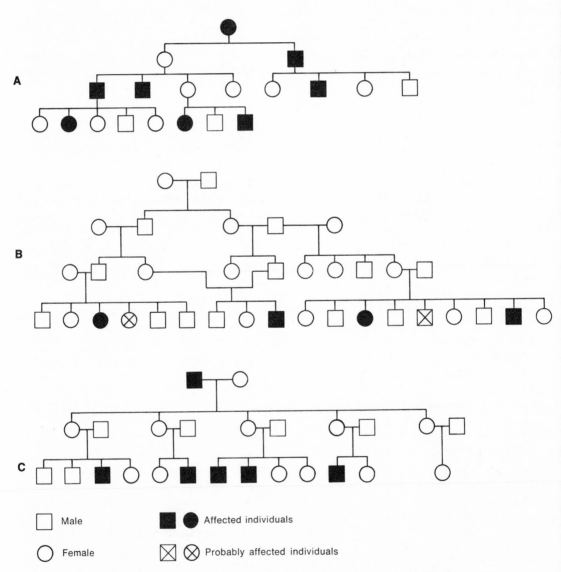

Fig. 12-1. **A**, Huntington's chorea. **B**, Phenylketonuria. **C**, Color blindness.

are dealing with discrete, unitary traits or with continuous (or quasicontinuous) complex characters. The genetic principles underlying each are, of course, the same as explained earlier; however, the statistical methodologies and procedures are somewhat different in each case. We will first consider unit characters and their inheritance.

UNIT CHARACTERS IN HUMAN POPULATIONS
Pedigrees

The most obvious and immediate way of examining the possibility that some unit character is inherited is by assessing its pattern of occurrence and nonoccurrence in family pedigrees. Pedigree analysis was, in fact, the method first put forward in systematic fashion by Galton (1869), although, undoubtedly, it must have been used by many people before him. We will see some examples of his use of it in respect to intellectual and personality traits in chapters that follow. Today, pedigree analysis is still widely used by medical geneticists for purposes of genetic counseling and for obtaining at least provisional notions regarding the heritability and mode of transmission of some deleterious character. Three family pedigrees representing, respectively, autosomal dominant, autosomal recessive, and sex-linked transmission are shown in Fig. 12-1. A casual inspection of these will indicate that several general guidelines may be used to make a genetic assessment of such data.

Case 1. Huntington's chorea (autosomal dominance)

This usually fatal disease has a late age of onset (after 35) and involves marked choreic movement often preceded by a neurotic or psychotic episode. In around 7% of cases, it terminates in suicide. In a typical pedigree we find the following indicators of dominant autosomal inheritance: (1) the trait appears in each generation without skipping; (2) the trait is transmitted to half of an affected person's children on the average because over many families the most common type of mating will be between a normal homozygote and a heterozygote carrying the abnormal gene; (3) unaffected parents do not transmit the trait to their children; and (4) no sex differences are present. Note that in

some instances, carriers may die before showing the illness themselves. Likewise, some individuals may outlive most of the risk period. These facts account for some seeming anomalies in the pedigree shown in Fig. 12-1, *A*.

Case 2. Phenylketonuria (autosomal recessivity)

This disorder shows up at birth and produces a number of physical abnormalities, usually with gross mental defects. It is due mainly to a deficiency in the liver enzyme phenylalanine hydroxylase, which normally metabolizes the amino acid phenylalanine. Typically, (1) the trait appears only among siblings and not parents because reproduction is highly unlikely in an affected individual; (2) on the average, one fourth of the siblings of an index case are affected; (3) there is an increased incidence of consanguinity between parents of an affected child because the most likely mating type is between two carriers (genes for a rare disorder are much more likely to be found in relatives than in the general population; thus for a condition carried by a gene with a frequency of 0.002, the nonrelative mating risk is 0.000004; with two cousins, it is increased to 0.0001, a risk over thirty-two times higher [Stern, 1973]); and (4) no sex differences are present.

Case 3. Color blindness (sex-linked recessivity)

Of the several forms of color blindness, a deficiency at the red-green end of the spectrum is carried by a sex-linked recessive gene (Chapter 13). Mode of transmission is indicated by (1) a markedly higher incidence of the anomaly in the male component of the population; (2) transmission of the trait from an affected male through his daughters to half of their sons; and (3) transmission never occurring directly from an affected father to his son, since he passes on to his sons only a Y chromosome and not the X chromosome with the color-blind gene.

Case 4. Blood group X_g (sex-linked dominance)

(1) Affected males transmit the trait to all their daughters and to none of their sons; (2) carrier females transmit the trait to half their offspring of either sex; and (3) homozygous affected females transmit the trait to all their children of both sexes. Note that this pattern of transmission can only be distinguished from that of autosomal dominance by reference to the progeny of affected males.

The preceding may be considered classical cases in which analysis is relatively straightforward. However, variations in the manner in which genes are expressed may be produced by a large number of factors, including penetrance and expressivity, pleiotropy, consanguinity, variable age of onset, genetic heterogeneity (production of same trait by different gene loci), and gene interaction and method of ascertainment. All or any of these will act to complicate the task of pedigree analysis considerably. Procedures for reducing ambiguity have been worked out for a number of these factors. Let us look at two such procedures relating to ascertainment and age of onset.

Bias of ascertainment. A general inspection of one or two family pedigrees may certainly provide an investigator with some reasonable hunches as to the mode of inheritance of the disorder or character under study. However, unusual and atypical concentrations of a disorder in a family may attract more attention than isolated cases and hence lead to premature conclusions. Likewise, reporting on all the members of a family may be very incomplete. Secondhand reports of the presence (and even more so, of the absence) of traits are often very unreliable. Neel and Schull (1954) described the appearance of a pedigree of diabetes mellitus before and after the siblings of the propositus or index case were given sugar-tolerance tests. Before the tests, the propositus claimed that she was the only diabetic among eleven living siblings. After tests were given to all available subjects, the score stood at five diabetic, four tested and found normal, and two untested. Nonprofessional medical or psychiatric diagnoses are also untrustworthy. Certainly the most valuable data on human inheritance have come from studies in which a qualified investigator and assistants have examined all cases with their relatives and verified personally (and, if possible, *independently*) the presence or absence of a trait according to a uniform procedure. Blind diagnosis is likewise desirable.

Accurate and complete ascertainment is a minimal requirement. But, particularly in the case of rare, recessive traits, an additional statistical difficulty arises. When a trait is rare in a population, a general survey of all families is of little usefulness, since most of them will not contain a trait bearer and hence throw no light on the mode of inheritance of the trait. Instead, it is the practice to study only families that do contain a trait bearer or propositus. If a rare recessive gene

is involved, then it is most likely that the parents will be heterozygotes and that, therefore, the expected proportion of affected offspring will be one fourth. However, it is also true that there will be many such parents who will have no affected children, and these, of course, will not be located for analysis even though their offspring should be counted in order to secure a correct genetic ratio. This is particularly true when families are small.

Several methods for overcoming such a sampling problem are available. We will consider only one of them, the Weinberg proband method, as in the following example.

Suppose an investigation is dealing with a disorder carried by a single recessive gene. In sixteen cases of marriage between two heterozygotes each having two offspring, the most likely distribution of the disorder among these thirty-two children will be as shown in Table 12-1.

There are eight affected cases. If the investigator studied only the families in which these occurred and included them in his count, he would get an incidence of 8 out of 14, or 57.1%. This would be too high. On the other hand, if he excluded the index cases, then he would emerge with a figure either of 0 (if both members of family 1 were counted as index cases) or 1 out of 14, or 7.1% (if only one member of family 1 was an index case). This would clearly be too low in the case of

Table 12-1. Most likely distribution of affected cases among thirty-two offspring from sixteen matings between two carriers*†

Families‡							
1	AA	5	AN	9	NN	13	NN
2	AN	6	AN	10	NN	14	NN
3	AN	7	AN	11	NN	15	NN
4	AN	8	NN	12	NN	16	NN

*From Genetic theory and abnormal behavior by David Rosenthal. Copyright © 1970 by McGraw-Hill, Inc. Used with permission of McGraw-Hill Book Co.
†Recessive inheritance is assumed.
‡A, affected; N, nonaffected.

the types of families being considered and the type of inheritance being postulated. What have been left out of consideration, of course, are the nine families, which might have had affected offspring but, by chance, did not. The Weinberg method attempts to correct for the loss of these families from the computation

$$P = \frac{\sum\limits_{1}^{n} [x(x-1)]}{\sum\limits_{1}^{n} [x(s-1)]}$$

where

P = probability that a child of heterozygous parents will be affected

x = number of affected children in family, including proband

s = total number of children in family

n = number of families with particular values of x

The actual computation for the data being considered is shown below. Entailed in the formula are (1) the rejection of the proband, leaving only one case of an affected offspring, and (2) the counting of a family as many times as there are affected offspring in it. Thus family 1 is counted twice to give a total of eight families. This also increases the number of affected offspring (nonproband) to two. Hence, we arrive at the expected ratio of 0.25, or one fourth.

Besides the Weinberg proband method, several others have been put forward by human population geneticists. These include Dahlberg's later-sib method, Hogben's a priori method, Haldane's a posteriori method, as well as others. However, since they have not found common usage among behavior geneticists, we will not attempt to present them here (see Fuller and Thompson, 1960).

Age. A second major category of problems is created by the fact that many diseases—and perhaps especially psychological disorders—have a variable age of onset. In some disorders, such as Huntington's chorea, symptoms may appear abruptly; in others, it is more gradual. For example, in schizophrenia the risk rises gradually to a maximum between ages 20 and 30 for men and between ages 30 and 40 for women. From a genetic point of view, this fact is exceedingly interesting, since it suggests either the operation of age-related cultural variables, age-related gene expression, or both. We will return to the substantive problem in Chapter 16. At this juncture, we wish only to consider the implication that such data have for estimation of correct incidence rates for such a syndrome.

It will be clear that if an age relationship exists for a character, then some relatives of a proband who are unaffected at the time a

---

Weinberg's propositus method for correcting observed incidence for comparison with expectancy for a Mendelian recessive character*

Number of children in family (s)	Number of affected in family (x)	Number of families (n)	x(x − 1)	x(s − 1)	n[x(s − 1)]
2	2	1	2	2	2
2	1	6	0	1	6
			$\Sigma[x(x-1)] = 2$		$\Sigma[x(s-1)] = 8$

$P = \dfrac{2}{8} = 0.25 = 25\%.$

*Modified from Rosenthal, D. 1970. Genetic theory and abnormal behavior. McGraw-Hill Book Co., New York.

study is made may yet contract the disease sometime in the future. Furthermore, their likelihood of doing so will be greater or less, depending on whether they have passed the age of maximum risk. Several methods have been developed to cope with the problem (Strömgren, 1950; Larsson and Sjögren, 1954). We will discuss only the one most commonly used, the Weinberg short method (Weinberg, 1927). Briefly, it involves weighting observations made according to the incidence of the illness at different age ranges. Thus if the risk is considered to extend from age t_1 to age t_2 years, then *unaffected* subjects below t_1 are weighted 0, unaffected subjects between t_1 and t_2 are weighted 0.5, and unaffected subjects beyond t_2 are weighted 1. This is simply a numerical way of saying that a person who is very young cannot be considered not to have the illness, even though he does not yet show it; a person within t_1 and t_2 still has a 50% chance (on the average) of showing it later; and a person who is beyond t_2 and is still unaffected never will be.

Weinberg's formula is

$$q = \frac{\Sigma f_t}{\Sigma[W_t \cdot n_t]}$$

where

q = corrected morbidity risk
f_t = frequency in each age class
W_t = weighting factor
n_t = number of individuals in each age class

An example is worked out in the box below.

The standard error of q can also be calculated by the formula

$$\sigma_q = \frac{q(1-q)}{N}$$

where
$N = \Sigma[W_t \cdot n_t]$

In the example, σ_q works out to be ±0.013.

Application of Hardy-Weinberg law

So far, we have been considering the genetic analysis of rare discrete traits that can be found in only a very few families in a population. Some traits of interest to the behavioral sciences, however, are very common and may be found in as many as 50% of families. In such cases we can examine a random sample of the population to compare the observed incidence of the trait with the incidence expected under a particular genetic model. Some aspects of the method have already been discussed in earlier chapters and need not be repeated here. We shall describe Snyder's (1932) study on the inheritance of ability to taste phenylthiocarbamide. Preliminary results suggested the hypothesis that nontasting corresponded to the genotype tt, and tasting, to the genotypes Tt or TT. The problem was how to test a Mendelian hypothesis when the genotypes of the taster parents were unknown. The solution is

---

Computation of morbidity risk by Weinberg's short method*

Age class	Affected (f_t)	Age class frequency (n_t)	Weighting factor (W_t)	Corrected population size $[W_t \cdot n_t]$
0-14	0	100	0	0
15-45	10	200	0.5	100
46+	20	300	1.0	300
Sums	30	600		400

$$q = \frac{\Sigma f_t}{\Sigma[W_t \cdot n_t]} = \frac{30}{400} = 0.075 = 7.5\%.$$

*Modified from Rosenthal, D. 1970. Genetic theory and abnormal behavior. McGraw-Hill Book Co., New York.

based on application of the Hardy-Weinberg law. We shall use the symbolism already adopted:

$$p = \text{frequency of } T$$
$$q = (1 - p) = \text{frequency of } t$$

At equilibrium the proportions of each genotype are:

$$p^2TT, \ 2pqTt, \text{ and } q^2tt$$

The results of marriages between tasters and nontasters may be summarized as follows:

Type of marriage	Frequency	Offspring Tasters	Offspring Nontasters
$TT \times tt$	$2p^2q^2$	$2p^2q^2$	0
$Tt \times tt$	$4pq^3$	$2pq^3$	$2pq^3$

Marriages between tasters and tasters may be of three types:

Type of marriage	Frequency	Offspring Tasters	Offspring Nontasters
$TT \times TT$	p^4	p^4	0
$TT \times Tt$	$4p^3q$	$4pq^3$	0
$Tt \times Tt$	$4p^2q^2$	$3p^2q^2$	p^2q^2

Summarizing, we find that the proportion of nontaster offspring from taster-by-nontaster marriages is simply

$$\frac{q}{1+q}$$

and from taster-by-taster marriages

$$\frac{q^2}{(1+q)^2}$$

The results of the survey are summarized in Table 12-2.

The fact that five children from nontaster-by-nontaster marriages did not fit the prediction may be explained by illegitimacy, incomplete gene penetrance, or faulty classification on the test. A few individuals have taste sensitivities intermediate to the taster-nontaster groups.

Counting both parents and offspring, 2556 tasters and 1087 nontasters are listed in Table 12-2. Thus the fraction of nontasters, q^2, is 0.298, and $q = 0.545$. From this value the expected proportions of offspring from each type of marriage were calculated, and the observed results are seen to be in excellent agreement with the theory.

The study, although it represents a classic example of the application of the Hardy-Weinberg rule to a character, has little direct relevance to behavioral scientists. So far, attempts to link the PTC gene to any psychological traits of interest have not been markedly successful (Kaplan, 1968). We will discuss some of this work in the next chapter.

It has been pointed out already that exact

Table 12-2. Numbers of tasters and nontasters classified according to parentage*

Marriages (total)	Offspring Tasters	Offspring Nontasters
Tasters × tasters (425)		
Number observed	929	130
Percent observed	87.7 ± 0.7	12.3 ± 0.7
Percent calculated	87.6 ± 0.1	12.4 ± 0.1
Tasters × nontasters (289)		
Number observed	483	278
Percent observed	63.5 ± 1.2	36.5 ± 1.2
Percent calculated	64.6 ± 1.2	35.4 ± 1.2
Nontasters × nontasters (86)		
Number observed	5	218
Percent observed	2.2	97.8
Percent calculated	0.0	100.0
TOTAL (800)	1417	626

*From Snyder, L. H. 1932. The inheritance of taste deficiency in man. Ohio J. Sci. **32:**436-440.

application of the Hardy-Weinberg rule to the solution of genetic problems depends on a number of assumptions that hold true only for a population at equilibrium. These have been set out in Chapter 4. In the case of behavioral traits, these assumptions seldom hold true. Apart from this difficulty, an additional complication arises from the fact that most behavioral traits are very plastic or fluid and have a highly variable penetrance, depending on the influence of environmental forces. Furthermore, we can expect some genotypes to be more responsive than others under a given set of environmental conditions; for example, a specified trait may be more completely expressed in homozygotes than in heterozygotes. Ideally, it would be desirable to set up models that incorporated, in an exact way, such genotype-environment interactions and then test them against observed data. In practice, this does not often prove to be possible. Hence, most investigators have been forced to fit their models to the data after the fact by making suitable ad hoc adjustments.

As an example of this kind of procedure, we may consider the work of Trankell (1955) on handedness in human populations. There has recently been a strong resumption of interest in this trait from both a physiological and a genetic point of view. We will discuss some more recent work on it in Chapter 13.

Trankell specifically treated handedness as a character for which one type of expression, dextrality, is favored by the environment. He hypothesized that left-handedness is carried by the genotype aa, and right-handedness by AA or Aa. Because the world has a bias toward dextrality, a certain proportion of aa individuals are prevented from expressing their genotype. Thus the aa constitution is necessary but not sufficient for expression of left-handedness.

With the preceding model in mind, Trankell accordingly modified the Hardy-Weinberg equation for the frequency of various phenotypes as related to gene frequency by hypothesizing that the proportion (k) of trait bearers is less than the total proportion (q^2) of homozygous recessives, aa. Accordingly,

the relation between phenotypes, genotypes, and frequencies for a population at equilibrium may be set out as follows:

Recorded phenotype	Genotype	Proportion
Trait absent, i.e.	AA	P^2
right-handed	Aa	$2pq$
	aa	$q^2 - k$
Trait present, i.e.,	aa	k
left-handed		

Multiplying the frequencies of each genotype with every other genotype, adding, simplifying, and summarizing leads to the following predictions with respect to the proportion of aa offspring in the progeny of various types of marriages.

Type of marriage	Proportion of recessive homozygotes in progeny
I Trait bearer × trait bearer	1
II Trait bearer × nontrait bearer	$\dfrac{(q - k)}{(1 - k)}$
III Nontrait bearer × nontrait bearer	$\dfrac{(q - k)^2}{(1 - k)^2}$

Let

N_1, N_2, N_3 = total number of children from each type of marriage

X_1, X_2, X_3 = number of children with trait recorded in each type of family

b = proportion of trait bearers in total filial population; that is, b corresponds to k in parental population

then

$$\frac{b}{q^2} = \frac{X}{q^2 \cdot N} = \frac{X}{N_{aa}} = \text{proportion of recessive}$$

homozygotes in filial generation recorded as trait bearers under prevalent conditions of penetrance

Multiplying this proportion by the expected number of recessive homozygotes from each type of marriage yields an estimate of the number of trait bearers from each family type as follows:

Family type	Predictive equation
I LH × LH	$X_1 = \dfrac{b}{q^2} \cdot N_1$
II LH × RH	$X_2 = \dfrac{b}{q^2} \dfrac{(q - k)}{(1 - k)} \cdot N_2$
III RH × RH	$X_3 = \dfrac{b}{q^2} \dfrac{(q - k)^2}{(1 - k)^2} \cdot N_3$

Table 12-3. Familial occurrence of handedness in three studies as related to Trankell's incomplete-penetrance hypothesis*

	Source of data		
	Ramaley (1913)	Chamberlain (1928)	Rife (1940)
Parental generation			
N	610	4354	1374
Left-handed	49	155	72
k	0.0803	0.0356	0.0524
Filial generation			
N	1130	7714	2178
Left-handed	177	367	191
b	0.1566	0.0476	0.0877
Type I marriages			
LH $\times$ LH $\quad N_1$	7	25	11
X_1	6	7	6
q_1	0.427	0.412	0.401
Type II marriages			
LH $\times$ RH $\quad N_2$	170	464	174
X_2	55	53	34
q_2	0.427	0.393	0.414
Type III marriages			
RH $\times$ RH $\quad N_3$	953	7225	1993
X_3	116	307	151
q_3	0.425	0.401	0.439
Best estimate of q	0.427	0.402	0.410

*From Trankell, A. 1955. Aspects of genetics in psychology. Am. J. Hum. Genet. **7**:264-276.

The values of k and b can be determined from randomly selected families, and those of N and X for each type of marriage separately. The result is three separate estimates of q.

Applying this method to three published accounts of the transmission of handedness and comparing observed and predicted values of chi-square showed substantial agreement between them. A summary is displayed in Table 12-3.

It is clear from these data that although the final estimates of the gene frequency (q) of the a gene are fairly consistent across the three studies, the estimates of k and b are not very close. Were we to average them, we would emerge with the general conclusion that about 16% of people in a population carry the aa genotype, and, of these, some-

what less than half actually express the trait of left-handedness. This does not seem an unreasonable conclusion, but it must be emphasized that many other hypotheses could also account for the empirical data, and making a final choice between alternatives is liable to be difficult. It would therefore seem desirable, in the case of such complex traits, to use the Hardy-Weinberg method only as a general frame of reference, within which an investigator proceeds cautiously, asking one empirical question at a time. For example, he may ask whether, in fact, there is assortative mating for handedness. Is expression of the trait encouraged or discouraged by some parents? If so, how? Many other such basic questions can be generated by reference to the Hardy-Weinberg law, and perhaps solutions to them will prove more interesting than the tentative confirmation of any highly generalized genetic model. Some of the work to be discussed in subsequent chapters seems to have been following such a prescription.

CONTINUOUS CHARACTERS IN HUMAN POPULATIONS

So far, we have been dealing with the genetic analysis of discrete and readily identifiable traits carried by major genes. However, most traits of interest to behavioral scientists are continuous rather than all-or-none, and their genetic variation is probably attributable to polygenic systems. To deal with them satisfactorily thus requires some extensions of the kinds of methods previously outlined.

Let us start by considering what may, at first sight, seem like a relatively simple case described by Li (1971). Some characters that are actually continuous may have, in a few individuals, an expression that is of an unusually striking nature. Such individuals we may designate as being talented, brilliant, or geniuses. Often these appear to run in families; an example is the Bernoulli pedigree, set out in Fig. 12-2. As indicated, we find four or five generations containing outstanding mathematicians. This seems impressive evidence for a genetic component in a quantitative trait, at least in its extreme form. However, as Li points out, we can readily find many families in which there are only one or two or perhaps no mathematicians, a fact that does not seem to fit with a hereditary hypothesis. Thus there seems to be some confusion as to which view is valid. But actually the confusion is more apparent than real and is simply a result of the fact that the unusual pedigree of the Bernoullis represents a very rare event. For the most part, with continuous traits, even when dependent on only a few genetic factors, like does not always beget like, and the variation among offspring in a

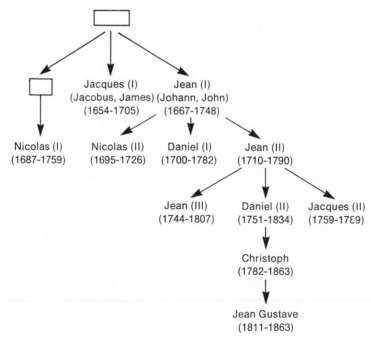

Fig. 12-2. Simplified pedigree of the Bernoulli family. Only mathematicians are shown in the diagram. (From Li, C. C. 1971. A tale of two thermos bottles: properties of a genetic model for human intelligence. In R. Cancro, ed. Intelligence: genetic and environmental influences. Grune & Stratton, Inc., New York. By permission.)

family is likely to be quite large. Thus a very dull father may produce a very bright child, or a very bright father, a dull child. In fact, as Li states (1971, p. 172), "only very strong social and environmental forces can perpetuate an artificial class; heredity does not. From this point of view, social forces are more conservative than hereditary ones."

Thus it is necessary to be very wary of using single pedigrees, even though sometimes the hypotheses they suggest may be true. But this can be confirmed only by sampling widely from the population under study. This was not, of course, done by Galton in his original pedigree studies.

Use of relatives and families

The fundamental social unit in many mammalian and most primate groups, including humans, is the family. Thus it is logical that most of the work done on human behavioral traits has centered on the similarities and differences shown by members of a family related in various ways and to varying degrees. The two major statistics used are the correlation coefficient and the variance. In the case of each of these, the broad goals are the same: partitioning of variance and covariance into a number of environmental and genetic components (Chapter 5). Without the use of special procedures, this is not possible, since a family unit supplies its members not only with certain kinds of genes but with certain kinds of environments. In a previous chapter, we showed how the average correlations between individuals in a family can be computed from Mendelian principles. The correlation between parent and offspring is 0.5, between siblings 0.5, between half-siblings 0.25, and so on. Extensive tables of genetic correlations have been set out by Charles (1933), Hogben (1933), and, even earlier than this, by Fisher (1918). However, it is considered possible, in the case of human groups, that such correlations might at least be approximated by environmental influences. It is true, of course, that genetic models are able to predict more or less exactly the order and size of familial correlations according to degree of kinship. There

are, today, no environmental models sophisticated enough to do this. For example, in many extended families there are no precise reasons why cousins or grandparents and grandchildren should be alike at all for purely environmental reasons, though ad hoc hypotheses may always be invented to suit the data (Urbach, 1974). Nevertheless, given the fairly large error factors generally associated with estimates of familial likeness, it is difficult to establish unqualified proof for or against a general environmental hypothesis.

Another problem associated with the use of family correlations in estimating heritability of a trait is the very large sample sizes needed to obtain any degree of precision. This difficulty has been discussed by Eaves (1972) and by Klein, DeFries, and Finkbeiner (1973), among others. Thus a modest family study may suggest that some trait has some degree of heritability, but it will probably not be able to do more than set broad limits on the estimate it supplies.

Correlations and covariances computed on conventional family groups are therefore by themselves of little genetic value. However, when these statistics are obtained for special cases in which there is some fairly definite control over both genetic and environmental variables, they may prove to have usefulness. In subsequent chapters, we will consider various studies in which such control has been exerted. At this point, however, it is appropriate to discuss the two most critical methods by which some degree of control is achieved: *fostering studies* and *twin studies*. Following this discussion, we will attempt to show how the data so obtained may be fitted to biometric genetic models.

Adoption studies. One mode of obtaining some control over environmental influences is afforded by studying children who have been adopted. Ideally, we should be able to sample children coming from a wide range of known genotypes and growing up in a wide range of known types of environments. Such conditions are virtually never achieved, since adoptive procedures are not governed by scientists. More often than not, welfare agencies attempt to place their wards in homes appropriate to their most probable capacities as judged by agency personnel. Fortunately for the behavioral scientist, such selective placement is imperfect, and thus there are usually available fairly good samples of adoption cases involving a minimum of genotype-environment correlations.

In using the adoption method, we may start either with the characteristics of adoptees themselves and then make predictions about their natural and adopting parents, or, conversely, we may start with the characteristics of the natural and adopting parents and then make predictions about the adoptees. In this way, we can, in theory, pit genetic hypotheses against environmental hypotheses. To give an example: we may locate, through a register, a sample of adopted children who have schizophrenia. These constitute our index cases. We then examine the incidence of schizophrenia among all biological relatives (e.g., parents, siblings, and half-siblings) of these cases and among all their adopted relatives. Other things being equal, a higher incidence of the illness in the former groups would indicate a genetic rather than an environmental etiology. As we shall see in Chapter 16, exactly this design has been used by Kety et al. (1971) in Denmark.

With more obviously continuous traits, for example, intelligence, the method is basically the same. However, we are also able to inquire not only into the relative similarity of children to their adopted and natural parents, as given by correlation coefficients, but we can also ask about mean values of the trait in children with different combinations of parents, for example, children of low-IQ natural parents but high-IQ adopting parents. These two questions are not the same. Thus at least one study (Skodak and Skeels, 1949) demonstrated that although the IQs of adopted children rose, on the average, as much as 20 points above the IQ level of their biological mothers, the correlation between these children and their biological mothers became larger with time; the corresponding correlation for IQ between these children and their adopting parents, however, remained insignificantly different from zero. What this means is simply that although a good environment may increase the intellectual level of a sample of children, it will probably do so differentially according to their genotypes. Confusion on this point has led to much futile debate. Since it bears very importantly on any discussion of the heritability of behavior traits, we will return to it in several of the ensuing chapters.

It is appropriate to make several points before leaving the adoption study method. First, it is probably true to say that neither adopting parents nor adopting children can be considered to be typical of a population of children and parents living in a conventional family structure. As Pringle (1966), in a review of the adoption method, has emphasized, they constitute socially deviant groups. This fact may not affect such characters as intelligence or achievement, but it may well influence personality variables. The children, especially, may suffer from what has been called by one writer (Sants, 1964) "genealogical bewilderment," that is to say, a lack of precise knowledge about their origins. Often, they may engage in extended fantasies about their real parents—that they were of royal blood, were millionaires, and so forth. To some extent, the adopting parents themselves may also have some curiosity about the same question.

Apart from this, it is sometimes claimed that adopting families are special in respect to more concrete characteristics, such as levels of education and income and similarity of husband and wife (Kamin, 1974). It cannot be definitely stated that clear differences have been firmly established. Nevertheless, we should emphasize that in any adoption study, care must be taken to look for such differences and, if possible, to control for them if the data obtained are to be generalized.

A second point is that the procedure of adoption may be becoming less common because of the development of reliable birth-control methods and more permissive attitudes toward abortion (DeFries, 1975). Consequently, it may happen that before long the behavior geneticist will have to look elsewhere for such data. Some major sources might be the more than 200 Israeli kibbutzim or Polynesian and Eskimo groups in which adoption is a cultural practice. As far as we are aware, these have not so far been used as a means of unraveling the kinds of problems we are considering. These and other methodological problems of the adoption method have been thoroughly discussed by Munsinger (1975).

The adoption method allows us a greater degree of control over genetic and environmental variables than is possible by using conventional family groups. Control is still far from perfect, however, and a further improvement can be achieved by use of the next method to be considered, the twin method.

Twin studies. The unique advantage afforded to human behavior geneticists by the use of twins is that they allow the study of identical genotypes. Monozygotic twin pairs are thus somewhat analogous to the isogenic lines of mice that have been produced by inbreeding over many generations, although, unlike such strains, they are not necessarily homozygous at the majority of genetic loci. Since many of the major conclusions about the inheritance of behavior characters have been based on data from twins, it is important to consider them carefully. Three excellent reference works have been published by Scheinfeld (1967), Bulmer (1970), and Mittler (1971).

Twin types. Generally speaking, there are two kinds of twins, *monozygotic (MZ)* and *dizygotic (DZ)*. The former are derived from the splitting, at a very early developmental stage, of a fertilized ovum or zygote. From this event, two embryos develop. Dizygotics, on the other hand, derive from two different and independently fertilized ova. Consequently, they are no more alike genetically than ordinary siblings. They may also, of course, be of unlike sex. Supertwins—triplets, quadruplets, or quintuplets—may be monozygotic, multizygotic, or combinations of these types. Members of one of the most famous twin groups, the Dionne quintuplets, were diagnosed as all monozygotic; likewise

with the Morlock quadruplets and the Genain quadruplets. Two other quintuplet sets, the Fischers of Aberdeen, S.D., and the Prietos of Venezuela, combined identicals and fraternals (Scheinfeld, 1967).

Little is known about the exact circumstances surrounding the formation of twins. The females of species of lower animals often produce several offspring at each birth. The production of identical twins, however, is a relatively rare event. A remarkable exception to this is the nine-banded armadillo, which regularly produces identical quadruplets. Some cattle breeds also show a relatively high incidence of twinning (Bulmer, 1970). In humans, dizygotic or fraternal twins may result when more than one egg is released from the maternal ovaries at ovulation. It is thought that unusually active and fertile sperm from the father at this time may constitute a necessary condition. It is not clear what factors lead to the production of monozygotics. However, genetic influences may be at work. This is especially suggested by the kinds of variables associated with DZ twinning. We will look at some of these in a moment. Before doing so, however, it is necessary to qualify the broad statement that twins are of two types only. It is at least theoretically possible that there is a range in respect to identity of genotypes.

In some instances, fraternal twins may actually have different fathers. This can happen if two eggs are released, and each fertilized successively by different fathers. In a case described by Scheinfeld (1967), a woman who ran a boarding house in Chicago had sexual relations with two boarders within a few hours. Each later laid claim to being the father of the twins to which she gave birth. However, blood tests indicated that each man had, in fact, sired one member of the pair. We must grant that such a circumstance is most unusual. However, the results of it would be fraternal twins who are only like each other genetically as half-siblings.

It is also theoretically possible that there are other types of twins (Mijsberg, 1957; Bulmer, 1970). Were the ovum to divide prior to fertilization and then each of the two resulting eggs be fertilized by a different sperm, the result would be twins identical on the maternal side and alike as sibs in the paternal complement of genes. These *uniovular dispermatic* twins would then be less similar than MZs but more similar than DZs. Again, equal division of the primary oocyte, giving rise to two ova, would, if fertilized, result in twins less alike than DZs though more alike than unrelated individuals. Such twins would be termed *primary oocytary*. Finally, equal division of the secondary oocyte during meiosis II would result, on the average, in twins *(secondary oocytary)* more alike than DZs but less alike than uniovular dispermatic twins (Lehmann and Huber, 1944). Thus the gradation of likeness would be primary oocytary, DZs, secondary oocytary, uniovular dispermatic, and MZs. Exact amount of likeness would depend on amount of crossing-over during meiosis I and vary according to distance of genes from the centromere. These are interesting possibilities. However, the actual evidence that they occur is very slender and is confined to rather primitive animals (Bulmer, 1970). But given the irregular concordance rates commonly found for various characters in twins, it is possible such cases may exist in humans.

Further variation in respect to degree of genetic similarity may be produced by chromosomal changes occurring in the zygote such that one member of a twin pair could be a mosaic or chimera, or that possibly both members could be, but differently constituted. One of the most striking examples of such a change is found in the rare case of opposite-sex identical pairs. Normally, at the initial cleavage of a genetically male zygote, each of the cells receives a Y together with an X chromosome. But in some instances the Y does not appear in one of the cells, giving that zygote an XO constitution. This individual will grow up to be an abnormal female with Turner's syndrome. The other will, of course, become a normal boy (Turpin et al., 1961). A similar case of XO/XX mosaicism has also been reported (Mikkelsen, Frøland, and Ellebjerg, 1963).

During the last twenty years, we have learned a good deal about the causes and effects of major chromosomal abnormalities. Perhaps as the methods of cytogenetics improve, they will show that less obvious minor changes in chromosome structure may also have important consequences on the physiological and biochemical makeup of organisms.

Apart from the kinds of genetic events just described, innumerable environmental forces both before and after birth may act to increase or decrease phenotypic similarity between the members of MZ or DZ twin pairs. We will return to them when we consider the basic assumptions involved in MZ-DZ comparisons. First, let us consider the incidence of twinning in different populations and its relation to maternal characteristics.

Incidence of twins. For whites in North America, about 1 birth in every 87 is a twin birth. In other words, 2 out of every 88 births are twins, that is, 2% of the population. Triplets occur about once in every 87^2 births, quadruplets once in 87^3 births. This mathematical relation, known as Hellin's law, is only approximated, however, at least partially because twins and supertwins have a much higher mortality rate than singletons.

Blacks in North America have a somewhat higher rate of twinning than whites: about 1 pair in 73 births. This may conceivably reflect the remote ancestry of members of this ethnic group, since African blacks show rates varying from 1 in 40 births up to as high as 1 in 22 among the Yorubas of Nigeria (Scheinfeld, 1967; Bulmer, 1970). At the other end of the scale, we find one of the lowest rates among the Japanese: 1 in about 165 births (Inouye, 1957).

The preceding differences relate mainly to the proportionate numbers of fraternal twins produced in these groups. For example, in the United States, black fraternal twins are 15% to 20% more frequent than white fraternal twins. The relative MZ-DZ rates per 100 are for whites, 65 DZ:35 MZ, and for blacks, 71 DZ:29 MZ. In Japan, on the other hand, MZs outnumber DZs by a 2:1 ratio. Such data as these strongly suggest an inherited basis, at least for the production of DZ twins.

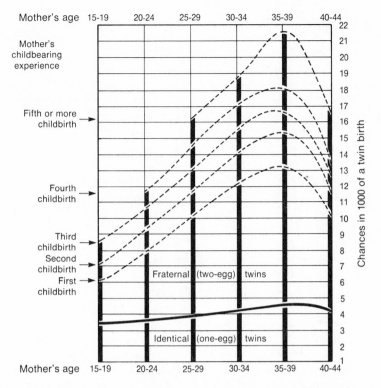

Fig. 12-3. Relationships between maternal age, parity, and twinning rate. (From Twins and supertwins by Amram Scheinfeld. Copyright © 1967 by Amram Scheinfeld. Reproduced by permission of J. B. Lippincott Co., Philadelphia, and Chatto & Windus Ltd., London.)

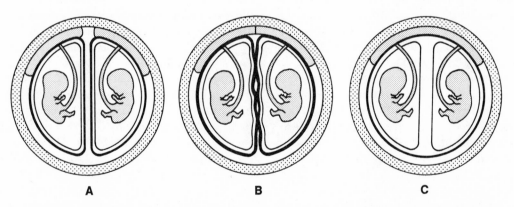

Fig. 12-4. Diagramatic representation of some common arrangements of the placenta and fetal membranes in twins. Uterine wall is lightly stippled; fetus, umbilical cord, and placenta are heavily stippled; chorion is shown by heavy line; and amnion is shown by light line. **A,** Twins with separate placentas and separate membranes. This arrangement occurs in either DZ or MZ twins. **B,** Twins with separate but secondarily fused placentas and chorions. The two halves of the placenta have separate circulations. This arrangement occurs in either DZ or MZ twins. **C,** Twins sharing a single placenta with common circulation and a single chorion, but having separate amniotic sacs. This arrangement occurs only in MZ twins and is diagnostic of MZ twinning. (From Thompson, J. S., and M. W. Thompson. 1967. Genetics in medicine. W. B. Saunders Co., Philadelphia.)

Maternal characteristics and twinning.
Number of previous births, maternal age, and parity bear an orderly relation to probability of twinning. The relationships are shown in Fig. 12-3. The more children the mother has borne previously, the more likely the next birth will be twins, and, regardless of parity, the older the mother—up to a peak of 35 to 39 years—the greater the chances of a twin birth. These relationships do not apply, however, for monozygotic twinning. Again, such data suggest, indirectly, a genetic basis for DZ but not for MZ twinning.

We will now discuss the diagnosis of twin types. Since the central feature of the twin method is a comparison between likeness of MZs and likeness of DZs, it is crucial that we be able to separate the two major twin types.

Zygosity diagnosis. There are two basic approaches to determination of the zygosity of twins. The first of these relates to the characteristics of the maternal membranes that surround and sustain the fetuses during the period of prenatal development. The second relates to various morphological and physiological traits with respect to which the members of a pair may be similar or dissimilar.

Any fetus developing in the maternal uterus is enclosed by two membranes. The inner and more delicate one is called the *amnion*. The outer thicker membrane is the *chorion*. The latter is attached to the *placenta*, the part of the uterine wall through which nutrients are supplied to the fetus. For the most part, the process of fraternal twinning as described earlier will result in the two zygotes becoming implanted at quite

separated uterine sites. Since members of monozygotic pairs originate from the same single zygote, they are likely to be implanted very close together. As a consequence of these two different developmental patterns, the two types of twins differ in respect to the manner in which they are enclosed in the chorionic and amniotic membranes. The basic differences are shown in Table 12-4 and Fig. 12-4.

It is clear that the occurrence of a single placenta and a single chorion (whether there are one or two amniotic sacs) is a sufficient condition for the diagnosis of monozygosity. The presence of two chorions is a reasonably good prediction of dizygosity, though not an absolutely reliable one; however, coupled with two placentas, it is an almost infallible indicator that the twins are not identical. Note that two chorions exclude the possibility of a single amnion.

It is also clear from the preceding discussion, however, that there is an area of ambiguity in the case of a single, fused placenta with two chorions. Since there is a considerable possibility of error in such a circumstance, additional confirmation is demanded. This can be supplied by postnatal examination of various characteristics of the members of a twin pair. The earliest methods, as used by Galton (1875) and developed later by Siemens (1924), relied on assaying a large number of physical characteristics. This has been called the *polysymptomatic method*, or *similarity method*. Although it is liable to some error, on the whole it is remarkably successful. Thus if two members of a pair are obviously unlike in such features as ear

Table 12-4. Relation between twin types and fetal membranes*

	Percentage distribution			
	Monochorionic		Dichorionic	
	Monoamniotic	Diamniotic	Single placenta (secondary fusion)	Two placentas
MZ	Rare	75%	~25%	Rare (~1%)
DZ	—	—	~50%	~50%

*From Thompson, J. S., and M. W. Thompson. 1967. Genetics in medicine. W. B. Saunders Co., Philadelphia.

shape, eye shape, color of eyes, eyebrows, eyelashes, mouth and lip shape, chin and jaw structure, and hair form (e.g., curly or wavy), then they probably are not monozygotic. Note, however, that MZ twins may still differ, and usually do at birth, in respect to more plastic characters such as weight. However, these initial differences tend to reduce as the twins grow up. Further, mirror imaging sometimes appears for characters liable to asymmetry. Thus hair whorling, which is usually clockwise in direction, may be counterclockwise in one member of a pair (Rife, 1933). Likewise, mouths may tilt in opposite directions, the right eye may be larger in one twin but smaller in the other, and one may be right-handed and the other left-handed. Mirror imaging does not, however, disturb the general impression that the members of the pair are still identical. In fact, a test of zygosity that is 95% accurate is given by asking twins the simple questions, "When growing up, were you as alike as two peas in a pod or of a family likeness only?" (Cederlöf et al., 1961).

Nevertheless, there still occur some pairs of identicals who do not look very much alike and some pairs of "look-alike" fraternals. Consequently, most investigators resort to additional methods of zygosity diagnosis to reduce still further any uncertainty. One of these is blood typing. Ten or more blood groups may be used, for example, ABO, MNS, P, Rhesus, Lutheran, and Duffy. If members of a pair are discordant on any of these, they are fraternal twins. The importance of blood grouping analyses has been emphasized by Carter-Saltzman and Scarr (1977). Another method is by the use of dermatoglyphics, that is, the patterns of the skin ridges on fingers, palms, toes, and soles of the feet. This was first put to scientific use by Galton. Fingerprints are classified into three types: whorls, loops, and arches. Palm and foot characters are similarly classified. Again, we look for identity if the twins are monozygotic.

Both of these methods can clearly exclude the possibility that twins are identical. However, identity for blood groups or dermatoglyphics does not necessarily exclude dizygosity. If parents are identical for all the blood types, for example, it is likely that members of a fraternal pair will also be identical and hence may be misdiagnosed as monozygotic. To get around this problem, it is possible to make a computation regarding the probability of monozygosity that takes into account the parental genotypes. This is done in such a way that if, for example, both parents are alike for several of a number of blood-group systems used, then the probability that concordant twins are monozygotic is reduced (Smith and Penrose, 1955; Race and Sanger, 1968; Wilson, 1970). This makes intuitive sense. The final and possibly most certain method of diagnosis is the skin graft. Grafts are normally accepted between identical twins but are rejected between fraternals (Bain and Lowenstein, 1964). However, this procedure is usually not feasible for practical reasons. Furthermore, some authorities have argued that it may not always be reliable on account of occasional mixture of fetal circulation in nonidentical twins (Bulmer, 1970).

Besides the preceding, some less conventional methods are sometimes used relating to less obvious physiological characters. PTC tasting is one example. Another more exotic method has been reported by Kalmus (Scheinfeld, 1967) who found that police dogs could quite readily distinguish the scents of fraternal twins but became confused when confronted by identicals. This confusion could, however, be overcome by greater familiarization.

Before turning to the statistical methods used in establishing heritability of a behavioral character by use of MZ and DZ twins, we will offer a few cautions about their general usefulness in this regard. Many of the problems have been summarized by Allen (1965).

In the first place, as we have already indicated, there is the possibility of multiple twin types. If these are rather rare, as is thought by most workers, most conclusions educed from twin data would not be seriously affected. However, if they are more common, then they could be of consequence.

Table 12-5. Proportions of twins and singletons born at different stages of gestation (weeks)*

Weeks of gestation	Percentage born at this stage	
	Twins	Singletons
Under 28 weeks	4.0	0.5
28-31 weeks	5.2	0.8
32-35 weeks	10.8	1.9
36 weeks	12.7	8.4
37-39 weeks	14.9	8.8
40 weeks	51.4	76.4
41 weeks and over	1.1	3.2

*From Twins and supertwins by Amram Scheinfeld. Copyright © 1967 by Amram Scheinfeld. Reproduced by permission of J. B. Lippincott Co., Philadelphia, and Chatto & Windus Ltd., London.

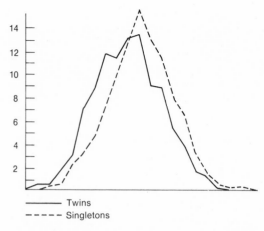

——— Twins
- - - - - Singletons

Fig. 12-5. Comparison of reading scores of twins and single children (girls). (From Husén, T. 1960. Scand. J. Psychol. **1:**125-135.)

Secondly, it should be stressed that twins represent a somewhat atypical sample of the population. To the extent that they are atypical, we must not be too ready to generalize conclusions obtained from twin data to the ordinary populations of singletons. To start with, the crowded uterine conditions in which twins spend the early part of their lives may well have profound effects. Twins usually have a shorter gestation period, as shown in Table 12-5. Furthermore, the period of birth is much more variable than in singletons, most of the latter being clustered in the fortieth week. Only 51.4% of twins are born at this time, however, almost 50% are born prematurely. One consequence of this is that twins tend to have a much lower birth weight. Scheinfeld (1967), summarizing a large amount of data, concludes that more than half of all twins weigh no more than 5½ pounds at birth, almost eight times the number among singletons. Furthermore, as many as 1 twin in 20 weighs as little as 2 pounds 3 ounces; only 1 in 2000 singletons has such a low birth weight. The tendency to low birth weight and short gestation period is greatly accentuated in black as against white twin populations and in same-sex as opposed to opposite-sex pairs (Howard and Brown, 1970).

The preceding factors, plus the additional complication of a much higher incidence of problems associated with delivery, probably make twins rather vulnerable to pathology. Indeed, a very high rate of mental subnormality has been noted for twins by Rosanoff, Handy, and Plesset (1937) and Allen and Kallman (1955). Allen and Kallman found, in fact, that twins constitute 3.1% of all admissions to New York State institutions for the mentally retarded, though only 1.9% (1 out of 88) of the general population are twins. Again, even within the normal IQ range, both Husén (1960) and Zazzo (1960) have found consistent differences for reading scores and IQs, respectively, in favor of singletons. The relevant data are summarized in Figs. 12-5 and 12-6.

On the other hand, it also seems to be true, as pointed out by Vandenberg (1968a), that twins are not necessarily worse on all particular abilities. Thus, while singletons did better in one study on verbal, quantitative, and spatial tests, they were poorer than twins on perceptual tasks (Koch, 1966). It is not clear why this should be, assuming such a finding has some generality.

A final point should be made. It would appear that if twins are different from singletons, this is probably because of the fact that twin samples contain an elevated incidence of individuals who have suffered some pre-

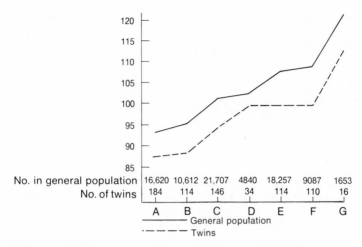

No. in general population	16,620	10,612	21,707	4840	18,257	9087	1653
No. of twins	184	114	146	34	114	110	16
	A	B	C	D	E	F	G

——— General population
----- Twins

Fig. 12-6. Distribution of IQs, by occupation of father, for twins and singletons. *A*, Farm workers; *B*, workers in towns of less than 2000; *C*, workers in larger towns and cities; *D*, salaried employees in towns of less than 2000; *E*, salaried employees in larger towns and cities; *F*, managers and merchants; *G*, professional men. (From Zazzo, R. 1960. Les jumeaux, le couple et la personne. Presses Universitaires de France, Paris.)

natal or perinatal insult. One way of identifying such individuals is by inspection of birth weight records. Pencavel (1976) has done this with 34 MZ pairs reared apart from Shields' (1962) sample, for whom reliable birth data were reported. Out of these, in 22 cases the firstborn had a higher IQ than the second born. Likewise, in 21 pairs for whom qualitative or quantitative data on birth weight were also available, in 11 pairs the firstborn was also the heavier. Putting these results together, Pencavel was able to fit a linear probability function predicting higher or lower IQ in terms of birth order and birth weight, plus an error term. The net result was that, with weight held constant, the firstborn twin had a 15% probability of having the higher IQ. Being the heavier twin, however, yielded a 65% probability of being the more intelligent.

Munsinger (1977) has extended this general finding using all available separated MZ twin samples. He showed that when twins with markedly different birth weights are omitted, similarity of IQs in remaining pairs is markedly elevated. In addition, mean IQ of these is within the normal range.

Furthermore, problems associated with the twin method have less to do with their representativeness as a biological sample than with the representativeness of the environments in which they grow up and especially with the nature of early MZ as opposed to DZ environments. We will consider these problems in the larger context of biometric genetic analysis of continuous characters in human populations.

STATISTICAL ANALYSIS IN HUMAN POPULATIONS

The years after our first book, *Behavior Genetics* (1960), saw a tremendous increase in the development and use of sophisticated methods of analyzing family and twin data for genetic effects. Fundamentally, however, the basic problem has remained the same. The starting point in all analyses relates to the specification of how much the similarities or differences on a trait are caused by genetic and how much by environmental factors. This is the basic division. We may schematize it very simply as

$$P = G + E$$

where
P = phenotypic score on some trait
G = genetic influences
E = environmental influences

This simple division was not accomplished, however, by Galton or by many of those following him who simply presented, as data, correlations between relatives. There is no doubt that it was clear to all of them that families shared the same environments as well as the same genes. Certainly, many of them offered reasonable qualitative arguments as to why a genetic model fitted their data better than an environmental one. However, it is obvious that such arguments are not very convincing. As a consequence, it became necessary to develop quantitative techniques for studying the relative influence of nature as against nurture.

The simplest expression for this has always been the heritability coefficient, h^2. As defined in Chapter 5, this is simply a statistical way of stating the extent to which the variation of a character is determined by genetic influences. Thus:

$$h^2 = \frac{G}{G + E}$$

This appears to be straightforward enough. However, we must bear in mind two major points. In the first place, as we will describe in more detail shortly, both G and E may be separated into between- and within-family influences. Furthermore, these may both be broken down into more specific components. Second, it must be emphasized that a heritability estimation has no absolute value. It is simply a population statistic, like an estimate of mortality rate or birthrate. It is unlikely that we will ever be able to put forward, for some human characters, a universal heritability estimate. Neither is it likely that such a number would, in any case, serve any useful or desirable purpose. We will expand on these two major reservations as our discussion proceeds. First, let us look at some of the methods that have been used to estimate the heritability of a trait in human beings by reference to twin data.

Twin data

A great variety of mathematical expressions have been used to analyze data from MZ and DZ twins, with the goal of quantify-ing the relative effects of nature and nurture. Many of the formulations have no real foundation in biometric theory but, nevertheless, do provide rough and ready ratios that can be compared for different traits studied.

One of the earliest methods, put forward by Lenz and Von Verschauer (1928), involved a single comparison of within-pair MZ differences and within-pair DZ differences, each being first divided by the summed scores of the pair to correct for scale effects. Dahlberg (1926) compared standard deviations of like-sex fraternal and identical twins, taking into account variation due to sex and errors of measurement. Gottschaldt (1939) used the expression

$$MD_{DZ} - MD_{MZ} = \frac{E + H + HE}{E}$$

where

MD_{DZ} = mean difference between DZ co-twins

MD_{MZ} = mean difference between MZ co-twins

E, H, and HE = functions of environment, heredity, and interaction

When discrete traits are involved, the percentage of concordance (agreement between members of a pair) in MZ and DZ pairs may be used to give a measure of heritability.

$$\text{Heritability} = \frac{C_{MZ} - C_{DZ}}{100 - C_{DZ}}$$

with C_{MZ} and C_{DZ} being the percentage of MZ and DZ pairs classified as alike with respect to a trait. Some important refinements of this general method are discussed by Allen, Harvald, and Shields (1967) and by Smith (1974).

With continuous data, analysis of variance leading to the calculation of an intraclass correlation is generally employed. As discussed in Chapter 5, the intraclass correlation is equivalent to the ratio of the between-groups component to total variance (Haggard, 1958) and is defined as

$$r_i = \frac{\Sigma (x_i - \bar{x})(y_i - \bar{x})}{n\sigma^2} = \frac{\Sigma x_i y_i - n\bar{x}^2}{n\sigma^2}$$

where

$\bar{x}$ = mean of all measurements

x_i and y_i = measurements on the ith pair of twins
n = number of pairs
σ^2 = variance of total sample

Note that in this computation, each individual in a pair is entered both as x and y, and the variance is computed for all individuals around $\bar{x}$.

In the present context, the intraclass correlation is interpreted as the proportion of the total variance which arises from the fact that the members of twin pairs differ from each other. The correlation would be 1 if co-twins always received the same score, and 0 if co-twins were no more likely to receive the same score than two individuals selected at random. Minus values would indicate that co-twins were more dissimilar than randomly chosen pairs.

The formal relationship of r_i to variance is stated in the equations

$$1 - r_i = \frac{n\sigma^2 - \Sigma xy + n\bar{x}^2}{n\sigma^2} = \Sigma \frac{(x - y)^2}{2n} \times \frac{1}{\sigma^2}$$

where

$\dfrac{\Sigma (x - y)^2}{2n}$ = Mean square deviation between twin pairs

$\dfrac{1}{\sigma^2}$ = Reciprocal of total variance

Then

$$1 - r_{MZ} = \frac{\sigma^2_{MZ}}{\sigma^2}$$

$$1 - r_{DZ} = \frac{\sigma^2_{DZ}}{\sigma^2}$$

where

σ^2 = population variance
σ^2_{MZ} = within-pair MZ variance
σ^2_{DZ} = within-pair DZ variance

Holzinger's (1929) well-known heritability coefficient, H, is:

$$H = \frac{r_{MZ} - r_{DZ}}{1 - r_{DZ}} = \frac{\sigma^2_{DZ} - \sigma^2_{MZ}}{\sigma^2_{DZ}}$$

This coefficient is not the same as that which we symbolized as h^2 previously. Holzinger's ratio gives the proportion of differences produced only by genetic differences *within families*. Accordingly, H and other comparable indices derived from a comparison of MZ and DZ intrapair differences underestimate the effects of the genetic component in the general population by a factor that approaches 2 as heritability decreases.

In view of this latter fact, Nichols (1965) has used a formula that corrects for this underestimation as follows:

$$HR = \frac{2 (r_{MZ} - r_{DZ})}{r_{MZ}}$$

It is usually essential to evaluate the significance of the difference between MZ and DZ variances. This can be readily accomplished by use of an F test. Thus:

$$F = \frac{\sigma^2_{DZ}}{\sigma^2_{MZ}}$$

Vandenberg (1966) has shown the relationship between F and H when the latter is expressed in variance terms:

$$H = \frac{\sigma^2_{DZ} - \sigma^2_{MZ}}{\sigma^2_{DZ}}$$

$$H = \frac{\sigma^2_{DZ}}{\sigma^2_{DZ}} - \frac{\sigma^2_{MZ}}{\sigma^2_{DZ}} = 1 - \frac{\sigma^2_{MZ}}{\sigma^2_{DZ}}$$

Since

$$F = \frac{\sigma^2_{DZ}}{\sigma^2_{MZ}}$$

$$H = 1 - \frac{1}{F}$$

or

$$F = \frac{1}{1 - H}$$

The relation between H, HR, and F and genetic and environmental components of variance has been shown by Jinks and Fulker (1970) to be

$$H = \frac{r_{MZ} - r_{DZ}}{1 - r_{DZ}} = \frac{G_1}{G_1 + E_1}$$

$$HR = \frac{2 (r_{MZ} - r_{DZ})}{r_{MZ}} = \frac{2G_1}{G_1 + G_2 + E_2}$$

$$F = \frac{1}{1 - H} = \frac{G_1 + E_1}{E_1}$$

where
G_1 = within-family genetic component
G_2 = between-family genetic component
E_1 = within-family environmental component
E_2 = between-family environmental component

All three formulae involve several basic assumptions that can now be stated.

1. Environmental and genetic variances are additive. Another way of stating this is to say that environmental and genetic components neither interact nor covary. It is not always easy to test this assumption, although, as we shall see, several attempts have been made to do so in the domain of human traits.

2. MZ and DZ pairs have equivalent means and between-pair variances on the character studied. This can readily be checked on the sample of twins used.

3. MZ and DZ pairs are treated nearly enough alike so that environmental differences between co-twins are equal in both types. This is perhaps the most crucial assumption and has been widely debated. A number of factors may operate to reduce σ_{MZ}^2 as compared with σ_{DZ}^2 and thus lead to an overestimation of the importance of heredity (Östlyngen, 1949; Scheinfeld, 1967). MZ co-twins may be treated more alike and are often confused by parents and associates (Jones, 1955; Scarr, 1969). They may model their behavior on each other to a greater extent than DZ co-twins (Smith, 1965). Furthermore, when comparisons involve a subjective element, the obvious physical similarity of MZ twins may induce an underestimation of psychological differences *(halo effect)*. Likewise, the expectation of psychological differences associated with physical differences may lead to an overestimation of σ_{DZ}^2.

These possible sources of bias in twin studies are difficult to handle. It may well be that the real genetic similarity of MZ co-twins simply serves to homogenize the environments in which they grow up. Thus parents who treat them alike may do so because of their likeness, but without, however, actually increasing this likeness. On the other hand, being treated alike and exposed to similar environments may increase the likeness beyond what it would be if they had been raised apart. This would be a case of genotype-environment covariance. Finally, it may be that their genetic identity has no relation to their likeness on some measured character that is caused solely or mainly by parents or other agents in the environment. This would, of course, render trivial the meaning of any heritability estimates derived from twin data. So far as the writers are aware, no general decision can be made between these alternatives. It is likely that each twin study must be examined separately on its own merits. However, one promising approach to the problem has been put forward by Scarr (1969a), who outlined two hypotheses: one is that the belief of the parents is irrelevant; that however they may classify their twins, they treat them according to their true zygosity. The second is that the similarities of MZ and DZ twins will be governed mainly by parental beliefs about their zygosity. In other words, if parents think two co-twins are monozygotic, they will treat them more alike; if they think they are dizygotic, they will treat them as less alike and accentuate their differences.

Scarr gathered data on 19 MZ and 22 DZ twin pairs correctly classified by parents, and 4 MZ and 7 DZ pairs incorrectly classified. These were compared with respect to five simple rated variables and two personality scales. On the whole, the data support the first hypothesis. Thus MZ twins are said by parents to be treated as more alike than DZ twins regardless of the correctness of their classification. At the same time, the disparity between σ_{MZ}^2 and σ_{DZ}^2 is a good deal greater for MZs and DZs correctly classified than for incorrectly classified twins, a finding in line with the second hypothesis.

Since the twin samples used were very small, these conclusions can only be regarded as tentative. It is probable that, regardless of blood typing (the criterion for zygosity determination), misclassified MZs looked less alike than correctly classified MZs and, likewise, that incorrectly classified DZs looked more alike than correctly classified DZs. In fact, the data themselves indicate this. Thus the main determinant of parental treatment is probably appearance rather than actual zygosity (a rather academic matter for most parents). If, in general, MZ co-twins appear more similar than DZ co-twins, as is certainly true on the average, then it seems likely that parental treatment will at the very least sustain this similarity, if not increase it.

However, as indicated previously, it is difficult to formulate any conclusion that will be true for all twins. Some parents appear to like having "two of the same" and will actively promote likeness, whereas others will dislike such redundancy and do their best to magnify differences. An indication that the latter course of action is often taken is supplied by the twin data of Wilde (1964) who found that DZs reared in different homes were more alike on a number of personality traits than DZs reared in the same families.

Two later studies have directly attacked the problem of whether perceived similarity makes for increased likeness in twins. The first of these by Plomin, Willerman, and Loehlin (1976) involved the use of a "confusability" index, which is essentially a scale given to parents who rated how often their twins were confused with each other. This, in turn, was correlated with measures of activity, sociability, and impulsivity in two independent studies. The major result was that in MZ pairs confusability did not correlate with similarity in personality. In fact, five of the eight correlations were slightly negative, suggesting a contrast effect. Two correlations were significantly positive for DZ twins, however. The authors explained this result by suggesting that DZ twins who look more alike are, in fact, genetically more alike and, therefore, more similar in personality. This is possible. However, it is perhaps more likely that the relative sizes of the correlations are determined by the magnitudes of variances in the two groups. Confusability shows a variance of around 1.7 in the MZ sample, but between 4 and 5 in the DZ sample. Furthermore, if we consider the combined sample of MZs and DZs, variance is between 5 and 6, and, coordinately, all but one of the confusability × personality trait correlations are significant. Consequently, the results must be treated with some caution.

The study by Matheny, Wilson, and Dolan (1976) yielded results essentially in agreement with those of Plomin, Willerman, and Loehlin. On the Porter and Cattell Chil-

dren's Personality Questionnaire, there were significant correlations for only three dimensions out of fourteen for MZ pairs, and none for DZ pairs. Likewise, no significant correlations were found for either type of twin between similarity score and two IQ tests, two perception tests, an achievement test, and a speech accuracy test. The authors therefore concluded that "perceived similarity of same-sex twins is not a significant determiner of behavioral outcome."

As matters stand, then, there is little support for the simple environmentalist assumption that MZ co-twins are more alike because they look more alike. Note, however, as we suggested before, that data on one link in the chain of events still appears to be lacking: the relation between perceived similarity of twins and their actual treatment by authority figures, by peers, and by each other.

Some possible causes of increased σ^2_{MZ} relative to σ^2_{DZ} have been suggested (Östlyngen, 1949; Price, 1950; Allen, 1965; Bulmer, 1970). The prenatal conditions of monochorionic twins, particularly mutual blood circulation, may often be unfavorable for one co-twin and may result in an increased environmental variation unique to MZ pairs. MZ twins have more reversed asymmetries than DZ twins, as is shown by the larger proportion of discordance of handedness in MZ pairs (Nagylaki and Levy, 1973). Again, MZ twins have been observed to adopt complementary roles in their outside contacts, one serving as spokesman while the other is quiet (von Bracken, 1936). Sometimes rebellion against identification with an identical co-twin leads to the adoption of a different role. When persons are rating behavior, a halo effect can work in reverse if the obvious resemblance between MZ co-twins leads to the exaggeration of minor differences (*contrast effect*). Finally, errors of measurement are more serious when the true difference is small. For example, even if the true σ^2_{MZ} were zero, a test of low reliability could often yield different scores for MZ co-twins. On the other hand, the same unreliable test might serve to distinguish quite adequately

DZ co-twins who were more different from each other. The problem of intrusion of error measurement into twin studies is of obvious importance (Loehlin, 1965a).

It is probable that some of the effects just listed will balance each other out. Yet others are almost impossible to control. For example, little control is possible over the primary prenatal and natal biases that must inevitably affect the makeup of twins. Likewise, it is impossible to keep the development of genetically different DZ co-twins as closely in step as it is in a pair of MZ twins. The different genotypes of DZ pairs must interact differently with the environment, and their responses lead to further differentiation accumulated on a genetic base. If MZ twins are placed in objectively different environments, it is conceivable, even probable, that they will select similar parts of these environments for their attention and effectively reduce the psychological consequences of environmental variability. Such environment self-selection must be a process of great importance. Yet behavior geneticists have hardly begun to study this experimentally.

Another variant in the usual twin design has been the method of co-twin controls for specific experimental procedures. Specifically this involves the imposition of some experimental variable on one MZ co-twin with the other member of the pair serving as an untreated control subject. For example, one co-twin may be given some learning experience, but not the other. The two are then compared at the end of the training period. Such deliberate manipulations of environment can tell us much about the plasticity of genotype. However, although the usefulness of the method was argued by us in 1960 and later by Thompson and Wilde (1973), it has not begun to be used on any large scale. We will consider in subsequent chapters a few examples of the application of this approach.

In summary, the use of twins in human behavior genetics involves many hidden assumptions and many sampling difficulties. For more information on the statistical treatment of these, the reader should consult Kempthorne and Osborne (1961), Haseman

and Elston (1970), and Christian, Kang, and Norton (1974), in addition to some of the sources already listed.

Twin-family methods

We have indicated previously some of the limitations of twin methods. We may obtain a great deal more information if we include in our designs relatives of other kinds, for example, full sibs reared together and apart, half-sibs, foster children, and others. This allows us to set up sets of equations in which various empirical values representing resemblances (correlations) or differences (variances) are expressed in terms of theoretical genetic and environmental components. Estimated values for the latter can then be found by application of various procedures deriving from the general method of simultaneous equations. We will present here three such approaches, each of which handles the problem somewhat differently.

Cattell's MAVA. *Multiple abstract variance analysis,* developed by Cattell (1960, 1973), attempts to apportion the variance of a trait between the following components: within- and between-family genetic, within- and between-family environmental, and the covariances between all of these. Cattell is careful to acknowledge the fact that gene-environment interactions also may occur but submits that the assumption of uncorrelated heredity and environment, which a formal analysis of variance would require, is not tenable for most behavior traits (Cattell, 1960).

The complete expression for the total variance of a trait in some society contains ten terms.

$$\sigma_{so}^2 = \sigma_{wg}^2 + \sigma_{we}^2 + \sigma_{bg}^2 + \sigma_{be}^2 +$$
$$2r_{wg,we}\sigma_{wg}\sigma_{we} +$$
$$2r_{wg,be}\sigma_{wg}\sigma_{be} + 2r_{wg,bg}\sigma_{wg}\sigma_{bg} +$$
$$2r_{we,be}\sigma_{we}\sigma_{be} + 2r_{we,bg}\sigma_{we}\sigma_{bg} +$$
$$2r_{be,bg}\sigma_{be}\sigma_{bg}$$

The last six terms simply represent the covariances between the two genetic and two environmental components in all possible combinations. Some of these may be dropped on the grounds that they may not correspond plausibly to any real situation. For example, there is no reason why a genetic deviation from the family mean should be correlated with a genetic deviation of the family from the population mean. Hence, $r_{wg,bg}$ can be taken as zero. The same should apply to the term $r_{we,be}$, which can then also be dropped. The other covariance terms, however, may assume importance through *autogenic* mechanisms, by which certain environments are selected by, or themselves select, certain varieties of genotypes. For example, a family whose members are, on the average, genetically bright will likely create for these members an environment conducive to increasing brightness even more. Thus the term $r_{bg,we}$ is probably not zero. A firm decision as to each term must, in the end, be an empirical matter. But given the large number of unknown or abstract variances that are possible, it would seem advisable to exclude those which are less plausible on a priori grounds.

The critical step in MAVA is the selection of various types of families for which variances are expressible in terms of the unknown variances. In fact, as Loehlin (1965b) has pointed out, there are basically two equations, each with ten possible terms: one representing the within-family, the other, the between-family variance. These equations are modified according to the type of family used simply by omitting whichever correlation terms are not relevant. For example, the complete within-family variance equation is given by

$$\sigma_w^2 = \sigma_{wg}^2 (1 - r_{wg_1wg_2}) + \sigma_{we}^2 (1 - r_{we_1we_2}) + \\ \sigma_{bg}^2 (1 - r_{bg_1bg_2}) + \sigma_{be}^2 (1 - r_{be_1be_2}) + \\ 2\sigma_{wg}\sigma_{we} (r_{wg_1we_1} - r_{wg_1we_2}) + \\ 2\sigma_{wg}\sigma_{bg} (r_{wg_1bg_1} - r_{wg_1bg_2}) + \\ 2\sigma_{wg}\sigma_{be} (r_{wg_1be_1} - r_{wg_1be_2}) + \\ 2\sigma_{we}\sigma_{bg} (r_{we_1bg_1} - r_{we_1bg_2}) + \\ 2\sigma_{we}\sigma_{be} (r_{we_1be_1} - r_{we_1be_2}) + \\ 2\sigma_{bg}\sigma_{be} (r_{bg_1be_1} - r_{bg_1be_2})$$

where the subscripts *1* and *2* are used to designate the two individuals in a pairing (for example, siblings)

By setting to 1 all correlational terms in this equation, except $r_{we_1we_2}$, which is set at

0, one comes out with the expression for variance between identical twins reared together, that is, σ_{we}^2. What hypotheses are made about the covariation between genotype and environment in any instance are thus crucial.

In his complete multiple variance design, Cattell puts forward seventeen such equations. His "limited resources design" involves only ten equations by omitting the use of identical twins reared apart and half-siblings reared apart—material not so easily available. One set of family equations and its solutions are given on p. 250 as an example of Cattell's procedure. Others may be found in his original articles (Cattell, 1953, 1960, 1973).

The solution of this group of equations involves setting the values of each noncanceling correlation coefficient at intervals of 0.1 from −1 to +1 and finding which estimate gives the greatest internal consistency among variances that depend and do not depend on it.

The MAVA method is capable of extensions to include many variables of psychological importance: order of birth, sex and ages of siblings and of parents, cultural subgroups, foster-home characteristics, and the like. Cattell himself (1973) has used it in a number of ingenious ways. However, it is still true (as it was in 1960) that few others have done so. This is probably mainly a result of the very large samples that are necessary because of the magnitude of standard errors generated by the sheer number of linear combinations of terms used (Loehlin, 1965b; Eaves, 1972). Furthermore, the method tells us nothing about gene action, a lack that is rectified in alternative methods. In spite of these failings, however, MAVA has been useful in generating some most interesting ideas about gene-environment relationships in human behavior. We will take up some of these in later chapters.

Path coefficient analysis. We have pointed out that a straightforward correlational analysis of family data is not likely to yield very useful results, unless the study is designed in special ways to allow partitioning of the co-

Components of family variance

Identical twins together	$\sigma^2_{ITT} = \sigma^2_{we'}*$
Siblings together	$\sigma^2_{ST} = \sigma^2_{wg} + \sigma^2_{we} + 2r_{wg,we}\sigma_{wg}\sigma_{we}$
Siblings reared apart	$\sigma^2_{SA} = \sigma^2_{wg} + \sigma^2_{we} + \sigma^2_{be} + 2r_{wg,we}$
	$\sigma_{wg}\sigma_{we}\ (+\ 2r_{wg,be}\sigma_{wg}\sigma_{be})$†
Unrelated children together	$\sigma^2_{UT} = \sigma^2_{wg} + \sigma^2_{we} + \sigma^2_{bg} + 2r_{wg,we}$
	$\sigma_{wg}\sigma_{we} + 2r_{we,bg}\sigma_{we}\sigma_{bg}$‡
Unrelated children apart	$\sigma^2_{UA} = \sigma^2_{wg} + \sigma^2_{we} + \sigma^2_{bg} + \sigma^2_{be} +$
	$2r_{wg,we}\sigma_{wg}\sigma_{we} + 2r_{bg,be}\sigma_{bg}\sigma_{be}$§

Solutions for the unknown variances are as follows:

$$\sigma_{we'} = \sqrt{\sigma^2_{ITT}}$$

$$\sigma_{be} = \sqrt{\sigma^2_{SA} - \sigma^2_{ST}}$$

$$\sigma_{wg} = -r_{we,wg}\sqrt{\sigma^2_{ITT}} \pm \sqrt{\sigma^2_{ST} + (r^2_{wg,we} - 1)\sigma^2_{ITT}}$$

$$\sigma_{bg} = \sqrt{\sigma^2_{UT} - \sigma^2_{ST}}$$

$$r_{be,bg} = \frac{\sigma^2_{UA} - \sigma^2_{UT} - \sigma^2_{SA} + \sigma^2_{ST}}{2\sqrt{\sigma^2_{SA} - \sigma^2_{ST}}\sqrt{\sigma^2_{UT} - \sigma^2_{ST}}}$$

---

*$\sigma^2_{we'}$ acknowledges the possibility that twins may occupy a different kind of family environment than nontwins.

†The term in parentheses can be used if selective placement is considered to operate.

‡The term $r_{wg,bg}$ is dropped.

§The terms $r_{wg,bg}$ and $r_{we,be}$ are dropped.

variance at least into genetic and environmental components. Few studies done before 1960 were of this type and therefore contributed not much more than a glimpse of the complexities that lay ahead. Path analysis represents a more sophisticated type of correlational analysis. Although it has been used only in a few instances by behavioral geneticists, it does seem to have great potential usefulness, as already demonstrated in agricultural genetics. It was, in fact, a distinguished worker in the latter field, Sewell Wright, who first put forward the method in a series of articles dating back as far as 1918 (Wright, 1931, 1934a). Some examples of its application to behavioral science problems are furnished by Duncan (1966, 1968), Morton (1974), and Rao, Morton, and Yee (1974).

Correlations express simply the degree of relationship between two or more variables. By using partial correlations, it is possible to specify the relationship between any set of these with the influence of some others removed. For example, we may start with the three correlations between (1) IQ, (2) school achievement, and (3) socioeconomic class. We may state these as r_{12}, r_{13}, and r_{23}. If, however, we wish to examine, for example, the relation of IQ and school achievement, with the influence of the mutually correlated variable socioeconomic class removed, then we may compute a partial correlation, expressed as:

$$r_{12\cdot3} = \frac{r_{12} - r_{13}r_{23}}{\sqrt{(1 - r^2_{13})\ (1 - r^2_{23})}}$$

This sometimes can provide very useful information. However, with a large number of variables it becomes an unmanageable task. Furthermore and related to this is the fact that any conventional correlation tells us nothing about causality, only association.

However, it would be quite reasonable to suppose in the preceding example, that on a priori grounds it is more likely that a high IQ produces good school marks than the converse. This could be argued both on the grounds that an IQ has temporal priority over school achievement (a child possesses intelligence before he enters school) and also on the grounds of generality (an IQ relates to many other activities besides school performance). This causal direction between two variables is indicated conventionally as follows:

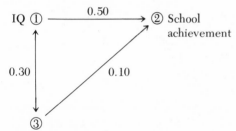

If we consider IQ to causally influence school achievement but to be indeterminate in respect to socioeconomic class, we would represent the situation as follows:

The two-headed arrow between variables 1 and 3 indicates bidirectionality of relationship. The numbers inserted give the values (fictitious) of the two path coefficients, p_{12} and p_{32}, and the correlation, r_{13}. The former are essentially partial regression coefficients in standard forms of school achievement on IQ and on socioeconomic class. The path coefficients indicate the extent to which variance on the achievement variable is influenced independently by the other two.

The basic theorem of path analysis may be stated as follows:

$$r_{ij} = \sum_q p_{iq} r_{jq}$$

where

r_{ij} = correlation between any two variables i and j

q = index representing set of variables from which paths lead directly to X_i

p_{iq} = path coefficient from any of the set of q variables to X_i

r_{jq} = correlation between any q variable and variable j

The actual path values are calculated as *beta coefficients* by a regression procedure. These simply represent the values by which other variables should be weighted in order to yield together the best possible prediction of some main variable of interest. In a complex system, it is necessary to compute beta coefficients for variables independently, with correlated variables partialed out. For example, a five-variable system may be so arranged that the independent path from variable 4 to variable 5 must be calculated as $p_{54 \cdot 123}$. The resulting value may well turn out to be negligible in magnitude. When this happens, one can seek to eliminate some pathways from the original diagram on the grounds that these do not have any real significance. Such a procedure acts as a check on the validity of an a priori causal series (Duncan, 1966). This is of great value, since too fine an analysis with the consequent inclusion of a great many variables may yield the impression that the system contains little more than noise, whereas the use of fewer variables with fewer postulated relationships between them may reveal a basic order. Sometimes a macroscopic approach may have advantages over a microscopic one.

Biometric genetic analysis. Like MAVA, the biometric genetic approach put forward by Jinks and Fulker (1970) also attempts to partition the variance on a trait in a population into genetic and environmental components. Deriving as it does, however, from the earlier formulations of Fisher (1918) and Mather (1949), it goes well beyond MAVA in its dissection of the genetic component. Moreover, it places much greater stress on the major components than on the various interactions between them, such interactions being eliminated by scaling, if possible.

The basic components of interest may be designated as follows:

G_1 = within-family genetic component
G_2 = between-family genetic component

Table 12-6. Expectations of variance components for three kinds of families according to a simple genetic model[*]

Monozygotic twins reared together (MZT)

$$\sigma_W^2 = E_1$$
$$\sigma_B^2 = G + E_2$$
$$\sigma_T^2 = G + E_1 + E_2$$

Monozygotic twins reared apart (MZA)

$$\sigma_W^2 = E_1 + E_2$$
$$\sigma_B^2 = G$$
$$\sigma_T^2 = G + E_1 + E_2$$

Dizygotic twins reared together (DZT) or full sibs reared together (FST)

$$\sigma_W^2 = G_1 + E_1$$
$$\sigma_B^2 = G_2 + E_2$$
$$\sigma_T^2 = G_1 + G_2 + E_1 + E_2$$

Dizygotic twins reared apart (DZA) or full sibs reared apart (FSA)

$$\sigma_W^2 = G_1 + E_1 + E_2$$
$$\sigma_B^2 = G_2$$
$$\sigma_T^2 = G_1 + G_2 + E_1 + E_2$$

---

[*]From Jinks, J. L., and D. W. Fulker. 1970. Comparison of the biometrical genetical, MAVA, and classical approaches to the analysis of human behavior. Psychol. Bull. 73:311-349. Copyright 1970 by the American Psychological Association. Reprinted by permission.

E_1 = within-family environmental component
E_2 = between-family environmental component

These components are estimated from empirical findings on the within-group variance (σ_W^2) and between-group variance (σ_B^2) from different types of families. Expectations concerning these components in various family groups are shown in Table 12-6. The first three sets in the table supply the minimum data for estimating G_1, G_2, E_1, and E_2. MZA, DZT, and DZA (or FST and FSA), however, can yield about as much information, with the exception of the G × E interaction. This first step of obtaining estimates of G_1, G_2, E_1, and E_2 allows calculation of heritabilities by means of the usual formulae described previously.

The decomposition of G_1 and G_2 into components representing different types of gene action is the next step. If we could assume that all gene action was additive and that mating was random, then the observed value of G would estimate half the additive genetic variance. That is:

$$V_G = \tfrac{1}{2}V_A{}^*$$

In this expression

$$\tfrac{1}{2}V_A = \Sigma p_i q_i a^2$$

where

p_i = frequency of allele at ith locus increasing score

q_i = frequency of allele at ith locus decreasing score

a = genotypic value given by each allelic pair

Thus the whole expression represents the case for which it is possible to add the effects of all genes contributing to the score on a trait. It is assumed that p and q add to 1. Where $p = q = 0.5$:

$$V_A = 2\Sigma(\tfrac{1}{2} \times \tfrac{1}{2})a^2$$
$$V_A = \tfrac{1}{2}\Sigma a^2$$

The last expression has already been presented in Chapter 5 in a discussion of quantitative genetics in experimental populations. Similarly, it will be recalled that if we include the possibility of dominant gene action, then

$$G = \tfrac{1}{2}A + \tfrac{1}{4}D$$

and breaking this down for twins and siblings

$$G_1 = \tfrac{1}{4}A + \tfrac{3}{16}D$$
$$G_2 = \tfrac{1}{4}A + \tfrac{1}{16}D$$

Complete derivations of these expressions may be found in Mather and Jinks (1971). It should be noted, however, that the two equations do not apply in the case of other kinds of family groups such as half-siblings. When these are used, other types of G's are necessary.

The occurrence of assortative mating for a trait whereby, for example, spouses tend to be alike tends to alter G_2 but not G_1. If G_2 is larger than G_1 and other things being equal, we may suspect positive assortative mating or homogamy. Its presence may then be

---

[*]Jinks and Fulker and others of the Birmingham group have preferred to use the symbol D for additive gene action and H for dominance. Here we will use A for additive and D for dominance.

tested for directly if the empirical data are available.

As indicated previously, fitting the preceding model to observations is readily feasible only if no genotype-environment interactions or covariances are present. Consequently, it is necessary to test for these in the initial part of any study. The differential reactions of genotypes to the same and different environments is easily testable in populations of animals. But it is not so easy with humans. The method proposed by Jinks and Fulker involves, basically, an examination of the relation between the sums of scores of pairs of MZ twins on a trait and the differences between them. If an interaction between G and E exists, then we should find some correlation between sums $(t_{11} + t_{12})$, $(t_{21} + t_{22})$, . . . $(t_{n1} + t_{n2})$ and differences $(t_{11} - t_{12})$, $(t_{21} - t_{22})$, . . . $(t_{n1} - t_{n2})$, since the former is a measure of genotype, and the latter is a measure of environment. Jinks and Fulker have elaborated this technique for different situations, according to whether the twins are together or separated and whether dominance deviations are present. Both linear and curvilinear covariance may also be detected by means of scatterplots of sums against differences.

Although this procedure is certainly a step in the right direction, it can only be considered imperfect. The sums of scores of MZ co-twins represent not only genotypes but also between-family environments. In other words, both members of a pair may have high or low scores because of environmental reasons. Hence a correlation between sums and differences may also reflect a between-environment and within-environment ($E_2 \times E_1$) interaction. Jinks and Fulker refer to this possibility, which can be covered at least partly by using identical twins reared apart.

A test for genotype-environment correlation can be made in a similar manner. Jinks and Fulker suggest a comparison of the two basic MAVA equations for biological families reared together and reared apart. These are the same, except that the expression for the former involves the term $r_{bg \cdot be}$ and the latter, the term $r_{bg \cdot we}$. If one σ_T^2 is larger than the other σ_T^2, as indicated by an F test, then we may suspect that this is a result of the fact that one of these correlations is appreciably higher.

If either correlated environments or genotype-environment interactions are found to be present, it is difficult to proceed beyond the estimation of G_1, G_2, E_1, and E_2 unless the data can be appropriately rescaled to get rid of them.

If these complications are not present or are removable, analysis can be carried out somewhat along the lines of MAVA. That is to say, observed variances are computed for the different family groups, and these are equated to the different unknown components as outlined before, that is, G_1, G_2, E_1, E_2, A, and D, plus an assortative mating factor (μ), if any. Values are chosen that give the best fit. Standard errors for these can also be calculated. For further details of the statistical procedures, we refer the reader to Jinks and Fulker (1970). We will present later on some of the applications made by these authors to data on personality and intelligence. Variants on the preceding method have been put forward in a number of articles by Burt and colleagues (e.g., Burt and Howard, 1956; Burt, 1972) and by Jensen (1973a). Fundamentally, they involve similar assumptions and procedures.

QUASICONTINUOUS CHARACTERS IN HUMAN POPULATIONS

Most of the characters with which behavior genetics is concerned are complex and probably not inherited in any simple manner. For some of them, as we have shown, evidence for their heritable basis has come from observations of incidence of the character in relatives of affected individuals or probands. However, as Falconer (1965) has pointed out, an increased incidence in relatives tells us little about how strong the hereditary factor is, and this is surely the main issue.

To attempt to solve this problem, Falconer (1965) has suggested that the methods of quantitive genetics developed to deal with *threshold characters* may be applicable. In addition, he draws heavily on the basic con-

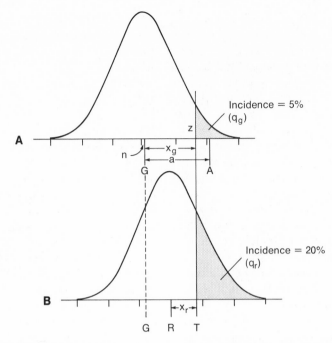

Fig. 12-7. Distributions for general population (**A**) and relatives of affected probands (**B**) compared with reference to the fixed threshold of liability, *T*. *G*, Mean liability of general population; *A*, mean liability of affected individuals in general population; *R*, mean liability of relatives; *q*, incidence, that is, proportion of individuals with liabilities exceeding the threshold; *x*, deviation of threshold from mean, that is, normal deviate; *z*, height of the ordinate at the threshold; *a*, mean deviation of affected individuals from the population mean (*z*/*q*); *n*, mean deviation of normal individuals from the population mean, *z*/(1 − *q*); subscript *g* refers to the general population, subscript *r* to the relatives. (From Falconer, D. S. 1965. Ann. Hum. Genet. **29:**51-76:)

cept of *liability* as first proposed by Carter (1961) in connection with congenital pyloric stenosis. The basic analytical operation involved in Falconer's method is the conversion of the information contained in incidence data into estimates of family correlations. Let us now look at the theory behind the model.

It is assumed that underlying an illness which has an all-or-none character there is a graded attribute designated as liability to the disease. Below a certain value of liability—the *threshold*—the individual is normal; beyond this value he is affected. It is important to emphasize that this attribute includes not only the individual's innate tendency to contract the illness *(susceptibility)*, but also the whole set of environmental circumstances

that contribute as well. The relationships between variation of liability, threshold, and incidence are shown in Fig. 12-7. It is assumed, as shown, that liability is normally distributed. This allows us to use units (standard deviations) for the attribute. It is also an assumption that excludes characters carried by major genes, since such characters would show discontinuous liabilities. The distributions shown in Fig. 12-7 illustrate the way in which incidence data relate to liability and threshold. The upper curve represents the general population with an incidence of the illness of 0.05. The bottom curve has a higher mean liability and hence shows a higher incidence of 0.20. This distribution is taken to represent relatives of affected probands. The two populations have, respectively, mean liabilities of -1.6 and -0.8 below threshold. This assumes, of course, equal variance of liability in the two groups.

Now by reference to tables of the normal distribution, it is possible to establish, for any given incidence, the distance of the threshold from the mean in standard deviation units. These distances may be labeled x_g and x_r. The difference in liability between the general population and relatives of probands is then $R - G = x_g - x_r$.

Individuals in the general population who are affected have a mean liability, A, which deviates from the general mean by the amount a in σ units. This mean is above the threshold from which it deviates by an amount $a - x_g$, where $a = z/q$. Z is the height of the ordinate of the normal curve at the threshold corresponding to the incidence q.

We may also wish to specify the mean liability of normal individuals in the general population. Unless the incidence of the illness is high, this will deviate only slightly from the population mean. It is given as n in Fig. 12-7. Note that since $a = z/q$ and $n = z/p$ where $p = 1 - q$, then $n = aq/p$.

We are now ready to estimate heritability. The data in the form just given are analogous to those used in a selection experiment. What we have done, essentially, is to "select" out of the general population a number of persons having a trait (the sample with mean A) and then look at their relatives who have a mean R. The difference between the general population and the selected group is $A - G$ and represents the selection differential. The difference between the means of the relatives and the mean of the general population represents the selection response, $R - G$. The ratio of these two differences in mean liability is equivalent to the regression (b) of relatives on probands in respect of liability. Thus:

$$b = \frac{R - G}{A - G}$$

We have already discussed the relation between selection differential and selection responses in Chapter 5. It will be recalled that

if we graph R against A, the slope of the resulting line is the regression of relatives on probands. If we transform incidences into liabilities, we get:

$$b = \frac{x_g - x_r}{a}$$

These may be found, as already indicated, by reference to normal distribution tables. Various methods can provide us with a standard error of estimate for b.

Once the regression is found, this leads to an estimate of heritability. The two are related as follows: let P be the phenotypic value (i.e., liability) of any individual, R be the phenotypic value of a relative, and r be the coefficient of genetic relationship. The regression of R on P is

$$b_{RP} = \frac{COV_{RP}}{V_P} = \frac{rV_A}{V_P} = rh^2$$

where

COV_{RP} = covariance of probands and relatives
V_A = additive genetic variance
V_P = total phenotypic variance

Thus:

$$h^2 = \frac{b}{r}$$

The value of r will vary according to the degree of relationship, being ½ in the case of first-degree relatives (children, parents, sibs), ¼ for second-degree (uncles, aunts, nephews, nieces), and so on. For MZ twins, of course, $r = 1$; hence $h^2 = b$.

Since incidences in the various groups bear a fixed relation to the quantities x_g, x_r, and a, it is possible to construct a nomograph from which h^2 may be read off directly for first-degree relatives. This is depicted in Fig. 12-8. It will, of course, supply an approximate rather than an exact solution. We will present a number of examples of its application in later chapters.

Falconer is careful to point out several basic assumptions that must be met if the model is to have validity. First, the general population samples and the probands sampled should both be representative samples of the same population. Second, the sample

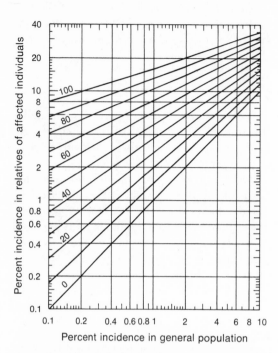

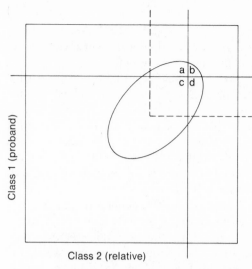

Fig. 12-8. Graph for estimating the heritability of liability from two observed incidences, when the relatives are sibs, parents, or children. (Explanation in text.) (From Falconer, D. S. 1965. Ann. Hum. Genet. 29:51-76.)

Fig. 12-9. Distribution of a dichotomized variable in two genetically related populations: for example, class 1 = normal and affected probands; class 2 = normal and affected relatives of individuals in class 1, such as siblings. Broken lines designate the means of each class. Solid lines separate affected from nonaffected individuals. Thus quadrant *a* would represent, for example, normal sons of affected fathers; *b*, affected sons of affected fathers; *c*, normal sons of normal fathers; and *d*, affected sons of normal fathers. (From Edwards, J. H. 1969. Br. Med. Bull. 25:58-64.)

from which the relatives are drawn should also belong to the same population. Essentially, this means that the three groups should have equal variances of liability and should be equated in other ways as well. One of the best ways of making such equations is by using a series of controls. The main control would be simply a group chosen from the population so as to be matched with affected probands for critical variables deemed to be important, such as age and sex. Relatives of these controls are then located, and incidence of the character established for both groups. Falconer discusses various extensions of the method and also the errors that can arise from its use. However, these need not concern us at present. We refer the reader directly to the articles of Falconer (1965, 1967).

The basic method just outlined has been extended and refined by various workers af-

ter Falconer. These include Edwards (1969), C. Smith (1970, 1971), James (1971), and Reich, James, and Morris (1972). The complexity of most of these treatments is beyond the scope of this book. However, it is perhaps worthwhile to discuss briefly one of them which seems fundamental, that put forward by Edwards (1969). In general, this model attempts to deal more satisfactorily with a problem fully recognized by Falconer: the possibility of reduced variance and skewness of liability in samples of probands' relatives. The procedure involves the estimation of a correlation between a sample of normal and affected individuals drawn from the general population and their normal and affected relatives. This can be computed from a bivariate surface divided into four quadrants by dichotomies (equivalent to thresholds) *x* and *y* units from the mean. Such a surface is illustrated in Fig. 12-9. Note that ratios of frequencies in different quadrants or combinations of quadrants yield the values we are interested in; in particular, $b/(b + d)$ gives incidence of the character among relatives of probands. Likewise, $(b + d)/N$ (where $N = a + b + c + d$) gives incidence of the trait in the population. All relevant information can, in fact, be summarized by means of a tetrachoric correlation. The latter can be computed or found directly from appropriate tables. Edwards (1969) has constructed a nomograph showing the relationship between population, incidence, incidence in first-degree relatives, and heritability up to 80%. Beyond the latter value, heritability may be derived by the application of a formula given by Edwards.

We may conclude this discussion with a word of caution. There are many hidden assumptions in models of the kind presented. Depending on which are accepted and which are qualified or rejected, it is possible to educe almost any kind of genetic model from data. Furthermore, depending on the model accepted, heritability estimates will differ radically. Consequently, it may be that further developments in statistical methodology will not add appreciably at the time to our knowledge of the genetics of behavior traits. There still remains much to be done at a simpler and more basic level. We will attempt to outline some possibilities in Chapter 18.

SUMMARY

In summary, we may state that the methods of human behavior genetics will vary somewhat depending on whether the trait studied is discrete, continuous, or quasicontinuous. However, all methods are derived from basic Mendelian principles, and all involve assessment of trait incidences or trait scores in individuals genetically related to probands. Twins and adoptees constitute particularly useful material, though the use of these involves many special problems, some purely methodological and others of some substantive interest. Among the great variety of methods available for partitioning genetic and environmental components, we have discussed in particular those of Cattell, Jinks and Fulker, Falconer, and Edwards. Few, if any, of these allow us to arrive at uniquely valid conclusions. Consequently, caution must be exercised in their application to human behavioral traits.

13

Neurobiological characteristics

An organism's capacity to behave adaptively is ultimately dependent on the basic sensory and response systems it can utilize. These must work in conjunction with each other. Thus the superb raptorial equipment of a hawk would be useless in capturing prey if it were not coupled with extraordinary visual acuity. Likewise, the scavenging responses of a catfish work mainly through chemoreceptors on its body surface. Many of the psychological differences between species are finally understandable in terms of their basic input and output processes.

This chapter will attempt to survey the genetics of relatively simple sensory-perceptual and response systems in human beings. It is possible that information about these will, in time, prove to be invaluable for an understanding of the more complex forms of behavior considered in other chapters.

SENSORY PROCESSES

Anomalies of sense organs are quite common in man (Sorsby, 1970; Wolstenholme and Knight, 1970). Many of these are associated with gross deformity or even the absence of particular organs. In severe conditions such as amaurotic idiocy, there are concurrent anomalies of the central nervous system that directly impair intelligence.

We shall deal only briefly with sensory variations that result from major structural defects. Blindness and deafness have important psychological consequences, but in educational or therapeutic work, the genetic or nongenetic etiology of the condition is somewhat less relevant than the management of the problem. The question of hereditary origin is, however, of great importance to the genetic counselor who may have to advise on

the possibility of the trait appearing in the siblings or offspring of an affected individual. Major emphasis will be placed here on variations that are known only by their behavioral manifestations and that are most conveniently studied by psychological and psychophysical techniques. For the most part, we will be dealing with capacity to discriminate stimuli in the different sensory modalities.

Taste

In 1931 Fox reported on the phenomenon of "taste-blindness." He used this term to describe the inability of some members of his laboratory staff to characterize as "bitter-tasting" the substance phenylthiocarbamide (PTC). Blakeslee and Salmon (1931) and Snyder (1931) corroborated Fox's finding and announced independently that the inability to taste PTC and related compounds was transmitted by a single recessive gene. In Chapter 12, we presented Snyder's actual data bearing on this conclusion, so we will not report them again here. Generally speaking, the model has stood up well in numerous other studies (e.g., Pons, 1960). Distributions of taste thresholds for most chemical compounds are Gaussian. Taste of quinine is an example (also bitter), and there appear to be substantial correlations between at least high thresholds for this substance and various others, including sodium chloride, sucrose, and chlorpromazine (Kaplan, 1968). However, taste thresholds for the phenylthiourea (the correct name for PTC) type of antithyroid compounds is distinctly bimodal in nature. Members of this group contain the characteristic:

$$H-N-C=S$$

Sensitivity to all compounds containing this is strongly correlated with ability to taste PTC.

Various methods have been used for assaying PTC tasting. These include the use of dry crystals, impregnated filter paper, and test solutions sampled through straws. The general technique of Harris and Kalmus (1949) appears most adequate, although it is too complex and time consuming for the large-scale type of field investigation usually desirable in population genetics. Subjects are given a few cubic centimeters of test solution in a glass. The concentration is gradually raised until a positive response is given. A confirmatory test requires that the subject separate correctly eight glasses, four containing water and four with the test concentration. Some examples of the bimodal distributions found with this method are shown in Fig. 13-1 (Barnicot, 1950). These are quite typical of findings in a great variety of populations and racial groups (e.g., Saldanha and Nacrur, 1963). The wider distribution shown by tasters includes, of course, both heterozygotes and dominant homozygotes.

The nontaster, it should be noted, is not deficient in general taste acuity, which may vary greatly between different compounds. To Blakeslee, this has meant "we live in different taste worlds." PTC tasting ability itself turns out to be extraordinarily specific. Thus

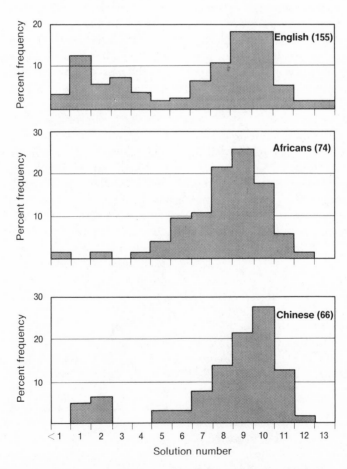

Fig. 13-1. Taste thresholds in populations of English, Africans, and Chinese. Solution 1 is 0.13% PTC; each succeeding solution is one half the concentration of the preceding. Bimodality of taste acuity is clearly shown in the English and Chinese groups. (From Barnicot, N. A. 1950. Ann. Eugen. **15**:248-254.)

tasters are apparently not able to detect PTC if it is dissolved in the saliva of nontasters or even of other individuals who are themselves tasters. Only when PTC is dissolved in the subject's own saliva can a response be elicited. The reasons for this specificity are still not clear (Cohen and Ogden, 1949).

The function of the taster gene (beyond being a useful marker trait for geneticists) is unknown. Attempts to demonstrate an association between its presence and various other characteristics have not proved very successful. A rather intensive study by Kaplan and colleagues (Kaplan, 1968) using the related compound 6-n-propyl-2-thiouracil (PROP) showed the following: sensitive tasters reported a larger number of food dislikes, preferred mild-tasting food, contained a relatively high proportion of nonsmokers, and displayed greater introversion (as inferred from WAIS results) and greater "general systemic reactivity." There was some association, also, between tasting and incidence of duodenal ulcers and between nontasting and incidence of gastric ulcers.

Many, if not most of these associations are correlational. However, it is possible that in some cases the taste threshold may be significantly altered by some of its correlates. For example, it is not unlikely that prolonged smoking may have such an effect. The whole problem of threshold variability as a function of environmental variables has not been sufficiently explored.

In general, then, although PTC tasting is clearly genetic, its function is not entirely clear. Boyd (1950) has suggested that it might protect against intake of substances high in antithyroid compounds. However, there is no definite evidence for this. More research explicitly aimed at this problem is desirable.

Auditory function

Many types of inherited deafness have been identified. Probably the majority of these appear to be secondary effects of some primary disorders, as, for example, in the case of Pendred syndrome (goiter) or of Down syndrome. Pedigree analyses have implicated both autosomal and sex chromosomes and both recessive and dominant inheritance. The psychological consequences of deafness are obvious enough and have been discussed by Anastasi (1958a).

Of more interest to the student of behavior are the studies concerned with auditory discrimination. Many of these were initiated in order to look for possible genetic bases for musicality. Pitch discrimination is obviously useful to a musician, although many other capacities are needed to make a fine performer or composer. The tests have often been criticized as predictors of musicality, but this does not necessarily detract from their value as indicators for genetic studies.

Stanton (1922) administered the Seashore tests of pitch, intensity, time, and interval discrimination to 85 members of the families of 6 well-known American musicians. She concluded that a child from musical stock had a better chance of being musically gifted, but the basis of her sampling was too narrow, and no statistical analyses were possible.

Mjoen (1925) gave a pitch discrimination test to a group of parents and offspring and found a high correlation between midparent and offspring scores. He proposed that the trait had a simple genetic basis, but his evidence for this view cannot be regarded as adequate.

Friend (1939) computed familial correlations for parents and 42 kindergarten children on performance on the Seashore tests of pitch, intensity, and consonance. Results are shown in Table 13-1. Correlations were uni-

Table 13-1. Parent-offspring correlations for pitch, intensity, and consonance as measured by the Seashore test*

	Midparent × child	Father × child	Mother × child
Pitch	0.14	0.02	0.09
Intensity	0.46	0.16	0.28
Consonance	−0.11	−0.04	−0.08

*From Friend, R. 1939. Influences of heredity and musical environment on the scores of kindergarten children on the Seashore measures of musical ability. J. Appl. Psychol. **23**:347-357.

formly low, although highest for intensity and higher for mother-child than for father-child. The latter presumably reflects maternal influence in the home.

A critical review of "absolute pitch" was carried out by Neu (1947), who cited much evidence for the acquisition of pitch discrimination by training. He justifiably criticized the notion that this ability was an inherent faculty or quality determined strictly by genotype and developing in vacuo. However, the tone of Neu's review implied that he thought of heredity as a static factor whose work was finished when the organism was born. Equal training in pitch discrimination might produce greater individual differences in pitch discrimination than no training. The possible genetic determinants of auditory discrimination should be sought by methods that parcel out experiential and genetic influences.

A series of studies by Fry (1948) and by Kalmus (1949) focused on "tune deafness." They used a test featuring a number of well-known tunes played either correctly or else distorted by the insertion of several wrong notes. Rhythm and tempo were held constant. They measured performance in terms of two types of errors, A and B. The former was given by judging a correct tune as wrong ("miss"); the latter by judging an incorrect tune as right ("false-alarm"). Total error scores were calculated as 3B-A. The experimenters found a clearly bimodal distribution. Defined in this manner, the trait appeared to segregate in families and sib pairs in such a way as to suggest that much of the variation might be a result of a single locus, possibly involving dominant gene action (Kalmus, 1952). Some support was given to this by the findings of Ashman (1952), who traced ability for simple musical memory through four generations, including members of three families. His data do not clearly fit any obvious genetic model, though they suggest polygenes and some degree of autosomal dominance.

Kwalwasser (1955), using his own ten-test battery of musical ability, computed correlations on 255 pairs of sibs. The general sib correlation was 0.48. Brothers' scores correlated 0.56, and sisters' scores 0.46. These figures are close to that obtained by Shuter (1966) using the Wing Standard Tests of Musical Intelligence on samples of children and parents. She obtained a sib correlation of 0.475. The individual parent-offspring correlation, however, was only 0.29 (slightly higher when a "selected" group of children was omitted). The assortative mating coefficient was 0.331. It is of some interest that the strongest familial resemblance was between father and child (0.627), in spite of the apparent fact that the mother characteristically seemed to "set the musical environment." This result is at odds with much of the work on family patterns in personality dimensions (Chapter 15), for which the maternal influence appears much stronger.

A number of twin studies on musical ability have also been carried out. One of the more extensive investigations by Vandenberg (1962) involved 33 MZ and 43 DZ pairs who were given Seashore's tests of pitch, loudness, and rhythm and Wing's tests of pitch and memory. Heritability values were in excess of 40% for loudness, rhythm, and memory, but very low for pitch (both tests). It was of interest that acuity of hearing, which is unrelated to pitch discrimination, was found to be highly heritable, but only for the right ear. Curiously enough, a later study by Stafford (1965) found significantly different intrapair variances between 48 MZ and 54 DZ twin pairs on pitch discrimination. Rhythm also showed significant heritability.

Shuter (1966) administered the Wing tests to 28 MZ and 32 DZ pairs, some adults and some children. Intraclass correlation for "musical quotients" for the total MZ sample was 0.794; for the whole DZ sample, 0.721, yielding a heritability of only 0.262 (Holzinger's statistic). However, if only the 10 MZ and 9 DZ boy pairs were used, heritability rose to 0.617. Heritability for pitch separately was 0.45. Shuter also tested 5 pairs of identical twins reared apart (from Shields' sample). Intrapair differences ranged from 2 points (2 pairs) to 20 points (1 pair). The 2 other pairs differed by 12 and 15 points.

On the whole, the family and twin studies just reviewed suggest that some limited components of musical ability are heritable. However, musical talent in general is obviously a highly complex character depending on both genetic and environmental support. In spite of this, a number of investigators have put forward specific models of genetic transmission. These include recessivity for high ability (Hurst, 1912), single-gene dominance (Reser, 1935), double-gene dominance (Scheinfeld, 1956), and polygenic inheritance (Ashman, 1952). The data base for their divergent views is weak at best. Consequently, it would not be wise to educe any definite conclusions at this time.

Certainly, the problem of hereditary factors in musical ability is a most interesting one, not only from a practical point of view. It is a complex skill clearly involving many components. Apart from peripheral factors such as pitch, rhythm, and loudness discrimination, it is probable that, at a more central level, both serial and parallel information processing must be simultaneously involved. This unusual demand on the resources of the central nervous system may well account for the relative rarity of highly talented musicians. However, these more theoretical aspects of musical ability have not been explored within the context of behavior genetics.

Basic visual functions

An organ whose functioning is dependent on precise correlated growth of many parts might be expected to be highly susceptible to both environmental and genetic influences. An entire book has been devoted to a consideration of genetic factors in ophthalmology (Sorsby, 1970). The author makes the point that as standards of communicable disease control have improved, the importance of heredity as a factor in blindness has increased. In Liverpool in 1791, two thirds of the applicants for admission to an institution for the blind were victims of smallpox. By 1951, smallpox had been virtually eliminated from England. At the same institution, 21% of the patients at that time had blindness attributable to hereditary defects, as contrasted with 44% associated with degenerative changes of age. However, the hereditary conditions accounted for more expected years of blindness and presented more serious educational problems.

We shall not attempt a catalog of the varieties of heritable eye anomalies of clinical importance. The interested reader is referred to Sorsby (1970) as an authoritative source. Provided there is no brain damage, intellectual retardation in properly educated blind children is not exceptionally large, and the average IQ seems to be about 90. The partially sighted child often has a more severe handicap than the visually uneducable child. However, many such problems become increasingly correctable with the improvement in ophthalmological technology.

Undoubtedly, the aspect of this modality most thoroughly studied by geneticists has been defective color vision. Genetic studies have, in fact, played an important role in the development of theories of color vision.

Basically, the rods and cones that constitute the vertebrate retina contain photolabile pigments which absorb light waves within the visible spectrum. Simply put, these pigments involve, in part, carotenoids built up of isoprene units. The stereoisomeric configuration of the carotenoid is such that it may exist in a *cis* or *trans* state. When struck by photons, an 11-*cis* molecule essentially straightens out into a stable all-*trans* configuration. It is this photochemical event that provides the basis for visual experience (Wald, 1966; Hubbard and Kropf, 1967; Brindley, 1970). One compound importantly involved in the vision cycle is the enzyme alcohol dehydrogenase, a fact that, as we shall see, may have wider significance. A series of experiments carried out both with normal and color-blind subjects has now established fairly clearly that there are probably three types of cones containing pigments that differentially absorb different wavelengths of light. Rushton (1966) has labeled these pigments "erythrolabe" (red-catching), "chlorolabe" (green-catching), and "cyanolabe" (blue-catching). It is presumably de-

Table 13-2. Simple types of color vision defects*

Trichromats		Dichromats	Monochromats
Normal	**Anomalous**		
No defects	Protanomalous	Protanopes	Cone monochromats (at high illumination)
	Deuteranomalous	Deuteranopes	
	Tritanomalous	Tritanopes	Rod monochromats (at low illumination)

*Modified from Wright, W. D. 1957. Diagnostic tests for colour vision. Ann. R. Coll. Surg. Engl. **20**:177-191.

fects in or absence of one or more of these pigments that are responsible for the various forms of color blindness.

A large number of tests have been developed to detect color-vision abnormalities. The details of some of them have been discussed by Kalmus (1965), Cruz-Coke (1970), and by a variety of experts in the volume edited by Verriest (1974). As a result of the application of these, there is now a fair degree of agreement in the major forms of color-vision defects. So-called red-green color blindness (daltonism) was for many years regarded as a unitary defect. Now it is divided into at least four phenotypes. In addition, several other less common types of deficiency have also been identified. The major types are set out in Table 13-2. The main principle involved in the classification derives from Rayleigh's equation. The latter simply specifies the proportions of monochromatic red (671 nm) and monochromatic green (535 nm) light needed by an observer to match a monochromatic yellow light (589 nm). Using an anomaloscope, the proportionate amounts of red and green tend to be relatively constant for normal trichromats. However, anomalous trichromats, though needing the same components, require them in different proportions. Thus some individuals, given the wavelengths just specified, find the resulting mixture reddish yellow. Such individuals are categorized as deuteranomalous, being less sensitive to green. Other individuals, however, judge the yellow mixture as greenish, since they are less sensitive to red. These are called protanomalous. To look at the mat-

ter from the angle of the normal observer, the mixture matched by protanomalous subjects to yellow will appear too red; that matched by deuteranomalous will appear too green.

Dichromats are more difficult to detect with the anomaloscope procedure. In general, however, they may match any setting of yellow with a different mixture of green and red. Thus protanopes ("red blind") will match yellow to green alone or reduced yellow to red alone. Deuteranopes, on the other hand, can match yellow with red or green over a range of intensities. These people are "green-red blind," since they have a deficiency of the green retinal pigment, chlorolabe. However, it is also possible that they have no differential sensations of red and green but see only yellows of varying intensity (Kalmus, 1965). The same may be true of protanopes. In any case, the luminosity functions of both cover a narrower range than normal, being truncated at the red end for protanopes and at the green end for deuteranopes (Pitt, 1944).

Not all subjects fit the preceding classification, however. Some individuals falling between anomalous trichromats and dichromats are usually designated as "extreme" anomalous subjects. These behave, in a matching task, very much like anomalous individuals, except that they accept matches over a wider range of red-green mixtures. A convenient schematic summary of the diagnostic ranges for the various categories on the Nagel anomalscope is shown in Fig. 13-2. The normal person, when asked to match a red-green

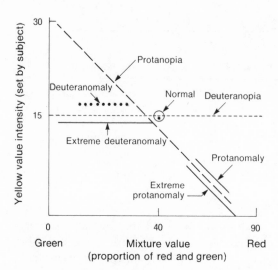

Fig. 13-2. Nagel anomaloscope results. Ranges of matching of red-green mixtures with yellow intensities for normals and for different categories of color-vision defectives. (Modified from Cameron, R. G. 1967. Aerospace Med. **38:**51-59.)

mixture at a setting of 40, sets the yellow at a scale point of 15. Deuteranomales find a region of equivalence at lower mixture values, covering a range between 15 and 30. Extreme deuteranomales match much in the same way, except over a wider range of values (e.g., 5 to 40). Deuteranopes match the yellow to red-green mixtures over the full range of 0 to 90. Protanomales equate a yellow of around 5 to a predominantly red mixture value of around 60. Extreme protanomalous subjects extend this range (yellow values, 5 to 12, with mixture values 45 to 60). Finally, protanopes match yellows in the range 5 to 30 with red-green mixtures 0 to 70 (Cameron, 1967).

Other tests besides the anomaloscope can be used to detect the types of defects just mentioned. Two well-known examples are the Ishihara and the Farnsworth-Munsell 100-hue tests. Comparisons between measures have been made by Kalmus (1965) and by Cruz-Coke (1970), among others. Probably all have particular problems and limitations associated with them. Cameron (1967) has concluded that for simple, rapid detection of color defect, the Ishihara plates are

to be preferred. The use of anomaloscopes, particularly of the Nagel type, is best for identifying the particular defect.

So far we have dealt only with protan and deutan defects. The third tritan category includes the defects of tritanomaly and tritanopia. Both involve a deficiency in the blue pigment, cyanolabe. In the normal eye, foveal vision is, in fact, tritanopic. Thus it is possible that, in tritans, the retinal area free of blue cones is simply more extended (Kalmus, 1965; Wald, 1966). However, since tritan defects are rare and difficult to detect, their exact nature is not well known. An additional defect, possibly falling into the tritan group, is tetartanopia, involving the inability to see yellow (580 nm), blue (470 nm), or violet (420 nm) (Willmer, 1946). However, authorities do not all agree on the existence of this type of dichromasy.

The genetics of color blindness are reasonably well understood. In an early survey of over 18,000 school children in Oslo, Waaler (1927) found that the same types of color defects were found repeatedly in related individuals. This led him to postulate that protanomaly and protanopia were results of alleles recessive to the normal allele which contributes something to the red-seeing system. Deuteranopia and deuteranomaly were considered by Waaler to be expressions of alleles at separate but closely linked loci. In addition, Waaler's data, as well as those of many other investigators, have confirmed that most types of color blindness are carried by genes on the X chromosome. This conclusion may be educed not only from the relative incidence of the defects in males and females, but also from linkage studies, which we will consider shortly. The exception appears to be tritanopia (unlike tritanomaly), which is thought to be carried by an autosomal dominant gene with incomplete manifestation. However, this conclusion is based on rather minimal data (Kalmus, 1955).

The critical test for the separation of two allelic systems (protan and deutan) requires an examination of the progeny of a female heterozygous for both. Such a woman should have normal vision, but half of her sons

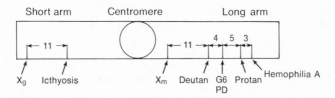

Fig. 13-3. Locations of protan and deutan genes in relation to genes for other X-linked characteristics. Distances between loci are in centimorgan units. (Modified from Cavalli-Sforza, L. L., and W. F. Bodmer. 1971. The genetics of human populations. W. H. Freeman & Co. Publishers, San Francisco; and McKusick, V. A., and F. H. Ruddle. 1977. Science **196**:390-405.)

should have a deutan and half a protan defect. Quite a number of such women have been located but they have not always turned out to have normal color vision (Walls, 1955; Franceschetti and Klein, 1956; Kalmus, 1965). Special theories of gene action must be invoked to explain such anomalies. According to the single-locus hypothesis, protan-deutan compounds should not be found in the offspring except perhaps under very special conditions of codominance. However, if separate loci exist, we should find at least four types if we assume two grades for each defect and nine types if we assume three grades (plus normal-abnormal compounds). Unfortunately, extensive data on this matter are not available. Walls and Matthews (1952) and Pickford (1965) have reported instances of such compounds, but, because of difficulties of precise diagnosis, their status is somewhat uncertain. At present, however, opinion appears to favor the existence of two loci, each with three abnormal alleles, the less abnormal dominant over the more abnormal.

Since most color-blindness genes are sex-linked, they afford an excellent opportunity for chromosome mapping studies. The major results of some of these have been usefully summarized by various authors and are schematically depicted in Fig. 13-3. It should be emphasized that the map distances indicated are highly tentative, being based on recombination frequencies in fairly small samples. Some other problems have also been pointed out by Kalmus (1965).

A final empirical aspect of color blindness of interest relates to manifestation in the het-

erozygous female. According to the Lyon hypothesis, one X chromosome is inactivated in the somatic cells of the female. This may be of maternal or paternal origin and hence may carry a particular gene or not. Depending on the time during development at which such inactivation occurs and for which chromosome, the female may be more or less a mosaic in respect to the gene involved. We may thus expect to find, in some carrier females, some expression of the trait. One of the most compelling examples of this in the present context is afforded by cases of women, described in the literature, who have been color blind in one eye but of normal vision in the other. At least some of these can be explained by the Lyon hypothesis. Others may involve a different etiology, since monocular color blindness also shows up occasionally in males and cannot, of course, be due to X-chromosome inactivation (Kalmus, 1965).

Walls (1955; Walls and Matthews, 1952) has employed data on the chromatic and brightness vision of protans and deutans to construct a comprehensive color-vision scheme. His point of view is that each kind of hereditary color-vision defect represents a normal system minus something, and that each may be considered as an experiment from which one can deduce the properties of the normal system. This approach is common in physiology, which could hardly exist as a science without the procedure of surgical extirpation. Walls' hypothesis is likewise derived from a consideration of the effects of biochemical "extirpations" produced by mutant genes.

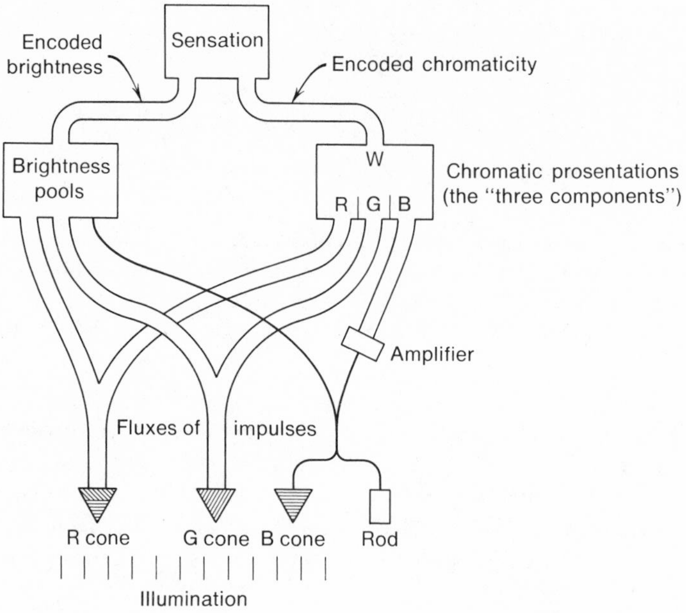

Fig. 13-4. Schema for normal color vision based on data from color-blind subjects: an example of using genetic lesions to interpret normal functions. (From Walls, G. L. 1955. Am. J. Ophthalmol. **39**:8-23.)

Walls' scheme for normal color vision is shown in Fig. 13-4. His original articles deal with the experimental argument for this theory. He has claimed it provides an explanation for the different effects of color blindness (and various experimental procedures) on brightness and chromaticity. A protanope, for example, is assumed to lack the R-cone system (sensitive to long wavelengths) and their associated paths to the brightness and chromaticity centers. A deuteranope has only G and R cones (or perhaps a common cone type) sensitive to longer, but not to shorter, wavelengths. Other less common anomalies of color vision can also be fit into the scheme by making appropriate deletions, including monochromasy, both photopic and scotopic.

Walls' model is, of course, highly conjectural and allows no exact predictions. Consequently, it is useful only as a general framework for guiding future research on the topic of color vision.

Visual perception

Remarkably little work has been carried out on this topic since the early studies previously reported by us (Fuller and Thompson, 1960). The major findings of these studies are summarized in Table 13-3. Perhaps the most that can be said about them is that they are suggestive of hereditary influences at work in the determination of some aspects of visual perception. However, all suffer from methodological problems to some degree or other and, in addition, have at best a tenuous relation to modern work in this field (e.g., Dodwell, 1970).

Little more seems to have been done during the last decade or so. One exception is a twin study by Matheny (1971) on the Ponzo illusion. The test involves a comparison between the lengths of two lines placed on an arc of lines radiating from an apex. One test line is closer to the apex, the other farther away. When the lines are objectively equal, most observers judge the one nearer the apex

Table 13-3. Summary of results of early studies on visual perception

Author	Character	Comparison		
von Bracken (1939)	Afterimage size	No difference between 7 MZ and 12 DZ pairs		
	Perpendicularity judgment	No difference between 7 MZ and 12 DZ pairs		
	Müller-Lyer illusion magnitude	No MZ-DZ difference		
Malan (1940)	Spatial orientation in four "blindfold" tests	Percent of pairs with "large" difference*		
		Test	MZ	DZ
		1	10	55
		2	15	52
		3	15	50
		4	22	48
Smith (1949)	Afterimage size in four tests	MZ-DZ interclass correlations		
		Test	MZ	DZ
		1	0.71	0.08
		2	0.68	0.00
		3	0.98	0.22
		4	0.75	0.23
Smith (1953)	Müller-Lyer illusion magnitude under four instructional conditions	MZ-DZ intraclass correlations†		
		Test	MZ	DZ
		1	0.53	0.39
		2	0.55	0.05
		3	0.51	0.37
		4	0.57	0.28

*The definition of "large" varied from test to test and is not precisely specified.
†MZ-DZ differences tended to increase with repeated exposures according to further work by Smith.

to be the longer. Comparisons were made of 21 MZ and 15 DZ (9 to 11 years) pairs. Magnitudes of intrapair MZ differences were found to be significantly smaller, thus implicating hereditary factors. One ancillary finding of interest was that in 2 MZ pairs showing a large intrapair difference, the member showing the greater illusion susceptibility was the one with a higher score on a digit-span subtest of an intelligence test. Matheny suggests that this may relate to the notion that illusions like the Ponzo, in which susceptibility increases with age (type II), involve a component of serial processing and hence might be expected to relate to digit-

span tasks. This hypothesis is perhaps worth further examination.

Two studies have explored possible genetic factors in critical flicker frequency (CFF). Murawski (1960) found that 4 MZ pairs showed greater similarity than pairs of unrelated individuals. Likewise, Klein and DeFries (1973) found differences between sexes and between four ethnic groups (Caucasians, Hawaiian, Japanese, and Chinese) in respect to CFF measures taken under various conditions.

A final study worth summarizing is another by Matheny (1972) at the University of Louisville. This examined the pattern of visual exploration in 70 MZ and 50 DZ twins between the ages of 5 and 11 years. The test involved a series of six cards, on each of which were printed pictures arranged according to some ordering principle, for example, in the shape of a triangle. On one card the arrangement was random. Verbal responses of subjects to the cards were recorded, thus yielding an index of where they started their "exploration," where they stopped, and the actual order they followed. Matheny was able to generate a total "pattern of exploration" score for which intrapair differences were computed. For children 5 to 7 years and for those 7 to 11 years, MZ pairs were significantly more alike than DZ pairs. Thus he concluded that this type of visual search had some genetic components.

It will be clear from the preceding summary that there is a great deal of room for important behavior-genetic work in the area of perception in general and visual perception in particular. Indeed, the study of perception has traditionally been one of the mainstreams of psychology, and thus it seems a pity that behavior geneticists have virtually ignored it.

RESPONSE PROCESSES

In a strict sense, the studies to be described in this section deal with the genetic basis of physiological rather than behavioral characters. There can be little doubt, however, that the functional activity of the nervous system is correlated with and must underlie all behavior, even though psychology has not yet become a branch of neurophysiology. Thus the area being considered is one of great importance.

Various indices of the activity of the nervous system have been used. Although some of them, at least, have achieved a high level of technical sophistication, it is still by no means certain what all of them are really measuring. In this sense, they are not much ahead of the purely psychological measures discussed in other chapters. We will consider them under two headings, one relating to measures of central nervous system activity, the other relating to measures of autonomic functioning.

Central nervous system

The neurophysiological trait whose genetic aspects have been studied most extensively is the electroencephalogram (EEG). Apart from the ambiguity as to its real meaning, however, some difficulty has related to the judgment of similarity between records and which electrode placements and record parameters are the most satisfactory. Thus there is not a marked consistency between the studies that have been carried out.

Work on the topic appears to have started with the studies of Davis and Davis (1936) and of Loomis, Harvey, and Hobart (1936). The former investigators reported that waking alpha activity was much more similar between members of MZ than of DZ pairs, and, in fact was as alike as between successive readings on the same individual. Loomis, Harvey, and Hobart, within a broad sample, included two pairs of identical twins, one of 27 months, the other between 3 and 4 years. They claimed that there was a marked intrapair similarity. Gottlober (1938), on the other hand, reported failure of four judges to group correctly records of fifteen families, each including parents and two or more children. He concluded that hereditary effects were absent, or of minor importance. His method, however, seems unsuitable for studying the heritability of a quantitative trait. In contrast, Raney (1938) reported EEGs of members of 17 MZ pairs were more alike than those of randomly paired age-matched unrelated individuals, in respect to percentage of alpha wave activity, amplitude of alpha, and frequency of alpha. This was true for both central and occipital electrode placements using age-corrected correlations (Spearman rank order). One interesting feature of the data is that there appeared a tendency for one member of each pair to show a "significantly larger relative difference between the two sides of the head, especially in amplitude measurements." Generally, one twin showed a bilateral EEG asymmetry, the reverse of that shown by his co-twin. If they can be considered reliable, these findings are of great interest and very relevant to current work on laterality, whose etiology, as we shall show, is still something of a mystery.

One of the larger studies was carried out by Lennox, Gibbs, and Gibbs (1945) on 55 MZ and 19 DZ pairs. In this case, judges attempted to diagnose zygosity by inspection of EEG tracings. Results are as shown in Table 13-4. They clearly indicate a very high success rate and suggest an additional and useful method of establishing twin type.

Table 13-4. Judgments of zygosity of MZ and DZ twins from EEG tracings*†

	Judgments		
Actual type of twin	MZ	DZ	?
MZ	47	2	6
DZ	1	18	0

*From Lennox, W. G., E. L. Gibbs, and F. A. Gibbs. 1945. The brain-wave pattern: an hereditary trait: evidence from 74 "normal" pairs of twins. J. Hered. 36: 233-243.
†$\chi^2 = 49.4$; $p < 0.001$.

Further work appears to have supported the results of these early studies (Vogel, 1957, 1970; Juel-Nielsen and Harvald, 1958; Tangherani and Pardelli, 1958; Dumermuth, 1968; Hume, 1973; Surwillo, 1977). A study by Juel-Nielsen and Harvald (1958) is worth special mention, since it is the only one involving the use of separated MZ twins. The latter were 1 male and 7 female pairs ranging in age from 22 to 72 years. The following EEG measures were taken: frequency of dominant activity, amplitude of dominant activity, and distribution in time of dominant activity calculated as a percentage of the total recording time. In 7 pairs, the effects of hyperventilation and of visual flicker ("provocation" tests) were also examined. The results were compelling. Frequency and amplitude measures showed intrapair variation only in one pair of twins. Members of other pairs showed near identity. However, less similarity was found for the distribution measure. These data are summarized in Table

Table 13-5. EEG measures in 8 pairs of MZ twins reared apart*

Pair number	Sex	Age (years)	Co-twin	Frequency	Dominant activity	
					Amplitude (microvolts)	Distribution (percent)
I	M	22	1	11	50	25-50
			2	11	50	25-75
II	F	37	1	9	75	75-100
			2	9-10	75	75-100
III	F	42	1	9-10	75	25-50
			2	9-10	75	25-50
IV	F	46	1	10	120	75-100
			2	10	70	75-100
V	F	49	1	8-9	120	25-50
			2	10	120	75-100
VI	F	56	1	10	75	25-50
			2	10	75	75-100
VII	F	71	1	9-10	50	0-25
			2	9-10	50	25-50
VIII	F	72	1	9-11	50	50-75
			2	9-11	50	50-75

*Modified from Juel-Nielsen, N., and B. Harvald. 1958. The electroencephalogram in uniovular twins brought up apart. Acta Genet. 8:57-64.

13-5. Reactions to the provocation tests were also remarkably alike within pairs. Thus the authors concluded that their results constitute unique proof that the EEG pattern is primarily determined by genetic factors. It must be pointed out, nevertheless, that the measures used are rather coarse ones that might have given similar results in DZ or sib pairs.

Three studies have taken a step beyond using simple EEG tracings and have examined cerebral evoked responses in MZ and DZ twins (Dustman and Beck, 1965; Osborne, 1970; Lewis, Dustman, and Beck, 1972). These found greater MZ than DZ similarity in form of evoked response. We can illustrate by reference to the data of Lewis, Dustman, and Beck (1972). These investigators studied 44 MZ, 46 DZ, and 46 age-matched pairs of unrelated individuals. They measured both waveform and wave amplitudes in response to visual (100 10 μsec flashes), auditory (200 0.25 msec clicks presented binaurally), or somatosensory (100 25 msec electrical pulses at threshold voltage) signals. Waveforms were measured by converting the analogue readouts to digital values sampled over time periods after the signal. Temporally corresponding digital values were then correlated between members of each twin pair. An amplitude measure was computed by summing the absolute differences between successive digital values over a time period and dividing the resulting cumulative sum by a constant to convert the sum to voltage. Again, these cumulative voltage changes (CVCs) were correlated between members of the twin and nontwin pairs. A summary of the results reported by Lewis, Dustman, and Beck is given in Table 13-6 and Fig. 13-5. Several comments may be made. First, average MZ correlations are of uniformly greater magnitude than both DZ and UR correlations for all three modalities. For VERs and AERs over both time periods, MZ similarity was statistically higher than that in DZ and UR samples ($p < 0.05$). However, only for AERs were the DZ-UR comparisons significantly different. Second, no real differences between the groups were found for SERs. Using 2 ($r_{MZ} - r_{DZ}$), the following heritability estimates were computed for the two time periods: for VER, 0.38 and 0.28; for AER, 0.16 and 0.36; and for SER, 0.04 and 0.10. Thus the locus of placement apparently makes a large difference. If reliable, this interesting finding suggests that the neural functioning involved in different parts of the brain may have varying amounts of genetic determination. Or to put it another way, some parts of the brain may be much more plastic to environmental influence than other parts. Third, concordances for amplitudes of evoked potentials are, in general, less than those for waveform. The authors regard this largely as a function of the greater variability of the waveform measure. Fourth, in the comparisons made there were no obvi-

Table 13-6. Correlations for waveform of evoked responses in MZ, DZ, and unrelated pairs of individuals*

| Modality† | Time epoch following signal | | | | | |
| | 0-88 msec | | | 90-300 msec | | |
	MZ	DZ	UR	MZ	DZ	UR
Visual (VER)	0.67	0.48	0.27	0.80	0.66	0.51
Auditory (AER)	0.83	0.75	0.60	0.74	0.56	0.37
Somatosensory (SER)	0.50	0.48	0.38	0.63	0.58	0.42

*Modified from Lewis, E. G., R. E. Dustman, and E. C. Beck. 1972. Evoked response similarity in monozygotic, dizygotic and unrelated individuals: a comparative study. Electroencephalogr. Clin. Neurophysiol. 32:309-316.
†Correlations shown are averages of correlations from three electrode placements for the visual and somatosensory and two for the auditory modality.

ous differences between the evoked potential measures taken at different epochs. This finding is of interest, since it is thought that the early phases of evoked potential reflect input from specific thalamic pathways, whereas later phases reflect extralemniscal (reticular or nonspecific thalamic) input. However, the lack of difference found by Lewis, Dustman, and Beck does not necessarily mean that some difference might not be found with some other measure. Finally, the authors examined the possibility that their results might be due merely to the relatively trivial fact that twins are highly similar in morphological cranial characteristics. They ruled out this possibility by the fact that evoked response correlated only −0.04 with cephalic index (Dustman and Beck, 1965).

The preceding study suggests that the evoked potential in man is under some degree of genetic control. However, as we have indicated, the extent appears to vary with

sensory modality. It may also turn out that it will be found to vary with technique of measurement. There is a rich field for research here.

Another approach to the problem of gene-neuron relationships has involved examination of the alpha attenuation response (AAR) or alpha blocking. This describes the shift in the EEG from the slower and regular high-voltage alpha rhythms to low-voltage fast activity when some stimulus is suddenly presented to a resting subject. Psychologically, it reflects an orienting, alerting, or attentional response. Two studies have used this measure in twin samples. Young and Fenton (1971) found that intraclass correlations for habituation of AAR in a sample of 17 MZ twins were significantly higher than the correlations for 30 unrelated individuals (all negative correlations) but not different from those for 15 DZ pairs. Heritabilities (computed by Thompson) for the AAR taken at various time periods during 60 signal trials

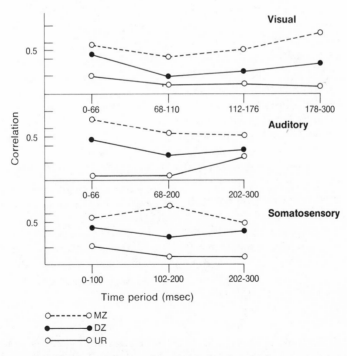

Fig. 13-5. MZ, DZ, and UR correlations for amplitude of evoked responses (CVCs) for three modalities over time. (Based on data from Lewis, E. G., R. E. Dustman, and E. C. Beck. 1972. Electroencephalogr. Clin. Neurophysiol. **32:**309-316.)

and over the whole trial period ranged between 0.36 and 0.82. The second study by Hume (1973), however, found no difference between MZ and DZ in recovery of alpha activity after a block induced by a rotating Archimedes spiral. On the other hand, a varimax factor analysis carried out on the twenty-one physiological measures used in the study yielded four interpretable factors, on one of which the measure of alpha blocking had a significantly high loading (factor 3, "EEG component"). Both MZ and DZ intraclass correlations for this factor were found to be significantly different from zero, but not significantly different from each other. Consequently, no real conclusion can be firmly put forward in regard to the heritability of the AAR.

A final study to be reviewed is unique (to the knowledge of the authors) in studying the electrical activity of the brain in twins during sleeping and dreaming. Zung and Wilson (1967) recorded sleep waves in 4 MZ and 2 DZ pairs over 4 consecutive nights. The obtained data were averaged and plotted out over time by computer as shown in Fig. 13-6. It is clear—qualitatively, at least—that there is strong concordance for MZs but not for DZs in respect to the major shifts in sleep stages during the nights of testing. Much the same applies to REM (paradoxical sleep) patterns. The authors also present data on percent of time occupied by the five sleep stages (*B, C, D, E,* and *REM*) and the presleep stage (*A*) and on time to fall asleep and total sleep time. However, presumably because of their very small sample size, they performed no correlation analyses on these data; neither does it seem worthwhile to do so. Consequently, their conclusion favoring genetic determination of sleep patterns seems overly strong. A larger-scale replication of their study would be most desirable.

Work on the heritability of EEG patterns has also been carried out with abnormal populations. Gottlieb, Ashby, and Knott (1947) and Knott et al. (1953) took records of a large sample of patients with primary behavior disorder and psychopathic personality. These patients had character defects such as ego-

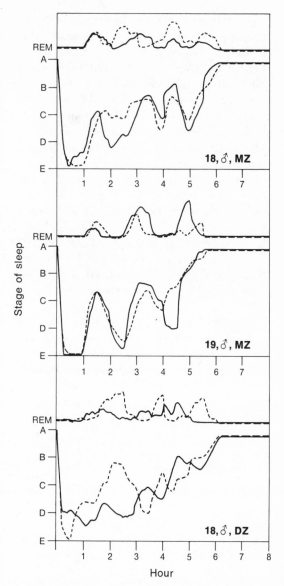

Fig. 13-6. Mean all-night sleep EEG and REM patterns in two pairs of MZ and one pair of DZ twins. (From Zung, W. W. K., and W. P. Wilson. 1967. Sleep and dream patterns in twins: Markov analysis of a genetic trait. In J. Wortis, ed. Recent advances in biological psychiatry, vol. 9. Plenum Publishing Corp., New York.)

centric motivation and poor judgment of the outcomes of deviant behavior. The EEGs of 86 patients had significantly more fast and slow rhythms than were shown by individuals in a normal "standard" sample. Also found to deviate significantly from normal, though not as much, were 172 parents of the probands. Data are summarized in Table 13-7. Kennard (1949) reported a similar finding.

The weight of evidence that we have just summarized suggests some degree of hereditary determination of EEG pattern both in its normal and abnormal aspects. However, most of this research has been carried out on small and specialized samples so that generalization is not very feasible.

Table 13-7. EEGs of patients and parents*

		Patient's EEG	
Parent's EEG	N	Normal	Abnormal
Normal × normal	22	15 (68%)	7 (32%)
Normal × abnormal	17	4 (24%)	13 (76%)

*From Gottlieb, J. S., M. Ashby, and J. R. Knott. 1947. Studies in primary behavior disorders and psychopathic personality. II. The inheritance of electrocortical activity. Am. J. Psychiatry **103**:823-827. Copyright 1947, the American Psychiatric Association. Reprinted by permission.

Autonomic nervous system

Homeostatic regulation in the higher vertebrates depends on a system of reflexes involving the sympathetic and parasympathetic divisions of the autonomic nervous system. When an organism is aroused psychologically, some of these reflexes may be activated. Thus many psychologists have attempted to assess emotionality in terms of heart-rate changes, respiration, galvanic skin responses, and other measures. Generally speaking, although there is some specificity between systems, individuals can be shown to differ from each other reliably in both resting levels and response amplitudes. Furthermore, such differences appear to show considerable stability over time (Lacey and Lacey, 1962). These facts have led some authors to suggest that eventually multiple measures of autonomic functioning may supply us with a kind of basic physiological "fingerprint" of an individual (Sargent and Weinman, 1966), by reference to which we may be able to differentiate various personality types (Claridge, 1967; Eysenck, 1967; Claridge, Canter, and

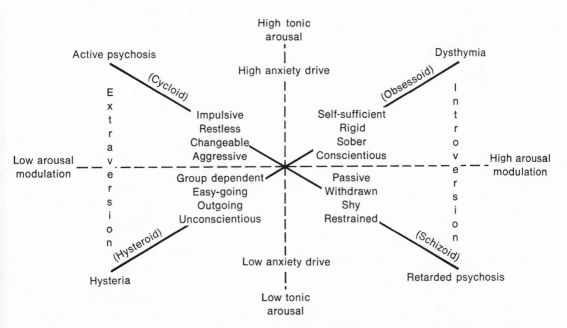

Fig. 13-7. Claridge's model relating personality types to levels of tonic arousal and arousal modulation. (From Claridge, G. S. 1967. Personality and arousal. Pergamon Press Ltd., Oxford.)

Hume, 1973; DiCara, 1974). An example of the kind of personality models that may be educed is set out in Fig. 13-7 from Claridge (1967). Here two psychophysiological dimensions are used: tonic arousal, which is mainly associated with autonomic measures, and arousal modulation, which mainly relates to CNS variables such as those discussed in the previous section of this chapter. It is difficult to assess the validity of such a model, although the same may be said about almost any available model of personality. We will indicate more fully in Chapter 15 the nature of some of the problems. However, conceptualizations of the kind put forward by Claridge at least have the virtue of building on more secure biological foundations. Let us now examine some of the studies that have attempted to show the influence of genotype on measures of autonomic functioning.

Summaries of the literature have been made by Lader and Wing (1966), by Block (1967), and by Hume (1973). In general, it may be said that there is fairly strong evidence for the genetic control of autonomic patterns. Some early data were put forward by Jost and Sontag (1944) at the Fels Institute. Measurements of the following variables were taken on twin, sibling, and unrelated pairs: volar conductance, pulse pressure, salivation, heart period, respiration rate, vasomotor persistence time (skin reddening after pressure), and palmar conductance. An index of autonomic balance was

computed from the weighted sum of scores on the separate measures. The correlations for this measure taken over 3 successive years showed some familial effect, as indicated in Table 13-8. However, if we consider that DZ twin correlations, had they been taken, would have been at least as high as those found for sib pairs, then heritability estimates would probably have an upper limit of 40%. In addition, Jost and Sontag found considerable variation in correlations between the specific measures they used.

The results of more recent studies are summarized in Table 13-9. It will be clear that there is not a great deal of consistency between them. This is perhaps not surprising, since they involved often quite different techniques of measurement and different scoring procedures. Generally speaking, it seems as though the cardiovascular measures tend to show more heritability than the indices of electrodermal activity. However, even this result does not always appear. What seems to be obviously needed is a large-scale parametric study aimed at establishing which measurements are reliable and which yield maximal heritability coefficients. Until this is accomplished, we will be able to say rather little either about genetic factors in autonomic activity or about the relationship of the latter to personality variables.

LATERALITY

It is a compelling feature of human beings that they are predominantly right-handed defined in terms of such indices as skill and preference. This bias to the right appears to hold not only in all known present-day cultures (Annett, 1972), but also in very ancient, extinct ones. Thus Yakovlev (1964) has suggested that Stone Age men of the Upper Paleolithic were probably dextral, as inferred from their artifacts and cave paintings. Likewise, Dart (1949), on the basis of examination of Pleistocene baboon skulls (*Parapapio africanus*) with punctured depressions and fracturing, has concluded that these were inflicted by right-handed tool-using predators, presumably some form of *Australopithecus*. Data such as these are interesting but must

Table 13-8. Correlation of autonomic balance measure in pairs of twins, sibs, and unrelated individuals over a 3-year test period*

	r		
Year	MZ twins	Sibs	Unrelated
1940	0.434	0.255	0.164
1941	0.470	0.406	0.017
1942	0.489	0.288	0.080

*From Jost, H., and L. W. Sontag. 1944. The genetic factor in autonomic nervous system function. Psychosom. Med. **6:**308-310.

Table 13-9. Summary of studies on heritability of measures of autonomic arousal

Author	Subjects	Genetic determination*	
		Low	High
Eysenck (1956)	26 MZ and 26 DZ twins	GSR latency	Systolic and diastolic pulse rate
Rachman (1960)	35 MZ twins	GSR habituation	Latency of GSR
Mathers, Osborne, and DeGeorge (1961)	34 MZ and 19 DZ twins		Resting heart rate
Kryshova et al. (1962)	11 MZ and 2 DZ twins		Blood pressure response to cold stimulus
Osborne, DeGeorge, and Mathers (1963)	34 MZ and 19 DZ twins	Basal blood pressure†	Casual blood pressure†
Vandenberg, Clark, and Samuels (1965)	22 MZ and 16 DZ twins	GSR	Respiration rate; heart rate (during stimulus presentation)
Lader and Wing (1966)	11 MZ and 12 DZ twins	Most GSR measures	"Spontaneous fluctuations" in GSR; habituation of GSR; pulse rate
Block (1967)	21 MZ twins		Resting heart rate; stimulus-induced change in heart rate and in respiration
Miall et al. (1967)	First-degree relatives of normal probands		Systolic and diastolic blood pressure
Shapiro et al. (1968)	12 MZ and 12 DZ twins	Pulse rate and catecholamine response to pain stimulus	Blood pressure response to pain stimulus
Downie et al. (1969)	50 MZ and 31 DZ same-sex and 28 DZ opposite-sex twins	Casual systolic and diastolic blood pressure	
Barcal, Simon, and Sova (1969)	39 MZ and 51 DZ twins	Casual blood pressure	Average of several blood pressure readings
van den Daele (1971)	6 MZ and 3 DZ twins (infants)	Activity level: low proprioceptive and no auditory stimulation or high proprioceptive and auditory stimulation	Activity level: low proprioceptive and auditory stimulation or high proprioceptive stimulation and no auditory stimulation
Hume (1973)	44 MZ and 51 DZ twins	Most measures of blood pressure and GSR; some heart rate measures	Blood pressure and GSR change in response to cold, GSR habituation, resting heart rate, and heart-rate change during orienting response

* Decisions regarding high or low genetic determination are mostly based on criteria used by the authors of each study. They are not necessarily the same between studies.

† Basal blood pressures were measured in the morning in fasting subjects under standardized conditions. Casual readings were taken during routine examinations and generally in the evening. They are thought likely to reflect daily activities of subjects as well as genetic factors.

be treated with some caution. Nevertheless, there is little question that human beings and many lower animals are quite asymmetrical in their anatomy, their physiology, and their sensory systems and performance. Handedness is perhaps the most obvious and striking example. But, as we shall describe shortly, there are many others for which laterality is equally pronounced, though its skewness is often not as marked.

General interest in laterality clearly goes back a very long way, and much folklore and mythology surrounds it. In psychology, attention was first focused on it early in the century as a result of theories that speech, reading ability, and personality were deleteriously affected by attempts to change "naturally" left-handed children into right-handers. Such ideas at least implicitly involved the assumption that there was genetic control of the trait. Considerable opposition to this view was voiced by those favoring an experiential etiology of laterality. Unfortunately, it is fair to say that even today this issue has still not been resolved. One reason for this, as we have just implied, is that laterality has turned out to be a surprisingly complex character with many facets and levels. One such aspect has, in fact, been mainly responsible for a striking resurgence of interest in it during the last decade. This is the problem of differential function between the two cerebral hemispheres studied by such workers as Zangwill (1967), Hecaén (1969), Sperry and Gazzaniga (1967), Kinsbourne and Smith (1971), Harnad et al. (1976), and Hardyck and Petrinovitch (1977). Indeed, a brief perusal of such journals as *Cortex* and *Neuropsychologia* will indicate to the reader the strength of interest in brain asymmetry during the last decade or so. It is difficult to assert at the time of writing how much of basic significance will come out of all this work. But whatever the case, the problem of laterality represents an almost ideal model system for behavior genetic analysis. The phenotype is relatively simple and readily measurable. It can be defined in terms of a binary or a continuous distribution, and, like most behaviors, it is clearly plastic to

change, at least within limits. We will now consider definition and measurement of laterality.

Measurement

As we suggested previously, laterality is a trait that can be defined in a great many ways. Most of the early work took it as being a binary character expressed in terms of which hand an individual used for writing. A variant on this—one which is still used—is the manner in which individuals fold their arms or clasp their hands; that is to say, right arm or hand over left arm or hand, or vice versa. As we shall see shortly, the binary classification of intergenerational populations into dextrals and sinistrals can be made to fit (with certain assumptions) quite simple genetic models. On the other hand, it is obvious that such models may not be appropriate when the character is defined on a continuum. Thus the problem of measurement is, as usual, a critical one if we are to understand fully the etiology of laterality.

A summary of some of the major methods used to establish laterality is offered in Fig. 13-8. According to this schema, three major dimensions are always involved. Thus we may use either a self-report form or actual behavioral observation. Probably most genetic studies have relied on the former. Likewise, we may ask for simple preference, or attempt to establish some estimate of dextral or sinistral speed or skill. Finally, either output or input or central functions may be tapped. The latter category is not really exclusive of the other two, of course, in the sense that we know that, for example, right visual field inputs or right ear auditory inputs travel immediately to the left cerebral hemisphere and only indirectly to the right through the corpus callosum. Conversely, we know that in most subjects, at least, damage to one hemisphere incurred naturally, by surgery, or by some anesthetic agent such as sodium amytal injected in the carotid artery will produce both input and output deficits, mostly on the contralateral side. Nonetheless, there is perhaps some virtue in separating them, since the relations between cen-

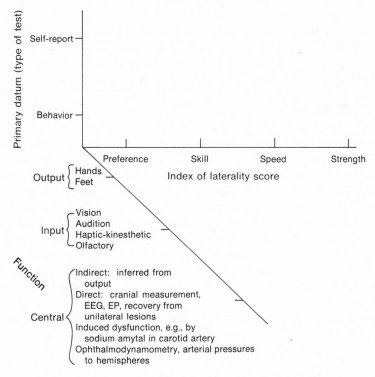

Fig. 13-8. Measurement of laterality: summary of methods.

tral and peripheral laterality are by no means simple or absolute (Hardyck and Petrinovitch, 1977).

Beyond the three dimensions specified, other minor subdivisions could be added. Thus a self-report of lateral preference could involve simply the question, "Which hand do you write with?" or be elaborated into a fairly lengthy questionnaire like the Edinburgh Inventory (Oldfield, 1971). This involved, initially, twenty items, later reduced to ten, relating to hand usage, for example, writing, throwing, cutting with scissors, sweeping, and brushing teeth. An even lengthier questionnaire has been evolved by Berman (1971) involving input (auditory and visual) as well as output functions.

Another kind of breakdown might be made, in terms of details of the subject's behavior observed. For example, on the output side, a simple measure of speed on some task may be obtained, such as moving pegs from one row on a board to an adjacent row (An-

nett, 1970b, 1972). Likewise, strength of grip as measured by a dynamometer or force displacement transducer has been used both in human beings (Woo and Pearson, 1927) and in animals (Collins, 1968b). It is easy to see that such measures can be made much more complicated. On the input side, a competition situation can be used, as in a dichotic listening task with information presented to both ears simultaneously (Kimura, 1967), or, again in the visual modality, relative performance in the left and right fields can be compared on a masking task (McKeever, Van Deventer, and Suberi, 1973). The interested reader will find a great number of variants on the preceding methods described in the literature.

As is often the case with psychophysical measurement, there is sometimes a negative correlation between degree of assessed laterality and task difficulty. Harshman and Krashen (1972) have proposed, on this account, the use of an "unbiased laterality coefficient" defined as "Percent of Error (POE)" and estimated by number of errors on the left divided by total errors; that is,

$$POE = \frac{Le}{(Re + Le)} \times 100.$$ In a survey of data

from a number of studies, Harshman and Krashen found that this laterality measure was only slightly correlated with accuracy (r = 0.21). It may also yield different conclusions from those derived from the more conventional measure of difference between correct on right and correct on left (Rc − Lc). The latter, for example, shows children to become *less* lateralized between ages 5 and 10 years. POE, on the other hand, shows no change in lateralization during this period. Thus the measurement problem on which Krashen and Harshman have focused is an important one.

However, there are difficulties with the POE measure. For example, let us compare

two subjects, both of whom make 50% errors on the left but score respectively 70% and 30% errors on the right. The measure (Rc − Lc) would yield scores of −20% and +20%; that is, the two would be equally lateralized, though in opposite directions. On the other hand, use of the POE measure would generate scores of 0.41 and 0.62, suggesting much less lateralization in one individual than in the other and a clear correlation with total accuracy. These and other such technical problems have been considered in some detail by Marshall, Caplan, and Holmes (1975). However, their exposition is beyond the scope of this present section.

A somewhat different approach to the problem of measuring laterality has been put forward by Annett (1970a) using "association analysis." This is a method previously developed for use in plant ecology and aimed at

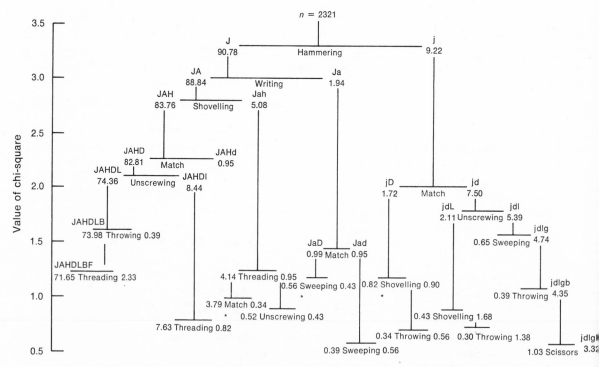

Fig. 13-9. Association analysis. Letters refer to the questionnaire item. Capitals indicate a response of "right" or "either"; lowercase letters indicate a response of "left." Numbers refer to percentage of subjects responding each way. (* Further subgroups involving less than five subjects are omitted.) (Modified from Annett, M. 1970. Br. J. Psychol. **61**:303-332.)

describing features of a given environment that affect the distribution of the species of plants occupying it. In the present context, the method attempts essentially to order questionnaire items according to the degree that they correlate with (or are associated with) all other items as estimated by correlation coefficients. For example, of all items used by Annett, "hammering" turns out to correlate most highly with all twelve other items. "Unscrewing the lid of a jar," on the other hand, correlates the lowest. Accordingly, the items may be arranged in a hierarchy as shown in Fig. 13-9. The particular data shown relate to the breakdown "left" or "not-left" in response to the items. "Not-left" includes both "right" or "either." The ordinate indicates, in terms of chi-square, the degree of association between adjacent items. Only significant associations are shown. All in all, the data indicate 23 patterns of association, a figure arrived at simply by counting the horizontal lines in the graph, starting with "hammering" (90.78% "not-left" and 9.22% "left"). However, there are still a large majority of pure right-handers (72%), a small proportion of pure left-handers (3%), and a remainder of mixed (25%). Perhaps the major points to emerge from Annett's analysis are, first, that

laterality, whether measured in terms of skill, speed, or preference, is a continuous rather than a dichotomous trait and, second, that left-handedness is not necessarily the mirror image of right-handedness, although this depends, as Annett put it, "on where the mirror is placed" in Fig. 13-9.

Distribution: humans

It will be clear from the preceding discussion that the form of the laterality distribution in any population studied will depend heavily on the method by which it is assessed. This form will, in turn, place constraints on the kind of genetic (or environmental) model chosen. However, it is still true to say that whether we take laterality as a dichotomous, a continuous, or a quasi-continuous trait, we will invariably observe, in human beings, a strong bias to the right. This fact can easily be demonstrated by looking first at three early sets of data, using a simple division into dextrals and sinistrals. Each study involved both a parental and a filial generation. Results are shown in Table 13-10. It is clear that the three surveys are quite consistent with each other and show, when pooled, unweighted mean percentages of 6.1 for incidence of sinistrals and 93.9 for

Table 13-10. Three early studies of the distribution of laterality in parental and filial generations

	Parental		Filial	
Study	L*	R†	L	R
Ramaley (1913) (no criterion of laterality stated)	49	561	177	953
Chamberlain (1928) (writing hand)	155	3199	367	7347
Rife (1940) (L = 1 or more acts out of 10 performed with left hand)	72	1302	191	1987
TOTALS	276 (5.2%)	5062 (94.8%)	735 (6.7%)	10,287 (93.3%)
	L		**R**	
Combined parental and filial	6.1%		93.9%	

*L, numbers of sinistrals.
†R, numbers of dextrals.

dextrals. It is notable, also, that there is a fairly consistent increase in left-handedness in the filial generations. This also appears in some later surveys by Annett (1973). These are derived from questionnaires requesting subjects simply to report handedness as well as that of various relatives. As indicated in Table 13-11, her incidence figures are quite close to those of earlier investigators. An additional feature of the table worth noting is

the higher frequency of dextrality in females, both mothers and daughters.

Besides questionnaire data, Annett (1967) has also used direct observation of university students and school children of 5 to 15 years while engaged in five actions. Incidence of pure sinistrals for the two groups were, respectively, 2% and 4.6%. The second of these figures, at least, is in agreement with most of the self-report data; the first, for reasons unknown, is a little low.

Among other laterality traits for which incidence figures are available are hand clasping and arm folding. These differ from handedness, of course, inasmuch as they are purely expressive gestures and as such are not involved in the direct relation of a person to the world. However, it is a fact that most people show a distinct preference in respect to which thumb is on top in clasping hands or which arm is on top in arm folding. A good deal of data on the incidence of these characters is available from quite widely dispersed regions of the world. Some examples are shown in Table 13-12. Viewing these samples as well as others not shown would lead us to conclude with Freire-Maia, Quelce-Salgado, and Freire-Maia (1958) that there is signifi-

Table 13-11. Incidence of laterality in surveys by Annett (1973)*

Type of relative	Percent L	R
Fathers	4.4	95.6
Mothers	3.7	96.3
Sons	11.3	88.7
Daughters	9.8	91.2
Total sample (main plus special left-handed families)		
Parental	4.1	95.9
Filial	10.6	89.4

*Based on data from Annett, M. 1973. Handedness in families. Ann. Hum. Genet. 37:93-105.

Table 13-12. Incidence of right hand clasping and arm folding in various ethnic groups

Source	Group	Incidence of trait (%) Hand clasping (R)	Arm folding (R)
Freire-Maia, Quelce-Salgado, and Freire-Maia (1958) (Brazil) Hand: N = 7462	Caucasians	55.2	41.4
	Mulattoes	61.5	40.4
	Negroes	68.7	41.8
	Indians	54.7	44.1
Quelce-Salgado, Freire-Maia, and Freire-Maia (1961)* (Brazil) Arm: N = 4592	Mongolians (Japanese)	60.6	43.9
COMBINED GROUPS		58.9	41.9
Freire-Maia and de Almeida (1966) (Angola and Portuguese West Africa) N = 1431	African Negroes {Males	62.3	57.1
	Females	58.1	50.0
Falk and Ayala (1971) (United States) N = 2000	Caucasians	47.0	46.0

*Two groups from 1961 article omitted; totals corrected accordingly.

cant heterogeneity between ethnic groups. Indeed, the full range for right hand clasping is probably from around 45% to 60% or higher and for arm folding from about 40% to as high as 91% in a small sample of Russians in Brazil. On the average, however, there is a slight bias to the right for hand clasping. Arm folding appears to be more symmetrically distributed. Some data are available on changes in trait incidences with age, but these are contradictory.

On the whole, then, although there is little doubt that lateral preferences in the two traits do exist, there is some ambiguity regarding their etiology. The equality of incidence between dextrals and sinistrals certainly does not readily fit any simple genetic model.

Another common laterality character is eye dominance. A study by Merrell (1957), using a sighting test, indicated that 69.5% of 426 subjects shared right ocular dominance. Various writers have attempted to distinguish between different types of eye dominance, but a study by Gronwell and Sampson (1971) found high correlations between twelve different tests, suggesting only one dimension is involved. For at least some of the tests they used, they found an incidence of around 65% of right-eyed subjects, a figure close to that just listed and to ones put forward by Duke-Elder (1949) and Spong (1962). There appears to be some relation between ocular and hand dominance. Merrell (1957) tested 497

individuals with four handedness tests and a measure of ocular dominance. He concluded that no association was present. However, as Levy (1976) has pointed out, the relation is, in fact, statistically significant. We have verified this revaluation. The nature of this relationship could be of some interest, since the two traits are rather different. Dominance of one hand usually implies dominance of the contralateral hemisphere; however, it is not certain how true this is for the dominant eye.

Distribution: lower animals

A good deal of data has been gathered on paw preference in various lower animal species. These include chimpanzees (Finch, 1941), Rhesus monkeys (Warren, 1953; Ettlinger, 1961; Milner, 1969), cats (Cole, 1955), rats (Harper, 1970), and mice (Collins, 1968b, 1969). Table 13-13, taken from Annett (1972), summarizes the main results. It is clear that with the criteria used, the percentages of dextrals, sinistrals, and mixed almost all agree with binomial proportions (Annett, 1972). That is to say, most animals use either hand or paw in reaching for food, but a few show strong left or right preferences. A graphic representation of typical data for mice was shown in Fig. 7-3. This suggests the operation of chance factors rather than genes. Additionally, Peterson (1934) was unable to select for pawedness in rats, and Collins (1969) found no differences in incidence between the inbred strains he

Table 13-13. Distribution of handedness (paw preference) in nonhuman animals*

Animal group	Criteria of R or L	Percent			
		N	Right	Mixed	Left
Chimpanzee (Finch, 1961)	90	30	30.0	40.0	30.0
Rhesus monkey					
(Warren, 1953)	90	84	27.4	45.2	30.0
(Ettlinger, 1961)	90	42	14.3	45.2	27.4
(Milner, 1969)	90	58	15.5	55.2	40.5
Cat (Cole, 1955)	75	60	20.0	38.3	41.7
Rat (Harper, 1970)	100	149	20.1	53.0	26.9
Mouse (Collins, 1969)	~88	858	26.3	43.8	29.8

*From Annett, M. 1972. The distribution of manual asymmetry. Br. J. Psychol. 63:343-358.

used and no obvious familial patterns in crosses. It is due at least in part to this work with lower animals that some investigators have begun to suspect that the good fit of some genetic models that have been put forward to account for handedness in humans may be mainly a function of the way the trait is measured. Let us now look again at the human data when a category of mixed handedness is included.

Binomial distribution of handedness

Annett (1967, 1972), after initially treating handedness as a binary trait, finally concluded that it can be measured so as to include a third category of mixed or inconsistent handers. In 1967, she published data on seven groups of university students, enlisted men, and schoolchildren using two questionnaires (8 acts and 12 acts, respectively) and observation (5 acts). Subjects were classified as dextrals (left hand for no acts), sinistrals (right hand for no acts), or mixed (inconsistency between acts). In none of the seven samples did the distribution depart from binomial expectation. We present a summary of the data in Table 13-14. Expected numbers calculated by Thompson on the hypothesis of a binomial distribution do not differ significantly from observed (chi-square = 1.20, not significant). It should be noted, as Annett points out (1967), that this result is quite compatible with a simple monofactorial model. With the data in Table 13-14, the frequency of the gene for dextrality would be 0.81, and sinistrality, 0.19. However, there are a number of reasons to be considered later why such simple genetic explanations are probably inadequate.

It will be clear from the preceding discussion that one can go beyond a tripartite division of handedness and treat it as a continuous dimension. This is readily done in the case of measures involving speed and skill rather than preference. As long ago as the late 1920s, Woo and Pearson (1927) and Woo (1928) showed that differences in left and right hand strength vary over subjects in a unimodal normal distribution in which the mean favors the right hand. That is to say, most subjects, or the average subject, have somewhat greater strength in the right hand. A similar situation is found with measures of skill. Distribution of the latter has been examined again by Annett (1970b, 1972), using speed of moving ten doweling pegs from one row on a pegboard to another row 8 inches below. The measure was time with left hand minus time with right hand. Distributions from males and females separately are shown in Fig. 13-10. As with binary measures of handedness, we again find that females are more asymmetrical to the right. The female mean falls about 1.1 above the point of equality between hands, as compared with 0.5 for males.

Annett (1972) has suggested that the relation between the distribution of lateral skill,

Table 13-14. Observed incidence of right-, mixed, and left-handedness in seven samples compared with incidences expected in a binomial distribution*

	Observed	Expected
Right	827	820
Mixed	352	366
Left	47	40
	$P_R = 0.812$	
	$P_L = 0.188$	

*From Annett, M. 1967. The binomial distribution of right, mixed, and left handedness. Q. J. Exp. Psychol. **19**:327-333.

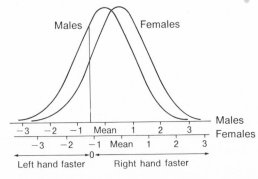

Fig. 13-10. Lateral distributions of skill (peg-moving) in males and females. Note that both distributions are biased to the right. (Modified from Annett, M. 1972. Br. J. Psychol. **63**:343-358.)

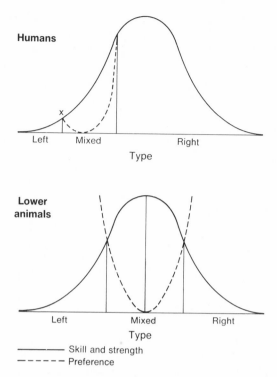

Fig. 13-11. Relationships between skill, strength, preference, and laterality type. (From Annett, M. 1972. Br. J. Psychology **63**:343-358.)

lateral preference, and the categories of left, right, and mixed are as shown in Fig. 13-11 for humans and lower animals separately. In both cases, the distributions conform to binomial proportions, but with humans showing the usual dextral bias.

Distribution in special groups

As we have indicated previously, in most populations studied, there is a marked predominance of dextrality. At the same time, the range is considerable, going from around 75% to almost 99% (Hecaén and de Ajuriagheria, 1964). Consequently, it is of some interest to look for possible systematic differences in groups marked by same special characteristic.

Twins. If handedness was completely determined by genotype, we could expect 100% concordance in monozygotic twins. However, it is clear that this is not the case. In Table 13-15, a summary is presented of eleven sets of data for MZ and nine sets of

Table 13-15. Handedness in members of MZ and DZ twin pairs: summary of data*

		Both R	Mixed	Both L
MZ twins	N	650	197	28
(11 sets of data)	p	0.743	0.225	0.032
DZ twins	N	697	170	11
(9 sets of data)	p	0.794	0.193	0.013

*Based on data from Collins, R. L. 1970. The sound of one paw clapping: an inquiry into the origin of left-handedness. In G. Lindzey and D. D. Thiessen, eds. Contributions to behavior-genetic analysis: the mouse as a prototype. Appleton-Century-Crofts, New York; Nagylaki, F., and J. Levy. 1973. "Sound of one paw clapping" isn't sound. Behav. Genet. **3**:279-292.

data for DZ twins. Incidences of the three types of pairs (both dextral, mixed, both sinistral) are shown. The first point to notice is that left-handedness has a rather high frequency in twins: around 23% in MZs and 18% in DZs. This suggests that twins are an abnormal group, that left-handedness is somewhat of an abnormal condition, or both. A third point, stressed by Collins (1970b), is that the proportions of R-R, R-L, and L-L pairs obtained correspond very closely to those expected under a binomial distribution with left = $q \cong 0.14$, and right = $p \cong 0.86$. A final point, related to the last, is that, for at least part of these data, the MZ and DZ correlations for handedness are not significantly different from zero: $\phi_{MZ} = -0.026$, and $\phi_{DZ} = -0.027$. Thus the twin data seem to show a complete lack of heritability for handedness, a conclusion rather strongly asserted by Collins (1970b). However, Nagylaki and Levy (1973) have argued by reference to the very high incidence of sinistrality in twin groups that they are an abnormal sample from which no valid generalization to nontwin populations can be made. Their assumption is basically that some proportion of left-handedness in twins is due to environmental factors—perhaps associated with crowding or mirror imaging—and that if such cases could be removed, concordance rates would be higher in twins than those reported. There is something to be said for this argument. Certainly, it is one that has been frequently used by critics of the work in other areas, such as IQ and mental illness. It is valid, however, only in cases where the major statistical parameters (e.g., mean and variance) of the experimental sample do not agree with those of the population to which generalization is being made. Sometimes when this happens, steps can be taken to remove individual subjects who show clearly aberrant features, for example, an abnormally low birth weight.

Age groups. Indirect evidence for the heritability of a trait is provided by its appearance early in life before cultural and social learning factors have had sufficient time to operate. Some of the data on early asymmetries has been reviewed by Levy (1976). In general, the neonate shows both structural and functional asymmetries. Thus sizeable hemispheric differences have been found both in adult and neonate brains, with the left being usually larger than the right (Witelson and Pallie, 1973). Corresponding differences in EEG and arterial blood supply have also been reported.

On the functional side, it has been found that neonates usually lie with the head turned to the right and more frequently show eye deviation to the right and right-hand fistedness when asleep or awake. A study by Cohen (1966) examined the association between hand preference as measured by direct grasping, and maturity as measured by the Bayley scale in 7½- to 8½-month-old infants. Of the total sample, 58% shared a definite preference. Of these, 52 were in the "advanced" maturity category, 34 in the "normal," and 6 in the "suspect" category. This difference is statistically significant and presumably indicates that extent of lateralization is an indicator of developmental maturity. Further, in the "normal or above" category, 40 out of 52 showed a right preference, that is, about 77%. Thus the data suggest that handedness develops quite early and reflects maturity at a given age level. The latter conclusion is perhaps not surprising, since the Bayley scale incorporates tests for which a stable hand preference would give an advantage.

Another study by Annett (1970b) encompassed a rather wider age range of 3½ to 15 years. A pegboard test was used as well as a record of the preferred hand in five simple acts. Distribution of handedness for the latter test in younger and older male and female age groups is shown in Table 13-16. Three categories are used: pure dextral, pure sinis-

Table 13-16. Handedness in younger and older children according to sex*

| Age in years | Percent | | | | | |
| | Males | | | Females | | |
	L	Mixed	R	L	Mixed	R
3½-8	6	30	64	2	19	79
9-15	6	34	60	4	22	74

*From Annett, M. 1970. The growth of manual preference and speed. Br. J. Psychol. 61:545-558.

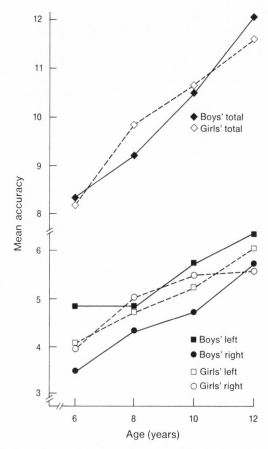

Fig. 13-12. Mean accuracy scores for recognition of nonsense shapes according to hand and sex. (From Witelson, S. F. 1976. Science **193**:425-427. Copyright 1976 by the American Association for the Advancement of Science.)

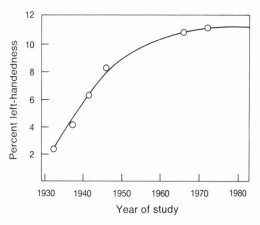

Fig. 13-13. Secular trend in incidence of sinistrality (left-handed writing) in the limited states. (From Levy, J. 1976. Behav. Genet. **6**:429-454.)

tral, and mixed. It is clear that there is very little change occurring as a function of age. However, there is a strong suggestion of a sex difference, with females showing a stronger dextral bias than males, particularly at the younger age level. This is a fairly characteristic tendency and has, so far, not been fully explained. The pegboard test results showed initially no change in right-left differences with age. Average right-hand performance was consistently better than average left-hand performance, and both improved between 3½ and 15 years.

A final paper involving both age and sex as variables was reported by Witelson (1976), one intriguingly titled "Sex and the Single Hemisphere." The laterality measure used was a dichhaptic stimulation test in which subjects were first required to palpate two nonsense shapes, one with each hand, and then to choose these from a visual display of six shapes. Two hundred boys and girls ages 6 to 12 years were subjects. The main results are shown in Fig. 13-12. No overall difference appeared between males and females, and both improve with age. However, there was a significant hand-by-sex interaction indicating a marked lateralization in boys from the age of 6 on, but none for girls even at the oldest age. On a verbal dichotic listening task, on the other hand, both sexes showed right-ear (i.e., left-hemisphere) superiority. Witelson concludes that the brains of boys and girls are differentially organized during development and that this neural dimorphism may be related to sex chromosomes and androgen levels. She further suggests that the difference has important implications for the teaching of verbal and spatial skills to the two sexes, as well as to the study of early pathologies such as dyslexia, aphasia, and autism. Assuming Witelson's results are generally valid, they would seem to have considerable basic significance.

Generations. In the surveys conducted by Annett (1973) and others just discussed, there is a suggestion that incidence of sinistrality increased from the parental to the filial generation. This appears to represent a genuine secular trend. Using a number of sources, Levy (1976) has plotted data for American populations over a 46-year period. Her graph, as reproduced in Fig. 13-13, suggests a monotonic increase in frequency of left-handed writing. As she projects the curve, it appears to have reached an asymptote in the middle 1970s. Accordingly, if this reflects the "relaxation of cultural pressures to be right-handed," then we may now have an incidence of sinistrality of about 11% that is uncontaminated by environmental factors and, in some sense, is a true estimate.

Heritability and genetics of laterality

In general, the demographic data we have presented are quite compatible with a genetic model. However, they are also consonant with a model involving the operation of random environmental factors coupled with a strong cultural pressure to dextrality. As we shall show, both points of view have been put forward by different authorities. Let us first consider some of the evidence that indirectly bears on choice of a general model.

It does appear to be true that many asymmetries are present in human beings from a very early age. As summarized by Levy (1976), they include functional and structural asymmetries of the brain, behavioral, dermatoglyphic, and motor, facial, and perceptual asymmetries. Taken together and considering that many of them are already present at or before birth, they strongly suggest the operation of genetic factors.

Another line of evidence has been provided by the results of a single case study published by Fromkin et al. (1974). They report on a young girl, Genie, kept in conditions of extreme isolation from about 20 months to over 13 years of age. Since her contact with other people was very limited and since her hands were strapped down for a good deal of the time, her opportunities to learn a hand preference were minimal. Nevertheless, when finally released from her unfortunate situation, she showed clear evidence of right-hand preference and right-hemispheric dominance, as indexed by di-

Table 13-17. Proportions of left-handed offspring in three types of family

	Parental phenotype combinations		
	RR	RL	LL
Average of three early studies (Trankell, 1955)	0.056	0.176	0.442
Annett (1973)			
Male	0.104	0.211	0.250 to 0.409*
Female	0.088	0.211	0 to 0.474*

*Range over "main sample" and two other sets of "special families" of both sinistral parents (see Annett, 1973).

chotic listening task performance. She also made rapid progress in language acquisition. We take the case of Genie as strong proof for both the innate capacity for language acquisition and for lateralization. This is not to say, of course, that environment had no effect here, since the combination of right-handedness and ipsilateral hemispheric language dominance is considered somewhat rare (Levy, 1976).

More direct evidence for the genetic basis of lateralization would normally be supplied by twin data. However, as we have indicated already, there appears to be a very low concordance for this trait both in MZ and DZ pairs. Whether this means no heritability, as Collins (1970b) has thought, or that twins are simply an atypical group, as Nagylaki and Levy (1973) have argued, is problematical. Whatever the case, we must turn our attention elsewhere—to family studies.

The incidences of sinistrality in three mating types from some older studies and a more recent one are shown in Table 13-17. In both sets of data there is an increase between incidence of left-handedness in children as we go from two-dextral to one-dextral to no-dextral parents. However, this trend varies somewhat with the two sexes; in addition, the incidence of left-handers is as high in families of left-handed mothers and right-handed fathers as in families of two sinistral parents.

Annett (1973) has gone on to estimate heritability values for handedness with Falconer's procedure for quasicontinuous variables (Chapter 12). Her results are as follows: for mothers of left-handed males, $h^2 = 57\%$; mothers of left-handed females, $h^2 = 51\%$; sisters of left-handed females, $h^2 = 44\%$ (just short of statistical significance); estimates for brothers, sisters, and fathers of males and for brothers of females are similar but nonsignificant (16% to 20%); for mothers and fathers of left-handed mothers, $h^2 \cong 100\%$ and 56%, respectively. The figure for fathers of females is only 4%. Annett also analyzed data on grandparents and some other family data gathered by Chamberlain (1928) and Rife (1940). These yielded similar results.

It is difficult to draw any definite conclusions from the preceding studies. However, there is some indication that degree of likeness is greater when one or two females are involved than where none is involved. This point, as well as several others, has been made by Annett (1973).

The twin and family data, taken at face value, do not seem to offer particularly strong support for genetic determination. As we have already shown, this ambiguity is even more pronounced in the animal studies. Nevertheless, several genetic models have been put forward, some of which fit at least selected portions of the total data quite well, though not uniquely so. We will now discuss these.

Basically, four types of model have been put forward. The most popular has been a single-gene model of which two variants have been explicitly suggested. Rife (1940), Merrell (1957), and Annett (1967) have argued for recessivity of sinistrality, with variable penetrance of the heterozygotic genotype, which is specified only by Rife as being 50%. This investigator attempted to fit this model to actual data with a moderate degree of success. However, his sample sizes were small, and, in any case, according to a further analysis by Collins (1970b), the model does not fit other data. Merrell and Annett did not attempt any Hardy-Weinberg analyses. The other variant, favored by Trankell (1955), posits variable penetrance in the double-re-

cessive genotype. He found a good and consistent fit over three independent sets of data (Ramaley, 1913; Chamberlain, 1928; Rife, 1940). The frequency of the left-handedness gene was between 40% and 43%, giving an incidence of phenotypic left-handedness in the population of about 5%. Since we have reproduced Trankell's analysis in Chapter 12, we will not present it again here. Collins (1970b) also fitted the model successfully to several sets of twin data. However, the values obtained from the fit for p (incidence of dextral gene), q (incidence of sinistral gene), and a parameter of manifestation in the double recessive seem implausible. For example, they would imply that more than 90% of the population are genotypic left-handers and 85% of these become right-handed for environmental reasons. Furthermore, as Collins further points out, the model converse to Trankell's—that is, dominance of sinistrality with variable penetrance in the homozygotic dextral genotype—fits the data equally well. It also involves some unlikely implications: for example, genotypic left-handers would constitute only about 2% of the population; a very high proportion of left-handedness would thus be of environmental etiology.

The single-gene models, therefore, do not seem to be adequate for a variety of reasons. As a consequence, Levy and Nagylaki (1972) have put forward a more complex theory that takes account of not only the readily observable phenotype of handedness but also the less readily ascertainable dimension of hemispheric dominance. Thus two pairs of genes are involved: L,l, LL and Ll, producing left-hemisphere language control with right-hemisphere control in ll individuals; and C,c, the dominant producing hemispheric control of contralateral hand with ipsilateral control in cc individuals. The nine possible combinations are shown in Table 13-18. Levy and Nagylaki tested their model with three parameters: α, the frequency of left-handers with right cerebral dominance; β, the frequency of left-handers with left cerebral dominance; and γ, the frequency of dextrals in the population. The data to which it was

Table 13-18. Genotypic constitution of dextral and sinistral phenotypes in the Levy-Nagylaki model

	Phenotype	
	Dominant hemisphere	Hand
Sinistral genotypes		
ll CC	Right	Left
ll Cc	Right	Left
LL cc	Left	Left
Ll cc	Left	Left
Dextral genotypes		
LL CC	Left	Right
Ll CC	Left	Right
LL Cc	Left	Right
Ll Cc	Left	Right
ll cc	Right	Right

applied consisted of Rife's family data and the incidence figures for recovery from aphasia in left-handers of Goodglass and Quadfasal (1954). They estimated that 47% of left-handers having brain damage in the language area recovered from aphasia and that 53% did not. Thus the ratio α/β was set at 47/53. γ was allowed to vary so as to minimize χ^2 for the fit to the breeding ratios and aphasia data separately and combined. For the latter, χ^2 was minimal (1.61) with γ set at 0.893. For the family data alone, γ was 0.891 for minimal χ^2 (1.07). Both of these values are close to the empirical value of γ (0.912) observed by Rife. However, it should be noted that all these figures are considerably lower than estimates put forward by most workers (Table 13-9).

Hudson (1975) independently tested the Levy-Nagylaki model on three other sets of family data, two from Annett (1973) and one from Chamberlain (reworked) (1928). χ^2 was minimized in terms of both γ and the ratio α/β. Unfortunately, none of these samples could be made to fit the model. Consequently, the validity and usefulness of the latter must be considered unproven. In spite of this fact, its conceptualization of laterality is surely an advance over previous models, particu-

larly in its strong emphasis on hemispheric dominance. This must surely be the level to which future genetic studies should be directed.

The third and fourth models to be discussed are somewhat similar. Collins (1970b) has favored a completely "nongenetic" model, according to which laterality arises either from cultural inheritance or from some "unknown extrachromosomal biologic predisposition to asymmetry." In view of the very early manifestation of laterality, both behaviorally and anatomically, the latter possibility might be more likely than the former. The final model, put forward by Annett (1973) argues that right-handedness is inherited presumably as any basic, universal human character but that sinistrality arises from various environmental causes. This would perhaps imply that left-handedness (and perhaps right cerebral dominance) are abnormal. The evidence, on the whole, does not seem to concur with such a notion.

In conclusion, it is clear that the problem of lateralization and its etiology is far from being solved. Since it is probably one of fundamental importance, further work on it would be very desirable.

MISCELLANEOUS RESPONSE PROCESSES

The response systems discussed earlier in the chapter are perhaps the most important of those studied. However, a great many other miscellaneous characters have also been given some attention by various research workers. They include a number of normal (Mittler, 1969; Bruggemann, 1970) and abnormal speech characteristics, such as stuttering and speech defects (see summary by Ehrman and Parsons, 1976), motor skills (McNemar, 1933; Brody, 1937), mutual imitation behavior (Wilde, 1970), and hypnotic susceptibility (Morgan, Hilgard, and Davert, 1970). As with many of the traits on which we focused earlier, heritability estimates vary with type of measurement, statistical analysis, type of sample, and time (when the response is assessed over trials). It is therefore difficult to educe any major conclusions from all this work, and, for that reason, we will not discuss it in detail.

SUMMARY

This chapter has focused on the genetic bases of simple sensory and response processes. The study of these has been relatively neglected by behavior geneticists to date, although many aspects of them have been of central interest to psychologists.

Data covering some simple gustatory, auditory, and visual functions were summarized. It is clear that many of these are under appreciable hereditary control. The same applies to relatively simple measures of central and autonomic nervous system activity, for example, EEG, evoked potential, GSR, and heart rate.

A final section of the chapter was devoted to a topic of great current interest in psychology and neurophysiology: lateralization. Although a number of genetic models fit the data on handedness quite well, none appears to do so uniquely. Furthermore, animal and twin data have suggested to some workers that genetic factors may be only minimally involved. No firm conclusions can be drawn at present.

14

Cognitive and intellectual abilities

In this chapter, we shall consider evidence relating to the inheritance of intellectual and cognitive ability in human beings. We shall examine first the meaning of the term "intelligence" and the ways in which it is commonly measured. Second, we shall attempt to assess critically data bearing on its heritability; third, we will outline and discuss various theories that have been put forward concerning the genetic transmission of intellectual abilities.

NATURE OF INTELLIGENCE AND INTELLIGENCE TESTS

Although intelligence is a commonly used term, it is one that most psychologists have found difficult to define satisfactorily. Interest in it can be traced, as with many other problems, at least as far back as the Greek philosophers. Until the eighteenth or nineteenth century, intelligence was studied largely as a special attribute of human beings and as a power enabling them to know reality. Thus for Aristotle, and for later medieval thinkers such as Aquinas, intelligence, or, more precisely, intellect, was a special faculty pertaining to the essence of humans as opposed to lower animals and was the means by which they could abstract from sensory data and arrive at concepts or ideas. No attempts were made to examine the differences between individuals in this respect, since all persons possessed it simply by virtue of being human. The approach is exemplified very aptly in the following excerpt from the writing of a philosopher of the seventeenth century, René Descartes (1637):

Good sense is, of all things among men, the most equally distributed; for everyone thinks himself so abundantly provided with it, that those even who are the most difficult to satisfy in every-

thing else, do not usually desire a larger measure of this quality than they already possess. And in this it is not likely that all are mistaken; the conviction is rather to be held as testifying that the power of judging aright and of distinguishing Truth from Error, which is properly what is called Good Sense or Reason, is by nature equal in all men; and that the diversity of our opinions consequently does not arise from some being endowed with a larger share of Reason than others, but solely from this, and we conduct our thoughts along different ways, and do not fix our attention on the same objects.

Descartes modestly deprecated even his own mind as no better than "those of the generality," although through his writings he was destined to influence profoundly the whole course of Western thought during the centuries following. It is of some interest to note from Descartes' statement at least a tacit recognition that there might be individual differences in reasoning ability but that if they do exist, they arose only because people fixed their attention on different parts of the environment, a circumstance that produced only "diversity of opinions." This might have been a modern environmentalist writing.

Descartes perhaps marks the beginning of a new way of looking at the world and at people. Until the Renaissance, the individual in society was virtually unrecognized, a fact that is strikingly demonstrated in the structure of medieval thought and society. Thus during the Middle Ages, the individual was so much part of some group that his individuality was virtually eclipsed by it (Huizinga, 1956). In the eyes of the church, for example, theologically at least, all men were equal before God. In the eyes of a feudal lord, a serf was a serf and nothing more or less. Likewise, a knight was completely defined by his knighthood. Stripped of it, he would be completely alienated from society. Thus roles and tradition superseded individuality.

With the economic and political changes that marked the decline of the medieval way of life, with the rise of nationalism and especially of Protestantism, and with the shift in emphasis in philosophy away from the relation of men to God toward the relation of men to each other, a realization of the worth of individual enterprise began to grow. By the beginning of the nineteenth century, the idea of individual differences and their importance came to be considered paramount. This was more apparent in some areas than others. It was certainly clear enough in nineteenth-century biology. Darwin's major concern was variation, and it is precisely this emphasis that distinguished his approach to systematics as against that of his predecessor, Carolus Linnaeus. In psychology, however, the most influential figures, such as Wundt, Fechner, Ebbinghaus, and Pavlov, were mostly not oriented in this way. It was only a few men on the margin of this newly developing field—notably Galton and Binet—who gave the testing of individual differences its start. Sir Francis Galton, indeed, has usually been designated as the "father of mental testing." He did not go to any lengths to define intelligence exactly, much less measure it adequately. But he did deal with it in a general way in his studies of the inheritance of abilities, eminence, and genius. His view of the nature of intelligence was, in many ways, a completely operational one. For example, to belong to Galton's more select groups, a man "should have distinguished himself pretty frequently either by purely original work or as a leader of opinion" (Galton, 1883, p. 9). Binet's interest in intelligence was rather more practical than scientific in that one of his main concerns was the problem of dealing with subnormal children in Paris schools. Though his tests were largely derived from empirical grounds, he put forward a definition of intelligence as the ability to maintain a definite direction, the ability to make adaptation leading to a desired end, and the ability to criticize one's own behavior.

Under the influence of Galton and Binet, the mental testing movement grew rapidly, especially in America, and it became a popular pastime to generate definitions of intelligence. Spearman reduced it to the ability to educe relations and correlates. Thorndike regarded it as the power of making good responses from the standpoint of truth or fact, whereas Terman, in more classical style, defined it as the ability to abstract.

Eventually, American behaviorism engineered a moratorium on definitions of intelligence by suggesting that intelligence should be regarded simply as that which an intelligence test measures (Goodenough, 1949). This definition is not as circular as it seems if the test is first constructed on the basis of some clear postulational definition.* Given such a test, we may then proceed to analyze in detail the nature of the functions it involves and the pattern of their relationships. In this way, a hypothesis may be generated that can be examined and corrected until a satisfactory understanding of intellectual ability is achieved. Such an a posteriori approach is implicit in most of the work of factor analysis.

The originator of this latter technique was Charles S. Spearman (1927). He was the first to explore and develop the idea implicit in Pearson's coefficient of correlation that if two traits or abilities vary together, then they have something in common. Spearman found that practically all intelligence tests, when corrected for their lack of reliability (correction for attenuation), intercorrelated to a high degree. From this fact he reasoned that all intellectual tasks must have some basic common element, an element he called g, or general mental ability. He explained the fact that correlations between tests were never perfect (i.e., $r < 1.0$) by suggesting

---

*Kempthorne (1978) has claimed that use of the term "intelligence" in reference to so-called mental tests is as arbitrary as use of the term "dog" in reference to the taxon to which it refers. This simplistic view is clearly erroneous. Mental tests were initially constructed to assess what was considered to be intelligence.

that each test had a specific as well as a general component. He conceived of these specifics as being the particular "engines" through which *g* was expressed. Spearman's work was seminal, and from it emerged a variety of factor-analytic methods, as well as a number of important factorial theories of intelligence. We will discuss some of these shortly.

Today, the meaning of the concept of intelligence and the value of IQ tests are still being debated. In fact, because of the rapidly accumulating data concerning its genetic determination, the argument during the late 1960s and early 1970s has become at times quite passionate. It is difficult to define exactly the pros and cons in this debate. But, roughly speaking, opinion ranges along a continuum with some at one end who see the IQ test as a kind of "in game" forced on society by a self-protecting establishment and, at the other end, those who see intelligence as a meaningful term and the IQ test as a sophisticated and objective scientific instrument that allows rapid and efficient management of the variety of talents in a society. In between, there are still others who are willing to accord to tests some merit but still insist that what they test primarily reflects cultural values. Something of the flavor of these views can be obtained from such sources as Butcher (1968), Dockrell (1970), Cancro (1971), Hunt (1972), and Block and Dworkin (1976).

The extreme position that intelligence test scores are no more than arbitrary "credentials," much like manner of dress or accent, has been put forward by a number of critics, one of the more influential being McClelland (1973). Basically, he has argued that IQ tests relate mostly to achievement in school, that such achievement is decided on social and economic grounds, and that therefore test scores may be expected to predict success in life when this success depends on the same kinds of social values that produce good school grades. McClelland's argument seems to bear not so much on whether intelligence is something real or not, but rather on the question of the relation between IQ and capacity to perform in various work situations. Since he apparently regards the latter as having the greater reality, he can then conclude that anything which fails to relate perfectly to these must be, to that extent, fictitious. This is not a logical conclusion. One could equally well assert that a height measurement also has a mythical quality if it fails to predict perfectly athletic performance. What we do with any measurement of anything is largely an arbitrary matter. However, this does not render the measurement trivial or arbitrary. The dimension of temperature is not given validity wholly by the fact that it may be used in giving us comfortably heated houses. So it is with intelligence. As long as we are dealing with a reasonably stable property of human beings, we are on perfectly respectable and scientific grounds in seeking to map its structure and to explore its correlates.

A less extreme position than that of McClelland has been put forward by Humphreys (1971). He has defined intelligence as "the totality of responses available to the organism at any one period of time for the solution of intellectual problems. *Intellectual is defined by a consensus among psychologists*" (italics ours). Humphreys thus appears to advocate that intelligence is not merely a fanciful construct and that tests are more than merely political devices designed to preserve privilege. However, it is also clear that he regards intelligence as having an arbitrary definition. It is simply whatever responses happen to be called "intellectual" by a certain group of people—hopefully experts, at least —at a particular time.

The final position we wish to delineate is probably the one held by most people working in the field of individual differences and human behavior genetics. Since it is also our position, it is appropriate to articulate it a little more closely.

1. In the first place, we consider that so-called intelligence tests do measure some stable character in human beings. Most IQ tests are highly reliable, with the standard error on the order of ±5.0, that is, about one third of the standard deviation. Changes in

IQ greater than 5 points occur with a frequency inversely proportional to their magnitude (Wechsler, 1971).

2. An IQ score is a relative rather than an absolute measure. That is to say, it reflects the ability of an individual in relation to a random sample of individuals taken from his age group and growing up in comparable environmental conditions.

3. Following from 2, it is recognized that special environmental conditions (e.g., enriched nursery school training) may alter IQ scores. However, the relative ranks of individuals exposed to such environments do not necessarily change much.

4. The many IQ tests presently available tend to show, on the average, appreciable correlations with each other. This fact favors the concept of general mental ability.

5. The intelligence of an individual is a function of his genotype and his environment, particularly his early environment.

6. Stability of intelligence, that is, its resistance to change, may be as much a result of early experience as of genotype.

7. Measured intelligence relates significantly to a number of other variables, notably to school achievement, occupational level, socioeconomic status, and, most importantly, to age. Little is known, however, about its relation to fitness in the biological sense, though this is a question of the most basic interest. Likewise, little is known about the physiological correlates of intelligence in humans, although at least one promising attempt has been made in this direction (Halstead, 1947). It has been argued by some writers that there is some basic substratum of intelligence which underlies its expression through any particular cultural medium. If local sociocultural elements could then be eliminated, there should remain only this essential substrate. What this would involve is problematical. Vernon, in an early article (1954), suggested that the ability to follow "oral directions" of a greater or lesser complexity represents a basic intellectual ability that is demanded by all societies and that is little affected by differences in educational background. Obviously, this approach to de-

fining intelligence can be almost as a priori as that followed by early workers in the field. Furthermore, it is probably true to say that there is really no such thing as a "culture-free" test but that there can be tests which are relatively "culture-fair," in the sense that they minimally involve the use of information restricted to particular sectors of the population (e.g., upper socioeconomic level). Tests such as the Raven Progressive Matrices, the Cattell Culture-Fair Test, and some of the performance tests of the Stanford-Binet or the Wechsler-Bellevue are examples. It is, of course, often argued that such tests are only seemingly culture fair and may actually be still biased in some tacit way. This problem has been considered in a penetrating discussion by Jensen (1973a). He has urged that an empirical approach be taken whereby test items are given to groups known by some definite criteria to differ in cultural acquisitions and that then items be designated as culture free or culture bound according to whether they distinguish the groups.

GENERAL INTELLIGENCE AND SPECIAL ABILITIES
Studies of natural families

Pedigrees. The oldest method used to study the heritability of a trait is the pedigree method. Although many people before him had traced family pedigrees, Galton (1883) was the first to do so in a way that can be called scientific. He was particularly impressed with what he called *eminence*—unusually high intellectual endowment—and considered this trait so rare as to have an incidence in the population of only 0.025%. On the basis of reputation and available records, he constructed pedigrees of distinguished statesmen, commanders, writers, scientists, poets, musicians, painters, divines, and even university oarsmen and "North Country wrestlers." An example of one of his pedigrees is shown in Fig. 14-1. To allow a more quantitative examination of pedigrees for intellectual ability, he used a rough scale based on a normal curve distribution with eighteen intervals ranging from eminent to idiot and imbecile. Since Galton found from his data

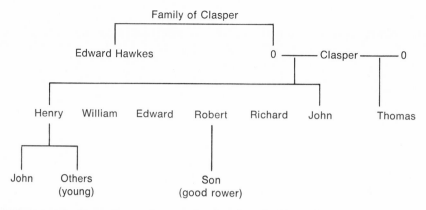

Fig. 14-1. Example of a pedigree of rowing families described by Galton. All those mentioned in the pedigree except one were able or excellent oarsmen. (From Galton, F. 1883. Hereditary genius. Appleton-Century-Crofts, New York.)

that the chances of an eminent man having an eminent relative were rather high, he inferred that intellectual ability was clearly inherited. Furthermore, after a careful examination of the environmentalist hypothesis, he concluded that true capacity could not be altered by lack of opportunity and would always surmount social barriers and unfavorable circumstances. It is interesting and perhaps amusing, depending on one's viewpoint, to note that one of his arguments for this involved a comparison of British and American eminence. According to his judgment, the number of really able men was very much higher in England in spite of the more repressive social life in that country. If the more democratic form of society in America in which education and opportunity, more available to the masses, did not produce proportionately more ability, then, obviously environment could not make much difference. This general conclusion is, of course, obviously open to many criticisms. Be this as it may, Galton's analysis of pedigree data, as well as his biometric and psychometric contributions stand as milestones in the development of human behavior genetics.

Since Galton's day, the pedigree method has been used widely, although for quantitative traits it has been refined by the use of more sophisticated statistical methodology. Such workers as van Bemmelen (1927), Gun

(1930a,b), and Bramwell (1944) have presented pedigrees of royal families, holders of the Order of Merit, and various other similar classes that cannot be easily described in exact terms but are nonetheless worth examination. In the realm of special abilities, we have a great many pedigree studies devoted to musical talent. One of the most famous of musical families is, of course, the Bach family, first studied by Galton (1883) and later by many others. Although the problem of separating nature from nurture complicates matters, the amount of musical ability that consistently appears in the Bachs for a number of generations is suggestive of hereditary factors at work.

Inasmuch as pedigree studies represent a first step in psychological genetics, they have considerable historical importance. Their value is reduced, of course, by the fact that they do not always show the systematic operation of heredity independent of environmental influences. At the same time, when the traits studied are uncommon ones and if their appearance in family lineages conforms closely to genetic expectations, the pedigree method can be very useful (Chapter 12). Even with common characters showing a normal distribution, the patterns of variance and covariance for different classes of kinship may often be found to fit genetic models or at least *not* to fit any obvious environmental models. Consequently, we can still learn

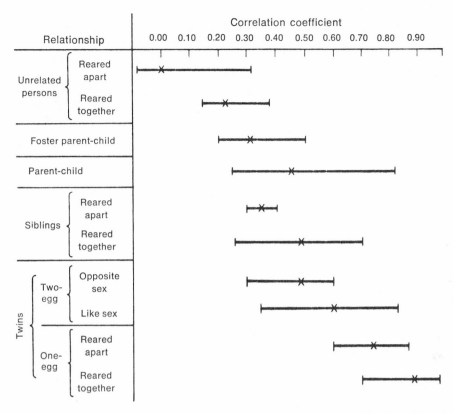

Fig. 14-2. Summary of family correlation studies: medians and ranges of correlations from fifty-two studies. (From Erlenmeyer-Kimling, L., and L. F. Jarvik. 1963. Science **142**:1477-1478. Copyright 1963 by the American Association for the Advancement of Science.)

much from examination of natural families. Their usefulness has been clearly shown, for example, by the work of Jinks and Fulker (1970), McCall (1970), and Eaves (1973). Two of the best examples of pedigree studies that are gold mines of information are those by Terman on gifted children (Terman and Oden, 1959) and by Reed and Reed (1965) on mental retardation.

Familial correlations in intelligence. The first systematic examination of familial correlation with respect to mental ability was made by Karl Pearson around the turn of the century. Using teachers' rankings based on a 7-point scale of intellectual capacity, he found that the correlation between brothers was 0.52, between sisters, 0.51, and between brother and sister, 0.52. Since these values were approximately the same as those obtained with physical characteristics such as

height, hair color, and cephalic index Pearson concluded that psychic traits are inherited in the same way and to the same degree (1904). Woods (1906), using a rating scale of 1 to 10 for "intellectual and moral excellence," analyzed some royal family pedigrees. He obtained a parent-child correlation of 0.3. Schuster and Elderton (1907), using the records of academic performance at Charterhouse, Harrow, and Oxford, were able to establish a parental correlation of 0.31 and sibling correlations of around 0.4.

One of the first to use standardized mental tests was Gordon (1919). She found a correlation of 0.61 with the first Stanford revision of the Binet-Simon test given to 216 sibling pairs. Both Pearson (1918) and Elderton (1923) reworked her data (correcting for age and using all possible sib pairings) and obtained correlations between 0.5 and 0.6.

Table 14-1. Correlations between relatives*

	Burt		Other investigators		
	Number of pairs	Correlation	Number of investigations	Median correlation	Theoretical value
Direct line					
With parents (as adults)	374	0.49	13	0.50	0.49
With parents (as children)	106	0.56	—	—	0.49
With grandparents	132	0.33	2	0.24	0.31
Collaterals					
Between monozygotic twins					
Reared together	95	0.92	13	0.87	1.00
Reared apart	53	0.87	3	0.75	1.00
Between dizygotic twins					
Same sex	71	0.55	8	0.56	0.54
Different sex	56	0.52	6	0.49	0.50
Between siblings					
Reared together	264	0.53	36	0.55	0.52
Reared apart	151	0.44	3†	0.47	0.52
Between uncle (or aunt) and nephew (or niece)	161	0.34	—	—	0.31
Between first cousins	215	0.28	2	0.26	0.18
Between second cousins	127	0.16	—	—	0.14
Unrelated persons					
Foster parent and child	88	0.19	3	0.20	0
Children reared together	136	0.27	4	0.23	0
Children reared apart	200	−0.04	2	−0.01	0

*From Burt, C. 1966. The genetic determination of differences in intelligence: a study of monozygotic twins reared together and apart. Br. J. Psychol. **57:**137-153.
†The actual figure given by Burt (N = 33) is an error.

Around the same time, Pintner (1918) and Madsen (1924) contributed by computing correlations on unrelated control pairs of children. Since these turned out to be rather low (between 0.19 and −0.4), the case for the inheritance of intellectual ability was thought to be further strengthened.

Since this early work on the problem, numerous other studies have been done. To give the reader some idea of the magnitude of this work and its degree of consistency, we present in Fig. 14-2 and Table 14-1 two summaries compiled up to 1963 and 1966, respectively. The first, by Erlenmeyer-Kimling and Jarvik (1963), consists of the results for various kinship classes of 52 studies involving, altogether, over 30,000 correlational pairings. Since some of the studies reported on several categories and several samples, the total number of separate observations comes to 99. Excluded from the list were (1) studies in which the type of test was other than a conventional IQ test, (2) studies in which atypical subjects were used (e.g., mental defectives), (3) studies in which zygosity information was inadequate (for twin studies), and (4) studies involving too few special subjects (e.g., one or two twin pairs, or triplets).

Several comments about Table 14-1 are in order. In the first place, it is clear from inspecting only the bottom ends of the ranges that as genetic resemblance increases, so intellectual resemblance also becomes more pronounced. The relevant categories are unrelated pairs, siblings or parent-offspring, and MZ twins. This also applies if we use either the median or mean correlations. Furthermore, the median correlations closely approximate those expected on purely genetic grounds. The observed correlations are 0.5 for parent-child, 0.49 for siblings reared together, 0.53 for DZ twins, and 0.87 for MZ twins reared together. Corresponding genetic correlations (under the simplest assumptions) are 0.5 for the first three groups, and 1 for MZ twins.

A second point to note is that, for each group, the range is considerable. This is particularly true for the 12 reported parent-offspring correlations, which vary from around 0.2 to 0.8. The reason for this is not clear, but it perhaps reflects the fact that the data reported came from eight countries and cover a span of two generations.

Third, it is obvious that environment must play some role in determining intelligence level. This shows up in every category involving a reared-together versus reared-apart comparison.

Finally, we should emphasize that the matter of deciding which value in each category is the most representative is not easily solved. Erlenmeyer-Kimling and Jarvik used simple unweighted medians. However, it can be argued that weighted medians or weighted means might be more appropriate, with the weighting factor based on the size of the sample involved. Likewise, corrections might be made for restriction of range and for test reliability (Jencks, 1972; McAskie and Clarke, 1976). However, at least some of these corrections may be expected to balance each other out.

The second summary presented in Table 14-1 by Burt cites 95 studies plus 10 of Burt's own observations. Median correlations for the various groups are compared with genetic correlations expected on the basis of Mendelian inheritance with incomplete dominance and some assortative mating. The formulae for these expectations were worked out by Burt and Howard (1956) and are ultimately based on those of Fisher (1918). Note that for the several new categories included—grandparents, uncles and aunts,

and cousins—observed correlations are not far from those expected on the basis of the genetic model. This applies to the more familiar categories as well.

There have been fewer family studies reported since the two preceding summaries appeared, probably because most investigators have relied more on the use of twin data. However, one large-scale study has been published by Reed and Reed (1965) on 545 families involving over 80,000 individuals. Although the study was aimed mainly at tracing the familial patterns of mental retardation, it also recorded IQ scores for normal individuals in some families for as many as five generations, plus every variety of collaterals. This is a major piece of work, much of which still remains to be analyzed. Two examples of such analyses are afforded by the articles of Jinks and Fulker (1970) and of Eaves (1973). The Reeds' data do not allow the exact separation of environmental and genetic effects. However, they can be analyzed for degree of fit to various genetic models. Jinks and Fulker and Eaves, using somewhat different methods, were successfully able to do this, emerging with the conclusion that IQ involves a large additive component of genetic variation, dominance of high IQ, and marked assortative mating. Apart from this, Higgins, Reed, and Reed (1962), using a sample of 1016 parental pairs and 2039 children, found a sibling correlation of 0.52 and father-child and mother-child correlations of 0.43 and 0.45, respectively. Their figure for husband-wife correlation was 0.33, which is considerably lower than has usually been found. Waller (1971b) has also derived heritability estimates from the Reeds' data. He used the following procedures: (1) regression of midchild on midparent for 258 2- to 5-child families weighted according to family size, yielding $b_{\overline{O}\overline{P}} = 0.631$ ($b = h^2 = 0.631 \pm 0.046$); (2) regression of midchildren on father's score, also with a weighting procedure; $b_{\overline{O}P} = 0.334$ ($b_{\overline{O}P} = \frac{1}{2}h^2$ and $h^2 = 0.668 \pm 0.025$); (3) regression of midchild on mother; $h^2 = 0.654 \pm 0.025$; and (4) intraclass correlation between siblings; $r_i > \frac{1}{2}h^2$ and $h^2 \leq 0.666$. These estimates are highly consistent and agree closely with those put forward by some of the other workers cited previously.*

All in all, familial correlational data up to 1966 are reasonably consistent on the average. By themselves, however, they are too limited to allow either firm estimates of heritability or elaboration of any genetic model of intelligence. Nonetheless, as we shall see, they can contribute significantly in these directions when pooled with other kinds of data on foster children and especially separated and nonseparated twins.

In fact, both the summaries presented contain sufficient data to allow rough specification of some of the components of variance for IQ. Thus Morton (1972), using the Erlenmeyer-Kimling and Jarvik median correlations, has estimated a heritability of 0.675, with 0.139 due to common environment, 0.016 due to common environment specific to twins, and 0.17 to random environmental effects. Different combinations of familial correlations can also supply heritability estimates. For example, twice the difference between sibs and foster sibs, $h^2 = 2 \ (r_S - r_{FS}) = 0.52$; likewise, $h^2 = 2 \ (r_{MZ} - r_{DZ}) = 0.68$. Corresponding estimates from Burt's data are 0.52 and 0.74. Since these methods involve untested assumptions about how environmental influences operate in the different groups, however, they cannot be given too much weight.

Setting aside these problems for the time being, we may note that certain of the studies reviewed reveal some interesting aspects of the operation of heredity and environment and of mating systems in human populations. We will now discuss some of these briefly.

Effects of age. Quite a number of studies have examined the effects of age disparities on family resemblances. It is assumed that since children of different age groups occupy different environments, the intellectual similarities between siblings of different ages should be less marked than between sibs of the same age if environmental effects are correlated with age. If environment is of little importance, on the other hand, sibling correlations should not be greatly affected by age differences. On the whole, the evidence seems to favor the second alternative. In Burt's table, the DZ twins reared together and sibs reared together (presumably of variable ages) correlate 0.56 and 0.55, respectively. Erlenmeyer-Kimling and Jarvik report medians of 0.53 and 0.49 for the two categories. The problem has been examined more specifically by several writers. Conrad (1931) found no difference in magnitude of intellectual resemblance (Stanford-Binet and Army Alpha) between sibs less than 3 years apart as against sibs more than 3 years apart. Likewise, Finch (1933), using 1023 pairs, found a correlation of 0.49, although within this sample, within-pair age differences ranged up to 11 years. A correlation computed between size of age difference and size of score difference was not significantly different from zero. Again, artificially "twinning" sibs by using only their IQ scores taken at identical ages had no effect on resemblance, as shown by Richardson (1936). One study that appears to indicate an opposite conclusion is that by Sims (1931), who artificially paired unrelated children according to socioeconomic background, age, and school. The correlation between these unrelated "sibs" was 0.35, a figure not much lower than that obtained by many investigators with real siblings.

An interesting variant on this problem has been put forward by McCall (1970), who examined parent-offspring correlations for IQ with the test scores of parents taken at the same ages as when their children were tested. A similar procedure was used in the case of sib correlations. In all, eleven ages were examined, between 42 months and 132 months. The general method allowed assessment of heritability of general level and heritability of sequential pattern. The median correlation for siblings turned out to be 0.55, but for parent-offspring, only 0.29. For nei-

---

*Note that the regressions estimate heritability only if some simplifying assumptions are accepted, for instance, no role of common environment. This is probably not tenable.

ther of the two groups, however, was there a significant "pattern" effect. In general, sizeable changes in IQ did occur with age, but trends in related individuals were not more similar than in unrelated pairs. This suggests, of course, that environmental influences are not systematically related to a person's age—or if they are in any single individual, the relationship is not generalizable to other individuals. This is perhaps not surprising from a commonsense point of view. Certainly, it is unlikely that parents will accord to their second child, say at the age of 2, the same kind of treatment they accorded their first child at this age. Presumably, parents learn from their mistakes and successes. Or, failing this, they may just change their minds as to what is "good" for a 2-year-old or a 6-year-old. In comparison, however, cotwins might very well encounter more uniform environments at each age level and might be thus expected to show a more significant patterning effect. Some interesting work by Wilson, Brown, and Matheny (1971) bearing on this point will be discussed shortly.

Correlations in different populations. One line of work on familial correlations has dealt with degree of resemblance in differently constituted populations. Hart (1924) attempted to compare sibling correlations in three groups: a random selection of schoolchildren, a rural group of lower mean intellectual level, and a group selected for high intelligence. The three correlations were, respectively, 0.45, 0.46, and 0.4. Another study by Hildreth (1925) tested subjects in three schools, two of which differed from the other in intellectual level. Differences in size of sib correlations were found to be a function of group homogeneity rather than of intelligence level. Hildreth's data are summarized in Table 14-2. It is of some interest to note that in the two "homogeneous" schools, sib correlations were very low when age was partialed out. However, the overall "best estimate" put forward by Hildreth was 0.47, a figure not far from the median correlation of 0.49 of Erlenmeyer-Kimling and Jarvik.

Outhit (1933), using the Stanford-Binet and the Army Alpha, examined parental and sibling correlations in 51 families drawn from the city, small towns, and the country. Parental subjects covered the full range of occupational status, ranging from unskilled laborer to professional, an age range between 27 and 67 years, and an educational range from grade 3 to postgraduate university level. Ages of children were between 3 years 2 months and 39 years, with a median age of 10 years 4 months. Most of the correlations Outhit obtained were higher than others have found previously (or later). All sib correlations (computed by various pairing methods) were in excess of 0.5; the midparent-midchild correlation turned out to be 0.802. It is likely that these high figures are the result of using a rather heterogeneous sample. There did not appear to be any differences in the sizes of correlations between the various groups; neither was there any correlation between parental ability and variability of offspring. In addition, Outhit found a clear offspring regression effect and also reported a phenotypic correlation for IQ between

Table 14-2. Resemblance in intelligence of siblings in three different schools*

School type	Number	Sib correlation	Sib correlation age constant
Average, heterogeneous	450	0.63	0.47
Superior, homogeneous	325	0.32	0.08
Average, homogeneous	253	0.27	0.13
Composite	1028	0.68	0.42

*From Hildreth, G. H. 1925. The resemblance of siblings in intelligence and achievement. Teachers Coll. Columbia Univ. Contrib. Educ. **186**:1-65.

spouses of 0.741. It should be mentioned in passing that Outhit has an excellent review of the literature prior to 1933.

Lawrence (1931) obtained IQ test data in a large population of orphanage children. At least two important points emerged. First, even for such children, living in what might be considered to be a relatively homogeneous environment, variability of intelligence was as high as that of children living in their own homes. Second, the correlation between intelligence of the orphanage children and their parents was low at the time the children entered the orphanage but tended to increase as the children grew older. This would mean, of course, that homogeneity of environment does not necessarily produce homogeneity of intelligence and that the influence of social class may operate primarily through genetic rather than through educational factors.

These general conclusions will be explored in more detail in a later chapter. Suffice it for now to state that they are generally supported by a large body of data not only involving social class but also rural-urban comparisons. Much of the early work on the topic has been summarized by Jones (1954).

Culture-fair tests. A somewhat different approach to the problem we are considering has been to attempt to hold cultural or environmental differences constant by use of culture-fair or culture-free tests. This was done by Cattell and Willson (1938) in a study of 100 families. After correcting for age, attenuation, scatter, and skewness, these authors obtained a parent-offspring correlation of 0.91 and a sibling correlation of 0.71. They concluded on the basis of these data that parents and children have nine tenths of their respective intellectual levels in common and that four fifths of the variance of intelligence between families is due to heredity. Whether the test was really culture free or even culture fair is problematical. The fact that their correlations are higher than those found by any other workers suggests that cultural differences, when allowed to be expressed in test scores, tend to de-

press similarities and thus produce artificially low estimates of family resemblance. This is, of course, a hypothesis rather than a necessary and logical conclusion, since it could be argued that culture can make people either alike or different. However, it should also be noted that for polygenically controlled traits, the parent-offspring correlation is greatly increased by assortative mating—anywhere up to 0.5 plus one half of the assortative mating coefficient. Cattell and Willson report a coefficient of 0.811. Half of this added to a theoretical expectation of 0.5 would, in fact, give 0.905, a figure almost identical with the obtained one of 0.91. It is of some interest that Jones (1928) and Outhit (1933), who both reported rather high parent-offspring correlations, also found relatively strong assortative mating.

Whether justifiable or not, the Raven Progressive Matrices Test has often been considered to be a culture-fair test, since it involves exclusively figural material and logic. One family study by Guttman (1974) used this test to obtain parent-offspring, sib-sib, and first cousin correlations. These were found to vary somewhat with different parts of the test. For the total scores, heritability estimates ranged between 0.3 and 0.7, depending on the method of computation.

Twin-sib pairings. As Kamin (1974) has pointed out, an environmentalist might well be interested in another kind of correlational analysis, one not commonly reported: the resemblance between one member of a twin pair and a sibling from the same family. It could be argued that if environment is very important, twins should form their own microculture from which singleton sibs would be excluded. Hence twin-sib correlations should be smaller, on the average, then ordinary sib-sib correlations. Snider (1955) gathered such data in 329 twin-sib pairs for the Iowa Basic Skills Vocabulary Test. The correlation was 0.32 if the twin in the pairing was DZ and only 0.19 if the twin was MZ. In the same study, the conventional DZ correlation was 0.5, and the MZ correlation was 0.79. The twin-sib correlations reported by Partanen, Bruun, and Markkanen (1966) for two tests of verbal ability were 0.41 and 0.34. However, for a vocabulary test, Huntley (1966) reported a twin-sib correlation of 0.58. Consequently, the data must be regarded as ambiguous at the best and offer very weak support for a purely environmental model.

Special abilities. Most studies concerned with familial resemblance in special abilities have examined scholastic variables. Two early studies (Earle, 1903; Pearson, 1910) obtained fraternal and parental-child correlations about as large as those commonly found for general intelligence. Another study by Starch (1915) yielded a mean sibling correlation of 0.52 for general scholastic ability. One notable finding in this and a subsequent study (Starch, 1917) was that correlations for abilities supposedly not much affected by school work, such as memory or cancelation, were about the same magnitude as those which were specifically encouraged by it, such as reading, writing, and arithmetic. Starch concluded that heredity rather than training causes resemblances in families. Several subsequent studies by Cobb (1917), Huestis and Otto (1927), Willoughby (1927), and Banker (1928) supported this conclusion. Griffits (1926) obtained a much lower sib correlation for modal grades, however: 0.299.

One of the most exhaustive of the early family studies was done by Carter (1932a), who examined resemblances in respect to verbal and numerical abilities. His main data for these two classes are shown in Table 14-3. There is a tendency for members of like-sexed pairs to resemble each other more strongly than unlike-sexed pairs. On the whole, specific correlations obtained were rather low. He attributed this to the homo-

geneity of the group studied and also to the absence of much assortative mating between parents. Coefficients were 0.21 for vocabulary and −0.4 for arithmetical ability. It is of some interest that in respect to both types of abilities a clear regression effect showed up. That is, children of superior parents were less superior, and those of inferior parents were less inferior.

In summary, the work just discussed shows that strong familial resemblances can be found both for general intelligence and for some special abilities such as scholastic aptitude. By itself, such a body of data can be explained by invoking either genetic or environmental causation. In our view, however, environmental models need to be better worked out than they are at present if they are to fit the data as well as genetic models seem to do. This point has also been made by McAskie and Clarke (1976) in a thorough review of studies on parent-offspring resemblance.

Studies of adoptive families

It will be recalled that Gordon (1919) found that sibling pairs raised in orphanages in California showed as much similarity ($r = 0.53$) as sibs reared in their own homes. The rather few fostering studies that have been carried out mostly confirm this conclusion that likeness of genetic relatives is not eradicated by rearing in a different kind of environment. Some of these data have been summarized and critically discussed by Jencks (1972), Kamin (1974, 1978), and Munsinger (1975, 1978).

In 1928 Burks (1928) reported correlations for foster children's intelligence with foster fathers' intelligence, with foster mothers' in-

Table 14-3. Family resemblance in verbal and numerical abilities*

	Child with				
	Like sib	Unlike sib	Midparent	Like parent	Unlike parent
Average verbal	0.38	0.31	0.22	0.54	−0.11
Average numerical	0.24	0.17	0.20	0.64	−0.28

*From Carter, H. D. 1932. Family resemblances in verbal and numerical abilities. Genet. Psychol. Monogr. **12**:1-104.

telligence, and with home environment. These were, respectively, 0.07, 0.19, and 0.21. Corresponding correlations for children reared with their biological parents (matched to the foster parents) were 0.45, 0.46, and 0.42. By using certain correction procedures, both sets of correlations were elevated, without, however, altering very much the absolute difference between them. Much smaller differences were obtained in a study by Freeman, Holzinger, and Mitchell (1928). These authors found a correlation of 0.39 for Otis IQ between adopting midparent and foster child. This compared with a correlation of only 0.35 between child and true midparent. Such results may well have been due to selective placement, as suggested by Leahy (1932). In fact, mean IQs of foster children placed in "good," "average," and "poor" homes were, respectively, 111, 103, and 91. Leahy herself at a later date (1935) obtained results more like those of Burks. Snygg (1938), on the other hand, in a study of 312 foster children, reported a low correlation of 0.13 between true mothers' IQs and those of their fostered children. Since the children were young at the time of testing (3 to 8 years), the resemblance might have increased with age, though Snygg was not able to discern any trend in this direction.

A convenient summary of the major data has been made by Jencks et al. (1972), as modified in Table 14-4. Shown are weighted means, derived from four American studies with weights assigned according to number of subjects used in each study. It is clear from the table that the degree of resemblance generated between unrelated persons occupying the same environment is appreciably less than that commonly found between relatives, even when these are living apart. Perhaps the best demonstration of this point is made by the two important studies of Skodak and Skeels (1949) and of Honzik (1957). Let us look at these in more detail.

The Skodak-Skeels work was carried out over a period of almost 15 years, with the first report appearing in 1936 (Skeels, 1936) and the final one in 1949 (Skodak and Skeels, 1949). The initial sample consisted of 390 children placed in foster homes by adoption agencies. Over a follow-up period, IQ data were collected on four separate occasions. Because of difficulties in locating all members of the original group each time, the sample was successively reduced to a final size of 100 children, all of whom had been tested four times at various ages. Skodak and Skeels (1949) concluded from an examination of the IQ means and variances of these successive samples that these 100 subjects were representative of the original starting population. Data on various characteristics of true mothers and fathers and foster parents of adoptees were also obtained. At least four major points emerged from the data:

Table 14-4. Correlations between characteristics of adopted children and their adopting relatives*

Child IQ and	Studies	Correlation	N
Adopting father's IQ	3	0.21	536
Adopting mother's IQ	3	0.24	645
Adopting father's education	3	0.09	466
Adopting mother's education	3	0.18	486
Adopting father's occupation	2	0.28	588
Family income	1	0.23	181
Other adopted child in same home	4	0.42	165
Other natural child in same home	3	0.26	94
All unrelated children in same home	7	0.32	259

*Modified from Table A-3 in Inequality: a reassessment of the effect of family and schooling in America, by Christopher Jencks et al., © 1972 by Basic Books, Inc., Publishers, New York. (For a later summary see Munsinger [1978].)

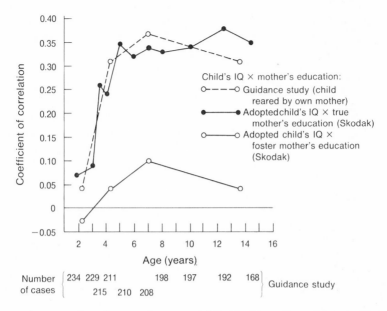

Fig. 14-3. Education of mother in relation to child's IQ. (From Honzik, M. 1957. Child Dev. **28**:215-228. Copyright 1957, The Society for Research in Child Development, Inc.)

1. It is clear that the foster children's IQs correlate increasingly with true mothers' IQs ($N = 63$) and education ($N = 100$) but not with foster parents' education. The relevant data, shown in Fig. 14-3, are taken from Honzik (1957). IQ data were available on only 13 true fathers and hence are too limited to analyze. However, the net impression given by the three graphs is clear enough. The foster children correlate with parental education and intelligence to an increasing extent over the period of time during which they were studied. But the similarity of children's IQ level to foster parent education remains close to zero and, if anything, becomes less as the children grow older. This seems strong grounds for invoking the operation of heritable factors. Note, however, that it can be argued (Kamin, 1974) that the differences in these correlations could be a function of (a) some peculiarity of adopting families that typically results in low parent-offspring resemblance, both for natural and adopted offspring, or (b) homogeneity of variance of scores (IQ or education) in either or both of the adopting or natural parents. Thus, if the adopting parents turn out to be a particularly

homogeneous group, the net effect of this could be to reduce the correlation between themselves and their children, both adopted and natural. On the first point, the evidence in favor of postulating something atypical about adopting families is very slender. As just indicated, Freeman, Holzinger, and Mitchell (1928) reported a correlation of only 0.35 between true children and parents in adopting families and a correlation of 0.39 between adoptive child and adoptive midparent (0.37 with fathers and 0.28 with mothers). The authors acknowledge that the true child-parent correlation was unusually low and, furthermore, suggest that the higher figure of 0.39 resulted from selective placement. The only comparable correlations from Leahy (1935) are 0.18 for adoptive child –midparent Otis IQ and 0.36 ($N = 20$, nonsignificant) natural child–midparent Otis. Consequently, these are flimsy grounds for supposing that, for some unknown reasons, adopting parents somehow make their children unlike themselves.

Data bearing on point b are to be found in the Skodak-Skeels study. They report the following standard deviations for years of educa-

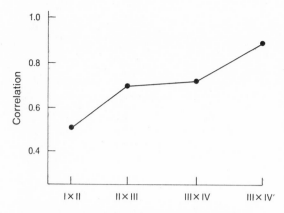

Fig. 14-4. Correlations between successive tests given to 100 adopted children. *IV'*, 1937 Stanford-Binet; other data points based on 1916 Stanford-Binet and on Kuhlman-Binet. (Based on data from Skodak, M., and H. M. Skeels. 1949. J. Genet. Psychol. **75:**85-125.)

tion of foster and true parents, respectively: foster fathers, 3.54; foster mothers, 2.89; true fathers, 2.73; and true mothers, 2.31. Thus, if anything, the correlations between children and true parents are probably smaller than they should be, rather than larger. Likewise, the variabilities of the 1945 sample of foster parents (in respect to education and occupation) are almost identical with those of the smaller 1949 sample. Therefore this particular criticism also appears to be poorly supported by the data.

2. It is of some interest that the correlations between successive tests given to the 100 children steadily rise with age. This fact is shown in Fig. 14-4. The graph means, essentially, that despite the variation in foster homes, the rank order of individuals is becoming progressively more stable. At the same time, it increasingly agrees with the rank order of true parents. A similar effect has appeared in data by Skeels (1966) on orphanage children fostered into a home for mental defectives and by Englemann (1968) on disadvantaged children exposed to a program of experimental enrichment for 2 years. We will refer to these studies later.

3. The mean IQ of all fostered children at final testing was very much higher than the mean IQ of their true mothers (106 as compared with 86) using the 1916 Stanford-

Binet. This has often been interpreted as strong evidence for the operation of environmental forces. As Jensen (1973b) has shown, however, this result is not unexpected, assuming a broad heritability as high as 80% and an assortative mating coefficient of around 0.39. We might logically expect, however, that there would be some association between "goodness" of home and resulting final IQ level. In actual fact, if we use years of education as an index, we find that for children with IQs over 120 ($N = 21$) the mean amount of education of their foster parents is 12.05 years as compared with 12.2 years for all foster parents. This is a puzzling result from an environmentalist point of view.

4. A point not particularly emphasized by Skodak and Skeels is that across ages and across four testings, mean IQs drop steadily. This effect is shown in Fig. 14-5. Thus the foster children become more like their natural mothers with time, not only as far as relative rank order of IQ but also in respect to mean level. It is of some interest, also, that the effect is about as pronounced for children placed in upper-class as for those placed in lower-class homes, a point illustrated by the data in Fig. 14-6. A related point is made by the data in Fig. 14-7, based on Leahy (1935). It is clear that the intelligence of an adoptee does not follow closely the social class of the parents, as occurs with control children raised in their own home. This result is in line with data of Burks (1928), who reported the following correlations: for foster child IQ and Whittier index,* $r = 0.15$, and for foster child and culture index, $r = 0.11$. For control children, corresponding correlations were $r = 0.46$ and $r = 0.37$. Further confirmation has recently been added by Claeys and Nuttin working in Belgium (Claeys, 1973). Their results are shown in Fig. 14-8 and make the same point. Further work by Munsinger (1975) offers additional verification.

---

*The Whittier index is an estimate of home quality and is based on five items: necessities, neatness, size of home, parental conditions, and parental supervision.

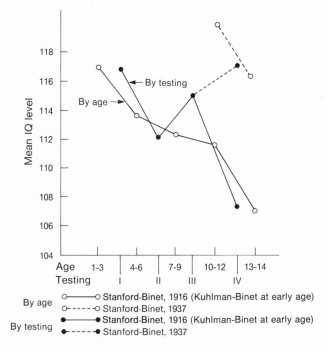

Fig. 14-5. Changes in IQs of adopted children over five ages and over four testings. Data are given for Stanford-Binet 1916 and 1937 revisions separately. (Based on data from Skodak, M., and H. M. Skeels. 1949. J. Genet. Psychol. **75:**85-125.)

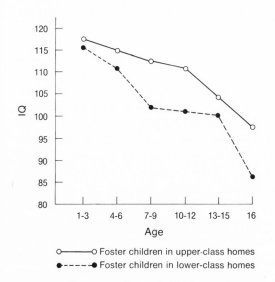

Fig. 14-6. Changes in foster-child IQ over age as a function of class of adopting home. (Based on data from Skodak, M., and H. M. Skeels. 1949. J. Genet. Psychol. **75:**85-125.)

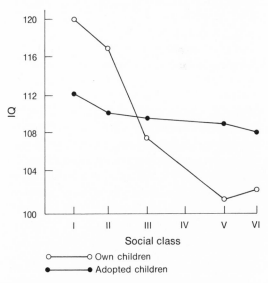

Fig. 14-7. Stanford-Binet (1916) IQs of "own" and adopted children by social class of parent. (From Leahy, A. 1935. Genet. Psychol. Monogr. **17:** 236-308.)

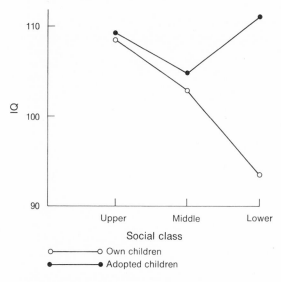

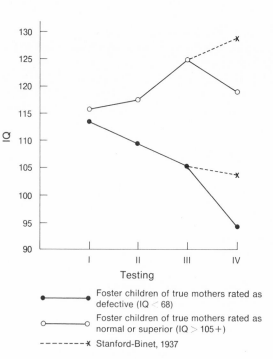

Fig. 14-8. PMA IQs (mean of factors V, Q, P, and S) of "own" and adopted children by social class. (Based on data from Claeys, W. 1973. Behav. Genet. **3**:323-338.)

Fig. 14-9. Stanford-Binet (1916) IQs of foster children born of normal or defective mothers; four testings. (Based on data from Skodak, M., and H. M. Skeels. 1949. J. Genet. Psychol. **75**:85-125.)

It is difficult to offer environmentalist explanations of such data. On the face of things, if environment is all important, then a "good" home should elevate intelligence, and a "poor" home reduce it. This clearly does not happen in the studies reviewed. One exception should be mentioned, however: the study of Freeman, Holzinger, and Mitchell (1928). These workers reported on sibling pairs with one member placed in a "better" home and one placed in a "poorer" home. The former turned out to be brighter. However, placement in better homes also occurred earlier, a factor that tended to elevate IQ. When Freeman et al. corrected for this and for age of test, the difference was only around 5 IQ points. We must add, also, that even this difference may have been a result of selective placement, a possibility seriously considered by the authors themselves (Freeman, Holzinger, and Mitchell, 1928). In fact, as we showed previously, comparable comparisons calculated from the Skodak-Skeels data show little association between high IQ and excellence of foster home.

5. A final datum of interest relates to the preceding discussion. It appears that regardless of social class of adopting family, the IQs of foster children appear to decline with age toward the level of their natural parents. We may thus fairly ask whether this decline also applies to children whose biological parents tested at higher IQ levels. Skodak and Skeels (1949) reported such data, which are shown in Fig. 14-9. It is clear that foster children whose real mothers had IQs of 105 or higher do not show a decline with age and, except perhaps at test I, show considerably higher IQs than children from so-called defective mothers. There was no difference between these groups of children in respect to either foster midparent educational level or foster father occupational level. In spite of this, Skodak and Skeels claim that the homes of the brighter children were nevertheless superior in terms of income level and interest of parents in providing their children with

enriched educational opportunities. However, since the authors offer only qualitative information on this point, their interpretation must be viewed with caution if not with skepticism.

The aspects of the adoption studies just discussed seem to us clearly to favor the operation of hereditary factors in determining intelligence. One must grant, of course, that each separate study is open to criticism on methodological grounds, as emphasized especially by Kamin (1978) and by Goldberger (1976). This includes the Freeman, Holzinger, and Mitchell (1928) study, the only one favorable to an environmentalist point of view. However, as we have tried to emphasize, although the fit of the data to a simple genetic model is undoubtedly imperfect, their fit to an environmental model is even worse. Naturally, readers should satisfy themselves on this point by directly consulting the primary sources.

Several attempts have been made to provide heritability estimates on the basis of the adoption data. Burks, using a simple path analysis, put forward a figure of "not far from 75 or 80 per cent" (Burks, 1928, p. 308). Wright (1931), extending Burks' analysis, suggested a lower bound of 0.49 and an upper bound of 0.81. Jencks (1972), also using path analysis, offered an estimate of around 0.5, but with a high covariance term of close to 0.2. Again, there seems little merit in arguing about which figure is the "right" one. Perhaps all are valid, depending on the operation of the usual parameters in the particular population studied.

Twin studies

General intelligence. Galton must again receive major credit for introducing the twin method to the scientific study of inheritance. In his book *Inquiries into Human Faculty* (1883) he describes the results of questionnaire data obtained from a large number of pairs of twins, for about eight of which there was close similarity between the members of the pair with respect to many different characters. Some reports that suggested dissimilarity Galton tended to discount on the grounds of "the tendency of relatives to dwell unconsciously on distinctive peculiarities and to disregard the far more numerous points of likeness that would first attract the notice of a stranger." His final conclusion was that "nature prevails enormously over nurture when the differences of nurture do not exceed what is commonly to be found among persons of the same rank of society and in the same country" (Galton, 1883, p. 241).

Galton's data were rather meager and lacked objectivity, but the inferences he drew from them have been in most part substantiated by ensuing work up to the present.

Basically, as we have pointed out earlier, two kinds of studies have been carried out. The first type simply makes a comparison between monozygotics and dizygotic pairs reared together. Such material is, of course, relatively easy to obtain but always encounters the major problem of possibly reduced environmental variance within MZ pairs as compared with DZ pairs. Other difficulties were outlined in Chapter 12. The second type of study compares separated MZs (MZAs) with MZs reared together (MZTs) and/or DZs apart (DZAs) or together (DZTs). The ideal design, for which very few examples are available, employs all four categories. As will be obvious, such cases are not so readily come by. At the time of this writing, workers in the field have together accumulated only around 122 MZA pairs, plus a number of individual pairs presented as case histories. Let us look at the first of these two types of study.

Before techniques of establishing zygosity became well standardized, most investigations dealt with comparisons between like-sexed and unlike-sexed twins. Typical of these is the study by Merriman (1924), who published one of the first incisive monographs on intellectual resemblances of twins. He used three standardized tests, the Stanford-Binet, the Army Beta, and the National Intelligence Test, as well as teachers' estimates. Merriman concluded that (1) environment has little effect on the size of correlations, since older and younger pairs were not consistently different in this respect;

(2) twins, as a group, do not show any obvious intellectual handicap; (3) correlations between like-sexed pairs are not significantly higher than correlations between unlike-sexed pairs; and (4) the resemblance of members of like-sexed pairs is close to that between members of ordinary sibling pairs.

Lauterbach (1925) confirmed and added to Merriman's work in a study involving 149 like-sexed and 63 unlike-sexed twin pairs. On the Stanford-Binet IQ test, he obtained correlations of 0.77 and 0.56, respectively, for these two groups. The correlations tended to be lower at a younger age. Lauterbach also examined resemblances on various scholastic achievement tests, but we shall consider these separately later.

One of the first to deal specifically with twins whose zygosity was established was Tallman (1928). He found a mean intrapair difference in IQ of 5.08 for identicals, 7.37 for like-sexed dizygotics, and 13.14 for sibling pairs. His findings agreed with those put forward in the same year by Kramer and Lauterbach (1928).

Further studies in the late 1920s and 1930s were carried out by Wingfield and Sandiford (1928), Holzinger (1929), Hermann and Hogben (1933), Stocks and Karn (1933), and Byrns and Healy (1936). All of these arrived at essentially the same conclusions as those of earlier workers. One of these studies, however, yielded slightly anomalous findings. Stocks and Karn (1933) reported an age-corrected correlation of 0.843 for 68 MZ pairs on the Stanford-Binet and 0.868 for 56 like-sexed DZ twins. This would yield a negative heritability estimate. The figure for all 119 DZ pairs, however, was 0.651. The authors concluded that this was probably the more accurate estimate, but, nonetheless, suggested that environment was perhaps much more influential than other workers had supposed.

Table 14-5. Summary of major twin studies of IQ

Study	Test	MZ		DZ	
		Correlation	N (pairs)	Correlation	N (pairs)
Holzinger (1929)	Otis IQ	0.92	25	0.63	26
	Binet IQ	0.88	25	0.62	26
Newman, Freeman, and Holzinger (1937) (raw, uncorrected)	Binet IQ	0.91	50	0.64	50
Wictorin (1952)	Otis IQ	0.92	50	0.62	50
	Simplex General Intelligence Test	0.85	141	0.70	128
	C-test, a general intelligence test	0.91	141	0.73	128
Blewett (1954)	PMA Composite	0.75	26	0.39	26
Husén (1960)	Swedish Military Induction Test	0.90	215	0.70	416
Nichols (1965)	NMSQT Composite	0.87	687	0.63	482
Schoenfeldt (1968) (in Jencks, 1972)	Project Talent "IQ Composite"	0.85	335	0.54	156
Halperin, Rao, and Morton (1975)	Composite of Stanford Achievement, Arithmetic subtest, and Knox Cube	0.83	146	0.59	155
Loehlin and Nichols (1976)	General ability	0.86	>500	0.62	>300
	Special abilities	0.74	>500	0.52	>300

One final point should be made. Jarvik, Blum, and Varma (1972) studied 13 MZ and 6 DZ twins over a 20-year period. They were between 77 and 88 years of age at final testing. Seven tests were used to measure level of intellectual functioning and intrapair similarity. DZ pairs tended to become more alike on five of the tests, while MZs became slightly less alike on all except one test. However, MZ intraclass correlations remained higher on six of the tests. Thus level of heritability of intelligence, as estimated in this way, appears to be maintained stably during the lifetime of the individual.

Work done since our earlier book, *Behavior Genetics* (1960), has been considerable in quantity and has focused rather more on special abilities. Before turning to a consideration of the latter, however, we will attempt to summarize the results of major studies done on general intelligence. Table 14-5 is based on various sources, notably Nichols (1965), Vandenberg (1968a), and Jencks (1972). The studies include populations from America, England, Sweden, and the Union of Soviet Socialist Republics. Jencks (1972) has attempted to compute heritability estimates for some of these. These include corrections for age, unreliability, and attenuation and range between 0.55 and 0.91. Halperin, Rao, and Morton (1975) estimated an h^2 of 0.492 for the Russian data.

Special abilities. Two of the earliest studies using factor scores rather than test scores were carried out by Blewett (1954) and by Strandskov (1954). In the first of these, 26 MZ and 26 DZ pairs, divided equally by sex, were tested on Thurstone's Primary Mental Abilities Test plus 3 composite scores. Results indicated heritabilities ranging from 0.508 to 0.680 (verbal), with number being close to zero (0.073) and Composite 1, a weighted average of the five factors, being the next lowest (0.339). There was no indication, contrary to Blewett's hypothesis, that a general factor was more strongly heritable. Strandskov's data, based on 48 MZ and 55 DZ pairs, showed significantly higher MZ concordance for space, verbal, fluency, and memory factors, but not for number and rea-

soning. Thus the two studies are mostly in agreement but strikingly different in respect to the reasoning factor. The latter is sometimes considered to be a rather poorly defined factor when tested in different populations (Vandenberg, 1959), and this may account for the discrepancy.

During the 1950s and 1960s, a great deal of work on special abilities was carried out by a number of workers, including Wictorin (1952), Husén (1963), Partanen, Bruun, and Markkanen (1966), Nichols (1965), and especially Vandenberg (1968). In many of these studies, MZ-DZ comparisons have been reported in terms of the F statistic and its probability level. This simply tells us whether the within-pair variances for MZs are significantly less than those of DZ pairs. Some representative results are shown in Tables 14-6 to 14-8.

Table 14-6 shows results obtained by Wictorin on 128 MZ and 141 DZ pairs for a variety of psychological tests. F ratios and intraclass correlations are shown. Note that none of the memory tests yielded significant Fs. This result is in agreement with Vandenberg (1968), using the PMA memory test of Thurstone. In one of Wictorin's tests, in fact, the intraclass correlation is larger for DZs than for MZs, although the MZ within-pair variance is still smaller. Presumably, this seeming anomaly is a function of the between-pair variances in the two sets of twins.

Husén's data, also gathered in Sweden on varying numbers of twin pairs, are shown in Table 14-7. All F ratios are significant.

Some of Vandenberg's data on special abilities are summarized in Table 14-8. It is not clear why some subtests are significant and others are not. Neither is it obvious why the same tests sometimes turn out to be heritable in one study and not heritable in another population. It does seem to be true, however, that the vast majority of special abilities examined by the twin method show less variation between members of MZ pairs than between members of DZ pairs. Thus the general data are supportive of those obtained for general intelligence (see also Loehlin and Nichols, 1976).

Table 14-6. *F* ratios and intraclass correlations for like-sexed DZ twins and MZ twins for twelve psychological tests*

	F	Intraclass correlations	
		MZ	DZ
Verbal analysis	1.12	0.63	0.57
Form perception (paper form board)	1.34†	0.65	0.53
Form perception (perceptual speed)	1.36†	0.64	0.61
Number perception (clerical checking)	1.59†	0.83	0.69
Numerical reasoning (series)	2.01†	0.74	0.49
Numerical reasoning (number analysis)	1.63†	0.69	0.57
Numerical reasoning (classification)	1.57†	0.70	0.55
Numerical reasoning (verbal arithmetic)	2.18†	0.87	0.73
Simple arithmetic	1.68†	0.81	0.74
Memory (recall)	1.24	0.62	0.58
Memory (recognition)	1.17	0.49	0.45
Memory (paired associates)	1.16	0.43	0.53

*From Wictorin, M. 1952. Bidrag till Räknefärdighetens Psykologi en Tvillingundersökning, Elanders, Gothenburg, Sweden; modified from Vandenberg, S. G. 1968. The nature and nurture of intelligence. In D. Glass, ed. Genetics. Rockefeller University Press, New York.
†$p < 0.05$ or 0.01.

Table 14-7. *F* ratios between MZ and DZ twins for five psychological tests*

	N_{MZ}	N_{DZ}	F
Following verbal instructions	215	415	2.62†
Finding synonyms	269	532	2.08†
Choosing odd word among five presented	269	532	1.77†
Raven Progressive Matrices	269	532	1.37†
Number series	54	117	1.54†

*From Husén, T. 1953. Tvillingstudien. Almqvist & Wiksell, Förlag AB, Stockholm; modified from Vandenberg, S. G. 1968. The nature and nurture of intelligence. In D. Glass, ed. Genetics. Rockefeller University Press, New York.
†$p < 0.05$ or 0.01.

A good deal of work has been done on the heritability of scholastic abilities. Three of the largest studies were undertaken by Husén (1963), Nichols (1965), and Schoenfeldt (1968). The first of these workers tested a large number of Swedish schoolchildren over a period of more than 10 years in reading, writing, and arithmetic skills. Among twins tested in grade 4, Husén found significantly higher intraclass correlations for MZs as compared with DZs in all three subjects for both sexes, with one exception being the comparison between MZ girls and DZ girls on arithmetic. By grade 6, however, intrapair similarities had all risen appreciably, being more marked for identicals and somewhat more marked in boys. Husén interpreted this as resulting from a genotype-environment interaction. This was expressed as a tendency for MZs, because of their "primary likeness," increasingly to share a common environment and hence to grow more alike. However, DZs by virtue of their primary dissimilarity should tend to become

Table 14-8. Abilities showing significant or nonsignificant F ratios in MZ-DZ comparisons*

Test	F ratio	
	Significant	Nonsignificant
PMA (two studies)	Verbal	Reasoning
	Space	
	Number	
	Word fluency	
Differential abilities test (two studies)	Verbal reasoning	Numerical ability
	Space relations (one study)	Space relations (one study)
	Clerical speed and accuracy	Abstract reasoning
	Language use: spelling	Mechanical reasoning
	Language use: sentences	
Wechsler Adult Intelligence Test	Information	Picture completion
	Comprehension	Object assembly
	Arithmetic	
	Similarities	
	Digit span	
	Vocabulary	
	Digit symbol	
	Block design	
	Picture arrangement	

*From Vandenberg, S. G. 1968. The nature and nurture of intelligence. In D. Glass, ed. Genetics. Rockefeller University Press, New York.

more and more unalike. Since they do not, this hypothesis, as Husén recognizes, is not entirely tenable.

Nichols used two sets of twin samples. The first was made up of 315 male and 372 female MZs and 209 male and 273 female DZs. The second was a subset of these which excluded twins reporting differences between them in environmental experience, such as a major illness, prolonged separation, or special training for one and not the other. Data for the complete set on the National Merit Scholarship Qualification Test (NMSQT) are shown in Table 14-9. The subtests are correlated and hence reflect a general factor of scholarly ability plus a factor specific to the area (residual subtest). Tests of significance of differences between intraclass correlations and heritability estimates with Nichols' HR are shown. It is clear that most t tests are significant and that heritability ratios are generally high. Nichols concluded that around 70% or more of the variance both in the general and specific components of the subtests

was attributable to hereditary factors. Results using the second sample agree with those obtained on the complete set. Correlations are slightly higher, as might be expected, but this is true for both MZ and DZ pairs. The exclusion rate, it may be noted, was 18% for the MZ and 25% for the DZ sample. If anything, this should have lowered heritability of the tests and subtests, but in fact these did not change appreciably according to Nichols.

The final study, "Project Talent," by Schoenfeldt (1968) used a sample of 335 MZ and 156 DZ twins. The tests included "Vocabulary," "Social Studies," "Mathematics," "Information," "Verbal," "Quantitative," and "Mechanical-Technical," plus an "IQ Composite" and a "General Academic Aptitude Composite." In every case, correlations for males and females together, corrected for attenuation, were higher for MZs than for DZs. On the specific tests, heritability estimates were notably low only on vocabulary and mechanical-technical.

Although studies of the preceding kind are

Table 14-9. Intraclass correlations and HRs for NMSQT subtests and residual subtests*

	Males				Females			
	Intraclass correlation				Intraclass correlation			
	MZ	DZ	t	HR	MZ	DZ	t	HR
Subtests								
English usage	0.71	0.64	1.46	0.22	0.77	0.49	6.16†	0.99
Mathematics usage	0.74	0.42	5.69†	1.16	0.70	0.47	4.48†	0.87
Social studies reading	0.76	0.50	5.08†	0.92	0.79	0.52	6.01†	0.92
Natural sciences reading	0.69	0.52	2.96†	0.60	0.66	0.48	4.48†	0.68
Word usage	0.85	0.64	5.32†	0.60	0.64	0.64	6.78†	0.62
Residual subtests								
English usage	0.40	0.42	−0.25	−0.10	0.48	0.25	3.37†	0.96
Mathematics usage	0.48	0.21	3.47	1.14	0.43	0.23	2.74†	0.92
Social studies reading	0.33	0.16	2.09	1.06	0.29	0.17	1.60	0.83
Natural sciences reading	0.27	0.32	−0.61	−0.36	0.31	0.10	2.77†	1.37
Word usage	0.55	0.37	2.50	0.65	0.55	0.26	4.48†	1.07

*From Nichols, R. C. 1965. The inheritance of general and specific ability. Natl. Merit Scholarship Corp. Res. Rep. 1:1-10.
†$p < 0.05$ or < 0.01.

still under way in many centers, no startling breakthroughs in data or methodology have yet occurred. We have already referred both in this and in a previous chapter to some of the problems associated with the use of MZ-DZ comparisons. Before leaving this section, we will consider some of the main ones more explicitly. Kamin (1974), in particular, has stressed the not uncommon criticism that MZ twins are usually treated more alike than DZ twins and hence are bound to be more similar on any given measure, including intellectual ability. He has cited two studies by Wilson (1934) and Smith (1965) in support of this contention. On the whole, both sets of data indicate that MZs spend more time together, are more likely to have the same friends, and study together more than do DZs. However, this is by no means prima facie evidence for an environmentalist model. Being more closely associated may *result* from being more alike genetically in the first place (Chapter 12). In any case, we simply do not know the extent to which such an en-

vironmental circumstance acts to produce likeness. In some instances, it could easily be argued that proximity could produce polarization or complementarity, thereby magnifying small initial differences.

A second problem follows from the first. As suggested in Chapter 12, twins, in many ways, may be considered as different from the general population of individual children. Consequently, it may be argued that conclusions obtained from MZ-DZ comparisons may not be able to be very widely generalized. There is, of course, some truth to this contention. Again, however, it is a difficulty that can easily be overemphasized. If the statistical properties of the trait distribution in twins are empirically not very different from those found for the general population, then there is no compelling reason that forbids generalization. Indeed, it is hardly uncommon in psychology and biology to find generalizations often being made from one species to another, for example, from one breed of rat to human beings at large.

Caution in science is certainly a virtue, but one can run the risk of discarding hypotheses that are in fact true.

A final related problem concerns ascertainment bias. A sample of twins may be atypical of twins in general simply because more of a particular type are more easily located. Thus most twin studies involve more female than male pairs. If the different sex pairs are then lumped together, this may distort our conclusions, a point made forcibly by Kamin. However, in practice, most investigators have reported data separately by age, sex, and other pertinent variables, allowing readers to draw their own conclusions.

These hazards are always present in work with twins. However, at least the most serious of them—the problem of environmental similarity—is somewhat reduced by studying separated twins. We will consider next the few studies that have relied on this method.

Separated monozygotic twins. Up to the time of writing, only four major studies using separated monozygotic twins have appeared. These supply data on the cognitive abilities and personality characteristics of 122 pairs. Beyond this sample, we also have a fair number of case studies on individual pairs as well as a few on supertwins such as the Dionne quintuplets (Blatz, 1937) and the Genain quadruplets (Rosenthal, 1963). General summaries of results in some of these may be found in Scheinfeld (1967), Bulmer (1970), and Mittler (1971). We will restrict our discussion here to the four major studies. One was carried out in the United States by Newman, Freeman, and Holzinger (1937), two in England by Shields (1962) and Burt (1966), and one in Denmark by Juel-Nielsen (1965).

Broadly speaking, the results of these have been in agreement, and it has been considered appropriate by one writer, at least (Jensen, 1970), to consider their pooled results. We will present these later in the chapter. First, it would seem best to examine the studies one at a time, since each entails specific and unique problems.

Newman, Freeman, and Holzinger (1937). The MZA sample consisted of 12 female and 7 male pairs ranging in age from 11 to 59

years with a median age of 26 years. In addition, 50 monozygotic pairs reared together (MZTs) and 50 dizygotic pairs reared together (DZTs) were also studied. The MZAs were obtained largely through newspaper and radio appeals, coupled (in the case of the final 9 pairs) with an offer of a free trip to the Chicago Fair, with all expenses paid. The procedure used by Newman et al. in initial selection of cases was fairly stringent due to the financial outlay their inducement entailed. This may have led to a bias in favor of more phenotypically similar as against less phenotypically similar twins. Confirmation of zygosity was carried out by the "similarity" procedure of Siemens and von Verschuer.

Besides the various physical indices, a number of psychometric tests were given to the three twin samples. For MZTs and DZTs these included Stanford-Binet (1916) MA and IQ, Otis score and IQ, the Stanford Achievement Test, the Woodworth-Matthews Personal Data Sheet, the Downey Will-Temperament Test, and three tests of Tapping Speed. The MZA sample was given all of these and, additionally, the Psychological Examination of Thurstone, the Otis Self-administering Test, the International Test (a nonlanguage ability test), the Pressey Test of Emotions, and the Kent-Rosanoff Free Association Test. Apart from these formal tests, extensive and detailed case histories were gathered for every individual pair of separated twins. We will focus here mainly on the tests of cognitive ability. The main results for the three groups of twins are presented in Table 14-10. Commenting on these data, Newman et al. (1937, p. 116) remarked, "it appears that from 75 to 90 per cent of the difference variance is attributable to nature for physical traits. In the case of intelligence, the values of h^2 range from .65 to .80, so that on the average nearly three-quarters of the variance in intelligence is attributable to nature." Notice, however, that the heritability estimates would have been lower had MZA pairs been used instead of MZT pairs. It is clear that separation does indeed lower similarity. This applies even to height, weight, and head width measures. Because of this

Table 14-10. Twin correlations (intraclass) and heritabilities in MZA, MZT, and DZT samples for various tests of ability, achievement, and some physical characteristics*

	MZA raw correlation	MZT		DZT		h^2‡
		Raw correlation	Age-corrected† correlation	Raw correlation	Age-corrected correlation	
Stanford-Binet Mental Age	0.637	0.922	0.861	0.831	0.599	0.65
Stanford-Binet IQ	0.670	0.910	0.881	0.640	0.631	0.68
Otis IQ	0.727	0.922	0.621	0.621	0.618	0.80
Stanford Achievement	0.507	0.955	0.883	0.883	0.696	0.64
Woodworth-Matthews	0.583	0.562	0.371	0.371	0.365	0.30
Height	0.969	0.981		0.934		0.81
Weight	0.960	0.965		0.901		0.78
Head width	0.880	0.908		0.654		0.75

*Modified from Newman, H. H., F. N. Freeman, and K. J. Holzinger. 1937. Twins: a study of heredity and environment. University of Chicago Press, Chicago.

†The procedure of age correction by partialing out correlations can be criticized, since the relation between IQ and age is not linear. MZA correlations were not corrected by Newman et al. However, McNemar (1938) later did so, thereby raising the Binet IQ correlation to 0.77. Jensen (1969) later added a further correction for unreliability, raising it to 0.81.

‡h^2 estimates are based on Holzinger's formula, $H = \dfrac{r_{MZT} - r_{DZT}}{1 - r_{DZT}}$. Age-corrected correlations were used.

fact, Newman et al. conclude at the end of their monograph that no determinate solution can be given to the problem of nature versus nurture. For example, if the comparison is made between MZAs and MZTs for, say, Otis IQ, the share of environmental determination turns out to be 0.64, but if the MZT-DZT comparison is used, environmental contribution is estimated as only 0.16 (Newman, Freeman, and Holzinger, 1937). The same kind of results were obtained for other tests. Thus Newman et al. express their agreement with Jenning's dictum that what heredity can do environment can also do.

This is perhaps a fair conclusion. There is no question that MZAs are less alike than MZTs. Nevertheless, it must be emphasized that in spite of often very different environmental circumstances, the members of the 19 pairs were still very similar in respect to intellectual ability and certainly much more similar than pairs of individuals drawn randomly from the population. It is certainly no surprise to find that environment can change IQ, and such a fact by no means contradicts a hereditarian model. Let us now look at some

of the causes of the MZA differences. Newman, Freeman, and Holzinger (1937) used five judges to rate differences between pair members in respect to the educational, social, and physical environments in which they were raised. A 10-point scale was used, and, for each variable, pooled reliabilities were 0.961, 0.907, and 0.913, respectively. Furthermore, the three variables shared only low correlations with each other. Next, twin differences on the various tests used were correlated with environmental difference ratings. Statistically significant relations are shown in Table 14-11.

Some comments are in order. It is clear, in the first place, that disparities in educational and social background correlate with differences in IQ. This is especially marked in the case of the Binet Test. In the second place, it is equally interesting, and perhaps surprising, that only one of the four personality tests is affected by environmental differences, and this by only one of the environmental dimensions, the physical. Finally, it should be noted that most of the relationships between environmental differences and IQ differences

Table 14-11. Statistically significant relationships between MZA intrapair differences and differences in rearing environments*

Trait	Environmental difference rating		
	Educational	Social	Physical
Weight			0.599
Binet IQ	0.791	0.507	
Otis IQ	0.547	0.533	
International Test	0.462	0.534	
Thurstone Psychological Examination	0.570		
Stanford Educational Age	0.908		
Downey Will-Temperament total score			0.465

*From Newman, H. H., F. N. Freeman, and K. J. Holzinger. 1937. Twins: a study of heredity and environment. University of Chicago Press, Chicago.

are accounted for by four extreme pairs. When these are omitted, *none of the correlation turns out to be significant, and the joint contribution of educational and social differences to variance of IQ differences drops from 72% to 20%* (Newman, Freeman, and Holzinger, 1937).

In the opinion of the writers, the Newman et al. study makes a strong case for the importance of genetic factors in the determination of intellectual ability, much stronger, perhaps, than its authors realized. No doubt one can quibble with some of the statistical analyses performed, particularly those carried out by others later, and it is easy to find some eccentricities and inconsistencies in the presentation of the individual MZA case studies. However, the data considered in toto certainly do not readily fit an environmental model. The case of Gladys and Helen is frequently cited in support of environmental influences, in view of the fact that a striking difference in rearing environments was associated with a difference of 24 IQ points on the Stanford-Binet. However, it is not so frequently noted that they differed by

only 8 points on Otis IQ and by only 4.1 points on the Otis S-A test. Only four cases showed a smaller difference than this last figure. By contrast, we may consider case IV, Mabel and Mary, who were both brought up on farms in Ohio and visited each other frequently. They differed by 17 points on the Binet IQ, 14 points on the Otis IQ, and 18.7 points on the Otis S-A test. Overall, in fact, for the 19 cases, we find that there is no association between superior IQ (Binet and Otis combined) and superior placement ($\chi^2 = 0.57$). Thus if environment works in some systematic manner, it is difficult to specify it on the basis of the data of this study.

Shields (1962). Shields used a total sample of 15 male and 29 female pairs of separated monozygotic twins. They were between 8½- and 59-years-old at time of testing. Of these pairs, 21 had been separated at or very near birth, and 3, 4, and 2 pairs at 3, 6, and 9 months, respectively. Another 6 were separated at between 12 and 24 months, 2 at 4 years, and 1 pair at each of the ages of 5, 7, 8, and 9 years. Shields notes that only in 14 cases were the members of a pair raised in unrelated families. A group of 44 MZT pairs, matched by age and sex to the MZA sample, was also used. In addition, 28 DZ pairs were given a limited number of tests. This sample consisted of 11 separated pairs (1 male and 10 female) and 17 pairs (1 male and 16 female) reared together. Detailed case histories and test data are supplied for all separated twins. The complete sample was originally obtained through a BBC television program entitled "Twin Sister, Twin Brother." The appeal yielded responses from some 5000 twins, from which population the final samples were selected. Final confirmation of zygosity was established by means of the similarity method, eight blood groups, PTC testing, color blindness, fingerprints, and handedness.

Intelligence was tested by means of (1) the Dominoes Test, a 20-minute nonverbal test developed during World War II that correlates 0.74 with the Raven Progressive Matrices, has a reliability of 0.92, and is considered to have a high g saturation; and (2) the Raven Mill Hill Vocabulary Scale,

Set A (Synonyms), Form B. Personality was assessed by means of the Self-Rating Questionnaire (SRQ). This test is made up of 38 items, some of which yield a score on extraversion, and others on neuroticism. Some additional items are common to both dimensions. Here we will consider results only for the two intelligence tests.

To obtain a combined estimate of intelligence, Shields added the Dominoes score to twice the Mill Hill score, this weighting being used to allow for the differential ranges of the two tests. Main results are shown in Table 14-12. As indicated in the footnote to the table, results on dizygotic twins are based on very small numbers and cannot be considered reliable. However, it is nonetheless true that a correlation of around 0.5 would be expected for a dizygotic twin or sibling sample.

These basic data certainly seem to support the notion of a hereditary component in intelligence. Nevertheless, they involve some problems (Kamin, 1974). In the first place, members of all except 5 MZA pairs and 1 MZT pair were both tested by Shields. If we compare the degree of similarity of these with the similarity of those not both tested by Shields, we find that the latter show a mean difference of 18.5 (combined score, excluding 1 pair with unreliable test scores), as compared with 7.1 for the other group. This may

Table 14-12. Correlations for combined intelligence scores for MZA, MZT, and DZ twins*

Twin type	N	Mean intrapair difference	Intraclass correlation
MZT	34	7.38	0.76
MZA	37	9.46	0.77
DZ†	7	13.43	0.51

*From Shields, J. 1962. Monozygotic twins brought up apart and brought up together. Oxford University Press, London.
†The DZ sample is made up of 4 DZA pairs and 3 DZT pairs. Correlation for the former group is 0.05, and for the latter, 0.71 (Kamin, 1974). Neither of these is statistically significant, nor is the total correlation of 0.51. Hence the DZ data are irrelevant.

represent tester bias. It may also represent the fact that the 5 pairs were, in fact, raised in rather more widely varying environments than the pairs that Shields tested. It may also reflect the fact that of these 5 pairs, 4 were reared by unrelated families.

In fact, members of pairs reared in unrelated families correlated only 0.51, compared with 0.83 for those reared in related families (Kamin, 1974). Mean intrapair differences in the two groups, however, are not remarkably different. For those in related families, the mean difference was 11.6, for those raised by unrelated families, it was 12.3. This small difference mostly involves the Mill Hill test.

Nevertheless, it has been argued that Shields' data only demonstrate the power of correlated environments. Thus Kamin (1974) selected 7 of the MZA pairs judged by him to have been reared in very similar circumstances. He computed the intelligence correlation between them to be 0.99, as compared with 0.66 for the remaining 33 MZA cases. Again, however, we must comment that this kind of finding is hardly surprising and does not in any way confirm an environmentalist position and disconfirm a hereditarian one.

A critique of Kamin's analysis has been made by Fulker (1975). As it turns out, both the greater similarity of twins reared in related families and of twins both tested by Shields is dependent largely on a very large difference shown by two pairs (both from other groups). These are extreme cases, and, as Fulker points out, inspection of the histories of these two pairs suggests that they are quite atypical in respect both to their rearing conditions and to their response to the test situation.

Burt (1966). The study carried out by Burt in England over a period of more than 40 years involved a final sample of 53 MZA pairs together with MZT, DZT, and sibling groups. His final report appeared in 1966, although earlier articles described the work as it progressed over this long period of time.

It must be stated at the outset that Burt's mode of presentation of methods and data was imprecise. It is difficult to know exactly what tests were used, how they had been standardized, and how they were administered. Likewise, no information is given as to the sex and age composition of the twin sample. However, setting aside these ambiguities for the moment, we present the basic results in Table 14-13. On the basis of the results for "final assessment," Burt (1972) later computed the total genetic contribution as 87.4%. Unfortunately, there are at least two major difficulties involved.

In the first place, it is not entirely clear what Burt meant operationally by "final assessment." It is obvious enough that the general procedure involved the adjustment of raw test scores upward or downward on the basis of consideration of circumstances that might have biased a subject's initial score, but such a procedure was so clinical and intuitive that it would be impossible to replicate by other researchers. We may also note that if we take the group test results as the most objective, the adjustment procedure has the net effect of increasing the correlation between MZAs and lowering it for MZTs. This is in line with Burt's starting definition of intelligence as "that part of the general cognitive factor which is attributable to the individual's genetic constitution" (Burt and Howard, 1956). Thus clearly Burt was

Table 14-13. Intraclass correlations for MZA, MZT, and DZT samples on various tests of intelligence and achievement*

	MZA	MZT	DZT
N (pairs)	53	95	127
Intelligence			
Group test	0.771	0.944	0.552
Individual test	0.863	0.918	0.527
Final assessment	0.874	0.925	0.531
Educational			
Reading and spelling	0.597	0.951	0.919
Arithmetic	0.705	0.862	0.748
General attainments	0.623	0.983	0.623

*Modified from Burt, C. 1966. The genetic determination of differences in intelligence: a study of monozygotic twins reared together and apart. Br. J. Psychol. 57:137-153.

attempting to, as Jensen (1974) puts it, "read through" environmental influences to pure, innate ability. Such a research strategy seems somewhat circular.

A second problem arises from a consideration of the sequence of reports made by Burt on his gradually accumulating samples between 1943 and 1966. There is a remarkable and unlikely consistency in some of the results. One striking instance relates to the MZA correlation for the group test. In four reports, sample sizes increase as follows: 21, "over 30," 42, and 53. MZA correlations for these samples are, respectively, 0.771, 0.77, 0.778, and 0.771. In an earlier paper (Burt, 1943) what appears to be a correlation for final assessment scores for 15 MZA pairs was also reported as 0.771. Likewise, final assessment correlations between 1955 and 1966 vary only in the third decimal place. Both Kamin (1974) and Jensen (1974) have drawn attention to these and other anomalies. It is difficult to know what they mean.

Nevertheless, we may consider several points raised by Jensen (1974). First, the samples are not independent but cumulative. Thus some consistency at least would be expected. Second, MZA correlations of 0.77 have been reported both by Shields (1962) and by Newman, Freeman, and Holzinger (1937) (as corrected by McNemar, 1938). Third, according to calculations made by Jensen himself, the distribution of twin differences on final assessments does not differ significantly from that obtained in the other three separated twin studies. However, this last point has been challenged by Kamin (1974).

Besides these two major problems, there are others with which we need not deal here. Thus, as they stand presently, the bulk of Burt's data cannot be used by themselves for testing genetic or environmental hypotheses about intelligence. Possibly, the initial data, reported in 1955 for 21 MZA pairs (Burt, 1955), might be regarded as reliable, since most of the anomalies we have mentioned crept in after this point. However, even these should be viewed with caution.

Juel-Nielsen (1965). The final study was

Table 14-14. Intellectual resemblance (correlations) between members of 12 MZA pairs*

	Wechsler-Bellevue	
	Full sample	*1 doubtful case omitted (Pair III)*
Verbal IQ	0.78	0.81
Performance IQ	0.49	0.56
Total IQ	0.62	0.68

	Raven Progressive Matrices
Raw score	0.79
Speed	0.84

*Based on data from Juel-Nielsen, N. 1965. Individual and environment: a psychiatric-psychological investigation of monozygous twins reared apart. Acta Psychiatr. Neurol. Scand. Monogr. Suppl. **183**.

carried out in Denmark with 3 separated male and 8 female monozygotic pairs. The males ranged in age from 22 to 77 years, the females 35 to 72 years. Age of separation varied from 1 day to 5¾ years. The sample was obtained from the Danish Twin Registry. Zygosity determination was carried out by conventional procedures, including blood typing. No MZT or DZT pairs were studied.

The tests used were the Wechsler-Bellevue Intelligence Scale (W-B), Form I; the Raven Progressive Matrices; the Rorschach inkblot test; and Rapoport's Word Association Test. The subjects were also given intensive interviews, in most cases between fifteen and twenty times. Testing was done blind by independent psychologists. The amount of detailed information so obtained was sufficient to fill 450 printed pages.

Results for intelligence are shown in Table 14-14. It can be seen that the correlations are of the same order of magnitude as those reported in the other studies reviewed. Juel-Nielsen also compared 6 completely separated pairs with 6 incompletely separated pairs. Intrapair variances on the two tests were slightly greater in the former group, but not to a statistically significant degree. Likewise, differences in amount of education were found to be associated with intrapair differences in Wechsler verbal but not performance IQ. No association at all was found for the Raven test. Birth weight and birth order were not found to relate to intrapair similarity. Juel-Nielsen concludes that, since the twin partners resemble each other more than do persons selected randomly (the total proband material), genetic factors play an important role in intellectual ability.

Again, however, this conclusion has been rejected by Kamin (1974) on a number of grounds, notably the small size and atypicality of the MZA sample, the poor standardization (or lack of it) of the tests in Danish populations, and what appears to be a strong association between IQ and age—positive for females but negative for males. His criticisms must be taken seriously and should be referred to directly by the interested reader.

Total MZA sample. Granting the problems associated with the four individual studies, we may nevertheless consider the total picture that emerges when their results are combined together. Jensen (1970) has attempted to do this, using Burt's final assessment scores, Shield's Dominoes and Mill Hill scores (transformed), Newman, Freeman, and Holzinger's Stanford-Binet scores, and Juel-Nielsen's Wechsler scores. The combined distribution has a mean of 96.8 and a standard deviation of 14.2, values close to those found in the general population. It may be noted in passing, however, that the variance of the Juel-Nielsen sample is significantly smaller than the variances of the other three studies; the means are also different from each other (Schwartz and Schwartz, 1976). The general appearance of the combined distribution of IQs of the 244 twins is as shown in Fig. 14-10. According to Jensen it does not deviate significantly from normality.

A distribution may also be plotted for intrapair difference scores. This also is normally distributed and represents purely environmental effects on IQ. Allowing for test error ($\sigma = 3.35$), the standard deviation of this distribution is 4.74. Knowing that the variance of a random sample of IQs in a normal population is 225, we may then compute, by sub-

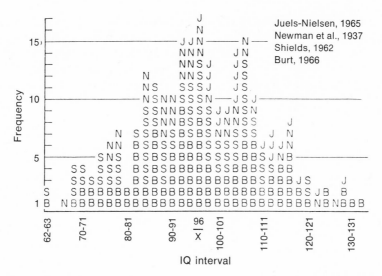

Fig. 14-10. Combined IQ distribution of 244 MZA twins from four studies. (From Jensen, A. R. 1970. Behav. Genet. **1:**133-148.)

traction, the hereditary variance. This turns out to be 191.25 (or $\sigma = 13.83$). In other words, genetic influences contribute 85% of the variance, environment 10%, and test error 5%. It is of interest to note that the range of environmental influence is close to 30 IQ points ($6 \times 1\sigma$); this estimate is about the same as that put forward by Burks (1928) on the basis of her adoption study data. Such a large difference would, of course, be an unlikely event and only arise from rearing two genetically like individuals in the most disparate of known environments. But such cases undoubtedly can occur, particularly as a result of deliberate and well-planned intervention procedures.

The kinship, adoption, and twin studies that we have reviewed constitute the bulk of the evidence for supposing that intellectual ability is at least to some extent determined by genotype. Although it does seem to hang together and, as well, agrees with the results of experimentation with animals (Chapter 9), it is by no means free of problems, as we have continually emphasized. Consequently, we would urge caution on the part of the reader and encourage consultation of the primary material as well as criticisms of it before forming conclusions.

We will next consider some ancillary data

bearing on the topic and then turn to a consideration of the heritability of abilities other than those measured by standard IQ tests. We will close the chapter with a discussion of the genetics of intelligence and a general comparison of the capacity of biological as against environmentalist models to account for the data.

ANCILLARY DATA ON THE HERITABILITY OF INTELLIGENCE
Stability of intelligence

Generally speaking, IQ scores remain relatively constant over part of the age range with, however, correlations between successive ages increasing in magnitude (Jensen, 1972). This of course, partly reflects the way IQ tests are standardized. Nonetheless, the fact that a variable can be defined in this way may mean that it relates to a basic biological fact about human organisms. It may also mean, of course, that it merely relates to cumulating environmental experience. This latter interpretation has been favored by Kamin (1974), who has suggested that much of the similarity between twins can be explained by virtue of their identity of age and sex, which are taken as *environmental variables*. Of course, it may be true that were we to pick pairs of individuals of like age and

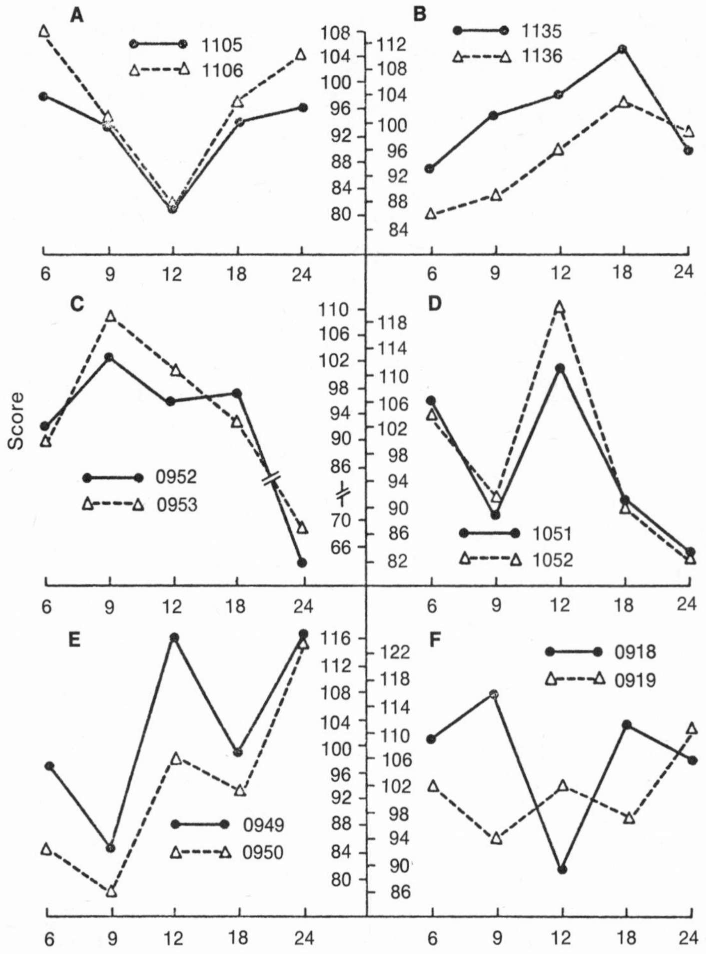

Fig. 14-11. Profiles of mental development scores for MZ twins at ages 6 through 24 months. All pairs except pair **F** are highly congruent. (From Wilson, R. S. 1972. Science **175**:914-917. Copyright 1972 by the American Association for the Advancement of Science.)

sex randomly from the population, they would probably be more alike in mental ability than pairs of differing age and sex to the extent that mental age or IQ correlated with age. However, this fact by no means uniquely supports an environmentalist position. Indeed, were the variable in question height rather than intelligence, we would be more inclined to view the greater similarity of like-sexed and -aged individuals as evidence for the genetic determination of height.

Further data bearing on this point have been put forward by Wilson (1972), who compared the intrapair correlations of MZ and DZ samples (more than 50 pairs of each) for mental ability at six successive ages during early development. MZ correlations ranged from 0.76 to 0.87, DZ from 0.61 to 0.75. The MZ correlations exceeded those of DZs at each age level. What is of greater interest, however, is the kind of fluctuation in scores shown by particular twin pairs. Fig. 14-11 shows developmental profiles for 6 pairs. It is clear that 5 of these show concordant fluctuations. One pair *(F)* is discordant. We should also note that some DZ pairs also show high developmental congruence (Wilson, 1977). Such data suggest that some transient changes in intelligence may

Table 14-15. Changes in IQ in experimental and control groups*†

Group	Before transfer		After transfer		Follow-up	
	Age (months)	IQ	Age (months)	IQ	Age (months)	IQ
Experimental	18.3	64.3	38.4	91.8	71.4	95.9
Control	16.6	86.7	47.2	60.5	83.3	66.1

*Modified from Skeels, H. M. 1966. Adult status of children with contrasting early life experiences: a follow-up study. Child Dev. Monogr. 31(3), Serial No. 105.
†Computations carried out by Thompson. See text for explanation.

be under genetic control. But even if they are not, they do not very much disturb the overall intrapair similarity.

Effects of systematic intervention

Wilson's results suggest that mental ability is certainly subject to fluctuation but in a rather orderly way. That is, like genotypes tend to show changes of like magnitude and like direction at around the same ages. Wilson's results do not tell us, of course, whether this is, in fact, a result of genes or of fortuitous environmental events occurring coincidentally. In discussing the adoption study of Skodak and Skeels (1949), we pointed out that the rank order of adoptees, in respect to IQ, tended to remain invariant across ages. This suggests some underlying facts or factors that confer on phenotypes a certain degree of stability relative to each other. Similar findings may be educed from data published by Skeels (1966) and by Englemann (1970), though neither of these workers examined this aspect of their results.

Skeels (1966) studied an experimental group of 13 children who had been transferred from an orphanage into a home for mentally retarded girls as "house guests." Their intelligence was compared with a control group of 12 children who remained in the orphanage. Three complete testings were carried out: one before transfer, another after transfer, and a third follow-up test. The Kuhlmann-Binet was used in the first two, the Stanford-Binet (1916) in the final testing. General results are shown in Table 14-15. It is clear that the experimental group shows a dramatic rise of over 30 IQ points over the

Table 14-16. Correlations between successive testings in experimental and control groups in Skeels' (1966) study*

	Testing			
	II		III	
	E	C	E	C
I	0.465†	0.302	0.464†	−0.411
II			0.838‡	0.562†

*Computations carried out by Thompson. See text.
†$p < 0.05$ (one tailed).
‡$p < 0.001$ (one tailed).

period of about 4½ years. Likewise, the controls show an equally dramatic loss of over 20 IQ points during approximately the same period of time. However, in spite of these shifts in mean levels, there is a strong tendency for rank order of children to be preserved; this is much more marked in the experimental group. Data are shown in Table 14-16. In both groups, however, this stability is higher at the end of the experimental period. Another computation from these data is of interest. This is the relation between initial IQ and amount of change over the 4½ years. The relevant correlations are, for the experimental group, −0.522 ($p < 0.05$, one tailed) and, for controls, −0.809 ($p < 0.001$, one tailed). These indicate, respectively, that the lower the initial IQ, the greater the gain, and the higher the initial IQ, the greater the loss. Although based on very small numbers, these results are of some interest and importance.

The second relevant study by Englemann

Table 14-17. Gains in IQ as result of special training program*

	Testing		
	I (initial)	II (after 1 year in program or school)	III (after 2 years in program or school)
Experimental	95.33	112.47	121.08
Comparison	94.50	102.57	99.61

*Based on data from Englemann, S. 1970. The effectiveness of direct verbal instruction on IQ performance and achievement in reading and arithmetic. In J. Helmuth, ed. Disadvantaged child. Brunner/Mazel, Inc., New York.

Table 14-18. Correlations between testings for experimental and comparison groups*

	Testing			
	II		III	
	E	C	E	C
I	0.846†	(no test)	0.702†	0.674†
II			0.749†	

*Computations by Thompson. See text.
†$p < 0.001$ (one tailed).

(1970) involved a group of 14 disadvantaged children exposed to an intensive training program in "rapid attainment of basic academic concepts." Their Stanford-Binet IQ was taken at entering (4.3 years), after 1 year in the program, and after 2 years. They were compared with a control group of 28 also disadvantaged subjects with the same initial mean IQ but given only traditional classroom instruction. Again we find large gains shown by the experimental group, as shown in Table 14-17. These data seem to leave little doubt that IQ can be markedly changed. Again, however, as indicated by the correlations in Table 14-18, the rank order is preserved in both groups over the 2 years. In neither group is there any significant change in the variance. Such results place a very serious qualification on Englemann's conclusion that "genetic influence seems to be a minor factor in the determination of intelli-

gence." Something must be acting to preserve the initial rank order of IQ. One would hardly suppose it would be the teachers, particularly those involved in the experimental program. In fact, it is explicitly stated by Englemann (1970, p. 343) that "the basic goal was to bring all of the children to 'average' on some of the more common measures of achievement, such as IQ measures." Of course, many other ad hoc environmental explanations might be invoked. Nevertheless, on their face value, the data, as we have presented them, are immediately compatible with a genetic model.

Regression

Regression describes the well-documented fact that bright parents tend to have children who, on the average, are duller and that dull parents tend to have children who, on the average, are brighter than the parents (Waller, 1971b). Genetic theory explains regression as arising from the improbability of a particular combination of genes responsible for brightness or dullness occurring again in a second generation. The greater the deviation of the parents from the mean of the population, the greater the improbability. It is a phenomenon, however, that is by no means immune from environmental qualification. Thus height is clearly a strongly heritable character. Yet over several generations there has been a secular trend for children to be taller, on the average, than their parents (Damon, 1974). This has presumably been a result of better standards of health and nutrition and, at least among more well-to-do groups, may be stabilizing. Such a trend, in a sense, masks the appearance of regression. Nevertheless, as environmental conditions approach an optimal level, we could expect to see its recurrence for height and such characters.

Apart from the preceding, it may also be argued that regression can as readily be explained by reference to improbable environments instead of improbable genetic combinations. This is, indeed, possible, but it does not make much intuitive sense if we consider the fact that quality of rearing environment is strongly correlated with social position and IQ level. That is to say, good environments are much more probable at the upper end of the intelligence distribution than at the lower end. But in spite of this, regression tends to be quite symmetrical, occurring as much in families below the mean as in families above the mean.

Inbreeding effects

Traits under genetic control often are depressed in offspring from the mating of related genotypes. This is commonly referred to as *inbreeding depression*. Thus animals and plants that are inbred may show less vigor, lowered fertility, and generally less adaptive capacity. Thus it may be argued that if we find such a depression for a trait in the offspring of related parents, this is evidence for that trait being heritable.

In human beings (and in animals) there are grades of inbreeding ranging from incestuous matings—usually brother-sister or father-daughter—to consanguineous pairings, as between first cousins, for example. Sometimes writers also make a distinction between effects occurring in offspring of consanguineous unions, called *consanguinity effects*, and effects in offspring of parents who are unrelated but one or both of whom are the product of a consanguineous mating, called *inbreeding effects* (Schull et al., 1970). Incestuous matings are relatively uncommon. However, in cases that have been studied, offspring have been found to show greatly increased mortality and morbidity, which includes severe mental retardation (Adams and Neel, 1967; Carter, 1967; Seemanová, 1971). Effects appear to be more marked in offspring of father-daughter than of brother-sister matings (Seemanová, 1971).

One of the first studies done on cousin marriages was by Böök (1957) in Sweden. Basically, he found that cousin marriages yielded a significantly greater proportion of children (>40%) classified as "below normal" than did marriages in a group of matched, unrelated pairs (<15%). Likewise, the incidence of mental deficiency (IQ range 0 to 70) was higher in the inbred group than in the general population, regardless of whether the cousin marriages involved normal parents or not. As a corollary of these findings, Böök found significantly fewer "gifted" children among the children of cousin marriages. However, as Böök acknowledged, it is difficult to know whether these were really direct effects of inbreeding or arose from the fact that in cousin families there were more parents with below normal intelligence and more with psychiatric illness.

Consanguinity and inbreeding effects have also been studied by Neel, Schull, and various collaborators in a number of communities in Japan, where cousin marriages are more common than in the West. In an initial monograph, Schull and Neel (1965) reported the effect of varying degrees of consanguinity on intelligence (Wechsler Intelligence Test for Children) and other characters in populations in Hiroshima and Nagasaki. In this study, a significant effect was found. Taking means of outbred offspring as standard, depression effects ranged from 5% to almost 12% for the eleven WISC subtests. Thus the effects were small but statistically significant. Socioeconomic status was found to relate to consanguinity and also to intelligence. When this variable is allowed for, the depression effect still remains. However, it is still not certain whether, in fact, their analysis totally removed the social class variable. In fact, Neel et al. (1970) later acknowledged that the depression effect reported earlier might have been overestimated. In this latter paper, data were presented on the effects of both inbreeding and consanguinity. In this case, the former appeared to have little or no influence on intelligence (Tanaka-Binet) or on a number of other measured traits. With socioeconomic status controlled, some effects of consanguinity on intelligence were found, but these were not statistically significant.

A final study was carried out by Bashi (1977) in the Arab population of Israel, a group in which cousin marriages are permitted and even considered desirable for economic reasons. On three intelligence and some achievement tests, 970 children from first-cousin and 125 children from double–first cousin matings were compared with control children. The order of performance in most of the tests was consistent with an inbreeding effect. However, differences were very small and, in most instances, not statistically significant. Consequently, it would be unwise to put too much faith in Bashi's conclusion that "the cause [of the alleged deficit] is indeed genetic."

In general, then, the data on consanguinity or inbreeding effects on IQ are rather meager and ambiguous. Few firm conclusions can be educed from them. If real effects do exist, they are probably rather slight, except in the case of incestuous matings. We should also note, however, that this does not disconfirm the heritability of IQ. It only implies that IQ, as measured, may have little relation to biological fitness (Falconer, 1960; McClearn and DeFries, 1973).

CREATIVITY AND DIVERGENT THINKING

It is widely recognized (Butcher, 1968) that there is a domain of cognitive functioning that is to some extent untapped by conventional IQ and ability tests. This domain at least includes creativity and divergent thinking. These capacities are thought to correlate fairly strongly with standard intelligence up to a certain level (perhaps IQ 120) but less strongly if at all beyond this level. This relationship would, in fact, be represented by a scatterplot in the shape of a "bent pear."

To date, very little is known about the inheritance of such abilities. Some distinguished lineages in the fine arts, such as the Bach pedigree, suggest that it is possible. Clearly Galton thought so. But the evidence is slim. One major reason for this, of course, is that the domain is exceedingly difficult to measure. If there are problems in knowing what intelligence really is, there are even greater ones associated with defining creativity.

At the time of this writing, only about five studies have been carried out. All of these used some of Guilford's measures of

divergent thinking or variants of them (Vandenberg, 1968a; Barron, 1969; Olive, 1972; Canter, 1973; Reznikoff et al., 1973). Barron (1969) used, in addition, the Barron-Welsh Art Scale. To give the reader a flavor of the kinds of data collected, we present descriptions of a sample of the kinds of subtests commonly used.

1. *Remote Associates Test.* The subject is given three words and is asked to supply a fourth word related to all three of the stimulus words.

2. *Alternate Uses Test.* The subject is asked to list as many unusual uses as possible for various common objects such as a brick or a paper clip.

3. *Gottschaldt Figures Test.* The subject is asked to locate a simple geometric shape hidden in a more complex design.

4. *Word Association Test.* The subject is given a list of 25 words with several meanings. He is asked to write down as many meanings as he can think of for each.

For such tests, scoring is necessarily rather loose. It is usually derived from total productivity, speed, and rated originality.

Let us now consider some of the studies that have been done. Vandenberg (1968b) obtained virtually no differences in intrapair variances between 67 MZ twins and 24 like-sexed DZ twins. Barron (1969) suggested that adaptive flexibility (Gottschaldt Figures Test) and aesthetic judgment (Barron-Welsh Art Scale) might be inherited, but Olive (1972) reported no differences between MZ and DZ twins in any aspect of divergent thinking. Reznikoff et al. (1973) used a subject pool of 28 male MZ, 35 female MZ, 19 male DZ, and 35 female DZ between 13- and 19-years-old. They were given a battery of eleven tests of creativity, including five drawn from Guilford's work. Testing was carried out blind. Intraclass correlations were high and significant on almost all tests and in both MZ and DZ groups. However, F ratios comparing MZ and DZ variances were significant for only three tests in the male and two tests in the female sample. One test was common to both, the Remote Associates Test (described before). In general, however, the authors concluded that they had no solid evidence for genetic factors in creativity.

The final study by Canter (1973) was reported as part of a larger twin study carried out in Glasgow in which 44 MZ and 51 DZ pairs were used, ranging in age from 16 to 55 years. As is so often the case in twin samples, there was a predominance of female pairs in the ratio of approximately 3 to 1. Five tests of divergent thinking were administered, yielding sixteen scores altogether. Since the within-test indices correlated, they were combined to give five scores plus a total divergent thinking score. Basically, only one of these yielded a significant MZ-DZ difference, this being the Word Association Test (described before). However, this test also showed the highest correlations with two tests of general intelligence: the Mill Hill vocabulary and the Raven Progressive Matrices Test. Thus Canter concluded that it was only this component of the Word Association Test that gave rise to its hereditary variance. In fact, a general trend emerged, indicating that the higher the loading of each of the five tests on general intelligence, the greater its hereditary variance.

In general, then, we may conclude that whatever creativity is, it is heritable probably only to the extent it reflects general intelligence. Alternately, it may be that psychologists presently have no tests that properly measure the functions underlying creative activity.

GENETIC TRANSMISSION OF INTELLIGENCE

Up until the early 1950s, only two major attempts had been made to construct a theory of the genetics of human intelligence. The first of these was made by Hurst (1932, 1934), who based his model on two sets of data: (1) the Woods royal family data, consisting of intelligence ratings of 212 European royal families with 424 parents and 558 offspring, and (2) Hurst's own Leicestershire data, consisting of IQ scores of 194 families with 388 parents and 812 offspring. The Stanford-Binet and the Healy Picture Completion tests were used with the children and some

lower-grade adults. Rating was used in the case of mediocre or higher-grade adults. His final scale of intelligence consisted of eleven grades, each grade being roughly the equivalent of 20 IQ points. Reliability of rating was checked whenever possible.

The total of 406 families appeared to fall into a number of different types. In the first place, they could be broken down into those which produced either all or half mediocre offspring and those which produced only a few mediocre offspring. There were 334 families that produced all or half mediocre offspring (incidence, about 75% or higher in the population), and of these families, 124, or one third, produced all mediocre offspring. The remaining 210, or two thirds, produced offspring of all grades, half being mediocre and half high or low grade. In other terms, of the type of family tending to produce an abundance of mediocre offspring, one third are nonsegregating, and two thirds are segregating. The 72 families of the second main type tended to produce only about one fourth mediocre and three fourths high- or low-grade offspring. None of the parents in these families was rated as mediocre. Hurst's data are schematized in Table 14-19. From these results, Hurst concluded that mediocrity was a dominant character dependent on a single major gene (N), while high or low intelligence was recessive. Thus any individual with the gene either in homozygous or heterozygous form was mediocre or normal. Lacking it, the individual would be either high or low in intellectual level. At the same time, he noted that some offspring (about 26%) from matings between high- or low-grade parents were also mediocre. This being so, Hurst found it necessary to postulate the presence of five minor modifiers, Aa, Bb, Cc, Dd, and Ee, which acted only when N was absent. In such a case, A, B, C, D, and E would act as unit increasers (10 IQ points), whereas a, b, c, d, and e would act as unit decreasers. Thus some individuals who carried the double recessive nn could also carry the modifiers in heterozygous form. These individuals would then also be mediocre.

This theory of genetics of intelligence is

Table 14-19. Summary of Hurst's data on the genetic transmission of intelligence*

Intelligence of parents	Number of families	Intelligence of offspring
Mediocre × mediocre	85	All mediocre
Mediocre × mediocre	77	½ mediocre, ½ high or low grade
Mediocre × high or low grade	39	All mediocre
Mediocre × high or low grade	133	½ mediocre, ½ high or low grade
High or low grade × high or low grade	72	¼ mediocre, ¾ high or low grade

*Based on data from Hurst, C. C. 1934. The genetics of intellect. Eugen. Rev. **26:**33-45.

undoubtedly ingenious and also fits rather well the data on which it is based. However, it is difficult to explain why, in the 77 families involving both mediocre parents, half the offspring should be above or below normal. If a simple dominant is involved, nonnormal children will appear only when both such mediocre parents are heterozygous for N, and in such a case the incidence of nonmediocrity should be only one fourth at the very most. Allowing for the action of the minor modifiers that may also produce normality, the proportion presumably would be, in practice, less than this figure. In a broad way, the theory affords some explanation of the phenomenon of exceptional children appearing in otherwise quite homogeneous families. Furthermore, as Hurst points out, it explains a broad range of intellectual level, from idiot (*nn:* aa, bb, cc, dd, ee) to "illustrious" (*nn:* AA, BB, CC, DD, EE). It does, of course, lack an independent test, as Conrad and Jones (1940) have pointed out. An analysis of the third generation, though perhaps beyond the resources of Hurst to obtain, would have been desirable, as would the application of some of the statistical techniques suggested by Fisher and others.

A second theory has been proposed by Pickford (1949). This formulation rests on a simple multifactor hypothesis. According to Pickford, the distribution of Stanford-Binet IQ scores in the general population is such as to justify the action of ten equal and additive gene pairs. Since this model is backed by no facts other than the population distribution of IQ, not a great deal can be said

about it. It is partially supported by the work of Burt and Howard (1956). A deficiency suggested (Conrad and Jones, 1940) is the inability of the theory to account for regression of offspring means to the population mean, a phenomenon well documented by empirical studies (Outhit, 1933; Hurst, 1934; Cattell and Willson, 1938). Regression cannot occur in a very simple multifactorial model such as Pickford's (McAskie and Clarke, 1976).

Since the time these two models were put forward, not much new has been added. It is fair to say that most workers in the field have favored a polygenic theory or some variant of this. In 1955, Burt published a preliminary attempt to fit empirical kinship data to a theoretical model derived from Fisher's classical paper (Fisher, 1918). On the basis of a multifactor theory involving many genes with equal frequencies and additive effects, it can be educed that the incidence of the three classes of intelligence—bright, average, and dull in offspring of bright, average, and dull parents—will approximate the proportions to be expected from the action of a single pair of genes (i.e., *AA, Aa,* or *aa*). Starting with a group of 954 children rated according to the three grades of intelligence, Burt then estimated the intelligence of their parents to yield a 3 x 3 bivariate table. The simplest expectation is set out in Table 14-20, together with the empirical data for parents and offspring. We may note that, although the data deviate slightly from the theoretically expected frequencies, the raw and column totals add up exactly, in parents and children,

Table 14-20. Bivariate distribution of intelligence in parents and their offspring: theoretical expectations and empirical data*

Parents	Theoretical frequencies children				Empirical frequencies children			
	Bright	Average	Dull	Total	Bright	Average	Dull	Total
Bright	12.5	12.5	0	25.0	10.8	12.3	1.9	25.0
Average	12.5	25.0	12.5	50.0	13.4	26.5	10.1	50.0
Dull	0	12.5	12.5	25.0	0.8	11.2	13.0	25.0
TOTAL	25.0	50.0	25.0		25.0	50.0	25.0	

*From Burt, C. 1955. The evidence for the concept of intelligence. Br. J. Educ. Psychol. **25:**158-177. Reprinted by permission of Scottish Academic Press Ltd.

to the proportions 0.25, 0.50, and 0.25. Since the figures appear to be based on adjusted assessments, however, this may be no surprise. A table for sibling data (not reproduced here) showed the same result. The small deviations from a simple polygenic model are explained by Burt as being results of (1) test unreliability, (2) environment, (3) dominance, and (4) assortative mating. In this particular article, Burt did not attempt firm separation of the relative contributions of these components.

This initial attempt at a biometric genetic analysis of intelligence data was elaborated further by Burt, particularly in articles in 1956 (Burt and Howard) and 1972.*

Basically, the method attempts to predict what correlations between relatives of varying degree ought to be, given certain assumptions about degree of assortative mating, degree of genetic dominance, and relative influence of all genetic as against all environmental influences. The basic formula for the parent-offspring correlation has been derived by Fisher (1918) as

$$r_{po} = C_1 C_2 \frac{(1 + \mu)}{2}$$

where

C_1 = effects of heredity, with maximum value of 1 if no environmental influence; that is:

$$C_1 = \frac{V_A + V_D}{V_p}$$

C_2 = degree of additive gene action with parameter values between 0.5 (complete dominance) and 1 (no dominance) that is:

$$C_2 = \frac{V_A}{V_A + V_D}$$

μ = marital coefficient, given by coefficient of assortative mating (i.e., correlation between spouses for a trait), which estimates parameter A = genetic correlation between spouses

It can be seen from the equation that all three variables act to increase the parent-offspring correlation. It is also clear that, given this expression alone, we can only guess at values for C_1 and C_2, though r_{po} and μ can be measured.

Similar correlations have been worked out by Fisher and by Burt for a range of relatives. For example, the sibling correlation is as follows:

$$r_{oo} = \frac{1}{4}(C_1 + C_1 C_2 + 2C_1^2 C_2^2 \mu)$$

Burt and Howard, in a first step, offered guesses as to the values of C_1 and C_2: 0.95 and 0.75, respectively. The former appears to be simply a provisional estimate, the latter based on the assumption of partial dominance. The empirical value of μ is 0.386. Fitting these values to the equations for r_{po} and r_{oo}, they give $r_{po} = 0.495$ and $r_{oo} = 0.514$. The empirical correlations are, respectively, $r_{po} = 0.489$, and $r_{oo} = 0.507$. The fit is a good one.

Since this procedure was based on guesses as to the values of C_1 and C_2, Burt and Howard next attempted to estimate them empir-

*We already touched briefly on Burt's application of Fisher's theorem to kinship data.

ically from observed data. This can be done by using r_{po}, r_{oo}, and μ. We have two equations and two unknowns, C_1 and C_2, that can then be deduced. Although Burt and Howard do not present the computed values, these turn out to be as follows: $C_1 = 0.937$ and $C_2 = 0.753$. Thus we have, in other words, estimates fairly close to those arrived at by a priori guesses, the first representing a large additive genetic component, the second an intermediate dominance effect. As indicated earlier in the chapter, in 1966 Burt published observed and theoretical correlations for twelve types of lineal and collateral relatives. It is not stated what parameter values he used for C_1 and C_2.

Burt and Howard carried their analysis further, using a slightly different method and applying it to Burt's adjusted assessments and also to unadjusted group test results. Although his twin data using these scores are suspect, we have no immediate reason to suspect his sibling and parent-offspring results. Using the latter, they partitioned the total variances into genetic and environmental components as shown in Table 14-21. Their estimate of environmental influence is derived from separated MZ twin data. In the same table, we have also presented results of computations by Cavalli-Sforza and Bodmer (1971) based simply on the correlations taken at face value without a separate estimate of environmental effects. In addition, column three shows similar calculations performed by Thompson for the unadjusted group test

results. Again, parent-offspring and sibling correlations are taken at face value, and no allowance is made for unreliability. Were the latter included, environmental effects would increase up to around the estimate made by Cavalli-Sforza and Bodmer. Other values would, of course, decrease.

A rather more complex genetic model has been applied by Jinks and Eaves (1974) to Burt's complete family data and also to the kinship data surveyed (and corrected) by Jencks (1972). The model has five components, four of which are additive genetic component, dominance component, marital correlation, and additive genetic deviation of spouses. The fifth is a common environmental component, that is, the contribution to covariance between relatives produced by living in the same environment. A sixth, the specific environmental component, may be obtained by subtraction. Analysis of Burt's data yielded a broad heritability of 83%; for Jenck's data, 68%. This latter figure is 50% higher than the one educed by Jencks using path coefficient analysis and inserting a term for genotype-environment covariance. Jinks and Eaves found, however, no significant improvement in the fit of their model by adding such a parameter. A second point emerging was that, for both sets of data, a strong dominance component emerged, being more marked in the Jenck's data. This agrees with the previous analysis by Burt and Howard (1956) and also with another analysis of a different kind done before by Jinks and Fulker

Table 14-21. Components of variance for intelligence as measured by Burt's adjusted assessments and group test*

Variance component	Adjusted assessments (percent)		Group test (percent) (raw correlations) Thompson (face value)
	Burt and Howard (corrected)	Cavalli-Sforza and Bodmer (face value)	
Environmental	12.4	6.1	1.56
Additive	47.8	51.2	48.28
Dominance	21.9	23.4	27.44
Assortative mating	17.9	19.2	22.72

*From Cavalli-Sforza, L., and W. F. Bodmer. 1971. The genetics of human populations. W. H. Freeman & Co. Publishers, San Francisco.

(1970) on largely independent sets of data. We will consider the latter in a moment.

Before doing so, however, it is important to comment that genetic models of the preceding kind do little to accommodate systematic effects of environment; nor do they take much account of all the possible interactions and covariances between types of environment and types of genetic influences. This fact has been noted by Kamin (1974), Layzer (1974), and Goldberger (1975), among others. Furthermore, it is also true that small changes in observed kinship correlations can produce quite strikingly different estimates of additive as against dominance variance components. Thus Kamin (1974) has estimated, for example, that if $r_{po} = 0.52$ and $r_{oo} = 0.48$, then $C_1 = 0.73$ and $C_2 = 1.02$. However, if $r_{po} = 0.46$ and $r_{oo} = 0.54$, then $C_1 = 1.16$ and $C_2 = 0.57$. This should make us cautious if we are working only with a limited number of kinship categories, particularly if the number of families is relatively small.

It is partly because of problems of this type that Jinks and Fulker (1970) have put forward a more exactly derived model involving special kinds of data. We have discussed its methodology in Chapter 12 and will now briefly show its application to the genetics of intelligence.

The minimum data are supplied by MZT, MZA, and DZT twins. Provided no genotype-environment interactions exist, analysis of these sets can supply us with estimates of within- and between-family genetic and environmental components of variance (G_1, G_2, E_1, and E_2, respectively). Under the simplest assumptions, G_1 and G_2 can be used to supply additive and dominance components of variance.

Jinks and Fulker applied the model to Shields' Mill Hill and Dominoes test data, and to Burt's Final Assessments and Educational Attainments results. Supplementary analyses were also carried out on the family data of Reed and Reed (1965), the inbreeding data of Schull and Neel (1965), and the Stanford-Binet and Otis scores of the MZA twins of Newman, Freeman, and Holzinger (1937).

Their analysis is lengthy and complex, and only a brief summary of it will be presented here, as shown in Table 14-22. Several points are worth emphasizing. In the first place, a significant negative skewness of distribution of IQ within families suggests dominant gene action. This finding is confirmed by the detection of a small inbreeding depression effect found in the consanguinity data of Schull and Neel (1965). Second, common family environment seems to be very important for educational attainment but not for IQ. Third, genotype-environment covariance or interactions are by no means as common for IQ as is sometimes supposed (e.g., Jencks, 1972; Layzer, 1974). Fourth, as noted earlier, the general results of the analysis are in good agreement with the Jinks-Eaves model fitted to the kinship data of Burt and of Jencks. Finally, we must again remind readers that whatever the merit and sophistication of the statistical analysis, its success is, in the end, limited by the quality of the primary data.

Before leaving this section, we may refer briefly to work by a number of investigators suggesting that one type of cognitive function, spatial ability and perhaps general intelligence itself, may involve sex-linked genes (Lehrke, 1972; Bock and Kolakowski, 1973). This idea is based mainly on two types of data. One of these relates to the range of IQ and spatial ability in males as against females. The female distribution appears to be much more truncated at both ends. The second relates to the relative magnitude of the four types of parent-offspring correlations. In general, these conform to a sex-linkage model that predicts father-daughter and mother-son to be higher than mother-daughter correlations.

GENETIC AND ENVIRONMENTAL MODELS AND INTELLIGENCE

It will be clear that the large amount of data we have just summarized is of varying quality and reliability. However, almost all studies report a strong tendency for correlations on many forms of IQ tests to be higher the more closely the groups are related; these data must be explained either by refer-

Table 14-22. Biometric genetic analysis of several sets of IQ data: summary of major results*

Test	Correlated environments	Genotype-environment interaction	Heritability ± SE		Genetic model
			Broad	**Narrow**	
Mill Hill Vocabulary (Shields, 1962)	None	Low-scoring genotypes more affected by environment	73 ± 12%	—	Poor sampling prevented complete analysis
Dominoes Test (Shields, 1962)	None	None	71 ± 7%	—	Very simple polygene model adequate; common family environment (E_2) unimportant
Group Test, Adjusted Assessments (Burt, 1966)	None	—	86 ± 1%	71 ± 1%	Simple model adequate, assortative mating
Educational attainments (Burt, 1966)	Positive correlation between families	—	< 30%	—	Simple model not adequate; assortative mating; common family environment (E_2) very important
Stanford-Binet and Otis (Newman, Freeman, and Holzinger, 1937)	—	None	—	—	Common environment (E_2) not important
Several Tests (Reed and Reed, 1965)	—	—	—	—	Additive and dominant gene action, with level of dominance 0.74
Wechsler Intelligence Scale for Children (Schull and Neel, 1965)	—	—	—	—	Dominance for high IQ, with many genes (~100) and gene frequencies equal on the average

*Modified from Jinks, J. L., and D. W. Fulker. 1970. Comparison of the biometrical genetical MAVA, and classical approaches to the analysis of human behavior. Psychol. Bull. **73**:311-349.

ence to some genetic model, some environmental model, or some combination of both. Urbach (1974), in a perceptive analysis of this problem, has put forward the view that the hereditarian-environmentalist debate is not so much between competing theories but between competing research programs, each involving a *series* of falsifiable theories relating to anomalies in the data that must be explained. He has further suggested that the merit of a program must be gauged according to whether it deals with anomalies in a *progressive* or *degenerating* manner. If the

"hard core" theory of a program is altered in a way that not only deals with anomalies but also makes new predictions, it can be called progressive. If it deals with them merely in an ad hoc way, it can be called degenerating. By this criterion, Urbach has concluded from a general survey of the data on heritability of intelligence that the hereditarian program has been, in fact, progressive, and the environmental program, degenerating.

We would be hesitant to agree completely with this conclusion. However, in general, it does seem to be true that environmentalists have primarily played the role of critics, quick to point up flaws in the basic data or in their analysis but usually unable to suggest testable alternate explanations of such order as the data possess. One reason for this is that whereas the genetic model may have less precision than is sometimes supposed, environmental models have almost no precision at all. Their central theme appears to be that, in respect to intelligence, all individuals are equipotential and that they will be similar or dissimilar as the environments they occupy are similar or dissimilar. However, no definition of "environment" is commonly put forward except post hoc and then only in the crudest terms. Thus Skodak and Skeels, for example, having confirmed that foster parents' education and foster fathers' occupational level apparently have nothing to do with producing higher intelligence, invoke the operation of a set of factors underlying "the dynamic aspects" of the home, but presently not measurable. Likewise, Skeel's attributes the gain in IQ in his experimental group to "developmental stimulation and the intensity of relationships between the children and mother-surrogates." The latter, it will be recalled, were mentally retarded girls. One pair of Burt's MZA twins, George and Llewellyn, brought up, respectively, by the widow of an Oxford don and by an elderly couple on an isolated farm in North Wales, were found to differ by 1 IQ point (136 and 137) (Conway, 1958). Kamin (1974), commenting on this seeming anomaly for the environmentalist model, suggests that, in fact, the stimulation provided by "comfortable farmers" should be equal to that provided by "slender-pursed and unmaternal widows of deceased dons." This is possible, of course, but somewhat counterintuitive, the more especially so because Kamin neglects to mention that George, unlike his twin, had a brilliant career at school and obtained a first-class degree in modern languages (Conway, 1958). If two such environments are to be

considered identical, we can only ask about the specifications of those an environmentalist would consider different.

Such examples as those just cited are by no means atypical of the environmentalist program. The term "environment" can mean literally anything, and its definition can be altered to fit the case. The basic unit of the genetic model, the gene, on the other hand, is not simply a construct. Its chemical structure is known, as are the rules governing its transmission across generations. This permits explanations that are not post hoc and predictions of a precise sort. For these reasons, the genetic model must be considered to be, at least provisionally, superior to the environmentalist. We may only hope that, in time, investigators in this area will develop conceptualizations and measures of environmental influence that are as precise and testable as those relating to genetic determination.

SUMMARY

This chapter has examined views about intelligence and the ways in which it is presently tested. It is argued that although IQ tests are fallible, they are not as arbitrary and unreliable as is sometimes supposed.

Family, adoption, twin, and some ancillary data were critically assessed. In our opinion, these suggest a substantial genetic determination of intelligence level in the populations studied. However, since many of the individual studies surveyed have flaws of one kind or another, it is doubtful that different authorities will reach agreement in respect to the problem in the near future.

Studies on creativity indicate very little, if any, hereditary influences except insofar as the measures of it correlate with general intelligence.

In closing, we suggested that there is an onus on environmentalists to develop competing models that fit the existing data as well as or better than do genetic models. This matter will be taken up in more detail in Chapter 18.

15

Personality and temperament

This chapter will summarize work relating to the inheritance of personality and temperament. As with the study of intelligence, data in this area have been accumulating rapidly, particularly during the last ten years, and now cover a wide range of personality dimensions. Before examining this material, we shall discuss briefly the various meanings of personality and temperament and attempt to outline some of the major problems associated with their genetic analysis.

Theories of personality have a long history (Lindzey, Hall, and Manosevitz, 1973). All through the ages people have asked the question, "What am I?" But it was not until the late nineteenth century that the problem came to be attacked in an empirical manner. Freud and his followers were among the first to open up the field of inquiry, though it should be emphasized that Galton played an important role. Since this early work, a great deal has been written on the subject, much more, in fact, than can be summarized here. The reader is referred to some of the large number of source books available on the subject. It is nonetheless pertinent, however, to establish an orientation for this chapter by exemplifying some of the kinds of behavioral dimensions with which we shall deal. Personality tests range from measures of verbal responses to Rorschach inkblots to measures of tapping rates. Almost any reaction may be looked on as reflecting some aspect of personality, insofar as it relates to the affective or emotional responsiveness of an individual. By "emotional" we refer not only to violent changes or disturbances in behavior, but also to the relatively mild affective states involved in likes or dislikes of particular objects or events. Such states are not observed directly but are usually inferred from what a subject does or says he does in particular situations. Thus the words given by a subject in a free-association experiment, his verbal response to a Thematic Apperception card, or the way he answers a questionnaire may all be regarded as relating to personality and temperament.

Obviously, the domain of personality takes in a lot of ground. It is for this reason that many workers have felt that the major task in personality theory relates to developing a sensible and manageable taxonomy of its fundamental aspects. This kind of endeavor has had some success, as we shall see. However, there is still rather little agreement among personality theorists as to which taxonomy, by some criterion, is the best. In fact, the disagreement is probably greater than it is in respect to the domain of intelligence, though perhaps it is less emotional.

We will not attempt to take any particular stand in this chapter but will merely deal with the empirical evidence gathered with methods commonly considered to measure personality. From a strictly scientific point of view, this may sound unsatisfactory. Nevertheless, it is also true that the topic is one of great interest and is of central importance to the clinician and perhaps also to the biologist. In a sense, it is through personality that the intelligence of an individual interfaces with real-life situations. Indeed, as many authorities have pointed out (Burt, 1959; Jencks, 1972), high intelligence by itself is no guarantee of success. Additional characteristics are necessary as well, and these have to do with personality variables. In lower animals, certainly, it is not known to what extent abstract intelligence relates to survival and reproductive success. However, at least in some species, social ability and temperament may be much more important and hence subject to natural selection. An interesting exposition of this point has been put forward by the primatologist Jolly (1966). We will return to it in a later chapter. All that is necessary to emphasize at this point is that personality variables may have great significance for biological fitness in human beings.

Another aspect of the domain of personality that is of some special significance relates to the great range of variables it includes. Some of these, for example, particular social or political attitudes, would seem, on an a priori basis, to be almost certainly determined by environment. Others, however, such as introversion-extraversion, may quite likely, on the same basis, be heritable. To examine patterns of similarity in family groups in respect to such different characters may thus shed much light on the problem of cultural versus genetic transmission. The latter has been articulated along mathematical lines by Cavalli-Sforza and Feldman (1973a,b) and by Morton (1974), but data bearing on it have not been put forward in any systematic way. Yet, as 'we shall attempt to show, information does exist that may help us in specifying the difference between the two modes of transmission. As an example, we may cite some data of Nichols (1966) relating to personality resemblances between members of MZ as against DZ twin pairs. In respect to particular items on one test, Objective Behavior Inventory, we find the following:

Large zygosity differences (i.e., high heritability)	Small zygosity differences (i.e., low heritability)
Took cough syrup	Took a laxative
Picked up a hitchhiker	Hitchhiked
Sang in a glee club	Took voice lessons
Played tennis	Played table tennis
Rode in a car	Rode in a sports car
Played in a concert orchestra	Attended an orchestra concert

From a similar personality test we find other seeming anomalies. Thus, with the California Psychological Inventory (CPI):

Large zygosity differences (i.e., high heritability)	Small zygosity differences (i.e., low heritability)
I dislike to have to talk in front of a group of people.	In school I find it hard to talk before the class.
I admit I am a high-strung person.	A windstorm terrifies me.

Results like these may, of course, be quite spurious and merely be the outcome of taking seriously the results of "one-item tests," as it were. Since the reliability of any test is a function of its length—that is, the number of items making it up—it is clear that the responses of MZ and DZ twins to separate single questions may not mean very much. On the other hand, they may mean something and may be valuable in formulating hypotheses about the relation between environmental experience and test taking in twins. This, as we shall see later, is the position taken by Nichols. We will also discuss other such curious findings from other studies. They may tell us a good deal more than simply the difficulty of measuring personality.

PEDIGREE AND FAMILY STUDIES

As with other traits in human beings, the inheritance of personality has been studied by four main methods: pedigrees, adopted children, familial resemblances, and twin similarities. We will deal with each of these, as well as with some ancillary data.

Pedigree data

Raw pedigree data of a nonquantitative type have some general historical interest but little scientific value. In his book, *English Men of Science: Their Nature and Nurture*, Galton (1874) managed to obtain letters from eminent scientists and their kin reporting on many characteristics that may be included in the category of personality. Most relevant here are energy, perseverance, practical business habits, independence of character, religious bias, truthfulness, and taste for science. As we noted before, Galton was inclined to place great weight on heredity in the determination of eminence. In respect to personality, his views were not different. Thus he regarded even a "taste for science" as innate, though recognizing that it might be suppressed by the bias in society against science. "A love of science might be largely extended by fostering and not thwarting innate tendencies" (Galton, 1874, p. 225). Of interest is that most of the personality traits studied by Galton were later studied by his successors, using more exact empirical tests. The conclusions reached by many of them were substantially the same, if more moderate.

Following Galton, many more pedigree studies of temperament and personality have been carried out. Reference has already been made to the work of Gun (1930a,b) on hereditary traits in several royal family lineages. For example, "efficiency" is described as being the key character of the Tudors, while "tactless obstinacy" is attributed to the Stuarts. These and other data (Finlayson, 1916; Davenport and Scudder, 1918; Gun, 1928) appear to indicate transmission of some kind. Whether this is genetic or cultural or, for that matter, merely a reflection of the historical biases of the writers is a problem that the pedigree method cannot adequately solve. In view of its evident limitations we will not discuss pedigrees further.

Family correlations

More exact data are supplied by the family-correlation method. A variety of tests have been used, covering roughly three areas: personality traits, attitudes, and interests. Work on the general area appears to have commenced with Starch (1917). It has continued sporadically up to the 1950s, but not a great deal has been done after that time with this method. Reviews of the early work have been made by Schwesinger (1933), Crook (1937), Roff (1950), and others (Sen Gupta, 1941; Eysenck and Prell, 1951).

Personality tests. For the measurement of personality and adjustment, one of the most popular tests has been the Bernreuter Personality Inventory (Super, 1949; Anastasi, 1954). This test, though apparently quite reliable, is used much less today than prior to 1960. One reason for this lies in the fact that it is not altogether certain what it predicts;

neither is it established that the responses given to the 125 questions reflect anything at all in the subjects' real behavior (Anastasi, 1954). If we are content merely to measure reliably the individual's phenomenological world, then, of course, these problems do not matter a great deal. In any case, it is of some interest to consider the work done with the Bernreuter. The test comprises scales to measure neuroticism, introversion, dominance, self-sufficiency, and solitariness. Not all need be scored. Although these dimensions are considered to be independent (on a priori grounds), they are probably correlated. Flanagan (1935) suggested that they were reducible to two factors: confidence and sociability. In spite of this alleged fact, family correlations turn out to be rather different for the scales. Hoffeditz (1934) found that in a sample of 100 fathers, 100 mothers, 111 sons, and 145 daughters, the mean-parent × mean-child correlations were as follows: neuroticism, 0.278; self-sufficiency, 0.2; and dominance, 0.294. Of interest is that daughters correlated more highly with both parents, particularly the mother, than did sons, at least on the neuroticism and dominance scales. However, all correlations for the individual scales are low and mostly nonsignificant. Consequently, only the overall results can be taken seriously. Crook and Thomas (1934) also computed parent-child and sibling correlations on the Bernreuter Scales. On the whole, their data agreed well with those of Hoffeditz for the same three scales used. The highest correlations were between sisters and between mothers and daughters. The lowest was between father and son, followed by brother and brother. The other combinations—brother-sister, mother-son, and father-daughter—were intermediate. However, again, all correlations were low and probably mostly nonsignificant. There did not appear to be any systematic differences between the three scales. Sward and Friedman (1935), in a study of "Jewish temperament," also obtained consistently low correlations for the neuroticism scale of the Bernreuter. Mother-child correlations again tended to be higher than those between fa-

Table 15-1. Family correlations (corrected for attenuation) for three of the scales of the Bernreuter Personality Inventory*

	Number of pairs	Scales		
		Neuroticism	Dominance	Self-sufficiency
Brother-brother	50	0.25	0.09	−0.08
Brother-sister	56	0.15	0.10	0.28
Sister-sister	51	0.36	0.33	0.36
Father-son	62	0.06	0.05	−0.03
Father-daughter	64	0.24	0.26	0.39
Mother-son	68	0.32	0.22	0.20
Mother-daughter	73	0.62	0.43	0.13
Husband-wife	79	0.07	−0.06	0.01

*From Crook, M. N. 1937. Intra-family relationships in personality test performance. Psychol. Rec. 1:479-502.

thers and offspring. Perhaps the most complete study has been that of Crook (1937) whose data for the three Bernreuter scales are displayed in Table 15-1. Only correlations corrected for attenuation are shown. Since age was found not to relate to scores, no correction for this variable was used. It will be noted that, as before, all relationships involving a female (mother-daughter, father-daughter, mother-son) tend to be higher than those involving males only. Since the coefficients are all small, it is difficult to say whether this trend is significant. Nonetheless, the fact that it appears in all of the four studies cited is worthy of mention. As we shall show shortly, similar trends have also been found in more recent work. If actually valid, an interpretation in terms of cultural rather than genetic transmission would seem more plausible.

Most early tests like the Bernreuter have been difficult to validate. Because of this, attempts were later made to key items to some empirical criteria. An outstanding example of this procedure is afforded by the Minnesota Multiphasic Personality Inventory (MMPI) (Hathaway and Meehl, 1951; Dahlstrom and Welsh, 1960). This was first developed to measure basic traits associated with disabling psychological abnormality. It consists of 550 affirmative statements, to which the subject responds: "True," "False," or "Cannot say." Two examples are "I do not tire quickly,"

and "I believe I am being plotted against." Scores in terms of ten scales are provided. Eight of these are constituted by items that were previously found to differentiate a specific clinical group and a normal control group. Another scale was derived from items on which males and females differed, and another from items that correlated with an independent test of introversion-extraversion. Thus the ten scales are as follows:

1. Hs: Hypochondriasis
2. D: Depression
3. Hy: Hysteria
4. Pd: Psychopathic deviate
5. Mf: Masculinity-femininity
6. Pa: Paranoia
7. Pt: Psychasthenia
8. Sc: Schizophrenia
9. Ma: Hypomania
10. Si: Social introversion

Four so-called validity scales are also involved. One assesses amount of uncertainty of subjects, another gets at stereotypic responding, a third at carelessness in responding or possible malingering, and a fourth attempts to measure faking, either in a "good" or "bad" direction. Scores on the substantive scales are assessed in light of the subject's performance on the validity items.

Because of its wide popularity, the MMPI has been used in a number of behavior genetic studies, especially with twins. Two examined familial correlations. Thus Gjerde

(1949), using nine of the test scales, reported the four types of parent-offspring correlations that turned out to be uniformly low and nonsignificant. This may have been because the author worked with a highly selected group of children drawn from the Laboratory School of the University of Chicago. Hill and Hill (1973a,b) later repeated the study starting from a more representative group of subjects. They also corrected for age differences by using parents and children who had been tested at approximately the same ages. This, and other constraints they imposed, reduced their sample size from several thousand down to 28 mothers, 28 fathers, and 28 children. Only one single family correlation turned out to be significant. Since sixty were computed, it is doubtful if much weight can be given to this single positive result. The authors went on to give heritability estimates based on both random and assortative mating models. Again, these are uniformly low and probably nonsignificant. The statistical data were further checked by three clinical psychologists who assessed, in an overall way, the general resemblances of parents and offspring for MMPI profiles. These judgments confirmed the lack of significant scale-by-scale correlations.

It is clear that the results of the study are unremarkable except in a negative way. It is somewhat counterintuitive to find that parents and children do not resemble each other at all in respect to what are commonly thought of as basic personality dimensions. This poses less of a problem for genetic than for environmental models. Furthermore, the results of Hill and Hill disagree rather sharply with those obtained by several investigators using twins. They also appear to be discrepant with the MAVA data of Cattell, Stice, and Kristy (1957), emphasizing the importance of within-family as opposed to between-family influences on personality. It may well be, of course, that the Hill and Hill findings are spurious and result merely from their use of highly selected and homogeneous samples.

A personality dimension considered by many to be of great importance and measurable by a number of tests (including the MMPI) is introversion-extraversion. This was given great theoretical emphasis originally by Jung on the basis of clinical observation, and later by Eysenck (Eysenck, 1971), who, with his students and colleagues in London, has produced a massive amount of empirical and theoretical work on it. Most of the behavior genetic studies have used twins; however, there are a few family studies, of which we will discuss two.

The first of these was done by Coppen, Cowie, and Slater (1965). They used the Maudsley Personality Inventory (MPI), a test purporting to measure two aspects of personality: neuroticism (N), defined as "the general emotional lability of a person, his emotional over-responsiveness and his liability to neurotic breakdown under stress", and extraversion (E) defined as "outgoing social proclivities." Validity and reliability of the test are discussed by Coppen, Cowie and Slater (1965). The test was administered to 224 patients at the Belmont Hospital and to 735 of the first-degree relatives of these. Correlations were computed between the patients and relatives as well as within groups of relatives of the patient as proband. Several findings of interest emerged.

1. There was a striking difference between correlations for families of male and female patients. Thus the families of male patients showed significant correlations, on the average, both for N and for E scores, whereas the families of female patients did not. This difference was largely accounted for by the resemblance of all relatives of the patient to the mother but not to the father. These results are shown in Table 15-2.

2. Relatives of all patients tended to score about the same on extraversion. However, they showed *less* neuroticism on the MPI than a standard normal level. Coppen et al. suggest that this may be due to a "defensive reaction to having a close relative in hospital." Hence they may have falsified their responses to give a stronger impression of normality.

3. Statistically significant assortative mating for neuroticism occurred among parents

Table 15-2. Correlations for neuroticism *(N)* and extraversion *(E)* within families of male and female patients*

	N		E	
	Male patients	**Female patients**	**Male patients**	**Female patients**
Average family *r*	0.187†	0.043	0.236†	0.081
Average *r* between all family members and mother	0.322†	0.059	0.336†	0.089
Average *r* between all family members and father	−0.027	0.057	0.083	0.155

*From Coppen, A., V. Cowie, and E. Slater. 1965. Familial aspects of "neuroticism" and "extraversion." Br. J. Psychiatry **111**:70-83.

†$p < 0.01$.

of male patients but not of female patients. In neither group was there assortative mating for extraversion.

The first of these three conclusions is perhaps the most interesting. No obvious genetic explanation presents itself. The authors suggest that an environmental explanation is more meaningful; in particular, they put forward the view that there is "something special about the affective relationship between the mother of a male neurotic and her children, such as does not obtain between the mother of a female neurotic and her children." Specifically, they hypothesize that some males who have "over-influential" mothers fail to make an appropriate transfer, after age 10, from the mother to the father as a model. This effect apparently does not hold in the case of females with overinfluential fathers. This is an interesting explanation but may be oversimplified. If we take particular correlations at face value (ignoring their significance levels), we find that for both N and E the father-son relationships are stronger than the father-daughter relationships. However, for N, the mother-son is higher than the mother-daughter correlation, but the reverse holds for E. If these relationships are real, they would require some rather complex explanations in terms of family dynamics. The chances are probably good, however, that they are not real. Some indication of this is afforded by another larger-scale study carried out by Insel (1974). Insel used a sample of 98 families consist-

ing of 589 subjects representing three generations. These were drawn mainly from middle and lower middle classes in the greater London area. Grandparents' ages ranged from 49 to 94, parents' from 32 to 59, and children's from 9 to 32. All subjects were given Eysenck and Eysenck's (1969) Psychoticism-Neuroticism-Extraversion (PEN) Inventory and Junior Personality Inventory. In addition, two attitude tests of conservatism were also administered. Since the results are complex, we have summarized only the statistically significant correlations in order of magnitude, as shown in Table 15-3. It is difficult to make sense of the complete arrays of correlations. However, from the table several points of interest seem to emerge. First, there appears to be a fairly strong maternal effect for all three of the PEN variables. In no case does the male of one generation appear to exert a significant influence on a relative in the two succeeding generations. Second, at least in the case of the variables E and N, one might infer that the mother's influence on her daughter and her son generates a significant correlation between them. Third, one curious exception to the maternal effect is given by the *negative* correlation (−0.334) between paternal grandmother and father for neuroticism. Insel does not attempt to explain this, and it may well be spurious. Curiously enough, the highest of all the correlations reported is between paternal grandmother and father for psychoticism. Fourth, what has been stated

Table 15-3. Statistically significant correlations for different family relationships on three personality dimensions and a social attitude scale*

Psychoticism		Neuroticism		Extraversion		Conservatism	
Relation†	Corre-lation	Relation	Corre-lation	Relation	Corre-lation	Relation	Corre-lation
PGM × F	0.701	M × D	0.521	M × D	0.498	MGF × MGM	0.663
M × S	0.594	M × S	0.353	MGM × M	0.452	M × D	0.658
MGF × MGM	0.359	D × S	0.265	D × S	0.344	M × F	0.657
D × S	0.354			M × S	0.343	F × S	0.603
MGM × M	0.287‡			MGM × D	0.295	M × S	0.597
MGM × S	0.279			PGM × F	−0.334	PGF × PGM	0.594
						MGM × M	0.587
						PGF × F	0.516
						F × D	0.435
						D × S	0.422
						PGM × S	0.388
						MGF × D	0.387
						MGF × PGF	0.381
						MGF × S	0.376
						PGM × MGM	0.364
						PGF × F	0.353‡
						MGF × M	0.338
						MGM × F	0.323

*Based on data from Insel, P. 1974. Maternal effects in personality. Behav. Genet. 4:133-144.
†*PG*, paternal grand-; *MG*, maternal grand-; *F*, father; *M*, mother; *D*, daughter; *S*, son.
‡$p > 0.05$.

previously does not seem to apply at all in the case of the social attitude scale. In respect to this, the male line appears to play as significant a role as the female, though a slightly less influential one. Furthermore, the whole network of relationships is clearly more salient for conservatism, as represented by seventeen significant family correlations as against about six for P and E and three for N. It may well be that we are seeing here the operation of purely environmental factors and, in the case of P, E, and N, the outcome of some kind of complex gene-environment covariations or interactions. The data do not permit any conclusions as to this possibility. However, one point seems important: the strong maternal effect in respect to variables that might seem, on a priori grounds, to have some genetic components. This agrees with the results of studies using the Bernreuter scale, reviewed previously. Coordinately, Insel's data show a

lack of strong maternal effect in respect to data from tests measuring dimensions that could hardly be other than environmental in etiology. Thus whatever else Insel's study may show, it is one that must give us some food for thought. We will now discuss briefly some of the work done specifically on social attitudes and interests. Most of this predates Insel's study by several decades and is perhaps mainly of historical interest. However, we consider it here to emphasize the point that behavior genetics must deal not only with traits whose transmission over generations follows a Mendelian pattern, but also with the transmission of traits through a cultural mode.

Attitudes and interests. Most of the early work done on attitudes and interests shows much the kind of family uniformity found by Insel for social conservatism. Kulp and Davidson (1933), for example, found sibling correlations for social attitudes ranging from

Table 15-4. Family correlations for several attitude scales*

	Attitude to		
	Church	War	Communism
Parent-child	0.63	0.44	0.56
Sibling	0.60	0.37	0.48
Father-son	0.59	0.40	0.40
Mother-son	0.57	0.44	0.61
Father-daughter	0.64	0.44	0.62
Mother-daughter	0.71	0.46	0.51
Father-mother	0.76	0.43	0.58

*From Newcomb, F., and G. Svehla. 1937. Intra-family relationships in attitude. Sociometry 1:180-205.

0.29 to 0.6, with a mean of 0.4. Kirkpatrick and Stone (1935) found for religious attitudes the following correlations: mother-daughter, 0.53, and father-son, 0.33. Children tended to be more similar in general to the mother than to the father. This tendency was found again by Kirkpatrick (1936), using an attitude-to-feminism scale, although all correlations were somewhat lower for religious attitudes. Curiously enough, the author found that greater intimacy between a parent and a child actually tended to lower the correlation between them for the attitude-to-feminism scale. Correlations between mothers and children of 0.4 as opposed to father-children correlations of less than 0.3 have also been reported by Peterson (1936) using the Purdue Attitude Scales. Newcomb and Svehla (1937), with the Thurstone scales for attitudes to the church, to war, and to Communism, obtained correlations of a fairly large order of magnitude. These data are shown in Table 15-4. Again, this is clear evidence for the cultural transmission of attitudes, although the specific pattern is not at all clear. One interesting aspect of these data is the fairly strong degree of assortative mating for the three attitudes. Given this degree of parental solidarity, it is perhaps no surprise that children come to think the same way.

Several studies have been done on family resemblances in interest patterns. Forster (1931) used twenty-five scales given to 122 father-son pairs. The median correlation turned out to be 0.35, with a range of 0 to 0.49. The highest correlations were obtained on the scales for farmer, advertiser, real-estate salesman, physicist, chemist, and YMCA secretary. Those for personnel manager, accountant, artist, and city school superintendent yielded the lowest coefficients. Another study by Strong (1943) obtained a range of correlations similar to that of Forster's: 0.11 to 0.48, with an average of 0.29 for eleven vocational interest scales. As we shall see later, there is little concordance between the results of these two studies and analogous ones using the twin method.

ADOPTION STUDIES

In several of the adoption studies reviewed in Chapter 14, some data were gathered on some aspects of personality as well as intelligence. Thus Freeman, Holzinger, and Mitchell (1928) located 32 cases out of the 401 in their Home (or fostered) group who had serious behavior problems. The latter ranged from disobedience and stealing to masturbation and property destruction. Their mean IQ was 89. The ratings of the homes into which they had been adopted were not significantly below average. However, Freeman et al. found that 72% of the 32 had "morally defective" true parents. Some examples (condensed) are as follows:

Parent	Child
1. Father a day laborer, worthless, deserted his family; mother is weak in mind and body.	1. Girl, IQ 86, steals, lies, defiant of authority, stubborn, boisterous.
2. Father deserted his family; mother of bad reputation; accomplice in a robbery; poor housekeeper.	2. Girl, IQ 96, steals money, deceitful, destructive of property, mean-spirited, unreliable.
3. Father in jail for bootlegging; one-half Indian; mother immoral; home dirty; general moral tone of home bad.	3. Girl, IQ 71, steals, lies, masturbates, walks streets, unmanageable.

The authors emphasize that not all children with "morally defective" parents had behavior problems. However, they concluded that at least in the case of the group of 32, "heredity played some part in [their] behavior" (p. 207). Freeman, Holzinger, and Mitchell, curiously, seem rather less cautious about genes underlying immorality than about genes underlying intelligence.

Burks (1928) also examined the personality traits of her foster and control groups. Three traits were chosen as reliable from a larger list: cheerfulness and optimism, sympathy and tenderness, and conscientiousness. These were correlated in the two groups with midparental rating by field visitors on "kindliness," "sympathy," and "tact." In neither group was any of the three correlations significant. The Woodworth-Cady questionnaire —a test of 85 items designed to sift out psychotic tendencies and emotional instability —was also administered to some of the children and correlated with a number of home variables, including parental kindliness and culture index. Only one correlation approached statistical significance, that between emotional stability and high score of the home on the Whittier index for the control group. Since this is only one correlation out of eleven, it probably does not reflect anything of importance. Thus Burks concluded that environment was "possibly much more potent" in determination of personality than of intelligence.

The third major adoption study by Leahy (1935) also involved a test of emotional stability, the Woodworth-Matthews, given to control and adopted children. Scores were correlated with midparent Otis, midparent vocabulary, cultural index of home, child training index of home, and occupation of father. Although the pattern of correlations was similar between groups, all of them were low (highest 0.18) and probably nonsignificant. The test scores bore no relation in either group to occupational status of the home.

TWIN STUDIES
Pairs reared together

At least some of the ambiguities in the data obtained by the methods just described are somewhat reduced in the twin studies of personality inheritance. We will first briefly review some of the work done in the 1930s and 1940s. This has involved a wide variety of tests of varying reliabilities and validities. Many of the studies are inconclusive and often contradictory. For a fuller account, the reader is referred to Fuller and Thompson, 1960.

Miscellaneous tests. Mainly three types of study have been carried out: those using self-report inventories, those using performance tests, and those using so-called projective tests. In the first category, the Bernreuter has been used in several studies. Carter (1933, 1935), in two experiments, tested 133 pairs of twins, 55 MZ and 78 DZ pairs. His results for the six scales of the test are shown in Table 15-5. Without exception, MZ pairs showed greater similarity than did DZ pairs. The effect was largely independent of age. However, like-sexed DZ pairs showed much greater similarity than unlike-sexed pairs. Raw intraclass correlations are shown in the table. If corrected for range and attenuation, these rise somewhat in magnitude. The MZ correlation for dominance, for example, goes up to 0.86. The like-sexed DZ coefficient rises to 0.38. Estimates of heritability were not computed by Carter. However, with the formula $2(r_{MZ} - r_{DZ})$ (like-sexed DZ only), these would range from 0.2 for introversion to 0.88 for self-sufficiency. The very low figure for introversion is somewhat surprising, since in the majority of ensuing studies heritability of this dimension has been estimated as a good deal higher.

A partial replication of Carter's study by Portenier (1939) yielded rather different overall results. He compared 12 pairs of sibs with 12 pairs of twins, of whom, however, only 2 pairs were monozygotic. His data indicated that, on five of the six scales, sibs were more alike than twins. However, Portenier's sample size was very small, and, in fact, only two of the twelve correlations reported were statistically significant. Consequently, not much weight can be accorded to his data. Much the same holds true for additional data he reported using the Maller Character Sketches Test, the Allport Ascendance-Sub-

Table 15-5. Twin similarities on scales of the Bernreuter Personality Inventory*

	Number of pairs	Scale					
		Neuroticism	Self-sufficiency	Intro-version	Domi-nance	Self-confidence	Socia-bility
MZ	55	0.63	0.44	0.50	0.71	0.58	0.57
DZ							
Like sex	44	0.32	−0.14	0.40	0.34	0.20	0.43
Unlike sex	34	0.18	0.12	0.18	−0.18	0.07	0.30

*From Carter, H. D. 1935. Twin similarities in emotional traits. Char. Pers. 4:61-78.

mission Scale, and the Meier-Seashore Art Judgment Scale. One possibly exceptional result was obtained on the Strong Masculinity-Femininity Scale. Twins correlated 0.92, siblings, 0.67. These figures accord somewhat with some more recent studies. One point of general interest perhaps arises from Portenier's work. This is that being the same age, genetically related, and living together does not necessarily make individuals alike, even in respect to characters that would seem to be fairly plastic.

Newman, Freeman, and Holzinger (1937) examined the heritability of emotional adjustment as measured by the Woodworth-Matthews test. They found only a small difference between MZT and DZT pairs, yielding a heritability estimate of 0.3 of only borderline significance.

A popular early performance test designed to measure personality in a more objective way was the Will-Temperament Test of Downey. The test emerged from a series of studies on handwriting and muscle recording and is based on the theory that temperament depends on two fundamental factors: (1) the amount of "nervous energy" an individual has and (2) the pattern of discharge, direct or indirect, of such energy. The scale thus attempts to measure three groups of traits: (1) speed and fluidity of reaction, (2) forcefulness and decisiveness of action, and (3) carefulness and persistence of reaction. Each of these is measured by a number of simple tests (Freeman, 1926). Although at the time it appeared, the test appeared to have promise (being fairly objective and derived from at least some kind of theory) it

had poor reliability and validity and is now out of use. Nevertheless, in our view, some of the concepts and methods involved in the test sound surprisingly modern, and it would not seem too difficult to devise some similar test directly related to contemporary theories of stress and informational load. However, the few studies that have explored the heritable basis of temperament as measured by the Downey test have not found evidence for the operation of genetic factors (Newman, Freeman, and Holzinger 1937; Tarcsay, 1939).

In this connection, it is worth noting that there is an extensive older literature on the inheritance of the various components of handwriting, a behavior featuring importantly in the Downey test. At this time, no definite conclusions can be stated. It is clear that results must depend in large measure on the particular handwriting dimensions being studied. Thus Carmena (1935) and Miguel (1935) stressed genetic factors in the "pressure component," though von Bracken (1940) disagreed and favored the inheritance of speed. Writing angle was emphasized by Hermann (1939) and Nicolay (1939). Vandenberg (1966a), in surveying some of this literature, concluded that there was a general trend in the data supporting some degree of heritability of handwriting.

The problem of speed or tempo of reaction is an interesting one and was studied more specifically by a number of early investigators. Frischeisen-Köhler (1933a,b) concluded that it was definitely conditioned by genetic factors. Tests included tapping with finger, foot, and hand at a rate "most agreeable to the subject," and, second, choosing a preferred metronome speed. MZ twins were more alike than DZ pairs, whose members deviated from each other about the same as sibs. Unrelated pairs were most different. Other familial resemblance data supported the twin results. If both parents were "quick," 4% of the children were rated as "slow," 56% as "quick," and 40% as "moderate." If both parents were "slow," 71% of the children were also "slow," none "quick," and 29% "moderate." With both "moderate," 17% of the children were "slow," 17% "quick," and 66% "moderate." Frischeisen-Köhler concluded that this was strong evidence for hereditary transmission and suggested a simple genetic model involving two series of multiple alleles, with dominance of a gene or genes for "quick" tempo.

Indirect confirmation of these results was attempted several years later by Newman, Freeman, and Holzinger (1937). Their procedure involved only measures of maximal rather than "agreeable" tapping rate. Hence their scores may have reflected physiological capacity rather than temperament. Mean intrapair difference in tapping rate for 50 MZ pairs was 19.3 ($r = 0.814$) and for 51 DZ pairs, 29 ($r = 0.689$). However, the difference between these values was not statistically significant.

Along similar lines, but at the verbal rather than at the motor level, Carter (1939) found, for speed of word association, correlations of 0.53 for MZ and 0.44 for DZ twins. Sorensen and Carter (1940) subsequently obtained corrected correlations of 0.52 for 38 MZ pairs and 0.3 for 34 DZ pairs. Cattell and Malteno (1940), studying associational fluency, obtained similar results. Needless to say, although such results are in the right direction for a genetic hypothesis, the heritability estimates that emerge are quite low and would probably be nonsignificant.

Somewhat related to the traits of tempo and association fluency is perseveration. This refers to the tendency for individuals to continue in some repetitive performance. Usually, simple motor tasks are used such as writing letters or numbers in a prescribed

order that is changed from time to time by the experimenter. Perseverative tendencies are usually measured by the degree to which a preceding task interferes with a subsequent one. The greater the lag in shifting, the higher the perseveration score. The two studies that have examined this trait in MZ and DZ twins have, however, obtained directly contradictory results (Yule, 1935; Cattell and Malteno, 1940).

The final aspect of personality studied by early behavior geneticists is that assessed by projective tests. The latter involve some indeterminate stimulus, such as a picture or an inkblot, which subjects are asked to describe. The assumption is that they will project their own personality dispositions into their descriptions. Various elaborate scoring systems are used for assessing the latter. The most popular projective test has been the Rorschach Inkblot Test, which is still used fairly widely by clinicians more than fifty years since its initial development. A number of studies have explored the possibility of Rorschach dimensions being heritable. However, on the whole, the data are quite ambiguous and certainly offer no real support for a genetic hypothesis. Much the same applies in the case of studies using another projective test, the Szondi test, in which a proclivity to a certain form of mental disorder is supposedly indexed by the preference a subject has for a picture of a patient with that disorder. Primary references to this literature may be found in Fuller and Thompson (1960) and Vandenberg (1966a).

It will be clear from the preceding discussion that the early twin work on personality yielded information of very limited value. Especially during the 1950s, however, there occurred some important advances in personality test construction and theory, particularly in respect to the methods of factor analysis. These advances generated a large number of sophisticated and elaborate twin studies that we will now consider. The reader is warned in advance, however, that the degree of ambiguity has not been much reduced. For the most part, the studies carried out have focused on a relatively small number of well-standardized tests. We will center our discussion around these.

Cattell's personality factors. One of the most prolific workers in the area of personality study has been Cattell. He has used three types of assessments to locate and identify basic dimensions of personality: (1) life-record behavior in situ (L data), (2) questionnaire responses (Q data), and (3) objective test behavior (T data). To indicate the scope of his work, it may be noted that for T data alone, he and colleagues have developed well over 2,000 separate tests (Cattell and Warburton, 1967). From correlations both within and between the three kinds of data, Cattell has developed "purified" measures that assess most directly the basic factors. He and others have used these instruments to study the inheritance of personality. We will attempt to summarize the results of four major studies. Although they involved somewhat different measures and samples of slightly different ages, all dealt

with all or some of the sixteen personality factors derived from Q data. A summary of these factors is supplied in Table 15-6. The descriptions of each are taken from a number of authors (Cattell, Blewett, and Beloff, 1955; Gottesman, 1963; Canter, 1973). Table 15-7 presents F and h^2 values obtained in the four studies referred to previously. Values of h^2 were calculated by the equation

$$h^2 = \frac{\sigma^2_{DZ} - \sigma^2_{MZ}}{\sigma^2_{DZ}}$$

or

$$h^2 = \frac{r_{MZ} - r_{DZ}}{1 - r_{DZ}}$$

It is clear that the results of the four studies do not yield any very coherent clues as to what dimensions of personality are heritable. Out of 52 F values, we find only 14 significant; of 52 h^2 estimates, only 5 are above 0.5. Even factor B, general intelligence, shows moderate predominance of environmental

Table 15-6. Cattell's personality factors derived from questionnaire data*

Symbol	Description of trait measured	
	Low score	High score
A	Stiff, aloof, reserved, schizothymic	Warm, sociable, cyclothymic
B	Mental defect	Intelligent
C	Neurotic	Stable, ego strength
D	Phlegmatic	Excitable, impatient
E	Submissive	Assertive, dominant
F	Sober, serious, desurgency	Enthusiasm, surgency
G	Casual, undependable, expedient	Superego strength
H	Shy, sensitive, schizothymic	Adventurous, cyclothymic
I	Tough minded, realistic, poised	Tender minded, sensitive, emotional
J	Liking group action, energetic conformity	Individualistic, quiet eccentricity
K	Socialized, trained mind	Boorish, rejecting of education
L	Suspicious, paranoid	Trusting
M	Unimaginative	Imaginative
N	Naive, gullible	Shrewd
O	Confident, adequacy, placidness	Guilt proneness
Q1	Radical	Conservative
Q2	Group dependent	Self-sufficient
Q3	Uncontrolled, lax	Controlled, willpower
Q4	Relaxed, composed	Tense, excitable, somatic anxiety

*Tests are available for different age levels and come in different versions. Some do not involve all factors listed above.

Table 15-7. Summary of F and h^2 values obtained for Cattell's factors by four different investigators*

Factor	Cattell, Blewett, and Beloff (1955)[†]		Vandenberg (1962)		Gottesman (1963)		Canter (1973)[‡]	
	F	h^2	F	h^2	F	h^2	F	h^2
A	1.08	0.08	1.30	0.23	1.11	0.10	1.02	0.02
B	1.91§	0.47			1.05	0.05	1.84§	0.46
C	1.60§	0.37	3.20§	0.69	1.03	0.03	1.03	0.02
D	1.35	0.26	0.93	0	0.62	0		
E	0.90	0	0.97	0	1.44	0.31	1.10	0.10
F	1.47	0.32	1.45	0.31	2.29§	0.56	0.98	0
G							0.77	0
H	1.34	0.26	0.93	0	1.62	0.38	1.48	0.38
I	1.47	0.32	0.97	0	1.07	0.06	2.89§	0.66
J	1.57§	0.36	1.54	0.35	1.41	0.29		
K	1.39	0.28	1.06	0.06				
L							1.74§	0.43
M							1.47	0.33
N							0.84	0
O					1.85§	0.46	1.89§	0.47
Q1							0.89	0
Q2					2.28§	0.56	1.26	0.21
Q3	1.08	0.07	1.87§	0.47	1.53	0.12	0.80	0
Q4	1.56§	0.36	2.08§	0.52	0.53	0.06	0.77	0
N								
MZ	104		45		34		39	
DZ	30		37		35		44	

*Based on data from Gottesman, I. I. 1963. Heritability of personality: a demonstration. Psychol. Monogr. **77**:1-21; Vandenberg, S. G. 1966. Contributions of twin research to psychology. Psychol. Bull. **66**:327-352; and Canter, S. 1973. Personality traits in twins. In Canter, S., ed. Personality differences and biological variations: a study of twins. Pergamon Press Ltd., London.
[†]h^2 estimates based on raw variances and calculated by Vandenberg (1966a).
[‡]h^2 estimates based on intraclass correlations and calculated by Thompson.
§Statistically significant F ratios.

contribution in two studies and almost complete environmental determination in the third study. Other notable features we may mention are the following:

1. There occur some sex differences within the studies in respect to heritabilities. However, these are inconsistent and impossible to interpret.

2. In one study (Canter, 1973) it was possible to divide the twin sample (MZ and DZ) into pairs separated more than five years or less than five years. In the case of MZ pairs, those longer separated show significant intraclass correlations for five factors; those sepa-

rated for shorter times, significant correlations for nine factors. The corresponding figures for DZ twins are zero and two, respectively.

3. In the same study, three second-order factors were derived from the correlations between the first-order factors, neuroticism, anxiety, and extraversion. F values were computed separately for these. None was statistically significant, however.

4. Three of the samples were drawn from U.S. populations, one from Scotland. However, there does not appear to be any more consistency within the U.S. samples than be-

tween any of these and the Scottish sample.

5. Cattell, Blewett, and Beloff's study represented an illustration of how Cattell's MAVA technique (Chapter 12) could be applied to the study of personality. Consequently, he used, in addition to twins, samples of sibs reared together, unrelated individuals reared together, and unrelated reared apart (i.e., the general population). For all twelve factors, there was a linear increase in variance as genotypic similarity and communality of environment decreased. Because of problems of unreliability of G_1, G_2, E_1, and E_2, Cattell et al. were not able to shed much light on the relative contribution of these components, separately or in interaction, to the total variance of each factor.

The preceding results relied on questionnaire data. In a second study, Cattell, Stice, and Kristy (1957) attempted to examine the heritability of ten factors derived from "objective test data" (T data). These factors do not necessarily match the factors previously described, though it is assumed that they might bear some relationship to each other. The names of each of the factors are indicated in Table 15-8. Each is measured by three or four short tests. The specifics of the latter take some effort in tracking down, since in Cattell's program tests and factor names appear to change somewhat over the years. However, an example can be given as follows (Cattell and Warburton, 1967):

Universal Index No. 22: corticalertia

Defined as "speed in basic, simple, nervous responses resembling hyperthyroidism . . .," this index predicts "pilot success."

Measures
1. Ideomotor speed
   a. Fast psychomotor speed on simple tasks (M.I. 6): for example, S is shown a sequence of letters and instructed to write an X under every P and a Z under every A; speed is stressed.
2. High ratio speed to accuracy in open-pencil mazes (M.I. 120): for example, S must trace a pencil through a simple line maze without touching or crossing the sides; time limit of 15 seconds; errors scored for touching or crossing sides or breaks in the tracing.

MAVA was again used, samples consisting of MZ and DZ twins, sibs together and apart, unrelated together, and a group representing the general population. Reliabilities (corrected for test length) for the ten factors ranged between 0.1 and 0.77, with the majority falling below 0.4. Results for the MZ and DZ twins only are shown in Table 15-8. Three of the ten F values are significant, with heritabilities of 0.67, 0.47, and 0.45.

Table 15-9 shows complete results from the MAVA applied to each factor. The correlational and variance terms are not absolutely derived but are the estimates that seemed to yield the greatest degree of "internal consistency." Four consistency principles were explicitly stated by Cattell et al., and sets of MAVA equations chosen that best satisfied them. Other such choices must be made, based mainly on inspection rather than definite statistical criteria. These ambiguities somewhat weaken the force of conclusions drawn from the data by the authors, but let us look at them.

Table 15-8. F and h^2 values for ten personality factors as measured by objective tests*

Universal Index number and name	F	h^2
16. Assertiveness	1.10	0.04
17. Inhibition	1.34	0.26
19. Critical practicality	0.51	0
20. Comention (gregariousness)	0.88	0
21. Exuberance	2.77†	0.64
22. Corticalertia (see text)	1.88†	0.47
23. Neural reserve vs. neuroticism	1.82†	0.45
26. Self-sentiment control	1.04	0.04
28. Asthenia	0.82	0
29. Immediate overresponsiveness	1.04	0.04

N

MZ	104 pairs
DZ	30 pairs

*From Vandenberg, S. G. 1966. Contributions of twin research to psychology. Psychol. Bull. **66**:327-352; based on data from Cattell, R. B., G. F. Stice, and N. F. Kristy. 1957. A first approximation to nature-nurture ratios for eleven primary personality factors in objective tests. J. Abnorm. Soc. Psychol. **54**:143-159.
†Statistically significant F values.

Table 15-9. Within and between genetic and environmental variances for ten personality factors*

Components of variance	Personality factors									
	16	17	19	20	21	22	23	26	28	29
r_{bebh}	—†	−0.51	−0.28	−0.53	0.23	−0.8	−0.56	−0.98	−0.91	—†
r_{wewh}	−0.85	0.10	−0.20	−0.60	−0.10	0.5	−0.50	−0.60	−0.30	0.50
σ^2_{we}	10.50	2.60	1.30	16.70	5.20	5.8	2.20	2.20	2.60	3.10
σ^2_{wh}	0.06	1.80	1.10	24.00	3.10	1.2	1.60	1.20	1.10	0.50
σ^2_{be}	3.20	3.20	1.60	4.60	0.40	2.3	5.80	3.10	3.70	3.40
σ^2_{bh}	0	1.00	0.50	21.10	2.40	0.9	1.00	0.60	0.71	0

*Modified from Cattell, R. B., G. F. Stice, and N. F. Kristy. 1957. A first approximation to nature-nurture ratios for eleven primary personality factors in objective tests. J. Abnorm. Soc. Psychol. 54:143-159. Copyright 1957 by the American Psychological Association. Reprinted by permission.
†No solution.

In the first place, out of ten personality factors, four (16, 20, 21, and 22) show within-family environmental variances decidedly higher than between-family components. The remaining six factors show the opposite, although it is doubtful if they are, in fact, significantly different. In the case of No. 16, assertiveness, both hereditary variances are of negligible size. The same applies, perhaps surprisingly, to No. 22, corticalertia. For the other two, No. 20 (comention) and No. 21 (exuberance), hereditary influences are of comparable magnitude to environmental. Thus the authors' conclusion that their results favor the importance of within-family rather than between-family happenings is perhaps overly strong.

Second, it may be seen from Table 15-9 that out of the eighteen correlational terms shown, fourteen are negative, ranging from −0.1 up to −0.98. Cattell et al. educe from this a "law of coercion to the biocultural mean." By this they mean that hereditary deviants for any personality trait are constrained by environmental influence toward more socially acceptable moderate behavior. If we examine the nature of the traits showing the highest r_{bebh} values (No. 26 and 28), this has some plausibility. Thus No. 26, self-sentiment control, is described as being "associated with strength of will and self-examination . . . it appears the most important factor in determining school achievement."

Likewise, No. 28, asthenia, "combines sociability and emotionality with some evasiveness of reality. It has complaisant 'social climber' qualities. . . ." However, it is a little puzzling why for No. 21, exuberance, hereditary variation should be extended by societal influences.

A third point relates to the fact that, in most cases, r_{wewh} values are less than r_{bebh} values. Cattell et al. take this to mean that, although the family also tends to discourage deviants, it does this much less than society does. That is, the family is more permissive. In fact, for two factors, inhibition and corticalertia, the family and society appear to work in strongly opposite directions. In another, exuberance, the situation appears to be reversed, so that society encourages variation, and the family constrains it. It is hard to know exactly what such data mean.

The final conclusion is the simplest and perhaps the strongest one. The data show fairly convincingly that environment has a much stronger influence on personality than heredity. There appears to be one curious exception. This is No. 20, comention, interpreted variously as "gregariousness," "honesty," and "acceptance of social and ethical values." On the face of it, it seems like one of the least likely candidates for traits with a genetic basis. Because of this anomaly, the authors suggest that a reinterpretation of it will now be necessary.

Table 15-10. *F* ratios and heritability estimates in three twin studies of the MMPI*

Scale	Gottesman (1963)		Gottesman (1965)		Reznikoff and Honeyman (1967)		Combined data	
	F	*h*²†	*F*	*h*²†	*F*	*h*²†	*F*	*h*²†
Hypochondriasis	1.19	0.16	1.01	0.01	2.33‡	0.57	1.21	0.17
Depression	1.81‡	0.45	1.82‡	0.45	1.62	0.38	1.53‡	0.35
Hysteria	0.86	0	1.43	0.30	2.70‡	0.64	1.27	0.21
Psychopathic deviate	2.01‡	0.50	1.63‡	0.39	1.54	0.35	1.39‡	0.28
Masculinity-femininity	1.18	0.15	1.41	0.29	2.37‡	0.58	1.10	0.09
Paranoia	1.05	0.05	1.61	0.38	1.78	0.44	1.27	0.21
Psychasthenia	1.58	0.37	1.46	0.31	0.82	0	1.52‡	0.35
Schizophrenia	1.71	0.42	1.49‡	0.33	1.40	0.27	1.36‡	0.27
Hypomania	1.32	0.24	1.15	0.13	1.65	0.39	1.21	0.17
Social introversion	3.42‡	0.71	1.49‡	0.33	2.02	0.51	1.59‡	0.37
N								
MZ	34		68		18		120	
DZ	34		82		18		132	

*Modified from Thompson, W. R., and G. J. S. Wilde. 1973. Behavior genetics. In B. Wolman, ed. Handbook of general psychology. Prentice-Hall, Inc., Englewood Cliffs, N.J.

†$h^2 = 1 - \dfrac{1}{F}$.

‡Statistically significant *F* values.

We will leave the work of Cattell et al. without further comment. It will be clear to the reader that the MAVA method involves many problems, not the least of which relates to the statistical criteria used in drawing inferences from the data.

Minnesota Multiphasic Personality Inventory (MMPI). As we noted earlier, the MMPI is considered by authorities to be one of the better devices for measuring personality mainly because its items have been developed by empirical criterion keying. Three major studies have examined heritability of MMPI performance using twins. Two of these were carried out by Gottesman (1963, 1965) in Minneapolis and Boston, the third by Reznikoff and Honeyman (1967). Table 15-10 summarizes the results of these three studies, showing *F* ratios and heritability estimates obtained in each case. In addition, *F*'s for the combined data have also been computed (Thompson and Wilde, 1973). It is clear that the congruence between studies is not very obvious, except insofar as all three

find most scales to have very low heritabilities. (We will explicate more fully this lack of congruence later in the chapter.) The differences may, of course, be real and represent different gene pools and environments being sampled. Again, they may be partly a function of sample sizes. To take account of the latter possibility, we have presented results for data combined over the three studies. As shown, five scales turn out to have statistically significant *F* ratios: depression, psychopathic deviate, psychasthenia, schizophrenia, and social introversion. It will be noted, however, that the heritabilities of these are all low, the highest being 0.37 for the introversion scale. As we shall shortly, this particular trait has come in for special study, but the results have not always been consistent.

A second point emerging, at least from the Gottesman studies, is that there are differences between the sexes. In the Minnesota study, females show a significant *F* on only one scale, social introversion. But three

Table 15-11. Statistically significant MZ-DZ differences (either variance differences or intraclass correlation differences) for scales of the California Psychological Inventory*

	Nichols (1966)		Gottesman (1966)	
	Scale	h^2	Scale	h^2
Males	Dominance	0.52	Dominance	0.51
	Capacity for status	0.32	Self-acceptance	0.48
	Responsibility	0.39		
	Socialization	0.44		
	Tolerance	0.42		
	Achievement-via-independence	0.35		
	Intellectual efficiency	0.43		
	Acquiescence	0.42		
	Social desirability	0.32		
Females	Good impression	0.25	Dominance	0.42
	Communality	0.32	Sociability	0.56
	Psychological-mindedness	0.22	Self-acceptance	0.42
			Responsibility	0.50
	Flexibility	0.40	Self-control	0.47
			Achievement-via-independence	0.46
			Intellectual efficiency	0.46
Total sample†	Sociability	~0.34	Dominance	0.49
	Social presence	~0.39	Sociability	0.49
	Self-acceptance	~0.30	Self-acceptance	0.46
	Sense of well-being	~0.28	Social presence	0.35
	Self-control	~0.36	Socialization	0.32
	Achievement-via-conformance	~0.36	Good impression	0.38
	Value orientation	~0.35	Value orientation	‡
	Extraversion-introversion	~0.39	Extraversion-introversion	0.49
	Rigidity	~0.35		
TOTAL N				
MZ		498		79
DZ		319		68

*Based on data from Nichols, R. C. 1966. The resemblance of twins in personality and interests. Natl. Merit Scholarship Corp. Res. Rep. **2**:1-23; and Gottesman, I. I. 1966. Genetic variance in adaptive personality traits. J. Child Psychol. Psychiatry **7**(36):199-208.
†h^2 is given here as simple average of male and female h^2 values and is thus only an approximation (Nichols' data only).
‡No single h^2 computed.

scales have significant F's for males. This is reversed in the Harvard study, in which only one scale is significant for males and four for females.

Gottesman (1963), in one study, also performed an analysis of profile similarity for MZ and DZ twins. Profiles were slightly more alike for MZ pairs, but the trend is a very weak one.

California Psychological Inventory (CPI). This is another of the better personality inventories available (Anastasi, 1968). It is, in fact, derived from the MMPI, from which half of its 480 items are taken. However, it was developed specifically for use with normal populations and involves, in its eighteen scales, a predominance of normal personality dimensions. Some examples are "dominance," "self-acceptance," "responsibility," "achievement-via-independence," and "introversion-extraversion." Three of the scales are validity scales. Some of the criteria used

in developing the majority of the scales were course grades, social class membership, and participation in extracurricular activities.

Two workers, Nichols (1966) and Gottesman (1966), have used the CPI in twin studies (see also Loehlin and Nichols, 1976). Their basic results are presented in Table 15-11. Considering males first, we find that the two studies agree on only one trait: dominance. The heritability estimates are almost identical. In females, there is no agreement at all, except in respect to scales neither author found to be significant. For the total sample, however, there is agreement for five scales, positively, as well as agreement on a number of scales neither author found to have significant heritable components. Thus, for the CPI also, we do not find very consistent results. In fact, Nichols calculated correlations between h^2 values in his and Gottesman's studies, for males and females separately on the eighteen scales. The correla-

Table 15-12. Some of the variables that reflect the personality dimension of introversion-extraversion*

Variable	Introversion	Extraversion
Neurotic syndrome	Dysthymia	Hysteria, psychopathy
Body build	Leptomorph	Euryomorph
Intellectual function	Low IQ/vocabulary ratio	High IQ/vocabulary ratio
Perceptual rigidity	High	Low
Persistence	High	Low
Speed	Low	High
Speed/accuracy ratio	Low	High
Level of aspiration	High	Low
Intrapersonal variability	Low	High
Sense of humor	Cognitive	Orectic
Sociability	Low	High
Repression	Weak	Strong
Social attitudes	Tender-minded	Tough-minded
Rorschach test	M% High	D High
T.A.T.	Low productivity	High productivity
Conditioning	Quick	Slow
Reminiscence	Low	High
Figural-aftereffects	Small	Large
Stress reactions	Overactive	Inert
Sedation threshold	High	Low
Perceptual constancy	Low	High

*From Classification and the problem of diagnosis, by H. J. Eysenck, in the Handbook of abnormal psychology: an experimental approach, edited by H. J. Eysenck, © Pitman Medical Publishing Co. Ltd., 1960, Basic Books, Inc., Publishers, New York.

Table 15-13. Heritability estimates for the trait of introversion-extraversion in a number of representative studies*

Author	Test	N MZ	N DZ	r_{MZ}	r_{DZ}	h^2
Carter (1935)	Bernreuter	55	44	0.57	0.41	0.22
Eysenck (1956)	Factor II (52 tests)	26	26	0.50	−0.33	(0.62)
Shields (1962)	Test based on Maudsley Personality Inventory	43	16	—	—	0.67†
Vandenberg (1962)	Thurstone F Sociable	45	35	0.50	−0.06	0.47
Freedman and Keller (1963)	Bayley IBP responsive to people			All r_{MZ}'s higher		
Gottesman (1963)		34	34			
	MMP1 O			0.55	0.08	0.71
	HSPQ Q2			0.60	0.15	0.56
	HSPQ F			0.47	0.12	0.56
	HSPQ H			0.38	0.20	0.38
Wilde (1964)	Amsterdam Biographical Questionnaire (E score)	88	42	0.37	0.35	0.03
Gottesman (1966)	CPI sociability	79	68	$F = 1.97$		0.49
	CPI self-acceptance			$F = 1.85$		0.46
	CPI social presence			$F = 1.55$		0.35
Nichols (1966)	CPI factor II (extraversion)	498	319		males	0.46
					females	0.32
Partanen, Bruun, and Markkanen (1966)	Bruun Scale	157	189	0.51	0.26	0.41
Vandenberg (1966b)	Myers-Briggs	40	27	$F = 1.84$ ($p < 0.05$)		0.46
Vandenberg (1966b)	Comrey Shyness	111	90	$F = 1.94$		0.48
	Stern need affiliation	50	38	$F = 1.54$		0.35
Scarr (1969)		24	28			
	ACL need affiliation			0.83	0.56	0.61
	ACL counseling readiness			0.56	0.03	0.55
	Fels friendliness			0.86	0.36	0.78
	Fels social apprehension			0.88	0.28	0.83
	Observer rating likableness			0.93	0.82	0.61
Canter (1973)	Eysenck Personality Inventory (E scale)	40	45	0.34	0.29	0.07
	Cattell's 16 personality factors (E factor)	39	44	0.43	0.08	0.38
Eaves and Eysenck (1974)	Eysenck Personality Inventory	451	257	—	—	0.48

*Modified from Scarr, S. 1969. Social introversion-extraversion as a heritable response. Child Dev. **40**:823-832; Thompson, W. R., and G. J. S. Wilde. 1973. Behavior genetics. In B. Wolman, ed. Handbook of general psychology. Prentice-Hall, Inc., Englewood Cliffs, N.J.
†As analyzed by Jinks and Fulker (1970).

tions were for males, -0.22, and for females, -0.24. However, such correlations may overestimate the disagreeement if we consider the fact that a large number of h^2 estimates are derived from nonsignificant MZ-DZ differences and hence may be regarded as no different from zero. In other words, although there is only slight agreement in locating heritable personality characters, there is a good deal of agreement in respect to those which have no heritability at all.

The discussion to this point perhaps suggests that workers in the area should look more closely at two aspects of personality that may be related. These are introversion-extraversion and what might be called sociability. A number of studies have focused directly on these.

Introversion-extraversion. We have already described briefly the dimension of introversion-extraversion in connection with the family-resemblance studies that have examined it. We noted also that first Jung, on largely clinical grounds, and later Eysenck, mainly on the results of factor analytic work, considered it to be a major personality dimension, somewhat equivalent in importance to abstract intelligence on the cognitive side. To give readers who are biologists some impression of the variety of empirical variables it is supposed to underlie, we refer them to Table 15-12, taken from Eysenck (1960). Other variables besides these have also been implicated, notably "social popularity" and "general social liking." What is perhaps most impressive is the heterogeneity of behaviors involved. It seems unlikely that environmental influences could operate so as to produce communalities between such seemingly different variables. It is perhaps for this reason that a number of workers have attempted to explore its hereditary basis using twins. Apart from Nichols and Gottesman, at least ten studies have been published on the topic. We will first offer a general summary (Table 15-13) of most of the results obtained to date and then explore in more detail some of the more interesting aspects of individual studies. A wide variety of tests and subject samples is represented. The

range of heritabilities runs from a low of 0.03 to a high of 0.83. Out of the 23 h^2 estimates given, 12 are below 0.5. However, only 6 are below 0.4. It should also be mentioned that the estimate of 0.03 given by Wilde (1964) is based on separated and nonseparated MZ and DZ pairs (see Chapters 12 and 14). It was a curiosity of his data that while separated MZs were much less alike than MZs reared together in respect to extraversion, separated DZs were much more alike. Thus lumping all MZs and DZs together reduces the difference between them. However, if we derive a heritability estimate only from MZTs and DZTs, we obtain a figure of 0.48, which is more in line with the results of other workers. This is not, of course, to say that Wilde's results are wrong. But they are rather unusual, since they imply, in effect, that living in the same family makes MZs *more* alike, but makes DZs *less* alike. Wilde (1970) later attempted to study conformity behavior experimentally in MZ and DZ twins but was not able to establish a well-defined difference between them. However, the possibility that his initial findings have some generality should make us cautious in interpreting twin data for any personality traits. Such an admonition is reinforced by the only other study of the group involving separated twins, that of Canter (1973). In this case, on the EPI scale, both MZ and DZ twin pairs separated for more than five years were more alike than MZs and DZs separated for less than five years. This is partly similar to Wilde's data. Thus h^2 for all MZs was 0.17; for all DZs, zero. But, for Cattell's 16 PF scale, MZs and DZs separated less than five years correlated 0.29 and -0.65, respectively, yielding $h^2 = 0.57$ or 0.29 (depending on whether we take the negative correlation at its face value or as zero). MZs and DZs separated for more than five years correlated 0.85 and 0.5, respectively, yielding $h^2 = 0.70$.

Sociability. We will deal with this dimension only briefly, since, although intuitively it sounds sensible, it is as hard to define exactly as introversion-extraversion. In fact, if we look back at the data dealing with the latter, we can readily see that the two may overlap to a large degree. An extravert is traditionally thought of as someone who is highly sociable, and an introvert as someone rather asocial. The work we have already cited appears to bear this out. MZ twins are more alike than DZ twins on a number of defined social dimensions, defined, at least, in terms of various questionnaire items. It must be recognized, however, that the manner in which responses to such items translate into real behavior in the real world is not very well known.

The reader may consult some of the tables previously discussed to locate sociability traits showing some heritability. There are, of course, many negative results as well. Thus Vandenberg, Stafford, and Brown (1968) reported no significant MZ-DZ differences on four measures of social intelligence involving interpretation of facial expressions. However, Canter (1973) found significant heritability for sociability and for impulsivity as measured by items generally reflecting components of extraversion. This finding applied, however, exclusively to females. Furthermore, for twins separated more than five years, h^2 for sociability was a high 0.88; for those less than five years, only 0.35. Impulsivity, on the other hand, had zero heritability for the same two groups.

If aspects of sociability are heritable, we might expect its apparent opposite—hostility—to be so as well. Canter (1973) has been one of the few workers to explore this question, using the Foulds Hostility Scale. Nine characteristics comprising the general domain of hostility are assessed. In this study, for the whole group of twins, only two yielded significant F ratios and heritabilities. For "self-criticism," h^2 was approximately 0.63, and for the presumably closely associated trait of intropunitiveness, h^2 was 0.58. In the male sample, only intropunitiveness showed significant heritability, and self-criticism only in the female sample. In twins separated less than five years, both measures plus "general hostility" show moderate heritability. In twins separated more than five years, only for self-criticism are MZ and DZ

pairs different. However, on the whole, twins separated longer also tended to be older. Consequently, the two variables, as acknowledged by Canter, are somewhat confounded.

On the whole, the kind of evidence just summarized must be considered weak as far as a hereditarian position toward sociability. Only a few aspects of this general trait (if it is such) appear to have any genetic basis, and these may well be only random sampling effects. Canter is fully aware of this possibility but rightly emphasizes that the information on MZ and DZ similarity is valuable from either a hereditarian or an environmentalist viewpoint.

Attitudes and vocational interests. Work on familial resemblances in attitudes has been going on since the 1930s. On the whole, we find a good deal of uniformity between relatives. We have already reviewed these data. One study of more recent vintage by Eaves and Eysenck (1974) used twins to examine the possible heritability of two attitude dimensions described by Eysenck: "radicalism" versus "conservatism" and "toughmindedness" versus "tender-mindedness." Compared were 450 MZ pairs with 257 DZ pairs in respect to their responses to a 60-item Public Opinion Inventory and an 80-item Personality Inventory. The authors attempted to test between two broad models, a "simple environmental" model and a "simple genetic" model. The expectations derived from each are set out in Table 15-14. Analysis of the variance components showed a significant departure from the expectations generated by the environmental model. The genetic model, on the other hand, fit reasonably well. About half the individual variation in respect to the traits studied was estimated as being genetic. There was a significant tendency for "tough-mindedness" to be associated with high scores on the psychoticism factor. The authors speculated that a tendency to psychopathy might possibly underlie this relationship.

We will now consider the matter of vocational choice. It may strike the reader as unlikely that occupational preferences might be

Table 15-14. Expectations of mean squares on two simple models*

| | Expectation | | | |
| | Environmental | | Genetic | |
Model	E₁ (within family)	E₂ (between family)	G₁ (additive genetic)	E₁ (within family)
Mean square/product				
Between MZ pairs	1	2	1	1
Within MZ pairs	1			1
Between DZ pairs	1	2	¾	1
Within DZ pairs	1		¼	1

*From Eaves, L. J., and H. J. Eysenck. 1974. Genetics and the development of social attitudes. Nature **249:**288-289.

thought of as having a heritable base. Nonetheless, it is also obvious that genetically determined dispositions, perhaps accentuated by environment, might well incline an individual to prefer and excel in certain kinds of activities rather than others. To give a simple example, a boy who is physically strong and well coordinated could well develop a liking for sports, a category choice that, in turn, could exclude certain other kinds of activities. Likewise, other preferences, both positive and negative, could well come about through the more subtle interactions of personality and cognitive variables. Hence such a line of research is not altogether implausible. We will consider several major studies that have attempted to explore the problem.

Two of the studies by Carter (1932) and Vandenberg and Kelly (1964) may be considered together, since the latter authors have explicitly done this. Both used the Strong Interest Inventory administered to sets of MZ and DZ twins. Basically this test requires the subject to express his liking of, disliking of, or indifference to several hundred discrete items dealing with various activities, which are not confined to occupational categories but encompass virtually all aspects of daily living. They were chosen on the basis of criterion keying to different occupational groups. Thus if an individual shows the same pattern of preferences as, say, doctors are known to have, then it is assumed that this individual is disposed to this occupational category, other things being equal.

Table 15-15. Heritability of four groups of occupational interest scales as measured by the Strong Interest Inventory*

| | Average h^2 estimates | |
Category of scale	Carter	Vandenberg
8 science scales	0.42	0.39
4 language scales	0.20	0.10
4 people scales	0.41	0.20
5 business scales	0.24	0.26
All 21 scales	0.36	0.28
N		
MZ	43	43
DZ	43	34

*Modified from Vandenberg, S. G., and L. Kelly. 1964. Hereditary components in vocational preferences. Acta Genet. Med. Gemellol. **13:**266-277. Based on data from Carter, H. D. 1932. Twin similarities in occupational interests. J. Educ. Psychol. **23:**641-655.

The test has been revised over the years and in the version used by Vandenberg and Kelly included forty-seven occupational scales. The version used by Carter involved only twenty-three, however, of which only twenty-one could be directly compared between the studies. This comparison is shown in Table 15-15. Significance levels are not available for the Carter data. However, in Vandenberg's study, twelve individual scales yielded significant MZ-DZ differences. Of these, the four showing highest heritability values were "physicist," "mathematician," "osteopath," and "dentist." The high value for the average "science" scale in both studies reflects the same trend. However, whereas Carter and Vandenberg appear to agree in general, they do not agree very well in particular. The rank-order correlation for h^2 values for all twenty-one scales across studies was only 0.16. Thus the evidence is not exceptionally strong for inferring a hereditary basis for vocational choice.

Another study was attempted by Vandenberg with a different inventory, the Minnesota Vocational Interest Inventory (Vandenberg, Stafford, and Brown, 1968). This covers rather more particular nonprofessional occupations. The author found significant MZ-DZ differences for the following: baker, carpenter, hospital attendant, IBM operator, retail salesclerk, truck driver, truck mechanic, warehouse worker, and for a general factor score called "machine repairs."

Vandenberg (Vandenberg et al., 1968), in surveying some of these results, appears to be rather sanguine as to the importance of genes in determining vocational preferences. He may be right. Nevertheless, we should consider the results of the last study by Nichols (1966). This used a different measure, the Vocational Preference Inventory developed by Holland (1968) and intended to represent the personality through occupational preference rather than vice versa. Some of his more bizarre findings are set out in Table 15-16. It is not intended to be implied that all pairs of occupations are equivalent. Nonetheless, the differences are rather subtle in some cases, particularly considering the testees were all of high school age. Nichols' conclusions about the validity of the MZ-DZ difference (apart from what they really indicate about genotypic influences) are rather less optimistic than Vandenberg's.

The general results of most of the studies on personality have been validated in the large-scale study of Loehlin and Nichols (1976). They used more than 800 MZ and DZ twin pairs whose zygosity was mostly established by questionnaire data. Some of their major conclusions are worth presenting here.

1. MZs tend to be more alike than DZs over a wide range of personality variables. "Typical intraclass correlations" for MZ

and DZ pairs were as follows: for inventory scales, 0.50 and 0.28; for self-concept clusters, 0.34 and 0.10; for ideals, goals, and vocational interest clusters, 0.37 and 0.20; and for activities clusters, 0.64 and 0.49.

2. MZ-DZ differences did not appear consistently greater for some personality domains than others.

3. Few, if any, sex differences were found.

4. Data concerning early experiences of twins were gathered both from their parents and from the twins themselves. On the whole, all twins were treated more alike than nontwin siblings, and MZs more alike than DZs. This tendency held across homes in which "impact" of home environment was high or low. That is to say, active attempts to impose standards of behavior did not make twins any more alike.

5. Similarity of early treatment did not, however, predict similarity of personality when measured in adolescence either in MZ or DZ twin pairs or in the combined sample. Thus we have the paradox that although environment must contribute to personality similarities in twins, and, in fact, MZs are treated more alike than DZs by parents, degree of later similarity in personality is not predicted by similarity of early environment.

Table 15-16. Vocational Preference Inventory items with large and small zygosity differences (i.e., MZ versus DZ intraclass correlations)*

Large MZ-DZ differences (t ratio > 2)	Small MZ-DZ differences (t ratio < 1)
Police judge	Judge
Juvenile delinquency expert	Director of welfare agency
Poet	Composer
Post office clerk	Bank teller
Free-lance writer	Newspaper editor
Crane operator	Tool designer
N	
MZ	516
DZ	334

*Based on data from Nichols, R. C. 1966. The resemblance of twins in personality and interests. Natl. Merit Scholarship Corp. Res. Rep. **2**:1-23.

Thus, as Loehlin and Nichols put it (1976, p. 94), environment must be operating "in remarkably mysterious ways."

Monozygotic twins reared apart

We have already discussed some work on twins separated for various periods of time (Canter, 1973). However, these ranged in age from 16 to 55 years. It appears that few, if any, of these, had been separated so early that they had not been exposed to a common environment for a good portion of their lives. For more relevant information, we must then turn to the four major MZA studies reviewed in Chapter 14. Of these, three directly involved measures of personality.

Newman, Freeman and Holzinger (1937). It may be recalled that besides intelligence tests, these investigators used the following measures of personality for their MZA sample of 19 pairs: the Woodworth-Matthews, the Kent-Rosanoff Free Association Test, the Pressey Test of the Emotions, and the Downey Will-Temperament Test. None of these, however, was very well standardized. Only the Woodworth-Matthews and the Downey were given to the MZT and DZT pairs. As we saw earlier, the differences in pair similarity between these two groups were inconsequential. The only real comparison made between these and the MZAs was in respect to Woodworth-Matthews performance. Correlations (raw intraclass) were as follows: MZA, 0.583; MZT, 0.562; and DZT, 0.371. It seems unlikely that there is a significant effect here, although an estimate of her-

itability would fall around 0.3, a figure not atypical in this area. It is not possible to interpret results for the other tests given.

Shields (1962). Shields used a test devised by Eysenck, the Self-Rating Questionnaire, designed to measure mainly extraversion and neuroticism. In addition, the author attempted to make a general overall assessment of personality resemblance in the separated group. The major results are shown in Table 15-17. It can be seen that again, as in the results of Newman, Freeman, and Holzinger (1937), of Wilde (1964), and of Canter (1973), MZAs are more alike than MZTs. All correlations for MZ pairs are statistically significant. Those for DZ pairs are not. It is of interest that there is virtually no difference between separated and nonseparated DZ pairs as indicated in the table. It is also surprising to find that the DZ pair separated latest, at nine years (Ds 7), showed the largest difference in extraversion and neuroticism of any of the DZA group. An extensive biometric genetic analysis of Shield's extraversion and neuroticism data has been carried out by Jinks and Fulker (1970). We will return to this at the end of the chapter.

Juel-Nielsen (1965). The 12 pairs of twins in this sample were given the Rorschach Test, the Rapaport Word Association Test, and a number of intensive psychiatric interviews. Unfortunately, no detailed statistical analysis of these results was attempted. Juel-Nielsen concludes that, in the main, "environmental factors play a decisive role in the development of personality," although

Table 15-17. Extraversion and neuroticism in MZA twins as compared with MZT, DZT, and DZA pairs*

	Number of pairs	Extraversion		Neuroticism	
		Mean intrapair difference	Intraclass correlation	Mean intrapair difference	Intraclass correlation
MZA	43	2.5	0.42	3.1	0.53
MZT	42	2.7	0.61	3.0	0.38
DZA	8	3.9	−0.17	4.6	0.11
DZT	17	5.0		3.7	

*Modified from Shields, J. 1962. Monozygotic twins brought up apart and brought up together. Oxford University Press, London.

Table 15-18. Factors from tests with high- or low-hereditary variance*

Factor	Description	Loading
From high-heredity clusters		
I	Optimistic, poised	0.61
	Socially outgoing	0.42
	Has own opinion	0.42
	Quick thinking	0.40
	Socially dominant	0.40
II	Likes to take things slow	0.38
	Gets going easily	0.36
	Adventurous, self-confident	0.29
	Socially dominant	0.29
III	Controls impulses	0.55
	Gets angry, frightened, upset	−0.52
	Good social adjustment	0.37
	Likes to work with tools	0.33
IV	Likes to work with tools	0.36
	Intellectual interests	−0.33
	Likes physical work	0.28
	Impatient, impulsive	0.28
From low-heredity clusters		
I	Impulsive, outgoing	0.56
	Enjoys group activity	0.54
	Seeks social stimulation	0.53
	Shy	−0.42
	Good memory for recent events	0.41
	Vigorous, active	0.40
	Considers self fortunate	0.37
II	Feels restricted by adults and rules	0.44
	Nervous, suspicious, jumpy	0.42
	Shy	0.31
III	Enjoys team sports	0.55
	Vigorous, active	0.37
	Likes racing, boxing, betting	0.35
	Considers self fortunate	0.33
IV	Likes school and teachers	0.44
	Good behavior	0.39
	Gets along well with parents	0.31

*Modified from Loehlin, J. C. 1965. A heredity-environment analysis of personality inventory data. In S. G. Vandenberg, ed. Methods and goals in human behavior genetics. Academic Press, Inc., New York.

"there are remarkable points of similarity between the personality structures of the twins" (p. 135). Such a conclusion seems modest enough; the interested reader is referred directly to Juel-Nielsen's extensive case histories.

HEREDITARY VERSUS ENVIRONMENTAL PERSONALITY FACTORS

An interesting study by Loehlin (1965a) has attempted to separate out hereditary from environmental personality factors. Basically, the procedure involves two factor analyses, one of variables known to be high in genetic variance, the other of variables known to be high in environmental variance. Two tests were used, the Thurstone Temperament Survey (TTS) and the Cattell Junior Personality Quiz (JPQ). Whether an item was high or low in genetic variance was determined by a separate twin analysis. In the end, fifteen clusters of items showing the highest MZ-DZ difference and fourteen clusters with little or no MZ-DZ difference were chosen for factoring. Four factors were extracted from the correlations of clusters in each of the two matrices. The results are shown in Table 15-18. As Loehlin points out, there is a good deal of similarity between the two sets of factors. Both the factor I's seem to have to do with extraversion-introversion. Factor III in the high-heredity and factor II in the low-heredity groups seem to reflect an emotional adjustment dimension; factors IV and III both have to do with physical activities. The other two factors, however, do not match particularly well. In general, Loehlin suggested that there was a difference in flavor in the two sets. Thus factors of the first set seem to have to do more with the individual himself, those in the second set with how he reacts to his environment. The hereditary extraversion factor, for example, involves what a person *brings* to group activities. The environmental extraversion factor emphasizes what the person *gets* from group activities. Somewhat the same applies to the other two matching sets. It is difficult to interpret Loehlin's conclusions. Inferences drawn from tables of factor loadings are based more on personal judgment than on definite scientific criteria. However, the kind of methodology used in this study seems generally promising and could well be extended much further.

A later twin study was carried out on this problem by Horn, Plomin, and Rosenman (1976) using the California Psychological Inventory. From the eighteen scales that make up the test, they eliminated all items scored on more than one scale and all items that were unreliable. An item was then classified as genetic if the MZ correlation for it exceeded the DZ by at least 0.1 and as environmental if the MZ-DZ correlation difference was smaller than 0.1. Heritabilities computed on these purified scales were found to range from 0 to 0.78. Such a result is in contrast to the earlier results of Nichols (1966) who found moderate-sized heritabilities for most of the eighteen unpurified scales. Thus the procedures of Horn et al. represent a gain in precision in assessing genetic influences on personality dimensions. The authors next went on to apply factor analysis separately to the genetic items and to the environmental items. The factors they educed for the two sets were quite different, a result in marked contrast to that of Loehlin (1965) described previously. The first two genetic factors (together accounting for 38.8% of the variance) were "Conversational Poise" and "Compulsiveness." None of the environmental factors extracted accounted for much of the variance among environmental items. The first two were "Confidence in Leadership" (4% of variance) and "Impulse Control" (2.9% of variance). Horn et al. interpret the high dimensionality of the environmental matrix to mean that "there are no broad systematic environmental influences operating to make the children within a family similar in personality" (p. 26). This conclusion appears very similar to that of Loehlin and Nichols (1976) in respect to early experiential effects as discussed previously.

PERSONALITY DEVELOPMENT

We discussed in a previous chapter the strategy of analyzing characters at an early age, before environment has had much of a chance to have an impact. A number of workers have applied this strategy to personality study with the hope of finding some basic dimensions of temperament, perhaps genetically based, out of which the adult personality gradually emerges.

An important study along these lines was carried out by Rutter, Korn, and Birch (1963). These investigators obtained ratings by parents of 3 MZ and 5 DZ twin pairs and 26 sib pairs on seven "primary reaction patterns" (PRPs): (1) activity level, (2) rhythmicity or regularity, (3) approach or withdrawal, (4) adaptability, (5) intensity of reaction, (6) threshold of responsiveness, and (7) quality of mood. The highest genetic loadings were found for 1, 3, and 4 in this list, the lowest for rhythmicity or regularity. Evidence of genetic influences for any PRP was stronger in the first year than in either the second or third. Over the three years, not all PRPs were equally stable. Curiously enough, the two seemingly most heritable were also the most unstable. Thus as we pointed out earlier, particularly in our discussion of intelligence, high heritability need not by any means imply unchangeability.

Clearly, in view of the very small sample sizes, the work of Rutter *et al.* is interesting and suggestive but hardly definitive. More explicit attempts to establish hereditary precursors of personality have been made by a number of subsequent workers. Freedman (1965) used the Bayley Infant Behavior Profile to study 9 MZ and 11 DZ pairs in the first year of life. In twelve major areas, represented by twenty-two variables, significant MZ-DZ differences were found on ten of these as follows: (1) social orientation, (2) object orientation, (3) goal directedness, (4) attention span, (5) activity, (6) reactivity, (7) fearfulness, (8) looking, (9) sound producing: banging or other, and (10) manipulation. Of interest also is that in at least two indicators of social orientation, onset and frequency of smiling and fearfulness of strangers, MZ twins were much more concordant than DZ twins. Smiling, at least, appears to be a strongly endogeneous response, since it appears in blind infants (rubella cases) in response only to auditory and tactile stimuli.

Further studies by Brown, Stafford, and Vandenberg (1967), by Wilson, Brown, and Matheny (1971), and by Owen and Sines (1971) confirm the impression that at least two factors in early temperament are important, one having to do with general vegetative functions—for example, feeding, sleeping problems, or irritability—and the other

having to do with sociability. It is, of course, difficult to pin these down exactly. Neither are the variables supposedly reflecting them equally stable over time. However, some general proclivities appear very early and do seem to predict later behavior fairly well. Thus Cohen et al. (1972) have presented suggestive evidence that rating on a First Week Evaluation Scale (FES) measuring such behaviors as "vigor," "calmness," "attention," and others predicts, rather well, later developmental maturity in test taking and natural situations and also IQ (Stanford-Binet) at around 4 years. Again, unfortunately, this study involved very small samples and cannot be considered conclusive.

In general, the line of research just outlined seems a most promising one. The evidence so far obtained suggests quite strongly that there are very basic differences in temperament between individuals at least as early as birth and that these dispositions, in interaction with environmental influences, come to constitute the structure of adult personality. More work in this important area is presently being carried on, particularly at the University of Louisville.

A developmental approach has also been taken by Dworkin et al. (1976), although relating to somewhat older age levels. In a longitudinal assessment of a sample at adolescence and later at adulthood, they found that, for MMPI and CPI scales, patterns of heritability change markedly. That is to say, various scales showed significant heritabilities at the two ages, but these were not the same ones. In fact only one, CPI dominance, showed definite stability across both ages. Thus we again find that the manner in which genotype and environment interact during development must be highly complex. Since the Dworkin et al. study strikes at some fundamental issues in behavior genetics, it will be discussed again in Chapter 18.

GENETIC TRANSMISSION OF PERSONALITY

In view of the ambiguity of the data on the simple heritability of personality, it is not surprising that almost nothing is known or has even been suggested about possible genetic mechanisms involved. There is only one notable exception that we will discuss: the biometric genetic analysis by Jinks and Fulker (1970) of the Shields' data on extraversion. We have referred already to the latter in our discussion of MZA studies.

The basic samples consisted of MZA, MZT, and DZT twin pairs of both sexes. Analysis of the variances of these groups indicated the following: (1) no evidence of correlated environments; (2) some evidence of genotype-environment interaction, indicating that more introverted genotypes are more susceptible to environmental influence, a finding somewhat consonant with the greater conditionability of introverts; (3) a simple polygenic model does not fit the data due to peculiarities in the samples; and (4) common family environment is of little importance for this trait.

Jinks and Fulker also analyzed Shields' neuroticism data to which a fairly simple genetic model could be adequately fitted. However, we will discuss this work in Chapter 16.

SUMMARY

Compared to the work on inheritance of intelligence, the study of human personality has been somewhat ambiguous in its conclusions. This is mostly the fault of the rather poorly standardized procedures so far developed for assessing personality. The endless proliferation of tests and the continual invention of more exotic trait dimensions are hardly likely to take us very far. However, if we were to name one character that seems to hold special promise, we would have to single out the dimension of extraversion-introversion. Most studies have found it to be strongly heritable. Likewise, the large volume of experimental work done on it by Eysenck and colleagues and others suggests that it may well have some fundamental importance. Again, of course, as with intelligence, one may wonder if it is unitary or multidimensional, and, if the latter, whether some dimensions have a better genetic definition than others. No doubt such problems will gradually be solved by future workers. For a fruitful theoretical and empirical analysis of these, the reader is referred to Loehlin and Nichols (1976).

16

Mental illness: major psychoses

In this chapter and Chapter 17, we will present data on the heritability and genetic transmission of some of the more common forms of mental illness, particularly the major psychoses and psychoneuroses. We will also consider some miscellaneous categories such as alcoholism and psychopathy and touch briefly on a few types of mental deficiency. Since our 1960 text, *Behavior Genetics*, the field has burgeoned, and there are now available a number of excellent summarizing works in English and other languages. For treatments much fuller than we can offer, the reader is referred to some of the following: Zerbin-Rüdin (1967, 1969), Rosenthal (1970), Slater and Cowie (1971), Kaplan (1972), Gottesman and Shields (1972, 1976), and Bleuler (1974).

Let us first examine briefly the nature of mental illness in general. The problem of definition continues to be difficult. From a purely operational point of view, we can follow Scott (1958) in listing the following criteria for mental illness: (1) exposure to psychiatric treatment, (2) maladjustment, (3) labeled mentally ill, (4) awareness of illness and seeking assistance, (5) diagnosed by tests as mentally ill, and (6) absence of mental health. It will be clear that these criteria either separately or collectively are not entirely satisfactory. For example, a person may seek help from a priest or a minister rather than from a psychiatrist. Again, being "labeled" mentally ill may say more about the predilections of the labeler than the labelee. Likewise, there are clearly degrees of assistance seeking; often this may not reflect anything seriously wrong. Finally, there are probably many individuals who, although, in fact, seriously mentally ill, are not themselves aware of it. The community of clinical

psychologists and psychiatrists is, of course, fully aware of such problems and divides rather sharply as to what kinds of basic models are most appropriate for conceptualizing mental illness. Some of these we will consider in a moment.

Accepting hospital admission and residence as a simple criterion, we illustrate the general magnitude of the problem of mental illness by noting that, in the United States in one year (1965), 135,476 first admissions to state and county mental hospitals were recorded. Of these, the largest group was composed of schizophrenics, with alcoholics running a close second (Sahakian, 1970). If we take into account the number already resident in these hospitals, we can easily see that the sheer monetary cost, direct and indirect, is extremely high. One source has estimated that for schizophrenia alone, it amounts to as much as $14 billion annually* (Mosher and Feinsilver, 1971). There is therefore a serious need to direct resources toward the understanding of how mental illness comes about and what kinds of treatment procedures are most appropriate. How this can best be accomplished constitutes no small problem. As noted before, there have been a number of different models of mental illness put forward, and some of these differ rather radically. Let us now look at some of them.

MODELS

Sahakian (1970) has delineated five types of model that have been formulated by various workers. The first of these, and perhaps the most conventional, is the so-called *medical model*. This viewpoint takes mental illness to be a complex disease entity with one or many somatic bases. These may or may not be genetically determined. Treatment and cure lie properly with the profession of medicine and the various paramedical disciplines. A second point of view, put forward most forcibly by Szaz (1960), appears to be grounded in an extreme belief in *cultural relativism*. The model holds that there is no such entity as mental illness, which is simply a term used by the establishment to describe the behavior of persons who are, in some sense, outside this establishment. Such people, because they are, for various reasons, in opposition to the dominating customs and mores, face serious *problems of living.* Thus they are deemed to be sick, not because they actually are sick (as with someone having pneumonia, for example), but because their viewpoints and attitudes are deviant. Szaz has a good many followers. However, we cannot count ourselves among them.

A third model that still claims many adherents is the *dynamic model* derived originally from Freud. The central features of this viewpoint are a belief in human instinctual drives whose aims can come into conflict with the dictates of culture as transmitted through the parents and, second, a strong emphasis on the importance of experiences encountered during early development. On the face of them, these two propositions are quite biological and, in fact, find a good deal of support in scientific data. Nevertheless, in practice, the methods and orientation of psychoanalysis have not been framed in a manner that is scientific and open to proof or disproof. The concepts and language used by its proponents have often been obscure and mystical, and the therapy of overly long duration and limited in its success.

A fourth model of mental illness is the *learning theory model.* This also has many adherents, notably Eysenck, Wolpe, and others (Eysenck, 1959). Basically, the model posits that mental illness consists simply in any behaviors that are regarded by the individual or by society as undesirable. These behaviors have been acquired by learning and hence can be removed by the standard procedures involved in elimination of any response. Note that the disease and the symptoms are syn-

onymous. Note also that there is very low priority given to predisposing genetic factors.

The final model may be called *sociocultural*. It is somewhat similar to the second one we listed, insofar as it suggests that mental illness is largely relative to cultures, just as are (supposedly) law, morality, and all notions of what is good or bad, true or false, beautiful or ugly. However, it recognizes pragmatically that the members of a society must not deviate too widely from its behavioral norms for that society to function well. This viewpoint is, of course, a favorite of many social anthropologists and sociologists. Again, it completely omits any reference to the place of genetic factors in mental illness.

It will be obvious, perhaps, that the preceding models listed are not necessarily exclusive of each other. The differences between them lie mainly in the degree of priority each assigns to various aspects of mental illness. Thus we would agree that the symptomatology of a disorder like schizophrenia may be conditioned by the peculiarities of the society in which it occurs. Likewise, we would agree that early experiences are probably of critical importance and that such experiences must be stored by some types of learning mechanisms. However, our main commitment in this book is to examine the part played by genes, and we will, accordingly, emphasize this aspect.

CLASSIFICATION

If one allows that mental illness is a real, albeit complex, entity, then it is clear that there is a necessity for developing some kind of classificatory scheme or taxonomy. Two such schemes are currently available. One is that of the World Health Organization, the other that of the American Psychiatric Association (Sahakian, 1970). Though differing slightly in detail, both separate the following: (1) disorders caused by or associated with impairment of brain function resulting from infections, trauma, or toxic agents; (2) mental deficiencies or mental retardation, both endogenous or exogenous; and (3) psychogenic disorders without any clearly defined structural causes. This last group includes involutional psychoses, the various schizophrenias, the affective psychoses, paranoid states, various ill-defined psychotic reactions, psychoneuroses, personality disorders, alcoholism, sexual deviations, and sundry others. We shall be primarily concerned with this last group, although we shall also touch on some of the other categories. It should be noted that the specific types of disorders within the broad classes are delineated both nosologically and etiologically. The difference between schizophrenia and manic-depression, for example, is defined largely by the characteristic symptoms involved in each entity. On the other hand, a disorder like alcoholism, often grouped under "personality disorders," is defined in terms of an agent that may produce abnormal symptoms, although it is also at least implicit that the propensity to consume this drug is, itself, dependent on some ulterior cause. Although we suggested in 1960 that classification was likely to become more and more etiological, this does not appear to have happened to any great extent. If anything, the opposite has occurred, at least in treatment, as epitomized by the popularity of behavior modification techniques with their emphasis on removal of present symptoms regardless of their cause.

Many readers will be aware that, although the available classificatory schemes are as explicit as possible, in practice their application to individual patients is often precarious. This is one major source of difficulty in psychiatric genetics. Thus in estimating the heritability of schizophrenia, for example, the value generated may go up or down depending on the looseness or strictness of the diagnosis of this disorder. National differences are thought to exist in this respect. Thus in the United States, the diagnosis of schizophrenia is made much more often than in the United Kingdom. However, in recent years some gains have been made in respect to uniformity. Wing et al. (1967) have developed a procedure known as the Present State Examination (PSE) on which independent diagnosticians are able to achieve a very high degree of agreement. In addition, Shields and Gottesman (1972) compared diagnoses of a con-

secutive series of twins at the Maudsley-Bethlem hospital made by six clinicians from three countries (United Kingdom, United States, and Japan) and with rather widely differing backgrounds. They found that even between the two judges with the most disparate views, there was good agreement on what was normal, what was nonschizophrenia, and what was schizophrenic or schizophrenic-like. The main source of disagreement was the cut-off point used in the latter category before a case was called definitely schizophrenic rather than doubtful or borderline. A six-judge consensus diagnosis following "middle-of-the-road" lines was easily reached and gave highly reliable results. Shields and Gottesman thus express considerable optimism about the future of diagnostic procedures and conclude not only that their present unreliability is greatly exaggerated but also that they can readily be improved still further.

METHODS

The methods that are commonly used to study the inheritance of mental illness are essentially similar to those used to study any other kind of behavior. We have discussed them most explicitly in Chapter 12. They are the familial resemblance or proband method, the twin resemblance method, and the adoptees method. An additional method, not commonly used in other behavior genetic studies, is the long-term follow-up of children at "high risk," for example, children born of one or two schizophrenic parents. This latter approach has great merit but, like any long-term follow-up study, is costly and time consuming. In general, insofar as mental disorders represent discrete phenotypic categories rather than regions on a continuous scale, the previous methods are applied somewhat differently than those dealt with in the two previous chapters. Since the various methods used here were discussed in Chapter 12, they will not be reviewed here at any length. We shall only consider briefly their development in the history of psychiatry and their applications to the study of mental illness.

The twin method, as noted earlier, traces to Galton and was used in the field of psychiatry by many workers after him. Most of these early reports, however, dealt with isolated instances of twins, both of whom showed mental illness of some kind. A number of critics rightly pointed out that such cases could not be combined to make an unbiased statistical sample. Luxenburger (1928) proposed, instead, the use of the so-called consecutive admissions procedure that involved the use of an uninterrupted series of hospital admissions over a period of time. All twins in such a series were located and constituted the final sample for which concordance rates would be computed. This procedure would give a truer picture of the place of hereditary factors, since there would be no bias operating in favor of describing concordant pairs only, as there was in the early case-report publications. An antiheredity bias would, of course, probably locate more discordant cases. In recent years, great care has been given to sampling, with efforts made to include not only concordant and discordant pairs, but also appropriate proportions of male and female pairs, of age groups, and of identical and fraternal pairs.

The second major method of psychiatric genetics is the proband method. This was first used by the Munich school under the direction of Rüdin. It has been used since the 1920s by investigators in the field in all parts of the world, notably Germany, Scandinavia, Iceland, the United States, and the British Isles. Essentially, this method involves the selection of a random or representative sample of cases (probands, or propositi) in a particular psychiatric category. All relatives of these cases are then located for diagnosis and study. When this has been done, observed incidence rates among them are computed, and these compared with rates in relatives of a randomly chosen control group from the general population. Various special problems such as changes in risk with age and the inclusion or exclusion of the probands themselves in the random population sample have required special attention. There are many pitfalls in this methodology; many of these have been carefully outlined by Rosenthal in two very important articles (Rosenthal, 1961, 1962).

The final two methods are clear enough. All variants of the adoptees method simply attempt to examine the fate of children born of mentally ill parents and raised by normal parents or, in the rather rare cases, of children of normal parents raised by abnormal foster parents. Likewise, the study of children at risk simply attempts to follow closely the development of children born of and raised by parents one or both of whom have been diagnosed as mentally ill.

In broad outline, these are the main methods employed in the field of psychiatric genetics. Their exact usage and specific details involve many more problems than can be dealt with adequately at this point. The reader is referred back to Chapter 12 on general methodology in human genetics. We will attempt in the course of our presentation here to point up particular problems of special interest.

SCHIZOPHRENIA

The term schizophrenia refers to a broad array of severely abnormal symptoms that have as their central core a disorganization of basic personality. This disorganization may involve hallucinations, a dissociation between ideas and emotional responses, and numerous other such symptoms. Historically, the syndrome was first delineated by Connolly in England and later designated by Morel in 1860 as *dementia praecox*, meaning, literally, premature loss of mind. Morel felt it was mainly determined by heredity. Kraepelin in 1896 extended but also specified the definition to include several other disease entities that he felt had basically the same common denominator, these being catatonia, hebephrenia, and paranoia. Finally, the term dementia praecox was replaced by the designation schizophrenia by Eugen Bleuler in 1911. His modifications to the diagnosis were influenced in part by the insights of both Freud and Jung. Following Kraepelin, Bleuler accepted four subtypes of schizophrenia: simple, hebephrenic, catonic, and paranoid. Although the first of these is somewhat of a "wastebasket" category, the others refer to fairly definite symptomatologies that underlie the common disintegration of personality. Hebephrenia is usually characterized by shallow and inappropriate affect, general silliness, and incongruous laughing or giggling. It has an early onset. Catatonia involves gross disturbance of motor behavior, ranging from stupor and catalepsy to marked excitation. Paranoid schizophrenia is characterized by delusions with megalomaniacal or persecutory content. Hallucinatory activity is frequently involved. These categories are still recognized today, though sundry other categories have been added, for example, residual schizophrenia, schizo-affective type, schizophreniform attack, plus some childhood varieties. For the most part, however, genetic work has dealt with the schizophrenic syndrome as a whole, although a few workers have attempted to analyze some of the subcategories separately. It

should be noted also that, although it has usually been treated as a discrete entity, some investigators have felt that it can appropriately be treated as a continuous or quasi-continuous variable.

Another distinction often made is between process and reactive schizophrenia. The former is thought to be more severe, more chronic, more progressive, and more genetically determined. The latter is considered to have more rapid onset, usually as the result of some environmental stress, and better prognosis. The usefulness of this distinction is not, however, universally accepted (see discussion by Abelin, 1972).

It is clear that we are dealing with a complex entity whose definition and etiology are still poorly understood. However, a synoptic view of schizophrenia that usefully summarizes the main aspects of the phenomenon has been put forward by Meehl (1962) and is summarized in Fig. 16-1. The main features of the schema are the following: an individual

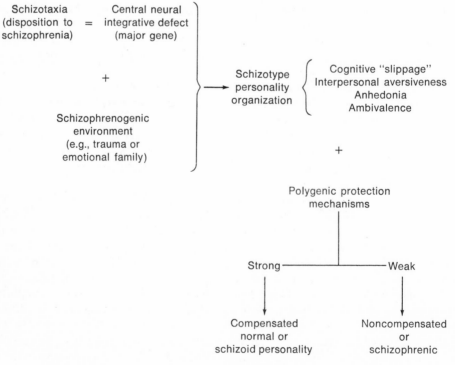

Fig. 16-1. Schizotaxia, schizotypy, and schizophrenia: Meehl's summary.

can carry a major gene for schizophrenia, but this is not a sufficient condition for developing schizophrenia, though it is a necessary one, and, second, even given the presence of the gene and a poor environment, an individual may still not develop the illness in its full-blown form if he has genes that confer some protection against the deleterious effects of the major gene. This viewpoint was not particularly new when Meehl presented it; Kallman had previously put it forward (Kallman, 1953). However, coming in America at the time it did and from a prestigious clinical psychologist, its impact was considerable. Let us now turn to some basic data relevant to the inheritance of schizophrenia.

Distribution

In the countries in which it has been most intensively studied, that is, the advanced or industrialized countries, there seems little doubt that the disorder of schizophrenia exists, though, as we shall see in a moment, with a variable incidence. However, there is still some doubt as to whether it is truly universal. An influential review by Benedict and Jacks (1954) concluded that it was, basing this conclusion mainly on five studies of Maoris in New Zealand, native Hawaiians, Bantu, native Kenyans, and Gold Coast bushmen. Torrey (1973), reviewing these studies, has concluded that they do not by any means make the case that Benedict and Jacks claimed. Not only were there serious problems of diagnoses (many were done "secondhand"), but there also appears to have been a strong correlation between incidence of schizophrenia and degree of Western acculturation. Thus the question of whether schizophrenia is universal is still an open one. This ambiguity, of course, does not argue against either a genetic or an environmental hypothesis. Conversely, were it established firmly that schizophrenia is universal, this, also, would not resolve the issue. However, if cultures could be found in which it was demonstrably absent, these would furnish most interesting research material. A prior question, of course, is whether there are any societies in which any forms of mental illness are totally absent. The evidence available suggests this is unlikely (Rosenthal, 1970; Dunham, 1976).

In any case, the association between incidence of schizophrenia and degree of civilization is an interesting one and deserves further exploration. The process of civilization does not by any means connote peace and harmony. It may, in fact, involve considerable social stress. Furthermore, the process of adjusting to a new culture may be difficult. In line with this idea, Murphy (1968) has presented data which shows that the incidence of schizophrenia was appreciably higher in immigrant populations to Canada (e.g., Irish, English, German, Italian, and Scandinavian) than in the Canadian-born offspring of these immigrants. Thus there is a good argument to be made for an association be-

Table 16-1. Morbidity risks of schizophrenia in different countries*

Date	Country	Number of studies	Mean risk (percent)	Range (percent)
1928-1950	Germany	8	0.42	0.35-1.40
1929-1962	Switzerland	3	2.38	0.98-2.40
1938-1951	Denmark	3	0.69	0.47-0.90
1935-1956	Sweden	4	1.25	0.68-2.85
1942	Finland	1	0.91	
1953	Formosa	1	0.59	
1964	Iceland	1	0.73	

*Based on data from Zerbin-Rüdin, E. 1967. Endogene Psychosen. In P. E. Becker, ed. Humangenetik, ein kurzes Handbuch, vol. 2. Georg Thieme Verlag KG, Stuttgart, West Germany; and Slater, E., and V. Cowie. 1971. The genetics of mental disorders. Oxford University Press, London.

tween problems of adjusting to the dictates of a culture and the incidence of mental illness in general and of schizophrenia in particular. Data showing a much higher incidence of schizophrenia in lower economic classes (Hollingshead and Redlich, 1957) lend further credence to this idea, particularly since the relation holds most firmly in larger (more than 100,000 population) than in smaller cities (Kohn, 1968).

The morbidity risks of schizophrenia have been calculated in at least twenty studies for seven or more countries. Table 16-1, modified from Zerbin-Rüdin (1967), summarizes the main results. There are at least two notable features of the data. One is that there is a rather large range in incidence figures across countries, going from a low of about three per thousand up to almost three per hundred. This is a tenfold difference. Second, the ranges within countries are also considerable. We are thus faced with an immediate problem at the most basic level of inquiry. Are these differences results of different diagnostic procedures, different sampling methods, or actually differences in the genetic structures of the various populations? No

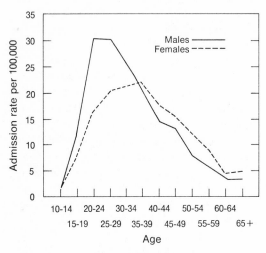

Fig. 16-2. Number of first admissions for schizophrenia to state mental hospitals in the United States. (Modified from Rosenthal, D. 1970. Genetic theory and abnormal behavior. McGraw-Hill Book Co., New York; based on data from Landis, C., and J. D. Page. 1938. Modern society and mental disease. Farrar & Rinehart, Inc., New York.)

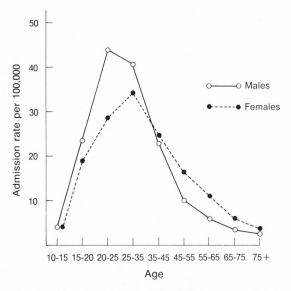

Fig. 16-3. First admissions to mental hospitals in England and Wales during 1952 to 1960. (Based on data from Slater, E., and V. Cowie. 1971. The genetics of mental disorders. Oxford University Press, London.)

definitive answer can be given to these questions. All three possibilities may well be involved. However, it is worth noting that the highest figure of 2.85%, found in Böök (1953), was for a population in a remote part of northern Sweden. This might suggest a possible stress effect elevating the rate. Counter to this is the fact that, in the same sample, Böök found almost no manic-depressive psychosis. This argues somewhat against an environmental hypothesis. Differences in age structures in the populations could be involved. The incidence figures shown have been corrected by Weinberg's procedure, but whether the latter is completely adequate is not entirely certain. A final point is that the studies cited have used three different methods in computing incidence: the normal proband method, the birth register method, and the census method. However, the ranges and median rates do not differ appreciably between them. Slater and Cowie (1971) estimate 0.85 as the best estimate of the general population risk.

Morbidity rates for schizophrenia differ by sex as well as by age. The relationships are shown in Fig. 16-2, based on data of Landis and Page (1938) for state mental hospitals in the United States. Fig. 16-3 presents the results of a more recent survey of first admissions to mental hospitals in England and Wales. Both sets of graphs make essentially the same points: risk increases up to a peak in the late twenties, and the peak risk period is earlier for males than for females. There is no appreciable difference between the sexes as far as overall risk, however.

Again, it is not definitely known why these sex differences are present. They could be results of some hormonal factors, to greater community acceptance of young female as compared with young male schizophrenics, or to differences in the kinds of stresses faced by males and females at different times of life. The matter remains to be resolved.

Risks in relatives

A large number of studies has been carried out on this topic since work was first initiated at Munich in 1916 by Rüdin. Rüdin himself studied 701 sibships with unaffected parents and found the age-corrected risk for schizophrenia in the brothers and sisters of a proband to be 4.48%. The risk for other psychoses was almost as high: 4.12%. If one parent was schizophrenic, the rates were elevated to 6.18% and 10.3%, respectively. If both parents were psychotic or alcoholic, the risk for schizophrenia rose to 22.72% (Rüdin, 1916).

Other major studies of considerable historical importance were carried out by Kallman and associates, initially in Berlin. Some main features of the data may be noted from Table 16-2. The rate among step-sibs of schizophrenics is somewhat higher than the incidence in the general population, but not very much. This indicates that simple association with a schizophrenic (or with the environment that he occupies) is not a very important precipitating factor. Second, the occurrence in sibs of probands is greater than in the parents of probands, a fact that suggests recessivity of any gene or genes involved. Third, having two parents diagnosed as schizophrenic gives a very poor prognosis for the children.

Table 16-2. Expectancy of schizophrenia in relatives of probands*

Relation to proband	Percent expectancy
Step-sibs	1.8
Half-sibs	7.0-7.6
Full sibs	11.5-14.3
Children	
One parent affected	16.4
Both parents affected	68.1
Parents	9.3-10.3
Grandparents	3.9
Grandchildren	4.3
Nephews and nieces	3.9

*From Kallman, F. J., and S. E. Barrera. 1942. The heredoconstitutional mechanism of predisposition and resistance to schizophrenia. Am. J. Psychiatry **98:**544-550. Copyright 1942, the American Psychiatric Association. Reprinted by permission; and Kallman, F. J. 1946. The genetic theory of schizophrenia. Am. J. Psychiatry **103:**309-322. Copyright 1946, the American Psychiatric Association. Reprinted by permission.

Kallman's estimates have been regarded as rather high. To put them in a broader context, we present the data of Table 16-3, which summarizes the results of over sixty-five sets of observations. At least twelve investigators from a number of countries were involved. Expectancies listed are age corrected. Of course, we are looking only at mean values. Actual ranges within each category tend to be rather large. For example, for parents of probands, the percentage expectancies in different studies go from a low of 0.2 to a high of 12; for sibs, from 3.3 to 14.3; and for children of a proband, from 7 to 16.9. Thus there is fairly wide variation,

Table 16-3. Summary of observations of expectancy rates of schizophrenia for relatives of proband cases*

Relationship	Mean percent expectancy of schizophrenia including	
	Certain cases	Probable cases
Parents	4.4	5.5
Sibs		
All	8.5	10.2
Parents normal	8.2	9.7
One parent schizophrenic	13.8	17.2
Children		
One parent schizophrenic	12.3	13.9
Both parents schizophrenic	36.6	46.3
Half-sibs	3.2	3.5
Uncles and aunts	2.0	3.6
Nephews and nieces	2.2	2.6
Grandchildren	2.8	3.5
Grandparents	0.7	—
First cousins	2.9	3.5

*Based on data from Zerbin-Rüdin, E. 1967. Endogene Psychosen. In P. E. Becker, ed. Humangenetik, ein kurzes Handbuch, vol. 2. Georg Thieme Verlag KG, Stuttgart, West Germany; Rosenthal, D. 1970. Genetic theory and abnormal behavior. McGraw-Hill Book Co., New York; Slater, E., and V. Cowie. 1971. The genetics of mental disorders. Oxford University Press, London; and Gottesman, I. I., and J. Shields. 1972. Schizophrenia and genetics: a twin vantage point. Academic Press, Inc., New York.

which may be due to sampling or to differing diagnostic criteria. This makes the data difficult to interpret. However, the *majority* of studies have reported elevated rates in sibs as compared with parents of probands. If the majority is correct (which it may not be), this again argues for a recessivity hypothesis, provided that the parental sample is a representative one. It is of some interest that studies reporting high rates for sibs also find high rates for parents. For fourteen studies, Rosenthal (1970) reported a correlation of 0.78 between these two sets of rates. This correlation is not a function of sample sizes, which are independent of rates in relatives. However, Slater and Cowie (1971) have argued that parents rarely produce children after becoming schizophrenic. This means, essentially, that until the age at which they produced the index case, their own risk was negligible. Taking this lowered risk into account and making a new age correction for parents, Slater and Cowie arrive at an expectancy for parents of 14.12%. This figure is not very different from those for sibs and children.

Assuming that environmental factors are not correlated with genotype, we can use the preceding data to compute heritabilities by means of Falconer's method (Chapter 12). For nine studies in which the general population risks are furnished, heritability values range from 0 to 0.85, using parents, and from 0.51 to 0.83, using sibs. Median values are, respectively, 0.45 to 0.73 (Rosenthal, 1970). These would represent upper bounds.

It will be noted that the rate of schizophrenia in children when both parents are schizophrenic is about three times the rate when only one parent is affected. However, the figure of 36.6% is a good deal lower than that found by Kallman. It was arrived at on the basis of four studies involving 53 cases. As one might suppose, the marriage of two schizophrenics is not a very common event. Thus, the data may or may not be reliable. In any case, if we take the lower figure as the more valid, it is clearly a long way from what would be predicted on the basis of a simple monogenic theory, even allowing for reduced penetrance. Equally, it would be hard to conceive of an environment more unfavorable to offspring normality than one provided by two schizophrenic parents. Thus the data are puzzling. A further point of interest is that the mean hospitalization age was 13.1 years earlier in children than in parents. This could be explained as a regression of children to age of peak risk (though rather a large one), or, alternately, one could suppose that the children had been exposed to rearing conditions much more unfavorable than those of their parents. A detailed analysis of these studies has been made by Rosenthal (1966).

A final point to be made about the family-risk data concerns the incidence in second-degree relatives of probands, that is, uncles, aunts, nieces, nephews, grandparents, and grandchildren. If we take the median risks for sibs or children as a standard, then these estimates for second-degree relatives should be higher than they are, closer to 4% or 5%. Possibly this is due to unreliable data.

Twins

Relatively few twin studies have been carried out on schizophrenia. The first work appears to have been done by Luxenburger over a period of years starting in the late 1920s (Luxenburger, 1928), after which there were reports by investigators from a number of countries carried out up to the present. A simple summary of the main data is given in Table 16-4. It is obvious that there is a good deal of variability between studies in methodology and results. Thus heritability estimates range from 0 to a high of 83% if we take simply the most extreme concordance rates.

Sample sizes are also variable, diagnostic procedures have been different, and sampling has sometimes been of consecutive admissions and sometimes of resident hospital populations. Some investigators have corrected for concordance estimates, and others have not. However, some salient aspects of the data are worth noting.

1. If we consider seven studies with sample sizes greater than 50 pairs (both MZ and DZ together), we find that the median heritability for these is 0.65. Indeed, except for one, all are in the range 0.63 to 0.66. For five studies with sample sizes under 50, the median heritability is 0.24.

2. In several continuing studies in which sample sizes were increased over the years, heritability likewise increased. This applied in the case of Kringlen (1964, 1968) and of Tienari (1963, 1971).

3. With one possible exception (Tienari, 1963), concordance rates have always been found to be higher in MZ than in DZ twins, regardless of diagnostic procedure.

4. As shown in Table 16-5, concordance rates tend to be somewhat higher for female twins, of whom there is a preponderance in most of the studies.

5. Like-sexed DZ pairs show somewhat higher rates than unlike-sexed pairs.

6. MZ samples obtained by sampling resident hospital populations, as opposed to sampling consecutive admissions over a period of years, usually yield higher concordance rates. One of the reasons for this lies in the fact that more severe, chronic cases are likely to be found by using the former method. These characteristically also show less discordance. This point is illustrated by data

Table 16-4. Summary of twin studies on schizophrenia, 1928 to 1972

Number of studies*	Countries	Range of sample sizes (pairs)		Range of concordance (percent)	
		MZ	DZ	MZ	DZ
11	9	16-174	17-517	0-86	2-17

*Some studies were carried out over a number of years and reported seriatim, for example, Luxenburger (1928, 1934) and Kringlen (1964, 1968). These are counted only as one study in the table.

Table 16-5. Concordance by sex and sampling: several studies*

Samples (MZ)	N		Percent concordance for schizophrenia	
	M	F	M	F
Based on consecutive admissions	59	53	46	47
Not so based	55	84	51	71
All studies	114	137	48	62

*Modified from Shields, J. 1968. Summary of the genetic evidence. In D. Rosenthal and S. S. Kety, eds. The transmission of schizophrenia. Pergamon Press Ltd., London.

Table 16-6. Relation between severity of schizophrenia in index case and concordance*

Severity of index case	Severity of co-twin's illness				Percent concordance
	Extreme	Medium	Little or no deterioration	No schizophrenia	
Extreme	29	10	9	0	100
Medium	0	33	20	0	100
Little or no deterioration	0	0	19	54	26

*Modified from Rosenthal, D. 1970. Genetic theory and abnormal behavior. McGraw-Hill Book Co., New York.

from Kallman (1946) as presented by Rosenthal (1961). These are shown in Table 16-6.

Beyond these relatively straightforward aspects of the data are many important points of detail relating to each individual study. Since we cannot cover these here, we refer the reader to some of the sources we have already mentioned. Gottesman and Shields (1972), in particular, have given excellent, short synopses of most of the twin studies carried out. Before leaving this section, however, it seems worthwhile to discuss a little more fully the largest of the programs carried out, one which, in addition, has also yielded perhaps the highest heritability estimates. This is the work of Kallman (1946, 1953).

In his initial publication in 1946, Kallman reported on 174 MZ and 517 DZ pairs. Later on, he added cases to raise the total to 953 pairs. This represents a massive amount of labor. Unfortunately, as some writers have pointed out (Rainer, 1966; Shields, Gottesman, and Slater, 1967), he reported on his

methodology and data in such a terse manner that he left himself open to criticism by a number of writers, notably Jackson (1960). This criticism has centered mainly around the following points: (1) overly loose and possibly idiosyncratic diagnosis of schizophrenia; (2) "contaminated" diagnosis, that is, allowing knowledge of the zygosity of a co-twin to influence his diagnosis; (3) atypicality of the sample; and (4) unorthodox procedures for arriving at age-corrected morbidity risks.

Shields, Gottesman, and Slater (1967) have carefully reexamined Kallman's data and procedures. As they indicate, in 1946 Kallman found an uncorrected concordance rate of 69% for MZ and 10.3% for DZ pairs. These figures, when corrected for age, become 85.8% and 14.7%, respectively. There is some doubt as to exactly how the age correction was carried out. However, Shields et al. are of the opinion that Kallman did not overcorrect for age and that had he included suspected schizophrenics as well as definite

Table 16-7. Incidence of schizophrenia in MZ twins reared apart*

Study	Age at separation	Concordant pairs	Discordant pairs
Kallman (1938)	Soon after birth	1	
Essen-Möller (1941)	7 years	1	
Craike and Slater (1945)	9 months	1	
Kallman and Roth (1956)	Not stated	1	
Shields (1962)	Birth	1	
Tienari (1963)	3 years and 8 years		2
Kringlen (1964, 1968)	3 months and 22 months	1	1
Mitsuda (1967)	Infancy	5	3
Inouye (1972)	Before 5 years	6†	3
TOTALS		17	9

*Based on data from Slater, E. 1968. A review of earlier evidence on genetic factors in schizophrenia. In D. Rosenthal and S. S. Kety, eds. The transmission of schizophrenia. Pergamon Press Ltd., Oxford; and Gottesman, I. I., and J. Shields. 1972. Schizophrenia and genetics: a twin vantage point. Academic Press, Inc., New York.
†Only three pairs showed "complete concordance."

schizophrenics in his computations, he would have arrived at an even higher MZ concordance rate than he did.

Kallman's diagnoses were in the majority of cases (73.4%) of individuals who had received already an independent hospital diagnosis of schizophrenia. An additional number had been inpatients in a mental hospital, and a few others (presumably not hospitalized) were diagnosed by him as definitely schizophrenic. The remainder (23 cases) were regarded as "suspected" schizophrenics. Thus it seems very unlikely that Kallman's definition was so broad and loose as to elevate concordance rates. In any case, it is notable that using only cases already diagnosed by the mental hospital as schizophrenic yields rates of 50% for MZ and 6% for DZ pairs. This is a ratio of 8.3:1. Using all Kallman's cases, the corresponding MZ:DZ ratio is only 7:1. This fact seems to indicate, in addition, that diagnosis was not contaminated. If it had been, we would expect, as criteria became looser and more independent, the MZ concordance rate to rise more than the DZ concordance rate. This does not appear to happen (Shields, Gottesman, and Slater, 1967).

In summary then, most of the methodological criticisms of Kallman's work are probably without much basis. It seems most likely that his high rates were mainly a result

of his use of samples from a resident hospital population. This may mean, of course, that his results are not widely generalizable. But it does not mean that they are untrue.

Separated MZ twins. Unfortunately, we still have very meager information on concordance rates for schizophrenia in twins reared apart. Available data have been summarized by Slater (1968) and by Gottesman and Shields (1972) and are presented in Table 16-7. Curiously enough, the concordance rate of about 65% in the 26 pairs is actually higher than the median rate for MZ twins reared together. It is very much in line, however, with the kinds of estimates found in most of the larger twin studies done (63% to 66%). Thus, assuming the sample is not atypical, the data provide strong support for the genetic model. Furthermore, they suggest that whatever factors make for discordance or concordance, these have little to do with the within-family environment. This brings us to a consideration of another research stratagem: the study of twins found to be discordant.

Discordant MZ twins. As we have shown in the previous section, concordance in MZ twins is far from perfect. Its rate also varies widely, notably between the United States and the Scandinavian countries. This has led some to argue that therefore genetic factors are absent in schizophrenia (e.g., Jackson,

1960). However, it is obvious that imperfect concordance by no means discredits a hereditarian hypothesis. In fact, in consideration of MZ twins discordant for schizophrenia, one can examine two possibilities: one is that there are genetic forms of schizophrenia and environmental forms, that is, phenocopies. The latter are hardly new in genetics, and it is well known now that environmental influences can produce behaviors that closely resemble behaviors otherwise produced by genes. The hypothesis that there might be two forms of schizophrenia, one genetically, the other environmentally, determined, was initially put forward by Rosenthal (1959), although he appears to have abandoned it now (Rosenthal and Van Dyke, 1970). The more cogent possibility, however, is that we distinguish between the schizophrenogenic genotype and its clinical manifestation. Such a distinction is essentially entailed by the genetic concept of expressivity. In building genetic models to account for schizophrenia, it has invariably been necessary to invoke expressivity or manifestation rate, sometimes to such an extent that it becomes a "fudge" factor. The study of discordant MZ twins, however, allows us to give some definite empirical meaning to the notion.

Generally speaking, investigators have focused on two kinds of possible differences in discordant MZ twins: extrapersonal variables relating to the life history of each and representing presumed triggering factors and, second, intrapersonal variables, for example, biochemical, that dispose one member of the pair to schizophrenia more than the other. The two must, of course, be considered to combine in some fashion to produce the discordance.

Thus what we search for are similarities and differences between members of discordant MZ pairs, it being assumed that the former represent genotypic and the latter environmental influences.

A good example of the general method is afforded by the study of Rosenthal and Van Dyke (1970) referred to previously. This attempted to follow up the hypothesis put forward by Essen-Möller (1941) that what

was inherited in schizophrenia was what he called a "characterological defect," which might or might not develop into clinical schizophrenia, depending on various circumstances. Rosenthal and Van Dyke compared 11 schizophrenic index cases with their non-schizophrenic co-twins on the Wechsler Adult Intelligence Scale. They found that co-twins scored higher than index cases on all except one (block design) of the eleven scales. This accords with the general notion put forward by Bleuler and others that schizophrenia involves thought disorder. However, a striking constancy between the index and co-twins was their profiles across scales. Thus, as Rosenthal and Van Dyke state, "the pattern of performance . . . remains essentially the same, despite the impairment of level." Their results are shown in Fig. 16-4. Striking as they seem to be, they are based on a very small sample, one typically involving a predominance of female pairs. Furthermore, there are no control groups representing either the normal general or twin population. Thus it could be that all people show such profiles, or all twins, both, or neither. Consequently, the study is

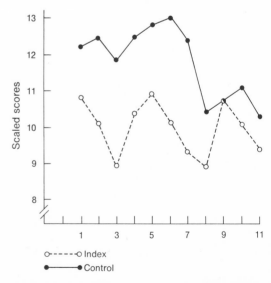

Fig. 16-4. Scores by index and control twins discordant for schizophrenia on the eleven WAIS subscales. (From Rosenthal, D., and J. Van Dyke. 1970. Acta Psychiatr. [Suppl.] **219:**183-189.)

by no means definitive, as the authors themselves have recognized. However, it usefully demonstrates the general research strategy.

In point of fact, it is only recently that intensive work involving adequate control groups has been instituted. Prior to the early 1960s, many individual case histories concerning discordant MZ twins were published, but, considered separately at least, these were not very illuminating. However, investigators gradually became aware of some of the basic methodological problems involved. Apart from the sampling difficulties to which we have just alluded, there is a problem in merely defining what a discordant pair is. Thus a pair of discordant MZ twins before, at, or shortly after the peak age of risk may well become concordant. On the other hand, pairs that are past the age of risk (e.g., 45 years plus) may not constitute very appropriate material for study. It is clear that the discordant MZ twin method is far from straightforward. Some of the problems have been discussed by Kringlen (1964).

In the early 1960s, an intensive program on discordant MZ twins was initiated by Pollin and colleagues at the Twin and Sibling Unit of the National Institute of Mental Health. In one of their earlier reports, Pollin and Stabenau (1968) gathered information on 100 pairs of discordant MZ pairs located through published reports from 1929 on and also through the National Science Foundation Twin Registry. Index cases and controls were compared on twenty-six life-history factors. Some of the factors discriminating best were as follows:

Index twin is:
1. more neurotic as a child
2. more submissive
3. more sensitive
4. more of a "serious-worrier"
5. more obedient and gentle
6. more CNS illnesses as a child
7. more likely to have had birth complications
8. more likely to have suffered asphyxia at birth
9. more dependent
10. less likely to be married
11. less intelligent
12. worse at school

The authors concluded that index cases usually had suffered more stress in childhood, this stress relating less to complex psychological problems and more to problems involving some central nervous system deficiency. Other data obtained by Pollin and colleagues (Pollin, Stabenau, and Tupin, 1965; Pollin et al., 1966) give credence to this idea. Thus the index twins (in their sample) tended consistently to be smaller at birth. This caused or was associated with a lowered physiological competence, which, in turn, produced an experience of the world different from that of their co-twin and, likewise, different relationships with their parents. All these factors would gradually accumulate to heighten the probability of manifestation of the schizophrenic genotype.

Again, however, we must emphasize that their findings are based on a very small sample and are of a rather clinical character involving no adequate control groups. Neither has a definite association between birth weight and schizophrenia been always found by other investigators. Consequently, we must consider their theory unconfirmed, though still highly plausible (Gottesman and Shields, 1976).

Other work by the same group has focused on biochemical variables. In one article, Stabenau et al. (1969), following up some leads in the literature, studied in 16 pairs of discordant MZ twins (age range, 16 to 45 years) the following variables: (1) lactate-pyruvate (L/P ratios), (2) rabbit red cell agglutination titer, (3) serum S_{19} macroglobulin urine 3,4-dimethoxyphenylethylamine (DMPE), and (4) serum protein-bound iodine (PBI). Evaluations of these variables were run blind. They found "normalized" L/P ratios and measures of antirabbit red cell hemagglutinin to be significantly higher in index cases. Unfortunately, at least some of the latter had been on phenothiazine medication at various times prior to the present study. The phenothiazines in fact elevate both of the two variables. Consequently, the validity of their findings in respect to these is in some doubt. PBI levels were found to be significantly lower in the index group. However, both in this and in a larger NIMH twin sample, PBI levels were found to correlate significantly with birth weight, which was lower in the schizophrenic subjects. The authors suggest that low birth weight is, in fact, the cause of the low PBI levels.

In a later article, Pollin (1971) reported on two other classes of biochemical variables with particular relevance to coping with stress: the catecholamines and the adrenal steroids. More importantly, in this study two control groups were added: a group of 4 normal MZA pairs and a group of 3 MZ pairs concordant for schizophrenia. The discordant group consisted of 11 pairs. Altogether, seven catecholamines, 17-OH steroids, and urine volume were measured. However, not all of them were measured in all pairs for various reasons.

Results were as follows: in the first place, for the whole sample, intraclass correlations for catecholamine excretion were positive and significant for all except one of the substances measured (4-hydroxy-phenyl-glycol). Steroid levels correlated zero. Among discordant pairs, however, only four of the cate-

cholamine levels were significantly correlated, and *none* in the concordant pairs. Three catecholamines correlated in the normal control group. Also, in this group, curiously enough, steroid levels correlated 0.910, $p < 0.01$. It is difficult to make much sense out of these data. However, if we follow Pollin and take the findings based on all 18 pairs as being the most robust, we might conclude that catecholamine levels are under genetic control, whereas 17-OH steroid levels are not. Further analysis of between-group means lends some weight to this conclusion. The main comparisons achieving statistical significance (by individual *t* tests only, however, rather than *F* tests) are shown in Fig. 16-5. At least taken qualitatively, the results are of some interest. Generally speaking, for the catecholamines we find that the individuals having the genotype whether manifesting or not manifesting the disorder show elevated levels. This suggests, as Pollin states, that a high level of catecholamines is related to the schizophrenic *genotype* but that elevated steroid values are related to the schizophrenic *phenotype*. This otherwise in-

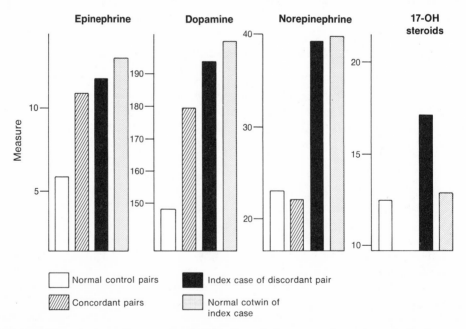

Fig. 16-5. Group means for excretion levels of catecholamines and 17-OH steroids. (Concordant sample not tested for 17-OH steroids.) (Modified from Pollin, W. 1971. Am. J. Psychiatry **128:**311-317.)

teresting idea is undercut, unfortunately, by a number of difficulties. One is the fact that, for norepinephrine, the concordant schizophrenic group is no different from the normal group. Another is the fact that no real differences were obtained for the four other catecholamines measured. Other difficulties are as follows: the ranges within each group are large, the sample sizes were small, the statistical analyses left much to be desired, and there is little information given concerning possible confounding effects of medication or treatment administered to the schizophrenic cases at the time. Consequently, we must view the data of this study with a great deal of caution. It is to be hoped that future studies using this important methodology of studying discordant twins will find ways of coping with the difficulties inherent in it.

Adoption studies

The basic methodology and types of designs falling under this general heading have been discussed in some detail by Rosenthal (1970, 1972). He has identified roughly four strategies which differ somewhat in detail. We will now discuss some of the attempts to apply these.

Extended family or adoptee's family. One of the first workers to use this method was Karlsson (1966) in his extensive investigations of schizophrenic pedigrees in Iceland. He located 8 cases of schizophrenics who had been reared in foster homes. Out of the 29 biological sibs of these index cases, 8 became schizophrenic. Out of the 28 foster sibs with whom the index cases were reared, none was schizophrenic.

The same approach has been used in a more systematic way by Kety et al. (1971), a joint team of Danish and U.S. researchers. Using the thorough registers maintained in Denmark, they were able to obtain a sample of about 5500 adoptees from which they located 33 definitely diagnosed schizophrenic index cases. Chosen next were 33 nonschizophrenic and normal adoptees to match the index cases on twenty variables, for example, time spent with biological parent, socioeconomic class of adoptive parents, and age. All

biological and adoptive relatives of index and control adoptees were located. From the Psychiatric Register, it was then ascertained how many in each group of relatives had at any time suffered from "schizophrenia-spectrum disorders." Results are shown in Table 16-8. The most relevant comparisons are within columns. Significantly more disorders were found in biological relatives of index cases than in those of control cases. The groups were not different in respect to incidences among adoptive relatives. This finding applies somewhat more strongly for the smaller subgroup of cases adopted out at a very early age. The data are clearly in accord with a genetic hypothesis. It would be difficult to explain them, granting their validity, in environmental terms.

Adoptees study. This method starts with parental index cases, diagnosed as schizophrenic, who have given up their children for adoption and then compares the incidence of the disorder in these foster children against a control group. The first systematic study

Table 16-8. Percent schizophrenia-spectrum disorders among biological and adoptive relatives of index and control adoptees*

	Relatives			
	Biological		Adoptive	
	N	Percent	N	Percent
Total				
Index cases	150	8.6	74	2.8
Controls	156	1.9	83	3.6
Separated from biological parents within one month of birth				
Index cases	93	9.7	45	4.5
Controls	92	0	51	1.9

*Modified from Kety, S. S., D. Rosenthal, P. H. Wender, and F. Schulsinger. 1971. Mental illness in the biological and adoptive families of adopted schizophrenics. Am. J. Psychiatry **128**:302-306. Copyright 1971, the American Psychiatric Association. Reprinted by permission.

was done by Heston (1966) who single-handedly undertook a search in fourteen states and in Canada for relevant cases. This yielded, finally, 47 experimental children and 50 control subjects. The latter were matched in terms of sex, age at placement, type of home placed in, and length of time in child-care institutions. Main results are presented in Table 16-9. All differences shown are statistically significant. In addition, the index group had a lower mean IQ (94 as against 103.7 for controls), had fewer members married, and had fewer total children. Thus it is clear that, on the whole, there was a good deal more general pathology in individuals born of schizophrenic mothers even though their exposure to her had been only a few days. This does not rule out the possibility of prenatal factors operating.

A later and rather more elaborate study by the NIMH-Copenhagen group (Rosenthal et al., 1968) yielded essentially the same results as those of Heston. The rate for schizophrenic-spectrum disorders in index adoptees was somewhat over 30%; in controls, only 15%. This latter figure is surprisingly high and sug-

gests some pathology in the biological or adoptive families of the so-called normal subjects. Interestingly enough, the two groups did not differ significantly on any scale of the MMPI, *including the schizophrenia scale*. Since this initial report, Rosenthal and colleagues have added further subjects (Rosenthal et al., 1971) to almost double the sample sizes. The original results have stood up well. Rates of schizophrenia in both experimental and control groups rose slightly to 31.6% and 17.8%, respectively.

A rather interesting variant on the adoptees method was carried out by Fischer (1971) in Denmark. She studied the incidence of schizophrenia in 47 children of MZ schizophrenic index cases. This group was compared with another group of 25 children of the normal (i.e., discordant) co-twins of the index cases. Thus both groups of children would have pathological genes, but the first, in addition, would have pathological rearing conditions. Age-corrected incidences were not significantly different between groups: 12.3% for children of index cases and 9.4% for children of unaffected co-twins. Thus Fischer concluded that hereditary factors were more important.

Adoptive parents. A third method starts with adopted schizophrenic index cases and then looks for pathology in their biological as compared with adoptive parents. This strategy has been used by Wender, Rosenthal, and Kety (1968) with three groups of parents: 20 fathers and mothers who had adopted children later diagnosed as schizophrenic, 20 fathers and mothers who had reared their own schizophrenic children, and a third group of 20 parents with adopted nonschizophrenic children. Incidence and severity of pathology was found to be significantly higher in the biological parents rearing their own schizophrenic children. However, the adoptive parents of schizophrenics were rated as showing significantly more psychopathology than the adoptive parents of normals. The authors explored several possible explanations of this last finding. In general, however, it would seem most likely to suppose that it would arise simply from the strain and

Table 16-9. Status of adult offspring of schizophrenic ($N = 47$) or normal ($N = 50$) mothers reared in foster homes from shortly after birth (3 days)*

Variable	Biological mother	
	Normal mother	Schizophrenic mother
Mean rating (Menninger Health-Sickness Scale)	80.1	65.2
Schizophrenia (age-corrected risk = 16.6%)	0	5
Mental deficiency	0	4
Sociopathic disorder	2	9
Neurotic disorder	7	13
> One year in penal or psychiatric institution	2	11
Felons	2	7

*Modified from Heston, L. L. 1966. Psychiatric disorders in foster home reared children of schizophrenic mothers. Br. J. Psychiatry 112:819-825.

depression engendered from finding that the children they adopted became schizophrenic.

Cross-fostering. The final method entails the kind of design used as standard practice in animal studies: that is, the comparison of offspring of schizophrenics raised by normals with the offspring of normals raised by schizophrenics. Wender et al. (1973) have reported a study using this design. Essentially, they have simply added a new group (the second listed before) to the two other groups already reported in previous publications. Thus three groups were compared. An adopted-index group ($N = 69$), having schizophrenic biological parents with (presumably) normal rearing parents; an adopted-control group ($N = 79$) with normal biological parents and normal foster parents; a cross-foster group ($N = 28$) with normal biological parents and schizophrenic foster parents. Basically, for the total sample, their results indicated that the largest percent of severe pathology ("upper pathology quartile") was found in the adopted-index group (18.8%). The other two groups were about equal (~10%). However, when the samples were "purified" by eliminating from the cross-fostered group all subjects who had been "deviant infants, difficult to place, or whose parents had been psychiatrically diagnosed," incidence in this group fell to under 5%. The statistical significance of most of the differences is borderline, and the authors are careful to point out a number of methodological limitations to their study. However, on the whole, it does seem that possession of a nonschizophrenic genotype (i.e., normal biological parent) is quite a powerful protection against a schizophrenogenic rearing environment.

In general, the adoption studies give support to a genetic hypothesis of the etiology of schizophrenia. However, we must advise caution, since the sample sizes in the relevant studies have been small and, in many ways, atypical. More research is still needed.

Postdiction and prediction studies

A strategy more time consuming and difficult than those we have so far examined is to try to follow closely the life history of some-one who is schizophrenic (Garmezy and Streitman, 1974; Garmezy, 1974). Thus, we may either start with an index case and attempt to reconstruct the past, or we may start with individuals considered to be at "high risk" and record minutely all events during the course of their development.

The first of these methods may be called postdictive or retrospective-clinical. It is an old strategy that has been used at least since the 1920s. Basically, it is no different from psychoanalytic therapy that attempts to find the roots of an illness in the patient's history. Used in research, however, it obviously becomes more complicated. In the first place, suitable controls are necessary. Thus index cases need to be matched with normal subjects and, preferably, with others having some organic or nonschizophrenic illness. Numerous other matching variables should, of course, also be employed. Even then, we can never be sure that any conditions which occur in a patient's history are uniquely necessary to producing the condition.

A second category of difficulty relates to the reliability of the reconstruction. Reliance on the memory of the patient or of associates is obviously not a good idea. Forgetfulness and distortion, conscious or unconscious, can and obviously do occur. The same applies to records, which may be kept with greater or lesser completeness by different agencies. Furthermore, the latter may often be reluctant, for obvious reasons, to give out the information to researchers. Other problems have been noted by Garmezy and Streitman (1974).

In any case, as Anthony points out in a review (1972), most of the studies using this method have been poorly designed and planned. Consequently, the conclusions we can draw from them are limited. Some have been of better quality, however. Thus Wahl (1956), studying the early histories of 568 male schizophrenic naval personnel, found that 41.4% had lost or been separated from a parent (not confined to the mother) for fairly extensive periods early in life. This is not an uncommon finding and has been found by a number of other investigators. Rutter (1966) claimed a similar relation between "mental disorder" in children and early maternal or paternal bereavement. However, he pointed out also that such bereavement (e.g., death) is often itself associated with parental mental disorder and/or chronic illness. Consequently, the relation, though plausible, must be a complex one and, furthermore, is probably by no means unique to the etiology of schizophrenia.

An extensive study carried out by Watt (1972) gathered data from the school records of a schizophrenic sample and matched controls. He was able to identify a number of psychological variables differentiating the two groups. Preschizophrenic boys had scored low on such variables as motivation, emotional stability, and agreeableness. Girls were also less emotionally stable and tended to be more introverted. Again, incidence of early parental deaths was higher in the schizophrenic sample (19%) as compared with controls (8%).

The second strategy, the prospective study of children at risk, has become rather popular in recent years. There are presently as many as twenty or more major studies going on in the United States, Denmark, Sweden, Norway, Israel, and Mauritius. Again, the investigator must face many problems. The main one is the sheer amount of time and effort that must be devoted to keeping track of the original risk and nonrisk individuals over long periods of time, at least up to the age of peak risk in the middle or late twenties. Another difficulty relates to the definition of risk and choice of controls. Generally, risk is defined as having a schizophrenic parent. However, one can subdivide further, as Erlenmeyer-Kimling (1968) and others have done, by adding a "high-risk" group of individuals with two schizophrenic parents. A "low-risk" group could also be added by picking individuals with a second-degree schizophrenic relative. As for controls, it has been usual to include, besides individuals from normal families, some whose parents have shown some other psychological or organic disability. Thus as the researcher plans more carefully, so the magnitude of the project increases. In the actual follow-up, on the other hand, the opposite happens; as time passes, it is usual to find the starting N's gradually eroded through loss of subjects. Clearly, the prospective method is not for the fainthearted.

One of the first workers was Fish, who, since 1952, has been following up, in an intensive clinical way, a few cases considered vulnerable to schizophrenia. They were chosen not only on the basis of having hospitalized schizophrenic mothers but also on the basis of showing abnormally uneven development at 1 month of age. In three cases —Peter, Frank, and Conrad—one eventually became schizophrenic, one psychopathic, and the other a "moderately well-compensated neurotic" (Fish et al., 1966; Fish, 1971). Fish's descriptions are very qualitative. Furthermore the erratic environment in which the boys grew up makes it hard to pinpoint key variables. Nevertheless, the work as a whole provides a valuable source of detailed information for other investigators.

One of the most important programs has been that carried out by Mednick and Schulsinger in Copenhagen, starting in 1968. They began with 207 index cases, the normal-functioning adolescent (mean 15.7 years) children of schizophrenic mothers. These children were paired in diads on the basis of a number of variables. A further group of 104 control subjects was used, each again being matched to one of the index diads. The use of the latter was to allow for the possibility that one member of a diad might become sick and the other not; in which case, there would be afforded an opportunity to look for critical precipitating factors. Unfortunately, since matching of members of diads was imperfect, they were not able to be used. A large number of dependent variables were measured. Some salient results were as follows:

1. There appeared to be significantly more pregnancy and birth complications (PBCs) associated with risk children. This was based mostly on midwives' reports. It is in accord with the data of Pollin et al. discussed earlier.

2. There was a definite relation between maternal absence in the first two years of life and later pathology.

3. Risk subjects had experienced a greater amount of conflict in the home.

4. A sizable group of risk subjects showed clearly deviant behavior, behavior that is sometimes designated as schizoid.

5. Subjects in the risk group appeared to react differently on a number of psychophysiological measures, chiefly galvanic skin resistance (GSR) (e.g., shorter latencies, greater amplitudes).

6. During the initial five years of the research, about 20 of the risk group were either hospitalized for mental disorder or diagnosed as alcoholic, delinquent, or schizoid.

Two of the preceding points are worth elaborating further. Mednick (1970) has attributed major importance to pregnancy and birth difficulties and has gone on to suggest that possible anoxia occasioned by these may produce early hippocampal damage in risk children. The latter notion has not been readily accepted (Kessler and Neale, 1974), since, although it may seem plausible, it goes rather far beyond the data and, for that matter, beyond feasible experimentation. However, Mednick and colleagues (Mednick, 1970; Mednick et al., 1971) did consider the possible involvement of PBCs important enough to pursue further. In general, the use of more exact procedures and methods has at least partly verified their previous findings. However, it must also be mentioned that most other studies have not been able to confirm the conclusions of the Mednick group (Garmezy, 1974). Consequently, the matter remains in doubt.

If PBCs are of importance, they may, of course, predispose to psychiatric illness in a number of ways. An obvious possibility stressed by some workers is the effect they may have on the ensuing social interactions between mother and child. Distress connected with birth and labor could well dispose a mother to behave negatively to her child. This, in turn, could set off a whole chain of events offering poor prognosis for the child. Mednick et al., however, as suggested previously, have preferred to emphasize a more direct causal route, the effect of PBCs on the body's stress-response mechanisms. It was this reasoning that led to a focus on the nature of psychophysiological responses in risk and nonrisk cases. Mednick

(1958) has, in fact, proposed a microtheory of the etiology of schizophrenia. He has theorized that, in coping with a stressful stimulus, preschizophrenics (or schizophrenics) exhibit a higher than normal reactivity, relative slowness of habituation and poor extinction of the conditioned GSR response (Mednick, 1970). Although Mednick was able to educe some support for his theory, other investigators have not, beyond the generality that risk cases are usually different from normal cases. However, the particular line of research is a promising one and certainly deserves more work. It represents a noteworthy departure from the quasiliterary manner in which schizophrenia and other mental disorders are usually discussed.

A final program we will discuss briefly is that initiated by Anthony (1968), a child psychiatrist at the University of Washington. His studies involved at least two features of interest: (1) intensive clinical and experimental investigation of index children and their families and (2) reliance on a Broadbentian information-processing model in dealing with the etiology of schizophrenia. His data are still being collected and assessed. However, preliminary reports (Anthony, 1972) indicate that high-risk subjects are overresponsive to stimuli, habituate abnormally slowly, and show poor discrimination at the stimulus and response level.

The studies just summarized can be taken as representative of work using the prospective strategy. It will be obvious to the reader that, in spite of enormous cost and labor, the results, so far, have not been that illuminating. There are few definitive agreements between programs (except trivial ones) and many disagreements. If all that finally emerges is that children at high risk have problems in early childhood, then we cannot consider the investment to have been very worthwhile.

Genetic models and schizophrenia

So far, we have merely summarized some data which appear to indicate that there is a heritable component in the etiology of schizophrenia. This represents only the first step. The second step involves attempting to fit genetic models to these data. Clearly, it is rare with behavioral characters that we immediately find a precise fit of one model. In the case of schizophrenia, depending on how it is defined, on which types of data are emphasized, and on various statistical machinations performed on the data, we find that a number of models can fit the data about equally well. Each of these has its adherents, and each points research in a certain direction.

In general, there are two types of models that are not necessarily mutually exclusive: monogenic models, which invoke the action of one or a few major genes, and polygenic, or multifactor, models, which postulate the action of many genes, each with small effects but which, in a particular combination, can produce a dramatic phenotype like schizophrenia. As we shall see, there are a number of variants of these two classes of theory and some models that involve aspects of both.

Monogenic models are more "optimistic" in the sense that they reflect a faith that a disease like schizophrenia will turn out to have a major, determinate cause, presumably biochemical in nature and possibly correctable. Multifactor theories, on the other hand, must necessarily place more reliance on environmental intervention. They have, for the same reason, at least the political advantage of appearing more palatable to environmentalists. But apart from these considerations, the models currently being put forward have somewhat different conceptual starting points regarding the entity with which they are dealing.

Perhaps one of the most fundamental problems to be solved first (it has not yet been solved) relates to the specificity of schizophrenia. Concretely the question has at least three parts: first, is schizophrenia a discrete, a continuous, or a quasicontinuous character? Second, is the schizophrenic genotype a unit or unitary dimension underlying the diverse phenotypic forms the syndrome may take? Third, does the genotype underlying schizophrenia also underlie other classified forms of mental illness such as manic-depression, melancholia, neuroticism, or psychopathy? These problems have received full discussion by many workers (Kaplan, 1972) and have been well reviewed by Rosenthal (1970) and by Slater and Cowie (1971). Let us consider each question separately.

There is little doubt that there are degrees of severity of schizophrenia. Many workers commonly accept this in their use of such terms as schizoid, schizophreniform, and schizotaxic. However, it also seems plausible that below a certain threshold an individual can be considered as not ill enough to warrant full hospital treatment, but that beyond this point he should get it. It is probable that such a threshold is not a value arbitrarily fixed by local psychiatric practice. Hence, some theorists have found it most plausible to treat schizophrenia as a quasicontinuous variable, that is, a dimension on which the normal population has a certain threshold of risk, and schizophrenic genotypes a much lower threshold of risk. This approach has been favored by Falconer (1965), Edwards (1969), and Kidd and Cavalli-Sforza (1973), among others. Conceptually, however, the difference between a simple continuity model

Table 16-10. Incidence of type of schizophrenia in probands and affected children*

Probands	Children			
	Hebephrenic	Catatonic	Paranoid	Total
Hebephrenic	34	8	14	56
Catatonic	6	18	10	34
Paranoid	5	9	7	21
TOTAL	45	35	31	111

*From Kallman, F. J. 1938. The genetics of schizophrenia. J. J. Augustin, Inc.—Publisher, Locust Valley, N.Y.

(Heston, 1970) and a quasicontinuity model is not great. Espousal of continuity usually calls for a major (though not exclusive) role to be given to polygenes.

Whether continuous or not, schizophrenia shows up in a bewildering variety of forms. Hence it is sensible to inquire whether these are fundamentally different. Kallman (1938) attempted to answer this question by examining, in probands and their affected children, the incidence of the four Kraepelinian categories of schizophrenia. His results are shown in Table 16-10. As calculated by Slater and Cowie (1971), the association between probands and children in respect to subcategory is significant ($\chi^2 = 15.32$, $p < 0.01$). Kringlen (1968) has also reported this conclusion. At the same time, all types of proband have all types of affected children. Hence, Kallman concluded in favor of a major schizophrenic genotype whose expression could be controlled by minor modifiers. An additional piece of information Kallman put forward related to the expectancy rates for subtypes in children: for hebephrenic, 20.7%; for catatonic, 21.6%; for paranoid, 11.6%; and for simple, 10.4%. It could be argued from these data that the first two were more closely related to the nuclear genotype than the last two. This finding has been confirmed by Hallgren and Sjögren (1959).

Other workers, however, using somewhat different taxonomies, have favored a heterogeneity model. Leonhard (1936) distinguished between what he called "typical" and "atypical" schizophrenics. By examining the incidence of these in probands and their affected children and sibs, he educed support for the conclusion that typical schizophrenia was carried by recessive genes with low penetrance, and the atypical group by dominant genes with much higher penetrance. Similar distinctions have been made by other workers: for example, between process and reactive forms (Abelin, 1972), chronic (progressive or transient) and relapsing (Inouye, 1961), and a number of others to be discussed later.

Obviously, some confusion still surrounds the problem of the unity of the schizophrenic genotype. The problem is not yet solved. Slater and Cowie (1971) are probably correct in concluding that, although much of the data collected point to heterogeneity of schizophrenia, this cannot yet be regarded as firmly established.

The final question is even more complex. There is no question that in the relatives of schizophrenic probands we may find a range of other disorders. In particular, it is quite common for manic-depression and schizophrenia to show up in the same families. Some illustrative data, compiled by Rosenthal (1970), are shown in Table 16-11. Note, however, that the ratios of manic-depressive to schizophrenic children are in the general order of 5:1. It is likewise true that if the proband is schizophrenic, the first-degree relatives are about five times more likely to have schizophrenia as manic-depression (Ödegaard, 1963). Thus there is some tendency for each disorder to "breed true" but not

Table 16-11. Expectancies of manic-depression and schizophrenia in children of a manic-depressive parent*

| Study | N (age corrected) | Percent | |
		Manic-depressive	Schizophrenic
Hoffman (1921)	94	13.8	2.5
Weinberg and Lobstein (1936)	94	6.3	0.8
Röll and Entres (1936)	82	9.7	2.0
Slater (1938)	204	12.8	3.1
Stenstedt (1952)	149	6.0	3.1

*Modified from Rosenthal, D. 1970. Genetic theory and abnormal behavior. McGraw-Hill Book Co., New York.

completely. No instance has been reported, according to Rosenthal, of an MZ pair with one co-twin showing clear-cut schizophrenia, the other, clear-cut manic-depression.

Thus the relationship between these two major psychoses is still something of a puzzle. Even less well understood are the relationships of schizophrenia with such disorders as neurosis, psychopathy, sexual deviations, and epilepsy.

At least a partial solution to the preceding difficulties has been offered by Eysenck (1972) on empirical and theoretical grounds. He has argued for dimensions rather than Kraepelinian categories and has put forward evidence that any personality, from normal to abnormal, can be located in a three-dimensional space defined by the three basic factors: psychoticism, extraversion-introversion, and neuroticism. These have broad heritabilities of 0.53, 0.37, and 0.65, respectively. The main point, in the present context, is the notion of a unitary dimension of psychoticism. This would underlie all psychotic reactions, which might, however, differ in their expression depending on the part played, in any particular case, by the two other variables. In many ways, this is an appealing model. It is similar conceptually to the rather old concept in German psychiatry of *Einheitpsychosen*—unitary psychosis. Eysenck's updated version has the advantages of being empirically derived through factor-analytic procedures and also subject to experimental analysis. Nevertheless, it can hardly be said to have found universal acceptance (Kaplan, 1972).

Most genetic models, either explicitly or implicitly, take some position in respect to the three problems listed before. Any model will have to take account of them eventually.

Turning now to the models themselves, we will first quickly review the kinds of data that are offered as supportive of different kinds of genetic transmission. Dominant gene action is indicated by the following: approximately equal incidence in parents, sibs, and children of probands; about equal likelihood of collateral and lineal relatives showing the disorder; parental consanguinity not particu-

larly frequent; often late onset; mild severity; and variability of clinical expression. Recessivity, on the other hand, is usually suggested by much higher incidence in sibs than in parents or children; collateral inheritance; higher than usual consanguinity rate; early onset, high degree of severity; and usually clear-cut expression. It is not easy to define exact tests for a polygenic model. Some have been suggested by Carter (1969). One of these has involved examining the incidence in families of single affected probands with the incidence in families with two (usually sibs) or more affected probands. In the latter case, the loading of unfavorable polygenes should be greater, and hence the familial incidence higher. Monogenic theories, however, would predict no differences in incidence between the two classes of family. A related method involves comparing the symmetry of incidence in ascendant, secondary relatives of a proband on the maternal and paternal sides. Given certain conditions, polygenic theory would predict a less unilateral and more bilateral distribution. The logic of this has been elucidated by Slater (1966). A second test between monogenic and polygenic theories devolves around the rather high frequency of schizophrenia that has been maintained, perhaps over the last century and a half. There is known to be some selection against schizophrenia. To explain why frequency has not gone down sharply is easier for a polygenic theory, from a purely statistical point of view. Any monogenic model, however, must invoke some notion of balanced polymorphism, giving a carrier some compensating advantage, to explain the apparent resistance to selection. This is more especially true in the case of a dominance theory (Gottesman and Shields, 1972).

Additional data relevant to genetic transmission may be obtained from twin data and from data on offspring of dual matings. These, as we have seen, have to do mostly with problems of penetrance and manifestation.

We will now turn to a discussion of the two major classes of theory and variants of them. Anticipating somewhat, we may point out that due to the imprecision of the tests listed previously and also to the variability of the data base, no single model has, as yet, found universal acceptance by workers in the field.

Monogenic theories. A simple recessivity hypothesis was originally put forward by Rüdin (1916) and others of the Munich school, for example, Luxenburger (1928) and Weinberg and Lobstein (1943). The conclusion was reached mainly on the apparently greater incidence in collateral than in lineal relatives. At the same time, it became clear that the incidence did not reach the approximately 25% expectancy rate in siblings of schizophrenic probands. Again, the incidence in sibs of a proband with one schizophrenic parent is appreciably higher than in sibs of a proband of normal parents (Rosenthal, 1970). Furthermore, not all MZ co-twins of probands are concordant, and not all children of two schizophrenics are also schizophrenic. Consanguinity rates have varied from study to study and, at least according to Rosenthal (1970), cannot be considered as being uniformly higher than normal expectation. Thus it is obvious that a simple recessivity theory cannot be made to fit the facts. This led Rüdin to modify his initial model by adding other pairs of genes.

Kallman (1938, 1953), on the basis of his extensive German family study and American twin study, also espoused a recessivity theory. He also was impressed by the tendency for collateral transmission and (at least in his own data) elevated consanguinity rates. To accommodate the model to the difficulties listed before, however, he suggested that manifestation was determined by a genetically nonspecific constitutional defense mechanism which was "unquestionably polygenic" in nature. The latter he related to body type, being weakest in asthenics and strongest in athletics and, physiologically, involving the reticuloendothelial system. This latter notion has not seemed plausible to some critics (Slater and Cowie, 1971). But others, for example, Burch (1964) and Hurst (1972), have felt that perhaps Kallman was on the right track. Burch, in fact, has put forward the view that Kallman's constitutional

defense mechanism is not polygenic but is, in fact, carried by a second pair of recessive genes. His general approach to the etiology of schizophrenia (and some other psychoses) is a highly novel one but, to date, has not commanded much attention.

It seems clear enough that a recessivity model has problems. The postulate of an additional system governing expressivity can make it credible up to a point. But, in view of the lack of perfect concordance between MZ co-twins, one must, at some point, invoke within-family environmental influences as well.

Perhaps the main problem, however, seems to relate to the choice of data. Thus using mainly the observations by Kallman and by Garrone (1962), Hurst has found a good fit for a simple major recessive gene with homozygous penetrance of 60% to 70 percent. A reanalysis by Garrone of Böök's (1953) data gathered in northern Sweden yielded the same conclusions, though they were not, as we shall see, the same as those educed by Böök himself.

Other exponents of a recessive model are Elston and Campbell (1970), who reanalyzed Kallman's family data by sophisticated mathematical methods. Basically, they used four parameters: the gene frequency (0.066) and penetrances for the *AA* genotype of 0%, the *aa* genotype of 100%, and the *Aa* genotype of 5.3%. A slightly different alternative set of estimates was also chosen but gave about the same fit to observed data with a maximum likelihood method. Their main results are shown in Table 16-12. The only large departure of expected from observed values involves MZ twins; Kallman's figures are much higher. However, it may be recalled that the most representative value found in all the twin studies carried out is probably closer to 60%. Allowing for some special influence of prenatal conditions common to MZ twins, this may not be too far away from the expected estimates given. Elston and Campbell therefore conclude that a monogenic recessive model fits the Kallman data very adequately, in fact, much better than a polygenic model. In a later publication, Elston, Kring-

Table 16-12. Observed and expected incidences of schizophrenia in various family groups*

Relation to proband	Sample size	Observed incidence (percent)	Expected incidence (percent)	
			I†	II†
MZ twins	174	85.8	43.0	40.5
DZ twins	517	14.7	15.1	15.3
Sibs	6453	12.7	15.1	15.3
Children	2000	11.9, 16.4	8.2	9.6
Parents	1191	9.2	8.2	9.6
Half-sibs	259	7.3	4.6	5.5
Grandchildren	1016	4.3	4.6	5.5
Nephews and nieces	2170	3.0	4.6	5.5

*Based on analysis of Elston, R. C., and M. A. Campbell, 1970. Schizophrenia: evidence for the major gene hypothesis. Behav. Genet. 1:3-10. Based on data from Kallman, F. J. 1938. The genetics of schizophrenia. J. J. Augustin, Inc. —Publisher, Locust Valley, N.Y.; and 1946. The genetic theory of schizophrenia. Am. J. Psychiatry **103**:309-322.
† I and II represent expectations based on two sets of parameter values (see text).

len, and Namboodiri (1973) present data indicating the possibility of a relationship between general psychosis and the *Gc* blood group locus and between schizophrenia and the *Gm* and/or Rhesus loci. This is a suggestive finding but, as the authors themselves point out, one to be treated with caution.

It will be clear from the preceding discussion that the hypothesis of major recessive genes can by no means be dismissed out of hand. However, its validity does appear somewhat to depend on selection of the right data. Let us now consider the alternative dominance model.

It is probably fair to say that no worker has seriously suggested that the simple operation of a dominant gene produces schizophrenia. As with the recessivity model, much weight must be placed on the parameter of penetrance. One of the first to present a sophisticated dominance model was Böök (1953). He sampled an isolate of about 9000 individuals in northern Sweden about 100 kilometers above the Arctic Circle. The population was relatively homogeneous, had a higher than usual rate of inbreeding, and existed under climatic conditions that would have to be regarded as stressful. Slater and Cowie (1971) have summarized some of his major findings as follows: (1) the incidence of schizophrenia was unusually high (2.92 for males and 3.2 for

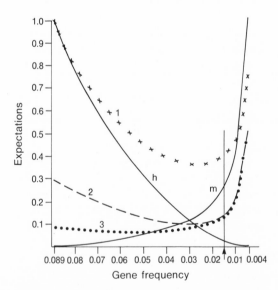

Fig. 16-6. Theoretical expectations of incidence of schizophrenia in relatives of schizophrenics *(broken lines),* with varying gene frequency and varying penetrance. *1,* Children of two schizophrenics; *2,* sibs of schizophrenics; *3,* children of one schizophrenic; *h,* proportion of all schizophrenics who are homozygous; *m,* manifestation rate of gene in heterozygote. From Slater, E. 1958. Acta Genet. Stat. Med. **8**:50-56. Reprinted by permission of S. Karger AG, Basel, Switzerland.)

females); (2) the incidence of manic-depression and other major psychoses was very low: no male and only four female manic-depressives were located; (3) the predominant form of schizophrenic illness was catatonia; and (4) morbidity risks in relatives of probands were for parents, 12%; for sibs, 9%; and for sibs of a proband of one schizophrenic parent, 12.7%. Böök analyzed his data using simple applications of the Hardy-Weinberg principle for simple recessive, simple dominance, and intermediate models. He concluded in favor of a single dominant gene pair with homozygous penetrance of 100% and heterozygous penetrance of about 20%. Gene frequency in the population was estimated at around 7%.

Since he felt that Böök's method of analyzing his data might be somewhat idiosyncratic, Slater (1958) undertook a more general treatment. Basically, his approach is as follows: if the frequency of the schizophrenic gene (a) is p, then the frequency of the normal allele (A) is q, or $1 - p$. Given random mating and a population at equilibrium, the proportions of the three genotypes AA, Aa, and aa are:

$$q^2AA + 2pqAa + p^2aa = 1$$

If we hypothesize a manifestation rate, m, in the heterozygote, then the incidence, s, of schizophrenia in the population will be:

$$s = 2mpq + p^2$$

Now the value of s has been estimated empirically to be around 0.8%. Knowing this, we can then plot values for m, p, and q that are, of course, interdependent. Slater has accordingly taken values of p from 0.089 to 0.004, from which may be derived the values of m. The values may then be fit to equations for expected incidence of schizophrenia among various relatives of probands. Slater plotted the results of these for eleven values of p and m for children of two schizophrenics, children of one schizophrenic, and sibs of schizophrenics. The results are shown in Fig. 16-6. At the extreme left of the diagram, we have the case of complete recessivity, m being zero at this point; expectations are for 100% incidence in offspring of dual matings, and

8.9% in offspring of one schizophrenic proband, and 30% in sibs of a proband. At the extreme right, we have simple dominance; that is, manifestation is complete in heterozygotes (Aa). Under these conditions, risks in sibs and in children of one schizophrenic are equal at 50%. For children of two schizophrenics, the rate is about 75%. Obviously, the empirical data fit best for some intermediate value of p. This turns out to be, approximately, $p = 0.015$. The values yielded are approximately 14% for children and sibs of probands and 39% for children of two schizophrenics. Manifestation in heterozygotes is about 26%. Slater matched these estimates to data from Kallman and Elsässer. In a later publication (Slater and Cowie, 1971), he used all the family data summarized by Zerbin-Rüdin (1967) and categorized so as to distinguish between diagnostically certain cases only and diagnostically certain plus "probable" cases. The relevant observed and expected values are set out in Table 16-13. For the most part, agreement is fairly good. However, risk in children of dual matings is higher than expected, and risk in second-degree relatives is about 25% less than expected. Slater and Cowie (1971) suggest that these deviations may be a result of the operation of unknown environmental factors.

It should be noted that Slater and Cowie omit parents in this table of data. It may be recalled that the observed incidence given by Zerbin-Rüdin (1967) for fourteen studies was the low figure of 4.38%. This would suggest a departure from the parent-sib-children equality expected by a dominance model. However, as noted before, Slater and Cowie regard the parental value as an underestimate due to lack of suitable age correction. Taking this into account, they arrive at a risk "in the neighbourhood of 10 per cent" (p. 16). It may also be true, as Rosenthal (1970) has suggested, that, since schizophrenics have a reproductive disadvantage (about 7%), those who do marry and have children represent a sample biased toward normality. This consideration would, of course, also raise the true risk estimate.

On the basis of the preceding sets of data,

Table 16-13. Comparison between observed and expected incidences of schizophrenia in relatives of probands*

Relation	N	Observed schizophrenic (percent)		Theoretical expectation (percent)
		Certain cases	Certain plus doubtful	
Sibs (all)	8505	8.5	10.2	10.2
Sibs (parents normal)	7535	8.2	9.7	9.4
Sibs (one parent schizophrenic)	675	13.8	17.2	13.5
Children	1227	12.3	13.9	8.8
Children of dual matings	134	36.6	46.3	37.1
Second-degree relative				
Half-sibs	311	3.2	3.5	4.7
Uncles and aunts	3376	2.0	3.6	4.7
Nephews and nieces	2315	2.2	2.6	4.7
Grandchildren	713	2.8	3.5	4.7

* Modified from Slater, E., and V. Cowie. 1971. The genetics of mental disorders. Oxford University Press, London. Based on data from Zerbin-Rüdin, E. 1967. Endogene Psychosen. In P. E. Becker, ed. Humangenetik, ein kurzes Handbuch, vol. 2. Georg Thieme Verlag KG, Stuttgart, West Germany.

Slater (1972) concludes in favor of a "major, partially dominant, gene model." Some findings supportive of his general position have been offered in the extensive work of Karlsson (1966). He traced the incidence of schizophrenia in the 967 descendants of the 12 children of an Iceland priest born in 1781. The 12 pedigrees were found to include 23 psychotic descendants. However, these were not distributed normally across the pedigrees but were clustered into only 6 of them; of these 6, 5 included 3 or more cases. Slater and Tsuang (1968) have also reported the same kind of asymmetry in the pedigrees of 53 schizophrenics in England. Karlsson (1972) suggests that this pattern is "highly suggestive of dominant transmission and it is inconsistent with polygenic inheritance." However, the reader should be reminded that the term "suggestive" is not equivalent to "scientifically proven." In fact, in consideration of his own data and those of other workers, Karlsson concludes in favor of the operation of two pairs of genes, one major pair being dominant, and the other pair also probably dominant but possibly recessive (Karlsson, 1972). He has found, in fact, that a two-locus dominant-recessive combination yields expectations that fit Kallman's data

well (including MZ twins). On the other hand his own recently gathered data reported in 1972 fit better with a double-dominant model.

Gregory, in a useful article in 1960, attempted to examine the fit of three hypotheses to some of the main empirical data then available from Kallman (1938, 1946, 1953), Elsässer (1952), Slater (1958), and a number of others. Specifically, he examined (1) hypothesis A: simple recessivity with incomplete manifestation in homozygotes; (2) hypothesis B: partial dominance with complete manifestation in homozygote but incomplete manifestation in heterozygote; and (3) hypothesis C: partial dominance with homozygous manifestation rate double that of heterozygous manifestation rate. Only the second of these hypotheses gave a good fit to the data. It is significant that Gregory's statistical procedures are almost identical to those of Slater (1958), though their choice of empirical data differs somewhat.

A final approach which should be mentioned is that of Heston (1970). He has suggested that so-called schizoidia and schizophrenia are "alternative expressions of the same genotype" (p. 252), the latter being, essentially, a simple dominant gene pair. Unlike other models, however, this one postu-

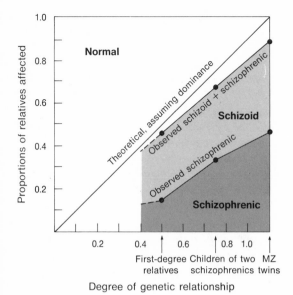

Fig. 16-7. Observed and expected proportion of schizoids and schizophrenics. (From Heston, L. L. 1970. Science **167**:249-256. Copyright 1970 by the American Association for the Advancement of Science.)

lates virtually complete manifestation, but with a continuum of expression ranging from mild (schizoid) to extreme (full schizophrenia). The degree of expression is hypothesized by Heston to depend on various other traits, themselves partly genetic and partly environmental. Heston's schematization of his model is shown in Fig. 16-7.

All preceding monogenic theories have the burden of explaining the high rate of schizophrenia that apparently has been maintained for many generations in spite of alleged negative selection. We offer this statement in such a provisional way because the data bearing on both the incidence of schizophrenia over a long time period and on its negative fitness are not at all strong. Clearly, the criteria by which schizophrenia has been diagnosed over the last hundred years have changed; or, if they have not, there is some onus of proof on those who take this position. Likewise, the claim that schizophrenics have a lowered reproductive advantage may or may not be true. Data on these two points are hardly incontrovertible. However, if we grant that schizophrenia (if not psychosis in

general) has perhaps remained at a rather high constant level over many years, we may look at some data bearing on the second point. One of the first to suggest that a schizophrenic gene in heterozygous form may give its bearer some advantage was Caspari (1961) in a review of our previous text (1960). Since then, other workers have put forward data relating to this. Thus Heston (1966) reported in his adoption study that about 50% of offspring born of schizophrenic mothers and adopted out were more "colorful" and "held more creative jobs" than normal control subjects. This general finding has been confirmed by other investigators. Karlsson (1972), for example, has offered data indicating that in Iceland the likelihood of being listed in the *Who's Who in Iceland* is twice as high for relatives of psychotics as it is for members of the general population. Juda (1949), Schaffner, Lane, and Albee, (1967), and McConaghy and Clancy (1968), among others, have reported similar findings. Of course, being colorful or creative does not necessarily confer a reproductive advantage, which is the point at issue. Empirical data bearing on this are slender. However, Erlenmeyer-Kimling and Paradowski (1966) found that, over a twenty-year period (between 1934 to 1936 and 1954 to 1956), not only did schizophrenics show considerable reproductive gains, but also their normal siblings, who, by the end of this period, were reproducing at a rate 140% higher than the rate of the general population (Erlenmeyer-Kimling et al., 1969). This evidence has much more direct bearing on the point at issue. Again, Kay and Lindelius (1970) have reported lowered mortality rates (birth to 15 years) in nieces and nephews of schizophrenics, but this finding was not verified by Buck, Simpson, and Wanklin (1977), who corrected for misreporting of deaths.

It has been suggested by Osmond and Hoffer (1966) that schizophrenics themselves, though perhaps disadvantaged in their capacity to get along in society, do actually derive some side benefits from their genotype. Thus they claim that schizophrenics are more tolerant of such drugs as insulin, thyroxine, and epinephrine (Adrenalin), less liable to rheumatoid arthritis, more resistant to surgical shock, have more stable blood pressure, have reduced incidence of allergies, and show, as a group, a lower death rate from pneumonia. These epidemiological data must be treated with caution. However, if they are valid, one might suppose that the possession of such physiological advantages but without the schizophrenic symptomatology—a combination most likely in normal relatives of schizophrenics—would confer considerable fitness.

In general, the problem must be regarded as a real one, but the data bearing on it are weak and provisional. Biological fitness is difficult to measure and, in the case of human beings, is heavily overlaid with variables of custom and culture. Obviously, many individuals might be capable of having 20 or more children but choose to have only 1. If the question is to be properly pursued, it will be necessary to obtain much more accurate estimates than we have now of reproductive patterns in schizophrenics and their normal relatives.

Polygenic theories. Some current proponents of some form of polygenic, or diathesis-stress, mode are Falconer (1965), Gottesman and Shields (1967), Rosenthal (1970), Eysenck (1972), and Ödegaard (1972). Basically, the theory assumes that a large number of genes, each with small individual effects, contribute to a disposition, or diathesis, to schizophrenia. Actual clinical manifestation is triggered by some environmental stress.

As indicated already, there are really two main classes of this model, one postulating a continuous variation for the phenotype, the other a quasicontinuous variation. A possible third (as suggested by Gottesman and Shields, 1972) is one in which the action of many genes with very unequal effect is posited. An example is afforded by bristle number in *Drosophila*. This is a continuous character (or can be treated so) under polygenic control, but 87.5% of the variation is controlled by only about five loci (Thoday, 1967).

The model put forward by Esysenck (1972) is basically a polygenic one. Its novel feature lies not so much on the genetic but rather on the taxonomic side. As previously indicated, his own and his colleagues' work over a period of years, using various psychometric and experimental tests, has indicated the presence of three orthogonal dimensions of personality: psychoticism, neuroticism, and introversion-extraversion. From this analysis, Eysenck has constructed the so-called PEN inventory, which we discussed briefly in Chapter 15. The main point to be stressed here, however, is that Eysenck argues for a unitary dimension of psychosis (as against heterogeneity theories), which is inherited polygenically. A twin study by himself and Eaves (1974) shows, in fact, high concordance for the psychoticism scale in MZ twins living together (r = 0.56). The similarity is only slightly reduced for MZ twins living apart (r = 0.52). This updated version of *Einheitpsychosen* theory is considered by some to be a retrograde step (Gottesman

Table 16-14. Morbidity risk in siblings of probands according to the number of psychotic relatives registered in the parent generation (parents, uncles, and aunts)*

Number of psychotic relatives in parental generation	Number of siblings	Number of psychotic siblings	Percent psychotic siblings
0	990	67	8.27 ± 0.97
1	521	55	14.88 ± 1.85
2 or more	284	45	20.92 ± 2.77
TOTAL	1795	167	11.97 ± 0.87

*From Ödegaard, O. 1972. The multifactorial theory of inheritance in predisposition to schizophrenia. In A. R. Kaplan, ed. Genetic factors in "schizophrenia." Charles C Thomas, Publisher, Springfield, Ill.

and Shields, 1972). However, Eysenck is certainly correct in stressing the fact that different forms of psychoticism are often found in the same family pedigrees. As we have suggested already, these facts require explanation. Furthermore, the extensive use of objective test batteries followed by the application of factor analysis does seem a sensible procedure. Of course, history has shown that general factors often turn out themselves to have factorial complexity. The same may turn out to be the case with Eysenck's psychoticism factor.

Ödegaard, a proponent of a simple polygenic model, has put forward two main arguments. One relates to the "heterogeneous diagnostic pattern which is generally observed in the families of schizophrenic probands" (1972, p. 274). We have discussed this problem already. A second argument relates to the increasing incidence of psychosis in families having a greater load of schizophrenia. This is illustrated by his own material from the Gaustad Psychiatric Hospital in Norway, the data from which are summarized in Table 16-14. They indicate that the morbidity risk increases steadily with the number affected in the parental generation. However, as Ödegaard admits, the differences are not statistically significant and hence can only be regarded as "suggestive." Additional data at least are consonant with the preceding. Furthermore, a comparison of risk in sibs of probands with either psychotic

inheritance (two or more) on one side only or on both paternal and maternal sides indicated no difference. This seems more in agreement with a polygenic than a monogenic model. However, these data may be atypical (as suggested by Slater and Cowie) and, in fact, seem not to be in accord with the findings of Slater and Tsuang (1968) referred to previously.

Gottesman and Shields (1972) have attempted to supply a similar test of the polygenic model, using the Zerbin-Rüdin data. They compare (1) risk to sibs of a proband, depending on whether neither or one parent is schizophrenic, and (2) risk to children, depending on whether one or both parents are schizophrenic. A simple monogenic theory predicts no increase in the case of sibs and a 50% to 75% increase in the case of children. Polygenic theory predicts a rise in both instances. The empirical data are set out in Table 16-15. Clearly, a simple monogenic model cannot explain these data. However, the expectations generated by the modified monogenic theory put forward by Slater are about as much in accord with the data as those of a polygenic theory. Certainly, if anything, the latter fits rather better, though this would be difficult if not impossible to demonstrate statistically.

The model put forward by Gottesman and Shields (1967, 1972) is somewhat more sophisticated. It assumes that schizophrenia is a quasicontinuous or threshold character and

Table 16-15. Schizophrenic risk as a function of parental status: empirical values and values expected from monogenic and polygenic theories*

	Risk (percent)			
	Probands' sibs		Probands' children	
	Neither parent affected	One parent affected	One parent affected	Two parents affected
Predicted, polygenic	6.5	18.5	8.3	40.9
Predicted, monogenic (modified)	9.4	13.5	8.8	37.1
Observed (Zerbin-Rüdin)	9.7	17.2	13.9	46.3

*From Gottesman, I. I., and J. Shields. 1972. Schizophrenia and genetics: a twin vantage point. Academic Press, Inc., New York.

is based on the derivations of Falconer (1965) and Edwards (1969) discussed in Chapter 12. Basically, a dimension of liability is assumed to be inherited. The threshold for schizophrenics, however, is much lower and will be shared by their relatives to a degree depending on closeness of relationship. Ignoring the details of statistical methodology, we may simply state that, depending on the parameter of population risk, estimates of heritabilities for various types of genetic relation to a proband may be predicted. Gottes-

man and Shields have proceeded to do this, assuming six population risk values over a range of 0.85% to 3%. The results of this analysis are shown in Fig. 16-8. The empirical data are taken from the Veterans Administration records of all white males who served in the United States armed forces between 1941 and 1955 and also from Zerbin-Rüdin's pooled data. Values are as follows: MZ twins, 50%; DZ twins, 9%; sib risk, 10.2%; offspring of dual matings, 46%; second-degree relatives, 3.3%.

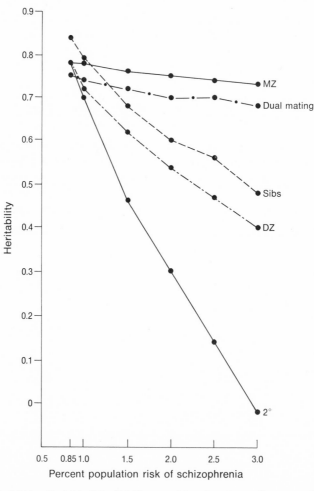

Fig. 16-8. Heritabilities (Smith) of the liability to schizophrenia as a function of varying population risks, estimated from risks in different classes of probands' relatives. (From Gottesman, I., and J. Shields. 1972. Schizophrenia and genetics: a twin study vantage point. Academic Press, Inc. [London] Ltd., London.)

It is clear from the graph that heritability estimated from MZ concordance is relatively insensitive to changes in population risk. Heritabilities estimated from second-degree relatives, sibs, and DZ twins are much more sensitive. The major point, however, is that at one level of population risk, all groups yield about the same heritability value. This is about 85%, with population risk at between 0.85% and 1%. The latter is, in fact, the most likely value for population risk from all empirical studies available. The consistency achieved between groups sharing quite different kinds and amounts of environmental communality is remarkable. As such, it represents strong evidence for the operation of multiple genes controlling liability for a threshold trait. However, whether the results of the Gottesman-Shields analysis are incompatible with all simple or monogenic theories is by no means certain. For that matter, these authors themselves speculate that in the polygenic system they hypothesize, there are a few genes of large effect. It may be that, in time, rapprochement will be achieved between the two major classes of theories.

Mixed models. Besides the models we have discussed, we may briefly discuss a few others that do not fit quite as readily into the two major classes.

A clinical breakdown was made by Leonhard (1936) between "typical" and "atypical" schizophrenics. On the basis of examination of family pedigrees of these, he concluded that the "typical" phenotype was an expression of recessive genes with low penetrance and that the "atypical" was carried by dominant genes with high penetrance. This theory has not, however, been accepted by most workers, since the distinction is difficult to make in practice, and, furthermore, when the types are distinguished by some criteria, they are often found to occur in the same families (Rosenthal, 1970).

Mitsuda (1967, 1972) has taken an approach that has used the typical-atypical distinction and extended it in some interesting ways. He has argued strongly for the avoidance, at this stage, of the complicated mathematical treatment of huge masses of data and, instead, for what he calls a clinicogenetic approach. This involves simply the detailed clinical and genetic analysis of individual cases. His work has led him to distinguish three forms of schizophrenia: typical, atypical, and intermediate, the latter two, however, probably being a single category. Pedigree studies indicate both monogenic and polygenic patterns of inheritance occurring for all three forms, but with typical most commonly being recessive. The relevant data are shown in Table 16-16. Mitsuda also reported finding no cases of typical and atypical schizophrenias occurring in a single pair of MZ twins, a result consonant with Inouye (1961). Significantly, however, he did find cases of schizophrenia and neurosis in cotwins. Going further, Mitsuda carried out a factor analysis of a battery of 53 items relating to objective and subjective features of schizophrenia. They were administered to 211 psychotic subjects. Four factors emerged, of which the first appeared to reflect the typical-atypical distinction. A few items with high loadings are shown in Table 16-17. It will be clear from the descriptive items that the typical-atypical distinction is not too dif-

Table 16-16. Modes of inheritance in three forms of schizophrenia*

Proband	N	Mode of inheritance		
		Dominant	Recessive	Intermediate
Typical	182	8.2%	72.5%	19.3%
Intermediate	32	37.5%	34.4%	28.1%
Atypical	102	42.2%	42.2%	15.6%

*Based on pedigree data from Mitsuda, H. 1972. Heterogeneity of schizophrenia. In A. R. Kaplan, ed. Genetic factors in "schizophrenia." Charles C Thomas, Publisher, Springfield, Ill.

Table 16-17. Examples of items loading on atypical and typical poles of factor I*

Item number	Atypical pole	Item number	Typical pole
12	Incoherence	48	Gradual onset
40	Amnesias as to pathological experience	14	Emotional blunting
52	Episodic, periodic course	53	Personality deterioration
45	Insight into illness at recovery stage	50	Chronic-progressive course
17	Disturbed orientation	3	Rigid, cold countenance
35	Delusional perception	2	Disturbed rapport
		51	Chronic-propulsive course

*From Mitusda, H. 1972. Heterogeneity of schizophrenia. In A. R. Kaplan, ed. Genetic factors in "schizophrenia." Charles C Thomas, Publisher, Springfield, Ill.

ferent from the reactive-process categorization used by other workers.

Mitsuda's final step has been to extend the so-called two-entities principle of Kraepelin (schizophrenia and manic-depression) to a three-entities principle by the addition of epilepsy. He has felt that this inclusion is demanded by the relatively high incidence of epilepsy in pedigrees with atypical schizophrenia. His final model is shown in Fig. 16-9. It depicts, essentially, a three-dimensional system for psychosis, with the apices of the triangle representing nuclear or typical schizophrenia, epilepsy, and manic-depression, and the sides representing peripheral forms that merge into each other.

Mitsuda's approach is an interesting one, but its heavy reliance on clinical methods makes it very hard to evaluate scientifically. Most certainly, he does not really commit himself to a definitive genetic model that is really testable. However, he is hardly unique in this respect in this difficult field.

We have already referred to the distinction made by many workers between process and reactive schizophrenia. A strong case for the usefulness of this particular categorization has been made by Abelin (1972). On the basis of family data from a variety of sources in Europe and the United States, he has concluded that so-called process schizophrenia, a psychosis with slow onset, chronicity, and low recovery rate, is carried by recessive genes with a high penetrance (20.9% in children), whereas reactive schizophrenia, an illness usually with acute onset but good prognosis,

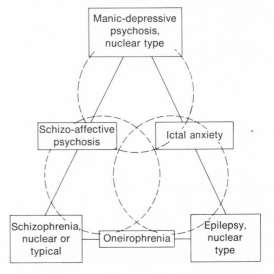

Fig. 16-9. Schematic representation of relationships between typical or nuclear major psychoses. (Modified from Mitsuda, H. 1972. Heterogeneity of schizophrenia. In A. R. Kaplan, ed. Genetic factors in "schizophrenia." Charles C Thomas, Publisher, Springfield, Ill.)

is manifested at a fairly low rate in heterozygotes (10.4% in children of probands). This model is clearly a variant of those of Kallman, Hurst, and others.

Finally, we may refer again to the model of Meehl presented earlier (Meehl, 1962). In a sense, this model is very much like Kallman's, though Meehl arrived at it seemingly on more intuitive grounds than Kallman, who, over the years, gradually modified his simple monogenic theory on the basis of data accumulated.

Comparison between models. At present, it seems hardly possible to rank the models presented in some kind of priority. Each one is largely based on a qualitative estimate of often highly selected data. Polygenic theories have been criticized by some (e.g., Moran, 1972) as deficient because they are not falsifiable in Popper's sense. This may be true. On the other hand, the same criticism may well be leveled at monogenic models, since all but the most simple (which cannot fit the data anyway) must invoke the risky concept of manifestation rate or penetrance. The latter is undoubtedly necessary and also valid. However, it can be used so loosely as to permit the fit of almost any model to any set of data.

The data are, in fact, very heterogeneous. Kidd and Cavalli-Sforza (1973) have applied chi-square tests to the estimates provided by Zerbin-Rüdin for various familial risks. They showed that for almost all such groups, chi-square values were significant, indicating that, at least from a statistical point of view, there is a real question as to whether the data from different studies should reasonably be lumped together. Nevertheless, Kidd and Cavalli-Sforza proceeded to analyze the data in terms of four parameter values: ϵ^2 = environmental variance, q = gene frequency, T = threshold on scale at which schizophrenic phenotype is manifested, and h' = position of heterozygote relative to that of the two homozygotes. Four sets of values were adopted for these parameters, most especially in reference to h', for which the range from complete recessivity to complete dominance was used. Results are summarized in Table 16-18. The best fit was yielded by a recessivity model with 50% penetrance in the homozygote. However, both heterozygotes and "normal" homozygotes would be expected to show some manifestation, the latter presumably for exclusively environmental reasons. Their conclusions are essentially in agreement with those of Elston and Campbell (1970) discussed earlier. At the same time, Kidd and Cavalli-Sforza also found that a polygenic threshold model can be fit to the data about equally well. Thus their results

Table 16-18. Incidence of schizophrenia in different genotypes under the assumption of a single-gene model*

	Parameter set†: values				Percent of each genotypic class affected		
	q	ϵ^2	T	h'	AA	Aa	aa
A	0.14	0.49	2.0	0	0.2	0.2	50.0
B	0.10	0.36	1.6	0.25	0.4	1.2	74.8
C	0.12	0.49	2.0	0.25	0.6	0.6	50.0
D	0.09	0.36	1.8	0.50	0.1	1.5	63.1

*From Kidd, K. K., and L. L. Cavalli-Sforza. 1973. An analysis of the genetics of schizophrenia. Soc. Biol. **20:**254-265.
†See text.

are inconclusive. Not only are the available data highly variable between studies but even when pooled do not fit any one model much better than another. The two models yield very different heritabilities, however, with h^2 from the single gene analysis on the order of 15% and from the polygenic model, about 80% or higher. Thus it would appear that if a single gene is controlling some critical biochemical pathway, the nature of the latter will be difficult to discover. On the other hand, under polygenic theory we would have to assume that, although such pathways might, in principle, be more readily discoverable (h^2 being very high), their sheer number may defeat attempts to do so.

As we shall see, however, the problems involved in the genetic analysis of schizophrenia are no greater than those involved in the study of other forms of mental illness. We have discussed schizophrenia in some detail because it represents perhaps the greatest social problem and therefore the greatest challenge. We will now turn to the category of affective psychoses and then consider in Chapter 17 some of the other deviations of personality.

AFFECTIVE PSYCHOSES

The syndrome of manic-depressive psychosis was designated in 1896 by Kraepelin. He suggested that its two main symptoms, depression and mania, were merely different phases in a single disease entity that depended primarily on some common organic pathology. More recently it has been usual to consider together a large number of severe mental disorders that involve mainly *affective disturbances*. Thus the American Psychiatric Association (Sahakian, 1970) lists a category of "affective reactions," within which are included "manic-depressive reactions, characterized by severe mood swings"; "manic-depressive reactions, manic type"; "manic-depressive reactions, depressed type"; "manic-depressive reactions, other"; and "psychotic depressive reaction." The World Health Organization nomenclature is slightly different and includes "involutional melancholia."

In general, it seems to be true that the disorder with which we are dealing encompasses a very wide range of symptomatology. According to Winokur, Clayton, and Reich (1969), little improvement has been achieved since the description given by Kraepelin in his 1921 monograph. He distinguished four types of mania: *hypomania, acute mania, delusional mania,* and *delerious mania*. These are specified in rather literary and imprecise ways but represent, roughly, a continuum from mild to severe. Likewise, depression may also be graded in steps as follows: *melancholia simplex, stupor, melancholia gravis, paranoid melancholia, fantastic melancholia,* and *delerious melancholia*. Often, manic and depressive periods follow one another cyclically. However, in some patients only one phase appears. Thus many workers (e.g., Leonhard, Korff, and Schulz, 1962) have felt it useful to make a distinction between bipolar forms of affective disorder and monopolar or unipolar forms, the latter involving, almost exclusively, bouts of depression rather than of mania. We shall discuss this distinction in more detail shortly.

However, the central feature of all kinds of illnesses we are considering is that they have a *primary affective* component which appears to arise from endogeneous causes. They are thus to be distinguished from so-called *secondary affective* disorders that may involve episodes of extreme excitation or melancholia triggered by some other causal factors such as alcoholism and family problems. Many attempts have been made, by means of personality tests, to separate the two classes, but it cannot be said that these have so far been very successful. Some of them have been reviewed by Winokur, Clayton, and Reich (1969). We will be mostly concerned with studies that have used genetic criteria.

Distribution

The incidence of manic-depressive psychosis appears to be somewhat variable between different populations. Some sample estimates are shown in Table 16-19. As Rosenthal (1970) has pointed out, the highest estimate is many times greater than the lowest. In Böök's sample, schizophrenia was very common, manic-depression very low. The highest—those of Essen-Möller in Sweden—are undoubtedly inflated by the inclusion of mild cases of affective disturbance. The same may have been true for von Tomasson's sample. On the other hand, there are probably many real differences, both genetic and environmental, between different populations. The median risk across all studies is probably somewhat below 1%, a figure usually accepted by most investigators.

Two other points in Table 16-19 are worth noting. First, manic-depressive illnesses appear to be less common in males than females, the ratio being about 0.69 (Slater and Cowie, 1971). This fact has led a number of investigators to postulate genetic models involving dominant sex-linked genes. Second, over the time period covered, the risk of this type of mental illness does not appear to have increased. This contrasts with schizophrenia. For the latter, between 1922 and 1946, number of first admissions doubled, probably reflecting, however, a growing interest on the part of psychiatrists rather than an actually elevated prevalence (Rosenthal, 1970). On the other hand, this may be changing. In England and Wales between 1960 and 1966, first-admission rates for manic-depression have increased markedly (46% to 49%), par-

Table 16-19. Incidence of manic-depressive psychosis*

Study	Country	Percent risk	
Early studies (Luxenburger, 1927-1936, and others)	Germany		<0.4
von Tomasson (1938)	Iceland		7.0
Mayer-Gross (1948)	Scotland		0.35
Sjögren (1948)	Sweden		0.6-0.8
Fremming (1951)	Denmark		1.2-1.6
Stenstedt (1952)	Sweden		~1.0
Böök (1953)	North Sweden		0.07
Slater (1953)	England		0.5-0.8
Larsson and Sjögren (1954)	Sweden	Males	0.9
		Females	1.2
Essen-Möller (1956)	Sweden	Males	1.7
		Females	2.8
Crombie (1957)	England		0.5-1.0
Kallman (1959)	United States		0.4
Norris (1959)	England	Males	0.8
		Females	1.4
Essen-Möller and Hagnell (1961)	Sweden	Males	8.5†
		Females	17.7†
Ödegaard (1961)	Norway	Males	0.4
		Females	0.6
Primrose (1962)	England		0.35
Helgason (1964)	Iceland	Males	1.8-2.18†
		Females	2.46-3.23†
Watts (1966)‡	England		0.6

*Modified from Rosenthal, D. 1970. Genetic theory and abnormal behavior. McGraw-Hill Book Co., New York.
†Mild cases included.
‡Cited by Rawnsley (1968).

ticularly in females (Slater and Cowie, 1971). But again, this more likely reflects changes in culture and in psychiatric practice rather than changes in gene frequency.

The age distribution is somewhat different between sexes. The peak for first admissions for men occurs rather late, around 55 years; for women, however, it peaks around 35 to 45, probably about a good ten years earlier. Data relevant to this point were first presented by Landis and Page (1938) in the United States almost three decades ago. Their sex-by-age curves are shown in Fig. 16-10. Judging from British statistics gathered for 1960 and 1966, the general picture still holds true. Note that it is almost the reverse of the sex-by-age distribution for schizophrenia. For the latter, age of onset is a

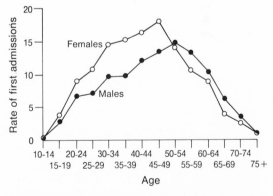

Fig. 16-10. Age and sex distributions for first admissions to mental hospitals for manic-depressive psychosis during 1933. (From Rosenthal, D. 1970. Genetic theory and abnormal behavior. McGraw-Hill Book Co., New York; based on data from Landis, C., and J. D. Page. 1938. Modern society and mental disease. Farrar & Rinehart, Inc., New York.)

Table 16-20. Risk in first-degree relatives of manic-depressive probands*

Incidence of deviations of manic-depressive type	Parents	Sibs	Children
Age-adjusted sample size	1772	2770	821
Manic-depressive, certain cases	7.7%	8.7%	11.8%
Manic-depressive certain, probable, and suicides	11.7%	12.3%	16.0%
Cycloid personalities	7.3%	7.8%	9.3%
Schizophrenia	0.4%	0.8%	2.3%

*Based on data from Zerbin-Rüdin, E. 1967. Endogene Psychosen. In P. E. Becker, ed. Humangenetik, ein kurzes handbuch, vol. 2. Georg Thieme Verlag KG, Stuttgart, West Germany; and Slater, E., and V. Cowie. 1971. The genetics of mental disorders. Oxford University Press, London.

good deal earlier for men than for women. No compelling explanations have been offered in either case, though some imaginative speculations have been offered (Burch, 1964). A further feature of the sex-by-age distributions appears to be that when the curves are corrected for the age structure of the base population, they both show a distinct bimodality. This is not obvious from the Landis and Page data, but is clearly evident in material gathered by Slater (1938) in Bavaria and corrected by the Strömgren method. Thus women show peaks at approximately 38 and 48, men at 36 and 54 years of age. Slater's curves also showed a narrower main risk period, extending mainly from 30 to 60 years for both sexes.

Family studies

The twelve main studies of risk in relatives of probands have again been collated by Zerbin-Rüdin (1967). Slater and Cowie (1971) have provided a convenient summation of the individual results as shown in Table 16-20. It should be mentioned that methods used vary fairly widely from study to study, and hence the ranges of risk for each category of relative are in excess of 20%. However, in terms of averages, parents and sibs do not differ markedly, but children are unaccountably high. This may reflect an underestimate of risk in parents (as appeared to be the case in parents of schizophrenic index cases) or an overestimate for children.

In general, most workers appear to take the view that the risk in first-degree relatives is roughly constant and somewhere around 10%. This suggests the action of a dominant gene with rather low penetrance.

Another interesting feature of the data in Table 16-20 is that the risk of schizophrenia in parents and sibs is at a level *below* expectation in the general population. In children, however, it is markedly elevated. This is the general picture. There are, however, more detailed aspects of the work done that should now be discussed.

Some work has been done on second-degree relatives of probands. From the figures summarized by Rosenthal (1970) for half-sibs, uncles, aunts, nieces, nephews, and grandchildren, the median risk is slightly over 2%. Thus morbidity risk goes down with decreasing consanguinity but is still above that in the general population. This is in accord with genetic theory.

In late onset cases of affective disorders (mostly involutional melancholia) the family risk seems appreciably greater than that in families of early onset cases. The work done on this dimension is somewhat ambiguous, particularly since there are grounds for supposing we may be dealing with different entities at different age levels. However, Hopkinson and Ley (1969) have reported risks in first-degree relatives of probands below and above 40 years. Risks found were, respectively, 28.9% and 12.5% (age corrected). The

Table 16-21. Morbidity risk in families of bipolar and unipolar probands: results of several investigations*

Study	Proband	Percent risk in first-degree relatives	
Leonhard, Korff, and Schulz (1962)	Unipolar	27.7	
	Bipolar	39.9	
Asano (1967)	Unipolar	34.5	
	Bipolar	39.3	
		Bipolar	*Unipolar*
Angst (in Angst and Perris, 1968)†	Unipolar	0.29	9.10
	Bipolar	3.70	11.20
Perris (in Angst and Perris, 1968)†	Unipolar	0.35	7.40
	Bipolar	10.80	0.58

*Modified from Winokur, G., P. J. Clayton, and T. Reich. 1969. Manic-depressive illness. The C. V. Mosby Co., St. Louis.
†Independent studies.

notion of two types of affective illnesses, each peaking at a certain age, generally fits with the bimodal character of the age-risk function. However, other workers have not agreed with the Hopkinson-Ley results (Price, 1968; Slater and Cowie, 1971).

As might be expected, if psychological or somatic precipitating factors are found to be present in the proband's history, risk in his family is commensurately less. Stenstedt (1952) found the reduction to be on the order of 50%.

One of the most significant lines of investigation has been that concerned with breaking down the global syndrome of manic-depression into more precise subcategories. Leonhard (1957) has been a pioneer in this respect in suggesting a distinction between bipolar, or classical manic-depressive psychosis, and unipolar or monopolar disorders involving predominantly episodes either of depression or of mania. The dichotomy is by no means an absolute one, as shown by family studies. However, it does seem to be useful. Some relevant data are summarized in Table 16-21. The data of Leonhard, Korff, and Schulz (1962) show, in the first place, a tendency for the families of bipolar patients to be at considerably higher risk. This was confirmed by Asano (1967). However, in both studies, the numbers were small. Winokur and Clayton (1967) used a reverse ap-

proach by examining the kind of affective disturbance in families with a generally poor history and in those with a good history. Bipolar psychoses occurred more frequently in the former.

Second, taking the lower half of the table, we can see that the data of both Perris and Angst confirm the preceding. In addition, the findings of Perris (reported by Perris in 1966 and then jointly with Angst in 1968) tend to show that polarity runs in families. Bipolar probands have a high percentage of bipolar psychotic relatives and a low proportion of unipolar relatives. The findings of Angst, however, are not quite consonant. As shown in Table 16-21, his data show that unipolar psychosis is more common than bipolar in relatives of both unipolar and bipolar probands. Price (1968) has examined some earlier data of Stenstedt (1952) and has concluded that these also do not fully support the Leonhard-Perris dichotomy. Consequently, the matter is still in some doubt. Clearly, unipolar and bipolar psychoses must overlap to a considerable degree, and the familial risk figures any worker obtains for either will depend greatly on the initial diagnostic criteria used (Price, 1968; Winokur, Clayton, and Reich, 1969).

Adding sex differences to the polarity category makes the whole problem even more complex. We have already noted that females

Table 16-22. Concordance rates for manic-depressive and affective disorders in MZ and DZ twins*

Study	MZ		DZ	
	N	Percent con-cordant (rounded)	N	Percent con-cordant (rounded)
Luxenburger (1930b)	4	75	13	0
Rosanoff, Handy, and Plesset (1935)	23	70	67	16
Kallman (1954)	27	93; 100†	55	24
Slater (1953)	8	50‡	30	23
Da Fonseca (1959)	21	75	39	38
Harvald and Hauge (1965)	10	50	39	3

*From Genetic theory and abnormal behavior by David Rosenthal. Copyright © 1970 by McGraw-Hill, Inc. Used with permission of McGraw Hill Book Co.
† Age-corrected rate.
‡ Slater's estimate is higher (80%) when only strict manic-depressive psychoses are included and rates are age corrected.

show a higher incidence of affective disorders. On the whole, this general difference does not appear to interact in any reliable manner with subtypes of manic-depressive disorder.

Twins

Only about six studies deal with twin concordance for affective disorders. These have been summarized by Zerbin-Rüdin (1969), Rosenthal (1970), and Slater and Cowie (1971). Major findings are shown in Table 16-22. Again, studies vary according to type of sampling method, diagnosis, and statistical procedures. Consequently, the range of MZ and DZ concordances is fairly large. Another point, emphasized by Rosenthal (1970), is that the number of MZ proband cases is much smaller than we might expect, given the population incidence of affective illnesses. The most probable answer to this is that prognosis for such disorders is relatively good compared to that for schizophrenia. Thus if sampling of resident hospital populations is carried out, many twins are liable to be missed, having been discharged back into the community.

Winokur, Clayton, and Reich (1969) have analyzed the details of the data summarized

in Table 16-22 and have shown the following: (1) for 34 MZ pairs involving a co-twin who had had a manic attack at least once, concordance for affective disorder was 82%; (2) 7 of these 28 concordant pairs showed dissimilar symptoms, 1 twin showing mania, the other depression; (3) for 19 manic DZ probands, concordance was 37%; and (4) of these 7, 5 showed dissimilar types of affective illness.

A somewhat simpler summation procedure has been carried out by Slater and Cowie (1971), who educe a 72% MZ and 19% DZ same-sexed concordance rate. However, in view of the heterogeneity of the studies, they suggest caution in accepting this.

A final question relates to twin concordance in respect to polarity of symptoms. Zerbin-Rüdin (1969) has attempted to assess studies in which such information is available. Table 16-23 summarizes the main results of this analysis. It seems clear that there is fairly high concordance for polarity. In fact, the relatively few cases where there is disparity may be so classified only because of once-in-a-lifetime episodes. The twin data are thus in line with the family data in pointing to a genetic distinctiveness in dispositions to unipolar or bipolar affective illness. A later study by Bertelsen, Harvald,

Table 16-23. Distribution of affective illness in twins according to clinical subtypes*

	Twin pairs			
	MZ		DZ	
Subtypes	N	%	N	%
Both unipolar depressive	22	26.5	8	14.3
Both bipolar	16	19.5	0	0
Both manic	5	6.0	0	0
One manic, one manic-depressive	0	0	1	1.8
One unipolar, one bipolar	5	6.0	3	5.4
One manic, one depressive	2	2.4	1	1.8
Incompletely concordant	9	10.8	7	12.5
Discordant (co-twin normal)	24	28.9	36	64.3
TOTAL	83		56	

*From Slater, E., and V. Cowie. 1971. The genetics of mental disorders. Oxford University Press, London; based on data from Zerbin-Rüdin, E. 1969. Zur Genetik der depressiven Erkrankungen. In H. Hippius and H. Selbach, eds. Das depressive Syndrom. Verlag Urban Schwarzenberg, Berlin.

and Hauge (1977) has confirmed this using 55 MZ and 52 same-sexed DZ twin pairs. In the 32 concordant (strict criteria) MZ pairs, 11 were concordant unipolar (all females), and 14 bipolar. In the remaining 7, one co-twin was unipolar, the other bipolar. In the 9 concordant DZ pairs, in 2 female pairs both members were bipolar, and in another 2 female pairs both were unipolar. Of the concordant DZ pairs, 5 were discordant with respect to polarity.

GENETICS OF AFFECTIVE DISORDERS

The matter of mode of inheritance of the affective illnesses is still an open question (Gershon et al., 1976). Many workers prior to 1960 (e.g., Slater, 1938; Merrell, 1951; Stenstedt, 1952; Kallman, 1954) were disposed to favor a major autosomal dominant with low penetrance. This view is supported fairly well by (1) the near equality of risk in parents, sibs, and children of probands; (2) the apparent absence of consanguinous mating; and (3) the relative risks in first-, second-, and third-degree relatives of probands.

Aside from the perennial problem of penetrance, another difficulty involves the fairly appreciable sex differences. This has been explained in different ways by different writers. Slater, Merrell, and Kallman, for example, suggested simply that penetrance varied between the sexes or, possibly, that there was a higher incidence of suicide among males, thereby removing them from a sample. Sainsbury (1968) has shown that, in fact, suicide rate is much higher in males than females, particularly in older age groups. More specifically, Pitts and Winokur (1966) estimated that among manic-depressives, suicide risk is four times greater for males. However, Perris and d'Elia (1966) found no difference between sexes in their sample of manic-depressive patients. Another possibility is sex limitation. Female hormones, especially those relating to the menstrual cycle, might dispose to affective disorders. Again, there appear to be no hard data for this suggestion, either.

An interesting variant on the autosomal dominant model was put forward by Rosanoff, Handy, and Plesset (1935). On the basis of their data (though going considerably beyond them) they postulated the operation of two genetic factors. The first, an autosomal dominant, is a gene for cyclothymia, labeled C. This is assumed to have a wide distribution in the population and to function in all phases of emotional disturbance as well as in severely pathological conditions. It does not

Table 16-24. Distribution by sex of affected relatives of bipolar and unipolar probands: empirical risks and risks expected by sex-linked dominant gene action*

Group	Number of affected relatives		Theoretical male:female ratio
	Males	Females	
Bipolar male probands			
Sibs	10	17	1:1
Children	6	7	0:1
Bipolar female probands			
Sibs	18	19	1:3
Children	1	2	1:1
Unipolar male probands			
Sibs	14	11	1:1
Children	1	9	0:1
Unipolar female probands			
Sibs	11	16	1:3
Children	3	2	1:1

*Based on data from Perris, C. 1968. The course of depressive psychoses. Acta Psychiatr. Scand. 44:238-248; and Slater, E., and V. Cowie. 1971. The genetics of mental disorders. Oxford University Press, London.

give rise to psychosis, however, unless the second gene is also present. This is a sex-linked activation gene, *A*, also dominant, which may also relate to normal emotional life. By itself it cannot give rise to affective illness. Only the combination of the two genes does this in the model. Thus there are two possible male genotypes in which psychotic manifestation can occur: *CCAY* and *CcAY*. In females, correspondingly, there are four vulnerable genotypes: *CCAA*, *CcAA*, *CCAa*, and *CcAa*. This theory is an interesting one but, not being very easily testable, has not won much favor among psychiatric geneticists.

Similar models have been put forward by other workers. Hoffman (1921) favored the operation of three major dominant genes: Rüdin (1923) argued for two recessive and one dominant gene. Clearly, there are many alternatives. At present, the main ones are between major gene and polygene models and between sex-linked as against autosomal transmission.

Data relating to the first point are ambiguous. Slater and Tsuang (1968) showed no difference in the distribution of ascendant secondary cases, thus suggesting polygenic in-

heritance. However, as Slater and Cowie (1971) emphasize, the question is still an open one.

The matter of sex-linkage is just as uncertain. Perris (1968) has put forward data broken down by polarity and by sexes. These may give us some test of dominant sex-linkage and are summarized in Table 16-24. It is clear that the approximation of empirical to expected values is poor if not nonexistent in the bipolar groups at least. The fit is somewhat better for the unipolar groups.

On the other hand, Winokur, Clayton, and Reich (1969) have presented data that they feel are fully consonant with a sex-linked dominant model, *applying, however, only to unipolar affective illness.* Their study sampled a series of consecutive admissions at a hospital in the United States. The final sample was made up of 35 females and 26 males all diagnosed as manic depressives. They were manic at the time of admission, but most had also had depressive attacks. Information on 56 of these 61 probands was available. Age-corrected rates are shown in Table 16-25. Winokur et al. note from the table the following points: (1) of the female probands, about twice as many had ill mothers as ill fa-

Table 16-25. Age-corrected morbidity rates for risks of affective disorder in first-degree relatives of manic probands*

Group	Number	Percent risk for affective disorder	SE
Relatives of female probands (N = 31)			
Fathers	31	23	±7.7
Mothers	31	50	±9.1
Brothers	33	23	±9.0
Sisters	39	46	±9.4
Relatives of male probands (N = 25)			
Fathers	25	0	—
Mothers	25	63	±9.9
Brothers	29	30	±9.5
Sisters	24	39	±11.5
Relatives of male and female probands combined (N = 56)			
Fathers	56	13	±4.6
Mothers	56	56	±6.8
Brothers	62	27	±6.6
Sisters	63	44	±7.3
All parents	112	34	±4.6
All sibs	125	35	±5.0

*From Winokur, G., P. J. Clayton, and F. Reich. 1969. Manic-depressive illness. The C. V. Mosby Co., St. Louis.

thers; (2) however, 63% of mothers of male probands showed some affective disorder; (3) female probands had fewer ill brothers than ill sisters (though this difference is probably not statistically significant); (4) about 35% of parents and 35% of sibs of probands had some affective disorder; and (5) of all first-degree relatives, females of probands had a significantly greater risk.

From these data, Winokur et al. rule out autosomal and sex-linked recessive transmission. In fact, neither of these possibilities has been seriously put forward by any authority. Second, they rule out a simple dominance model (though not a dominant type of inheritance) on the grounds of clearly reduced penetrance in sibs and parents and on the grounds of the sex differences found. They conclude, therefore, in favor of a model of sex-linked dominance with incomplete penetrance. The model clearly seems to fit their data well. However, it is by no means certain that their data base is typical of all

data (it does not seem to be); nor is it clear whether their model uniquely fits their data (Rosenthal, 1970; Slater and Cowie, 1971). However, as additional supportive evidence, Winokur, Clayton, and Reich (1969) present data from a single pedigree (the "Alger" family) in which are manifested a variety of affective disorders together with protan color blindness. These appear to be linked genetically. The authors estimate that such a linkage could occur by chance only at the 0.004 level. They therefore conclude that the dominant gene for manic-depression is located on the short arm of the X chromosome. It is to be hoped that more work will be done on this interesting model by independent researchers. It may well be wrong. But it still represents a quite definitive position that deserves to be examined further.

The preceding data on affective illness and the various genetic models fit to them by different investigators have received a thorough review by Gershon et al. (1976). We urge the interested reader to consult this valuable article.

SUMMARY

The survey we have presented on schizophrenia and the affective psychoses is by no means complete. However, it is perhaps sufficient to supply to the reader most of the types of strategies used in psychiatric genetics, some of the kinds of problems encountered, and some of the major conclusions that may be educed. It must be admitted that the latter are, at best, rather general and tentative and derive from a data base that most experimental scientists in psychology or biology might regard as shaky. Nonetheless, the problem of the etiology of psychosis is obviously one of great importance. We cannot ignore it simply because it is not easy to study with precision. In fact, much progress has been made during the last two decades, as will be clear from a comparison of this chapter with the corresponding treatment in our first book (Fuller and Thompson, 1960). Specifically, we may note significant advances in (1) statistical methodology, particularly in respect to estimations of risk from prevalence data; (2) diagnostic procedures, which appear to be resulting in more uniformity in hospitals and across national boundaries; (3) development of controlled designs, such as in adoption studies; (4) biochemical and cytological analyses; (5) fitting of genetic models to data; and (6) development of large-scale team research programs.

It is difficult to know what the future will bring. Really major breakthroughs have not so far occurred in the study of severe mental disorders. If we were to predict what approach might be most likely to produce one, we would perhaps be inclined to opt for biochemical or pharmacological genetics. However, a breakthrough may be some time in coming.

17

Mental illness: other forms of mental disorder

NEUROTIC DISORDERS

In the early days of psychoanalysis, many considered the psychoses and the neuroses to be on a continuum. Today, however, they are classified separately. The World Health Organization lists ten categories under the general heading of "Neuroses, Personality Disorders and other Nonpsychotic Mental Disorders (300-309)." The most important category, neurosis, includes anxiety neurosis, hysterical neurosis, phobic neurosis, obsessive compulsive neurosis, and depressive neurosis. Other categories are personality disorders, sexual deviations, alcoholism, drug dependence, physical disorders of presumably psychogenic origin, special symptoms not elsewhere classified, transient situational disturbances, behavior disorders in childhood, mental disorders not specified as psychotic associated with physical conditions (World Health Organization, 1965). The classification used by the American Psychiatric Association is essentially similar if perhaps less explicit (Sahakian, 1970).

In this section, we will not be able to deal adequately with all of the separate entities falling under the general heading. However, at the outset, it may be well to examine briefly the basis for separating the general neurotic disorders from the psychoses.

In the first place, the distinction between the two appears to have at least a face validity. On the average, at least, they appear as different to most psychiatrists (and perhaps to the layman) as a bee appears different from a wasp to an entomologist. This does not mean, of course, that they may not merge into each other.

Second, we have the factor-analytic work of Eysenck and colleagues (Eysenck, 1971) demonstrating a fairly clear separation between a psychoticism factor and a neuroticism factor.

Third, Cowie (1961) has gathered data on the incidence of neurosis in children of psychotics compared with the incidence of neuroses in relatives of nonpsychotic control patients also attending hospitals (general and neurological). The children were of the same mean age. The major result was that there was no difference in incidence of neuroticism in children of psychotic as against control probands. In fact, as judged by both the neuroticism scale of the Maudsley Personality Inventory and by a Teacher's Report Form, children of psychotic probands showed slightly less neuroticism than children of normals. Since Cowie was not able to interview her subjects personally and since the matching procedures may not have been very exact, we should treat these results with caution. Slater and Cowie (1971) may exaggerate in concluding that such data "go flatly against the hypothesis of any genetical connexion between psychosis and neurosis" (Slater and Cowie, 1971, p. 93). However, Cowie's results taken at face value do tend to support the independence of psychosis and neurosis. This is not to say that they cannot occur in the same families. Not only do MZ schizophrenic index cases often have neurotic co-twins (Kringlen, 1967), but, in addition, according to Gardner (1967), preschizophrenic males were significantly more prone to show obsessive-compulsive symptoms than controls. Phobias were also much more common. Preschizophrenic females, on the other hand, were not very different from controls in respect to neurotic traits, although the largest difference was found for obsessive-compulsive traits. Mitsuda, Sakai, and Kobayashi (1967) have reported a fairly high incidence of schizophrenia in the families and sibs of some categories of neurotic probands. Again, obsessional states appear to be most implicated, as well as anxiety reactions and "depersonalization." This led Mitsuda et al. to postulate a genetic relation between schizophrenia and neurosis.

Thus the picture is still somewhat confused. On the whole, the authors would tend to support a genetic distinction between psychosis and neurosis. The incidence of the latter is not well established, but it must be very common; therefore there is little reason why it cannot occur with fairly high frequency in the families of any proband case,

psychotic or normal. Furthermore, if the incidence of neurosis is really higher in families with psychoses, this elevation may well be purely for environmental reasons. Indeed, stress surely plays a large part in the development of neurosis. The extensive wartime studies of Slater (1943) on 2000 soldiers in England have shown a clear relationship between stress encountered and neurotic breakdown and also between number of abnormal personality markers (e.g., poor home life, poor school record, poor intelligence) and amount of stress needed to produce disorder. Furthermore, some high correlations were found between type of personality and type of symptoms manifested. Some of these data are shown in Table 17-1. Since the assessments of personality were taken post hoc, the data may involve a certain amount

Table 17-1. Correlations (tetrachoric) between personality traits and neurotic symptoms in a sample of 400 neurotic soldiers*

| Category | Examples of | | Correlation |
	Trait	Neurotic symptom	
Obsessional	Caution Orderliness Rigidity	Compulsive thoughts or actions	0.76
Hysterical	Egocentricity Excitability Emotional dependence Vanity	Conversion symptoms	0.51
Paranoid	Resentment Suspiciousness	Suspicion Hostility Ideas of reference	0.50
Anxious	Inferiority feelings Self-distrust Timidity	Somatic anxiety Tremor Sweating Mood of fear	0.40
Depressive	Ready pessimism Moodiness	Observed mood changes of consider- able duration	0.39
Hypochondriacal	Preoccupation with bodily health Interest in get-fit schemes Fondness for patent medicines	Somatic localization of illness and preoccu- pation with physical aspect of illness	0.19

*From Slater, E. 1943. The neurotic constitution: a statistical study of two thousand neurotic soldiers. J. Neurol. Psychiatry 6:1-16.

of confounding. However, the results are hardly counterintuitive and broadly suggest constitutional dispositions towards neurotic breakdown in general and the manifestation of specific types of symptoms.

Incidence and distribution

It is obviously very difficult to put forward any firm figures in respect to incidence and distribution of neuroses in the general population. Undoubtedly, most neurotics, or a large proportion, never seek help and hence go officially unnoticed. An early survey by Fremming (1947) gave 2% as the population risk. Later, Bille and Juel-Nielsen (1963) carried out a fairly complete survey of the population in a Danish county in Jutland, defining incidence of neurosis as number of individuals requiring contact with a public medical or psychiatric service over a 12-month period and receiving a diagnosis of neurosis. In the total population of more than 161,000 persons over age 15 years, 0.7% were so diagnosed, with about three times as many females as males. Peak risk for females was around 30, for males, between 40 and 50 years. Another Danish study by Nielsen, Wilsnack, and Strömgren (1965) found a general rate of 2.63% with again an excess (4:1) of females over males. Still higher figures were found by Hagnell (1966) in southern Sweden. He reported lifetime prevalence rates for males over age 10 of between 7.8% and 10%; his estimate for females was between 16.5% and 20.4%. The peak ages of risk and the sex differences found in these studies are generally consonant with statistics gathered in the United States by the National Institute of Mental Health (Sahakian, 1970).

The preceding figures are probably lower bounds. Hence we can probably be sure that neurotic disorders are very common indeed. It also seems that they are more common in women, although this difference may be due mostly to environmental and cultural factors.

Family studies

There have been around twenty or more family studies of the neurotic disorders. Some of the earlier attempts focused largely on obsessive-compulsive symptoms, perhaps because these are more clear-cut and dramatic. Luxenburger (1930a) reported the following incidences in relatives of obsessional probands: fathers, 15%; mothers, 6%; and sibs, 14%. It is not clear why the maternal risk figure is so low (perhaps only at population risk), especially in view of some of the work we reviewed in Chapter 15 on maternal influences in personality. It seems highly likely that Luxenburger's figure is an underestimate, possibly due to the cultural mores in the population he studied. Lewis (1935) found much higher risk figures for obsessional traits: 37% for parents and 21% for sibs.

One of the more thorough family studies in the early literature was carried out by Brown (1942). His main findings are summarized in Table 17-2. It will be noted that he attempted to study some of the main subcategories of neurosis: anxiety states, hysteria, obsessional states, and anxious personality. The first fairly compelling point is that the family risk (16.4% to 33.2%) is, in general, considerably elevated above population risk, even assuming the figures we presented for the latter are lower bounds. Thus between 36% and 41% of first-degree relatives of neurotic probands also have a disorder of some sort. Second, however, the risk in second-degree relatives is very low—if anything, a good deal lower than population risk. This is a rather surprising finding and hardly predictable from any simple genetic model. Third, there is a tendency (perhaps overestimated by Brown and others) toward a consonance between probands and relatives in respect to category of disorder. This is quite marked for probands with anxiety states, particularly if we lump together, for relatives, the diagnoses of anxiety states and anxious personality. Somewhat the same is true for hysteria. However, such an association seems lacking in the case of relatives of obsessional probands. Most of those who were neurotic suffered from anxious personality (24.8%). Only 6.9% were obsessional. Fourth, we may note the finding, not recorded in Table 17-2, that some psychosis was also found in relatives of probands. Fifth, data for first-degree relatives taken for parents and sibs separately do not make a great deal of sense. Thus, generally, risks are higher in parents than in sibs, a finding that does not support any kind of simple genetic model. Brown's study, though of interest, must therefore be judged to be rather inconclusive.

Rüdin (1953) focused on obsessive illnesses and found risks in parents, sibs, and children of probands of 5%, about 2%, and 1%, respectively. These rates are clearly lower than some rates we noted before.

At least three other studies besides that of Brown have found fairly high risk of anxiety neuroses in first-degree relatives of probands with this diagnosis (McInnes, 1937; Cohen et al., 1951; Coppen, Cowie, and Slater, 1965). This runs around 15% for parents and sibs together, although generally, as in Brown's study, parents of probands show higher incidences than do sibs. In at least two studies (Cohen et al., 1951; Coppen et al., 1965) there was found to be an unusually high rate of anxiety neurosis in mothers of chronic anxiety patients.

Brown (1942) showed that there was some tendency for hysteria to be transmitted as an entity. Later, Ljungberg (1957) studied hysteria specifically, starting with a proband sample of 381 cases with such symptoms as difficulties in walking, fits, paralyses, and

other problems commonly associated with hysteria. A number of these, however, had suffered brain damage. In first-degree relatives, risks for hysteria were 2.4% for males and 6.4% for females. Rates for parents, sibs, and children (sexes combined) were not appreciably different (4% to 5%), suggesting dominant gene action with low penetrance. Ljunberg himself, however, favored a polygenic model. However, it should also be noted that progeny of two unaffected parents were hysteric as often as offspring of one unaffected and one affected parent. This observation, to the extent it is valid and general, does not fit easily with any simple genetic model.

Hysteria was also studied by Guze and colleagues in a series of investigations (Arkonac and Guze, 1963; Guze, 1967). They attempted to develop a more precise definition of the disorder, abandoning the usual reference to conversion symptoms. Diagnosis was based mainly on an adjective checklist involving such signs as early onset, various somatic complaints (e.g., to do with menstrual cycle or headaches), anxiety, and repeated hospital admissions. In several publications, Guze and others (Arkonac and Guze, 1963; Guze, 1967; Woerner and Guze, 1968) have shown an increased prevalence of hysteria in first-degree relatives of female hysteria probands (about 24%). In a comparison

Table 17-2. Risks in first- ($N = 573$) and second-degree ($N = 1247$) relatives of probands with different categories of neurotic disorders*

Type of proband	Percent incidence in relatives†								
	Anxiety state		Hysteria		Obsessive state		Anxious personality		Normal first degree
	First degree	Second degree	First degree	Second degree	First degree	Second degree	First degree	Second degree	
Anxiety state	15.1‡	2.7	2.2	0.4	0.3	0	16.7	11.8‡	57.0
Hysteria	6.5	1.9	11.2‡	1.5	0	0	9.3	9.2	64.5
Obsessional state	3.0	0.4	0	0.8	6.9‡	0.8	24.8‡	16.8‡	59.4

*From Brown, F. W. 1942. Heredity in the psychoneuroses. Proc. R. Soc. Med. **35**:785-788.
†Under first-degree relatives, the remaining percentages are accounted for by the category "other conditions." This is omitted here. Much the same applies in the case of incidence figures for second-degree relatives.
‡Significantly different from control figures.

group of 167 women in a maternity hospital, only 3, or 1.8%, were found to be hysteric. Male relatives of probands were found to show an increased prevalence of alcoholism and sociopathy. Guze has concluded, like Brown, that hysteria, as he defines it, is inherited as an entity. However, the numbers in his studies were small and his definition perhaps somewhat arbitrary.

We saw before that some attention has been given to obsessive-compulsive neurosis by earlier workers. Sakai (1967) found that in 26 out of a total of 65 families of obsessive-compulsive probands, at least one member showed this disorder. However, the families also contained schizophrenics, depressives, psychopaths, epileptics, and mental defectives. Rosenberg (1967) also found a similar picture in the first-degree relatives of his 144 obsessional probands; this is approximately a 10% incidence of some form of psychiatric illness. However, they included only two cases of obsessional neurosis. Consequently, the information we have about this form of illness is ambiguous.

So-called reactive depression has been studied by Stenstedt (1966). He sampled 1242 first-degree relatives of 176 proband cases (54 male and 122 female). Incidence in the relatives for affective disorders was set by him at 2% and 7.5% for males and females, respectively. He felt that these figures were appreciably higher than population risk. This may or may not be so. Furthermore, it should be noted that much of his information on relatives was obtained second-hand.

Two miscellaneous family studies may also be mentioned. Oki (1967) studied the parents and sibs of probands with the diagnosis of "early childhood neurosis." Of the parents, 45% manifested neurosis or "nervousness." The corresponding figure for sibs was 33%. Rates in parents and sibs of a group of controls matched for age, sex, IQ, parental occupation, and physical state were 4% and 0%, respectively. A fairly high incidence was also found in second-degree relatives of probands, but none in those of controls. Tsuda (1967) has reported elevated rates of various illnesses in families of probands diagnosed as having "depersonalization neurosis." Disorders included neurosis, psychopathy, and schizophrenia. Some depressions occurred but at a low rate (3.7%).

A number of workers have collected family data on enuresis. Perhaps the most important of these is Hallgren (1957), who has also provided an excellent review of literature on this topic from the beginning of the century. He examined 423 parents and 262 sibs of enuretic probands. His major results are shown in Table 17-3. It is clear, in the first place, that risks are higher in males than in females. The differences are, in fact, statistically significant. Second, expectancies are about equal (~25%) in parent and sib groups. Excluding families in which determination appeared to be clearly nongenetic, Hallgren arrived at the following estimates: 38.5% for fathers, 23.4% for mothers, and 38.5% for male and 20.5% for female sibs. Rates for uncles and aunts were around 8%, and for grandparents, 2%. These last two values were no higher than the population risk, placed by Hallgren at 9.5%. He went on to test eight different genetic hypotheses, but,

Table 17-3. Incidence of enuresis in relatives of 215 probands*

	Parents		Siblings	
	Affected *N*	Not affected *N*	Affected *N*	Not affected *N*
Male	64	144	40	89
Female	38	177	25	108
TOTAL	102	321	65	197

*From Hallgren, B. 1957. Enuresis, a clinical and genetic study. Acta Psychiatr. Scand. [Suppl.] 114:1-159.

because of the complexity of his data, he was unable to arrive at any definite conclusion. However, either of the two following models seemed to him most plausible: determination by a single dominant major gene whose expression was determined by environment and/or polygenes or determination solely by the interaction of polygenes and environment. It is of interest that in spite of apparent sex differences for enuresis, genetic analysis failed to support either complete or partial sex-linkage, although they did not definitely disprove these possibilities.

Virtually all of the studies just reviewed have been clinical in orientation and carried out in a psychiatric setting usually with patient samples. However, in Chapter 15 we reviewed the work of a number of researchers who have used objective psychometric tests. These studies have treated neuroticism as a graded and measurable character. The reader may wish to refer back to these at this point. However, we may mention that at least some of these (e.g., Coppen, Cowie, and Slater, 1965; Insel, 1974) demonstrated significant familial effects for neuroticism as measured by personality inventories. They also indicated fairly strong maternal effects, a trend that is supported by at least some of the data on neuroticism discussed in this chapter. It would seem advantageous if, in the future, those working on neurotic and personality disorders would use not only the clinical methods of psychiatry but the more objective instruments of the psychologist. Progress would perhaps be more rapid than it has been up to now.

Twin studies

Not a great deal of work has been done on the neuroses using twins, and much of it is of dubious scientific value. A reasonably complete sample of the studies done on general neurosis is shown in Table 17-4. Again, as with other summaries we have presented, the reader should be warned that the studies vary considerably in quality of diagnostic procedures, zygosity determination, and sampling. In addition, the range for relative concordance rates is fairly wide. The study by Idha (1961), for example, suggests that heredity is relatively unimportant. On the other hand, perhaps the most well-executed study of the group by Shields (1954) yielded a concordance rate in MZ twins about six times that found in DZ twins. Even if partial concordance rates are included, the ratio is still about 3:1 in favor of MZ twins. For the whole table, in fact, the MZ:DZ concordance ratio is about 3:1 to 4:1. Consequently, the data in Table 17-4, taken as a whole, suggest fairly strong hereditary influences.

This conclusion is supported by a widely cited study by Eysenck and Prell (1951). It will be recalled that Eysenck has argued strongly for a unitary dimension of neuroticism that is definable in terms of various inventory items and experimental tests. With

Table 17-4. Concordance rates for neurosis in MZ and DZ twins: some representative findings

Study	MZ		DZ	
	N	Concordant	*N*	Concordant
Kent (1949)	6	6	9	2
Slater (1953)	8	2	43	5
Shields (1954)	36	25 (30)*	26	3 (8)*
Idha (1961)	20	10	5	2
Tienari (1963)	21	12	—	—
Parker (1966)	9	6	11	4
TOTAL	100	61 (66)*	94	16 (21)*

*Cases including "partial concordance."

Prell, he administered these tests to 20 MZ, 24 DZ, and 6 pairs of twins whose zygosity was doubtful. His results are summarized in Table 17-5. On the basis of Holzinger's heritability statistic, neuroticism proved to be strongly dependent on hereditary factors. Hereditary loadings for particular tests are also included in the table. It should be noted, however, that of the tests listed, only three have loadings over 0.4 for the neuroticism factor. These are static ataxia, body-sway suggestibility, and autokinetic movement. These also (apart from intelligence) have the highest hereditary determination. Finally, the heritability of the neuroticism factor as a unit is in excess of 80%. This is higher than most corresponding estimates given for intelligence. According to Rosenthal (1970), however, an attempt by Blewett to replicate the Eysenck-Prell study was not successful.

Two studies somewhat along the same lines, except using separated MZ twins as well as MZ twins reared together and DZ twins, were carried out by Newman, Freeman, and Holzinger (1937) and by Shields (1962). The major thrust of both had to do with intelligence, and they have been reviewed in an earlier chapter. However, the authors also administered to their twin samples a questionnaire designed to measure neuroticism. The main results of both studies are shown in Table 17-6. It is obvious that heritability as measured by the Bernreuter is not high. However, in the Newman et al. study, there is little difference between MZAs and MZTs. In the Shields study, on the other hand, MZAs are more alike than MZTs and in both of these samples the cotwins are more alike than members of DZ pairs.

Jinks and Fulker (1970) have carried out a biometric genetic analysis of Shields' data.

Table 17-5. The hereditary determination of neuroticism: MZ and DZ twin pair correlations on individual tests and on a neuroticism factor*

Trait	MZ pairs		DZ pairs		Hereditary determination, H'†
	Raw correlation	Correlation corrected for age	Raw correlation	Correlation corrected for age	
1. Intelligence	0.905	0.890	0.670	.0.660	0.676
2. Tapping area	0.193	0.164	−0.148	−0.144	0.269
3. Tapping speed	0.557	0.552	0.266	0.011	0.547
4. Level of aspiration	0.320	0.272	0.084	0.038	0.243
5. Motor-speed test	0.700	0.643	0.296	0.243	0.528
6. Speed of decision	0.340	0.339	−0.122	−0.122	0.193
7. Static ataxia	0.857	0.856	0.537	0.532	0.692
8. Body-sway suggestibility	0.737	0.734	0.128	0.110	0.701
9. Strength of grip	0.850	0.774	0.468	0.392	0.628
10. Word dislikes	0.512	0.510	0.394	0.380	0.210
11. Personality inventory	0.369	0.365	0.273	0.257	0.145
12. Lie scale	0.485	0.481	0.167	0.109	0.418
13. Flicker fusion	0.709	0.705	0.229	0.209	0.627
14. Autokinetic movement	0.734	0.722	0.228	0.210	0.648
15. Autokinetic suggestibility	0.534	0.534	0.141	0.135	0.461
16. Backward "S"	0.711	0.708	0.491	0.477	0.423
17. Fluency	0.357	0.353	0.118	0.114	0.270
NEUROTICISM FACTOR	0.851		0.217		0.810

*From Eysenck, H. J., and D. B. Prell. 1951. The inheritance of neuroticism: an experimental study. J. Ment. Sci. 97:441-465.
†Holzinger's heritability statistic.

Table 17-6. Results of two studies of correlations for neuroticism in MZ twins reared together and apart and in DZ twins

Study	Test	Correlations		
		MZA	MZT	DZ
Newman, Freeman, and Holzinger (1937)	Bernreuter neuroticism scale	0.58	0.56	0.37
Shields (1962)	Neuroticism questionnaire	0.53	0.38	0.11

Using several estimation procedures, they arrived at heritability estimates ranging from 0.37 to 1. The figure they put forward as the most likely is 54% ± 7% for both broad and narrow heritability of neuroticism. Shields' sample of twins was such as to allow some control over environmental sources of variance, unlike samples used in most other studies. It also involved larger samples than found in almost any other study. Consequently, the conclusion of Jinks and Fulker must be taken seriously.

Various twin studies have focused on particular types of neurotic disorders. Marks et al. (1969) have reviewed work on obsessive-compulsive disorders. Overall, they estimate MZ concordance to be around 75%. Such a result is typified by the study of Inouye (1965), who used samples of 21 MZ and 5 DZ pairs. Out of six categories of neurosis, concordance differences were found only for obsessive-compulsive reactions. Concordance rates were 80% and 50% for MZ and DZ pairs, respectively. Inouye's review of the literature up to 1960 suggested even more striking differences. On the other hand, Slater and Cowie (1971) report a lower figure of 50%, but based on only 6 MZ pairs. However, the general picture appears to fit reasonably well with the results of family studies.

Hysteria has been examined by Stumpfl (1937) using 9 MZ and 9 same-sexed DZ pairs. Concordance rates were around 55% and 0%, respectively. Slater (1961) studied 12 MZ and 12 DZ probands with the diagnosis of "hysteria 311," defined by the World Health Organization as "hysterical reaction without mention of anxiety reaction." By this strict definition, concordance was 0% for MZ and 8% for DZ pairs. With the broader category of "neurosis," 3 MZ and 2 DZ pairs could be regarded as concordant. Slater's negative conclusion was supported in his follow-up study (1965) and that of Slater and Glithero (1965). It is not in agreement with at least some of the work discussed before, especially that of Guze and co-workers.

Finally, anxiety states were studied by Slater and Shields (1969) in a series of 20 MZ and 40 DZ twins. Concordance rates were 65% and 13%, respectively. According to the authors, the various psychiatric deviations found to occur in MZ co-twins occurred only along the anxiety axis. In the same study, Slater and Shields reported large differences in MZ-DZ concordance rates for personality disorders but no differences for "other" neuroses, mainly reactive depressions.

The authors conclude in favor of a constitutional predisposition for becoming anxious that is basically adaptive and normally distributed in the population. Given a certain genotype and sufficient stress, however, maladaptive anxiety may occur and be manifested in a variety of forms. This simple model would seem to have considerable heuristic value.

To close this section, we may refer back to the twin work by a number of researchers using the MMPI (Chapter 15). This had to do, of course, mostly with a presumably normal population in which only propensities toward different kinds of disorder might be present. For the combined data, the scales showing significant MZ-DZ differences were depression, psychopathic deviate, psychasthenia, schizophrenia, and social introversion. The applicability of these results to psychiatric samples is not altogether clear, as Slater and Cowie (1971) suggest.

Genetics

Little information is available about the mode of genetic transmission of neurotic disorders. Most authorities would probably favor a polygenic model. Indeed, about the only formal analysis that has been carried out —that by Jinks and Fulker (1970) on Shields' data—has concluded that a "very simple genetical model is adequate to explain the data." Only additive gene action appeared to be operating, indicating that intermediate expression of neuroticism has been favored by natural selection. This seems consonant with the implied suggestion of Slater and Cowie (1971) that a moderate amount of neurosis (or, more strictly, anxiety) is adaptive and valuable for coping with the environmental stresses that typically occur during the lifetime of an individual. Too much or too little is maladaptive.

OTHER FORMS OF DEVIANT BEHAVIOR
Criminality

It was popular in the eighteenth and nineteenth centuries to attribute criminal behavior to some defective or perhaps atavistic biological constitution. Men like Lombroso in Italy, for example, gathered large amounts of data to back up this claim for a "criminal type" or genotype. The same orientation was taken by Hooton in the United States and by Galton in England. It is at least implicit in the pedigree studies carried out by Dugdale and Goddard on the Jukes and Kallikak families. More recently, Glueck and Glueck (1956) attempted to establish relationships between somatotypic characteristics as derived from anthropometric measurements and criminal or delinquent behavior. It is safe to say that today the weight of opinion favors the importance of environmental factors. Undoubtedly these play a major role. However, there is still a considerable body of data accumulated over the last half-century that implicates genetic predispositions. Such

data fall into two categories, which we will discuss separately. The first encompasses twin studies; the second, of more recent vintage, relates to the involvement of aneuploidies, particularly the 47, XYY karyotype, in criminal behavior.

Twin studies. The major results of the twin studies of criminal behavior are summarized in Table 17-7. Five countries are sampled with totals of 260 MZ and 487 DZ twin pairs. If the studies are lumped together, we find that, on the average, criminality and delinquency occur in both members of an MZ pair about three times as often as in only one member. Almost exactly the opposite holds true for DZ pairs. Taken at face value, such findings are certainly compatible with a hereditarian model. They are not, of course, incompatible with an environmental model either, though the difficulties of such an explanation are perhaps much greater than in the case of personality and intelligence. Criminal behavior and delinquency are much more explicitly defined by legal and moral systems and therefore should entail much more explicit teaching (by instruction or example) than variables of personality or intellect. Stealing, for example, is a fairly clearly delineated behavior. Introverted or intelligent behavior is not. It is difficult, then, to suppose that parents would instruct one family member that stealing is permissible but not another member of like age and sex unless he or she has an identical genotype. Thus one can hardly avoid the conclusion that genetic factors play some role. Let us examine a few of the studies listed in more detail.

The first study by Lange (1929, 1931) is of considerable historical interest. It appeared as a monograph, entitled *Crime as Destiny*, and attracted quite a large amount of attention. His operational definition of criminality was actual conviction and incarceration rather than psychopathic personality. However, it is clear from his clinical descriptions that many subjects suffered from a variety of physical and mental disorders. His zygosity determinations were largely based on appearance and fingerprints. Sampling was, of course, far from systematic, and it may be,

Table 17-7. Summary of twin studies of criminal and delinquent behavior*

Condition	Study	Country	MZ		DZ	
			Con-cordant	Dis-cordant	Con-cordant	Dis-cordant
Adult crime	Lange (1931)	Germany	10	3	2	15
Adult crime	Legras (1933)	Holland	4	0	0	5
Adult crime	Rosanoff, Handy, and Rosanoff (1934)	United States	25	12	6	54
Juvenile delinquency	Rosanoff, Handy, and Rosanoff (1934)	United States	39	3	28	37
Childhood behavior disorders	Rosanoff, Handy, and Rosanoff (1934)	United States	41	6	34	55
Adult crime	Kranz (1936)	Germany	20	11	30	63
Adult crime	Kranz (1937)	Germany	11	5	13	9
Adult crime	Stumpfl (1936)	Germany	11	7	9	38
Adult crime	Borgström (1939)	Finland	3	1	4	11
Psychopathy and neurosis	Slater (1953)	England	2	6	5	38
Adult crime	Yoshimasu (1965)	Japan	14	14	0	26
Juvenile crime	Hayashi (1967)	Japan	11	4	3	2
TOTAL			191	69	134	353
Percent concordance or discordance			74	26	28	72

*Modified from Shields, J. 1954. Personality differences and neurotic traits in normal twin school children. Eugen. Rev. 45:213-246; Fuller, J. L., and W. R. Thompson. 1960. Behavior genetics. John Wiley & Sons, Inc., New York; Rosenthal, D. 1970. Genetic theory and abnormal behavior. McGraw-Hill Book Co., New York; and Slater, E., and V. Cowie. 1971. The genetics of mental disorders. Oxford University Press, London.

mainly for this reason, that subsequent studies failed to yield results quite as dramatic as Lange's.

Kranz's two studies (1936, 1937) were a good deal more satisfactory, involving better sampling techniques and the use of blood typing for zygosity determination. In fact, his MZ and DZ concordance rates were not different at an acceptable level of statistical significance. However, there was a large difference between same- and unlike-sexed DZ pairs. The former showed a concordance rate of 54%; the latter, only 14%. This argued, Kranz felt, for a stronger environmental than genetic determination of criminality. In his later article, he attempted to divide his cases into endogenous and exogenous categories. The former included cases in which various mental or physical disorders appeared without unfavorable home environments; the latter included cases coming from homes that

were clearly poor. For the former category, MZ concordance was about 63% (7 out of 11 cases); for DZ twins, concordance was only about 18% (2 cases out of 11). Rates were about the same for MZ and DZ twins in the exogenous category. However, as Rosenthal (1970) points out, concordance rates for MZ and DZ twins together were higher, overall, for those in the exogenous than for those in the endogenous category. Kranz also attempted to examine the degree of similarity between members of concordant pairs for five dimensions of criminal behavior: frequency, severity, type, age at first conviction, and "global crime pattern." Three grades of similarity rating were used: "very," "somewhat," and "scarcely" similar. The scale is somewhat arbitrary. However, if the MZ twins are compared with DZs, we find differences at or approaching significance level for all except the category "type of

Table 17-8. MZ and DZ concordance rates for adult crime, juvenile delinquency, and childhood behavior problems*

	Percent concordant		
	MZ (N = 126)	**DZ (N = 214)**	**Ratio MZ : DZ**
Childhood behavior problems	87	40	2.3 : 1
Juvenile delinquency	93	44	2.1 : 1
Adult crime	70	10	7 : 1

*Modified from Rosenthal, D. 1971. Genetic theory and abnormal behavior. McGraw-Hill Book Co., New York; based on data from Rosanoff, A. J., L. M. Handy, and I. A. Rosanoff. 1934. Criminality and delinquency in twins. J. Crim. Law Criminol. 24:923-934.

crime." For the four other categories, MZ pairs tend to be disproportionately represented toward the "very similar" end of the scale. These are interesting findings. Unfortunately, since Kranz did not make blind assessments, his results must be viewed with some skepticism.

The study by Rosanoff, Handy, and Rosanoff (1934) is useful in providing concordance data on the development of criminal behavior. They used a concrete operational definition of the behavior in question, but their sampling procedures and zygosity determinations may not have been adequate. The comparison between MZ and DZ concordance rates for adult crime as against juvenile delinquency or early behavior disorders is made in Table 17-8. As indicated, MZ concordance relative to DZ concordance is very much higher for adult crime than for childhood behavior problems or juvenile delinquency. This suggests the interesting possibility that whatever genetic factors predispose to criminal activity, they may emerge only later in life, and that the criminal problems of childhood and adolescence may have a much more strongly environmental etiology.

Stumpfl's (1937) main contribution lies in his attempt to get at basic personality characteristics underlying criminal activity. His general concordance rates were about 61% in MZ and 37% in DZ twins, a difference that does not exceed chance expectation. However, in respect to "social orientation" and "essential personality traits," concordance was high in MZ but totally absent in DZ twins. It is very likely, however, that this difference may have been due to bias, since assessments were apparently not carried out blind.

The other studies listed in Table 17-7 do not add a great deal to the points we have already discussed.

We may summarize as follows:

1. The evidence taken as a whole does not invalidate a genetic liability of some kind to criminal activities.

2. At the same time, it is very clear that environment must play a very important part in crime. The case reports in the various studies suggest a pattern of chaotic home influences that could hardly fail to leave some mark.

3. Almost all studies carried out have found a much higher incidence of crime in males than in females. It is not clear whether this means lowered disposition to crime or simply less likelihood of incarceration in females. The former possibility appears the more likely.

4. Adult and juvenile crime probably involve different etiologies. However, it is likely (as suggested especially by the work of Yoshimasu [1965]) that an unfavorable record in early childhood and adolescence may well increase the probability of adult crime.

5. If genetic factors are implicated, they must act very indirectly. That is to say, genes may produce all kinds of mental or physical disorders which so handicap an individual emotionally that he turns to crime. It is not clear how this happens, particularly in view of the fact that mental and physical problems

are not reliably associated with criminal behavior.

The last point brings us to the second major part of our discussion, that is, the consideration of the association between criminality and clear-cut chromosomal damage.

Chromosomal anomalies. There is now a large literature on the possible involvement of chromosomal anomalies in criminal behavior. It has been well reviewed by Borgaonkar (1969), Kessler and Moos (1970), Owen (1972), Hook (1973), and Jarvik, Klodin, and Matsuyama (1973). We will attempt to provide only the basic essentials in this present section.

The association between the XYY genotype with tall stature, possible mental retardation, and aggressive behavior was first suggested in a report by Jacobs et al. (1965). This study involved 197 mentally retarded criminals. Among these, 7 XYY cases (3.6%) were found, an incidence considered well above that found in the normal population. Several famous individual murderers had previously been found to have the XYY complement, these including Robert Tait in Australia, Daniel Hugan in France, and John Farley in the United States. Consequently, the study of Jacobs and associates attracted considerable attention.

It will be clear that if such an association is to be firmly established, several requirements must be met: in the first place, it must be shown that the incidence of this karyotype is higher in populations considered to be deviantly aggressive than in normal populations; second, it must be shown that the extra Y chromosome produces behavior that other aneuploidies do not (e.g., XXY); and third, it should be shown that the deviant behavior is a direct rather than an indirect result of the extra Y. So far, most of the available data do not permit very firm conclusions in respect to these points. Let us consider each of them in turn.

Many estimates of incidence of the XYY karyotype have been made. Using normal adult males only, data from ten studies give a pooled rate of 8 cases for a total sample size of 6148, that is, 0.13%. Range was from 0% to about 0.2%. Surveys on newborn males

have also been carried out in the United States, England, Scotland, Canada, and Germany. Out of a pooled total N of 28,346, 29 XYY and 29 XXY cases were located. Thus the rate for both karyotypes in normal male populations appears to be about 1 to 2 per 1000 (Hook, 1973; Jarvik, Klodin, and Matsuyama, 1973), and the incidence of an extra Y seems to be about the same as the incidence of an extra X chromosome.

The next question concerns the incidence of the XYY in institutional settings, for example, mental hospitals. In about fourteen surveys carried out in a number of different studies involving a pooled total of close to 3000 cases, 13 cases of XYY and 14 cases of XXY were found. This represents an incidence in each case of about 4.6 cases per 1000, or 0.46%. The range is again considerable. In any case, it seems that in mental institutions, both aneuploidies show about the same elevation in rate.

Turning next to exclusively penal institutions, twenty or more studies have yielded 98 cases of XYY genotypes in a pooled total of approximately 5000 cases, or almost 20 per 1000 (1.9%). In the same total sample, however, the incidence of XXY was 0.9%, only half as frequent. Jarvik et al. (1973) conclude from these statistics that "criminals are the only group in which an extra Y chromosome occurs significantly more often than an extra X chromosome." Such a relation was also found by Witkin et al. (1976) as discussed later.

The previous data, then, suggest a statistical association between criminality and the extra Y chromosome specifically. In a sense, this association is unremarkable because it is derived from a comparison between rates that are very small indeed; 98 cases in 5000 can hardly be considered to be of great significance from the standpoint of the criminologist. However, from another point of view, we may rightly wonder if a study of these relatively rare cases may shed some light on the relation between maleness (conferred by the Y chromosome) and violence and aggression.

At least three general hypotheses may be put forward about the nature of the association (Hook, 1973). They apply in the case of any association alleged between two characters and are as follows:

1. The *associative hypothesis* focuses on the possibility that criminality and the extra Y are connected only by virtue of each being related to some common factor. Thus they could be associated if it were shown, for example, that some segments of the population in which the crime rate was high also yielded higher rates for the XYY genotypes. So far, no data support this hypothesis, but it is almost impossible to disprove.

2. The *social hypothesis* contends that certain outcomes of the extra Y produce difficulties in the social presentation of its possessor, this, in turn, leading to antisocial behavior. Specifically, XYY individuals tend to be exceptionally tall and often suffer from severe acne. It is not implausible that two such physical features might, in some cases, produce socially deviant behavior. At the same time, such a causal chain cannot be exclusive, since, obviously, there are many individuals of normal karyotype who have great stature and acne and yet are not criminals. In fact, Witkin et al. (1976), from chromosome analyses of over 4000 tall men (<184 cm) reported incidences of 0.29% for XYY and 0.39% for XXY. These seem within the normal range. However, within both these groups there was an elevation of criminality rate, though it was significant only for the XYY subjects.

3. The *biological hypothesis* postulates some direct but unknown hormonal physiological or neural effects of the aneuploidy, which, in turn, produce deviant behavior. Data relevant to this hypothesis have been thoroughly reviewed by Owen (1972).

It is fair to say, by way of summary, that there are very few biological sequelae of the XYY genotype that can be considered as definitively established. Some that have been examined include abnormalities of external genitalia, bone and joint abnormalities, dental irregularities, skin disorders, hormone levels, and cardiac anomalies. Owen's careful analysis of the data on stature shows that even the classical association between XYY and tallness is in fact, rather uncertain. It does appear to be true that aggression is stronger in males than in females. This suggests some involvement of the Y chromosome in aggressive behavior, but the biology of this involvement is very far from clear. A closer study of it in the various combinations of X and Y that occur should prove worthwhile.

Homosexuality

A variety of behaviors relating to sexuality are commonly considered to be deviant. The most important is perhaps homosexuality. The World Health Organization also lists under sexual deviation fetishism, pedophilia, transvestism, and exhibitionism, plus a variety of other more specialized abnormalities. Whether many of these so-called disorders are abnormal in the same sense as, say, schizophrenia is a moot point. Certainly within the last 10 to 15 years, there has been a fairly striking change in the orientation of psychiatrists and society in general to such sexual deviations. In fact, the American Psychiatric Association no longer classifies homosexuality as abnormal. Be this as it may, we may still consider the etiology of homosexuality to be of some interest, if not from a psychiatric, then from a purely biological point of view.

Some of the first work done on the problem did start from a biological model. Lang (1940, 1941) attempted to adapt Goldschmidt's theory of intersexes to human male homosexuality. If some homosexuals are, in fact, transformed females, the sex ratio in their sibs should deviate from the usual 106 males to 100 females. Lang gained access to the confidential records of the police departments of Munich and Hamburg and reported initially that the sex ratio in the sibs of 1015 cases of known homosexual probands was 121.1:100 ($\chi^2 = 13.54$, $p < 0.001$). Among sibs of married homosexual males, who were less likely in Lang's opinion to be biological intersexes, the sex ratio was more nearly normal. Jensch (1941), a colleague of Lang's, also worked on the problem in Breslau and Leipzig. A summary by Lang (1960) of the results of their researches is shown in Table

Table 17-9. Sex ratios in sibs of male homosexuals as compared with sibs of normal controls*

Study	Group	Proband (N)	Sibs (N) sexes		M/F × 100
			M	F	
Lang	Control	1296	3571	3349	106.6
	Homosexual				
	<25 years†	825	1166	1006	115.9
	>25 years†	952	1712	1281	133.6
Jensch	Homosexual				
	<25 years†	683	961	984	97.7
	>25 years†	1389	2833	2349	120.7

*Based on data from Lang, T. 1960. Die Homosexualität als genetisches Problem. Acta Genet. Med. Gemellol. 9:370-381; and Slater, E., and V. Cowie. 1971. The genetics of mental disorders. Oxford University Press, London.
†Age at time of ascertainment.

17-9. The data mostly seem to support Lang's prediction. However, the sex ratio for Jensch's under-25 group is in the opposite direction. In fact, if we pool these data with those of Lang, the sex ratio for sibs of 1508 homosexuals under 25 turns out to be 106.9, which is not different from the control group. Pooled results for the over-25 age group gives a sex ratio of 125.2, which is certainly different from normal expectation. This is a peculiar result for which no obvious explanation offers itself. One possibility is that, in the younger age group, families were not yet complete, but since homosexuals would hardly appear in police files much below 18 years of age, this does not seem plausible. Another possible explanation is that police assessment of what constitutes homosexuality may have become broader over a period of time. In this case, the older group might be more "feminine" than the younger group. This notion has some credibility in view of the fact that Jensch found that in the sibs of 244 homosexuals judged by their behavior to be more "feminine," the sex ratio was 157:100. A final possibility is that the results are quite spurious and are due to some kind of bias in ascertainment. This has some credibility in view of the fact that later workers have failed to confirm the Lang-Jensch data. Thus Darke (1948) analyzed the sex ratios in sibs of 100 known homosexuals in an Ameri-

can federal prison. He found evidence for a predominance of males mainly among sibs of younger rather than older probands. Likewise, Kallman (1952c) found no departure from normal sex ratios in the sibs of 145 homosexual index cases. These findings do not necessarily contradict those of Lang and Jensch but do place some limitation on their generality.

Two other findings in the Lang and Jensch research are worth noting. First, Lang (1960) presents data on sibs of female homosexuals reported to him from the Payne-Whitney Clinic in New York. Among the sibs of 150 women seeking help for homosexuality at that institution, the male-to-female sex ratio was 75.7. This is, of course, the inverse of the male case. Second, both Lang and Jensch obtained sex ratios for paternal and maternal half-sibs of some of their index cases. The sex ratio for paternal half-sibs (pooled data) was 128.9. For maternal half-sibs, it was 86.8. The difference between the two figures is highly significant. Again, these data are not easily accounted for by either environmental or genetic models. However, a more sociological explanation has been put forward by James (1971), who suggests, in essence, that such sex ratios could occur in half-sibs if (1) the index case is a child of the first union; (2) parents separated or divorced tend to keep their like-sexed children, that is, fathers keep sons, mothers keep daughters. This is an interesting theory but has not been confirmed by careful examination of relevant data.

Considering the Lang model broadly, it is true, as Slater and Cowie (1971) point out, that male homosexuals clearly do not have a 46,XX karyotype. In this simple form, the model is wrong. Nevertheless, it is still possible that the condition of homosexuality might be brought about by some anomalies in the chromosomes or in the complex set of pathways between them and behavior. Slater (1962) has obtained some data that bear indirectly on this possibility. He computed, for a sample of 401 male homosexuals, mean ordinal birth position and maternal age at birth. These are usually directly correlated. The former is expressed by the index

$$\frac{m - 1}{n - 1}$$

where

m = ordinal position in sibship
n = number of children in family

The expression has a range of value from 0 to 1 and a mean, for a random sample, approaching 0.5. For the homosexual sample, the index value was 0.58, indicating that homosexuals tend to be later in position of birth. It is of interest, parenthetically, that Slater found the opposite to be true for exhibitionists and transvestites. Mean maternal age for the general population was found to be 28.5 years (N = 632,408). For male homosexuals, it was 31.3 years. The maternal age distribution curves for these two categories and for Down syndrome cases are shown in Fig. 17-1. Not only is the curve for homosexuals displaced to the right, but it also has a significantly higher variance than the normal group. Slater concluded that these data suggested a "heterogeneous" etiology for homosexuality with a "chromosomal anomaly, such as might be associated with late maternal age" as a possible causative factor. As Slater and Cowie (1971) have pointed out, however, psychological explanations of these results are also feasible.

A re-analysis of Slater's cases by Abe and

Moran (1969) added another interesting piece of information: an even greater shift in paternal age. Mean age of fathers of homosexuals was 34.9 years, as compared with around 31.6 for a normal population. The authors suggested that possibly homosexual tendencies were transmitted from fathers. In the latter, such deviance was manifested in a "tendency to marriage at a later age than the norm." There is clearly an interesting though difficult area for further exploration here.

A number of twin studies have also been carried out in the study of homosexuality. Main results are shown in Table 17-10. For 55 MZ co-twins, pooled concordance was around 86%. For 39 DZ twins, it was between 13% and 30%, the rather wide range a result mainly of the manner of classification used in Kallman's study. All studies except Kallman's have involved very small samples from prison or hospital populations. Kallman's cases were perhaps more representative and were gathered by extensive searching in homosexual meeting places. They were evaluated as to degree of homosexuality (scale 0 to 6) on the Kinsey index system (Kinsey, Pomeroy, and Martin, 1948). Unlike DZ twins, Kallman's MZ co-twins tended to show fairly strong concordance as

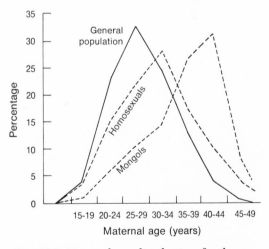

Fig. 17-1. Maternal age distributions for the general population, for a sample of homosexual patients, and for a sample of Down syndrome cases. (From Slater, E. 1962. Lancet **1**:69-71.)

Table 17-10. Twin studies of male homosexuality*

Study	MZ		DZ	
	N	Concordant	N	Concordant
Lange (1931)	2	1	—	—
Sanders (1934)	6	5	1	0
Habel (1950)	5	3	5	0
Kallman (1952c)	37	37	26	4-11
Heston and Shields (1968)	5	2-3	7	1

*Based on data from Heston, L. L., and J. Shields. 1968. Homosexuality in twins: a family study and a registry study. Arch. Gen. Psychiatry **18:**149-160.

to gradation of overt homosexuality. Interestingly, after a detailed study of individual case histories, Kallman commented that "index pairs have developed sexual tendencies independently and often far apart from each other. All deny any history of mutuality in sex relations."

Kallman's rather dramatic results need not be interpreted as final proof that homosexuality has a specific hereditary basis. Of his MZ pairs, 6 were concordant for schizophrenia as well as for homosexuality. Diagnoses for 22 additional cases were schizoid personality, obsessive-compulsive, or excessively alcoholic. Thus, given a common family environment during youth and assuming that some aspects of personality are heritable, the concordance of MZ twins in sexual orientation may be attributable to a similar pattern of experiences operating on similar substrates. Certainly, the appearance of other forms of mental disorders is common in the families of homosexuals. This was clearly shown earlier by Lang (1941), who carried out an intensive analysis of the families of 33 of his index cases. Mental illness, in some form or other, was found in all of them without exception. A homosexual orientation may be one kind of correlate of such an unfavorable combination of genotype and environment.

In conclusion of this section, we should emphasize, first, that whatever posture society may take toward homosexuality, it cannot be regarded as a biologically normal kind of behavior. Simply put, if heterosexuality were

not the norm, the human species would face extinction. Thus, in humans, as in other animal forms, there has been very strong selection for attraction between rather than within the sexes. Second, there must undoubtedly be fairly strong environmental causation in the etiology of homosexuality. But it seems likely that genetic factors are also operating, albeit by pathways as yet unknown.

Alcoholism

Except in acute cases where there is evident brain damage, it is difficult to define alcoholism exactly. The World Health Organization lists a spectrum from "acute alcoholic psychosis" to "other and unspecified alcoholism." Jellinek (1960) has distinguished forms of continuous drinking from what he calls "bout" drinking, but which forms, if any, have the greatest genetic loading is presently unknown.

It is clear that incidence and type of use of alcohol varies widely between races, countries, social classes, and probably between sexes. Individuals of Jewish and Chinese extraction are generally considered to use alcohol rather sparingly. The Irish, on the other hand, are traditionally considered to show a high level of alcohol intake. However, firm estimates are not readily available. One figure, put forward by Rosenthal (1970), for the United States is around 5% for chronic alcoholism. This may well be rather high. Goodwin and Guze (1974), on the basis of five surveys in various countries, put forward life-expectancy rates of 3% to 5% for males

Table 17-11. Incidence of chronic alcoholism in parents and siblings of alcoholic and normal probands*

	Fathers		Mothers		Brothers		Sisters	
	N	%	N	%	N	%	N	%
Alcoholic probands								
Brugger (1934)	70	24	72	5.0	83	28	107	3
Lemere et al. (1943)	500	11	500	0.4		14%		
Åmark (1951)	186	26	200	2.0	349	22	365	<1†
Bleuler (1955) United States	50	22	50	6.0	27	22	29	10
Bleuler (1955) Switzerland	49	33	49	8.0	49	12	61	3
Nagao (1967)	N = 281	13%			N = 238	9%		
Nonalcoholic probands (Bleuler, 1964)								
100 schizophrenics	100	3	100	1.0	54	4	59	2
200 surgical patients	200	13	200	3.0	192	13	185	3

*From Genetic theory and abnormal behavior by David Rosenthal. Copyright © 1970 by McGraw-Hill, Inc. Used with permission of McGraw-Hill Book Co.

†No alcoholics, only three cases of alcohol abuse.

and 0.1% to 1% for females. On the other hand, there is no doubt that the excessive use of alcohol is very common in virtually every part of the world and hence constitutes a problem of major importance. However, although there is now a sizeable and growing literature on the genetic factors in alcohol consumption in animals, relatively little has been accomplished with human subjects.

The main family studies carried out have been summarized by Rosenthal (1970) as shown in Table 17-11. It is clear that in most cases, rates in relatives of alcoholic probands are considerably elevated, this being a good deal more marked in males than in females, a fairly consistent finding in most work done on alcoholism in general populations. Åmark (1951) has provided some additional useful information. In a sample of 517 sibs of alcoholic probands with neither parent alcoholic, incidence was 17%. In a sample of 197 sibs of probands with one alcoholic parent, however, incidence was almost doubled—33%. There was also a marked incidence of psychopathy but not of the various forms of psychosis. It is also evident from the family studies that the environment provided by an alcoholic member of a family is not the best. Consequently, the results are suggestive but indeterminate.

One attempt to separate the two classes of causative factors was made by Roe (1944) with the adoption study method. She located a group of 36 subjects over 21 years old whose biological fathers had been classified as "heavy drinkers" with such accompanying problems as loss of job, disorganization, aggressiveness, and general disorderly conduct. These had been placed out for adoption at an early age (average of 5.6 years) with parents who showed no signs of inadequacy or alcoholism. These adoptees were then compared with a roughly matched control group of adoptees whose biological parents had shown no drinking problems. There were no significant differences in respect to use of alcohol by the two groups. For example, 63% of the index subjects used alcohol "occasionally," as did 55% of the control group. Of the index group, 7% was found to drink on a "regular basis"; of the control group, 9%. The main results, in other words, appeared to support an environmental etiol-

Table 17-12. Percentage of alcoholism in half siblings of alcoholic probands with different conditions of biological and adopting parentage*

	Half-sib	
	Alcoholic (percent)	Nonalcoholic (percent)
Biological parent alcoholic		
Rearing parent figure		
Alcoholic (N = 24)	46	54
Nonalcoholic	50	50
(N = 22)		
Biological parent nonalcoholic		
Rearing parent figure		
Alcoholic (N = 14)	14	86
Nonalcoholic	8	92
(N = 104)		

*From Schuckit, M. A., D. W. Goodwin, and G. Winokur. 1972. A study of alcoholism in half-siblings. Am. J. Psychiatry **128**:122-126. Copyright 1972, the American Psychiatric Association. Reprinted by permission.

ogy of alcoholism. However, there are problems with the study. The control group was placed at a significantly earlier age, fewer of them were placed in rural homes, fewer of them were married, and they had fewer children as a group. Thus the possibilities of general stress may have been greater among the control subjects, and this may have increased disposition to drinking.

A rather more sophisticated approach, using a similar methodology, was made later by Schuckit, Goodwin, and Winokur (1972). They examined the half-siblings of alcoholic proband cases. Some of these shared the same environment; some shared partly the same heredity. It will be obvious that many different designs can be incorporated under the half-sib method. In the Schuckit studies, however, most of the half-sibs studied shared a common biological mother rather than father. In general, the authors found that of half-sibs of alcoholic probands who had at least one alcoholic parent, 62% were alcoholic. Of those who were nonalcoholic, only

20% had an alcoholic biological parent. Strangely enough, among the alcoholic half-sibs, 0% had lived with an alcoholic foster parent, whereas among nonalcoholic half-sibs 5% had lived with an alcoholic parent surrogate.

Perhaps the results of major importance are those summarized in Table 17-12. These data separate out fairly clearly the effects of nurture and nature on alcoholism in half-sibs of alcoholic probands. Thus it is clear that if half-sibs share an alcoholic parent with their proband, about half of them turn out to be alcoholic regardless of whether raised by an alcoholic parent (biological or adopting) or a nonalcoholic parent. On the other hand, if biological parents are not alcoholic, then few half-sibs are alcoholic, though almost twice as many in this group are alcoholic if the rearing parent is alcoholic. Thus Schuckit et al. conclude that "the only consistent predictor of alcoholism in half-siblings was the presence of an alcoholic biological parent." They interpret these findings in terms of "genetic load." This term, as used here, seems to imply a polygenic or liability-threshold mode of inheritance. In general, the design used by Schuckit et al. seems a promising one that is especially well suited to cultures in which illegitimate births (and hence adoptions) have become less common but divorces and remarriages (and hence half-sib relations) have become much more common.

Another adoption study was carried out by Goodwin et al. (1973). They studied adoptees with at least one alcoholic biological parent and adoptees with nonalcoholic biological parents. Among 55 cases in the former group, 10 were diagnosed as alcoholic. Of these 10, 9 had, in fact, been treated for alcoholism. Among the 78 controls, however, only 4 were alcoholic, and none of these had received psychiatric treatment. Diagnosis was done blind, and social class was controlled.

Two later adoption studies have been reported by Goodwin et al. (1977a,b). They found elevated rates of alcoholism (~4%) in adopted-out females of both alcoholic and nonalcoholic biological fathers. Three of the four cases, however, had an alcoholic foster parent. Rates for female adoptees were far smaller than that for adopted-out sons of alcoholics, which was about 18%, as compared with a prevalence rate for Danish males of 3% to 4%. Thus the etiology and course of alcoholism may well be quite different between the sexes.

Several twin studies on alcoholism have been carried out, most of these of relatively recent vintage. A major contribution was made by Kaij (1960), who drew 174 pairs from two County Temperance Board registers in southern Sweden. Since in some pairs both members were used as probands, the final numbers were 58 MZ and 138 DZ pairs. Kaij used a 5-point scale to measure degree of alcoholism. This ranged from total abstention (0) to chronic alcoholism (4) involving pathological desire for alcohol, regular blackouts during intoxication, and withdrawal symptoms. Kaij then examined concordances in his twins for each degree of alcoholism. Several psychiatrists besides himself were involved in rating each individual. Interrater agreement was high. In general, 31 MZ pairs (53%) showed perfect concordance in respect to severity. The corresponding figure for DZ twins was 70 pairs (28%). Furthermore, with MZ twins, concordance appeared to increase as severity in the proband was higher. This was not the case with DZ pairs, however. Such a finding is reminiscent of the positive relationship between concordance and chronicity established for schizophrenia.

A second major twin study was carried out in Finland by Partanen, Bruun, and Markkanen (1966). They employed a series of 172 MZ and 557 same-sexed male DZ pairs from birth registries during the decade of 1920 to 1929. Thirteen drinking variables were measured. Analysis of the correlations between these yielded three identifiable factors: "density" (roughly, frequency of drinking per unit time), "amount," and "lack of control." Heritabilities of these factors were estimated as 0.39, 0.36, and 0.14, respectively. For the last factor, heritability was actually fairly high in younger twins but low in older. No such

age differences were found for "density" and "amount." To some extent, the general findings of Partanen et al. are consonant with those of Kaij in the sense that both "density" and "amount" must be factors that reflect severity of alcoholism.

Further evidence for this point has been provided by questionnaire data obtained by Loehlin (1972) as part of the National Merit Twin Study. Approximately 490 MZ and 317 DZ pairs responded to a large number of items, thirteen of which had to do either with drinking habits, drinking customs, or attitudes about drinking. Loehlin calculated heritability estimates for each of these items. They ranged from 0.62 for "had a hangover" to −0.36 for "women should not be allowed to drink in cocktail bars." The most significant finding was that five of the six items indicative of heavy drinking yielded the highest heritabilities. Curiously, the one of these six that did not was "become intoxicated" ($h^2 = 0.16$). It is odd that this "habit" should have virtually no genetic loading, whereas the items "had a hangover," "have never done any heavy drinking," and "have used alcohol excessively" should have high genetic loadings. We have noted in a previous chapter this kind of apparent paradox in connection with personality questionnaire items.

Drinking customs (e.g., wine or beer) had low heritabilities; drinking attitudes (e.g., "disapprove of women drinking in bars") had zero heritabilities. Loehlin is careful in interpreting these data to weigh environmental factors fairly heavily. Nevertheless, they are in accord with the Swedish and Finnish twin studies.

The final twin study to be reviewed here is of particular interest because it used a methodology not often found in human behavior genetic work. Vesell, Page, and Passanti (1971) measured rates of ethanol metabolism in 7 MZ and 7 DZ twin pairs. All were white, over 21 years of age, in good health, and members of pairs lived apart in fairly diverse environments. Each individual was given a single oral dose of 1 ml/kg of 95% ethanol diluted with ice water, following which five to six blood samples were

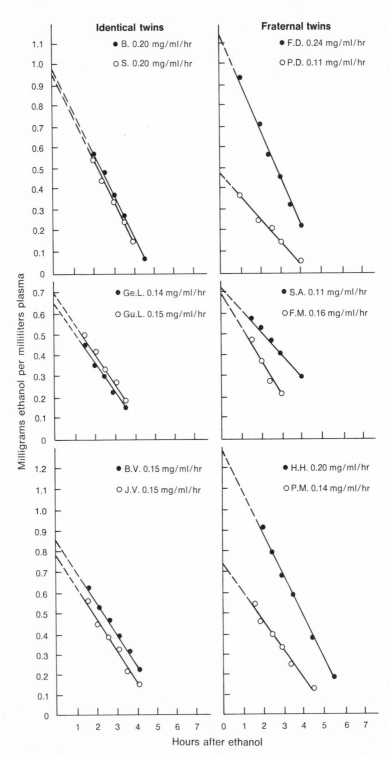

Fig. 17-2. Rates of ethanol removal in three pairs of MZ and three pairs of DZ twins. (From Vesell, E. S., J. G. Page, and G. T. Passanti. 1971. Clin. Pharmacol. Ther. **12:**192-201.)

taken at half-hour intervals. Plasma was later assayed for ethanol. For overall rates of ethanol metabolism the mean intrapair difference for MZ twins was found to be 0.004 mg/ml/hour; for DZ twins, the mean intrapair difference was 0.05 mg/ml/hour. The values differ by a factor of more than 12. Of the MZ twin pairs, 5 showed a zero difference. Vesell et al. computed heritability of ethanol metabolism to be 0.98, suggesting almost exclusive genetic control. Some sample data for 3 MZ and 3 DZ pairs are shown in Fig. 17-2. This is clearly quite a remarkable effect. It is still not certain, however, whether the high similarity of identicals is a direct result of genes or of various intermediate physical communalities possibly shared by them more than by fraternals, perhaps for environmental reasons. Convincing evidence on this point may be difficult to obtain. Nevertheless, the study is a most important and interesting one, not only by virtue of its dramatic results but also by virtue of the experimental methodology it involves. The latter is too rarely used in twin work.

A final point should be made. We know little about the relation between rate of metabolism of ethanol and alcoholism. Furthermore, as discussed in Chapter 8, there are reliable differences between mouse strains in alcohol preference and in alcohol dehydrogenase activity; there are no well-established differences between them in rate of alcohol metabolism. Consequently, the Vessel et al. data have only indirect bearing on the problem of alcoholism itself.

Little is known about the mode of genetic transmission of alcoholism. One seemingly significant fact, however, is the large difference in incidence between the sexes. Åmark (1951) found rates of 3.4% for males and 0.11% for females. Likewise, Marconi et al. (1955) reported incidences in Santiago, Chile, to be 8.3% for males and 0.6% for females. The South American figures are obviously much higher than the Scandinavian. However, as Cruz-Coke and Varela (1966, 1970) have noted, the female frequency is almost exactly the square of the male frequency. Thus $0.034^2 = 0.00116$; likewise, $0.083^2 = 0.0069$. Under appropriate equilibrium conditions, this relation is exactly what we would expect under recessive sex-linked transmission. Thus, if gene frequency is q, then incidence for males is also q, whereas, for females, it will be q^2, since both X chromosomes are present. Such a fit, for two sets of data in very different parts of the world, is impressive. Cruz-Coke and Varela have further claimed an association between cirrhosis of the liver, alcoholism, and defective color vision, the latter known to be carried by a sex-linked gene. Some relevant data are shown in Table 17-13. The comparison is between number of errors on the Farnsworth-Munsell 100-Hue Test in the sons and daughters of nonalcoholic versus alcoholic parent probands. It is clear that the two sets of parents are very different in the first place. In the second place, there is a significantly elevated rate in the daughters, though not the sons, of probands. These latter data agree with a sex-linkage model if we assume that defective color vision shows up in female carriers. Cruz-Coke and Varela conclude that alcoholism is a genetic polymorphism maintained at a high frequency in the population by a high fertility level among heterozygote females. Their data on fitness in the general Santiago population, of cirrhosis cases, and of alcoholic females appear to give some support to this notion. Smart (1963) has also

Table 17-13. Number of errors on the Farnsworth-Munsell 100-Hue Test in children of alcoholic and nonalcoholic male probands*

	N	Mean number of errors
Nonalcoholic probands	35	58.0 ± 5.3
Sons	21	45.9 ± 9.0
Daughters	21	36.3 ± 6.0
Alcoholic probands	21	152.2 ± 23.6
Sons	14	83.8 ± 21.8
Daughters	21	116.8 ± 17.5

*From Cruz-Coke, R., and A. Varela. 1970. Genetic factors in alcoholism. In R. E. Popham, ed. Alcohol and alcoholism: papers presented at the International Symposium in Memory of E. M. Jellinek, Santiago, Chile. University of Toronto Press, Toronto.

demonstrated that, in a Canadian population, an unusually high percentage of alcoholics come from large families. Under the given fertility conditions in Chile, equilibrium would be established at a frequency for the mutant gene of 0.243, much higher than the empirical estimate found by Cruz-Coke and Varela. Thus the authors suggest that this polymorphism is still in a state of transition in association with a cultural evolution from a nomadic to an industrial society. A survey of three types of populations in northern Chile, a seminomadic, several rural, and

several urban, showed an increasing incidence of color vision defects and alcoholism as extent of urbanization increased. This suggests a commensurate relaxation of selection pressures against both genes and an approach to equilibrium.

The preceding model is a very interesting one. However, it cannot at present be accepted without reservation. In the first place, a relation between alcoholism and color vision defects has not been found in all populations (Thuline, 1967). It has also been thought that the deficiencies found by Cruz-

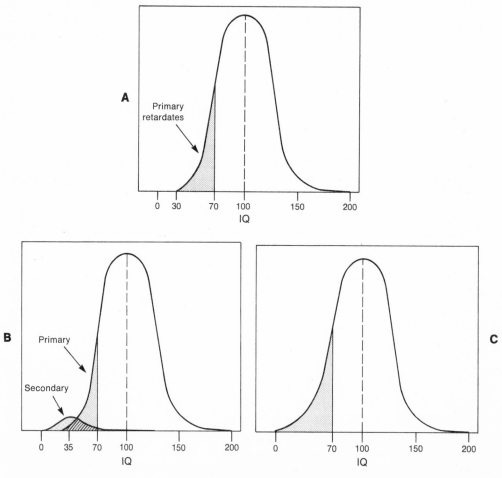

Fig. 17-3. The "two-group" approach to mental retardation. Curve **A** represents polygenic distribution of IQ and includes primary retardates. In **B** distribution for secondary retardates is added. When all retardates are lumped together, the IQ distribution is asymmetrical as shown in **C**. (Modified from Zigler, E. 1967. Science **155**:292-298. Copyright 1967 by the American Association for the Advancement of Science.)

Coke may, in fact, have been caused by alcoholism rather than correlated with it. Second, as Winokur (1967) has pointed out, under the sex-linkage model one would expect a very low risk for sons of alcoholic fathers, since the latter can transmit only the Y chromosome to their male offspring. Generally speaking, this relation does not seem to hold. Consequently, some modifications of the model would have to be made. It is also possible, of course, that it holds only for some populations or only for certain types of alcoholism. There is a need for much more work in this important area. For a useful summary of most of the data and theories bearing on alcoholism, the reader is referred to the volume by Goodwin (1976).

Mental retardation

Generally speaking, individuals with an IQ less than 70 are considered to be mentally retarded. The category includes "borderline," "morons," "imbeciles," and "idiots." Whatever causes their defect, such people usually find it difficult to get along in society and need specialized help and, in many cases, institutionalization. Precise classification is difficult because etiology is very diverse. However, it has been usual to divide the mentally retarded into two major groups. The first, sometimes labeled "primary," takes in perhaps as much as 75% of all retardates. Usually, no obvious major cause is discernible and we must therefore invoke the action of polygenes, unidentified environmental causes, or both. The second category, labeled "secondary" or "familial," includes cases involving some major neurological insult, whether caused by major genes, chromosomal damage, or serious environmental traumata.

Reed and Reed (1965) in their classic study of mental retardation have used a somewhat different classification according to presumed etiology. They divide retardates into the following groups: primarily genetic, probably genetic, primarily environmental, and "of unknown etiology." This is perhaps an overcautious taxonomy and somewhat blurs the distinction we have just made between primary and secondary mental deficiency. Let us look at this a little more closely.

It appears to be a fact that the lower tail of the distribution for intelligence departs significantly from a normal curve. This is shown in Fig. 17-3. This departure has usually been explained by the hypothesis illustrated in graph B. This postulates simply that individuals belonging to the primary group may be accommodated within the normal curve generated by the action of polygenes and, second, that the existence of a second group of individuals whose defects result from major causes is responsible for the skewness of the empirical distribution of IQ. Data bearing directly on this hypothesis are rather scarce. Many cases of secondary defect are, of course, readily identifiable and appear to occur in families of high, low, or average intelligence. Down syndrome is an obvious example, as is phenylketonuria. Also, the IQs of the affected individuals themselves may vary over a wide range, sometimes reaching normal levels. Consequently, it is not feasible simply to use a particular cut-off point on the IQ scale to differentiate the categories. However, Roberts (1952) has found that the fit of the empirical IQ distribution to normality is reasonably good down to about IQ 45. Below this, the numbers obtained by him in one survey of 3361 children were about eighteen times as many as expected (12.5 cases as against 0.7, respectively). Other workers, for example, Åkesson (1961) have set the cut-off point higher, in the 60–IQ point range. Thus it cannot be considered as absolute.

One way of exploring the two-category hypothesis has involved examining the IQ distribution of sibs in the respective groups and the proband sib correlations. Data using these methods have been somewhat ambiguous. Penrose (1939) found that the mean IQs of sibs of defective subjects of IQ below 30 was over 90, a figure higher than the mean IQ of sibs of cases between 70 and 100. Roberts (1940), in a survey in Bath, estimated that the mean IQ of 367 sibs of retarded subjects (IQs averaging 77.4) was 88.1, giving a regression of 0.53. For 17 sibs

Table 17-14. IQs (adjusted for age) of two proband groups*

	N	
IQ	Imbecile group	Feebleminded group
<53	66	21
53-60	40	54
>60	16	74
	122	149

*From Roberts, J. A. F. 1952. The genetics of mental deficiency. Eugen. Rev. **44:**71-83.

of 13 idiots and imbeciles, the mean IQ was 100. A larger-scale study carried out at Bristol and Colchester on 271 defective children (IQ 35 to 60) with 562 sibs yielded rather indeterminate results, however, mostly because of the author's handling of the data. Initial separation of the probands into high and low-IQ groups apparently did not produce two distinctive sib distributions, contrary to hypothesis. Faced with this problem, Roberts proceeded as follows: he arranged the probands in a table in order of IQ with their sibs alongside of each. Then, inspecting this table, he "divided the families by brute force into groups, doing the job quickly and ruthlessly with as little juggling as possible" (Roberts, 1952, p. 78). The resulting two proband groups were distributed as shown in Table 17-14. Next he plotted the IQ distributions of the sibs of these two groups. These are shown in Fig. 17-4. Roberts labels the distributions as representing "families." This might be thought to imply that they include the proband cases as well. However, in the text of his paper (p. 79) he specifies that they refer to "the sibs of the two arbitrary groups." Further comparison of the two groups showed that the imbecile families had fathers in considerably higher occupational levels than those of feebleminded families, and the quality of home was higher in the former group. Families of the feebleminded group were more fertile. Such differences were minimal for subject groups divided only according to IQ (60 IQ cutoff).

Roberts felt that his arbitrary separation corresponded to some "underlying reality," and his results are widely cited in leading textbooks. However, the criteria for categorizing individual probands as "imbecile" or "feebleminded" are not made clear. In response to a question on this point by Sir Godfrey Thomson in an ensuing discussion, he stated simply that his "brute force" division entailed "giving weight to two things simultaneously—namely, the subject's IQ and the distribution of the IQ's of its sibs" (p. 83). Such a procedure hardly seems objective enough to permit independent replication.

In summary, although there are some grounds for separating the two classes of mental retardation, the evidence is far from being impeccable. Whether both should be called "defective," however, is a related point. Zigler (1967) has argued strongly that a "defect" orientation should not be applied to individuals in the primary group, since it suggests a program of remediation, which is probably not applicable and may do positive harm. There is some cogency to his suggestions.

We will now review briefly some of the major forms of secondary mental defect. We will confine our discussion to the severe forms, since not a great deal can be said about the class falling within the normal range of intellectual variation.

Metabolic disorders. These may mainly be divided into three main categories: disorders of amino acid metabolism, disorders of carbohydrate metabolism, and disorders of lipid metabolism. We will consider an example of each.

Phenylketonuria (PKU). Perhaps the best known example of the first category, PKU, was first identified by Fölling in Norway in 1934. He originally labeled the syndrome phenylpyruvic oligophrenia, since the severe mental retardation he found in two sibs was associated with a high level of phenylpyruvic acid in their urine. Specifically, the biochemistry of the disorder is as follows: there is a basic deficiency in the liver enzyme L-phenylalanine oxidase, which normally hydroxylates the benzene ring of phenylalanine to form tyrosine. As a result, an alternative

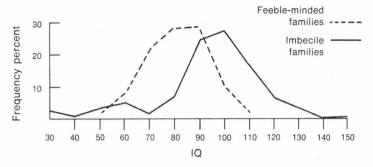

Fig. 17-4. Distribution of IQ in families (sibs) of "imbecile" and "feeble-minded" proband cases. (From Roberts, J. A. F. 1952. Eugen. Rev. **44:**71-83.)

metabolic pathway is used, involving the breakdown of phenylalanine into phenylpyruvic, phenyllactic, and phenylacetic acid. Since the renal threshold for these substances is low, they are excreted in large amounts in the urine. Use of this subsidiary pathway allows large amounts of phenylalanine to accumulate in the blood with various consequences. One is an inhibition of the hydroxylation of tryptophan and a decreased level of 5-hydroxytryptamine (serotonin) in the blood (Hsai, 1967). Another consequence is an interference with the normal metabolism of tyrosine causing a decreased melanin production. The observable effect of this is the characteristically light pigmentation of skin and hair found in phenylketonurics.

In general, the following symptomatology identifies the PKU syndrome:

1. High phenylalanine blood and phenylpyruvic acid urine levels.

2. Lack of pigmentation: subjects are usually blond and blue eyed with pale complexion and often a dry, rough eczematous skin.

3. Peculiar, musty odor resulting from high phenylacetic acid level.

4. Irritability, vomiting, and often epilepsy in early life.

5. Pithecoid (monkeylike) stance.

6. Reduction in brain weight with defects in myelination.

7. EEG abnormalities; spike or wave complexes of the petit mal type, even in absence of seizures.

8. Intellectual retardation.

The first seven of the list are physical or physiological characters. The last one is perhaps the most interesting from the standpoint of behavior genetics.

Most phenylketonurics have IQs below 50, possibly averaging around 25. The deficit is typically small at first, becoming more obvious with increasing age. In spite of this, there are thought to be some cases with IQs in the normal range despite strongly positive phenylalanine serum levels (Hsia, 1967). The deficit may be at least partly treated by keeping the PKU infant on a low-phenylalanine diet (e.g., Lofenolac or Ketonil). However, for children 6 years or older, this kind of diet appears to have no effect on the symptoms.

Curiously, Fuller and Shuman (1974) have shown that the correlation between normal sibs of PKU index cases is significantly below that found in the general population ($r = 0.13$, $N = 32$). This discordance is apparently attributable to a higher than usual incidence of high-IQ children in these families. A plot of the 78 sibs involved (252 PKU families) turned out to be significantly different from the empirical distribution for a normal sample as put forward by Terman and Merrill. Sibs of PKU cases fall in an IQ range of 78 to 148, with a mean of 112.7 and a standard deviation of 16.5. The explanation offered by Fuller and Shuman for their findings is rather complex. Roughly, they suggest that the PKU gene may "trigger mechanisms by which superior IQ's are produced." This will not occur, of course, in the double-recessive state but in at least a proportion of heterozygotes. The proportion in which such compensation does not occur will show only marginal intelligence and tend to be eliminated from the gene pool. The hypothesis is an interesting one but difficult to test, especially in view of the difficulty with carrier tests. However, a first step should be simply a large-scale replication of the Fuller-Shuman data to establish their generality and possibly to extend them to other kinship categories.

If relatives of PKU cases are unusual for intelligence, one can ask whether they are also unusual in respect to other characters. Numerous investigations have been undertaken to examine the incidence of psychosis and other psychiatric disorders. Some have obtained positive results, but on the whole the bulk of the evidence is negative. Slater and Cowie (1971) have provided an excellent review of this literature.

The final aspect of PKU to concern us is its genetic basis. Incidence has been found to vary fairly widely between different populations. Estimates range between 1 and 10 in 100,000. About 0.64% of all institutionalized mentally defective patients are phenylke-

tonuric (Menkes and Migeon, 1966; Hsia, 1967; Rosenthal, 1970; Slater and Cowie, 1971). It is rather rarely found among blacks and Jews, but elevated rates occur in populations with ancestral origins in Ireland and west Scotland (Carter and Woolf, 1961).

The most definitive study of the genetics of phenylketonuria was carried out by Jervis (1954) on 266 sibships. Since he found an equal sex distribution and a ratio of affected to normals of close to 25%, he concluded in favor of a single autosomal recessive gene. This hypothesis is generally accepted. There are two other points that should be mentioned. One is that there appears to be some impairment of phenylalanine metabolism in heterozygotes. Hsia et al. (1956) compared a sample of nonaffected parents of phenylketonurics (therefore presumably carrying the gene) with normal controls in respect to rate of uptake of phenylalanine after an overload of this substance. Results are shown in Fig. 17-5. It is very clear from these data that heterozygotes are somewhat abnormal metabolically even though not showing any other symptoms. Obviously, such carrier detection can be put to very useful purposes in

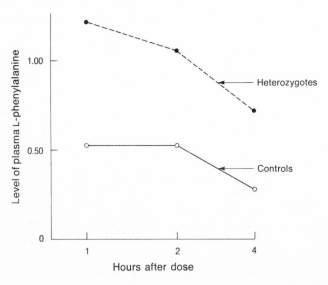

Fig. 17-5. Plasma L-phenylalanine levels after ingestion of 0.1 gram of L-phenylalanine per kilogram of body weight (micromoles per milliliter of plasma). (Heterozygotes show values that are significantly higher than controls at all postdosage hours.) (Based on data from Hsia, D. Y.-Y., K. W. Driscoll, W. Troll, and W. E. Knox. 1956. Nature **178:**1239-1240.)

genetic counseling. However, some reservations are in order, since not all workers have been convinced of the efficacy of presently available carrier tests (e.g., Bremer and Neumann, 1966).

A second point of interest arises from the study of children born of phenylketonuric mothers. Hsia (1970) reviewed a sample of 94 cases of such children. Of these, 8 were phenylketonuric, 79 were nonphenylketonuric, and 7 were classified as uncertain. Those in the first group were presumably homozygous for the recessive gene. Those in the second group were, in all probability, heterozygotes and hence should have been normal. Yet, 7 of them died very young, and at least 61 others showed mental retardation. Although data on the remaining 11 were incomplete, Hsia concluded that virtually all heterozygote offspring of phenylketonuric mothers are defective. Obviously, the uterine environment provided by the affected mother must have a large effect. However, a high phenylalanine level in fetal life is apparently not sufficient by itself, since offspring born of mothers who are hyperphenylalanemic only (i.e., do not carry the PKU gene) are apparently normal. Thus the presence of the gene in the child must presumably escalate the effects of high phenylalanine levels *in utero*. This is an interesting example of gene-environment-development interaction.

In general, we still have much to learn about phenylketonuria. However, since it has probably been more closely studied than any other biochemical error, the work done on it can serve as a very useful model for studying the large number of other comparable disorders.

Galactosemia. This is a disorder illustrating errors in carbohydrate metabolism. An autosomal recessive gene causes a defect in the enzyme galactose-l-phosphate uridyl transferase (Isselbacher, 1957). The symptoms have early onset involving vomiting, diarrhea, jaundice, and often anemia and death. Mental defect again is a symptom, although the deficit does not appear to be as great as in phenylketonuria. In one study, Hsia and Walker (1961) found a range of 40 to 100+ IQ in 45 patients, with 60% borderline or retarded but educable. Some homozygotes are apparently asymptomatic for reasons unknown, and the disorder responds well to dietary treatment. The latter involves chiefly the avoidance of milk, of which galactose is the main carbohydrate constituent. Incidence of galactosemia is estimated as 1 in 18,000 births (5.6×10^{-5}). Heterozygotes (parents of probands), as is the case with PKU, show a decreased ability to metabolize galactose after a loading dose, indicating a partial enzyme defect. However, they overlap a good deal with normal controls (Slater and Cowie, 1971).

Tay-Sachs disease. Infantile amaurotic idiocy, or Tay-Sachs disease, provides an example of lipid metabolism defect. The broad category of amaurotic idiocy takes in at least four types, distinguished according to age of onset. It is not yet clear whether these have a common pathogenesis or not. Tay-Sachs disease is thought by many workers to be caused by a recessive autosomal gene producing a defect in a component of the enzyme β-D-N-acetylhexosaminidase. However, the genetics of the disorder has not been definitively established (Slater and Cowie, 1971). The major symptoms include blindness, mental defect, and progressive muscular weakness. Perhaps the critical physiological cause behind many of the surface symptoms is the accumulation of abnormal amounts of lipoid substances in nerve cells.

There are an enormous number of gene-controlled biochemical defects besides these we have just discussed. On the whole, most of them are probably of more interest to physicians and biochemists than to behavior geneticists. At the same time, the study of at least some of them may offer basic insights about the physiological pathways between genes and behavior and, to this extent, are well worth some attention. The interested reader is referred to the excellent reviews by Slater and Cowie (1971) and by Omenn (1976) and to the comprehensive catalogue prepared by McKusick (1975).

Chromosomal anomalies. Changes in chromosome structure, morphology, or number occur fairly regularly in human beings and lower animals. Most of these produce death or drastic physical, physiological, and psychological abnormalities. Although they had been studied much earlier in such species as *Drosophila*, it was not until the late 1950s that their importance in the etiology of various disorders was fully appreciated. Two major types of chromosome anomaly may occur:

1. *Aneuploidy*, in which the number of chromosomes in cells is not an exact multiple of the haploid number *(n)*. This occurs as a result of an irregularity of the normal meiotic process. Thus either at the first or second reduction division, members of a homologous pair of chromosomes may fail to segregate into separate daughter cells. The result will be some zygotes with one too many chromosomes—trisomics $(2n + 1)$—and some with one too few—monosomics $(2n - 1)$. The latter are usually not viable.

2. *Chromosomal structure aberrations* may also occur. According to this type of change that takes place, these are usually classified according to the nature of the structural change. Thus *deletions* involve loss of part of a chromosome; *duplications* describe the addition of an extra segment of chromosome; *inversions* are produced by the looping and breaking of a chromosome in such a way that the normal sequence of genes on it is altered; and *translocations* involve the transfer of a piece of one chromosome to another nonhomologous chromosome. Most of these changes have serious and often lethal effects.

Aneuploidies and structural changes may occur in respect to either the autosomes or sex chromosomes. We will consider a few examples of each.

Down syndrome. One of the best known aneuploidies is *Down syndrome* or mongolism. This accounts for between 5% and 10% of all institutionalized mental defectives and has an incidence somewhere around 1 in 600 live births, probably on a worldwide basis. The syndrome is characterized by a number of physical abnormalities, in particular, slanted palpebral tissues with epicanthic folds, slanting eyes, and flatness of face, all

of which produce a superficial resemblance to members of the Mongolian race. However, apart from its pejorative quality, the term *mongolism* is perhaps not suitable because Mongolians (e.g., Chinese) with "mongolism" do not show any accentuation of their racial features. Furthermore, according to Slater and Cowie (1971), Japanese appear to think that patients with mongolism "look strikingly European."

For many years, Down syndrome was a puzzle to geneticists. On the one hand, there was abundant evidence that some kinds of prenatal agents were involved. This relates mostly to two facts. One is that there is a striking relation between incidence and maternal age. The ratio of percent Down to percent normal births rises from 0.39 for mothers 19 years or less to 14.33 for mothers 45 years and over, a thirty-sixfold increase (Penrose and Smith, 1966). This general pattern has been confirmed in a number of major, independent studies.

Again, birth rank is important. Data on this dimension with maternal age removed are shown in Table 17-15. The elevated incidence for primiparous infants is highly significant. However, the relationship may not be general; many workers have believed incidence to be higher in later births. If the latter were true, it would be consonant with the

Table 17-15. Observed and expected incidence of Down syndrome as a function of birth rank*

Birth rank	N		Difference (O − E)
	Observed	Expected	
1	61	42.7	+18.3
2	50	58.1	−8.1
3	34	34.0	0
4	26	27.9	−1.9
5+	46	54.3	−8.3
Σ	217	217.0	

*From Slater, E., and V. Cowie. 1971. The genetics of mental disorders. Oxford University Press, London; based on data from Smith, A., and R. G. Record. 1955. Maternal age and birth rank in the aetiology of mongolism. Br. J. Prev. Med. **9:**51-55.

maternal age effect in pointing to some kind of "mother-exhaustion" hypothesis. In any event, such data suggest environmental factors at work.

Emphasis had been placed on genetic factors by other workers, notably Kallman (1953) and Allen (1958), who observed several facts. In the first place, Down cases show strong physical similarity. Such uniformity could be taken to argue for a gene-controlled metabolic defect rather than for an exogenous cause, which would be expected to produce more variable effects. Second, according to some workers (e.g., Böök and Reed, 1950), there appears to be a moderately increased incidence among sibs of proband cases; third, the mothers of subjects have been found to show an elevated incidence of spontaneous abortions. If it were assumed that these aborted children were potential Down cases, this would suggest a familial incidence higher than generally computed. Fourth, Allen and Baroff (1955) showed that MZ twins are always concordant with respect to the syndrome, but DZ twins rarely so. This suggests either some heritable factors or else an environmental agent operating prior to fertilization.

Many of these seemingly discordant findings fell into place as a result of the rapid advances in cytogenetics taking place in the late 1950s. In 1959, Lejeune, Gautier, and Turpin analyzed the chromosome complement of three boys with Down syndrome and found the presence of a small extra chromosome associated with pair No. 21 (Denver classification). Hence the individuals concerned are usually designated as 21-trisomics. However, other findings have suggested that No. 22 of the same G group may be sometimes implicated (Slater and Cowie, 1971). In either case, the anomaly is produced by nondisjunction during either oogenesis or spermatogenesis, that is, the failure of the members of the chromosome pair to separate during meiosis and go to different daughter cells or polar bodies. This may occur either at the first or second reduction division. In the former case, two gametes will lack the chromosome. Fertilization will then yield 50% trisomics and 50% monosomics. In the latter case, 50% of the gametes will be monoploid, yielding normal zygotes; 25% will have No. 21 absent, yielding monosomic zygotes; and 25% will be diploid, leading to trisomic zygotes. The different possibilities are shown in Fig. 17-6.

Over 90% of Down cases have the preceding etiology. A few, however, involve a somewhat different underlying anomaly. These are the cases of so-called translocation mongolism. This involves an exchange of material between an arm of chromosome 21 with a part of a chromosome from group D (13 to 15) or group G (21 and 22). These are usually referred to as D/G or G/G translocations. A schematization of the zygotes produced in a translocation is shown in Fig. 17-7. Several points of interest may be mentioned. First, not all zygotes are abnormal. Besides the 25% who have a normal complement and no structural anomalies, another 25% are phenotypically normal, *even though they carry the translocation*. Both the G/G and D/G arrangements can apparently be present in some families for several generations before a Down case appears (Slater and Cowie, 1971). Second, data of Hamerton et al. (1961) suggest that carriers may actually have some biological advantage over noncarriers in the same family. Although such cases show some kind of "genic balance" (as shown in Fig. 17-7), it is not clear how this could convey additional fitness. In any case, the familial character of the translocation syndrome sharply separates it from the trisomic forms. Third, incidence of translocation mongolism is independent of maternal age. However, if a mother of 25 years or less has already had one Down child, her risk of having a second is fifty times the random risk. If she is 24 to 34 years old, the risk is five times random and is at random over 35 years (Carter and Evans, 1961). These figures are not based solely on translocation cases, but perhaps the latter constitutes a majority. Fourth, translocation and trisomy have been found to occur in the same families. Finally, translocation cases appear to have higher intelligence, on the average, than standard tri-

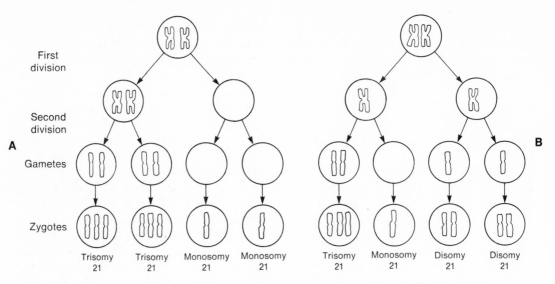

Fig. 17-6. Nondisjunction during meiosis. **A,** Occurring at first reduction division. **B,** Occurring at second reduction division (chromosome pair 21).

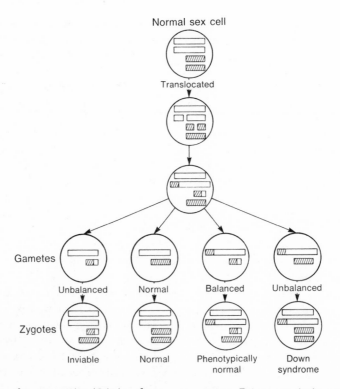

Fig. 17-7. Translocation 15/21 (G/D) and its consequences. Prior to meiosis, a part of one of chromosome pair 15 (unfilled rectangles) breaks off and exchanges with a part of one of chromosome pair 21 (dashed rectangle). Gametes and zygotes produced are as shown.

somics and to be more active and aggressive. They also manifest more "behavior problems" than do the usually rather placid and gentle trisomics (Johnson and Abelson, 1969).

Other disorders. Besides Down syndrome, there are quite a number of other disorders caused by chromosome aberrations. *Patau syndrome* (Patau et al., 1960) involves cleft palate, hand and finger abnormalities, and, again, severe mental retardation. Life expectancy is only about 4 months. It is produced by trisomy of chromosome 13 in the D group and is thus sometimes called *trisomy D syndrome.* Another autosomal trisomy is *Edwards syndrome* involving chromosome 18. It is also characterized by severe physical abnormality and very short life expectancy. Finally, Lejeune et al. (1963) reported a case involving a deletion of part of the short arm of a chromosome in the B group. Because infants suffering from this defect characteristically utter a peculiar weak cry like the mewing of a cat, it has been called *cri du chat syndrome.* Again, severe mental retardation is a central feature.

Anomalies of sex chromosomes, unlike those of the autosomes, do not always produce marked physical or mental abnormalities. There may be mild retardation, however, and, characteristically, defects associated with sex organs. Personality problems and sexually deviant behaviors may arise as secondary outcomes. The general incidence of sex chromosome anomalies is estimated as about 2.1 per 1000 for males and 1.6 per 1000 for females. The rate is somewhat higher in mental institutions (8.1 per 1000), though this reflects mostly the prevalence of XYY karyotypes (Menkes and Migeon, 1966).

One of the best known chromosomal anomalies is *Turner syndrome.* This involves an XO constitution, that is, an absence of the second X. Sometimes the patient is a mosaic, having one stem line of normal cells (XO/XX constitution), which is thought to attenuate the abnormality of the phenotype (Ferguson-Smith, 1965). Incidence is about 1 per 3000 female live births. A variety of physical symptoms are involved, although most of

these do not appear until puberty, when normal secondary sexual characteristics fail to appear. Such individuals are usually raised as girls and are recognizably female. However, the primary pathology relates to the ovaries, which are only streaks of connective tissue. Sometimes the karyotype may also involve isochrome X. This occurs as a result of division of the X chromosome perpendicular to its long axis rather than parallel to it. The end product is thus an imbalanced pair with each member having duplicate arms but the two members not duplicating each other.

Turner cases are chromatin negative in that they lack the presence of Barr bodies in cell nuclei. Sometimes, however, we find karyotypes with two, three, or even four Barr bodies, indicating a superabundance of X chromosomes. Such individuals are sometimes referred to as *superfemales.* Contrary to what this name suggests, such individuals have poorly developed secondary sexual characteristics and are usually infertile. Mental retardation is common.

Klinefelter syndrome is caused by an XXY constitution. Sometimes additional X or Y chromosomes may also be present, for example, XXXY or XXYY. The incidence is about 1 in 400 live male births. Most of the typically male characteristics are reduced, with some feminine characteristics present. Thus patients may show sparse pubic hair, absence of sperm, and exaggerated breast development. Mental retardation is common.

The final syndrome we may mention is the XYY constitution. However, we have discussed this earlier in connection with the topic of criminality.

This survey of chromosomal anomalies is very incomplete. However, it is hoped that it will give the reader at least a general familiarity with some of the most common disorders in this vast field. In respect to all of them, the major point of interest is, of course, their etiology. Little is known definitively. However, Thompson and Thompson (1967) have suggested that the following factors are implicated: late maternal age, genes predisposing to nondisjunction, autoimmune disease, radiation, and viruses. The challenge here is perhaps greater for the cyto-geneticist or biochemist than for the psychologist or the behavior geneticist. Nevertheless, a better understanding of the kinds of disorders just discussed may eventually shed light on the more general problem of the manner in which genic makeup determines behavior. In addition, a second general question relates to the modifiability of the behaviors characterizing these gross forms of retardation. This represents a useful point of contact between the hereditarian and the environmentalist.

SUMMARY

In this chapter, we have reviewed some of the behavior genetic work done on various forms of deviant behavior other than the two major psychoses. There is a fairly extensive literature on the etiology of the various forms of neurotic disorder with a good deal of data suggesting a genetic etiology. As with many psychiatric syndromes, however, precise figures on prevalence and risk are not available. Likewise, the taxonomy of this broad category is not very well worked out even with the help of psychometric evaluation. Consequently, there are, and will continue to be, serious difficulties standing in the way of genetic model fitting.

Somewhat the same applies to the other forms of deviant behavior with which we have dealt, for example, alcoholism and criminality. Studies of homosexuality were included here on the grounds that this sexual orientation, whether it be considered "abnormal" or simply "different," represents a departure from the biological norm and is, hence, a legitimate subject for genetic study. However, we would also reject the notion that homosexuality represents a psychological aberration in the same sense as does, say, schizophrenia. Clearly, many of the personality problems associated with this sexual orientation flow more from society's reaction to it rather than from the orientation itself.

We concluded the chapter with a brief survey of some of the more common forms of mental retardation, some of these arising from the biochemical action of single genetic loci, and others from various aneuploidies.

As will be clear, both from this and the preceding chapter, the general problem of psychological abnormality is one of great importance. Unfortunately, as we have continued to emphasize, in spite of the vast amount of time and money that have been devoted to its study, we still seem rather far away from arriving at definitive solutions.

IV
Summary and conclusions

18

Psychology, biology, and behavior genetics

The preceding chapters have summarized the evidence for supposing that there are heritable factors determining behavior. It will be clear to the reader that an enormous variety of methods and approaches are involved and, furthermore, that there is an extraordinary range in types of behavior studied. Thus the term *behavior* is used to encompass such phenomena as schizophrenia, intelligence, avoidance conditioning, color vision, audiogenic seizures, and laterality. It is therefore somewhat difficult to know what are the defining characteristics of the field of behavior genetics that give it some integrity as a unitary discipline, that set some limits to it, and that suggest major themes whose emphasis is likely to yield fruitful results in the future.

To tackle this major theoretical problem, we will begin first by examining some critical features of behavior and then proceed to delineate two major orientations that characterize the field and to show how these relate to the sciences of biology and psychology.

NATURE OF BEHAVIOR

Psychologists have struggled for a long time to give some precise definition to the primary datum of their field. It cannot be said that these attempts have been notably successful, as witnessed by the variety of "behaviors" in fact studied, as just pointed out. In the 1930s and 1940s, three of the great architects of learning theory, Tolman (1932), Skinner (1938), and Hull (1943), squarely confronted this problem. All agreed that behavior involves sequences of acts of an organism that have some effect on the outside world. Thus the depression of a bar by a rat

could be said to constitute behavior, although the specific movements entailed might well vary from one press to the next. In this sense, the definition was functionalist in nature, since it stressed that the uniformity of the *result* supplied coherence to the mechanism by which it was achieved. Tolman, Skinner, and Hull, as is well known, explicated this notion with rather different emphases, and, perhaps as a consequence, interest in the problem waned. Indeed, it may be that the need for the psychologist to define behavior is no greater than the need of the biologist to define "life." Nevertheless, it is still true that if the field we call behavior genetics is to have relevance to the two sciences it explicitly conjoins, then there is some advantage in at least pointing up some features of behavior that make it different from other characteristics studied by geneticists. The following have been specified earlier by Thompson (1968):

Behavior is continuous

Classical genetics evolved out of the study of simple discrete characters. It was not until half a century after Mendel put forward his basic formulations that the study of quantitative characters began and the notion of multiple factors and polygenes was proposed. It is now clear, of course, that the Mendelian model can readily be extended to encompass the transmission of continuous characters. However, it is not so clear under what circumstances it is desirable or valid to describe a trait as continuous or discrete. A good example, dealt with in Chapter 16, is schizophrenia. As we indicated, this phenotype may be taken either way, as well as something in between, or quasicontinuous; the consequences of such a choice may be quite different in respect to the genetic hypothesis accepted. Such ambiguity is by no means unique to the genetics of behavior but is perhaps most salient for this domain.

This might not matter very much except for the fact that monogenic theories suggest major biochemical pathways which can possibly be uncovered, whereas polygenic models suggest a complexity of chemical interactions probably intractable to exact study. Thus if most behavior traits must be fit to polygenic models, we may be left only with statistical analyses of such problems as how many genes are involved and the specification of the almost infinite number of interactions between them. Such mathematical exercises seem to us to have only trivial importance and, furthermore, to be of small interest to most biologists and psychologists. Let us turn to two other properties of behavior, which are of more potential fruitfulness.

Behavior is complex

The search for metabolic, hormonal, and neurological pathways through which genes can influence behavior has been successful. There are many mechanisms through which a gene substitution could produce changes in temperament, learning ability, or details of courtship behavior. In some instances the effects of a gene are so drastic and its influence on behavior so direct that the problem of looking for a pathway does not arise. The problems of interest lie in embryology rather than in ethology or psychology. Despite the abundance of potential pathways, many, perhaps most, behavioral differences clearly shown to be heritable have not been reduced to problems in biochemistry, electrophysiology, or embryology. Perhaps investigators have not looked in the right places. Or it may be that behavioral measures are the only reliable indicators of certain kinds of inherited organic characters. Biochemical, physiological, and anatomical techniques have limitations: they invade the integrity of the organism and, except for electrophysiology,

provide static rather than dynamic information.

Most experimental psychologists choose to study behaviors that are simple and readily controlled. The key-pecking response of a pigeon in a Skinner box is a good example. For the most part (there are some notable exceptions), the topography of the response is considered of little interest. Instead, attention is focused on rate of emission per unit time as a function of various environmental contingencies. Likewise, in the area of perception and information processing, researchers usually employ a highly complex input and, again, a simplified, often binary (e.g., yes-no) response. In both cases, the goal is to reduce interindividual variance (error) as much as possible.

In psychometrics, on the other hand, interest is precisely on the variation between behaviors and the variation between individuals in respect to these (Cronbach, 1957). Thus the enterprise is quite different, and one of its central concerns is the ordering of the complexity of traits like intelligence and personality into manageable units. The primary statistic used is, of course, the correlation coefficient with its extension into factor analysis.

This method begins with a matrix of intercorrelations between a number of measures and, by a series of statistical manipulations, extracts a smaller number of factors that can "explain" the variances of the original scores. There is no mathematically unique solution of such a matrix. Many psychologists have employed Thurstone's (1947) concepts of "simple structure" and "positive manifold." The first means that each test shall have loadings on as few factors as possible; the latter requires rotation of axes to eliminate significant negative factor loadings on all tests. This requirement is probably defensible in the area of intelligence testing in which Thurstone was particularly interested, but its validity in the realm of temperament is doubtful. Both these criteria are intrinsic to the original matrix; that is, they are applied to the relationships between the dependent variables as expressed in the test intercorrelations. With no definite relationship to causal factors (independent variables), they do not necessarily lead to factors that make biological sense. By itself factor analysis leads to more parsimonious description, not to hypothesis testing.

A possible method of accomplishing a rapprochement between factors and external criteria has been proposed by Eysenck (1950). Basically, his method involves rotating axes to maximal agreement of the first factor with some criterion test included in the matrix. For example, tetrachoric correlations might be employed to measure the success with which each of a set of tests discriminates between two genetically defined subgroups. Rotation of axes would continue to extract the factor that most nearly matched the criterion.

The same objective has been sought by the Taxonome program applied to differences between pure breeds of dogs (Cattell, Bolz, and Korth, 1973) (Chapter 10).

Another attempt to relate factor theory to genetics emphasizes the multiple-factor control of independent processes that can collectively be called intelligence (Royce, 1957, 1973). Royce's model (Fig. 18-1) assigns blocks of genes to various group factors. The relationship between the genotypes, for example, S and M, to their respective mental traits, such as space and memory, is shown as the area of psychophysiological genetics. Presumably the action is direct, since other genes are postulated to have indirect effects on intelligence through the nervous or endocrine system. The most notable features of

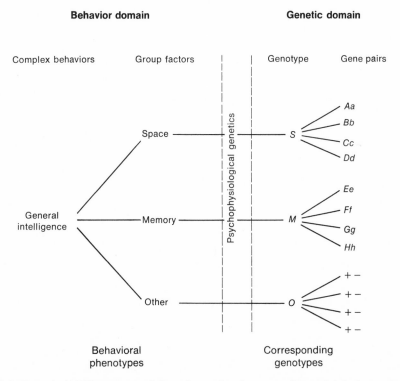

Fig. 18-1. Royce's (1957) concept of the relationship between the multiple-factor theory of psychology and the multiple-factor theory of genetics. Capital letters signify a plus effect on the trait or phenotype. Thus a high space-factor score would be given by *ABCD*, a low score by *abcd*. This model we have called "congruent," since there is a part-for-part correspondence between gene blocks and psychological factors. (From Royce, J. R. 1957. J. Educ. Psychol. Meas. **17**:361-376.)

the Royce model are the idea of congruence between genetic and psychological elements and the distinction between direct and indirect pathways between genes and intelligence.

The notion that dimensions derived from factor-analytic procedures should "make more sense" genetically than complex traits assessed by factorially impure tests is intuitively appealing perhaps, but it is neither logically nor empirically supportable at present. There are a great many ways of factoring correlation matrices, and, unless explicit steps are taken to assure this, there are no reasons to suppose that any particular set of factors will be more or less heritable or show simpler modes of genetic transmission than ordinary tests. Thus a comparison of heritability coefficients obtained for tests, first-order factors, and second-order factors (Thompson, 1968) shows no real differences among them. In fact, a correlation between two characters can be generated for a variety of reasons, both genetic and environmental (Fuller and Thompson, 1960). Since a factor is a kind of average intercorrelation, there are no grounds for supposing it will reflect more of one set of causes than the other. Some examples of breakdowns of several phenotypic correlations into genetic and environmental covariances are shown in Table 18-1. There are no grounds for thinking that the composition of factors is any more orderly. This is not to say that any particular factor or correlation is generated randomly, but rather that each one may follow quite different etiological rules.

A discussion of possible sources of correlation between behavioral traits was included in Chapter 5. Here we review this topic with explanatory diagrams, since it is central to the way in which we conceive of the genotype-phenotype-environment relationship.

Correlations between traits may arise from genic, chromosomal, gametic, or environmental communalities. A diagram of genic communality is shown on the left-hand side of Fig. 18-2. The correlation between traits ϕ and θ is a function of the contribution of physiological character l to each. This character is, in turn, controlled by gene D. Both ϕ and θ have genetic variances (from genes A, B, C, E, and F) which are either specific or shared with other traits. The *short arrows* extending from the physiological level boxes run to other behavioral traits omitted from the figure.

On the right-hand side of Fig. 18-2 is a diagram of chromosomal communality. The covariation between traits θ and Σ is dependent on the linkage of genes F and G. It will not be important in large random-breeding populations but may be significant in groups of closely related individuals.

Gametic communality is illustrated in Fig. 18-3. The associations of traits ϕ' and θ' and their alternates, ϕ'' and θ'', are maintained only as long as there is positive assortative mating for these combinations or there is selection against $\phi'\theta''$ and $\phi''\theta'$ phenotypes. The latter possibility is consistent with Dobzhansky's concept of the coadapted genotype (Dobzhansky, 1962). Selection operates on total phenotypes. Gene combinations that produce effective phenotypes are selected together and occur together in the gametes of

Table 18-1. Examples of phenotypic, genetic, and environmental correlations*

Traits	Correlations		
	r_P	$r_{G_A G_B}$	$r_{E_A E_B}$†
Milk yield × butterfat yield in cattle	0.93	0.85	0.96
Body length × back-fat thickness in pigs	−0.24	−0.47	−0.01
Fleece weight × body weight in sheep	0.36	−0.11	1.05
Body weight × egg weight in poultry	0.16	0.50	−0.05

*From Falconer, D. S. 1960. Introduction to quantitative genetics. Longman Group Ltd., Harlow, England.
†Environmental correlations also include nonadditive genetic components.

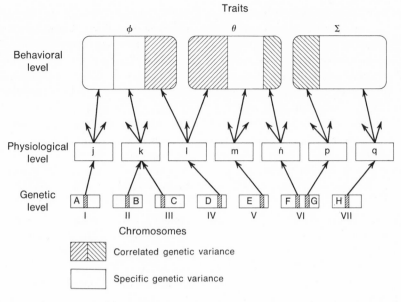

Fig. 18-2. Genic and chromosomal communality. The regulation of physiological process *l* by genes at locus *D* influences psychophenes ϕ and θ. Thus allelic differences at this locus result in covariation of ϕ and θ. The model is noncongruent because there is not a one-to-one correspondence between sets of genes and sets of behavioral traits. The short arrows extending upward from the middle row of boxes represent effects on other behavioral traits, here omitted for clarity. On the right side, chromosomal communality is depicted by the covariance between θ and Σ, which is dependent on the linkage between *F* and *G* in chromosome *VI*. The sign of the correlation can be + or − depending on the particular *F* and *G* alleles that are linked.

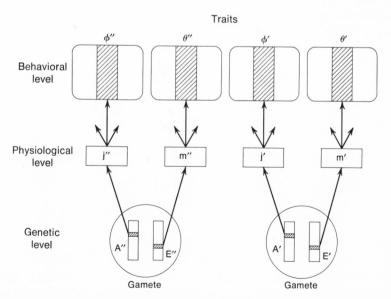

Fig. 18-3. Gametic communality. The covariance between ϕ'' and θ'' is maintained by the continuing association of chromosomes *I* and *II* bearing genes *A''* and *E''*. Covariance of ϕ' and θ' (lower ranking than ϕ'' and θ'') is similarly maintained either by inbreeding or by assortative mating between individuals of similar rank in ϕ and θ.

successful individuals. Adaptation to a changed environment may require not one but many changes in behavior. Thus selection pressure is applied to many loci simultaneously. This process combined with ecological and ethological barriers to interbreeding of individuals from different environments could lead to stable gametic communalities and the division of a species into subgroups characterized by behavioral differences. The best evidence for this outcome is found in *Drosophila paulistorum* (Chapter 11), but it must also occur in other wide-ranging species. Correlations of the same type will also turn up in matrices based on sets of inbred strains where the associations result more from random processes than from selection.

The diagrams of genetic communalities were drawn, for simplicity of exposition, without references to environmental effects. In Fig. 18-4 traits ϕ and θ are shown with both environmental and genetic contributions to variance. A portion of each type of variance is common to both traits; other portions are independent. If traits ϕ and θ are subsumed under a common factor, Z, because of their covariance, event *II* and gene *C* are both involved. If this model is representative of the true relationships between variables affecting behavior, one should not expect a factor analysis to produce purely biological or purely environmental factors unless special procedures are undertaken.

Attempts to do this systematically have been made by Loehlin (1965a) and Loehlin and Vandenberg (1968) in the realms of personality and intelligence, respectively. In the case of the former, as described in Chapter 15, separate factor analyses were carried out with tests having high or low heritability. As readers may recall, the four "hereditary" factors and the four "environmental" factors appeared very similar, though with slight shades of difference. Loehlin suggested that factors in the first set seemed "more focused on the individual himself; whereas factors in the second set were focused on 'his reaction to his environment.'" A slightly differ-

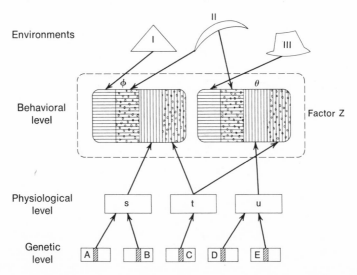

Fig. 18-4. Specific and common variances of genetic and environmental origin. Traits ϕ and θ may be subsumed under factor Z because of their intercorrelation. Factor Z exists because of common effects of gene *C* and event *II*. The factor provides a more parsimonious description of the traits, but it does not imply a single agent responsible for the covariation. *Horizontal lines*, Environmental variances; *vertical lines*, genetic variances; *dotted areas*, covariances of ϕ and θ.

ent approach was taken in regard to the domain of intelligence to compare factors derived from covariances of tests for within– *MZ and DZ twin-pair differences* and between–*MZ and DZ twin-pair differences* (i.e., pair sums). It was expected that factors for the first set might differ between MZ and DZ pairs, since differences in MZs reflect only environmental agencies, whereas differences in DZs reflect genetic variability as well. Factors obtained from MZ and DZ sums, however, should not differ. A final analysis was also carried out on a "difference matrix" obtained by subtracting the MZ variance-covariance matrix from the DZ variance-covariance matrix.

The general conclusions obtained from these analyses were as follows: (1) the same (or similar) cognitive ability factors were found in genetic and environmental covariation, and (2) a second-order, or general, factor was found in both but of somewhat different composition. The genetic second-order factor involved all five of the primary mental ability factors. However, the second-order environmental factor involved mainly verbal comprehension, word fluency, and reasoning, but not number and space, which were also independent from each other. Loehlin and Vandenberg suggest that the two factors perhaps correspond to Cattell's hypothesized factors of "fluid" and "crystallized" intelligence. The former might be thought of as representing general hereditary potential for *all* intellectual functioning; the latter, the effects of educational and cultural processes that produce communality mainly among various verbal types of tasks.

The data of Loehlin and Vandenberg, if valid, are intriguing, since they suggest that genetic and environmental agencies partially mimic each other in respect to the kind of covariation they produce between traits of personality and intelligence. Their results seem distinctly different from those found for morphological characters in domesticated animals (Falconer, 1960). It is not clear why this should be, although it may relate to the operation of the paths between genes and behavior, a topic we will consider shortly. It

may be recalled, also, that Horn, Plomin, and Rosenman (1976), following procedures similar to those of Vandenberg and Loehlin, failed to achieve the same results.

Another approach to the problem of units has come from ethology. In one sense the use of specific movements as units for behavior genetics stands at the opposite pole from the factor-analytic approach. The objective is not parsimony in the description of behavioral differences, but the analysis of gross general differences into a number of precisely defined acts. In another sense, the two approaches have something in common. Both methods seek to define basic units of behavior that can then be related to genetic and environmental determinants. The careful description of interspecific differences in behavior has been the concern of a number of biologists.

Behavior patterns are as useful in the taxonomy of some groups as are anatomical characters (Lorenz, 1950). Specific action patterns have served as psychophenes for a number of genetic analyses involving interspecific hybridization (Chapter 11). Most of these experiments have been done with insects, fish, and birds, groups in which fixed action patterns are particularly easy to identify. However, highly stylized motor responses are found in mammals, including humans, and more use might be made of them in genetic studies. Simple Mendelian models may not explain inheritance of such psychophenes any better than they explain that of complex psychophenes such as intelligence. In one of the best worked out systems, the calling song of a cricket, each discrete element of the song seems to be regulated by a polygenic system (Bentley and Hoy, 1972) (Chapter 11).

We have stressed the point that behavior is a process which involves the organism as a whole and that it must, therefore, have a polygenic basis. Single-locus gene substitutions can have significant effects on behavior by disrupting the development of receptors, motor organs, or central nervous system. A case of this type is convenient for genetic analysis, but we can only conclude from it that a certain gene is essential for normal functioning. We cannot say that the gene by itself organizes and controls a segment of behavior. Can we go further with polygenic systems than to describe them in terms of additivity and dominance? Can such genes be assigned to chromosomes and assigned specific functions? Something of this sort has been accomplished in *Drosophila melanogaster* with the polygenic system regulating the number of abdominal bristles (Thoday, 1961). Somewhat similar procedures were applied successfully in locating genes affecting geotaxis on the three large chromosomes of *D. melanogaster* (Hirsch and Erlenmeyer-Kimling, 1962). Unfortunately the techniques are impracticable for slow-breeding species with many chromosomes.

In mice the recombinant inbred strain (RIS) technique (Chapter 6) has been used in the search for individual loci affecting quantitative behavioral traits. There are limitations to the ingenious technique. To our knowledge all published material on the behavior of RIS mice has been, up to the time of writing, based on lines derived from two inbred strains; thus, only a small portion of the gene pool of this species has been sampled, the portion in which the two progenitor strains happen to have fixed different alleles. This limitation will be surmounted as additional RIS lines become available and are used in behavioral research. Also, locating a gene on a chromosome tells us nothing about its mode of action; neither does the demonstration that a gene affects the level of a particular form of behavior, such as aggression, mean that it is responsible for organizing agonistic patterns of response. Separating polygenic systems into individual components is only a first step toward understanding gene-gene and gene-phenotype relations. A start has been made (Messeri, Eleftheriou, and Oliverio, 1975; Simmel and Eleftheriou, 1977), and additional publications are awaited with interest.

Behavior is plastic

A third property of behavior, perhaps the most crucial, is its fluidity or changeability.

It is, in fact, this characteristic with which most of experimental psychology has been concerned. Thus one of the foremost such changes is that involved in *learning and memory*. As we have shown in an earlier chapter, animal behavior geneticists have done a great deal of work on these topics, though with perhaps limited success. On the other hand, experimental psychologists have only recently begun to accord a place to genetic factors in their models. Today such terms as "constraints on learning" (Hinde and Stevenson-Hinde, 1973) and "preparedness" (Seligman, 1970) have become standard terms in learning theory. Thus there are strong indications of a growing rapprochement between genetics and learning theory, as well as between both of these and neurophysiology.

In a very real sense, work on the genetics of changes involved in learning represents another way of dissecting complex behavior. In this case, the units tend to be temporal rather than structural, as with psychological factors previously described. Several illuminating examples to highlight this theoretical point have been touched on in earlier chapters.

One of the first studies to address itself to the problem in any systematic way was that of Wherry (1941) discussed in Chapter 10. By a factor analysis of maze errors made by maze-bright and maze-dull rats *over time,* he was able to extract three dimensions: forward-going tendency, food pointing, and goal gradient. The relative contribution of each of these to maze performance varied greatly from early to later trials. Thus in the initial period of learning, an animal's behavior was most influenced by a tendency simply to move forward in a straight line and least of all by the factor of goal gradient (fewer errors close to the goal box); however, after about one third of the total trials, this situation changed differentially for brights as against dulls. For brights, the goal-gradient factor now became strongly dominant, but for dulls, the food-pointing factor was most salient, with goal gradient catching up at the end of the learning session. It is difficult to put forward any watertight theoretical explanation of these results, though Wherry did make some conjectures. Nonetheless, they demonstrate convincingly that the changes which occur in behavior over time are governed by the relative salience of different psychological variables and that the expression of these is probably under genetic control.

A second example is afforded by the work of Broadhurst and Jinks (1966) involving a genetic analysis of open-field behavior over time in the Maudsley reactive and nonreactive strains. Basically, what they found using a diallel analysis was that heritability values changed with each trial and that the relative contribution to phenotypic variance of environmental, additive, and dominance components also altered. For elimination scores, there occurred over the several days of testing a proportional increase in dominance variation that, in turn, mainly underlay low scoring on this measure. On the other hand, ambulation scores over days showed a drift to intermediate values, this level also being mainly controlled by dominant genes. Such a finding is quite compatible with a large body of early experience work, suggesting that open-field activity is a complex trait involving quite different components whose relative contribution varies over time (Levine et al., 1967).

The same point is made more precise in an experiment by Wilcock and Fulker (1973). They employed a diallel cross design to analyze escape-avoidance conditioning in eight strains of rats. For successful avoidances in 30 trials, no evidence of directional dominance was found. However, analysis of early trials separately showed clear dominant gene action for low responding. After about trial 20, however, dominance for a *high* level of responding was found. This changeover suggests two processes operating, a conclusion very much in line with two-process models that explain avoidance learning by reference to a classical conditioning component (operating early) and an operant or instrumental component (operating later) (Mackintosh, 1974). Furthermore, it is reasonable to sup-

pose that inactivity is adaptive at first and high activity or "fleeing" adaptive later, as suggested by the genetic analyses. This consonance between genetic and behavioral architecture is of considerable interest (Broadhurst and Jinks, 1974).

A final example of a slightly different kind may be put forward. Genes can clearly control the processes involved in the temporal course of learning. But, more than this, they probably also control some of the basic parameters of learning as well. One such parameter in classical conditioning is the temporal distance between CS and US. Normally, strength of conditioning is an inverse function of this duration. Thus the greater the separation between the two, the weaker the conditioning; in fact, in most instances little will occur beyond a duration of a few seconds (Mackintosh, 1974). Thus it is noteworthy that rats injected with lithium chloride after flavored water ingestion develop strong aversion to the particular flavor, *even though the interval between ingestion and the actual sickness is several hours* (Garcia and Koelling, 1966). While such a result is apparently in violation of the usual principles of classical conditioning, it points up the fact that such a violation is highly adaptive for an animal in nature and has probably been brought about by natural selection. In other words, it is likely that animals who do not avoid food that later turns out to be toxic will not survive and reproduce.

There are many other studies dealing with the genetic analysis of learning and memory. Some have already been covered in Chapter 9. We will turn now to a second major type of behavioral plasticity, namely that involved in *developmental change*. This again is of fundamental interest both to psychologists and to biologists.

It is well known that particular genes express themselves at different times during development. A gene for baldness, for example, will not show itself until after adolescence. Likewise, more complex phenotypes like the major psychoses have varying probabilities of appearing at different age levels. With some characters like human body weight, heritability tends to be very low at first (Falconer, 1960) but rises gradually to around 0.5 (Vandenberg and Falkner, 1965). Likewise, barking in dogs shows quite different kinds of variation across breeds when measured at different ages (Scott, 1964). Again, and more remarkable, Henry and Haythorn (1975) have shown that in mice the albino gene, *c* (on a C57/BL background), influences seizure severity quite differently at 16 days as compared with 21 days of age. At 16 days, the normal allele is dominant. Thus no mice of the +/*c* genotype convulsed at 16 days, but showed at 21 days an incidence and severity of convulsion intermediate between the +/+ and *c/c* genotypes. Thus the interaction between the *c* locus alleles varies with age.

Results of the kind just described are difficult to interpret. It is possible that the same genes have different functions at different ages, perhaps because of modifying factors. It is also possible that behaviors thought to be the same at different ages are really quite different and are perhaps influenced by different genes. And, finally, it is possible

Table 18-2. MMPI and CPI scales with significant F ratios, or $R_{MZ} > R_{DZ}$, in adolescence and/or adulthood*

Personality inventory	Scales with significant F ratios, or $R_{MZ} > R_{DZ}$		
	Adolescence only	Adulthood only	Adolescence and adulthood
MMPI	Depression *(D)*	Hypomania *(Ma)*	Anxiety *(A)*†
	Psychopathic deviate *(Pd)*	Lie scale *(K)*	Dependency *(Dy)*†
	Paranoia *(Pa)*	Ego strength *(Es)*	
	Schizophrenia *(Sc)*		
	Masculinity-femininity *(Mf)*†		
	Hysteria *(Hy)*‡		
	Social introversion *(Si)*‡		
CPI	Sociability *(Sy)*	Sense of well-being *(Wb)*	Dominance *(Do)*
	Self-acceptance *(Sa)*	Psychological-mindedness *(Py)*	Self-control *(Sc)*§
	Tolerance *(To)*†		
	Social presence *(Sp)*‡		
	Socialization *(So)*‡		Good impression *(Gi)*‡
	Achievement through independence *(Ai)*‖		
	Intellectual efficiency *(Ie)*‖		

*Modified from Dworkin, R. H., B. W. Burke, B. A. Maher, and I. I. Gottesman. 1976. A longitudinal study of the genetics of personality. J. Pers. Soc. Psychol. **34:**510-518. Copyright 1976 by the American Psychological Association. Reprinted by permission.
†Significant in subsample only, that is, the part of whole initial sample that was retested in adulthood.
‡Significant in entire adolescent sample only.
§Significant in female subset of entire adolescent sample only.
‖Significant in subsample and female subset of entire adolescent sample only.

that the expressiveness of genes, or, conversely, the power of environment, varies with age. The problem is currently of great interest to molecular biologists concerned with the matter of how gene transcription and translation are regulated. Some of their work has led behavior geneticists to study this more directly at the phenotypic level. Some promising lines of work will now be discussed briefly.

It appears to be true that parent-child resemblance in IQ is quite low early in life, rises rather sharply around age 4, and then shows a moderate increase into adolescence. This might be thought to show that heritability increases with age, which may well be true. However, it should be noticed that if it does so increase, it may well decrease again going into adulthood. Thus Rao, Morton, and Yee (1974), applying a path analysis to published data on adopted children and parents, have estimated the genetic contribution to children's IQ as being 0.752 but to parents' IQ as being only 0.121 with 0.523 due to "common environment." If we ignore the fact that standard errors of variance components in parents are large, then these results would suggest that heritability shows a general decline with age. On the other hand, the analysis of Rao et al. is not in total agreement with the work of Jarvik, Blum, and Varma (1972) (Chapter 14), showing a maintenance of high heritability of IQ (MZ as compared with DZ twins) into advanced old age. However, any conclusions drawn from path analysis depend a good deal on the a priori arrangements of components and their causal structure. Consequently, it is rather difficult to educe any firm conclusions regarding the long-term developmental time course of heritability of IQ.

The short-term developmental changes in mental age have been studied by Wilson (1972, 1977) as discussed in Chapter 14. His critical findings were (1) at five ages of testing between 6 and 24 months, MZ pairs showed more similarity than DZ pairs at each age; (2) profile contours across ages of co-twins in MZ pairs were more similar than profiles of DZ co-twins; and (3) for most pairs, variability of score across ages is high. Thus

all individuals fluctuated greatly (as much as 30 mental development score points) over the testings, but whereas members of MZ pairs fluctuated in synchrony, members of DZ pairs tended not to do so. This is an interesting empirical finding. However, it does not necessarily reflect the action of genes over time (as in the operon model of Jacob and Monod [1961b], for example). The congruence may be a result of environmental rather than genetic events. In fact, computation of h^2 values from Wilson's intraclass correlations for the five tests $(2[r_{MZ} - r_{DZ}])$ yields values between 0.08 and 0.34, with the highest h^2 value of 0.5 obtained for profile contours over the early tests.

The general problem of "chronogenetics" has also been considered in some detail by Dworkin et al. (1976) (Chapter 15). They report extensive longitudinal data for MMPI and CPI scales given to an adolescent twin sample and to a subsample of the same group retested in adulthood. Results are shown in Table 18-2.

It is clear that remarkably few scales on either test show significant heritabilities at both ages. In fact, only one, CPI dominance, yields really unqualified results. Thus genetic determination of personality appears to shift markedly with age, a fact which suggests that prediction from an earlier age to a later age is difficult. Several hypotheses are put forward by Dworkin et al.: specifically, changes in gene regulation, changes in phenotypic variance, and changes in the traits actually measured by the scales at the two ages. However, they quite rightly refuse to opt for one of these explanations over the others.

An additional point of interest arises from an analysis of *difference scores* for the scales between the two ages. The question asked is whether MZ pairs show more agreement in amount and direction of change than DZ pairs. Significant results were obtained for five MMPI scales (*Hs, A, R, Es,* and *Dy*) and two CPI scales (*Wb* and *To*). It is somewhat paradoxical that two of these—*Hs* (hypochondriasis) and *R* (repression)—fail to show significant heritabilities at either cross-sectional assessment. The *change* is apparently under genetic control but not the scores themselves. This is a very curious result.

If we find some genetic basis for propensity to change, we might expect to see this occuring more exactly when specific and planned interventions are carried out. In fact, there is a voluminous literature on this topic, although it has not usually been discussed within the present frame of reference. Relevant portions of it have been usefully summarized by Erlenmeyer-Kimling (1972) under the general heading of "gene-environment interactions." Her survey deals mostly with the effects of early manipulations on various behaviors in different genetic lines of mice. A good example is afforded by data from Henderson (1970) on the effects of enrichment on mouse behavior. Major results are shown in Fig. 18-5, from which it is clear that different genotypes do, in fact, respond very differently to the same treatment. Three strains are quite strongly affected (BALB, C3H, and C57BL); three others (A/J, RF, and DBA), however, show very little change as a result of early treatment. In other words, some genotypes are much more plastic than others. Whether such plasticity is general across all treatments and all traits is not known, but, on the basis of other data presented by Erlenmeyer-Kimling, this seems unlikely.

An additional feature of great interest in Henderson's data is that the genetic contribution to phenotypic variance increases as a result of enriched rearing by a factor of almost ten. In other words, it appears that standard cage rearing inhibits the full expression of genotype. Note, also, that the order of strains is changed. C3H is the poorest under standard cage rearing but rises to third highest under enriched conditions. On the other hand, A/J and RF perform poorly under both conditions. These are fascinating and important results. Thompson (1968) suggested earlier that the ordinary notion of "heritability" does not fully cover such data and that what was needed was some conceptualization of *individual heritability*, referring to the plasticity of a given trait in a given genotype. A critic (Spuhler, 1968) rejected the view and asserted that the problem could be dealt with simply by treating performance in two environments as two characters and estimating the heritability of each. But such a treatment does not serve the purpose. In the case of Henderson's data, a genetic analysis of *difference scores* (between environments) would represent the critical step in assessing plasticity or buffering. Unfortunately, this was not done. That it is perfectly feasible, however, is clearly indicated by the study of Dworkin et al., dis-

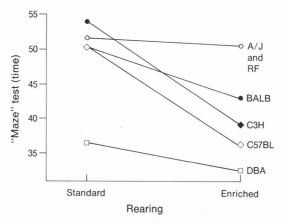

Fig. 18-5. Differences between six inbred mouse strains on a "maze" test as a function of early treatment. (From Erlenmeyer-Kimling, L. 1972. Gene-environment interactions and the variability of behavior. In L. Ehrman, G. S. Omenn, and E. Caspari, eds. Genetics, environment and behavior. Academic Press, Inc., New York; based on data from Henderson, N. D. 1970. J. Comp. Physiol. Psychol. **72:**505-511.)

cussed before. A thoughtful treatment of the general problem of plasticity and its statistical analysis has been put forward by Cavalli-Sforza and Feldman (1973a). We will return to this shortly.

Besides the plasticity involved in learning and development, there are other kinds. Circadian rhythms are an example; likewise, estrous and breeding cycles clearly modify the expression of a variety of behaviors. Finally, there is a large class of fluctuations that may be considered microenvironmental in origin and perhaps reflect merely "noise." These and other types of plasticity have been discussed fully by Thompson (1967) and more recently by Manning (1976).

In summary of this section, we may state that the three properties of behavior just outlined give it a status which calls for special kinds of conceptualizations and procedures. In general, however, two broad approaches encompass most work in behavior genetics. We will now discuss these in detail, emphasizing always the linkages between behavior genetics and its parent disciplines of biology and psychology.

APPROACHES TO INDIVIDUAL DIFFERENCES

As stated previously, there are two major approaches to the study of genetic contributions to individual differences. These have been termed by Dobzhansky and Spassky (1967) as the *Darwinian*, or *compositionist*, and the *Cartesian*, or *analytic*. Although these fully complement each other, they differ considerably in emphasis, as shown in Table 18-3. Both, in the end, operate within the general framework of evolutionary bi-

ology. Thus the latter focuses not only on changes in the genetic structure of populations related by descent, but also on the mechanisms by which genes act to produce a phenotype. The contrasts between the two approaches will be made sharper by examining some of the salient problems studied by each.

Population genetics and behavior

Adaptation and behavioral variation. As indicated in Table 18-3, one of the key concepts in a compositionist approach is adaptation. In order to survive, organisms must respond to stimuli in a way that results, on the average, in the satisfaction of tissue needs and the execution of reproductive functions. The usually accepted explanation for the correspondence between needs and behavior is the evolution of behavior mechanisms through natural selection.

Briefly, the natural selection theory of behavioral evolution postulates three related processes. First, genetic variation occurs within a population, presumably because of mutations, a small fraction of which are not lethal or deleterious. Second, this results in variable behavior, some forms of which are better adapted than others to the challenges that are encountered. Third, better-adapted individuals are more successful in reproduction, and the genes which are necessary for superior adaptation increase in frequency in that population. The process has no definite end, and evolution is a contemporary as well as a historical phenomenon.

Obviously, the evolution of behavior is explicable by these mechanisms only to the extent that behavior is heritable. However, su-

Table 18-3. Difference in emphasis between compositionist and analytic approaches

	Compositionist	Analytic
Level of study	Molar or organismic	Molecular
Type of explanation	Functionalist	Structuralist-mechanistic
Perspective	Historical	Ahistorical
Focus	Adaptive variation	Average or modal organism
Mode of data collection	Descriptive-naturalistic	Experimental-laboratory
Major framework	Evolution in populations	Evolutionary mechanisms

perior adaptations can be transmitted culturally as well as biologically; this is particularly true in the case of human beings. There is probably a continuum between these types of transmission. At the biological end, structures evolve that produce relatively stereotyped responses to critical stimuli impinging on an organism or parts of an organism. Through natural selection, each stimulus-response pattern is stabilized as the one most likely to permit survival and reproduction. Thus it is normal for most animals to "know" when they need food (by feeling hungry) and to undertake the necessary steps to reduce this need. In cases where the mechanism is out of kilter for some reason, as in aphagia or hyperphagia, the fitness of the organism is placed in serious jeopardy. At the cultural end of the continuum, we find adaptations involving the evolution of structures that become organized only in the course of their functioning to produce the most adaptive responses to particular circumstances. In this case, the stimulus-response chains are not stabilized by natural selection but by learning. Thus most children are born with the equipment for verbal communication. Many of the features of the latter, however, are clearly learned, for example, whether the language is German or English. It is difficult to know at what level the changeover occurs. Thus some (Skinner, 1957) have argued that virtually all features of language are learned. Others, however, have suggested that there are certain universal rules of language production and comprehension ("deep structures") that are innate (Chomsky, 1968).

Natural selection. It is often tacitly assumed that changes in behavior follow the selection of structures. Probably the opposite is true, since it is through behavior that the organism makes contact with the forces of selection. This was recognized by Darwin (1872) and has recently been emphasized by Wilson (1975). A good example is afforded by the classic studies of Kettlewell (1959, 1965) on the peppered moth, *Biston betularia*, in the United Kingdom. This species exists in two morphs, a dark and a light. What Kettle-

well observed was that there was a general increase in the proportion of dark morphs over time, this being mainly restricted, however, to industrial areas. In the latter, dark moths that alighted on trees blackened by industrial pollution were almost invisible to predatory birds, whereas light morphs were highly visible. The reverse held true in non-polluted areas in which trees were covered with white lichen. Thus gene frequencies for the two morphs changed drastically. What is important to note, in this case, is that a structural trait (melanism) was altered by natural selection by way of a behavior, the propensity of moths to settle on the vertical trunks of trees. Thus, to use Wilson's phrase, behavior is the "evolutionary pacemaker." It is for this reason that this author has given a very central place to behavior in his new sociobiology.

A second important point about natural selection is that it can operate at the individual or at the group level. Just as the fortunes of individuals can rise or fall, so also can those of groups, whether these be family units or whole populations (Wynne-Edwards, 1968; Wilson, 1975). Wynne-Edwards has, in fact, argued that selection at the group level is probably more common. Thus although it is undoubtedly factors like food supply, predation, climate, and disease that ultimately check the density of animal populations, it is interindividual social conventions which provide the proximate checks. These have to do with "rights" rather than with life and death. As Wynne-Edwards (p. 48) puts it, "promoting competition under conventional rules for conventional rewards appears to be the central biological function of society." In the sense that competition involves more than a single individual and that the value of an individual is not absolute but relative to the value of a competitor, it requires the invocation of a supraindividual selection process. This notion has aroused much argumentation and often confusion. Let us look at the major example given by Wynne-Edwards (1968).

It has been observed that the Scottish red grouse maintains local populations that seldom fall below or exceed a given density in spite of fluctuations in climate, food availability, and predation. The reason for this relative constancy appears to lie in ritualized social competitions, which are carried out every year beginning in August. During this time, territories are established, and the population comes to be sharply stratified into an establishment of owners and a residue of nonowners. The former are able to attract mates and reproduce; the latter, on the other hand, are evicted and suffer a very high mortality rate. Thus by the following August, when the young have been raised, the population size has been kept at a constant level. The idealized sequence of events is shown in Table 18-4.

Perhaps the critical problems arising out of the example given are the exact nature of the competition and the fitness dimensions on which selection is acting. Much has been written about these. However, perhaps this much can be stated with some certainty: first, the intragroup competition is largely ritualistic and hence not directly harmful to the competitors. This seems adaptive, since the continual selection of violent aggression would eventually eliminate the group. Second, the production of a hierarchy appears to be the essential feature of social selection; this depends on the "willingness" of some individuals to resign themselves to a lower

rank and on the capacity of the majority to be able to switch from a subordinate to a dominant role and vice versa, depending on circumstances. Wynne-Edwards and others have called the lower-rank group "altruists." However, note that the term as used in this context has a meaning rather different from that generally given to it. Certainly, any connotation of voluntary self-sacrifice is clearly inappropriate. As Wilson (1975, p. 120) has put it, "the theory of group selection has taken most of the good will out of altruism . . . [it] becomes just one more Darwinian enabling device."

A number of complex models have been put forward by population biologists to deal with the selection of altruism, notably the kin selection theory of Hamilton (1964) and the reciprocal altruism theory of Trivers (1971). Essentially, these both involve cost-benefit analyses of certain acts in terms of the survival and perpetuation of individual genes when the recipient of the act is a relative (Hamilton), or when there is some expectation that the recipient will reciprocate at some time in the future (Trivers). These models have been applied to a wide variety of phenomena ranging from the defensive behavior of "soldier" termites to sibling cooperation in wild turkeys (Wilson, 1975). The fit of the models to such data seems reasonable from the standpoint of population genetics. However, most psychologists will be somewhat perplexed to find in the sociobiological literature terms like "spitefulness," "cheating," "amnesty," and even "righteousness." Likewise, behavior geneticists may be wary of the easy assumption that all of these behaviors are carried by one or a few genes with certain mutation rates. Like the theory of evolution itself, such models represent grand schemata, whose conceptual dimensions are broad, whose propositions are largely unfalsifiable, and whose data base is very imprecise.

The writers would argue at this point that further elucidation of the phenomena of social behavior and group selection will not come (in contradiction to Wilson) either from population genetics or from neurophysiol-

Table 18-4. Idealized schema of recruitment and loss in a red grouse population*

August stock	37 old birds	63 recruits
	100	
Autumn territory contest	37 established (all ages)	63 surplus (all ages)
Winter mortality	7 die	56 die
	30 survivors	7 survivors
Spring stock	7 substitutes	
	37 breeders	0 surplus
August stock	37 old birds	63 recruits
	100	

*From Wynne-Edwards, V. C. 1968. Population control and social selection in animals. In D. Glass, ed. Genetics. Rockefeller University Press, New York.

ogy, but rather from the behavior-analytic procedures of psychology and ethology. These sciences are precisely equipped to reduce to experimentally manageable definitions such behavioral sequences as those referred to previously. Indeed, psychologists have been engaged in doing this for many years with such commonsense descriptive terms as "persistence," "impulsivity," "anxiety," and "relief." Just such analyses need to be made of the basic behaviors involved in group selection. This is largely a task for the future. However, to anticipate it somewhat, we will next turn to consider a concept that underlies much of the thinking about individual and group selection: *fitness* and its behavioral components.

Behavioral fitness in lower animals. Broadly speaking, biological fitness refers to the capacity of an individual to pass on genes to subsequent generations. It will be obvious that it must involve many different characters, ranging from simple fertility and fecundity to the ability to acquire territory or property and a mate. Less proximate characters, such as we will discuss next, will relate to these acquisitions and thus indirectly to fitness. Although the term is generally used in reference to an individual, it has been extended to cover the case of a collection of individuals who, by virtue of genetic relatedness, share genes in common. This has been called by Hamilton (1964) "inclusive fitness." It is a concept that occupies a central place in his model of kinship altruism.

Of major interest in the present context are the behavioral traits that relate to fitness. As we indicated earlier, at least some, and perhaps all, societies involve a degree of hierarchical arrangement whereby some members have greater access than others to the privileges of ownership and reproduction. Hierarchies are often rather difficult to specify in lower animals; they are even more difficult to define in human societies. In addition, they are likely to vary between species and between environments and probably vary with group size and density. Landau (1965) has developed a mathematical model which, in fact, predicts that as group size in-

creases and the number of uncorrelated ability components increases, the "hierarchy strength" (his index, "h") decreases. Wilson (1975) has taken this to mean that the more complex a society is, the more likely it is to be egalitarian. This may make mathematical sense; however, it does not seem to accord with at least some human data. For example, the income gap between rich and poor in the United States has increased appreciably during the last thirty years, though presumably the whole society has become more complex (Jencks, 1972).

We will now consider more closely the consequences in lower animal societies of being high or low in the hierarchy and the variables that relate to position. After this, we will turn to the human case.

According to Wilson (1975), evidence on the advantages of high social position is completely clear in nonhuman societies. It shows a direct relation to fitness. High-ranking animals have easier access to food and territory, supply better parental care to their young, are less vulnerable to predation, are exposed to less stress, and tend to rear more offspring. The number of species for which these generalizations appear to be true is fairly exhaustive and ranges from the social insects to primates. Although most of this work is descriptive, a number of controlled experimental studies have also been carried out. We may refer again to those of DeFries and McClearn (1970, 1972) on dominance and fitness in mice as an example (Chapter 11).

The apparatus they used was a unit made up of three mouse cages joined by a Y-shaped plastic manifold. Triads of males from several different strains were housed with females in these units for 2-week periods. Dominance was evaluated not only by behavioral observations but also by incidence of tail wounds. Paternity was established by means of coat-color markers. Results were compelling. In three separate studies, dominant males in the triads sired more than 90% of litters. In a few triads in which a dominance order was not well established, paternity was more evenly distributed. These interesting results were supplemented later by Horn (1974), who attempted not only to extend them to seminaturalistic environments but to clarify some ambiguities involved in the DeFries-McClearn work. One of these related to the baseline "fertilizing ability" of different strains. Horn found that this was, in fact, greatest (100%) in dominant RF mice and a good deal lower in submissive DBA, BALB, and C57 animals. The difference was increased when the mice lived in group conditions. A second ambiguity related to the mechanisms by which dominant mice achieved such high fitness in a competitive situation. Aggressiveness was shown clearly to be a factor. However, an additional factor was pregnancy blockage; that is, the presence of RF males caused females to lose a large proportion of their litters sired by other males. This applied more to some strains than to others.

The preceding kinds of data are of great importance, since they clearly support the results of the many field observations that have been made on the relation between social status and fitness. Nevertheless, they still do not delineate exactly what it is about a dominant animal that makes it successful. Before turning to this problem, however, we will offer a few comments about animals of low social status, "losers." Wilson (1975) has again offered some useful suggestions.

One is that subordinates which have no territory, as in the red grouse, may emigrate and thus disperse genes between populations. In this way, they may represent "the cutting edge of evolution." Another is that they may serve as an "aggression sink," that is to say, scapegoats which drain off wasteful aggression between high-ranking, reproducing animals. It must be admitted, of course, that these "compensations" directly benefit the group or species rather than the individual subordinates themselves. Perhaps the only real benefit the latter obtain from living within a group is physical protection. This applies, of course, mostly to non-territorial animals.

We may now consider the constitution of behavioral fitness itself. We have seen that it strongly relates to social position, which

in turn relates to dominance. In fact, social position and dominance might be regarded as synonymous. What then are the qualities underlying them? Wynne-Edwards (1968, p. 159) in consideration of this has remarked, "[the ability to succeed] depends not just on the sharpness of the teeth or the color of the scales, but on the total effect, which gives the eye its sparkle and spells confidence in action." Thus he appears to suggest that there are many routes to success, either by specialized narrow abilities or by some favorable combination of many of them. A partial list of such factors compiled by Wilson (1975) for a number of species includes size, age, caste, defensive ability, "persistence," previous experience of victories, mother's status, and "personality." Such attributes are rather vague and are also probably correlated. Thus age and size go together up to a point; it is unlikely that age per se could be of much importance, but rather the kinds of psychological and physiological attributes that relate to it. Just what these are is not clearly known.

One promising category of explanation is that of endocrine levels. It appears to be true that androgen levels, particularly testosterone, relate to aggression, which, up to a certain level, determines status. Thus red grouse cocks implanted with androgen become more aggressive, may double their territory size, and increase their courtship activity (Watson and Moss, 1971). Likewise, Rose, Holaday, and Bernstein (1971) have shown that in Rhesus monkeys plasma androgen levels are highest among animals in the first quartile of a dominance hierarchy and correlate 0.469 with "total aggressiveness." However, they also noted that the alpha male in their group was twelfth (out of 34) in order of aggression and had only a medium level of testosterone, whereas the most aggressive animal was only tenth in order of dominance but had a very high testosterone level. Thus the relations between hormones, aggression, and status are complex. But on the whole, it seems likely that they are somewhat curvilinear, possibly ∩ or Γ shaped, with moderate steroid levels producing moderate aggression and high dominance status.

Another putative fitness character listed by Wilson is "persistence." This also turns out to be influenced by steroid levels. An initial report by Andrew and Rogers (1972) showed that chicks injected with testosterone increased their "persistence" in a food-search task. The term was defined as a tendency to treat a newly encountered situation as if it were a familiar one. Their results were replicated by Andrew (1972) and Rogers (1974) and extended by Archer (1974), who demonstrated that testosterone reduced distractibility to stimuli "irrelevant" to an ongoing response but increased it for "relevant" stimuli.

In all these studies, the operational definition of "persistence" was imprecise. An attempt to reduce this imprecision was made by Thompson and Wright (1977),* who studied male rats in a discrimination procedure involving a shift from one dimension of responding to another. The two dimensions were presented in a double-bar Skinner box. One was position; the other was light. Thus, in one case, one bar (either left or right) always produced food; in the other, the bar over which a cue light was illuminated always produced reward. Groups injected with one of two levels of testosterone (high or medium) were compared with androsterone- and oil-injected controls in ability to shift from one cue to the other. Highly significant enhancement of persistence (inability to shift) was found in the medium-dose testosterone group compared with all other groups. In a second experiment, the same authors found that an antiandrogen, cyproterone acetate, significantly increased ability to shift. In no cases were the effects due to increased arousal level, which was measured by number of bar presses between trials or during trials when a bar press did not produce food. Consequently, the effects of these steroids on persistence, as defined, seem reasonably specific. Also, the fact that the steroids affect persistence in adult rats, as well as in chicks, suggests that the phenomenon has rather wide generality.

---

*Unpublished.

The functional utility of persistence has been discussed by Andrew (1972), who has argued that an ability to maintain such behaviors as territory acquisition and defense, mating, and food seeking in the face of changing circumstances should have a high fitness value. It is equally clear, however, that beyond a certain limit, persistence becomes "rigidity," which can hardly be thought to enhance survival. It is difficult to know exactly where the dividing line is. Furthermore, the relation between androgen level and degree of persistence is unclear. Thus in the Thompson-Wright study, the high-dosage group (55 mg/100 gram body weight) behaved exactly like controls, suggesting a U-shaped dose-response function. More work will be needed to establish this. Finally, a greater number of species should be studied. Analogous work done on human beings has been summarized by Klaiber et al. (1972).

Adrenal hormones may also be related to fitness, although information on this point is limited. Thus an article by Candland and Leshner (1971) showed that dominant males in a laboratory troop of squirrel monkeys had the highest corticosteroid levels but the lowest levels of catecholamines (epinephrine and norepinephrine). However, whether these were the causes or effects of dominance (and/or aggression) is problematical. It was noteworthy, in this study, that 17-ketosteroid levels were related to dominance by a L-shaped function, with the most dominant males having medium levels; the low-ranking, high; and middle-ranking, low levels.

The work on persistence just described is of great interest because it suggests a dimension of fitness-related behavior that may be close to what we earlier referred to as *plasticity* or *fluidity*. Although we supplied examples of the latter, we did not relate it specifically to fitness. At this point, we may refer to two studies that suggest such a relation. In the first of these, Henderson (1970) carried out a genetic analysis of the change in brain weight in mice produced by environmental enrichment. Six inbred strains and F_1 and F_2 hybrids were used. The major results were (1) only small increases in brain weight were found in inbreds, but (2) large increases (independent of body weight) with significant directional dominance were found in hybrids. Henderson therefore argued that capacity for an increase in brain size in response to enriched rearing is a fitness character.

Henderson's second study (1976) focused on developmental differences in the genetics of activity level in mice. He hypothesized that in infant mice a propensity *not* to leave the nest (i.e., low activity) should have fitness value and hence show directional dominance and low narrow heritability. The prediction was confirmed on over 2000 inbred 3- to 4-day-old mice from eight inbred and thirty-six hybrid crosses. In this case a capacity to show stable behavior (i.e., to stay in the nest) promoted survival.

Thus, although the two experiments yield, in a sense, opposite results, they both demonstrate that the same dimension of plasticity is of critical importance to fitness. Whether plasticity is a basic component of what we think of as animal "intelligence" is a matter of speculation. As discussed earlier, this term is notoriously difficult to define precisely. Perhaps to think of it as a capacity for optimal change may be to dilute it too much. Nevertheless, it is perhaps true that we may have restricted its definition so much to abstract types of problem solving that we have failed to take account of its applicability in other areas of living, particularly those to do with social interactions, motivation, and personality. We will enlarge on this point in our discussion of human behavioral fitness characters.

Behavioral fitness in humans. As in lower animals, almost all known human societies have some kind of ranking arrangement. We may ask again whether any of the attributes that characterize high- and low-status individuals bear some relation to biological fitness.

Undoubtedly the trait that has been studied most in this context has been IQ. In the literature relevant to the topic, several facts seem to stand out. In the first place, there appears to be a strong relation between IQ and occupational status. Some typical data gathered by Waller (1971a) are shown in Table 18-5. Studies by Burt (1961), Cliquet (1963), Scarr-Salapatek (1971), Claeys (1973), and numerous others are in agreement. It should be noted, however, that there is considerable overlap between classes such that some individuals in upper classes do appreciably worse than some individuals in lower classes. Thus in an open meritocratic society individuals and their families may rise and fall in social status depending on their cognitive abilities as measured in part by IQ tests.

Where little mobility exists, that is, in nonmeritocratic societies, we might expect to find that the relationship between cognitive measures and social class is smaller. According to a study by Johnson (1977), this indeed appears to be the case. In Hawaiian samples of American-European ancestry and American-Japanese ancestry, sizeable correlations were found between cognitive abilities, educational achievement, and occupation. For a sample of 209 Korean families, however, corresponding correlations were not significantly different from zero. Furthermore, although the Korean parent-offspring correlations for cognitive measures were all significant and appreciable in size, only four out of forty-two correlations between parental education and offspring cognitive abilities were significant. These ranged

Table 18-5. Relationship between IQ and social class in two generations*†

Social class of father	Father's mean IQ	Son's mean IQ by social class of father	Son's mean IQ by own social class
I	140‡	127‡	114
II	113	109	112
III	105	105	106
IV	93	101	97
V	81	91	88

*From Waller, J. H. 1971. Achievement and social mobility: relationships among IQ score, education, and occupation in two generations. Soc. Biol. 18:252-259.
†Total N = 131 fathers and 173 sons.
‡One case only.

in size from 0.12 to 0.17. Thus measured ability and social class membership are not the same thing (as sometimes supposed), neither do they always go together. They will do so only in meritocratic societies that allow upward movement according to ability. This agrees with the thesis put forward by Herrnstein (1973).

A second point to be mentioned is that the relation between IQ and social class is a good deal stronger than between IQ and income (Jencks, 1972). It is somewhat difficult to know which better reflects position in a social hierarchy.

Third, it has been claimed that there is a fairly clear inverse relationship between IQ (and social class) and biological fitness. The correlation reported by Higgins, Reed, and Reed (1962) was -0.3 ± 0.02 between IQ and family size. Similar findings have been put forward by a number of other workers (Cattell, 1974). In fact, the degree of consensus on this point led to the prediction that, over a number of years, the mean IQ for the whole population would fall. However, this eventuality did not come about. Most studies showed either no change or an increase in IQ over spans of between ten and twenty years. This rather "paradoxical" result was explained by Higgins, Reed, and Reed (1962), who showed that among low IQ groups a large proportion do not marry and hence have no legitimate children. They concluded that the zero fitness of these individuals balanced out the high fitness of those who did marry. In fact, when these childless persons are included, the relation between IQ and fertility reverses and becomes slightly positive. Both Bajema (1963, 1966) and Waller (1971a) have reported the same results. The data thus imply that intelligence or some component of it (possibly that related to social class) may have some fitness value. Note, however, that the effect is slight and is mostly a result of the low overall fitness in the below 70 IQ group. In the latter, fitness is distinctly bimodal, with a good many individuals not reproducing at all and a sizeable proportion reproducing at a very high rate. For this reason as well as others (emphasized by Cattell, 1974), we should be very cautious

about educing any firm conclusions about intelligence, social rank, and fitness. Of these three variables, probably the second, social position, is the most ambiguous in meaning. It is perhaps because of its imprecision that we find such sharply conflicting views as to its nature and importance. Thus Thorndike (1951) reported appreciable correlations between IQ and a number of social-class census variables, such as educational level of parents, home ownership, and quality and cost of housing. Likewise, Coleman et al. (1966) found that quality of schooling, as measured by factors like pupil expenditure or teacher-pupil ratios, had little influence on achievement and ability when "background and general social context" (particularly the socioeconomic level of home) were partialed out. On the other hand, writers like Jencks (1972) have emphasized the difficulty of locating the really critical variables that determine success.

Some evidence of a different kind has come from a study by Neel and associates of the Yanomama Indians of Brazil (Neel, 1970). Neel reports that in this highly polygynous society, there is open competition for a number of wives who tend to be distributed very unevenly. Headmen who are the most dominant also tend to have the most women and a disproportionately large number of surviving offspring. The "field impression" of Neel's group is that these men also tend to be more intelligent. Unfortunately, this impression is not supported by exact quantitative data. However, it is a very provocative suggestion, which should certainly be followed up.

The problem we are considering may be approached by another route. From an evolutionary point of view, fitness characters generally show the following: (1) low narrow heritability, given by the fact that directional selection has used up most of the additive variation, leaving only dominant gene action; (2) inbreeding depression; and (3) heterosis. We can then ask whether human intelligence qualifies as a fitness character according to these criteria.

In respect to the first, Thiessen (1972), accepting a high heritability estimate for IQ, has argued that it therefore reflects "genetic junk," that is to say, the expression of functionally equivalent polymorphic genes with no particular fitness value. Much the same suggestion was made earlier by McClearn (1968b), who speculated also that perhaps some components of IQ showing low heritabilities might have fitness value as compared with number, verbal ability, word fluency, and spatial ability. He further suggested that possibly many personality traits which characteristically show very low heritabilities might also have high fitness value and have therefore played a significant role in human evolution.

These are all provocative suggestions. But as Gottesman and Heston (1972) and Pollitzer (1972) have pointed out, the matter is fraught with difficulties. Thus we do not know empirically exactly what the heritability of IQ is. Neither do we know exactly the extent to which its variation is controlled by dominant genes (Eaves, 1973). Nor do we know, finally, the precise empirical relation between intelligence and biological fitness.

In respect to the other criteria for fitness characters, little can be said. Inbreeding depression is not well documented for IQ (Chapter 14), and heterosis not at all. This is not to say that more useful work cannot be carried out. However, it would need very extensive resources.

Having considered the evolutionary underpinnings of behavior, we will now turn to a rather different problem in population genetic analysis. This is the equally basic question of why relatives resemble each other (or do not) in behavioral traits and what factors, both genetic and environmental, contribute to these resemblances. The problem is perhaps at the very center of behavior genetics and is also the one that makes it a unique field.

Familial resemblance: genetic and cultural transmission. As we have seen in the preceding chapters, a key concept in most of the work discussed is *heritability*. However, its usage in behavior genetics has been unfortunate for at least two reasons. First, as pointed out by Feldman and Lewontin (1975), among others, the term is taken from animal and plant genetics, in which framework it was intended to supply guidelines for breeding programs. In its application to human beings in respect to such characters as IQ or personality traits, it therefore has little relevance in this sense. We do not have control over human breeding, neither would most of us wish to have any. Second, and following from the preceding, the term heritability focuses mainly on the *genetic* contribution to the obtained value. This has carried the tacit implication that the contribution of other factors may be lumped together under the general designation of "environmental," which, for the geneticist, is equivalent to noise. Furthermore, in cases where heritability is high, it is often assumed, even in the absence of proper controls, that only genetic forces are at work.

It is therefore understandable that wide use of the concept has led to a polarization in the scientific community, with the opposing camps using every statistical machination possible to push some heritability estimate (e.g., for IQ) as near to 1 as possible or as close to 0 as possible. This has led at least one authority (Morton, 1972) to suggest that we consider carefully our reasons for wanting to estimate heritability when no selection experiment is envisaged. Likewise, another writer (DeFries, 1972) has argued that we should separate out *measurable* from nonmeasurable environmental factors in the usual equation for partitioning components of variance. This would allow us to write, as an alternative to the heritability coefficient, a *coefficient of environmental determination*. Some other theorists (Cavalli-Sforza and Feldman, 1973a,b) have extended these ideas by attempting to write out explicit models that incorporate both genetic and cultural inheritance variables. Similar kinds of suggestions have been made by the various contributors to the text edited by Schaie et al. (1975) and also by McAskie and Clarke (1976).

Such ideas as these may seem to the reader to be unremarkable, if not obvious. To the writers, however, they appear of great importance, since they bring sharply into focus what we consider one, if not *the*, major thrust of behavior genetics: the *explanation of the behavioral phenotypes of individuals by reference to their relatives*. In a very real sense, this enterprise is a prelude to and serves as a foundation for the task of tracing biochemical and physiological pathways between genes and behavior. Let us examine it in more detail.

Perhaps the first person to understand the need for accommodating biometric genetics to psychology rather than vice versa was Cattell in his MAVA method. Although the latter is open to many criticisms of a technical type, nonetheless, it does attempt to incorporate specifically psychological notions and to deduce specifically psychological conclusions. We discussed some of these in Chapters 14 and 15. An example is the "law of coercion to the biocultural norm" educed from the predominantly negative between-family genetic-environmental covariances. The data base for such a "law" is no doubt shaky. Be this as it may, it hardly offends common sense, and the mere possibility that it might have even some validity under some circumstances is of great interest. Also, it can be explicated into numerous subquestions that are all of genuine psychological importance.

We may contrast Cattell's approach with that taken by Jinks and Fulker (1970), who attempt precisely to eliminate gene-environment interactions and covariances in an effort to reach more exact conclusions of a genetic kind: for example, how many genes are operating, how much dominance is present, and the extent of nonallelic interactions. Such questions are of interest to rather few people and certainly of almost no interest to most psychologists. The point has been emphasized by Cattell (1974, p. 174) as follows: "Jinks and Fulker attempt to assimilate human behavior genetics to standard notations and concepts of general genetics, and. . . . most geneticists seem unwilling to develop these restructurings from the standpoint of the needs of psychologists." Almost the same comment was made earlier by Thompson (1968). Let us now turn directly to the problem of the specific kinds of data that behavior

genetics is supposed to explain and the kinds of behavioral models that might supplement existing genetic ones.

McAskie and Clark (1976) have considered this problem in some detail. They suggest that genetic and environmental models (or some combined models) must be able to make predictions about (1) correlations between relatives of varying degree, including midparent-offspring, under conditions of random or assortative mating; (2) regression of offspring to the population mean; and (3) variances of offspring across different levels of parental IQ and total parental variances. Thus the basic data we have to consider are means, covariances, and variances in relatives. The latter may, of course, vary considerably in degree and kind, ranging from the simple case of parents rearing their own biological offspring to half-sibs reared apart and unrelated children reared together. As we indicated in Chapter 15, Cattell (1960) offered an acute psychological analysis of possible expectations about gene-environment covariances in such groups. More formal mathematical treatments of them have been more recently put forward by Cavalli-Sforza and Feldman (1973a,b) and Rao, Morton, and Yee (1974). However, some of the limitations of their approaches have been strongly emphasized by Lewontin (1974).

The kinds of predictions generated by the simpler genetic models are, of course, well known and are dealt with by McAskie and Clarke (1976). Such common occurrences as the mediocre family with one exceptional child, for example, are easily handled. In fact, it is a feature of population genetic models that they can be modified in various ways to fit almost any set of data. They have been rightly criticized on this account by many environmentalists. However, it is also true that these same critics have not usually been able to offer any alternative explanatory models involving well-defined environmental variables. McAskie and Clarke suggest two possibilities. One is an *exposure model*. It basically involves the simple hypothesis that duration or intensity of exposure of an individual (e.g., a child) to another (e.g., a parent) will generate a certain degree of likeness between them. By itself, this model would predict higher mother-offspring than father-offspring correlations in conventional families. However, this should not hold true in families in which the mother is often absent. There are probably no data that completely fulfill such a simple prediction. However, we have noted some instances in the area of personality that suggest slightly higher mother-daughter correlations than, say, father-daughter. This suggests the operation of the second model put forward by McAskie and Clark, an *identification model*. Parenthetically, it should be remarked that it is not really a second model but only a specification of type of exposure. Be this as it may, a factor of "identification" would predict higher like-sex parent-off-spring correlations than unlike-sex parent-offspring correlations. Again, no data convincingly support this expectation. Consequently, we might favor a model involving both "exposure" and "identification." This would make the more precise prediction about order of magnitude of parent-offspring correlation: $r_{md} > r_{ms} > r_{fs} > r_{fd}$, where the subscripts m, f, d, and s stand for mother, father, daughter, and son, respectively. A massive amount of data would be needed to test such an expectation.

One could readily add other modulating variables such as ages of parents, ages of children, and relative status of parents. Ultimately, this would take us into the whole area of *social learning and imitation*. Current models relevant to these processes would have to be assimilated to a framework of individual differences with due account given to the importance of genetic factors. *But it is certainly these aspects of social psychology that seem to represent the natural linkage between behavior genetics and the other behavioral sciences.*

We have not yet reached the point where we have exact information as to how different genotypes are transmuted by social contact into particular phenotypes. At present, work appears to be more at a general demographic level. To illustrate it, we will now discuss some studies dealing with the effects on intelligence of the two variables of birth order and family size.

Schooler (1972), in a review of the large amount of work done on these topics, expressed pessimism regarding the robustness of the findings reported up to that date. Nonetheless, one large-scale study since then by Belmont and Marolla (1973) has given rise to a resurgence of interest. Their major results are shown in Fig. 18-6. Since the sample size was very large (~400,000), most of the differences between data points are highly significant. It should also be noted that the largest difference, that between the older child in a two-child family and the youngest in a nine-child family, is only about 10 IQ points (about 4.3 transformed Raven scores). Thus the data can hardly be regarded

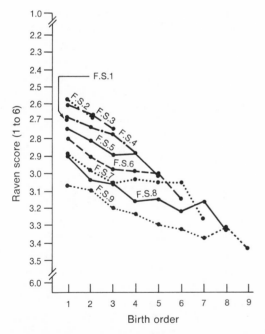

Fig. 18-6. Mean Raven class score by birth order within family size *(F.S.)* across the population $(N = 386,114)$. Class scores range from a high of 1.0 to a low of 6.0. (From Belmont, L., and F. A. Marolla. 1973. *Science* **182:**1096-1101. Copyright 1973 by the American Association for the Advancement of Science.)

as causing embarrassment to a genetic point of view. However, because of their orderliness and because of some theoretical postulates they suggest about family environmental effects, they still have considerable interest.

The basic findings of Belmont and Marolla agree with those of a number of investigators. Record, McKeown, and Edwards (1969) found the birth order effect and showed that it is independent of maternal age and socioeconomic status. However, Marjoribanks and Walberg (1975) reported some differences in respect to the latter variable when more precise statistical procedures were used. For example, in a "nonmanual" social class group (Belmont and Marolla data), successive decrements in intelligence became larger as birth order increased. These decrements became smaller however, in "manual"

and "farm" groups. Furthermore, they found that the intelligence test scores of children from small families within the manual group were often lower than test scores of children from large families in the nonmanual group. The author interpreted this to mean that family size makes less difference in high social groups than it does in lower ones.

A large-scale study by Davis, Cahan, and Bashi (1977) involving close to 200,000 Israeli subjects either of Asian-African or of European-American origin has also largely confirmed the Belmont-Marolla data. Their results are shown in Fig. 18-7. It will be noted that there appears to be a fairly strong cultural difference. In the Asian-African group, there is a tendency in families larger than five children for the curves to turn up with later children. This does not occur nearly so much in the European-American group.

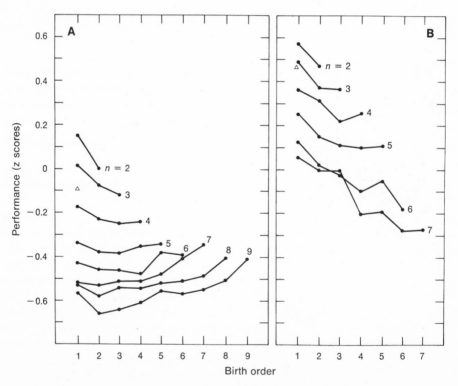

Fig. 18-7. Mathematics scores at age 14 as a function of birth order and family size. For Δ, $n = 1$. **A,** Israeli students of Asian-African origin ($N = 109,304$). **B,** Israeli students of European-American origin ($N = 82,689$). (From Davis, D. J., S. Cahan, and J. Bashi. 1977. Science **196:**1470-1472. Copyright 1977 by the American Association for the Advancement of Science.)

Results such as these are not readily explained by any known genetic or even biological model. Let us consider an environmental model put forward by Zajonc and Markus (1975), who designated it a *confluence model*, by which term they refer to the "mutual intellectual influences among children as they develop in the family context." The operating mechanism, which is assumed but not explicated in any detail, is *exposure*. Thus the IQ of any child is considered to be produced by the average intellectual level to which he is exposed, including his own. For example, the family of a first child born to a couple of average intelligence (IQ = 100) will have an intellectual exposure level of $(100 + 100 + 0)/3 = 67$. When a second child is born some time later, this level will then be $(100 + 100 + 40 + 0)/4 = 60$, the number 40 in the numerator being the intellectual level reached by the first child at the time of birth of the second. It can readily be seen that the model generates the predictions that average ability will be lower in large than in small families and that later children will be less able than earlier born. More precise predictions regarding particular children in particular families depend on specification of the preceding variables. Two important ones are intellectual growth rate of children and pattern of age gaps between births. By recourse to these, Zajonc and Markus explain the upswings we see in the curves for families of size eight and nine, as well as the usually sharp decline between the jth − 1 child and the jth child. Thus they argue that with a two-year gap between children, by the time the fifth child is born, the intellectual value of the older children will have increased to such a level that the total intellectual stimulation provided in the family commences to increase over that found in a four-child family. In other words, the relation between family size and richness of stimulation becomes quadratic rather than linear. When the age gap is only one year, however, this trend does not appear until eight-child families. This also means that, in general, in larger families, the birth order effect may be canceled out or reversed, with later children

being as able as earlier born, and those in the middle the worst. The Asian-African Israeli data of Davis, Cahan, and Bashi clearly conform to this pattern, although those for European-American Israelis do so only minimally. In these cases, however, age gaps between siblings do not seem to be a factor. What apparently is operating, according to the authors, is the intellectual level of the parents. This is notably lower in the Oriental-Israeli population, and the lower the parental contribution, the greater the relative gain achieved by later-born children in larger families.

Another feature of interest in most studies of this kind is that only-children generally fall below the ability level of firstborn in both two- and three-child families. One explanation of this phenomenon offered by Zajonc and Markus is that every child in a family, except last children and only-children, acts not only as a recipient of intellectual stimulation *but also as a transmitter of it*. This latter function, they argue, has the effect of raising ability.

There are many other interesting aspects of the Zajonc-Markus model that cannot be discussed here. Our intention is merely to indicate to the reader the manner in which an environmental model can be explicated. Obviously, it could be extended in many ways. For example, since, by itself, it explains a rather small proportion of the total variance of ability, the inclusion of some genetic parameters would be useful. Likewise, the fuller articulation of what psychological mechanisms are involved in "exposure" should also be carried out. Again, it is perhaps slightly embarrassing to the model that family size and birth order show about the same relationship to height as they do to IQ (Belmont, Stein, and Susser, 1975). It seems doubtful that the confluence model could easily explain both sets of data by reference to the same parameters. Obviously, some additions and modifications would have to be made. The critical point, however, is that all of them would relate to the general problem of familial structure and how members of a family influence each other genetically and environmentally. This seems to us to represent one of the most central themes in behavior genetics and certainly the one through which this field will commence to relate more closely to the mainstream of psychological and biological thought. In this, we are in agreement with Cattell (1974) and with Plomin, DeFries, and Loehlin (1977), among others.

PATHWAYS BETWEEN GENES AND BEHAVIOR

Although it is possible to demonstrate hereditary effects on a psychophene without understanding how a gene operates, there are reasons for probing more deeply. On the practical side, the modification of heritable defects is more likely to be successful if we understand what the causative alleles are doing. Furthermore, it is intellectually satisfying to explain a phenomenon in detail, even if control is neither possible nor desired. From the viewpoint of a geneticist, the ideal system is one in which an allelic difference at a single locus produces a reliable difference in a behavioral phenotype. Caspari (1963) has argued that the best course for behavior-genetic analysis is the study of the pleiotropic effects on behavior of well-known morphological and metabolic mutants, asserting that the common procedure of analyzing psychophenic strain differences genetically leads almost inevitably to a polygenic system whose further investigation is profitless. The conflict between the two approaches to behavior genetics is evident in the exchange of views between Wilcock (1969, 1971) and Thiessen (1971) discussed in Chapter 1. We shall argue here that both types of approach are necessary and that the two simply are directed at different questions.

The general problem of the relationship between genotype and phenotype was discussed in Chapter 3. When the phenotype of interest is more complex than the synthesis of a particular molecule—the so-called primary gene product—difficulty piles upon difficulty. Psychophenes, which are not structures but processes, are the most difficult of

all to relate functionally to a particular DNA molecule. True, a single mutation may disturb the orderly development of the nervous system and produce characteristic behavioral effects, as in the hyperkinetic *Drosophila* (Kaplan and Trout, 1969), the numerous neurological mutants of the mouse (Sidman, Green, and Appel, 1965), and the phenylketonuric human. Investigation of these mutants at many levels of organization may provide important information about neural development, and the functioning of a damaged nervous system can tell us something about how an intact one operates. However, there are severe limitations to this approach. Deviant mice and drosophilae exist only because they are sheltered in laboratories; information from them cannot be applied readily to the kinds of heritable behavioral variation that exist in Mendelian populations of wild-type individuals. In many mutants the pleiotropic effects on behavior are so all-embracing as to be of little interest; one can usually learn more about central nervous system functioning from a microlesion than from a mutant with extensive neurological abnormalities. On the other hand, the extent of behavioral pleiotropy emphasizes the integrational aspects of development and the havoc caused by one defective molecule or by failure of coordination of growth rates. As long as one does not expect that in complex organisms there will be one-to-one correspondence between a gene and a behaviorally defined trait, the *single-gene* approach can be useful.

Nature of genetic regulation

Complex organisms. In most of the organisms studied by behavior geneticists no cell produces all the products whose structures are encoded in its genome. Glands, muscles, and neurons of an individual have the same DNA but not the same RNAs, proteins, and other constituents. The relatively simple model of gene repression, depression, and activation discussed in Chapter 3 (Fig. 3-4) does not suffice to explain the complex coordinated syntheses that go on in the specialized cells of multicellular organisms. The Britten and Davidson (1969) model described in Chapter 3 (Fig. 3-5) integrates material from molecular genetics and developmental biology and outlines a system to account for gene regulation in the specialized cells of higher organisms. Basically it involves a process of intercellular and intracellular communication that turns synthetic activity on and off at appropriate periods of development. This model is particularly relevant to instances where stimulation of young developing organisms produces long-lasting behavioral effects. Strain differences in the effect of early experience could be caused by variation at various levels of the intragenomic communication system, which leaves permanent structural evidence at the molecular level. All this is at present theoretical. As we look at the attempts to define gene action on behavior in biochemical and structural terms, we shall find much less sophisticated models. The importance of the Britten and Davidson contribution for behavior genetics is that it points up the complexity of the gene-character relationship even when the characters are molecules rather than behavior patterns.

Simple organisms. One way of avoiding the complexities of tracking the pathways between genes and psychophenes in drosophilae, mice, and humans, species that dominate the behavior-genetic literature, is to work with simpler organisms. This approach is common in physiology, where much of our knowledge of nerve impulse conduction comes from experiments on the giant axon of the squid. It is reasonable to assume that the electrochemistry and membrane properties of squid axons are very similar to those of other axons which are not as convenient for study. Axons seem to be homologous structures in most species. No such homology can be assumed for behavior, which depends on integrated activity of organisms that differ tremendously in size, body plan, and complexity.

Nevertheless, since genetic effects on behavior are mediated through the soma in all species, the use of simple organisms has appealed to some behavior geneticists. Adler

(1969) has found mutant varieties of the motile colon bacillus *(Escherichia coli)* that deviate from wild-type in their degree of attraction for common sugars and amino acids. Since bacteria have no nervous systems, these differences must be associated with receptors for specific molecules associated with the regulation of flagellar movements. Because of their small size and lack of cellular differentiation, there are strict limits on the variability of bacterial behavior. Somewhat greater versatility is demonstrated by protozoa and roundworms, as will now be discussed.

All students of biology have observed the large unicellular ciliate, *Paramecium aurelia*, swimming in spirals and ingesting smaller microorganisms it encounters. By treating paramecia with a mutagenic agent, Kung (1971a,b) obtained variant forms that could be separated from wild-type by the absence of the usual negative geotactic response and by abnormal swimming patterns. *Paranoiac* mutants reversed cilia spontaneously, a type of behavior known to be associated with membrane depolarization. *Pawn* mutants were unable to reverse under ordinary circumstances and were unresponsive to changes in salt concentration of their medium. *Fast* mutants swam unusually rapidly, a behavior associated with hyperpolarization of the cell membrane. *Paramecium aurelia* has many advantages for genetic analysis, since it reproduces both by autogamy and by conjugation. It is also large enough for the insertion of microelectrodes, so that the membrane potentials in the mutants can be determined.

The behavior mutants apparently carry molecular lesions in their excitable membranes. The pawn mutants, some of which are temperature sensitive, have been most extensively studied and have great promise as material for the investigation of membrane structure and physiology (Chang and Kung, 1973; Chang et al., 1974). Genetic differences in excitable membrane properties could well be the basis of gene effects on the behavior of multicellular animals. Their membranes, however, are not accessible for the kind of experiments possible with paramecia. Furthermore, the behavior of an animal as small as a fruit fly involves actions of many neurons whose membranes may vary independently. Although the cytoplasm of *P. aurelia* is differentiated into organelles, it is a much simpler system with a more direct pathway between gene and psychophene than is found in metazoa. The genetic dissection of the properties of excitable membranes has implications beyond the species that are proving to be well suited for behavior-genetic analysis.

Caenorhabditis elegans is a free-living nematode worm about 1 mm in length that feeds on bacteria. Among its advantages for genetic experiments are its 3½-day life cycle and the fact that most individuals are self-fertilizing hermaphrodites. The latter characteristic makes the detection of recessive mutations easy and facilitates isolation of pure-breeding strains. Mutations are induced by chemical means (ethyl methanesulfonate). Several hundred mutations affecting chemotaxis, locomotion, and morphology have been detected and described (Ward, 1973; Brenner, 1974; Sulston and Brenner, 1974). Estimates of mutation frequency suggest that the genetic units of *C. elegans* are as large as in *Drosophila*. Research with this organism has been directed primarily at questions related to development. The following are typical questions: How is complexity represented in the genetic program? Is it the outcome of a global dynamic system with a very large number of interactions? Or are there defined subprograms that different cells adopt and execute independently? What controls the orderly temporal sequence of development?

Paramecium and *Caenorhabditis* have few chromosomes, self-fertilization as well as crossbreeding, short life cycles, and behavior so stereotyped that it is easy to detect deviants by mass screening procedures. These are characteristics ideal for subjects in genetic experiments. The limitations of these and similar species are on the behavioral side. Self-fertilizing hermaphrodites obviously cannot demonstrate sexual isolation;

neither has learning been clearly demonstrated in protozoans or nematodes. If these forms of behavior interest us, we must study their genetics in species that show them in a well-developed manner. Much of what can be learned from simple organisms can be applied to more complex ones, but complexity must be studied in its own right. In the following sections we shall discuss various ideas regarding the mechanisms through which genes in higher organisms influence behavior.

Chemical pathways

Logically all genetic effects on behavior involve biochemistry. There is no other way for genes to act. In this section we will consider cases in which a heritable general or localized biochemical difference unaccompanied by any marked structural modification is postulated as a prime factor in producing a behavioral difference. We will also discuss the field of psychopharmacogenetics, in which a metabolic difference is deduced on the basis of strain or family variability in reaction to a drug. For convenience, hormonal pathways, though also biochemical, will be considered separately.

Biochemical paths. Several strategies have been employed in the investigation of the gene-metabolism-behavior pathway. Research on the behavioral pleiotropy of the albino gene in mice is an example of one approach (Chapter 7). The genetics and biochemistry of albinism are well known, and the problems to be investigated are of this form: How does a deficit in tyrosinase result in less open-field ambulation? Once stated, the problem ramifies into questions regarding the influence of light intensity on an animal with no protective melanin pigment in its eyes and the behavioral implications of a reduced ipsilateral retinogeniculate tract.

Another strategy is to start with a behavioral syndrome that appears to be inherited in Mendelian fashion and look for some underlying metabolic error. Research on phenylketonuria described in Chapter 17 represents a highly successful application of this method. Not only have we gained some understanding of the way in which the damage to the nervous system occurs, but we have also been able to develop a rational treatment that ameliorates the severity of retardation and allows detection of heterozygous carriers who are not themselves retarded.

The search for a biochemical basis of behavioral variation is not always this successful. Thus the autosomal dominant mode of inheritance of Huntington disease has been known since the early twentieth century, but the primary mode of action of the gene is still not identified. Nevertheless, the PKU model has exerted a strong influence on behavior genetics, and the search for metabolic correlates of differences ascertained by purely behavioral criteria is a standard procedure. Neither has this been limited to behavioral traits inherited in simple Mendelian fashion. The method has often been used with pairs of strains measuring high and low (H and L) on some behavioral index. If measurements are taken on a biochemical characteristic, X, in these strains, there are three possibilities: $X_H > X_L$; $X_H = X_L$; $X_H < X_L$. The second type of outcome indicates that X has little to do with the difference between H and L. The other two outcomes could indicate that X is functionally related to the behavioral difference, although they could also indicate fixation within each strain of a different combination of functionally unrelated genes. Much of the literature on biochemical correlates of strain differences is, in fact, ambiguous in that no attempt has been made to distinguish between these interpretations.

There are several ways in which the ambiguity can be reduced. A behavior-biochemical correlation that holds up in a large set of unrelated inbred strains, in a heterogeneous stock, or in the segregating crosses between the strains is good evidence of a true functional relationship. Also, when a relationship between two variables already postulated from physiological experiments reappears in a genetic study, we are more confident in concluding that the biochemical difference is a mediator of the behavioral difference. For example, if it is known that

drugs reducing catecholamine levels in the brain increase susceptibility to audiogenic seizures, and if we find that strain H, which convulses readily, has less epinephrine than strain L, which is resistant, the case for genic control of susceptibility through neurotransmitters is supported. But even here data from a wider range of genotypes are desirable.

Gene-biochemical-behavior pathways in mice are nicely exemplified by the ascription of differences in free-choice alcohol consumption to aldehyde dehydrogenase activity (Chapter 8) and by differences in audiogenic-seizure susceptibility in response to the neurotransmitters norepinephrine, serotonin, and gamma-aminobutyric acid (Chapter 7). The possibility of these pathways is supported by considerable evidence, but it is premature to conclude that either system is fully understood or that the hypotheses will not be modified when additional data become available.

Interest in the genetic regulation of neurotransmitters in animals has been stimulated by the many clinical and experimental studies linking changes in catecholamines and indolamines with depressive reactions in humans (Akiskal and McKinney, 1973). These suggest that genetic variation in the rate of synthesis and metabolism of the biogenic amines plays a role in producing individual variation in affective responses, which in the extreme we label psychotic. Such data are difficult to obtain from humans, but they can be sought in lower animals whose breeding can be controlled and from whom tissue samples can be collected and analyzed as needed. An extensive series of experiments on genetic aspects of catecholamine synthesis in mice has been summarized by Barchas et al. (1975). There are two main sources of catecholamines (CAs) in animals, the adrenal medulla and the brain. In both of these, the biosynthetic pathways shown in Fig. 18-8 are similar. Since the adrenal gland is more accessible for metabolic studies, most of the genetic data are based on it rather than on the brain. Large and reliable differences in the activity of TH and PNMT (see legend for

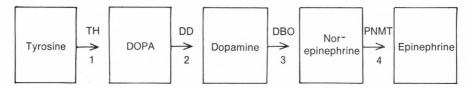

Fig. 18-8. Synthetic pathways for the biogenic catecholamines. The enzymes catalyzing the four steps are (1) tyrosine hydroxylase, *TH*; (2) DOPA decarboxylase, *DD*; (3) dopamine β-oxidase *(DOB)*; and (4) phenylethanolamine-N-methyl transferase, *PNMT*. The first three enzymes are active in brain and adrenal medulla; PNMT is active in the adrenal medulla.

Fig. 18-8) have been found in a selection of mouse strains. In F_1 hybrids, the mode of inheritance is variable, sometimes intermediate but sometimes dominant for moderate rather than for high or low activity (Kessler et al., 1972). The most interesting genetic finding has come from Mendelian crosses between two BALB/c strains separated for more than twenty-five years. These have diverged widely in CA levels (Ciranello et al., 1974). The difference between these two sublines in the level of activity of three enzymes, TH, DBO, and PNMT, appeared to be controlled by a single factor. This could be explained by a close linkage between the three genes coding for the enzymes or by the existence of a single regulator gene that activates all three producer genes as a group.

Evidence for genetic diversity in the regulation of synthetic enzyme activity by different modes of stimulation has been found for adrenal PNMT in mice (Ciranello, Dornbusch, and Barchas, 1972). As shown in Table 18-6, the pattern of responding to four different kinds of stimulation is unique for each of the three strains tested. Observations of this kind in animals encourage a search for similar genetic variability in our own species. The research discussed before suggests a change in the strategy of looking for biochemical correlates of serious psychiatric conditions and normal personality traits. Barchas et al. (1975) believe that instead of looking for "abnormal metabolites," which imply defective enzymes, we might better place more emphasis on study of the regulation of enzyme activities.

The approach to gene-behavior relationships through biochemistry has the advantages of being close to the gene and possibly involving a primary gene product. It has the disadvantages of being distant from behavioral events and always of being only one of the determinants of a particular action. One can employ genetic lesions to "naturally dissect" (Ginsburg, 1958) the nervous system, but the products of the dissection are not natural units of behavior as viewed by psychologists and ethologists. Rather they are a collection of functions that will be affected in common by an allelic substitution at the locus investigated. The genes that have been employed for such genetic dissections (Benzer, 1973) are incapable of surviving in natural populations because their disruptive effects are so great. Despite the power of the genetic dissection methods and their great contribution to the knowledge of development, it is doubtful that their results will be helpful in explaining the genetic basis of behavioral polymorphism in species living in a state of nature. Here the regulator approach seems more promising.

Psychopharmacogenetics. One of the liveliest areas in biochemically oriented behavior-genetic research has been named *psychopharmacogenetics*. At the simplest level, studies in this field are concerned with heritable differences in the metabolism of psychoactive drugs and in sensitivity of tissues to them as primarily determined by behavioral criteria. Many examples of such differences have been cited in previous chapters (7, 9, and 10). In a volume edited by Eleftheriou (1975), chapters on exploratory activity (Oliverio and Castellano), alcohol (Eriksson), audiogenic seizures (Schlesinger and Sharpless), and human studies (Omenn

Table 18-6. Comparison of effectiveness of four factors on PNMT activity in mice*†

Strain	Cold exposure	Glucocorticoid	Nerve stimulation	Direct ACTH‡
DBA/2J	+	+	+	0
C57BL/Ka	+	0	0	+
CBA/J	+	+	0	0

*From Barchas, J. D., R. D. Ciranello, S. Kessler, and D. A. Hamburg. 1975. Genetic aspects of catecholamine synthesis. In R. R. Fieve, D. Rosenthal, and H. Brill, eds. Genetic research in psychiatry. Johns Hopkins University Press, Baltimore.
†*PNMT*, Phenylethanolamine-N-methyl transferase.
‡*ACTH*, Adrenocorticotropic hormone.

and Motulsky) are particularly relevant to topics treated here. A similar emphasis on pharmacogenetic analysis of behavior may be found in several chapters of the text edited by van Abeelen (1974).

Most of this research goes well beyond the demonstration of strain differences. Thus the involvement of catecholamines in audiogenic seizures is deduced from the enhancing effects of reserpine and tetrabenzene (amine depleters) on seizure susceptibility and on the protective effects of monamine oxidase inhibitors (Schlesinger and Griek, 1970; Schlesinger and Sharpless, 1975).

The potential practical applications of psychopharmacogenetics in clinical medicine are important at a time when many behavioral conditions are treated with drugs and when drug addiction is a major public health problem. At the time of this writing, we do not have a good animal model of drug addiction suitable for genetic investigations and determination of its biochemical and psychological features. To develop one would require strains of animals that take drugs in a free-choice situation to the extent that removal of opportunity to obtain them results in withdrawal symptoms such as depression or convulsions. For genetic studies, strains differing in these traits would be needed. It may be that the strains are already available but that the techniques for demonstrating addiction have not yet been devised.

The pharmacogenetic analysis of behavior has its pitfalls. Although a drug may affect one specific chemical pathway, the physiological effects may ramify widely. For this reason, cerebral microinjections may be desirable to limit the region of activity. Furthermore, through metabolism, the injected compound may be greatly modified before it reaches the brain or other target organ. There are also problems in introducing many drugs into the brain because of the blood-brain barrier. The implication of these difficulties is that research in the field requires either a collaboration of specialists or training of individual researchers in both pharmacological and behavioral techniques. Despite these problems, psychopharmacogenetics shows much promise for the immediate future.

Hormones. Beach (1948) has formulated four statements regarding the relationship between hormones and behavior that are still valid today and relevant to behavior genetics:

1. Hormones may affect behavior by altering the organism's normal development and maintenance activities. Such effects, exemplified by the multiple deficiencies of the cretin, are relatively nonspecific.

2. Hormones may influence behavior through stimulation of structures employed in specific response patterns. For example, the growth of genital organs is dependent on hormones; adult sexual behavior cannot occur until these structures are fully developed.

3. Behavior may be altered through hormonal effects on peripheral receptors, sensitizing them to particular forms of stimulation.

4. Behavior may be influenced through

the effects of hormones on the integrative functions of the central nervous system.

Instances of all these possible mechanisms have been described. However, this does not mean that all of them are involved in the production of heritable differences in behavior. Hormonal effects that are demonstrable by surgical extirpation of a gland, by the injection of large doses of a hormone, or that accompany certain diseases are not necessarily involved in genetic regulation of metabolism and behavior.

Potentially, genes affecting the production, release, and metabolism of hormones could influence behavior through any of these four mechanisms. Genes could also act on target organs by influencing their sensitivity to specific chemical messengers. The number of hormonal pathways between genes and behavior is potentially large.

In fruit flies, domestic fowl, mice, and guinea pigs differences in the vigor of sexual behavior appear to be the result of variation in sensitivity of target organs rather than of differences in the amount of hormone produced (Chapter 11). Some of the target organs are presumably in the brain, probably in the limbic system, but effects on peripheral structures cannot be discounted. The same qualification applies also to the enhancement of aggressive behavior by androgens in castrated males and in neonatally androgenized females.

Genetic differences in sensitivity to hormones are sometimes more pronounced for behavioral than for somatic responses. Doses of four naturally occurring androgens —testosterone, dihydrotestosterone, androstenedione, and androstanedione—have been shown to have graded effects on mating and agonistic behavior in two strains of mice at dose levels that stimulate growth of male sex organs equally in both strains (Luttge and Hall, 1973a,b). Somatic responses were similar in the two strains; however, behavioral responses were more pronounced in Swiss-Webster than in CD-1 males. Strain differences in hormone response may be specific to particular target sites.

Shapiro and Goldman (1973) have re-ported a curious finding in genetically androgen-insensitive male rats that develop as sterile pseudohermaphrodites. These rats have a "feminine" type of high-saccharin preference that persists after gonadectomy. This suggests that insensitivity to androgen prevents "masculine imprinting of brain centers," which normally reduces saccharin preference in male rats. The importance of this somewhat esoteric piece of information is its implication that sex hormones influence more than mating and aggression, which are considered to comprise the core of functional sex-role differentiation.

In human beings, two genetic anomalies are instructive with respect to the hormonal link in gene-behavior relationships (Hamburg and Kessler, 1967; Money, 1973). The first of these is the adrenogenital syndrome, which is inherited as an autosomal recessive. In females the external genitalia are masculinized to a variable degree. Internal reproductive organs are female with the vagina opening internally near the neck of the bladder. Money has recommended surgical feminization and treatment with cortisone to compensate for the malfunctioning adrenal. Such women have no difficulty in developing a feminine gender identity and heterosexual orientation despite their early androgenization. A few female children with this syndrome have been reared as boys. After surgical masculinization, they generally acquired a masculine gender identity and were attracted sexually to females. In terms of chromosomal sex, such attraction would be designated as homosexual. In terms of psychological self-image, however, the relationship was heterosexual. Anatomy and rearing take precedence over the XX karyotype. Environment is not all, however. Girls with the syndrome, despite treatment and rearing as females, still tend to be tomboyish in their interests, again showing that the role of sex hormones is not restricted to reproductive processes. Boys with the adrenogenital syndrome, if untreated with cortisone, show precocious sexual maturity while still in grade school. Such precocity often leads to personality problems, since the child is a misfit in his own age group and is unlikely to be accepted by older individuals. Aside from the disruption of the normal rate of development, the effects of the hormonal imbalance on behavior are slight; these precocious males are not aggressive socially or sexually. Today the condition can be treated if discovered early enough, and the affected individual develops like other children.

The second example is the androgen insensitivity syndrome, which is inherited either as an X-linked recessive or an autosomal or X-linked dominant with expression limited to XY individuals. It is clearly impossible to distinguish between the three competing hypotheses. The testes of affected individuals function hormonally in a normal manner, although sperm are not produced. Genitalia are female in character, and breast development occurs at adolescence, although menstruation does not occur. Such "males" are phenotypically and psychologically female except for the absence of menstruation and of fertility.

In the first of these examples, the primary genetic effect is in the kind of hormone produced. In the second it is in the tissue response to a normal hormone. In both instances, hormonal and experiential factors override chromosomal sex. Such natural genetic experiments contribute to our knowledge of the manner in which humans develop a gender identity and the way hormones affect our personalities and interests.

Pituitary-adrenocortical influences. A decrease in the size of the adrenal glands accompanied the domestication of the Norway rat (Rogers and Richter, 1948). Hall's emotional rats had larger adrenals and thyroids than his nonemotional stock (Yeakel and Rhoades, 1941). Similar observations of line and strain differences have been made frequently since these early reports. Levine and Levin (1970) proposed that adrenocorticotropic hormone (ACTH) potentiates learning an avoidance response. A/J mice, which acquire passive avoidance more readily than DBA/2Js, also have higher plasma corticosterone levels. When treated with dexamethasone, a synthetic glucosteroid that blocks ACTH production, DBA mice remain unaffected, whereas the performance of A's falls to the DBA level. Where and how the hormone operates has not been specified.

The facilitating effect of adrenocorticoids on another acquired response, imprinting in ducks, has been reported (Martin, 1975). Mallards of recent wild origin are imprintable over a shorter, more sharply defined stage than are their domesticated descendants, Pekin ducks. Plasma corticosteroids are two to four times higher in mallards and peak sharply during the imprintable period. ACTH administration improved imprinting scores, thus supporting the hypothesis of a functional association.

The evidence is too meager to conclude that genetic effects on learning are mediated through the endocrine system and in particular through the pituitary and adrenal glands. These organs, however, continually interact in rhythmical cycles. They also reflect responses to stressful stimuli. It seems likely that they should affect learning processes, at least indirectly. Their effects may be small, however, in comparison with other sources of variability, both genetic and environmental. Fuller, Chambers, and Fuller (1956) found that neither adrenalectomy nor injections of exogenous cortisone in mice altered strain differences in activity and defecation in an open field where shock was administered.

In summary, the pathway from genetic variation to behavioral variation through the endocrine system is real but narrow. A number of mutations producing gross alterations in behavior have been noted previously. We can add to the list in humans. Extreme hypothyroidism, if untreated, produces extreme mental retardation, the well-known cretin. Hyperthyroidism can produce psychotic symptoms. But the hope that personality differences could be correlated with endocrine function or that the major psychiatric disorders would turn out to be subtle endocrinopathies that could be treated rationally has not materialized. Today the search for a physical basis for the genetic influence on psychoses is concentrated on the brain and its neurotransmitters. It was in the brain of animals that we found heritable variations in reactivity to hormones. We turn, therefore, to a consideration of the gene-behavior pathways that involve the nervous system.

Neural pathways

Behavior is integrated through the nervous system. It is here that genetic influences on the synthesis and metabolism of neurotransmitters and on the properties of excitable membranes could exert control over behavior. Also, since the functioning of the system is dependent on the interconnections of neurons, any gene-induced modification in the developmental pattern of the nervous system could produce permanent anatomical effects through which behavior might be affected over a lifetime. As with hormones, there is a plenitude of opportunities for genes to affect behavior by way of the nervous system.

Heritable defects of sense organs and of the central nervous system are found commonly in many species, including humans. Although in our species each type of defect is rare, the cumulative incidence of all types is high and places a heavy burden on the afflicted and those who care for them. For many conditions, supportive care for the affected and genetic counseling for relatives is the best that can be offered. When effects are limited to sense organs such as the eyes or ears, much can be done by special techniques of education. Without these aids the individuals will be severely retarded. The neurological mutants of *Drosophila* (Kaplan and Trout, 1969; Hotta and Benzer, 1970) and of mice (Sidman, Green, and Appel, 1965) provide study material for the analysis of the development and basic physiology of the nervous systems of these species. We will now turn to a consideration of quantitative variation in the nervous system that is not accompanied by gross structural or behavioral deviation from normal.

Central nervous system. The simplest quantitative attribute of the nervous system is brain size. Mere mass of the brain has been considered to be a measure of the psychological capacity of related species (Rensch,

1956), but the idea that bigger brains necessarily produce better adaptation to environment and greater prospects for species survival is erroneous (Horel, 1973). Brain weight by itself has not proved to be a reliable indicator of psychological differences within a species (Lashley, 1947). Despite this negative evidence the relationship of brain size to behavior continues to be a subject of investigation. Rats reared in impoverished environments behave less adaptively and have smaller brains than genetically similar rats reared in more stimulating conditions (Bennett et al., 1964). Perhaps larger brains are better after all, and the difficulty in finding reliable correlations between brain size and psychological capacities is attributable to an uncontrolled mix of genetic and environmental variation. Van Valen (1972) concluded on the basis of a considerable mass of data from diverse sources that the correlation between brain weight and IQ in humans might be as high as 0.3. The evidence is somewhat circumstantial, however, and better data are needed for verification of his estimate. There are obvious difficulties in obtaining accurate information in this area.

In two experiments, mice have been selected for high and low brain weight and compared with an unselected control line (Fuller and Herman, 1974; Roderick, Wimer, and Wimer, 1976). In both experiments the large-brained animals were also heavier overall, although Fuller and Herman tried to separate these two traits by selecting for large or small brains relative to body size. A number of behavioral comparisons between the selected lines have been made (reviewed by Roderick et al., 1976). In the earlier experiments heavy-brained mice made fewer errors in a water maze, but the interpretation of their superiority was made difficult by a high incidence of retinal degeneration in the low–brain weight line. Fuller and Geils (1973) found differences between the high and low lines in rate of development of a number of reflexes. Controls for maternal effects were made through outfostering and infostering. The line differences were not, however, related in any simple manner to brain size or rate of growth. Jensen (1977) demonstrated better performance in shuttle-box shock avoidance by heavy-brained mice, but accounted for it by a difference in level of activity rather than by a more rapid association of the CS with the US. In a heterogenous stock a moderate but reliable positive correlation of learning with brain weight was found even after adjusting for operant activity level (Jensen and Fuller, 1978). A conservative evaluation of these data is that increased brain size is associated with more rapid early development and a small advantage in certain kinds of learning tasks.

Genetic variation in brain structure extends to internal details as well as to size. Wimer, Wimer, and Roderick (1971) compared the ratio of hippocampal volume to total forebrain volume in a heterogeneous mouse stock. High values of this ratio correlated positively with general activity level and with modifiability of activity by reinforcement. No association was found between the ratio and habituation or discrimination reversal, two functions in which the hippocampus has frequently been implicated. In later articles, striking strain differences in connectivity between mossy and pyramidal cells of the hippocampus were shown (Barber et al., 1974; Vaughn et al., 1977). Genetic variation in fine structure may turn out to be a useful complement to lesion and stimulation methods for studying brain-behavior relationships.

The notion of a fixed standard pattern of neuroanatomy within a species is being eroded by genetic investigations. Lashley and Clark (1946) sharply criticized the architectonic method of dividing the cerebral cortex into small units based on common cell structure and arrangement. We quote from their account of quantitative comparisons between several specimens of macaques and spider monkeys: "Individual variations in cell size, density and arrangement (in homologous areas of different brains) exceed many interareal differences and make quantitative criteria, upon which the parcellations have been carried out, unreliable." Lashley (1947) cites Alexander's statement, "The myelo-

architectural pattern of the normal thalamus shows a surprising number of individual variations . . . human thalami are almost as different in appearance as human faces." It is reasonable that variation in human thalami, like that of human faces, is partly based on genes. Family resemblances in the shape of nose, ears, and chins are commonplace and are accepted as heritable. We cannot make direct observations on family likenesses in thalami and hypothalami in humans, but we can look for genetic variation in animals.

Wahlsten (1974b) reported four types of anomalous fiber tracts in the forebrains of six inbred, four hybrid, and two outbred strains of mice. The definition of "anomalous" was a pronounced variation from the representation in an atlas based on the C57BL/6J strain (Sidman, Angevine, and Taber, 1971). Variants included multiple bundles of the anterior commissure, deficient corpus callosum, stray bundles of the columns of the fornix, and unusual fiber bundles passing through the lateral portion of the septum. Some of these seemed to be found sporadically in all groups studied; others were restricted to one or a few strains and were transmitted to hybrids in an orderly fashion that permitted hypotheses of their mode of inheritance. Wahlsten (1975) also demonstrated large differences between strains in the rate of neural maturation based on anatomical criteria. Most striking was the strong heterotic effect manifested in F_2 hybrids and the high uniformity of the hybrids irrespective of the strains from which they were derived. On a scale of developmental age based on F_2 hybrids of a C57BL/6J $\times$ DBA/2J cross (Wahlsten, 1974a), inbred mice at 32 days postconception (about 13.5 days postpartum) ranged from 28.7 to 32.2 days in developmental age; F_2 hybrids from other crosses ranged from 31.5 to 32.7 days on the same scale. The difference is highly reliable. High variability within inbred lines suggests a strong environmental effect. The F_2 hybrids, although genetically more heterogeneous than an inbred strain, were less variable with respect to developmental age. Heterozygosis appears to favor developmental homeostasis. Inbreeding produces a disruption of the orderly processes of development with a consequent retardation in attaining a mature nervous system.

How important variation of the type reported by Wahlsten and by the Wimers and their collaborators is to behavior remains to be demonstrated. Nonetheless, the combination of genetic, neuroanatomical, and psychological techniques has the promise of providing new insights into the biological basis of behavioral individuality. We suspect that the observed genetic variations in structure are correlated with motivational and learning processes. Confirmation of this possibility is a task for the future.

Peripheral nervous system. Our attention to this point has been concentrated on genetic variation in the central nervous system because of its acknowledged role in the integration of information from external and internal receptors and the organization of appropriate responses. The role of the peripheral nerves seems by contrast to be prosaic, being limited to conduction between the center and the periphery. It is somewhat surprising to find that there are highly significant differences among six strains of mice in the velocity of nerve impulse transmission in the caudal nerve (Hegmann, 1972). A marked heterotic effect favoring rapid conduction was found in crosses between two of the strains. Mode of inheritance of conduction velocity was investigated in Mendelian crosses and by parent-offspring regression in a heterogeneous stock. The Mendelian analysis was complicated by strong heterosis and maternal effects; parent-offspring correlations showed a low heritability (Hegmann, White, and Kater, 1973). The behavioral significance of conduction velocity has not been proved. It is possible that peripheral effects are indicative of central conduction rates that could influence the speed and accuracy of information processing in brain and spinal cord.

CONCLUSION

The relationships between genotype and behavior are noncongruent but lawful (Fuller

and Thompson, 1960). The implications of this statement have been expressed by Fuller (1968) in an analogy that is repeated here in a slightly modified form:

A map of a city and the city itself are congruent because one can match any point on the map with a location in the city. In a similar way proteins and RNA molecules are congruent with DNA, and linkage maps of chromosomes with the position of mutant genes on a linear structure. As one moves from primary gene products to phenotypes arising from coactions and interactions of these products, it is no longer possible to pair each phene with a gene on a map. Yet all the characteristics of an organism, including its behavior, are completely dependent upon its genotype. The genotype is replicated in each cell; without it there is no organism.

The combination of the coadaptive and noncongruent approaches to behavior genetics makes the task of investigators in this field more difficult. If the effect of a gene substitution on a psychophene varies according to the remainder of the genotype, and if that substitution affects a variety of psychophenes, then our conclusions from any particular experiment must be specific to the subjects and variables used in it. There is therefore some danger that, in the search for causes of individual variation, behavior genetics could become a catalogue of special cases with no theory.

Although this danger is real, we are confident that it will not materialize. Instead genetics will make important contributions to the development of laws of behavior that integrate biological and psychological data. The quality of research in behavior genetics is improving. Classical Mendelian and biometric approaches are being applied to behavior that is more relevant to fitness in natural situations than were some of the older laboratory procedures. The genetics of regulatory processes in the endocrine and nervous systems are investigated actively. Strain comparisons are made under a variety of conditions rather than one standard procedure. More attention is being paid to developmental processes from a combined genetic-experiential view. In short, behavior genetics is being assimilated into the mainstream of psychobiological research. Switching from the typological view of a species to the populational view can only increase our understanding of how organisms function. There is no standard fruit fly, mouse, or human; a species is an aggregate of diverse individuals built on the same basic plan but rarely completely identical. The principles governing both the basic plan and the individual variability are the concern of behavior genetics.

In this chapter, we have attempted to focus on some major trends in the field and to show how these may serve to strengthen the bonds between psychology and biology, the two parent disciplines of behavior genetics. It is difficult to specify such trends exactly in summary form. However, the following may be listed as at least illustrative:

1. There is a growing interest in the manner in which phenotypic expression of genotype changes with time, whether the latter reflects repeated exposures to some situation or simply motivational change. Some of the kinds of changes found may reflect the operation of regulator genes that modulate the expression of the genes primarily underlying the character. On the other hand, other changes may simply reflect alterations in the components of the trait as measured. Again, the apparent fact that propensity to change may itself be heritable raises a host of interesting problems of fundamental interest to behavior genetics.

2. A second important line of work has been concerned with the behavioral correlates of biological fitness, both in lower animals and in human beings. The concept of fitness itself offers a fruitful link between behavior genetics and biology, particularly, of course, sociobiology. It may well be that there are numerous routes to "success," biological or psychological. However, even so, a thorough explanation of the characters that contribute to it and their genetic underpinnings would seem to be most worthwhile. Coordinately, it will be important to establish how the differential distributions of fitness in different groups may affect inter-

group selection. Such investigations will take us into an analysis of "selfishness" and "altruism," characters that now figure prominently in the writings of sociobiologists.

3. A third area of research concerns the manner in which groups of genetically related individuals impose on each other norms of behavior that produce varying degrees of similarity among them. This problem perhaps lies at the very basis of the entire nature/nurture controversy. Hereditarians have been able, by reference to Mendelian models, to explain both the similarities and differences that are regularly found in family groups. Environmentalists, on the other hand, though usually able to give ad hoc accounts of similarities, have not often come forward with explanations of radical differences. A few potentially fruitful models that we discussed earlier derive from the general area of developmental social psychology. As such, these models represent interesting areas of contact between behavior genetics and psychology.

4. A fourth trend is represented by the burgeoning research on pathways, both chemical and neural, between genotype and phenotype. The growth in this area is clearly due to the great advances in molecular biology and biochemical genetics that have occurred during the last two decades and, commensurately, to the increasing sophistication of training of workers in the field of behavior genetics. Though we have not discussed this in the same context, analysis of the final chain in such pathways—the behavioral—is of equal importance. Thus there will always be room for more precise dissection of the behavior being studied. This applies particularly to such characters as IQ, personality, and the mental disorders. To date, these are not precisely defined behaviorally but only in terms of fairly crude global assessments. Much can be accomplished here, particularly by reference to the methods of ethology and behavior analysis.

5. A final focus of interest appears to be in methodology. Like many areas in science, behavior genetics has tacitly adopted certain conventions of design that are not necessarily

acceptable to workers outside the field. The twin and adoption methods are good examples. Much work using these methods continues to be reported. However, during the last decade, when conclusions drawn from them have come under close public scrutiny, fundamental questions about their validity have been raised. It seems ironic that scientists should have to be reminded in this manner of the canons of logic. Nonetheless, at least some of the criticisms should, in the end, have a beneficial effect. None of us should attempt to defend methodology that is weak or ambiguous.

A different aspect of methodology in behavior genetics relates to the directions in which it should lead us. We contrasted before the approach taken by Jinks, Fulker, and others as against that taken by Cattell. The former, as primarily biometric geneticists, opt for methods suitable to uncovering more and more genetic rules. Cattell, on the other hand, as primarily a psychologist, has a much stronger interest in the laws that govern genotype-environment interactions. We favor Cattell's emphasis, since it seems to open up questions of much more direct psychological interest than does the biometric genetic orientation. The latter seems to have reached the limits of its usefulness as far as generating genuine research problems.

The five trends just identified are, of course, based largely on subjective judgment. It is always difficult to estimate the major activities that are occurring in some field and where they will lead. The progress of any science is not nearly as orderly as a student reading a textbook is led to see it. Consequently, we acknowledge that the directions that behavior genetics may take in the next decade may be quite different from those we have suggested. But what is fairly certain is that in the past two decades a great deal has been accomplished, and much remains to be done.

Bibliography

Abe, K., and P. A. Moran. 1969. Parental age of homosexuals. Br. J. Psychiatry 115:313-317.

Abelin, T. 1972. Etiological implications of the distinction between process and reactive schizophrenia: the monogenic model involving intermediateness. In A. R. Kaplan, ed., Genetic factors in "schizophrenia." Charles C Thomas, Publisher, Springfield, Ill.

Adams, M. S., and J. V. Neel. 1967. Children of incest. Pediatrics 40:55-62.

Adamson, R., and R. Black. 1959. Volitional drinking and avoidance learning in the white rat. J. Comp. Physiol. Psychol. 52:734-736.

Ader, R., S. B. Friedman, and L. J. Grota. 1967. Emotionality and adrenal cortical function: effects of strain, test, and the 24-hour corticosterone rhythm. Anim. Behav. 15:37-44.

Adler, J. 1969. Chemoreceptors in bacteria. Science 166:1588-1597.

Agranoff, B. W. 1972. Learning and memory: approaches to correlating behavioral and biochemical events. In R. W. Abers, G. J. Siegel, R. Katzman, and B. W. Agranoff, eds. Basic neurochemistry. Little, Brown & Co., Boston.

Åkesson, H. O. 1961. Epidemiology and genetics of mental deficiency in a southern Swedish population. Almqvist & Wiksell Förlag AB, Upsala.

Akiskal, H. S., and W. T. McKinney, Jr. 1973. Depressive disorders: towards a unified hypothesis. Science 182:20-29.

Allee, W. C., and D. Foreman. 1955. Effects of an androgen on dominance and subordination in six common breeds of Gallus gallus. Physiol. Zool. 28:89-115.

Allen, G. 1955. Comments on the analysis of twin samples. Acta. Genet. Med. Gemellol. 4:143-159.

Allen, G. 1958. Patterns of discovery in the genetics of mental deficiency. Am. J. Ment. Defic. 62:840-849.

Allen, G. 1965. Twin research: problems and prospects. In A. G. Steinberg and A. G. Bearn, eds. Progress in medical genetics, vol. 4. Grune & Stratton, Inc., New York.

Allen, G., and G. S. Baroff. 1955. Mongoloid twins and their siblings. Acta. Genet. 5:294-326.

Allen, G., B. Harvald, and J. Shields. 1967. Measures of twin concordance. Acta Genet. 17:475-481.

Allen, G., and F. J. Kallman. 1955. Frequency and types of mental retardation in twins. Am. J. Hum. Genet. 7:15-20.

Allport, G. W. 1966. Traits revisited. Am. Psychol. 21:1-10.

Alpern, H. P., and J. G. Marriott. 1972a. A direct measure of short-term memory in mice utilizing a successive reversal learning set. Behav. Biol. 7:723-732.

Alpern, H. P., and J. G. Marriott. 1972b. An analysis of short-term memory and conceptual behavior in three inbred strains of mice. Behav. Biol. 7:543-551.

Åmark, C. 1951. A study in alcoholism: clinical, social-psychiatric and genetic investigations. Acta Psychiatr. Neurol. Scand. [Suppl.] 70.

Anastasi, A. 1948. The nature of psychological traits. Psychol. Rev. 55:127-138.

Anastasi, A. 1954. Psychological testing. The Macmillan Co., New York.

Anastasi, A. 1958a. Differential psychology; individual and group differences in behavior. The Macmillan Co., New York.

Anastasi, A. 1958b. Heredity, environment and the question "how?" Psychol. Rev. 65:197-208.

Anastasi, A. 1968. Psychological testing, 3rd ed. The Macmillan Co., New York.

Anastasi, A., and J. P. Foley, Jr. 1948. A proposed reorientation in the heredity-environment controversy. Psychol. Rev. 55:239-249.

Anastasi, A., J. L. Fuller, J. P. Scott, and J. R. Schmitt. 1955. A factor analysis of the performance of dogs on certain learning tasks. Zoologica 40:33-46.

Anderson, E. E. 1937. The interrelationship of drives in the male albino rat. I. Intercorrelations of measures of drives. J. Comp. Psychol. 24:73-118.

Anderson, E. E. 1938a. The interrelationships of drives in the male albino rat. II. Interrelationships between 47 measures of drives and of learning. Comp. Psychol. Monogr. 14(8):1-119.

Anderson, E. E. 1938b. The interrelationships of drives in the male albino rat. III. Interrelationships among measures of emotional and exploratory behavior. J. Genet. Psychol. 53:335-352.

Andrew, R. J. 1972. Recognition processes and behavior with special reference to the effects of testosterone on persistence. In D. S. Lehrman, R. A. Hinde, and E. Shaw, eds. Advances in the study of behavior. Academic Press, Inc., New York.

Andrew, R. J., and L. J. Rogers. 1972. Testosterone, search behaviour and persistence. Nature 237:343-345.

Angst, J., and C. Perris. 1968. Zur Nosologie endogener Depressionen. Vergleich der Ergebnisse zweier Untersuchungen. Arch. Psychiatr. Nervenkr. 210:373-386.

Anliker, J., and J. Mayer. 1956. An operant conditioning technique for studying feeding-fasting patterns in normal and obese mice. J. Appl. Physiol. 8:667-670.

Annett, M. 1967. The binomial distribution of right, mixed, and left handedness. Q. J. Exp. Psychol. **19**: 327-333.

Annett, M. 1970a. A classification of hand preference by association analysis. Br. J. Psychol. **61**:303-332.

Annett, M. 1970b. The growth of manual preference and speed. Br. J. Psychol. **61**:545-558.

Annett, M. 1972. The distribution of manual asymmetry. Br. J. Psychol. **63**:343-358.

Annett, M. 1973. Handedness in families. Ann. Hum. Genet. **37**:93-105.

Anthony, E. J. 1968. The developmental precursors of adult schizophrenia. In D. Rosenthal and S. S. Kety, eds. The transmission of schizophrenia. Pergamon Press Ltd., Oxford, England.

Anthony, E. J. 1972. A clinical and experimental study of high-risk children and their schizophrenic parents. In A. R. Kaplan, ed. Genetic factors in "schizophrenia." Charles C Thomas, Publisher, Springfield, Ill.

Archer, J. 1973. Tests for emotionality in rats and mice: a review. Anim. Behav. **21**:205-235.

Archer, J. 1974. The effects of testosterone on the distractability of chicks by irrelevant and relevant novel stimuli. Anim. Behav. **22**:397-404.

Archer, J. 1975. The Maudsley reactive and nonreactive strains of rats: the need for an objective evaluation of differences. Behav. Genet. **5**:411-413.

Arkonac, O., and S. B. Guze. 1963. A family study of hysteria. N. Engl. J. Med. **268**:239-242.

Asano, N. 1967. Clinico-genetic study of manic-depressive psychoses. In H. Mitsuda, ed. Clinical genetics in psychiatry. Igaku Shoin Ltd., Tokyo.

Ashman, R. 1952. The inheritance of simple musical memory. J. Hered. **43**:51-52.

Averhoff, W. W., and R. H. Richardson. 1974. Pheromonal control of mating patterns in *Drosophila melanogaster*. Behav. Genet. **4**:207-225.

Avery, O. T., C. M. MacLeod, and M. McCarty. 1944. Studies on the chemical nature of the substance inducing transformation of pneumococcal types. J. Exp. Med. **79**:137-158.

Bagg, H. J. 1916. Individual differences and family resemblances in animal behavior. Am. Natur. **50**:222-236.

Bagg, H. J. 1920. Individual differences and family resemblances in animal behavior. Arch. Psychol. **6**:1-58.

Bailey, D. W. 1971. Recombinant inbred strains. Transplantation **11**:325-327.

Bain, B., and L. Lowenstein, 1964. Genetic studies on the mixed leucocyte reaction. Science **145**:1315-1316.

Bajema, C. 1963. Estimation of the direction and intensity of natural selection in relation to human intelligence by means of the intrinsic rate of natural increase. Eugen. Q. **10**:175-187.

Bajema, C. 1966. Relation of fertility to educational attainment in a Kalamazoo public school population: a follow-up study. Eugen. Q. **13**:306-315.

Banker, H. J. 1928. Genealogical correlation of student ability. J. Hered. **19**:503-508.

Barber, R. P., J. E. Vaughn, R. E. Wimer, and C. C.

Wimer. 1974. Genetically-associated variations in the distribution of dentate granule cell synapses upon the pyramidal cell dendrites in the mouse hippocampus. J. Comp. Neurol. **156**:417-434.

Barcal, R., J. Simon, and J. Sova. 1969. Blood pressure in twins. Lancet **1**:1321.

Barchas, J. D., R. D. Ciranello, S. Kessler, and D. A. Hamburg. 1975. Genetic aspects of catecholamine synthesis. In R. R. Fieve, D. Rosenthal, and H. Brill, eds. Genetic research in psychiatry. The Johns Hopkins University Press, Baltimore.

Barnett, S. A., and S. G. Scott. 1964. Behavioural "vigour" in inbred and hybrid mice. Anim. Behav. **12**:325-337.

Barnicot, N. A. 1950. Taste deficiency for phenylthiourea in African Negroes and Chinese. Ann. Eugen. **15**:248-254.

Barrett, R. J., N. J. Leith, and O. J. Ray. 1973. A behavioral and pharmacological analysis of variables mediating active avoidance behavior in rats. J. Comp. Physiol. Psychol. **82**:489-500.

Barron, F. 1969. Creative person and creative process. Holt, Rinehart & Winston, New York.

Bartlett, M. J. 1937. Note on the development of correlation among genetic components of ability. Ann. Eugen. **7**:299-302.

Bashi, J. 1977. Effects of inbreeding on cognitive performance. Nature **266**:440-442.

Bastock, M. 1956. A gene mutation which changes a behavior pattern. Evolution **10**:421-423.

Bastock, M., and A. Manning. 1955. The courtship of *Drosophila melanogaster*. Behaviour **8**:85-111.

Bauer, F. J. 1956. Genetic and experimental factors affecting social relations in male mice. J. Comp. Physiol. Psychol. **49**:359-364.

Beach, F. A. 1947. A review of physiological and psychological studies of sexual behavior in mammals. Physiol. Rev. **27**:240-307.

Beach, F. A. 1948. Hormones and behavior. Paul B. Hoeber, Inc. New York.

Beach, F. A. 1950. The snark was a boojum. Am. Psychol. **5**:115-124.

Beeman, E. A. 1947. The effect of male hormone on aggressive behavior of mice. Physiol. Zool. **20**:373-405.

Belmont, L., and F. A Marolla. 1973. Birth order, family size and intelligence. Science **182**:1096-1101.

Belmont, L., Z. A. Stein, and M. W. Susser. 1975. Comparison of associations of birth order with intelligence test score and height. Nature **255**:54-56.

Benedict, P. K., and I. Jacks. 1954. Mental illness in primitive societies. Psychiatry **17**:377-390.

Benjamins, J. A., and G. M. McKhann. 1972. Neurochemistry of development. In R. W. Albers, G. J. Siegel, R. Katzman, and B. W. Agranoff, eds. Basic neurochemistry. Little, Brown & Co., Boston.

Bennett, E. L., M. C. Diamond, D. Krech, and M. R. Rosenzweig. 1964. Chemical and anatomical plasticity of brain. Science **146**:610-619.

Bentley, D. R. 1971. Genetic control of an insect neuronal network. Science **174**:1139-1141.

Bentley, D. R., and R. R. Hoy. 1972. Genetic control of the neuronal network generating cricket *(Teleogryllus, Gryllus)* song patterns. Anim. Behav. **20:**478-492.

Bentley, D., and R. R. Hoy. 1974. The neurobiology of cricket song. Sci. Am. **231**(2):34-44.

Benzer, S. 1967. Behavioral mutants of *Drosophila* isolated by countercurrent distribution. Proc. Natl. Acad. Sci. U.S.A. **58:**1112-1119.

Benzer, S. 1973. Genetic dissection of behavior. Sci. Am. **229**(6):24-37.

Berendes, H. D. 1965. Salivary gland function and chromosomal puffing patterns in *Drosophila hydei.* Chromosoma **17:**35-77.

Berman, A. 1971. The problem of assessing cerebral dominance and its relationship to intelligence. Cortex **7:**372-386.

Bertelsen, A., B. Harvald, and M. Hauge. 1977. A Danish twin study of manic-depressive disorders. Br. J. Psychiatry **130:**330-351.

Bevan, W., G. W. Levy, J. M. Whitehouse, and J. M. Bevan. 1957. Spontaneous aggressiveness in two strains of mice castrated and treated with one of three androgens. Physiol. Zool. **30:**341-349.

Bielschowsky, M., and F. Bielschowsky. 1953. A new strain of mice with hereditary obesity. Proc. Univ. Otago Med. Sch. **31:**29-31.

Bielschowsky, M., and F. Bielschowsky. 1956. The New Zealand strain of obese mice. Their response to stilbestrol and to insulin. Aust. J. Exp. Biol. **34:**181-198.

Bignami, G. 1965. Selection for high rates and low rates of conditioning in the rat. Anim. Behav. **13:**221-227.

Bille, M., and N. Juel-Nielsen, 1963. Incidence of neurosis in psychiatric and other medical services in a Danish county. Dan. Med. Bull. **10:**172-176.

Birnbaum, A. 1972. The random phenotype concept, with applications. Genetics **72:**739-758.

Blakeslee, A. F., and M. R. Salmon. 1931. Odour and taste blindness. Eugen. News **16:**105-109.

Blatz, W. E. 1937. Collected studies on the Dionne quintuplets. University of Toronto Study, Child Development Series, University of Toronto Press, Toronto.

Bleuler, M. 1955. Familial and personal background of chronic alcoholics. In O. Diethelm, ed. Etiology of chronic alcoholism. Charles C Thomas, Publisher, Springfield, Ill.

Bleuler, M. 1964. Ursache und Wesen der schizophrenen Geistesstörungen. I. Forschungsergebnisse. Dtsch. Med. Wochenschr. **89:**1865-1870.

Bleuler, M. 1974. The offspring of schizophrenics. Schizophrenia Bull. **8:**93-108.

Blewett, D. B. 1954. An experimental study of the inheritance of intelligence. J. Ment. Sci. **100:**922-933.

Blizard, D. A. 1971. Autonomic reactivity in the rat: effects of genetic selection for emotionality. J. Comp. Physiol. Psychol. **76:**282-289.

Blizard, D. A., and C. K. Chai. 1972. Behavioral studies in mice selectively bred for differences in thyroid function. Behav. Genet. **2:**301-310.

Blizard, D. A., and R. A. Welty. 1971. Cardiac activity in the mouse: strain differences. J. Comp. Physiol. Psychol. **77:**337-344.

Block, J. D. 1967. Monozygotic twin similarity in multiple psychophysiologic parameters and measures. Recent Adv. Biol. Psychiatry **9:**105-118.

Block, N. J., and G. Dworkin, eds. 1976. The IQ controversy. Random House, Inc., New York.

Bock, R. D., and D. Kolakowski. 1973. Further evidence of sex-linked major-gene influence on human spatial visualizing ability. Am. J. Hum. Genet. **25:**1-14.

Boggan, W. O., and L. S. Seiden. 1973. 5-hydroxytryptophan reversal of reserpine enhancement of audiogenic susceptibility in mice. Physiol. Behav. **10:**9-12.

Böök, J. A. 1953. A genetic and psychiatric investigation of a North-Swedish population with special regard to schizophrenia and mental deficiency. Acta Genet. **4:**1-139.

Böök, J. A. 1957. Genetical investigations in a North-Swedish population. Ann. Hum. Genet. **21:**191-221.

Böök, J. A., and S. C. Reed. 1950. Empiric risk figures in mongolism. J. Am. Med. Assoc. **143:**730-732.

Borgaonkar, D. S. 1967. Bibliographie. 47, XYY (bibliography). Ann. Génét. **12:**67-70.

Borgström, C. A. 1939. Eine Serie von Kriminellen Zwillingen. Arch. Rassenbiol. **33:**334-343.

Bovet, D., F. Bovet-Nitti, and A. Oliverio. 1969. Genetic aspects of learning and memory in mice. Science **163:**139-149.

Boyd, W. C. 1950. Taste reactions to antithyroid substances. Science **112:**153.

Brace, C. L. 1961. Physique, physiology and behavior: an attempt to analyze a part of their roles in the canine biogram. Doctoral dissertation. Harvard University, Cambridge, Mass.

Brain, P. F., and A. E. Poole. 1974. The role of endocrines in isolation-induced intermale fighting in albino laboratory mice. I. Pituitary-adrenocortical influences. Aggress. Behav. **1:**39-69.

Bramwell, B. S. 1944. The order of merit: the holders and their kindred. Eugen. Rev. **36:**84-91.

Bray, G. A., and D. A. York. 1971. Genetically-transmitted obesity in rodents. Physiol. Rev. **51:**598-646.

Bray, G. A., and D. York. 1972. Studies on food intake of genetically obese rats. Am. J. Physiol. **223:**176-179.

Bremer, H. J., and W. Neumann. 1966. Tolerance of phenylalanine after intravenous administration in phenylketonurics, heterozygous carriers, and normal adults. Nature **209:**1148-1149.

Brenner, S. 1974. The genetics of *Caenorhabditis elegans.* Genetics **77:**71-94.

Brewster, D. J. 1968. Genetic analysis of ethanol preference in rats selected for emotional reactivity. J. Hered. **59:**283-286.

Brindley, G. S. 1970. Central pathways of vision. Ann. Rev. Physiol. **32:**259-268.

Britten, R. J., and E. H. Davidson. 1969. Gene regulation for higher cells: a theory. Science **165:**344-357.

Britten, R. J., and E. H. Davidson. 1971. Repetitive and non-repetitive DNA sequences and a speculation on the origins of evolutionary novelty. Q. Rev. Biol. 46:111-133.

Broadhurst, P. L. 1957. Determinants of emotionality in the rat. I. Situational factors. Br. J. Psychol. 48:1-12.

Broadhurst, P. L. 1958a. Determinants of emotionality in the rat. II. Antecedent factors. Br. J. Psychol. 49:12-20.

Broadhurst, P. L. 1958b. Determinants of emotionality in the rat. III. Strain differences. J. Comp. Physiol. Psychol. 51:55-59.

Broadhurst, P. L. 1960a. Analysis of maternal effects in the inheritance of behaviour. Anim. Behav. 9:129-141.

Broadhurst, P. L. 1960b. Experiments in psychogenetics: applications of biometrical genetics to the inheritance of behaviour. In H. J. Eysenck, ed. Experiments in personality, vol. 1. Psychogenetics and psychopharmacology. Routledge & Kegan Paul Ltd., London.

Broadhurst, P. L. 1967. An introduction to the diallel cross. In J. Hirsch, ed. Behavior-genetic analysis. McGraw-Hill Book Co., New York.

Broadhurst, P. L. 1969. Psychogenetics of emotionality in the rat. Ann. N.Y. Acad. Sci. 159:806-824.

Broadhurst, P. L. 1975. The Maudsley reactive and nonreactive strains of rats: a survey. Behav. Genet. 5:299-319.

Broadhurst, P. L. 1976. The Maudsley reactive and nonreactive strains of rats: a clarification. Behav. Genet. 6:363-365.

Broadhurst, P. L., and J. L. Jinks. 1966. Stability and change in the inheritance of behaviour in rats: a further analysis of statistics from a diallel cross. Proc. R. Soc. Biol. 165:450-472.

Broadhurst, P. L., and J. L. Jinks. 1974. What genetical architecture can tell us about the natural selection of behavioral traits. In J. H. F. van Abeelen, ed. The genetics of behaviour. North-Holland Publishing Co., Amsterdam.

Brody, D. 1937. Twin resemblances in mechanical ability, with reference to the effects of practice on performance. Child Dev. 8:207-216.

Brody, E. G. 1942. Genetic basis of spontaneous activity in the albino rat. Comp. Psychol. Monogr. 17(5):1-24.

Brody, E. G. 1950. A note on the genetic basis of spontaneous activity in the albino rat. J. Comp. Physiol. Psychol. 43:281-288.

Bronson, F. H., and C. Desjardins. 1968. Aggression in adult mice: modification by neonatal injections of gonadal hormones. Science 161:705-706.

Brown, A. M., R. E. Stafford, and S. G. Vandenberg. 1967. Twins: behavioral differences. Child Dev. 38:1055-1064.

Brown, F. A., Jr., and B. U. Hall. 1936. The directive influence of light upon *D. melanogaster* and some of its eye mutants. J. Exp. Zool. 74:205-221.

Brown, F. W. 1942. Heredity in the psychoneuroses. Proc. R. Soc. Med. 35:785-788.

Bruell, J. H. 1962. Dominance and segregation in the inheritance of quantitative behavior in mice. In E. Bliss, ed. Roots of behavior. Harper & Row, Publishers, New York.

Bruell, J. H. 1964a. Heterotic inheritance of wheel-running in mice. J. Comp. Physiol. Psychol. 58:159-163.

Bruell, J. H. 1964b. Inheritance of behavioral and physiological characters of mice and the problems of heterosis. Am. Zool. 4:125-138.

Bruell, J. H. 1967. Behavioral heterosis. In J. Hirsch, ed. Behavior-genetic analysis. McGraw-Hill Book Co., New York.

Bruell, J. H. 1970. Behavioral population genetics and wild *Mus musculus*. In G. Lindzey and D. D. Thiessen, eds. Contributions to behavior-genetic analysis: the mouse as a prototype. Appleton-Century-Crofts, New York.

Bruggemann, C. E. 1970. Twins: early grammatical development. Folia Phoniatr. 22:197-215.

Brugger, C. 1934. Familien untersuchungen bei Alkoholdeliranten. Z. Gesamte Neurol. Psychiatr. 151:740-788.

Buck, C., H. Simpson, and J. M. Wanklin. 1977. Survival of nieces and nephews of schizophrenic patients. Br. J. Psychiatry 130:506-508.

Bulmer, M. G. 1970. The biology of twinning in man. Clarendon Press, Oxford.

Burch, P. R. J. 1964. Involutional psychosis: some new aetiological considerations. Br. J. Psychiatry 110:825-829.

Burdette, W. J., ed. 1963. Methodology in mammalian genetics. Holden-Day, Inc., San Francisco.

Burks, B. S. 1928. The relative influence of nature and nurture upon mental development; a comparative study of foster parent-foster child resemblance and true parent–true child resemblance. In Twenty-seventh Yearbook of the National Society for the Study of Education, Part I. Public School Publishing Co., Bloomington, Ill.

Burrows, W. H., and T. C. Byerly. 1938. The effect of certain groups of environmental factors upon the expression of broodiness. Poultry Sci. 17:324-330.

Burt, C. 1943. Ability and income. Br. J. Educ. Psychol. 13:83-98.

Burt, C. 1955. The evidence for the concept of intelligence. Br. J. Educ. Psychol. 25:158-177.

Burt, C. 1959. Class differences in general intelligence. III. Br. J. Stat. Psychol. 12:15-33.

Burt, C. 1961. Intelligence and social mobility. Br. J. Stat. Psychol. 14:3-24.

Burt, C. 1966. The genetic determination of differences in intelligence: a study of monozygotic twins reared together and apart. Br. J. Psychol. 57:137-153.

Burt, C. 1972. Inheritance of general intelligence. Am. Psychol. 27:175-190.

Burt, C., and M. Howard. 1956. The multifactorial theory of inheritance and its application to intelligence. Br. J. Stat. Psychol. 9:95-131.

Butcher, H. J. 1968. Human intelligence: its nature and assessment. Methuen & Co. Ltd., London.

Byrns, R., and J. Healy, 1936. The intelligence of twins. J. Genet. Psychol. 49:474-478.

Cameron, R. G. 1967. Rational approach to color vision testing. Aerospace Med. 38:51-59.

Cancro, R., ed. 1971. Intelligence: genetic and environmental influences. Grune & Stratton, Inc., New York.

Candland, D. K., and A. I. Leshner. 1971. Formation of squirrel monkey dominance order is correlated with endocrine output. Bull. Ecol. Soc. Am. 52:54. (Abstr.)

Candland, D. K., K. D. Pack, and T. J. Matthews. 1967. Heart-rate and defecation frequency as measures of rodent emotionality. J. Comp. Physiol. Psychol. 64:146-150.

Canter, S. 1973. Personality traits in twins. In G. Claridge and S. Canter, eds. Personality differences and biological variations: a study of twins. Pergamon Press Ltd., Oxford.

Capretta, P. J. 1970. Saccharin and saccharin-glucose ingestion in two inbred strains of *Mus musculus*. Psychon. Sci. 21:133-135.

Carmena, M. 1935. Schreibdruck bei Zwillingen. Z. Gesamte Neurol. Psychiatr. 103:744-752.

Carr, R. M., and C. D. Williams. 1957. Exploratory behavior of three strains of rats. J. Comp. Physiol. Psychol. 50:621-623.

Carran, A. B. 1972. Biometrics of reversal learning in mice. II. Diallel cross. J. Comp. Physiol. Psychol. 78:466-470.

Carran, A. B. 1975. Biometrics of reversal learning in mice. III. Complete reciprocated mendelian analysis. J. Comp. Physiol. Psychol. 88:878-881.

Carran, A. B., L. T. Yeudall, and J. R. Royce. 1964. Voltage level and skin resistance in avoidance conditioning of inbred strains of mice. J. Comp. Physiol. Psychol. 58:427-430.

Carter, C. O. 1961. The inheritance of congenital pyloric stenosis. Br. Med. Bull. 17:251-254.

Carter, C. O. 1967. Risk to offspring of incest. Lancet 1:436.

Carter, C. O., 1969. Genetics of common disorders. Br. Med. Bull. 25:52-57.

Carter, C. O., and K. A. Evans. 1961. Risk of parents who have had one child with Down's syndrome (mongolism) having another child similarly affected. Lancet 2:785-787.

Carter, C. O., and L. I. Woolf. 1961. The birthplaces of parents and grandparents of a series of patients with phenylketonuria in south-east England. Ann. Hum. Genet. 25:57-64.

Carter, H. D. 1932a. Family resemblances in verbal and numerical abilities. Genet. Psychol. Monogr. 12:1-104.

Carter, H. D. 1932b. Twin similarities in occupational interests. J. Educ. Psychol. 23:641-655.

Carter, H. D. 1933. Twin similarities in personality traits. J. Genet. Psychol. 43:312-321.

Carter, H. D. 1935. Twin similarities in emotional traits. Char. Pers. 4:61-78.

Carter, H. D. 1939. Resemblance of twins in speed of association. Psychol. Bull. 36:641. (Abstr.)

Carter-Saltzman, L., and S. Scarr. 1977. MZ or DZ? Only your blood grouping laboratory knows for sure. Behav. Genet. 7:273-280.

Caspari, E. 1958. Genetic basis of behavior. In A. Roe and G. G. Simpson, eds. Behavior and evolution. Yale University Press, New Haven, Conn.

Caspari, E. 1961. Implications of genetics for psychology. Review of *Behavior Genetics* by J. L. Fuller and W. R. Thompson. Contemp. Psychol. 6:337-339.

Caspari, E. 1963. Genes and the study of behavior. Am. Zool. 3:97-100.

Caspari, E. 1964. Refresher course on behavior genetics: synthesis and outlook. Am. Zool. 4:169-172.

Caspari, E. 1967. Gene action as applied to behavior. In J. Hirsch, ed. Behavior-genetic analysis. McGraw-Hill Book Co., New York.

Caspari, E. W. 1972. Introductory remarks. In L. Ehrman, G. Omenn, and E. W. Caspari, eds. Genetics, environment and behavior. Academic Press, Inc., New York.

Castellion, A. W., E. A. Swinyard, and L. S. Goodman. 1965. The effect of maturation on the development and reproducibility of audiogenic and electroshock seizures. Exp. Neurol. 13:206-217.

Castle, W. E. 1941. Influence of certain color mutations on body size in mice, rats and rabbits. Genetics 26:177-191.

Cattell, R. B. 1953. Research designs in psychological genetics with special reference to the multiple variance analysis method. Am. J. Hum. Genet. 5:76-93.

Cattell, R. B. 1955. The chief invariant psychological and psycho-physical functional unities found by P-technique. J. Clin. Psychol. 11:319-343.

Cattell, R. B. 1960. The multiple abstract variance analysis equations and solutions: for nature-nurture research on continuous variables. Psychol. Rev. 67:353-372.

Cattell, R. B. 1973. Unraveling maturational and learning development by the comparative MAVA and structured learning approaches. In J. R. Nesselroade and H. W. Reese, eds. Life-span developmental psychology. Academic Press, Inc., New York.

Cattell, R. B. 1974. Differential fertility and normal selection for IQ: some required conditions in their investigation. Soc. Biol. 21:168-177.

Cattell, R. B., D. B. Blewett, and J. R. Beloff. 1955. The inheritance of personality. A multiple variance analysis determination of approximate nature-nurture ratios for primary personality factors in Q-data. Am. J. Hum. Genet. 7:122-146.

Cattell, R. B., C. R. Bolz, and B. Korth. 1973. Behavioral types in purebred dogs objectively determined by taxonome. Behav. Genet. 3:205-216.

Cattell, R. B., and E. V. Malteno. 1940. Contributions concerning mental inheritance. V. Temperament. J. Genet. Psychol. 57:31-47.

Cattell, R. B., G. F. Stice, and N. F. Kristy. 1957. A first approximation to nature-nurture ratios for eleven primary personality factors in objective tests. J. Abnorm. Soc. Psychol. **54**:143-159.

Cattell, R. B., and F. W. Warburton. 1967. Objective personality and motivation tests: a theoretical introduction and practical compendium. University of Illinois Press, Urbana.

Cattell, R. B., and J. L. Willson. 1938. Contributions concerning mental inheritance. I. Of intelligence. Br. J. Educ. Psychol. **8**:129-149.

Caul, W. F., and R. J. Barrett. 1973. Shuttle-box versus Y-maze avoidance: value of multiple response measures in interpreting active-avoidance behavior of rats. J. Comp. Physiol. Psychol. **84**:572-578.

Cavalli-Sforza, L. L., and W. F. Bodmer. 1971. The genetics of human populations. W. H. Freeman & Co. Publishers, San Francisco.

Cavalli-Sforza, L., and M. W. Feldman. 1973a. Cultural versus biological inheritance: phenotypic transmission from parents to children (a theory of the effects of parental phenotypes on children's phenotypes). Am. J. Hum. Genet. **25**:618-637.

Cavalli-Sforza, L., and M. W. Feldman. 1973b. Models for cultural inheritance. I. Group mean and within group variation. Theor. Pop. Biol. **4**:42-55.

Caviness, V. S., Jr., D. K. So, and R. L. Sidman. 1972. The hybrid reeler mouse. J. Hered. **63**:241-246.

Cederlöf, R., L. Friberg, E. Jonsson, and L. Kaij. 1961. Studies on similarity diagnosis with the aid of mailed questionnaires. Acta Genet. Stat. Med. **11**:338-362.

Chai, C. K., and M. M. Dickie. 1966. Endocrine variations. In E. L. Green, ed. Biology of the laboratory mouse. McGraw-Hill Book Co., New York.

Chamberlain, H. B. 1928. Inheritance of left-handedness. J. Hered. **19**:557-559.

Chang, S.-Y., and C. Kung. 1973. Genetic analysis of heat-sensitive pawn mutants of *Paramecium aurelia*. Genetics **75**:49-59.

Chang, S.-Y., J. van Houten, L. R. Guyer, S. S. Lui, and C. Kung. 1974. An extensive behavioral and genetic analysis of the pawn mutants in *Paramecium aurelia*. Genet. Res. **23**:165-173.

Charles, E. 1933. Collateral and ancestral correlation for sex-linked transmission irrespective of sex. Genetics **27**:97-104.

Chen, C. S. 1973. Sensitization for audiogenic seizures in two strains of mice and their F_1 hybrid. Dev. Psychobiol. **6**:131-138.

Chen, C. S., and J. L. Fuller. 1976. Selection for spontaneous or priming-induced audiogenic seizure susceptibility in mice. J. Comp. Physiol. Psychol. **90**:765-772.

Chen, C. S., G. R. Gates, and G. R. Bock. 1973. The effect of priming and tympanic membrane destruction on development of audiogenic susceptibility in BALB/c mice. Exp. Neurol. **39**:277-284.

Chlouverakis, C. 1972. Insulin resistance of parabiotic obese-*ob ob*. Horm. Metab. Res. **4**:143-148.

Chomsky, N. 1968. Language and the mind. Harcourt Brace Jovanovich, Inc., New York.

Christian, J. C., K. W. Kang, and J. A. Norton. 1974.

Choice of an estimate of genetic variance from twin data. Am. J. Hum. Genet. **26**:154-161.

Church, A. C., J. L. Fuller, and B. C. Dudek. 1976. Salsolinol differentially affects mice selected for sensitivity to alcohol. Psychopharmacology **47**:49-52.

Ciranello, R. D., J. N. Dornbusch, and J. D. Barchas. 1972. Regulation of adrenal phenyl-ethanolamine-*N*-methyltransferase activity in three inbred mouse strains. Mol. Pharmacol. **8**:511-520.

Ciranello, R. D., H. J. Hoffman, J. G. M. Shire, and J. Axelrod. 1974. Genetic regulation of the catecholamine biosynthetic enzymes. II. Inheritance of tyrosine hydroxylase, dopamine hydroxylase and phenylethanolamine-*N*-methyltransferase. J. Biol. Chem. **249**:4528-4534.

Claeys, W. 1973. Primary abilities and field-independence of adopted children. Behav. Genet. **3**:323-338.

Claridge, G. S. 1967. Personality and arousal. Pergamon Press Ltd., Oxford.

Claridge, G., S. Canter, and W. I. Hume. 1973. Personality differences and biological variations: a study of twins. Pergamon Press Ltd., Oxford.

Clark, E., L. R. Aronson, and M. Gordon. 1954. Mating behavior patterns in two sympatric species of Xiphophorin fishes: Their inheritance and significance in sexual isolation. Bull. Am. Mus. Nat. Hist. N.Y. **103**:135-226.

Clark, F. H. 1936. Geotropic behavior on a sloping plane of arboreal and nonarboreal races of mice of the genus *Peromyscus*. J. Mammal. **17**:44-47.

Cliquet, R. L. 1963. Bijdrage tot de kennis van het verband tussen de sociale status en een aantal antrobiologische ken merken. Verh. Kon. Vlaam. Acad. Wetensch. Lett. Schone Kunsten Belg. Kl. Wetensch. No. 72, Brussels.

Cloudman, A. M., and L. E. Bunker, Jr. 1945. The varitint-waddler mouse. J. Hered. **36**:259-263.

Cobb, M. V. 1917. A preliminary study of the inheritance of arithmetical abilities. J. Educ. Psychol. **8**:1-20.

Coburn, C. A. 1922. Heredity of wildness and savageness in mice. Behav. Monogr. **4**:1-71.

Cohen, A. I. 1966. Hand preference and developmental status of infants. J. Genet. Psychol. **108**:337-345.

Cohen, D. J., M. G. Martin, W. Pollin, G. Inoff, M. Werner, and E. Dibble. 1972. Personality development in twins: competence in the newborn and preschool periods. J. Am. Child Psychiatry **11**:625-644.

Cohen, J., and D. P. Ogden. 1949. Taste blindness to phenyl-thio-carbamide as a function of saliva. Science **110**:532-533.

Cohen, M. E., D. W. Badal, A. Kilpatrick, E. W. Reed, and P. D. White. 1951. The high familial prevalence of neurocirculatory asthenia (anxiety neurosis effort syndrome). Am. J. Hum. Genet. **3**:76-79.

Cole, J. 1955. Paw preference in cats related to hand preference in animals and men. J. Comp. Physiol. Psychol. **48**:137-140.

Cole, W. H. 1922. Note on the relation between the

photic stimulus and the rate of locomotion in *Drosophila*. Science **55**:678-679.

Coleman, D. L. 1973. Effects of parabiosis of obese with diabetes and normal mice. Diabetologia **9**:294-298.

Coleman, D. L., and K. P. Hummel. 1969. Effects of parabiosis of normal with genetically obese mice. Am. J. Physiol. **217**:1298-1304.

Coleman, D. L., and K. P. Hummel. 1973. The influence of genetic background on the expression of the obese *(ob)* gene in the mouse. Diabetologia **9**:287-293.

Coleman, J. S., E. Q. Campbell, C. J. Hobson, J. McPartland, A. M. Moody, F. D. Weinfeld, and R. L. York. 1966. Equality of educational opportunity. U.S. Government Printing Office, Washington, D.C.

Collins, R. L. 1964. Inheritance of avoidance conditioning in mice: a diallel study. Science **143**:1188-1190.

Collins, R. L. 1967. A general nonparametric theory of genetic analysis. I. Application to the classical cross. Genetics **56**:551. (Abstr.)

Collins, R. L. 1968a. A general nonparametric theory of genetic analysis. II. Digenic equations with linkage for the classical cross. Genetics **60**:169-170. (Abstr.)

Collins, R. L. 1968b. On the inheritance of handedness. I. Laterality in inbred mice. J. Hered. **59**:9-12.

Collins, R. L. 1969. On the inheritance of handedness. II. Selection for sinistrality in mice. J. Hered. **60**:117-119.

Collins, R. L. 1970a. A new genetic locus mapped from behavioral variation in mice: audiogenic seizure prone *(asp)*. Behav. Genet. **1**:99-110.

Collins, R. L. 1970b. The sound of one paw clapping: an inquiry into the origin of left-handedness. In G. Lindzey and D. D. Thiessen, eds. Contributions to behavior-genetic analysis: the mouse as a prototype, Appleton-Century-Crofts, New York.

Collins, R. L. 1975. When left-handed mice live in right-handed worlds. Science **187**:181-184.

Collins, R. L., and J. L. Fuller. 1968. Audiogenic seizure prone *(asp)*—a gene affecting behavior in linkage group VIII of the mouse. Science **162**:1137-1139.

Connolly, K. 1966. Locomotor activity in *Drosophila*. II. Selection for active and inactive strains. Anim. Behav. **14**:444-449.

Conrad, H. S. 1931. On kin resemblances in physique versus intelligence. J. Educ. Psychol. **22**:376-382.

Conrad, H. S., and H. E. Jones. 1940. A second study of family resemblances in intelligence: environmental and genetic implications of parent-child and sibling correlations in the total sample. In Thirty-ninth Yearbook of the National Society for the Study of Education. Public School Publishing Co., Bloomington, Ill.

Conway, J. 1958. The inheritance of intelligence and its social implications. Br. J. Stat. Psychol. **11**:171-190.

Cook, W. T., P. B. Siegel, and K. Hinkelmann. 1972. Genetic analyses of male mating behavior in chickens.

II. Crosses among selected and control lines. Behav. Genet. **2**:289-300.

Cooke, F., and F. G. Cooch. 1968. The genetics of polymorphism in the goose *Anser caerulescens*. Evolution **22**:289-300.

Cooke, F., G. H. Finney, and R. F. Rockwell. 1976. Assortative mating in lesser snow geese *(Anser caerulescens)*. Behav. Genet. **6**:127-140.

Cooke, F., and C. M. McNally. 1975. Mate selection and colour preferences in lesser snow geese. Behaviour **53**:151-170.

Cooke, F., P. J. Mirsky, and M. B. Seiger. 1972. Colour preferences in the lesser snow goose and their possible role in mate selection. Can. J. Zool. **50**:529-536.

Cooper, R. M., and J. P. Zubek. 1958. Effects of enriched and restricted early environments on the learning ability of bright and dull rats. Can. J. Psychol. **12**:159-164.

Coppen, A., V. Cowie, and E. Slater. 1965. Familial aspects of "neuroticism" and "extraversion." Br. J. Psychiatry **111**:70-83.

Cowie, V. 1961. The incidence of neurosis in the children of psychotics. Acta Psychiatr. Scand. **37**:37-87.

Craig, J. V., and R. A. Baruth. 1965. Inbreeding and social dominance ability in chickens. Anim. Behav. **13**:109-113.

Craig, J. V., D. K. Biswas, and A. M. Guhl. 1969. Agonistic behaviour influenced by strangeness, crowding and heredity in female domestic fowl *(Gallus gallus)*. Anim. Behav. **17**:498-506.

Craig, J. V., L. L. Ortman, and A. M. Guhl. 1965. Genetic selection for social dominance ability in chickens. Anim. Behav. **13**:114-131.

Craike, W. H., and E. Slater. 1945. Folie à deux in uniovular twins reared apart. Brain **68**:213-221.

Crombie, D. L. 1957. Coll. Gen. Pract. Res. Newsletter No. 16: 218.

Cronbach, L. 1957. The two disciplines of scientific psychology. Am. Psychol. **12**:671-684.

Crook, M. N. 1937. Intra-family relationships in personality test performance. Psychol. Rec. **1**:479-502.

Crook, M. N., and M. Thomas. 1934. Family relationships in ascendance-submission. Publ. Univ. Calif. Educ. Philos. Psychol. **1**:189-192.

Crow, J. F., and M. Kimura. 1970. An introduction to population genetics. Harper & Row, Publishers, Inc., New York.

Crowcroft, P., and F. P. Rowe. 1963. Social organization and territorial behaviour in the wild house mouse *(Mus musculus, L.)*. Proc. Zool. Soc. Lond. **140**:517-531.

Cruce, J. A. F., M. R. C. Greenwood, P. R. Johnson, and D. Quartermain. 1974. Caloric regulation in genetically obese and hypothalamic-hyperphagic rats. J. Comp. Physiol. Psychol. **87**:295-301.

Cruz-Coke, R. 1970. Color blindness: an evolutionary approach. Charles C Thomas, Publisher, Springfield, Ill.

Cruz-Coke, R., and A. Varela. 1966. Inheritance of alcoholism: its association with colour-blindness. Lancet 2:1282.

Cruz-Coke, R.,. and A. Varela. 1970. Genetic factors in alcoholism. In R. E. Popham, ed. Alcohol and alcoholism. University of Toronto Press, Toronto.

Da Fonseca, A. F. 1959. Analise heredo-clinica des purtubacoes affectivas. Doctoral dissertation. Universidade do Porto, Portugal.

Dahlberg, G. 1926. Twin births and twins from a hereditary point of view. Tidens Tryckery, Stockholm.

Dahlberg, G. 1953. Biometric evaluation of findings. In A. Sorsby, ed. Clinical genetics. Butterworth & Co., London.

Dahlstrom, W. G., and G. S. Welsh. 1960. An MMPI handbook: a guide to use in clinical practice and research. University of Minnesota Press, Minneapolis.

Damon, A. 1974. Larger body size and earlier menarche: the end may be in sight. Soc. Biol. 21:8-11.

Darke, R. A. 1948. Heredity as an etiological factor in homosexuality. J. Nerv. Ment. Dis. 107:251-268.

Dart, R. A. 1949. The predatory implemental technique of *Australopithecus*. Am. J. Phys. Anthropol. 7:1-38.

Darwin, C. 1872. The expression of emotions in man and animals. Philosophical Library, New York.

Darwin, C. 1927. The origin of species. Macmillan, New York. (First published in 1859.)

Davenport, C. B., and T. M. Scudder. 1918. Naval officers: their heredity and development. Carnegie Institution, Washington, D.C.

Davidson, E. H., and R. J. Britten. 1973. Organization, transcription and regulation in the animal genome. Q. Rev. Biol. 48:565-613.

Davis, B. D. 1954. Genetic and environmental control of enzyme formation. Proc. Assoc. Res. Nerv. Ment. Dis. 33:23-38.

Davis, D. J., S. Cahan, and J. Bashi. 1977. Birth order and intellectual development: the confluence model in the light of cross-cultural evidence. Science 196:1470-1472.

Davis, H., and P. Davis. 1936. Action potentials of the brain. Arch. Neurol. Psychiatry 36:1214-1224.

Dawson, W. M. 1932. Inheritance of wildness and tameness in mice. Genetics 17:296-326.

DeFries, J. C. 1964. Prenatal maternal stress in mice: differential effects on behavior. J. Hered. 55:289-295.

DeFries, J. C. 1967. Quantitative genetics and behavior: overview and perspective. In J. Hirsch, ed. Behavior-genetic analysis. McGraw-Hill Book Co., New York.

DeFries, J. C. 1969. Pleiotropic effects of albinism on open field behaviour in mice. Nature 221:65-66.

DeFries, J. C. 1972. Quantitative aspects of genetics and environment in the determination of behavior. In L. Ehrman, G. S. Omenn, and E. Caspari, eds. Genetics, environment, and behavior. Academic Press, Inc., New York.

DeFries, J. C. 1975. Commentary. In K. W. Schaie, V. E. Anderson, G. E. McClearn, and T. Money, eds. Developmental human behavior genetics. D. C. Heath & Co., Lexington,, Mass.

DeFries, J. C., and J. P. Hegmann. 1970. Genetic analysis of open-field behavior. In G. Lindzey and D. D. Thiessen, eds. Contributions to behavior-genetic analysis: the mouse as a prototype. Irvington Publishers, Inc., New York.

DeFries, J. C., J. P. Hegmann, and M. W. Weir. 1966. Open-field behavior in mice: Evidence for a major gene effect mediated by the visual system. Science 154:1577-1579.

DeFries, J. C., and G. E. McClearn. 1970. Social dominance and Darwinian fitness in the laboratory mouse. Am. Natur. 104:408-411.

DeFries, J. C., and G. E. McClearn. 1972. Behavioral genetics and the fine structure of mouse populations: a study of micro-evolution. In T. Dobzhansky, M. K. Hecht, and W. C. Steere, eds. Evolutionary biology, vol. 5. Appleton-Century-Crofts, New York.

DeFries, J. C., J. R. Wilson, and G. E. McClearn. 1970. Open field behavior in mice: selection response and situational generality. Behav. Genet. 1:195-211.

Del Solar, E. 1966. Sexual isolation caused by selection for positive and negative phototaxis and geotaxis in *Drosophila pseudoobscura*. Proc. Natl. Acad. Sci. U.S.A. 56:484-487.

Dempster, E. R., and I. M. Lerner. 1950. Heritability of threshold characters. Genetics 35:212-236.

Denenberg, V. H. 1959. Learning differences in two separated lines of mice. Science 130:451-452.

Denenberg, V. H. 1961. Comment on "infantile trauma, genetic factors and adult temperament." Psychol. Rep. 8:459-462.

Denenberg, V. H., S. Ross, and M. Blumenfield. 1963. Behavioral differences between mutant and non-mutant mice. J. Comp. Physiol. Psychol. 56:290-293.

Descartes, R. 1945. Discourse on the method of rightly conducting the reason and seeking truth in the sciences (1637). Penguin Books Ltd., Harmondsworth, England.

DiCara, L. V. ed. 1974. Limbic and autonomic nervous systems research. Plenum Publishing Corp., New York.

Dickerson, G. E., and J. W. Gowen. 1947. Hereditary obesity and efficient food utilization in mice. Science 105:496-498.

Dobzhansky, T. 1962. Mankind evolving. Yale University Press, New Haven, Conn.

Dobzhansky, T. 1967. Of flies and men. Am. Psychol. 42:41-48.

Dobzhansky, T., L. Ehrman, and P. A. Kastritsis. 1968. Ethological isolation between sympatric and allopatric species of the *obscura* group of *Drosophila*. Anim. Behav. 16:79-87.

Dobzhansky, T., and B. Spassky. 1967. Effects of selection and migration on geotactic and phototactic behavior of *Drosophila*. I. Proc. R. Soc. Biol. 168:27-47.

Dobzhansky, T., B. Spassky, and J. Sved. 1969. Effects of selection and migration on geotactic and phototactic behavior of *Drosophila*. II. Proc. R. Soc. Biol. 173:191-207.

Dobzhansky, T., and C. Streisinger. 1944. Experiments on sexual isolation in *Drosophila*. II. Geographic strains of *D. prosaltans*. Proc. Natl. Acad. Sci. U.S.A. **30**:340-345.

Dockrell, W. B., ed. 1970. On intelligence. Methuen & Co. Ltd., London.

Dodwell, P. C., ed. 1970. Perceptual learning and adaptation. Penguin Books Ltd., Harmondsworth, England.

Downie, W. W., J. A. Boyle, W. R. Greig, W. W. Buchanan, and F. P. Alepa. 1969. Relative roles of genetic and environmental factors in control of blood pressure in normotensive subjects. Br. Heart J. **31**:21-25.

Duke-Elder, W. S. 1949. Textbook of ophthalmology, vol. 4. Henry Kimpton Ltd., London.

Dumermuth, G. 1968. Variance spectra of electroencephalograms in twins. In P. Kellaway and I. Peterson, eds. Clinical electroencephalography of children. Grune & Stratton, Inc. New York.

Duncan, N. C., N. E. Grosen, and E. B. Hunt. 1971. Apparent memory differences in inbred mice produced by differential reaction to stress. J. Comp. Physiol. Psychol. **74**:383-389.

Duncan, O. D. 1966. Path analysis: sociological examples. Am. J. Sociol. **72**:1-16.

Duncan, O. D. 1968. Ability and achievement. Eugen. Q. **15**:1-11.

Dunham, H. W. 1976. Society, culture, and mental disorder. Arch. Gen. Psychiatry **33**:147-156.

Dunn, L. G. 1951. Genetics in the 20th century. Macmillan, Inc., New York.

Dustman, R. E., and E. C. Beck. 1965. The visual evoked response in twins. Electroencephalogr. Clin. Neurophysiol. **19**:570-575.

Dworkin, R. H., B. W. Burke, B. A. Maher, and I. I. Gottesman. 1976. A longitudinal study of the genetics of personality. J. Pers. Soc. Psychol. **34**:510-518.

Earle, E. L. 1903. The inheritance of ability to learn to spell. Columbia Univ. Contrib. Phil. Psychol. Educ. **2**:41-44.

Eaves, L. J. 1972. Computer simulation of sample sizes and experimental design in human psychogenetics. Psychol. Bull. **77**:144-152.

Eaves, L. J. 1973. Assortative mating and intelligence: an analysis of pedigree data. Heredity **30**:199-210.

Eaves, L. J., and H. J. Eysenck. 1974. Genetics and the development of social attitudes. Nature **249**:288-289.

Ebert, P. D., and J. S. Hyde. 1976. Selection for agonistic behavior in wild female *Mus musculus*. Behav. Genet. **6**:291-304.

Edwards, D. A. 1968. Mice: fighting by neonatally androgenized females. Science **161**:1027-1028.

Edwards, J. H. 1969. Familial predisposition in man. Br. Med. Bull. **25**:58-63.

Ehrman, L. 1964. Courtship and mating behavior as a reproductive isolating mechanism in *Drosophila*. Am. Zool. **4**:147-153.

Ehrman, L. 1965. Direct observation of sexual isolation between allopatric and between sympatric strains of the different *Drosophila paulistorum* races. Evolution **19**:459-464.

Ehrman, L. 1966. Mating success and genotype frequency in *Drosophila*. Anim. Behav. **14**:332-339.

Ehrman, L. 1969. The sensory basis of mate selection in *Drosophila*. Evolution **23**:59-64.

Ehrman, L. 1970a. Sexual isolation versus mating advantage of rare *Drosophila* males. Behav. Genet. **1**:111-118.

Ehrman, L. 1970b. A study of the mating advantage of rare males in *Drosophila*. Proc. Natl. Acad. Sci. U.S.A. **65**:345-348.

Ehrman, L. 1972. A factor influencing the rare male mating advantage in *Drosophila*. Behav. Genet. **2**:69-78.

Ehrman, L., G. S. Omenn, and E. Caspari, eds. 1972. Genetics, environment and behavior. Academic Press, Inc., New York.

Ehrman, L., and P. A. Parsons. 1976. The genetics of behavior. Sinauer Associates, Inc., Sunderland, Mass.

Ehrman, L., and C. Petit. 1968. Genotype frequency and mating success in the *willistoni* species group of *Drosophila*. Evolution **22**:649-658.

Elderton, E. M. 1909. The relative strength of nature and nurture. Eugenics Laboratory Lecture Series, vol. 3. Dulan, London.

Elderton, E. M. 1923. A summary of the present position with regard to the inheritance of intelligence. Biometrika **14**:378-408.

Eleftheriou, B. E. 1975. Psychopharmacogenetics. Plenum Publishing Corp., New York.

Ellingson, R. J. 1966. Relationship between EEG and test intelligence: a commentary. Psychol. Bull. **65**:91-98.

Elsässer, G. 1952. Die Nachkommen Geisteskranker Elternpaare. Georg Thieme Verlag, Stuttgart, West Germany.

Elston, R. C., and M. A. Campbell. 1970. Schizophrenia: evidence for the major gene hypothesis. Behav. Genet. **1**:3-10.

Elston, R. C., E. Kringlen, and K. K. Namboodiri. 1973. Possible linkage relationships between certain blood groups and schizophrenia or other psychoses. Behav. Genet. **3**:101-106.

Elston, R. C., and J. Stewart. 1973. The analysis of quantitative traits for simple genetic models from parental F_1 and backcross data. Genetics **73**:695-711.

Englemann, S. 1970. The effectiveness of direct verbal instruction on IQ performance and achievement in reading and arithmetic. In J. Helmuth, ed. Disadvantaged child. Brunner/Mazel, Inc., New York.

Eriksson, C. J. P. 1973. Ethanol and acetaldehyde metabolism in rat strains genetically selected for their ethanol preference. Biochem. Pharm. **22**:2283-2292.

Eriksson, K. 1968. Genetic selection for voluntary alcohol consumption in the albino rat. Science **159**:739-741.

Eriksson, K. 1969a. The estimation of heritability for the self-selection of alcohol in the albino rat. Ann. Med. Exp. Fenn. **47**:172-174.

Eriksson, K. 1969b. Factors affecting voluntary alcohol consumption in the albino rat. Ann. Zool. Fenn. **6**:227-265.

Eriksson, K., and K. Kiianmaa. 1971. Genetic analysis of susceptibility to morphine addiction in inbred mice. Ann. Med. Exp. Biol. Fenn. **49**:73-78.

Erlenmeyer-Kimling, L. 1968. Studies on the offspring of two schizophrenic parents. In D. Rosenthal and S. S. Kety, eds. The transmission of schizophrenia. Pergamon Press Ltd., Oxford.

Erlenmeyer-Kimling, L. 1972. Gene-environment interactions and the variability of behavior. In L. Ehrman, G. S. Omenn, and E. Caspari, eds. Genetics environment and behavior. Academic Press, Inc., New York.

Erlenmeyer-Kimling, L., J. Hirsch, and J. M. Weiss. 1962. Studies in experimental behavior genetics. III. Selection and hybridization analysis of individual differences in the sign of geotaxis. J. Comp. Physiol. Psychol. **55**:722-731.

Erlenmeyer-Kimling, L., and L. F. Jarvik. 1963. Genetics and intelligence: a review. Science **142**:1477-1478.

Erlenmeyer-Kimling, L., S. Nicol, J. D. Ranier, and W. E. Deming. 1969. Changes in fertility rates of schizophrenic patients in New York State. Am. J. Psychiatry **125**:916-927.

Erlenmeyer-Kimling, L., and W. Paradowski. 1966. Selection and schizophrenia. Am. Natur. **100**:651-665.

Essen-Möller, E. 1941. Psychiatrische Untersuchungen an einer Serie von Zwillingen. Acta Psychiatr. Neurol. Scand. [Suppl.] **23**.

Essen-Möller, E. 1956. Individual traits and morbidity in a Swedish rural population. Acta Psychiatr. Neurol. Scand. [Suppl.] **100**.

Essen-Möller, E., and O. Hagnell. 1961. The frequency and risk of depression within a rural population group in Scandia. Acta Psychiatr. Neurol. Scand. [Suppl.] **162**.

Ettlinger, G. 1961. Lateral preferences in monkeys. Behaviour **17**:275-287.

Ewing, A. W. 1963. Attempts to select for spontaneous activity in *Drosophila melanogaster*. Anim. Behav. **11**:369-378.

Ewing, A. W. 1969. The genetic basis of sound production in *Drosophila pseudoobscura* and *D. persimilis*. Anim. Behav. **17**:555-560.

Eysenck, H. J. 1950. Criterion analysis—an application of the hypothetico-deductive method to factor analysis. Psychol. Rev. **57**:38-53.

Eysenck, H. J. 1956. The inheritance of extraversion-intraversion. Acta Psychol. **12**:95-110.

Eysenck, H. J., ed. 1959. Handbook of abnormal psychology. Putnam & Co. Ltd., London.

Eysenck, H. J. 1960. Classification and the problem of diagnosis. In H. J. Eysenck, ed. Handbook of abnormal psychology. Sir Isaac Pitman & Sons Ltd., London.

Eysenck, H. J. 1967. The biological basis of personality. Charles C Thomas, Publisher, Springfield, Ill.

Eysenck, H. J., ed. 1971. Readings in extraversion-introversion. Staples Press, Ltd., London.

Eysenck, H. J. 1972. An experimental and genetic model of schizophrenia. In A. R. Kaplan, ed. Genetic factors in "schizophrenia." Charles C Thomas, Publisher, Springfield, Ill.

Eysenck, H. J., and D. B. Prell. 1951. The inheritance of neuroticism: an experimental study. J. Ment. Sci. **97**:441-465.

Eysenck, S. B. G., and H. J. Eysenck. 1969. Scores on three personality variables as a function of age, sex, and social class. Br. J. Soc. Clin. Psychol. **8**:69-76.

Falconer, D. S. 1960. Introduction to quantitative genetics. The Ronald Press Co., New York.

Falconer, D. S. 1963. Quantitative inheritance. In W. J. Burdette, ed. Methodology in mammalian genetics. Holden-Day, Inc., San Francisco.

Falconer, D. S. 1965. The inheritance of liability to certain diseases, estimated from the incidence among relatives. Ann. Hum. Genet. **29**:51-76.

Falconer, D. S. 1967. The inheritance of liability to diseases with variable age of onset, with particular reference to diabetes mellitus. Ann. Hum. Genet. **31**:1-20.

Falconer, D. S., and M. Latyszewski. 1952. Selection for size of mice on high and low planes of nutrition. In E. C. R. Reese and C. H. Waddington, eds. Quantitative inheritance. Her Majesty's Stationery Office, London.

Falconer, D. S., M. Latyszewski, and J. H. Isaacson. 1964. Diabetes insipidus associated with oligosyndactyly in the mouse. Genet. Res. **5**:473-488.

Falk, C. T., and F. J. Ayala. 1971. Genetic aspects of arm folding and hand clasping. Jpn. J. Hum. Genet. **15**:241-247.

Falk, J. L. 1961. Production of polydipsia in normal rats by an intermittent food schedule. Science **133**:195-196.

Farris, E. J., and E. H. Yeakel. 1945. Emotional behavior of gray Norway and Wistar albino rats. J. Comp. Psychol. **38**:109-117.

Feldman, M. W., and R. C. Lewontin. 1975. The heritability hang-up. Science **190**:1163-1168.

Fennell, R. A. 1945. The relation between heredity, sexual activity, and training to dominance-subordination in game cocks. Am. Natur. **79**:142-151.

Fenton, P. F., and M. T. Dowling. 1953. Studies on obesity. I. Nutritional obesity in mice. J. Nutr. **48**:319-331.

Ferguson-Smith, M. A. 1965. Karyotype-phenotype correlations in gonadal dysgenesis and their bearing on the pathogenesis of malformations. J. Med. Genet. **2**:142-155.

Festing, M. F. W. 1973a. Water escape learning in mice. I. Strain differences and biometrical considerations. Behav. Genet. **3**:13-24.

Festing, M. F. W. 1973b. Water escape learning in mice. II. Replicated selection for increased learning speed. Behav. Genet. **3**:25-36.

Festing, M. F. W. 1974. Water escape learning in mice. III. A diallel study. Behav. Genet. **4**:111-124.

Feuer, G. 1969. Differences in emotional behaviour and in function of the endocrine system in genetically different strains of albino rats. In E. Badjusz, ed. Physiology and pathology of adaptation mechanisms. Pergamon Press Ltd., Oxford.

Feuer, G., and P. L. Broadhurst. 1962a. Thyroid function in rats selectively bred for emotional elimination. I. Differences in thyroid hormones. J. Endocrinol. 24:127-136.

Feuer, G., and P. L. Broadhurst. 1962b. Thyroid function in rats selectively bred for emotional elimination. II. Differences in thyroid activity. J. Endocrinol. 24:253-262.

Feuer, G., and P. L. Broadhurst. 1962c. Thyroid function in rats selectively bred for emotional elimination. III. Behavioural and physiological changes after treatment with drugs acting on the thyroid. J. Endocrinol. 24:385-396.

Finch, F. H. 1933. A study of the relation of age interval to degree of resemblance of siblings in intelligence. J. Genet. Psychol. 43:389-404.

Finch, G. 1941. Chimpanzees' handedness. Science 94:117-118.

Finlayson, A. 1916. The Dach family: a study in the hereditary lack of emotional control. Eugen. Rec. Office Bull. 15.

Fischer, M. 1971. Psychoses in the offspring of schizophrenic monozygotic twins and their normal co-twins. Br. J. Psychiatry 118:43-52.

Fish, B. 1971. Contributions of developmental research to a theory of schizophrenia. In J. Hellmuth, ed. Exceptional infant: studies in abnormalities, vol. 2. Brunner/Mazel, Inc., New York.

Fish, B., R. Wile, T. Shapiro, and F. Halpern. 1966. The prediction of schizophrenia in infancy. II. A ten-year follow-up report of predictions made at one month of age. In P. H. Hoch and J. Zubin, eds. Psychopathology of schizophrenia. Grune & Stratton, Inc., New York.

Fisher, R. A. 1918. The correlation between relatives on the supposition of Mendelian inheritance. Trans. R. Soc. Edinb. 52:399-433.

Flanagan, J. C. 1935. Factor analysis in the study of personality. Stanford University Press, Stanford, Calif.

Fölling, A. 1934. Excretion of phenyl-pyruvic acid in urine as metabolic anomaly in connection with imbecility. Nord. Med. Tidskr. 8:1054-1059.

Forster, M. C. 1931. A study of father-son resemblance in vocational interests and personality traits. Ph.D. thesis, University of Minnesota.

Fowler, S. H., and M. E. Ensminger. 1960. Interactions between genotype and plane of nutrition in selection for rate of gain in swine. J. Anim. Sci. 19:434-449.

Fox, A. L. 1932. The relationship between chemical constitution and taste. Proc. Natl. Acad. Sci. U.S.A. 18:115-120.

Franceschetti, A., and D. Klein. 1956. Two families with parents of different types of color blindness. In First International Congress on Human Genetics, Part V. Copenhagen. S. Karger AG, Basel, Switzerland.

Fredericson, E. 1953. The wall-seeking tendency in three inbred mouse strains *(Mus musculus)*. J. Genet. Psychol. 82:143-146.

Fredericson, E., and E. A. Birnhaum. 1954. Competitive fighting between mice with different hereditary backgrounds. J. Genet. Psychol. 85:271-280.

Freedman, D. G. 1958. Constitutional and environmental interactions in rearing of four breeds of dogs. Science 127:585-586.

Freedman, D. G. 1965. An ethological approach to the genetical study of human behavior. In S. G. Vandenberg, ed. Methods and goals in human behavior genetics. Academic Press, Inc., New York.

Freedman, D. G., and B. Keller. 1963. Inheritance of behavior in infants. Science 140:196-198.

Freeman, F. N. 1926. Mental tests. Houghton Mifflin Co., New York.

Freeman, F. N., K. J. Holzinger, and B. C. Mitchell. 1928. The influence of environment on the intelligence, school achievement, and conduct of foster-children. In Twenty-seventh Yearbook of the National Society for the Study of Education, Part I. Public School Publishing Co., Bloomington, Ill.

Freire-Maia, A., and J. de Almeida. 1966. Hand clasping and arm folding among African Negroes. Hum. Biol. 38:175-179.

Freire-Maia, N., A. Quelce-Salgado, and A. Freire-Maia. 1958. Hand clasping in different ethnic groups. Hum. Biol. 30:281-291.

Fremming, K. H. 1947. Morbid risk of mental diseases and other abnormalities in an average Danish population. Munksgaard, International Booksellers & Publishers Ltd., Copenhagen.

Fremming, K. H. 1951. The expectation of mental infirmity in a sample of the Danish population. Eugenics Society Occasional Papers, No. 7. Cassell & Co., London.

Friend, R. 1939. Influences of heredity and musical environment on the scores of kindergarten children on the Seashore measures of musical ability. J. Appl. Psychol. 23:347-357.

Frings, H., M. Frings, and M. Hamilton. 1956. Experiments with albino mice from stock selected for predictable susceptibilities to audiogenic seizures. Behaviour 9:44-52.

Frischeisen-Köhler, I. 1933a. The personal tempo and its inheritance. Char. Pers. 1:301-313.

Frischeisen-Köhler, I. 1933b. Das personliche Tempo. Eine erbbiologische Untersuchung. Georg Thieme Verlag, Leipzig.

Fromkin, V., S. Krashen, S. Curtiss, D. Rigler, and M. Rigler. 1974. The development of language in Genie: a case of language acquisition beyond the "critical period." Brain Lang. 1:81-107.

Fry, D. 1948. An experimental study of tone deafness. Speech 1:4-11.

Fulker, D. W. 1966. Mating speed in male *Drosophila melanogaster:* a psychogenetic analysis. Science 153:203-205.

Fulker, D. W. 1970. Maternal buffering of rodent genotypic responses to stress: a complex genotype-environment interaction. Behav. Genet. 1:119-124.

Fulker, D. W. 1975. Review of L. J. Kamin, *Science and Politics of IQ.* J. Psychol. 88:505-519.

Fuller, D. W., J. Wilcock, and P. L. Broadhurst. 1972. Studies in genotype-environment interaction. I. Methodology and preliminary multivariate analysis of a diallel cross of eight strains of rat. Behav. Genet. 2:261-287.

Fuller, J. L. 1953. Cross-sectional and longitudinal studies of adjustive behavior in dogs. Ann. N.Y. Acad. Sci. 56:214-224.

Fuller, J. L. 1964. Measurement of alcohol preference in genetic experiments. J. Comp. Physiol. Psychol. 57:85-88.

Fuller, J. L. 1966. Transitory effects of experiential deprivation upon reversal learning in dogs. Psychon. Sci. 4:273-274.

Fuller, J. L. 1967. Effects of the albino gene upon behavior of mice. Anim. Behav. 15:467-470.

Fuller, J. L. 1968. Genetics and goal setting. In C. Bühler and F. Massarik, eds. The course of human life, Springer Publishing Co., Inc., New York.

Fuller, J. L. 1970. Strain differences in the effects of chlorpromazine and chlordiazepoxide upon active and passive avoidance in mice. Psychopharmacologia 16: 261-271.

Fuller, J. L. 1972. Genetic aspects of regulation of food intake. Adv. Psychosom. Med. 7:2-24.

Fuller, J. L. 1974. Single-locus control of saccharin preference in mice. J. Hered. 65:33-36.

Fuller, J. L. 1975. Independence of inherited susceptibility to spontaneous and primed audiogenic seizures in mice. Behav. Genet. 5:1-8.

Fuller, J. L., S. Brady-Wood, and M. F. Elias. 1973. Effects of retinal degeneration and brain size upon spatial reversal learning in mice. Percept. Mot. Skills 36:947-950.

Fuller, J. L., R. M. Chambers, and R. P. Fuller. 1956. Effects of cortisone and of adrenalectomy on activity and emotional behavior of mice. Psychosom. Med. 18:234-242.

Fuller, J. L., A. C. Church, and L. Dann. 1976. Ethanol consumption by mice selected for high and low ethanol sleep times. Behav. Genet. 7:59. (Abstr.)

Fuller, J. L., and R. L. Collins. 1968a. Mice unilaterally sensitized for audiogenic seizures. Science 162:1295.

Fuller, J. L., and R. L. Collins. 1968b. Temporal parameters of sensitization for audiogenic seizures in SJL/J mice. Dev. Psychobiol. 1:185-188.

Fuller, J. L., and R. L. Collins. 1970. Genetics of audiogenic seizures in mice: a parable for psychiatrists. Semin. Psychiatry 2:75-88.

Fuller, J. L., and R. L. Collins. 1972. Ethanol consumption and preference in mice: a genetic analysis. Ann. N.Y. Acad. Sci. 197:42-48.

Fuller, J. L., and C. W. Cooper. 1967. Saccharin reverses the effect of food deprivation upon fluid intake in mice. Anim. Behav. 15:403-410.

Fuller, J. L., C. Easler, and M. E. Smith. 1950. Inheritance of audiogenic seizure susceptibility in the mouse. Genetics 35:622-632.

Fuller, J. L., and H. D. Geils. 1973. Behavioral development in mice selected for differences in brain weight. Dev. Psychobiol. 6:469-474.

Fuller, J. L., and M. E. Hahn. 1976. Issues in the genetics of social behavior. Behav. Genet. 6:391-406.

Fuller, J. L., and B. H. Herman. 1974. Effect of genotype and practice upon behavioral development in mice. Dev. Psychobiol. 7:21-30.

Fuller, J. L., and G. A. Jacoby, Jr. 1955. Central and sensory control of food intake in genetically obese mice. Am. J. Physiol. 183:279-283.

Fuller, J. L., and F. H. Sjursen. 1967. Audiogenic seizures in eleven mouse strains. J. Hered. 58:135-140.

Fuller, J. L., and W. R. Thompson. 1960. Behavior genetics. John Wiley & Sons, Inc., New York.

Fuller, R. N., and J. B. Shuman. 1974. Genetic divergence in relatives of PKU's: low IQ correlation among normal siblings. Dev. Psychobiol. 7:323-330.

Galton, F. 1869. Hereditary genius: an inquiry into its laws and consequences. Macmillan & Co. Ltd., London.

Galton, F. 1874. English men of science: their nature and nurture. Macmillan & Co. Ltd., London.

Galton, F. 1875. The history of twins as a criterion of the relative powers of nature and nurture. J. R. Anthropol. Inst. 5:391-406.

Galton, F. 1883. Inquiry into human faculty. Macmillan & Co. Ltd., London.

Garcia, J., and R. A. Koelling. 1966. Relation of cue to consequence in avoidance learning. Psychon. Sci. 4:123-124.

Gardner, G. G. 1967. The relationship between childhood neurotic symptomatology and later schizophrenia in males and females. J. Nerv. Ment. Dis. 144:97-100.

Garmezy, N. 1974. Children at risk: the search for the antecedents of schizophrenia. II. Ongoing research, programs, issues, and intervention. Schizophrenia Bull. 8:55-125.

Garmezy, N., and S. Streitman. 1974. Children at risk: the search for the antecedents of schizophrenia. I. Conceptual models and research methods. Schizophrenia Bull. 8:14-90.

Garrod, A. E. 1923. Inborn errors of metabolism, 2nd ed. Oxford University Press, London.

Garrone, G. 1962. Étude statistique et génétique de la schizophrénie à Génève de 1901 à 1950. J. Génét. Hum. 11:89-219.

Gates, G. R., C. S. Chen, and G. R. Bock. 1973. Effects of monaural and binaural auditory deprivation on audiogenic seizure susceptibility in BALB/c mice. Exp. Neurol. 38:488-493.

Gates, W. H. 1934. Linkage tests of the new shaker mutation with other factors in the house mouse (Mus musculus). Am. Natur. 68:173-174.

Gauron, E. F. 1964. Nature of infantile shock traumatization, strain differences and adaptability to stress. Psychol. Rep. 14:775-779.

Gershon, E. S., W. E. Bunner, J. F. Leckman, M. Van Eerdewegh, and B. A. DeBauche. 1976. The inheritance of affective disorders: a review of data and of hypotheses. Behav. Genet. 6:227-261.

Ginsburg, B. E. 1954. Genetics and the physiology of the nervous system. Proc. Assoc. Res. Nerv. Ment. Dis. 33:39-56.

Ginsburg, B. E. 1958. Genetics as a tool in the study of behavior. Perspect. Biol. Med. 1:397-424.

Ginsburg, B. E. 1967. Genetic parameters in behavioral research. In J. Hirsch, ed. Behavior-genetic analysis. McGraw-Hill Book Co., New York.

Ginsburg, B., and W. C. Allee. 1942. Some effects of conditioning on social dominance and subordination in inbred strains of mice. Physiol. Zool. 15:485-506.

Ginsburg, B. E., J. S. Cowen, S. C. Maxson, and P. Y. Sze. 1969. Neurochemical effects of gene mutations associated with audiogenic seizures. In A. Barbeau and J. R. Brunette, eds. Progress in neurogenetics. Excerpta Medica Foundation, New York.

Ginsburg, B. E., and D. S. Miller. 1963. Genetic factors in audiogenic seizures. Colloq. Int. Cent. Natl. Rech. Sci. Paris 112:217-225.

Gjerde, C. M. 1949. Parent-child resemblance in vocational interests and personality traits. Doctoral dissertation. University of Minnesota, Minneapolis.

Glueck, S., and E. Glueck. 1956. Physique and delinquency. Paul B. Hoeber Inc., New York.

Goldberger, A. S. 1975. Statistical inference in the great IQ debate. Discussion Papers. Institute for Research on Poverty. University of Wisconsin, Madison.

Goldberger, A. S. 1976. Jensen on Burks. Educ. Psychol. 12:64-78.

Goldschmidt, R. 1938. Physiological genetics. McGraw-Hill Book Co., New York.

Goldstein, D. B. 1973. Inherited differences in intensity of alcohol withdrawal reactions in mice. Nature 245: 154-156.

Goldstein, D. B., and R. Kakihana. 1974. Alcohol withdrawal reactions and reserpine effects in inbred strains of mice. Life Sci. 15:415-425.

Goodale, H. D., R. Sanborn, and D. White. 1920. Broodiness in the domestic fowl. Data concerning its inheritance in the Rhode Island Red Breed. Mass. Agri. Exp. Stat. Bull. 199:93-116.

Goodenough, F. L. 1949. Mental testing. Rinehart & Co., New York.

Goodglass, H., and F. A. Quadfasal. 1954. Language laterality in lefthanded aphasics. Brain 77:521-548.

Goodwin, D. W. 1976. Is alcoholism hereditary? Oxford University Press, New York.

Goodwin, D. W., and S. B. Guze. 1974. Heredity and alcoholism. In B. Kissin and H. Begleiter, eds. The biology of alcoholism, vol. 3. Clinical pathology. Plenum Publishing Corp., New York.

Goodwin, D. W., F. Schulsinger, L. Hermansen, S. B. Guze, and G. Winokur. 1973. Alcohol problems in adoptees raised apart from alcoholic biological parents. Arch. Gen. Psychiatry 28:238-243.

Goodwin, D. W., F. Schulsinger, J. Knop, S. Mednick, and S. B. Guze. 1977a. Alcoholism and depression in adopted-out daughters of alcoholics. Arch. Gen. Psychiatry 34:751-755.

Goodwin, D. W., F. Schulsinger, J. Knop, S. Mednick, and S. B. Guze. 1977b. Psychopathology in adopted and nonadopted daughters of alcoholics. Arch. Gen. Psychiatry 34:1005-1009.

Gordon, K. 1919. Report on psychological tests of orphan children. J. Delinq. 4:45-55.

Gottesman, I. I. 1963. Heritability of personality: a demonstration. Psychol. Monogr. 77:1-21.

Gottesman, I. I. 1965. Personality and natural selection. In S. G. Vandenberg, ed. Methods and goals in human behavior genetics. Academic Press, Inc., New York.

Gottesman, I. I. 1966. Genetic variance in adaptive personality traits. J. Child. Psychol. Psychiatry 7: 199-208.

Gottesman, I. I., and L. L. Heston. 1972. Human behavioral adaptations: speculations on their genesis. In L. Ehrman, G. S. Omenn, and E. Caspari, eds. Genetics, environment, and behavior. Academic Press, Inc., New York.

Gottesman, I. I., and J. Shields. 1967. A polygenic theory of schizophrenia. Proc. Natl. Acad. Sci. U.S.A. 58:199-205.

Gottesman, I. I., and J. Shields. 1972. Schizophrenia and genetics: a twin vantage point. Academic Press, Inc., New York.

Gottesman, I. I., and J. Shields. 1976. A critical review of recent adoption, twin, and family studies of schizophrenia: behavioral genetics perspectives. Schizophrenia Bull. 2:360-398.

Gottlieb, J. S., M. C. Ashby, and J. R. Knott. 1947. Studies in primary behavior disorders and psychopathic personality. II. The inheritance of electrocortical activity. Am. J. Psychiatry 103:823-827.

Gottlober, A. B. 1938. The inheritance of brain potentials. J. Exp. Psychol. 22:193-200.

Gottschaldt, K. 1939. Erbpsychologie der elementar Funktionen der Begabung. In Handbuch der Erbbiologie des Menschen. Springer-Verlag, Berlin.

Goy, R. W., and J. S. Jakway. 1959. The inheritance of patterns in sexual behaviour in female guinea pigs. Anim. Behav. 7:142-149.

Goy, R. W., and W. C. Young. 1956-57. Strain differences in the behavioral responses of female guinea pigs to alpha-estradiol benzoate and progesterone. Behaviour 10:340-354.

Graves, E. A. 1936. Inter-relationships in performance in the albino rat. J. Comp. Psychol. 22:179-186.

Green, E. L. 1966. Breeding systems. In E. L. Green, ed. Biology of the laboratory mouse, 2nd ed. McGraw-Hill Book Co., New York.

Green, M. C. 1966. Mutant genes and linkages. In E. L. Green, ed. Biology of the laboratory mouse, 2nd ed. McGraw-Hill Book Co., New York.

Greenwood, M. R. C., D. Quartermain, P. R. Johnson, J. A. F. Cruce, and J. Hirsch. 1974. Food motivated behavior in genetically obese and hypothalamic-hyperphagic rats and mice. Physiol. Behav. 13:687-692.

Gregory, I. 1960. Genetic factors in schizophrenia. Am. J. Psychiatry 116:961-972.

Griffing, B. 1956. A generalized treatment of the use of diallel crosses in quantitative inheritance. Heredity 10:31-50.

Griffits, C. H. 1926. The influence of family on school marks. School Soc. **24**:713-716.

Gronwell, D. M. A., and H. Sampson. 1971. Ocular dominance: a test of two hypotheses. Br. J. Psychol. **62**:175-186.

Grüneberg, H. 1952. The genetics of the mouse, 2nd ed. Martinus Nijhoff, The Hague.

Grunt, J. A., and W. C. Young. 1952. Differential reactivity of individuals and the response of the male guinea pig to testosterone propionate. Endocrinology **51**:237-249.

Grunt, J. A., and W. C. Young. 1953. Consistency of sexual behavior patterns in individual male guinea pigs following castration and androgen therapy. J. Comp. Physiol. Psychol. **46**:138-144.

Guggenheim, K., and J. Mayer. 1952. Studies of pyruvate and acetate metabolism in the hereditary obesity-diabetes syndrome in mice. J. Biol. Chem. **198**:259-265.

Guhl, A. M., J. V. Craig, and C. D. Mueller. 1960. Selective breeding for aggressiveness in chickens. Poultry Sci. **39**:970-980.

Guhl, A. M., and D. G. Warren. 1946. Number of offspring sired by cockerels related to social dominance in chickens. Poultry Sci. **25**:460-472.

Guillery, R. W. 1974. Visual pathways in albinos. Sci. Am. **230**(5):44-54.

Guillery, R. W., C. S. Amorn, and B. B. Eighmy. 1971. Mutants with abnormal visual pathways: an explanation of anomalous geniculate laminae. Science **174**:831-832.

Guillery, R. W., G. L. Scott, B. M. Cattanach, and M. S. Deol. 1973. Genetic mechanisms determining the central visual pathways of mice. Science **179**:1014-1016.

Gun, W. T. J. 1928. Studies in hereditary ability. George Allen & Unwin Ltd., London.

Gun, W. T. J. 1930a. The heredity of the Stewarts. Eugen. Rev. **22**:195-201.

Gun, W. T. J. 1930b. The heredity of the Tudors. Eugen. Rev. **22**:111-116.

Guttman, R. 1974. Genetic analysis of analytical spatial ability: Raven's progressive matrices. Behav. Genet. **4**:273-284.

Guttman, R., I. Lieblich, and G. Naftali. 1969a. Variation in activity scores and sequences in two inbred strains of mice and their hybrids. Life Sci. **8**:893-899.

Guttman, R., I. Lieblich, and G. Naftali. 1969b. Variation in activity scores and sequences in two inbred strains, their hybrids and backcrosses. Anim. Behav. **17**:374-385.

Guze, S. B. 1967. The diagnosis of hysteria: what are we trying to do? Am. J. Psychiatry **124**:491-498.

Habel, H. 1950. Zwillingsuntersuchungen an Homosexuellen. Z. Sex. Forsch. **1**:161-180.

Hadler, N. M. 1964a. Genetic influence on phototaxis in *Drosophila melanogaster*. Biol. Bull. **126**:264-273.

Hadler, N. M. 1964b. Heritability and phototaxis in *Drosophila melanogaster*. Genetics **50**:1269-1277.

Haggard, E. A. 1958. Intraclass correlation and the analysis of variance. Dryden Press, New York.

Hagnell, O. 1966. A prospective study of the incidence of mental disorder. Svenska Boknorlaget Norstedts-Bonnier, Stockholm.

Hahn, M. E., S. B. Haber, and J. L. Fuller. 1973. Differential agonistic behavior in mice selected for brain weight. Physiol. Behav. **10**:759-762.

Hall, C. S. 1934. Emotional behavior in the rat. I. Defecation and urination as measures of individual differences in emotionality. J. Comp. Psychol. **18**:385-403.

Hall, C. S. 1938. The inheritance of emotionality. Sigma Xi Q. **26**:17-27.

Hall, C. S. 1941. Temperament: a survey of animal studies. Psychol. Bull. **38**:909-943.

Hall, C. S. 1947. Genetic differences in fatal audiogenic seizures between two inbred strains of house mice. J. Hered. **38**:2-6.

Hall, C. S. 1951. The genetics of behavior. In S. S. Stevens, ed. Handbook of experimental psychology. John Wiley & Sons, Inc., New York.

Hall, C. S., and S. J. Klein. 1942. Individual differences in aggressiveness in rats. J. Comp. Psychol. **33**:371-383.

Hallgren, B. 1957. Enuresis, a clinical and genetic study. Acta Psychiatr. Scand. [Suppl] **114**:1-159.

Hallgren, B., and T. Sjögren. 1959. A clinical and genetic-statistical study of schizophrenia and low-grade mental deficiency in a large Swedish rural population. Acta Psychiatr. Scand. [Suppl.] **140**.

Halperin, S. L., D. C. Rao, and N. E. Morton. 1975. A twin study of intelligence in Russia. Behav. Genet. **5**:83-86.

Halstead, W. C. 1947. Brain and intelligence. University of Chicago Press, Chicago.

Hamburg, D. A., and S. Kessler. 1967. A behavioural-endocrine-genetic approach to stress problems. In S. G. Spickett, ed. Endocrine genetics. University Press, Cambridge, England.

Hamerton, J. L., V. A. Cowie, F. Gianelli, S. M. Briggs, and P. E. Polani. 1961. Differential transmission of Down's syndrome (mongolism) through male and female translocation carriers. Lancet **2**:956-958.

Hamilton, W. D. 1964. The genetical theory of social behavior. I and II. J. Theor. Biol. **7**:1-52.

Hardy, G. H. 1908. Mendelian proportions in a mixed population. Science **28**:49-50.

Hardyck, C., and L. F. Petrinovitch. 1977. Left-handedness. Psychol. Bull. **84**:385-404.

Harnad, S. R., R. W. Doty, L. Goldstein, J. Jaynes, and G. Krauthamer. 1976. Lateralization in the nervous system. Academic Press, Inc., New York.

Harper. 1970. Cited in Annett, M. 1972. The distribution of manual symmetry. Br. J. Psychol. **63**:343-358.

Harrington, G. M., and J. R. Hanlon. 1966. Heart rate, defaecation and genetic differences in rats. Psychon. Sci. **6**:425-426.

Harris, H., and H. Kalmus. 1949. The measurement of taste sensitivity to phenylthiourea (PTC). Ann. Eugen. Lond. **15**:24-31.

Harris, V. T. 1952. An experimental study of habitat selection by prairie and forest races of the deermouse, *Peromyscus maniculatus*. Contrib. Lab. Vert. Biol. Univ. Mich. **56**:1-53.

Harshman, R., and S. Krashen. 1972. An "unbiased" procedure for comparing degree of lateralization of dichotically presented stimuli. U.C.L.A. Working Papers Phon. **23**:3-12.

Hart, H. 1924. Correlations between intelligence quotients of siblings. School Soc. **20**:382.

Harvald, B., and M. Hauge. 1965. Hereditary factors elucidated by twin studies. In J. V. Neel, M. W. Shaw and W. J. Schull, eds. Genetics and the epidemiology of chronic disease. U.S. Department of Health, Education, and Welfare, U.S. Government Printing Office, Washington, D.C.

Haseman, J. K., and R. C. Elston. 1970. The estimation of genetic variance from twin data. Behav. Genet. **1**:11-19.

Hathaway, S. R., and P. E. Meehl. 1951. An atlas for the clinical use of the MMPI. University of Minnesota Press, Minneapolis.

Hausberger, F. X. 1958. Parabiosis and transplantation experiments in hereditary obese mice. Anat. Rec. **130**:313. (Abstr.)

Hawkins, J. D. 1970. Single gene substitutions and behavior. In G. Lindzey and D. D. Thiessen, eds. Contributions to behavior-genetic analysis: the mouse as a prototype. Appleton-Century-Crofts, New York.

Hayashi, S. 1967. A study of juvenile delinquency in twins. In H. Mitsuda, ed. Clinical genetics in psychiatry. Igaku Shoin Ltd., Tokyo.

Hayman, B. I. 1954. The theory and analysis of diallel crosses. Genetics **39**:789-809.

Hayman, B. I. 1958. The theory and analysis of diallel crosses. II. Genetics **43**:63-85.

Hays, F. A. 1933. Characteristics of non-broody and intense broody lines of Rhode Island Reds. Mass. Agri. Exp. Stat. Bull. No. 311.

Hays, F. A. 1940. Inheritance of broodiness in Rhode Island Reds. Mass. Agri. Exp. Stat. Bull. No. 377.

Hays, W. L. 1963. Statistics for psychologists. Holt, Rinehart & Winston, Inc., New York.

Hebb, D. O., and K. Williams. 1946. A method of rating animal intelligence. J. Gen. Psychol. **34**:59-65.

Hecaén, H. 1969. Aphasic, apraxic and agnostic syndrome in right and left hemisphere lesions. In P. J. Vinken and G. W. Bruyn, eds. Handbook of clinical neurology: disorders of speech, perception and symbolic behavior, vol. 4. American Elsevier Publishing Co., Inc., New York.

Hecaén, H., and J. de Ajuriagheria. 1964. Left-handedness: manual superiority and cerebral dominance. Grune & Stratton, Inc., New York.

Hegmann, J. P. 1972. Physiological function and behavioral genetics. I. Genetic variance for peripheral conductance in mice. Behav. Genet. **2**:55-67.

Hegmann, J. P., J. E. White, and S. B. Kater. 1973. Physiological function and behavioral genetics. II. Quantitative genetic analysis of conduction velocity of caudal nerves of the mouse, *Mus musculus*. Behav. Genet. **3**:121-131.

Heinze, W. J. 1974. A genetic analysis of escape behavior in rats. Behav. Genet. **4**:125-131.

Helgason, T. 1964. The epidemiology of mental disorders in Iceland. Acta Psychiatr. Neurol. Scand. [Suppl.] **173**.

Henderson, N. D. 1967. Prior treatment effects on open field behaviour of mice: a genetic analysis. Anim. Behav. **15**:364-376.

Henderson, N. D. 1968. The confounding effects of genetic variables in early experience research: can we ignore them? Dev. Psychobiol. **1**:146-152.

Henderson, N. D. 1970. Genetic influences on the behavior of mice can be obscured by laboratory rearing. J. Comp. Physiol. Psychol. **72**:505-511.

Henderson, N. D. 1972. Relative effects of early rearing environment on discrimination learning in housemice. J. Comp. Physiol. Psychol. **79**:243-253.

Henderson, N. D. 1977. Genetics of locomotor behavior in infant mice. Behav. Genet. **7**:69. (Abstr.)

Henry, K. R. 1967. Audiogenic seizure susceptibility induced in C57BL/6J mice by prior auditory exposure. Science **158**:938-940.

Henry, K. R., and R. E. Bowman. 1970. Behavior-genetic analysis of the ontogeny of accoustically primed audiogenic seizures in mice. J. Comp. Physiol. Psychol. **70**:235-241.

Henry, K. R., and M. M. Haythorn. 1975. Albinism and auditory function in the laboratory mouse. I. Effects of single gene substitutions on auditory physiology, audiogenic seizures and developmental processes. Behav. Genet. **5**:137-149.

Henry, K. R., and M. Saleh. 1973. Recruitment deafness: functional effect of priming-induced audiogenic seizures in mice. J. Comp. Physiol. Psychol. **84**:430-435.

Henry, K. R., and K. Schlesinger. 1967. Effects of the albino and dilute loci on mouse behavior. J. Comp. Physiol. Psychol. **63**:320-323.

Henry, K. R., M. Wallick, and M. Davis. 1972. Inferior colliculus lesions: effects on audiogenic seizure and Preyer reflex. Physiol. Behav. **9**:885-887.

Hermann, E. 1939. Messungen an Handschriftproblem von Zwillingspaaren unter 14 Jahren. Z. Psychol. **147**:238-255.

Hermann, L., and L. Hogben. 1933. The intellectual resemblance of twins. Proc. R. Soc. Edinb. **53**:105-129.

Heron, W. T. 1935. The inheritance of maze learning ability in rats. J. Comp. Psychol. **19**:77-89.

Heron, W. T. 1941. The inheritance of brightness and dullness in maze learning ability in the rat. J. Genet. Psychol. **59**:41-49.

Heron, W. T., and B. F. Skinner. 1940. The rate of extinction in maze-bright and maze-dull rats. Psychol. Rec. **4**:11-18.

Heron, W. T., and S. Yugend. 1936. Basal metabolism and maze learning in rats. J. Genet. Psychol. **48**:471-474.

Herrnstein, R. J. 1973. IQ in the meritocracy. Little, Brown & Co., Boston.

Heston, L. L. 1966. Psychiatric disorders in foster home reared children of schizophrenic mothers. Br. J. Psychiatry 112:819-825.

Heston, L. L. 1970. The genetics of schizophrenic and schizoid disease. Science 167:249-256.

Heston, L. L., and J. Shields. 1968. Homosexuality in twins: a family study and a registry study. Arch. Gen. Psychiatry 18:149-160.

Heston, W. D. W., S. A. Anderson, V. G. Erwin, and G. E. McClearn. 1973. A comparison of the actions of various hypnotics on mice selectively bred for sensitivity to ethanol. Behav. Genet. 3:402-403. (Abstr.)

Heston, W. D. W., V. G. Erwin, S. M. Anderson, and H. Robbins. 1974. A comparison of the effects of alcohol on mice selectively bred for differences in ethanol sleep time. Life Sci. 14:365-370.

Heston, W. E. 1949. Development of inbred strains in the mouse and their use in cancer research. R. B. Jackson Memorial Laboratory's Twentieth Anniversary Lecture, Bar Harbor, Maine.

Higgins, J. V., E. W. Reed, and S. C. Reed. 1962. Intelligence and family size: a paradox resolved. Eugen. Q. 9:84-90.

Hildreth, G. H. 1925. The resemblance of siblings in intelligence and achievement. Teachers Coll. Columbia Univ. Contrib. Educ. 186:1-65.

Hildreth, P. E. 1962. Quantitative aspects of mating behavior in *Drosophila.* Behaviour 19:57-73.

Hildreth, P. E., and G. C. Becker. 1962. Genetic influences on mating behavior in *Drosophila melanogaster.* Behaviour 19:219-238.

Hill, M. S. 1973. Hereditary influence on the normal personality using the MMPI. II. Prospective assortative mating. Behav. Genet. 3:225-232.

Hill, M. S., and R. N. Hill. 1973. Hereditary influence on the normal personality using the MMPI. I. Age-corrected parent-offspring resemblances. Behav. Genet. 3:133-144.

Hinde, R. A., and J. Stevenson-Hinde. 1973. Constraints on learning. Academic Press, Inc., London.

Hirsch, J. 1959. Studies in experimental behavior genetics. II. Individual differences in geotaxis as a function of chromosome variations in synthesized *Drosophila* populations. J. Comp. Physiol. Psychol. 52:304-308.

Hirsch, J., ed. 1967a. Behavior-genetic analysis. McGraw-Hill Book Co., New York.

Hirsch, J. 1967b. Behavior-genetic analysis at the chromosome level of organization. In J. Hirsch, ed. Behavior-genetic analysis. McGraw-Hill Book Co., New York.

Hirsch, J., and J. C. Boudreau. 1958. The heritability of phototaxis in a population of *Drosophila melanogaster.* J. Comp. Physiol. Psychol. 51:647-651.

Hirsch, J., and L. Erlenmeyer-Kimling. 1962. Studies in experimental behavior genetics. IV. Chromosome analysis for geotaxis. J. Comp. Physiol. Psychol. 55:732-739.

Hirsch, J., and R. C. Tryon. 1956. Mass screening and reliable individual measurement in the experimental behavior genetics of lower organisms. Psychol. Bull. 53:402-410.

Hoenigsberg, H. F., and S. K. Santibanez. 1960a. Courtship and sensory preference in inbred lines of *Drosophila melanogaster.* Evolution 14:1-7.

Hoenigsberg, H. F., and S. K. Santibanez. 1960b. Observations on the sexual behavior of *Drosophila equinoxialis* and *Drosophila prosaltans.* Am. Natur. 44:382-384.

Hoffeditz, E. L. 1934. Family resemblances in personality traits. J. Soc. Psychol. 5:214-227.

Hoffman, H. 1921. Die Nachkommenschaft bei endogenen Psychosen. Springer-Verlag, Berlin.

Hogben, L. 1933. The limits of applicability of correlation technique in human genetics. J. Genet. 27:379-406.

Holland, H. C., and B. D. Gupta. 1966. Some correlated measures of activity and reactivity in two strains of rats selectively bred for differences in the acquisition of a conditioned avoidance response. Anim. Behav. 14:574-580.

Holland, J. L. 1958. A personality inventory employing occupational titles. J. Appl. Psychol. 42:336-342.

Hollingshead, A. B., and F. C. Redlich. 1957. Social class and mental illness. John Wiley & Sons, Inc., New York.

Holzinger, K. J. 1929. The relative effect of nature and nurture influences on twin differences. J. Educ. Psychol. 20:241-248.

Honzik, M. 1957. Developmental studies of parent-child resemblance in intelligence. Child. Dev. 28:215-228.

Hook, E. B. 1973. Behavioral implications of the human XYY genotype. Science 179:139-150.

Hopkinson, G., and P. Ley. 1969. A genetic study of affective disorder. Br. J. Psychiatry 115:917-922.

Horel, J. A. 1973. The brain and behavior in phylogenetic perspective. In D. A. Dewsbury and D. A. Rethlingshafer, eds. Comparative psychology: a modern survey. McGraw-Hill Book Co., New York.

Horn, J. M. 1974. Aggression as a component of relative fitness in four inbred strains of mice. Behav. Genet. 4:373-382.

Horn, J. M., R. Plomin, and R. Rosenman. 1976. Heritability of personality traits in adult male twins. Behav. Genet 6:17-30.

Hosgood, S. M. W., and P. A. Parsons. 1965. Mating speed differences between Australian populations of *Drosophila melanogaster.* Genetica 36:260-266.

Hoshishima, K., S. Yokoyama, and K. Seto. 1962. Taste sensitivity in various strains of mice. Am. J. Physiol. 202:1200-1204.

Hotta, Y., and S. Benzer. 1970. Genetic dissection of the *Drosophila* nervous system by means of mosaics. Proc. Natl. Acad. Sci. U.S.A. 67:1156-1163.

Hotta, Y., and S. Benzer. 1972. The mapping of behavior in *Drosophila* mosaics. Nature 240:527-535.

Howard, R. G., and Brown, A. M. 1970. Twinning: a marker for biological insults. Child. Dev. **41**:519-530.

Howells, T. H. 1946. The hereditary differential in learning—a reply to F. A. Pattie. Psychol. Rev. **53**: 302-305.

Hoy, R. R., J. Hahn, and R. C. Paul. 1977. Hybrid cricket auditory behavior: evidence for genetic coupling in auditory communication. Science **195**:82-83.

Hsia, D. Y.-Y. 1967. The hereditary metabolic diseases. In J. Hirsch, ed. Behavior-genetic analysis. McGraw-Hill Book Co., New York.

Hsia, D. Y.-Y. 1970. Phenylketonuria and its variants. In A. G. Steinberg and A. G. Bearn, eds. Progress in medical genetics, vol. 7. Grune & Stratton, Inc., New York.

Hsia, D. Y.-Y., K. W. Driscoll, W. Troll, and W. E. Knox. 1956. Detection by phenylalanine tolerance tests of heterozygous carriers of phenylketonuria. Nature **178**:1239-1240.

Hsia, D. Y.-Y., and F. A. Walker. 1961. Variability in the clinical manifestations of galactosemia. J. Pediatr. **59**:872-883.

Hubbard, R., and A. Kropf. 1967. Molecular isomers in vision. Sci. Am. **216**(6):64-76.

Hudson, P. T. 1975. The genetics of handedness: a reply to Levy and Nagylaki. Neuropsychologia **13**:331-339.

Huestis, R. R., and T. P. Otto. 1927. The grades of related students. J. Hered. **18**:225-226.

Huizinga, J. 1956. The waning of the Middle Ages. Anchor Books, New York.

Hull, C. L. 1943. Principles of behavior. Appleton-Century-Crofts, New York.

Hume, W. I. 1973. Physiological measures in twins. In G. Claridge, S. Canter, and W. I. Hume, eds. Personality differences and biological variations: a study of twins. Pergamon Press Ltd., Oxford.

Hummel, K. P. 1960. Pituitary lesions in mice of the Marsh strain. Anat. Rec. **137**:336. (Abstr.)

Humphreys, L. G. 1971. Theory of intelligence. In R. Cancro, ed. Intelligence: genetic and environmental influences. Grune & Stratton, Inc., New York.

Hunt, J. M. 1961. Intelligence and experience. The Ronald Press Co., New York.

Hunt, J. M., ed. 1972. Human intelligence. Transaction Books, New Brunswick, N.J.

Huntley, R. M. C. 1966. Heritability of intelligence. In J. E. Meade and A. S. Parkes, eds. Genetic and environmental factors in human ability. Oliver & Boyd, Edinburgh.

Hurst, C. C. 1912. Mendelian inheritance in man. Eugen. Rev. **4**:20-24.

Hurst, C. C. 1932. A genetic formula for the inheritance of intelligence in man. Proc. R. Soc. Lond. [Biol.] **112**:80-97.

Hurst, C. C. 1934. The genetics of intellect. Eugen. Rev. **26**:33-45.

Hurst, L. A. 1972. Hypothesis of a single-locus recessive genotype for schizophrenia. In A. R. Kaplan, ed. Genetic factors in "schizophrenia." Charles C Thomas, Publisher, Springfield, Ill.

Husén, T. 1953. Tvillingstudien. Almqvist & Wiksell Förlag AB, Stockholm.

Husén, T. 1960. Abilities of twins. Scand. J. Psychol. **1**:125-135.

Husén, T. 1963. Intrapair similarities in the school achievements of twins. Scand. J. Psychol. **4**:108-114.

Hyde, J. S. 1973. Genetic homeostasis and behavior: analysis, data, and theory. Behav. Genet. **3**:233-245.

Hyde, J. S. 1974. Inheritance of learning ability in mice: a diallel-environmental analysis. J. Comp. Physiol. Psychol. **86**:116-123.

Hyde, J. S., and P. D. Ebert. 1976. Correlated response in selection for aggressiveness in female mice. I. Male aggressiveness. Behav. Genet. **6**:421-427.

Hydén, H. 1970. The question of a molecular basis of the memory trace. In K. H. Pribram and D. E. Broadbent, eds. Biology of memory. Academic Press, Inc., New York.

Idha, S. 1961. A study of neurosis by the twin method. Psychiatr. Neurol. Jpn. **63**:861-892.

Ingalls, A. M., M. M. Dickie, and G. D. Snell. 1950. Obese, a new mutation in the house mouse. J. Hered. **41**:317-318.

Ingram, V. M. 1957. Gene mutations in normal haemoglobin: the chemical difference between normal and sickle cell haemoglobin. Nature **180**:326-328.

Inouye, E. 1957. Frequency of multiple births in three cities in Japan. Am. J. Hum. Genet. **9**:317-320.

Inouye, E. 1961. Similarity and dissimilarity of schizophrenia in twins. Proceedings of the Third World Congress on Psychiatry, Montreal. University of Toronto Press, Toronto.

Inouye, E. 1965. Similar and dissimilar manifestations of obsessive-compulsive neuroses in monozygotic twins. Am. J. Psychiatry **121**:1171-1175.

Inouye, E. 1972. Monozygotic twins with schizophrenia reared apart in infancy. Jpn. J. Hum. Genet. **16**:182-190.

Insel, P. 1974. Maternal effects in personality. Behav. Genet. **4**:133-144.

Isselbacher, K. J. 1957. Evidence for an accessory pathway of galactose metabolism in mammalian liver. Science **126**:652-654.

Iturrian, W. B., and G. B. Fink. 1968. Influence of age and brief auditory conditioning upon experimental seizures in mice. Dev. Psychobiol. **2**:12-18.

Jackson, D. D. 1960. A critique of the literature on the genetics of schizophrenia. In D. D. Jackson, ed. The etiology of schizophrenia. Basic Books, Inc., Publishers, New York.

Jacob, F., and J. Monod. 1961a. Genetic regulatory mechanisms in the synthesis of proteins. J. Mol. Biol. **3**:318-356.

Jacob, F., and J. Monod. 1961b. On the regulation of gene activity. Cold Spring Harbor Symp. Quant. Biol. **26**:193-209.

Jacobs, P. A., M. Brunton, M. M. Melville, R. P. Brittain, and W. F. McClemont. 1965. Aggressive behavior, mental subnormality and the XYY male. Nature **208**:1315-1352.

Jakway, J. S. 1959. The inheritance of patterns of mating behaviour in the male guinea pig. Anim. Behav. 7:150-162.

James, J. 1971. Incidence of an attribute in relatives of individuals with the trait. Ann. Hum. Genet. 35:47-49.

James, W. 1971. Sex ratios of half-sibs of male homosexuals. Br. J. Psychiatry 118:93-94.

Jarvik, L. F., J. E. Blum, and A. O. Varma. 1972. Genetic components and intellectual functioning during senescence: a 20-year study of aging twins. Behav. Genet. 2:159-171.

Jarvik, L. F., V. Klodin, and S. S. Matsuyama. 1973. Human aggression and the extra Y chromosome. Am. Psychol. 28:674-682.

Jellinek, E. M. 1960. The disease concept of alcoholism. Hillhouse Press, New Haven, Conn.

Jencks, C., et al. 1972. Inequality: a reassessment of the effect of family and schooling in America. Basic Books, Inc., Publishers, New York.

Jensch, K. 1941. Zur Genealogie der Homosexualität. Arch. Psychiatr. Nervenkr. 112:679-696.

Jensen, A. R. 1969. How much can we boost IQ and scholastic achievement? Harvard Educ. Rev. 39:1-123.

Jensen, A. R. 1970. IQ's of identical twins reared apart. Behav. Genet. 1:133-148.

Jensen, A. R. 1972. Genetics and education. Harper & Row, Publishers, Inc., New York.

Jensen, A. R. 1973a. Educability and group differences. Methuen & Co. Ltd., London.

Jensen, A. R. 1973b. Let's understand Skodak and Skeels finally. Educ. Psychol. 10:30-35.

Jensen, A. R. 1974. Kinship correlations reported by Sir Cyril Burt. Behav. Genet. 4:1-28.

Jensen, C. 1977. Generality of learning differences in brain-weight–selected mice. J. Comp. Physiol. Psychol. 91:629-641.

Jensen, C., and J. L. Fuller. 1978. Learning performance varies with brain weight in heterogeneous mouse lines. J. Comp. Physiol. Psychol. (In press.)

Jervis, G. A. 1954. Phenylpyruvic oligophrenia (phenylketonuria). Proc. Assoc. Nerv. Ment. Dis. 33:259-282.

Jinks, J. L. 1954. The analysis of continuous variation in a diallel cross of *Nicotiana rustica* varieties. I. The analysis of F_1 data. Genetics 39:767-788.

Jinks, J. L., and L. J. Eaves. 1974. IQ and inequality. Nature 248:287-289.

Jinks, J. L., and D. W. Fulker. 1970. Comparison of the biometrical genetical, MAVA, and classical approaches to the analysis of human behavior. Psychol. Bull. 73:311-349.

Joffe, J. M. 1964. Avoidance learning and failure to learn in two strains of rats selectively bred for emotionality. Psychon. Sci. 1:185-186.

Joffe, J. M. 1965. Genotype and prenatal and premating stress interact to affect adult behavior in rats. Science 150:1844-1845.

Joffe, J. M. 1969. Prenatal determinants of behaviour. Pergamon Press, Inc., New York.

Johnson, R. C. 1977. Degree of meritocracy and the association among abilities, educational attainment, and occupational status. Behav. Genet. 7:70. (Abstr.)

Johnson, R. C., and R. B. Abelson. 1969. Intellectual, behavioral, and physical characteristics associated with trisomy, translocation and mosaic types of Down's syndrome. Am. J. Ment. Defic. 73:852-855.

Johnson, W. H., R. A. Laubengayer, L. E. Delaney, and T. A. Cole. 1972. Biology, 4th ed. Holt, Rinehart & Winston, Inc., New York.

Jolly, A. 1966. Lemur social behavior and primate intelligence. Science 153:501-506.

Jones, H. E. 1928. A first study of parent-child resemblances in intelligence. In Twenty-seventh Yearbook of the National Society for the Study of Education. Public School Publishing Co., Bloomington, Ill.

Jones, H. E. 1954. Environmental influences on mental development. In L. Carmichael, ed. Manual of child psychology, 2nd ed. John Wiley & Sons, Inc., New York.

Jones, H. E. 1955. Perceived differences among twins. Eugen. Q. 2:98-102.

Jones, M. B. 1966. On the evaluation of variability in isogenic hybrids. Biometrics 22:623-628.

Jost, H., and L. W. Sontag. 1944. The genetic factor in autonomic nervous system function. Psychosom. Med. 6:308-310.

Juda, A. 1949. The relationship between highest mental capacity and psychic abnormality. Am. J. Psychiatry 106:296-307.

Juel-Nielsen, N. 1965. Individual and environment: a psychiatric-psychological investigation of monozygous twins reared apart. Acta. Psychiatr. Neurol. Scand. Monogr. [Suppl.]183.

Juel-Nielsen, N., and B. Harvald. 1958. The electroencephalogram in uniovular twins brought up apart. Acta Genet. 8:57-64.

Kaij, L. 1960. Alcoholism in twins: studies on the etiology and sequels of abuse of alcohol. Almquist & Wicksell Förlag AB, Stockholm.

Kakihana, R. Y. 1965. Developmental study of preference for and tolerance to ethanol in inbred strains of mice. Doctoral dissertation. University of California, Berkeley.

Kakihana, R. Y., D. R. Brown, G. E. McClearn, and I. R. Tabershaw. 1966. Brain sensitivity to alcohol in inbred mouse strains. Science 154:1574-1575.

Kallman, F. J. 1938. The genetics of schizophrenia. J. J. Augustin, Inc.—Publisher, Locust Valley, N.Y.

Kallman, F. J. 1946. The genetic theory of schizophrenia. Am. J. Psychiatry 103:309-322.

Kallman, F. J. 1952a. Genetic aspects of psychoses. In The biology of mental health and disease: Twenty-seventh Annual Conference, Millbrook Memorial Fund. Paul B. Hoeber, Inc., New York.

Kallman, F. J. 1952b. Twin and sibship study on the genetic aspects of male homosexuality. Am. J. Hum. Genet. 4:136-146.

Kallman, F. J. 1953. Heredity in health and mental disorder. W. W. Norton & Co., Inc., New York.

Kallman, F. J. 1954. Genetic principles in manic-depressive psychosis. In P. H. Hoch and J. Zubin, eds. Depression. Grune & Stratton, Inc., New York.

Kallman, F. J. 1959. The genetics of mental illness. In S. Areti, ed. American handbook of psychiatry, vol. 1. Basic Books, Publishers, Inc., New York.

Kallman, F. J., and S. E. Barrera. 1942. The heredoconstitutional mechanism of predisposition and resistance to schizophrenia. Am. J. Psychiatry 98:544-550.

Kallman, F. J., and B. Roth. 1956. Genetic aspects of preadolescent schizophrenia. Am. J. Psychiatry 112: 599-606.

Kalmus, H. 1949. Tune deafness and its inheritance. Proceedings of the Tenth International Congress on Genetics, Stockholm.

Kalmus, H. 1952. Inherited sense defects. Sci. Am. 186: 64-70.

Kalmus, H. 1955. The familial distribution of congenital tritanopia, with some remarks on some similar conditions. Ann. Hum. Genet. 20:33-56.

Kalmus, H. 1965. Diagnosis and genetics of defective colour vision. Pergamon Press Ltd., London.

Kamin, L. J. 1974. The science and politics of I.Q. John Wiley & Sons, Inc., New York.

Kamin, L. J. 1978. Comment on Munsinger's review of adoption studies. Psychol. Bull. 85:194-201.

Kaplan, A. R. 1968. Physiological and pathological correlates of differences in taste acuity. In S. G. Vandenberg, ed. Progress in human behavior genetics. Johns Hopkins University Press, Baltimore.

Kaplan, A. R. ed. 1972. Genetic factors in "schizophrenia." Charles C Thomas, Publisher, Springfield, Ill.

Kaplan, W. D., and W. E. Trout, III. 1969. The behavior of four neurological mutants of *Drosophila*. Genetics 61:399-409.

Karlsson, J. L. 1966. The biological basis of schizophrenia. Charles C Thomas, Publisher, Springfield, Ill.

Karlsson, J. L. 1972. A two-locus hypothesis for inheritance of schizophrenia. In A. R. Kaplan, ed. Genetic factors in "schizophrenia." Charles C Thomas, Publisher, Springfield, Ill.

Katz, J., and W. C. Halstead. 1950. Protein organization and mental functions. Comp. Psychol. Monogr. 20:1-38.

Katzev, R. D., and S. K. Mills. 1974. Strain differences in avoidance conditioning as a function of the classical CS-US contingency. J. Comp. Physiol. Psychol. 87: 661-671.

Kaufman, L. 1948. On the mode of inheritance of broodiness. Eighth World Poultry Congress, Copenhagen. Official Rep. 1:301-304.

Kay, D. W. K., and R. Lindelius. 1970. A study of schizophrenia. Acta Psychiatr. Scand. [Suppl.] 216.

Kekić, V., and D. Marinković. 1974. Multiple choice selection for light preference in *Drosophila subobscura*. Behav. Genet. 4:285-300.

Kempthorne, O. 1978. Logical, epistemological and statistical aspects of nature-nurture data interpretation. Biometrics 34:1-23.

Kempthorne, O., and R. H. Osborne. 1961. The interpretation of twin data. Am. J. Hum. Genet. 13:320-339.

Kennard, M. A. 1949. Inheritance of electroencephalogram patterns in children with behavior disorders. Psychosom. Med. 11:151-157.

Kent, E. 1949. A study of maladjusted twins. Smith Coll. Stud. Soc. Work 19:63-77.

Kerbusch, J. M. L. 1974. A diallel study of exploratory behaviour and learning performances in mice. In J. H. F. van Abeelen, ed. The genetics of behaviour. North-Holland Publishing Co., Amsterdam.

Kessler, P., and J. M. Neale. 1974. Hippocampal damage and schizophrenia: a critique of Mednick's theory. J. Abnorm. Psychol. 83:91-96.

Kessler, S. 1968. The genetics of *Drosophila* mating behavior. I. Organization of mating speed in *Drosophila pseudoobscura*. Anim. Behav. 16:485-491.

Kessler, S. 1969. The genetics of *Drosophila* mating behavior. II. The genetic architecture of mating speed in *Drosophila pseudoobscura*. Genetics 62:421-433.

Kessler, S., R. D. Ciranello, J. G. M. Shire, and J. D. Barchas. 1972. Genetic variation in catecholamine-synthesizing enzyme activity. Proc. Natl. Acad. Sci. U.S.A. 69:2448-2450.

Kessler, S., P. Harmatz, and S. A. Gerling. 1975. The genetics of pheromonally mediated aggression in mice. I. Strain differences in the capacity of male urinary odors to elicit aggression. Behav. Genet. 5:233-238.

Kessler, S., and R. H. Moos. 1970. The XYY karyotype and criminality: a review. J. Psychiatr. Res. 7:153-170.

Kettlewell, H. B. D. 1959. Darwin's missing evidence. Sci. Am. 200:48-53.

Kettlewell, H. B. D. 1965. Insect survival and selection for pattern. Science 148:1290-1296.

Kety, S. S. 1959. Biochemical theories of schizophrenia. Science 129:1528-1532, 1590-1596.

Kety, S. S., D. Rosenthal, P. H. Wender, and F. Schulsinger. 1971. Mental illness in the biological and adoptive families of adopted schizophrenics. Am. J. Psychiatry 128:302-306.

Kidd, K. K., and L. L. Cavalli-Sforza. 1973. An analysis of the genetics of schizophrenia. Soc. Biol. 20:254-265.

Kimura, D. 1967. Functional asymmetry of the brain in dichotic listening. Cortex 3:163-173.

King, J. A. 1957. Relationships between early social experience and adult aggressive behavior in inbred mice. J. Genet. Psychol. 90:151-166.

Kinsbourne, M., and W. L. Smith, eds. 1971. Hemispheric disconnection and cerebral function. Charles C Thomas, Publisher, Springfield, Ill.

Kinsey, A. C., W. Pomeroy, and C. E. Martin. 1948. Sexual behavior in the human male. W. B. Saunders Co., Philadelphia.

Kirkpatrick, C. 1936. A comparison of generations in regard to attitudes toward feminism. J. Genet. Psychol. 49:343-361.

Kirkpatrick, C., and S. Stone. 1935. Attitude measurement and the comparison of generations. J. Appl. Psychol. **19**:564-582.

Klaiber, E. L., D. M. Braverman, W. Vogel, G. E. Abraham, and F. L. Cone. 1972. Effects of infused testosterone on mental performance and serum LH. J. Clin. Endocrinol. Metab. **32**:341-350.

Klein, J., and D. W. Bailey. 1971. Histocompatibility differences in wild mice: further evidence for the existence of deme structure in natural populations of the house mouse. Genetics **68**:287-297.

Klein, T. W., and J. C. DeFries. 1970a. Similar polymorphism of taste sensitivity to PTC in mice and men. Nature **225**:555-557.

Klein, T. W., and J. C. DeFries. 1970b. Taste sensitivity in infrahuman species: use of a genetic model to test the validity of alternative measures. Behav. Res. Instr. **2**:106-107.

Klein, T. W., and J. C. DeFries. 1973. Racial and cultural differences in sensitivity to flickering light. Soc. Biol. **20**:212-218.

Klein, T. W., J. C. DeFries, and C. T. Finkbeiner. 1973. Heritability and genetic correlation: standard errors of estimate and sample size. Behav. Genet. **3**:355-364.

Knott, J. R., E. B. Platt, A. M. Coulson, and J. S. Gottlieb. 1953. A familial evaluation of the electroencephalogram of patients with primary behavior disorder and psychopathic personality. Electroencephalogr. Clin. Neurophysiol. **5**:193-199.

Knox, C. W., and M. W. Olsen. 1938. A test of crossbred chickens: single comb white leghorns and Rhode Island reds. Poultry Sci. **17**:193-199.

Koch, H. L. 1966. Twins and twin relations. University of Chicago Press, Chicago.

Kohn, M. L. 1968. Social class and schizophrenia: a critical review. In D. Rosenthal and S. S. Kety, eds. The transmission of schizophrenia. Pergamon Press, Inc., New York.

Kohn, M. L. 1973. Social class and schizophrenia: a critical review. Schizophrenia Bull. **7**:60-79.

Komai, T., J. V. Craig, and S. Wearden. 1959. Heritability and repeatability of social aggressiveness in the domestic chicken. Poultry Sci. **38**:356-359.

Kraepelin, E. 1921. Manic-depressive insanity and paranoia. E. & S. Livingstone Ltd., Edinburgh.

Kramer, E., and C. E. Lauterbach. 1928. Resemblances in the handwriting of twins and siblings. J. Educ. Res. **18**:149-152.

Kranz, H. 1936. Lebensschicksale kriminellen Zwillinge. Springer-Verlag, Berlin.

Kranz, H. 1937. Untersuchungen an Zwillingen in Fursorgeerziehunganstalten. Z. Induct. Abstam. Vererb. **73**:508-512.

Krechevsky, I. 1933. The hereditary nature of "hypotheses." J. Comp. Psychol. **16**:99-116.

Kringlen, E. 1964. Schizophrenia in male MZ twins. Acta Psychiatr. Scand. [Suppl.]178, **40**:1-76.

Kringlen, E. 1967. Heredity and environment in the functional psychoses: case histories. William Heinemann Ltd., London.

Kringlen, E. 1968. An epidemiological—clinical twin study on schizophrenia. In D. Rosenthal and S. S. Kety, eds. The transmission of schizophrenia. Pergamon Press, Inc. New York.

Kruse, M. 1941. Food satiation curves for maze-bright and maze-dull rats. J. Comp. Psychol. **31**:13-21.

Kryshova, N. A., Z. V. Beliaeva, A. F. Dmitrieva, M. A. Zhilinskiaa, and L. G. Perbov. 1962. Investigation of the higher nervous activity and of certain vegetative features in twins. Sov. Psychol. Psychiatr. **1**:31-41.

Kulp, D. H., and H. H. Davidson. 1933. Sibling resemblance in social attitudes. J. Educ. Sociol. **7**:133-140.

Kung, C. 1971a. Genic mutants with altered system of excitation. I. Phenotypes of the behavioral mutants. Z. Verg. Physiol. **71**:142-164.

Kung, C. 1971b. Genic mutants with altered system of excitation in *Paramecium aurelia*. II. Mutagenesis, screening and genetic analysis of the mutants. Genetics **69**:29-45.

Kuo, Z. Y. 1929. The net result of the anti-heredity movement in psychology. Psychol. Rev. **36**:181-199.

Kuppusawny, B. 1947. Laws of heredity in relation to general mental ability. J. Gen. Psychol. **36**:29-43.

Kutscher, C. L. 1974. Strain differences in drinking in inbred mice during ad libitum feeding and food deprivation. Physiol. Behav. **13**:63-70.

Kutscher, C. L., and D. G. Miller. 1974. Age-dependent polydipsia in the SWR/J mouse. Physiol. Behav. **13**:71-79.

Kwalwasser, J. 1955. Exploring the musical mind. Coleman-Ross Co., Inc. New York.

Lacey, J. I., and B. C. Lacey. 1962. The law of initial value in the longitudinal study of autonomic constitution: reproducibility of autonomic response and response patterns over a four year interval. Ann. N.Y. Acad. Sci. **98**:1257-1290.

Lader, M. H., and L. Wing. 1966. Physiological measures, sedative drugs and morbid anxiety. Oxford University Press, London.

Lagerspetz, K. M. J. 1961. Genetic and social causes of aggressive behaviour in mice. Scand. J. Psychol. **2**:167-173.

Lagerspetz, K. M. J. 1964. Studies on the aggressive behaviour of mice. Ann. Acad. Sci. Fenn. **131**:1-13.

Lagerspetz, K. M. J., and K. Y. H. Lagerspetz. 1971. Changes in the aggressiveness of mice resulting from selective breeding, learning and social isolation. Scand. J. Psychol. **12**:241-248.

Lagerspetz, K. M. J., and K. Y. H. Lagerspetz. 1974. Genetic determination of aggressive behaviour. In J. H. F. van Abeelen, ed. The genetics of behaviour. Elsevier North-Holland Inc., New York.

Lagerspetz, K. M. J., and K. Y. H. Lagerspetz. 1975. The expression of the genes of aggressiveness in mice: the effect of androgen on aggression and sexual behavior in females. Aggress. Behav. **1**:291-296.

Lagerspetz, K. M. J., and K. Wuorinen. 1965. A cross-fostering experiment with mice selectively bred for aggressiveness and non-aggressiveness. Rep. Inst. Psychol. Univ. Turku **17**:1-6.

Lagerspetz, K. Y. H., R. Tirri, and K. M. J. Lagerspetz. 1968. Neurochemical and edocrinological studies of mice selectively bred for aggressiveness. Scand. J. Psychol. **9**:157-160.

Landau, H. G. 1965. Development of structure in a society with a dominance relation when new members are added successively. Bull. Math. Biophys. **27**:151-160.

Landauer, W. 1945. Rumplessness of chicken embryos produced by injection of insulin and other chemicals. J. Exp. Zool. **98**:65-77.

Landis, C., and J. D. Page. 1938. Modern society and mental disease. Farrar & Rinehart, Inc., New York.

Lane, P. W., and M. M. Dickie. 1954. Fertile, obese male mice. J. Hered. **45**:56-58.

Lane, P. W., and M. M. Dickie. 1958. The effect of restricted food intake on the lifespan of genetically obese mice. J. Nutr. **64**:549-554.

Lang, T. 1940. Studies on the genetic determination of homosexuality. J. Nerv. Ment. Dis. **92**:55-64.

Lang, T. 1941. Untersuchungen an männlichen Homosexuellen und denen Sippen zwischen Homosexualität und Psychose. Z. Gesamte Neurol. Psychiatr. **171**:651-679.

Lang, T. 1960. Die Homosexualität als genetisches Problem. Acta Genet. Med. **9**:370-381.

Lange, J. 1929. Verbrecken als Schicksal. Thomas, Leipzig.

Lange, J. 1931. Crime as destiny. George Allen & Unwin Ltd., London.

Larsson, T., and T. Sjögren. 1954. A methodological, psychiatric, and statistical study of a large Swedish rural population. Acta Psychiatr. Neurol. [Suppl.]**89**.

Lashley, K. S. 1947. Structural variation in the nervous system in relation to behavior. Psychol. Rev. **54**:325-334.

Lashley, K. S. 1950. Functional interpretation of anatomic patterns. Proc. Assoc. Res. Nerv. Ment. Dis. **30**:529-547.

Lashley, K. S., and G. Clark. 1946. The cytoarchitecture of the cerebral cortex of *Ateles:* a critical examination of architectonic studies. J. Comp. Neurol. **85**:223-306.

Lauterbach, C. E. 1925. Studies in twin resemblance. Genetics **10**:525-568.

Lawrence, E. M. 1931. An investigation into the relation between intelligence and inheritance. Br. J. Psychol. Monogr. [Suppl.]**16**:1-80.

Layzer, D. 1974. Heritability analyses of IQ: science or numerology? Science **183**:1259-1266.

Leahy, A. M. 1932. A study of certain selective factors influencing prediction of the mental status of adopted children in nature-nurture research. J. Genet. Psychol. **41**:294-329.

Leahy, A. M. 1935. Nature-nurture and intelligence. Genet. Psychol. Monogr. **17**:236-308.

Legras, A. M. 1933. Psychose und Kriminalität bei Zwillingen. Z. Gesamte Neurol. Psychiatr. **144**:198-222.

Lehmann, A. 1967. Audiogenic seizures data in mice supporting new theories of biogenic amines mechanisms in the central nervous system. Life Sci. **6**:1423-1431.

Lehmann, A., and E. Boesiger. 1964. Sur le déterminisme génétique de l'épilepsie acoustique de *Mus musculus domesticus* (Swiss/Rb). C. R. Acad. Sci. **258**:4858-4861.

Lehmann, F. E., and W. Huber. 1944. Beobachtungen an *Tubifex* über die Bildung von Doppeleieren bei der zweiter Reifungsteilung und die Frage der Entstehung ovozytärer Zwillinge. Arch. Julius Klaus Stift. **19**:473-477.

Lehrke, R. 1972. A theory of X-linkage of major intellectual traits. Am. J. Ment. Defic. **76**:611-619.

Lejeune, J., M. Gautier, and R. Turpin. 1959. Les chromosomes humains en culture de tissus. C. R. Acad. Sci. **248**:602-603.

Lejeune, J., J. LaFourcade, R. Berger, J. Vialatte, M. Boeswillwald, P. Seringe, and R. Turpin. 1963. Trois cas de deletion partielle du bras court d'un chromosome 5. C. R. Acad. Sci. **257**:3098-3102.

Lemere, F., W. L. Voegtlin, W. R. Broz, P. O'Hallaren, and W. E. Tupper. 1943. Heredity as an etiologic factor in chronic alcoholism. Northwest Med. **42**:110-111.

Lennox, W. G., E. L. Gibbs, and F. A. Gibbs. 1945. The brain-wave pattern: an hereditary trait: evidence from 74 "normal" pairs of twins. J. Hered. **36**:233-243.

Lenz, F. and F. O. Von Verschauer. 1928. Zur Bestimmung des Anteils von Erbanlage um Umwelt an der Variabilität. Arch. Rassen. Ges. Biol. **20**:425-428.

Leonard, J. E., L. Ehrman, and M. Schorsch. 1974. Bioassay of a *Drosophila* pheromone influencing sexual selection. Nature **250**:261-262.

Leonhard, K. 1936. Die defekt-schizophrenen Krankheitsbilder. Georg Thieme Verlag, Leipzig.

Leonhard, K. 1957. Aufteilung der endogenen Psychosen. Akademie Verlag, Berlin.

Leonhard, K., I. Korff, and H. Schulz. 1962. Die Temperamente in den Familien der monopolaren und bipolaren phasischen Psychosen. Psychiatr. Neurol. Med. Psychol. **143**:416-434.

Lerner, I. M. 1950. Population genetics and animal improvement. University Press, Cambridge, England.

Lerner, I. M. 1954. Genetic homeostasis. John Wiley & Sons, Inc., New York.

Lerner, I. M. 1958. The genetic basis of selection. John Wiley & Sons, Inc., New York.

Lerner, I. M. 1968. Heredity, evolution, and society. W. H. Freeman & Co., San Francisco.

Les, E. P. 1968. Cage population density and efficiency of food utilization in inbred mice. Lab. Anim. Care **18**:305-313.

Lester, D. L. 1966. Self-selection of alcohol by animals, human variation, and the etiology of alcoholism. A critical review. Q. J. Stud. Alc. **27**:395-438.

Lester, D. L., and E. X. Freed. 1972. A rat model of alcoholism? Ann. N.Y. Acad. Sci. **197**:54-59.

Lester, D. L., and L. A. Greenberg. 1952. Nutrition and the etiology of alcoholism. The effects of sucrose, saccharin and fat on the self-selection of ethyl alcohol by rats. Q. J. Stud. Alc. 13:553-560.

Levin, B. H., J. G. Vandenbergh, and J. L. Cole. 1974. Aggression, social pressure and asymptote in laboratory mouse populations. Psychol. Rep. 34:239-244.

Levine, L. 1958. Studies on sexual selection in mice. I. Reproductive competition between albino and black-agouti males. Am. Natur. 92:21-26.

Levine, L. 1978. Biology of the gene, 3rd ed. The C. V. Mosby Co., St. Louis.

Levine, L., G. E. Barsel, and C. A. Diakow. 1965. Interaction of aggressive and sexual behavior in male mice. Behaviour 25:272-280.

Levine, L., G. E. Barsel, and C. A. Diakow. 1966. Mating behavior of two inbred strains of mice. Anim. Behav. 14:1-6.

Levine, L., C. A. Diakow, and G. E. Barsel. 1965. Interstrain fighting in male mice. Anim. Behav. 13:52-58.

Levine, L., and B. Lascher. 1965. Studies on sexual selection in mice. II. Reproductive competition between black and brown males. Am. Natur. 99:67-72.

Levine, S., and P. L. Broadhurst. 1963. Genetic and ontogenetic determinants of adult behavior in the rat. J. Comp. Physiol. Psychol. 56:423-428.

Levine, S., G. C. Haltmeyer, G. G. Karas, and V. H. Denenberg. 1967. Physiological and behavioral effects of infantile stimulation. Physiol. Behav. 2:55-59.

Levine, S., and R. Levin. 1970. Pituitary-adrenal influences on passive avoidance in two inbred strains of mice. Horm. Behav. 1:105-110.

Levy, J. 1976. A review of evidence for a genetic component in the determination of handedness. Behav. Genet. 6:429-454.

Levy, J., and T. Nagylaki. 1972. A model for the genetics of handedness. Genetics 72:117-128.

Lewis, A. 1935. Problems of obsessional illness. Proc. R. Soc. Med. 29:325-336.

Lewis, E. G., R. E. Dustman, and E. C. Beck. 1972. Evoked response similarity in monozygotic, dizygotic and unrelated individuals: a comparative study. Electroencephalogr. Clin. Neurophysiol. 32:309-316.

Lewis, W. L., and E. J. Warwick. 1953. Effectiveness of selection for body weight in mice. J. Hered. 44:233-238.

Lewontin, R. C. 1974. The analysis of variance and the analysis of cause. Am. J. Hum. Genet. 26:400-411.

Lewontin, R. C., and J. L. Hubby. 1966. A molecular approach to the study of genetic heterozygosity in natural populations. II. Amount of variation and degree of heterozygosity in natural populations of *Drosophila pseudoobscura*. Genetics 54:595-609.

Li, C. C. 1955. Population genetics. University of Chicago Press, Chicago.

Li, C. C. 1971. A tale of two thermos bottles: properties of a genetic model for human intelligence. In R. Cancro, ed. Intelligence: genetic and environmental influences. Grune & Stratton, Inc., New York.

Lindzey, G. 1951. Emotionality and audiogenic seizure susceptibility in five inbred strains of mice. J. Comp. Physiol. Psychol. 44:389-393.

Lindzey, G., C. S. Hall, and M. Manosevitz. 1973. Theories of personality, 3rd ed. John Wiley & Sons, Inc., New York.

Lindzey, G., D. T. Lykken, and H. D. Winston. 1960. Infantile trauma, genetic factors and temperament. J. Abnorm. Soc. Psychol. 61:7-14.

Lindzey, G., M. Manosevitz, and H. D. Winston. 1966. Social dominance in the mouse. Psychon. Sci. 5:451-452.

Lindzey, G., and H. D. Winston. 1962. Maze learning and effects of pretraining in inbred strains of mice. J. Comp. Physiol. Psychol. 55:748-752.

Lindzey, G., H. D. Winston, and M. Manosevitz. 1963. Early experience, genotype and temperament in *Mus musculus*. J. Comp. Physiol. Psychol. 56:622-629.

Ljungberg, L. 1957. Hysteria: a clinical prognostic and genetic study. Acta. Psychiatr. Scand. (Suppl.)112.

Loehlin, J. C. 1965a. A heredity-environment analysis of personality inventory data. In S. G. Vandenberg, ed. Methods and goals in human behavior genetics. Academic Press, Inc., New York.

Loehlin, J. C. 1965b. Some methodological problems in Cattell's multiple abstract variance analysis. Psychol. Rev. 72:156-161.

Loehlin, J. C. 1972. An analysis of alcohol-related questionnaire items from the National Merit twin study. Ann. N.Y. Acad. Sci. 197:117-120.

Loehlin, J. C., and R. C. Nichols. 1976. Heredity, environment and personality: a study of 850 sets of twins. University of Texas Press, Austin, Tex.

Loehlin, J. C., and S. G. Vandenberg. 1968. Genetic and environmental components in the covariation of cognitive abilities: an additive model. In S. G. Vandenberg, ed. Progress in human behavior genetics. Johns Hopkins University Press, Baltimore.

Loomis, A. L., E. N. Harvey, and G. Hobart. 1936. Electrical potentials of the human brain. J. Exp. Psychol. 19:249-279.

Lorden, J. F., G. A. Oltmans, and D. L. Margules. 1975. Central catecholamine levels in genetically obese (*ob ob* and *db db*) mice. Brain Res. 96:390-394.

Lorenz, K. 1950. The comparative method in studying innate behavior patterns. In Physiological mechanisms in animal behavior. Symp. Soc. Exp. Biol. 4:221-268.

Lucas, E. A., E. W. Powell, and O. D. Murphree. 1974. Hippocampal theta in nervous pointer dogs. Physiol. Behav. 12:609-613.

Lund, R. D. 1965. Uncrossed visual pathways of hooded and albino rats. Science 149:1506-1507.

Lush, J. L. 1945. Animal breeding plans, 3rd ed. Iowa State University Press, Ames, Iowa.

Luttge, W. G., and N. R. Hall. 1973a. Androgen-induced agonistic behavior in castrate male Swiss-Webster mice: comparison of four naturally occurring androgens. Behav. Biol. 8:725-732.

Luttge, W. G., and N. R. Hall. 1973b. Differential effectiveness of testosterone and its metabolites in the induction of male sexual behavior in two strains of albino mice. Horm. Behav. 4:31-43.

Luxenburger, H. 1928. Vorläufiger Bericht über psychiatrische Serienuntersuchungen an Zwillingen. Z. Gesamte Neurol. Psychiatr. 116:297-326.

Luxenburger, H. 1930a. Heredität und Familientypus der Zwangsneurotiker. Arch. Psychiatr. Nervenkr. 91:590-594.

Luxenburger, H. 1930b. Psychiatrische-neurologische Zwillingspathologie. Z. Gesamte Neurol. Psychiatr. 14:56-57, 145-180.

Luxenburger, H. 1934. Die manifestations wahrscheinlichkeit der Zwillingsforschung. Z. Psych. Hyg. 7:176-184.

Lyon, M. F. 1962. Sex chromatin and gene action in the mammalian X-chromosome. Am. J. Hum. Genet. 14:135-148.

MacArthur, J. W. 1949. Selection for small and large body size in the house mouse. Genetics 34:194-209.

McAskie, M., and A. M. Clarke. 1976. Parent-offspring resemblances in intelligence: theories and evidence. Br. J. Psychol. 67:243-273.

MacBean, I. T., and P. A. Parsons. 1967. Directional selection for duration of copulation in *Drosophila melanogaster*. Genetics 56:233-239.

McCall, R. B. 1970. Intelligence quotient pattern over age: comparisons among siblings and parent-child pairs. Science 170:644-647.

McClearn, G. E. 1959. The genetics of mouse behavior in novel situations. J. Comp. Physiol. Psychol. 52:62-67.

McClearn, G. E. 1961. Genotype and mouse activity. J. Comp. Physiol. Psychol. 54:674-676.

McClearn, G. E. 1968a. Genetics and motivation of the mouse. In W. J. Arnold ed. Nebraska Symposium on Motivation. Nebraska University Press, Lincoln.

McClearn, G. E. 1968b. Social implications of behavioral genetics. In D. Glass, ed. Genetics. Rockefeller University Press, New York.

McClearn, G. E. 1972. The genetics of alcohol preference. Int. Symp. Biol. Aspt. Alc. Consumpt. Finn. Found. Alc. Stud. 20:113-119.

McClearn, G. E., and J. C. DeFries. 1973. Introduction to behavioral genetics. W. H. Freeman & Co. Publishers, San Francisco.

McClearn, G. E., and R. Kakihana. 1973. Selective breeding for ethanol sensitivity in mice. Behav. Genet. 3:409-410. (Abstr.)

McClearn, G. E., and W. Meredith. 1964. Dimensional analysis of activity and elimination in a genetically heterogeneous group of mice. Anim. Behav. 12:1-10.

McClearn, G. E., and D. A. Rodgers. 1959. Differences in alcohol preference among inbred strains of mice. Q. J. Stud. Alc. 20:691-695.

McClearn, G. E., and D. A. Rodgers. 1961. Genetic factors in alcohol preference of laboratory mice. J. Comp. Physiol. Psychol. 54:116-119.

McClearn, G. E., J. R. Wilson, and W. Meredith. 1970. The use of isogenic and heterogenic mouse stocks in behavioral research. In G. Lindzey and D. D. Thiessen, eds. Contributions to behavior-genetic analysis: the mouse as a prototype. Appleton-Century-Crofts, New York.

McClelland, D. C. 1973. Testing for competence rather than for intelligence. Am. Psychol. 28:1-14.

McCollum, R. E., P. B. Siegel, and H. P. Van Krey. 1971. Responses to androgen in lines of chickens selected for mating behavior. Horm. Behav. 2:31-42.

McConaghy, N., and M. Clancy. 1968. Familial relationships of allusive thinking in university students and their parents. Br. J. Psychiatry 114:1079-1087.

McDaniel, G. R., and J. V. Craig. 1959. Behavior traits, semen measurements and fertility of white leghorn males. Poultry Sci. 38:1005-1014.

McDougall, W. 1927. An experiment for testing the hypothesis of Lamarck. Br. J. Psychol. 17:267-304.

McDougall, W. 1938. Fourth report on a Lamarckian experiment. Br. J. Psychol. 28:328-345, 365-395.

McEwen, R. S. 1918. The reactions to light and to gravity in *Drosophila* and its mutants. J. Exp. Zool. 25:49-106.

McEwen, R. S. 1925. Relative phototropism of vestigial and wild type *D. melanogaster*. Biol. Bull. 49:354-364.

McGaugh, J. L., and J. M. Cole. 1965. Age and strain differences in the effect of distribution of practice on maze-learning. Psychon. Sci. 2:253-254.

McGaugh, J. L., W. Westbrook, and G. Burt. 1961. Strain differences in the facilitative effects of 1757 I.S. on maze learning. J. Comp. Physiol. Psychol. 54:502-505.

McGill, T. E. 1962. Sexual behaviour in three inbred strains of mice. Behaviour 19:341-350.

McGill, T. E. 1970. Genetic analysis of male sexual behavior. In G. Lindzey and D. D. Thiessen eds. Contributions to behavior-genetic analysis: the mouse as a prototype. Appleton-Century-Crofts, New York.

McGill, T. E., and W. C. Blight. 1963a. Effects of genotype on the recovery of sex drive in the male mouse. J. Comp. Physiol. Psychol. 56:887-888.

McGill, T. E., and W. C. Blight. 1963b. The sexual behaviour of hybrid male mice compared with the sexual behaviour of males of the inbred parent strains. Anim. Behav. 11:480-483.

McGill, T. E., and C. M. Haynes. 1973. Heterozygosity and retention of ejaculatory reflex after castration in male mice. J. Comp. Physiol. Psychol. 84:423-429.

McGill, T. E., and G. R. Tucker. 1964. Genotype and sex drive in intact and in castrated male mice. Science 145:514-515.

McInnes, R. G. 1937. Observations on heredity in neurosis. Proc. R. Soc. Med. 30:895-904.

McKeever, W. F., A. D. Van Deventer, and M. Suberi. 1973. Avowed, assessed and familial handedness and differential hemispheric processing of brief sequential and non-sequential visual stimuli. Neuropsychologia, 11:235-238.

Mackintosh, J. H. 1970. Territory formation by laboratory mice. Anim. Behav. 18:177-183.

Mackintosh, N. J. 1974. The psychology of animal learning. Academic Press, Inc., London.

Mackintosh, N. J. 1975. Critical notice: Kamin, L. J., *The Science and Politics of IQ*. Q. J. Exp. Psychol. 27:672-686.

McKusick, V. A. 1972. Study guide in human genetics. Prentice-Hall, Inc., Englewood Cliffs, N.J.

McKusick, V. A. 1975. Mendelian inheritance in man, 4th ed. The Johns Hopkins University Press, Baltimore.

McKusick, V. A., and F. H. Ruddle. 1977. The status of the gene map of the human chromosomes. Science 196:390-405.

McLaren, A., and D. Michie. 1956. Variability of response in experimental animals: a comparison of the reactions of inbred, F_1 hybrid, and random bred mice to a narcotic drug. J. Genet. 54:440-455.

McNemar, Q. 1933. Twin resemblances in motor skills and the effects of practice thereon. J. Genet. Psychol. 42:70-99.

McNemar, Q. 1938. Newman, Freeman and Holzinger's twins: a study of heredity and environment. Psychol. Bull. 35:247-248.

Madsen, I. N. 1924. Some results with the Stanford revision of the Binet-Simon tests. School Soc. 19:559-562.

Mahut, H. 1958. Breed differences in the dog's emotional behavior. Can. J. Psychol. 12:35-44.

Mainardi, D. 1963a. Eliminazione della barriera etologica all'isolamento riproduttivo tra *Mus musculus domesticus* e *M.m. bactrianus* mediante azione sull'apprendiento infantile. Inst. Lombardo Acad. Sci. Lett. 97:291-299.

Mainardi, D. 1963b. Speciazione nel topo. Fattori etologici determinanti barriere riproduttivo tra *Mus musculus domesticus* e *M.m. bactrianus*. Inst. Lombardo Acad. Sci. Lett. 97:135-142.

Mainardi, D., M. Marsan, and A. Pasquali. 1965. Causation of sexual preferences in the house mouse. The behavior of mice raised by parents whose odor was artificially altered. Atti. Soc. Sci. Nat. Milano. 104:325-338.

Mainardi, D., F. M. Scudo, and D. Barbieri. 1965. Assortative mating based on early learning: population genetics. Acta Biomed. 36:585-605.

Malagolowkin-Cohen, C., A. S. Simmons, and H. Levene. 1965. A study of sexual isolation between certain strains of *Drosophila paulistorum*. Evolution 19:95-103.

Malan, M. 1940. Zur Erblichkeit der Orientierungsfähigkeit im Raum. Z. Morphol. Anthropol. 39:1-23.

Manning, A. 1959. The sexual isolation between *Drosophila melanogaster* and *D. simulans*. Anim. Behav. 7:60-65.

Manning, A. 1961. The effects of artificial selection for mating speed in *Drosophila melanogaster*. Anim. Behav. 9:82-92.

Manning, A. 1963. Selection of mating speed in *Drosophila melanogaster* based on behaviour of one sex. Anim. Behav. 11:116-120.

Manning, A. 1967a. Control of sexual receptivity in female *Drosophila*. Anim. Behav. 15:239-250.

Manning, A. 1967b. Genes and the evolution of insect behavior. In J. Hirsch, ed. Behavior-genetic analysis. McGraw-Hill Book Co., New York.

Manning, A. 1968. The effects of artificial selection for slow mating in *Drosophila simulans*. I. The behaviour changes. Anim. Behav. 16:108-113.

Manning, A. 1976. The place of genetics in the study of behaviour. In P. P. G. Bateson and R. A. Hinde, eds. Growing points in ethology. Cambridge University Press, Cambridge, England.

Manning, A., and J. Hirsch. 1971. The effects of artificial selection for slow mating in *Drosophila simulans*. II. Genetic analysis of the slow mating line. Anim. Behav. 19:448-453.

Manosevitz, M. 1965. Genotype, fear and hoarding. J. Comp. Physiol. Psychol. 60:412-416.

Manosevitz, M. 1967. Hoarding and inbred strains of mice. J. Comp. Physiol. Psychol. 63:148-150.

Manosevitz, M. 1972. Behavioral heterosis: food competition in mice. J. Comp. Physiol. Psychol. 79:46-50.

Manosevitz, M., R. B. Campenot, and C. F. Swecionis. 1968. Effects of enriched environment upon hoarding. J. Comp. Physiol. Psychol. 66:319-324.

Manosevitz, M., C. I. Fitzsimmons, and T. R. McCanne. 1970. Correlates of food competition behavior. Psychon. Sci. 19:141-142.

Manosevitz, M., and G. Lindzey. 1967. Genetics of hoarding: a biometrical analysis. J. Comp. Physiol. Psychol. 63:142-144.

Manosevitz, M., and R. J. Montemayor. 1972. Interaction of environmental enrichment and genotype. J. Comp. Physiol. Psychol. 79:67-76.

Marconi, J., A. Varela, E. Rosenblatt, G. Solaris, I. Marchese, R. Alvarado, and W. Enriquez. 1955. A survey on the prevalence of alcoholism among the adult population of a suburb of Santiago. Q. J. Stud. Alc. 16:438-446.

Mardones, R. J. 1952. On the relationship between deficiency of B-vitamins and alcohol intake in rats. Q. J. Stud. Alc. 12:563-575.

Mardones, R. J., N. M. Segovia, and A. D. Hederra. 1953. Heredity of experimental alcohol preference in rats. II. Coefficient of heredity. Q. J. Stud. Alc. 14:1-2.

Marinković, D. 1974. Light dependent matings of *Drosophila pseudoobscura*. Behav. Genet. 4:301-303.

Marjoribanks, K., and H. J. Walberg. 1975. Birth order, family size, and intelligence. Soc. Biol. 22:261-268.

Marks, I. M., M. Crowe, E. Drewe, J. Young, and W. G. Dewhurst. 1969. Obsessive-compulsive neurosis in identical twins. Br. J. Psychiatry. 115:991-998.

Marler, P. A. 1970. A comparative approach to vocal learning: song development in white-crowned sparrows. J. Comp. Physiol. Psychol. 71(2):1-25.

Marshall, J. C., D. Caplan, and J. M. Holmes. 1975. The measure of laterality. Neuropsychology 13:315-321.

Martin, J. T. 1975. Hormonal influences in the evolution and ontogeny of imprinting behavior in the duck. In W. H. Gispen, T. B. v. Wienersma Greidanus, B. Bohus, and D. deWied, eds. Hormones, homeostasis and the brain. Elsevier North-Holland, Inc., New York.

Matheny, A. P. 1971. Genetic determinants of the Ponzo illusion. Psychon. Sci. 24:155-156.

Matheny, A. P. 1972. Perceptual exploration in twins. J. Exp. Child Psychol. 14:108-116.

Matheny, A. P., R. S. Wilson, and A. B. Dolan. 1976. Relations between twins' similarity of appearance and behavioral similarity: testing and assumption. Behav. Genet. 6:343-352.

Mather, K. 1949. Biometrical genetics: the study of continuous variation. Dover Publications, Inc., New York.

Mather, K., and J. L. Jinks. 1971. Biometrical genetics: the study of continuous variation, 2nd ed. Chapman-Hall, Ltd., London.

Mathers, J. A., R. H. Osborne, and F. V. DeGeorge. 1961. Studies of blood pressure, heart rate and the electrocardiogram in adult twins. Am. Heart J. 62:634-642.

Mayer, J. 1953. Decreased activity and energy balance in the hereditary obesity-diabetes syndrome of mice. Science 117:504-505.

Mayer-Gross, W. 1948. Mental health survey in a rural area. Eugen. Rev. 40:140-148.

Maynard Smith, J. 1956. Fertility, mating behavior and sexual selection in *Drosophila subobscura*. J. Genet. 54:261-279.

Mayr, E. 1970. Populations, species and evolution. Harvard University Press, Cambridge, Mass.

Mednick, S. A. 1958. A learning theory approach to research in schizophrenia. Psychol. Bull. 55:316-327.

Mednick, S. A. 1970. Breakdown in individuals at high risk for schizophrenia: possible predispositional prenatal factors. Men. Hyg. 54:50-63.

Mednick, S. A., E. Mura, F. Schulsinger, and B. Mednick. 1971. Perinatal conditions and infant development in children with schizophrenic parents. Soc. Biol. 18(suppl):103-113.

Mednick, S. A., and F. Schulsinger. 1968. Some premorbid characteristics related to breakdown in children with schizophrenic mothers. In D. Rosenthal and S. S. Kety, eds. The transmission of schizophrenia. Pergamon Press Ltd., Oxford.

Meehl, P. E. 1962. Schizotaxia, schizotypy and schizophrenia. Am. Psychol. 17:827-838.

Meier, G. W., and D. P. Foshee. 1965. Albinism and water escape performance in mice. Science 147:307-308.

Mendel, G. 1865. Experiments in plant hybridization. Translated in J. A. Peters, ed. 1962. Classical papers in genetics. Prentice-Hall, Inc., Englewood Cliffs, N.J.

Mendlewicz, J., ed. 1975. Genetics and psychopharmacology. S. Karger AG, Basel, Switzerland.

Menkes, J. H., and B. R. Migeon. 1966. Biochemical and genetic aspects of mental retardation. Ann. Rev. Med. 17:407-430.

Merrell, C. 1957. Dominance of eye and hand. Hum. Biol. 29:314-328.

Merrell, D. J. 1949. Selective mating in *Drosophila melanogaster*. Genetics 34:370-389.

Merrell, D. J. 1951. Inheritance of manic-depressive psychosis. A.M.A. Arch. Neurol. Psychiatry 66:272-279.

Merrell, D. J. 1953. Selective mating as a cause of gene frequency changes in laboratory populations of *D. melanogaster*. Evolution 7:287-296.

Merrell, D. J. 1965. Methodology in behavior genetics. J. Hered. 56:263-266

Merriman, C. 1924. The intellectual resemblance of twins. Psychol. Monogr. 33(5):1-58.

Messeri, P., B. E. Eleftheriou, and A. Oliverio. 1975. Dominance behavior: a phylogenetic analysis in the mouse. Physiol. Behav. 14:53-58.

Messeri, P., A. Oliverio, and D. Bovet. 1972. Relations between avoidance and activity: a diallel study in mice. Behav. Biol. 7:733-742.

Meyer, D. R., D. A. Yutzey, and P. E. Meyer. 1966. Effects of neocortical ablations on relearning of a black-white discrimination habit by two strains of rats. J. Comp. Physiol. Psychol. 61:83-86.

Miall, W. E., P. Heneage, T. Khosia, H. Lovell, and F. Moore. 1967. Factors influencing the degree of resemblance in arterial pressure in close relatives. Clin. Sci. 33:271-283.

Miguel, C. 1935. Schreibdruck bei Zwillingen. Z. Gesamte Neurol. Psychiatr. 152:19-24.

Mijsberg, W. 1957. Genetic-statistical data on the presence of secondary oöcytary twins among non-identical twins. Acta Genet. 7:39-42.

Mikhail, A. A., and P. L. Broadhurst. 1965. Stomach ulceration and emotionality in selected strains of rats. J. Psychosom. Res. 8:477-479.

Mikkelsen, M., A. Frøland, and J. Ellebjerg. 1963. XO/XX mosaicism in a pair of presumably monozygotic twins with different phenotypes. Cytogenetics 2:86-98.

Miller, E. 1971. Handedness and the pattern of human ability. Br. J. Psychol. 62:111-112.

Milner, A. D. 1969. Distribution of hand preferences in monkeys. Neuropsychology 7:375-377.

Mitsuda, H., ed. 1967. Clinical genetics in psychiatry. Igaku Shoin Ltd., Tokyo.

Mitsuda, H. 1972. Heterogeneity of schizophrenia. In A. R. Kaplan, ed. Genetic factors in "Schizophrenia." Charles C Thomas, Publisher, Springfield, Ill.

Mitsuda, H., T. Sakai, and J. Kobayashi. 1967. A clinico-genetic study on the relationship between neurosis and psychosis. In H. Mitsuda, ed. Clinical genetics in psychiatry. Igaku Shoin Ltd., Tokyo.

Mittler, P. 1969. Genetic aspects of psycholinguistic abilities. J. Child Psychol. Psychiatry 10:165-176.

Mittler, P. 1971. The study of twins. Penguin Books Ltd., Harmondsworth, England.

Mjoen, J. A. 1925. Zur Erbanalyse der musikalischen Begabung. Hereditas. 7:109-128.

Money, J. 1973. Effects of prenatal androgenation and

deandrogenation on behavior in human beings. In W. F. Ganong and L. Martini, eds. Frontiers in neurochemistry. Oxford University Press, New York.

Moran, G. 1975. Severe food deprivation: some thoughts regarding its exclusive use. Psychol. Bull. 82:543-557.

Moran, P. A. P. 1972. Theoretical considerations on schizophrenia genetics. In A. R. Kaplan, ed. Genetic factors in "schizophrenia." Charles C Thomas, Publisher, Springfield, Ill.

Mordkoff, A. M., and J. L. Fuller. 1959. Variability in activity within inbred and crossbred mice. J. Hered. 50:6-8.

Mordkoff, A. M., K. Schlesinger, and R. A. Lavine. 1965. Developmental homeostasis in behavior of mice. Locomotor activity and grooming. J. Hered. 55:84-88.

Morgan, A. H., E. R. Hilgard, and E. S. Davert. 1970. The heritability of hypnotic susceptibility of twins: a preliminary report. Behav. Genet. 1:213-224.

Morgan, T. H., and C. B. Bridges. 1919. Contributions to the genetics of *Drosophila melanogaster*. I. The origin of gynandromorphs. Publ. No. 278:1-122. Carnegie Institution, Washington, D.C.

Morton, N. E. 1972. Human behavioral genetics. In L. Ehrman, G. S. Omenn, and E. Caspari, eds. Genetics, environment and behavior. Academic Press, Inc., New York.

Morton, N. E. 1974. Analysis of family resemblance. I. Introduction. Am. J. Hum. Genet. 26:318-330.

Mos, L., J. R. Royce, and W. Poley. 1973. Effects of postweaning stimulation on factors of mouse-emotionality. Dev. Psychol. 5:229-239.

Mosher, L. R., and D. Feinsilver. 1971. Special report: schizophrenia. Publ. No. (HSM) 72-9007. U.S. Department of Health, Education, and Welfare, U.S. Government Printing Office, Washington, D.C.

Munsinger, H. 1975. The adopted child's IQ: a critical review. Psychol. Bull. 82:623-659.

Munsinger, H. 1977. The identical twin transfusion syndrome: a source of error in estimating IQ resemblance and heritability. Ann. Hum. Genet. 40:307-321.

Munsinger, H. 1978. Reply to Kamin. Psychol. Bull. 85:202-206.

Murawski, B. J. 1960. Flicker fusion thresholds in control subjects and identical twins. J. Appl. Physiol. 15:246-248.

Murphree, O. D., R. A. Dykman, and J. E. Peters. 1967a. Genetically-determined abnormal behavior in dogs: results of behavioral tests. Cond. Reflex 2:199-205.

Murphree, O. D., R. A. Dykman, and J. E. Peters. 1967b. Operant conditioning of two strains of the pointer dog. Psychophysiology 3:414-417.

Murphree, O. D., and J. E. O. Newton. 1971. Crossbreeding and special handling of genetically nervous dogs. Cond. Reflex. 6:129-136.

Murphree, O. D., J. E. Peters, and R. A. Dykman. 1967. Effect of person on nervous, stable and crossbred pointer dogs. Cond. Reflex 2:273-276.

Murphy, G. 1947. Personality: a biosocial approach to origins and structure. Harper & Brothers, New York.

Murphy, H. B. M. 1968. Cultural factors in the genesis of schizophrenia. In D. Rosenthal and S. S. Kety, eds. The transmission of schizophrenia. Pergamon Press, Inc., New York.

Myers, A. K. 1959. Avoidance learning as a function of several training conditions and strain differences in rats. J. Comp. Physiol. Psychol. 52:381-386.

Myers, A. K. 1962. Alcohol choice in Wistar and G-4 rats as a function of environmental temperature and alcohol concentration. J. Comp. Physiol. Psychol. 55:606-609.

Nachman, M. 1959. The inheritance of saccharin preference. J. Comp. Physiol. Psychol. 52:451-457.

Nachman, M., C. Larue, and J. LeMagnen. 1971. The role of olfactory and orosensory factors in the alcohol preference of inbred strains of mice. Physiol. Behav. 6:53-59.

Nagao, S. 1967. Clinico-genetic study of chronic alcoholism. In H. Mitsuda, ed. Clinical genetics in psychiatry. Igaku Shoin Ltd., Tokyo.

Nagylaki, T., and J. Levy. 1973. "Sound of one paw clapping" isn't sound. Behav. Genet. 3:279-292.

Naik, D. V., and H. Valtin. 1969. Hereditary vasopressin-resistant urinary concentrating defects in mice. Am. J. Physiol. 217:1183-1190.

Nakamura, C. Y., and N. H. Anderson. 1962. Avoidance behavior differences within and between strains of rats. J. Comp. Physiol. Psychol. 55:740-747.

Neel, J. V. 1949. The detection of the genetic carriers of hereditary disease. Am. J. Hum. Genet. 1:19-36.

Neel, J. V. 1970. Lessons from a "primitive" people. Science 170:815-822.

Neel, J. V., and W. J. Schull. 1954. Human heredity. University of Chicago Press, Chicago.

Neel, J. V., W. J. Schull, M. Yamamoto, S. Uchida, T. Yanase, and N. Fujiki. 1970. The effects of parental consanguinity and inbreeding in Hirado, Japan. II. Physical development, tapping rate, blood pressure, intelligence quotient, and school performance. Am. J. Hum. Genet. 22:263-286.

Neu, D. M. 1947. A critical review of the literature on "absolute pitch." Psychol. Bull. 44:249-266.

Newcomb, T., and G. Svehla. 1937. Intra-family relationships in attitude. Sociometry 1:180-205.

Newman, H. H., F. N. Freeman, and K. J. Holzinger. 1937. Twins: a study of heredity and environment. University of Chicago Press, Chicago.

Nichols, J. 1950. Effects of captivity on adrenal glands of wild Norway rats. Am. J. Physiol. 162:5-7.

Nichols, J. R., and S. Hsiao. 1967. Addiction liability of albino rats: breeding for quantitative differences in morphine drinking. Science 157:561-563.

Nichols, R. C. 1965. The National Merit twin study. In S. G. Vandenberg, ed. Methods and goals in human behavior genetics. Academic Press, Inc., New York.

Nichols, R. C. 1966. The resemblance of twins in personality and interests. Natl. Merit Scholarship Corp. Res. Rep. 2:1-23.

Nicolay, E. 1939. Messungen an Handschriftproblem von Zwillingspaaren über 14 Jahren. Arch. Gesamte Psychol. 105:275-295.

Nielsen, J., W. Wilsnack, and E. Strömgren. 1965. Some aspects of community psychiatry. Br. J. Prev. Soc. Med. 2:1-23.

Nissen, H. W. 1958. Axes of behavioral comparison. In A. Roe, and G. G. Simpson, eds. Behavior and evolution. Yale University Press, New Haven, Conn.

Norris, V. 1959. Mental illness in London. Maudsley Monogr. No. 6, London.

Ödegaard, Ö. 1961. The epidemiology of depressive psychosis. Acta Psychiatr. Neurol. Scand. [Suppl.] 162.

Ödegaard, Ö. 1963. The psychiatric disease entities in the light of a genetic investigation. Acta Psychiatr. Scand. [Suppl.]169, 39:94-104.

Ödegaard, Ö. 1972. The multifactorial theory of inheritance in predisposition to schizophrenia. In A. R. Kaplan, ed. Genetic factors in "schizophrenia." Charles C Thomas, Publisher, Springfield, Ill.

Oki, T. 1967. A psychological study of early childhood neuroses. In H. Mitsuda, ed. Clinical genetics in psychiatry. Igaku Shoin Ltd., Tokyo.

Oldfield, R. 1971. The assessment and analysis of handedness: the Edinburgh inventory. Neuropsychology 9:97-113.

Olive, H. 1972. Sibling resemblances in divergent thinking. J. Genet. Psychol. 120:155-162.

Oliverio, A. 1971. Genetic variation and heritability in a measure of avoidance learning in mice. J. Comp. Physiol. Psychol. 74:390-397.

Oliverio, A., C. Castellano, and P. Messeri. 1972. Genetic analysis of avoidance, maze and wheel-running behaviors in the mouse. J. Comp. Physiol. Psychol. 79:459-473.

Oliverio, A., B. E. Eleftheriou, and D. W. Bailey. 1973a. Exploratory activity: genetic analysis of its modification by scopolamine and amphetamine. Physiol. Behav. 10:893-899.

Oliverio, A., B. E. Eleftheriou, and D. W. Bailey. 1973b. A gene influencing active avoidance performance in mice. Physiol. Behav. 11:497-501.

Oliverio, A., M. Satta, and D. Bovet. 1968. Effects of cross-fostering on emotional and learning behavior of different strains of rats. Life Sci. 7:799-806.

Omenn, G. S. 1976. Inborn errors of metabolism: clues to understanding human behavioral disorders. Behav. Genet. 6:263-284.

Opsahl, C. A., and T. L. Powley. 1974. Failure of vagotomy to reverse obesity in the genetically obese Zucker rat. Am. J. Physiol. 226:34-38.

Orenberg, E. K. 1975. Genetic determination of aggressive behavior and brain cyclic AMP. Psychopharmacol. Comm. 1:99-107.

Ortman, L. L., and J. V. Craig. 1968. Social dominance in chickens modified by genetic selection: physiological mechanisms. Anim. Behav. 16:33-37.

Osborne, R. H., F. V. DeGeorge, and J. A. L. Mathers. 1963. The variability of blood pressure: basal and casual measurements in adult twins. Am. Heart J. 66:176-183.

Osborne, R. T. 1970. Heritability estimates for the visual evoked response. I. Life Sci. 9:481-490.

Osmond, H., and A. Hoffer. 1966. A comprehensive theory of schizophrenia. Int. J. Neuropsychiatry 2: 302-309.

Östlyngen, E. 1949. Possibilities and limitations of twin research as a means of solving problems of heredity and environment. Acta Psychol. 6:59-90.

Outhit, M. C. 1933. A study of the resemblance of parents and children in general intelligence. Arch. Psychol. 149:1-60.

Owen, D. R. 1972. The 47, XYY male: a review. Psychol. Bull. 78:209-233.

Owen, D. R., and J. O. Sines. 1971. Heritability of personality in children. Behav. Genet. 1:235-248.

Owen, R. D. 1945. Immunogenetic consequences of vascular anastamoses between bovine twins. Science 102:400-401.

Padeh, B., D. Wahlsten, and J. C. DeFries. 1974. Operant discrimination learning and operant bar-pressing rates in inbred and heterogeneous laboratory mice. Behav. Genet. 4:383-393.

Pak, W., K. Grossfield, and N. White. 1969. Nonphototactic mutants in the study of vision in Drosophila. Nature 222:351-354.

Pare, W. P. 1969. Age, sex and strain differences in the aversive threshold to grid shock in the rat. J. Comp. Physiol. Psychol. 69:214-218.

Parker, N. 1966. Twin relationships and concordance for neurosis. Proceedings of the Fourth World Congress on Psychiatry. Excerpta Medica International Congress Series, No. 150, New York.

Parson, W., and K. R. Crispell. 1955. Studies of acetate metabolism in the hereditary obesity-diabetes syndrome of mice utilizing C^{14} acetate. Metabolism 4: 227-230.

Parsons, P. A. 1964. A diallel cross for mating speeds in Drosophila melanogaster. Genetica 35:141-151.

Parsons, P. A. 1965. Assortative mating for a metrical characteristic in Drosophila. Heredity 20:161-167.

Parsons, P. A. 1967a. The genetic analysis of behaviour. Methuen & Co. Ltd., London.

Parsons, P. A. 1967b. Variability within and between strains for mating behavior parameters in Drosophila pseudoobscura. Experientia 23:131-134.

Parsons, P. A. 1974. Male mating speed as a component of fitness in Drosophila. Behav. Genet. 4:395-404.

Partanen, J., K. Bruun, and T. Markkanen. 1966. Inheritance of drinking behavior. A study on intelligence, personality and the use of alcohol of adult twins. Finn. Found. Alc. Stud. 14:1-159.

Pastore, N. 1949. The nature-nurture controversy. Kings Crown Press, New York.

Patau, K., D. W. Smith, E. Thermon, S. L. Inhorn, and H. P. Wagner. 1960. Multiple congenital anomaly caused by an extra autosome. Lancet 1:790-793.

Pauling, L., H. A. Itano, S. J. Singer, and I. C. Wells.

1949. Sickle-cell anemia, a molecular disease. Science **110**:543-548.

Pearson, K. 1904. On the laws of inheritance in man. II. On the inheritance of the mental and moral characters in man, and its comparison with the inheritance of the physical characters. Biometrika **3**:131-190.

Pearson, K. 1910. Nature and nurture: Eugenics Laboratory Lecture Series, vol. 6. Dulan, London.

Pearson, K. 1918. Inheritance of psychical characters. Biometrika **12**:367-372.

Pelz, W. E., G. Whitney, and J. C. Smith. 1973. Genetic influences on saccharin preference of mice. Physiol. Behav. **10**:263-265.

Pencavel, J. H. 1976. A note on the IQ of monozygotic twins raised apart and the order of their birth. Behav. Genet. **6**:455-460.

Penrose, L. S. 1939. Intelligence test scores of mentally defective patients and their relatives. Br. J. Psychol. **30**:1-18.

Penrose, L. S. and G. F. Smith. 1966. Down's anomaly. J. & A. Churchill, London.

Perris, C. 1966. A study of bipolar (manic-depressive) and unipolar recurrent depressive psychoses. Acta Psychiatr. Scand. [Suppl.]194, **42**.

Perris, C. 1968. The course of depressive psychoses. Acta Psychiatr. Scand. **44**:238-248.

Perris, C., and G. d'Elia. 1966. A study of bipolar (manic-depressive) and unipolar recurrent depressive psychoses. Acta Psychiatr. Scand. [Suppl.]194:172-183.

Peterson, G. M. 1934. Mechanisms of handedness in the rat. Psychol. Monogr. **4**(46).

Peterson, T. D. 1936. The relationship between certain attitudes of parents and children. Purdue Univ. Stud. Higher Educ. **31**:127-144.

Petit, C. 1958. Le determinisme génétique et psychophysiologique de la competition sexuelle chez *Drosophila melanogaster*. Bull. Biol. Fr. Belg. **93**:248-329.

Petit, C., and L. Ehrman. 1969. Sexual selection in *Drosophila*. Evol. Biol. **3**:177-223.

Petras, M. L. 1967. Studies of natural populations of *Mus*. I. Biochemical polymorphisms and their bearing on breeding structure. Evolution **21**:259-274.

Pickford, R. W. 1949. The genetics of intelligence. J. Psychol. **28**:129-145.

Pickford, R. W. 1965. The genetics of color vision. In A. V. S. de Renck and J. Knight, eds. Ciba Foundation Symposium on Colour Vision. Little, Brown & Co., Boston.

Pintner, R. 1918. The mental indices of siblings. Psychol. Rev. **25**:252-255.

Pitt, F. H. G. 1944. The nature of normal trichromatic and dichromatic vision. Proc. R. Soc. Lond. [Biol.] **132**:101-117.

Pittendrigh, C. S. 1958. Adaptation, natural selection and behavior. In G. G. Simpson and A. Roe, eds. Behavior and evolution. Yale University Press, New Haven, Conn.

Pitts, F. N., and G. Winokur. 1966. Affective disorder. III. Diagnostic correlates and incidence of suicide. J. Nerv. Ment. Dis. **139**:176-181.

Plomin, R., J. C. DeFries, and J. C. Loehlin. 1977. Genotype-environment interaction and correlation in the analysis of human behavior. Psychol. Bull. **84**:309-322.

Plomin, R., L. Willerman, and J. C. Loehlin. 1976. Resemblance in appearance and the equal environments assumption in twin studies of personality traits. Behav. Genet. **6**:43-52.

Poley, W., and J. R. Royce. 1970. Genotype, maternal stimulation and factors of mouse emotionality. J. Comp. Physiol. Psychol. **71**:246-250.

Poley, W., and J. R. Royce. 1973. Behavior-genetic analysis of mouse emotionality. II. Stability of factors across genotypes. Anim. Learn. Behav. **1**:116-120.

Poley, W., L. T. Yeudall, and J. R. Royce. 1970. Factors of emotionality related to alcohol consumption in laboratory mice. Multivar. Behav. Res. **5**:203-208.

Pollin, W. 1971. A possible genetic factor related to psychosis. Am. J. Psychiatry **128**:311-317.

Pollin, W., and J. R. Stabenau. 1968. Biological, psychological and historical differences in a series of monozygotic twins discordant for schizophrenia. In D. Rosenthal and S. S. Kety, eds. The transmission of schizophrenia. Pergamon Press Ltd., Oxford.

Pollin, W., J. R. Stabenau, L. R. Mosher, and J. Tupin. 1966. Life history differences in identical twins discordant for schizophrenia. Am. J. Orthopsychiatry **26**:492-509.

Pollin, W., J. R. Stabenau, and J. Tupin. 1965. Family studies with identical twins discordant for schizophrenia. Psychiatry **28**:60-78.

Pollitzer, W. S. 1972. Discussion of paper of I. I. Gottesmann and L. L. Heston. In L. Ehrman, G. S. Omenn, and E. Caspari, eds. Genetics, environment and behavior. Academic Press, Inc., New York.

Pons, J. A. 1960. Contribution to the heredity of P.T.C. taste character. Ann. Hum. Genet. **24**:71-76.

Portenier, L. 1939. Twinning as a factor in influencing personality. J. Educ. Psyhcol. **30**:542-547.

Porter, R. H. 1972. Infantile handling differentially affects interstrain dominance interactions in mice. Behav. Biol. **7**:415-420.

Potter, J. H. 1949. Dominance relations between different breeds of domestic hens. Physiol. Zool. **22**:261-280.

Powell, B. J., and M. North-Jones. 1974. Effects of early handling on avoidance performance of Maudsley MR and MNR strains. Dev. Psychobiol. **7**:145-148.

Price, B. 1936. Homogamy and the intercorrelation of capacity traits. Ann. Eugen. **7**:22-27.

Price, B. 1950. Primary bias in twin studies: a review of prenatal and natal difference-producing factors in monozygotic pairs. Am. J. Hum. Genet. **2**:293-352.

Price, J. 1968. The genetics of depressive behaviour. In A. J. Coppens and A. Walk, eds. Recent developments in affective disorders. Br. J. Psychiatry Spec. Publ. No. 2.

Primrose, E. J. R. 1962. Psychological illness: a community study. Tavistock Publications Ltd., London.

Pringle, M. L. K. 1966. Adoption: facts and fallacies. Longman Group Ltd., London.

Pruzan, A., and L. Ehrman. 1974. Age, experience and rare-male mating advantage in *Drosophila pseudoobscura*. Behav. Genet. 4:159-165.

Punnett, R. C., and P. G. Bailey. 1920. Genetic studies in poultry. II. Inheritance of egg colour and broodiness. J. Genet. 10:277-292.

Quelce-Salgado, A., A. Freire-Maia, and N. Freire-Maia. 1961. Arm folding: a genetic trait? Jpn. J. Hum. Genet. 6:21-25.

Race, R. R., and R. Sanger. 1968. Blood groups in man. F. A. Davis Co., Philadelphia.

Rachman, S. 1960. Galvanic skin responses in identical twins. Psychol. Rep. 6:298.

Rainer, J. D. 1966. The contributions of Franz Josef Kallman to the genetics of schizophrenia. Behav. Sci. 11:413-437.

Ramaley, E. 1913. Inheritance of left-handedness. Am. Natur. 47:730-738.

Ramirez, I., and J. L. Fuller. 1976. Genetic influences on water and sweetened water consumption in mice. Physiol. Behav. 16:163-168.

Randall, C. L., and D. Lester. 1975. Social modification of alcohol consumption in inbred mice. Science 189:149-151.

Raney, E. T. 1938. Reversed lateral dominance in identical twins. J. Exp. Psychol. 23:304-312.

Rao, D. C., N. E. Morton, and S. Yee. 1974. Analysis of family resemblance. II. A linear model for family correlation. Am. J. Hum. Genet. 26:331-359.

Rawnsley, K. 1968. Epidemiology of affective disorders. In A. Coppen and A. Walk, eds. Recent developments in affective disorders. Br. J. Psychiatry Spec. Publ. No. 2.

Reading, A. J. 1966. Effects of maternal environment on the behavior of inbred mice. J. Comp. Physiol. Psychol. 62:437-440.

Record, R. G., T. McKeown, and J. H. Edwards. 1970. An investigation of the differences in measured intelligence between twins and single births. Ann. Hum. Genet. 34:11-20.

Reed, E. W., and S. C. Reed. 1965. Mental retardation: a family study. W. B. Saunders Co., Philadelphia.

Reed, S. C., C. M. Williams, and L. E. Chadwick. 1942. Frequency of wing beat as a character for separating races, species and geographical varieties in *Drosophila*. Genetics 27:349-361.

Reed, T. E. 1969. Caucasian genes in American Negroes. Science 165:762-778.

Reeve, E. C. R., and C. H. Waddington. 1952. Quantitative inheritance. Her Majesty's Stationery Office, London.

Reich, T., W. A. James, and C. A. Morris. 1972. The use of multiple thresholds in determining the mode of transmission of semi-continuous traits. Ann. Hum. Genet. 36:163-184.

Reimer, J. D., and M. L. Petras. 1967. Breeding structure of the house mouse, *Mus musculus*, in a population cage. J. Mammal. 48:88-99.

Rensch, B. 1956. Increase of learning capability with increase of brain size. Am. Natur. 90:81-95.

Reser, H. 1935. Inheritance of musical ability. Eugen. News 20:8-9.

Ressler, R. H. 1962. Parental handling in two strains of mice reared by foster parents. Science 137:129-130.

Ressler, R. H. 1963. Genotype-correlated parental influences in two strains of mice. J. Comp. Physiol. Psychol. 56:882-886.

Reznikoff, M., G. Domino, C. Bridges, and M. S. Honeyman. 1973. Creative abilities in identical and fraternal twins. Behav. Genet. 3:365-377.

Reznikoff, M., and M. S. Honeyman. 1967. MMPI profiles of monozygotic and dizygotic twin pairs. J. Consult. Psychol. 31:100. (Abstr.)

Rhine, J. B., and W. McDougall. 1933. Third report on a Lamarckian experiment. Br. J. Psychol. 24:213-235.

Richardson, J. S. 1936. The correlation of intelligence quotients of siblings of the same chronological age levels. J. Juven. Res. 20:186-198.

Richter, C. P. 1952. Domestication of the Norway rat and its implications for the study of genetics in man. Am. J. Hum. Genet. 4:273-285.

Richter, C. P. 1954. The effects of domestication and selection on the behavior of the Norway rat. J. Natl. Cancer Inst. 15:727-738.

Richter, C. P., P. V. Rogers, and C. E. Hall. 1950. Failure of salt replacement therapy in adrenalectomized recently captured wild Norway rats. Endocrinology 46:233-242.

Richter, C. P., and E. H. Uhlenhuth. 1954. Comparison of the effects of gonadectomy on spontaneous activity of wild and domesticated Norway rats. Endocrinology 54:311-322.

Rife, D. C. 1933. Genetic studies of monozygotic twins. I.-III. J. Hered. 24:339-345.

Rife, D. C. 1940. Handedness with special reference to twins. Genetics 25:178-186.

Riopelle, A. J., and C. W. Hill. 1973. Complex processes. In D. A. Dewsbury and D. A. Rethlingshafer, eds. Comparative psychology. McGraw-Hill Book Co., New York.

Roberts, E., and L. E. Card. 1934. Inheritance of broodiness in the domestic fowl. Proc. 5th World Poultry Cong. Rome 2:353-358.

Roberts, J. A. F. 1940. Studies on a child population. V. The resemblance in intelligence between sibs. Ann. Eugen. 10:293-312.

Roberts, J. A. F. 1952. The genetics of mental deficiency. Eugen. Rev. 44:71-83.

Roberts, R. C. 1967. Some concepts and methods in quantitative genetics. In J. Hirsch, ed. Behavior-genetic analysis. McGraw-Hill Book Co., New York.

Robins, E., and B. K. Hartman. 1972. Biochemical theories of mental disorders. In R. W. Albers, G. J. Siegel, R. Katzman, and B. W. Agranoff, eds. Basic neurochemistry. Little, Brown & Co., Boston.

Rockwell, R. F., and M. B. Seiger. 1973a. A comparative study of photoresponse in *Drosophila pseudoobscura* and *D. persimilis*. Behav. Genet. 3:163-174.

Rockwell, R. F., and M. B. Seiger. 1973b. Phototaxis

in *Drosophila:* a critical evaluation. Am. Sci. **61**:339-345.

Roderick, T. H. 1960. Selection for cholinesterase activity in the cerebral cortex of the rat. Genetics **45**: 1123-1140.

Roderick, T. H., R. E. Wimer, and C. C. Wimer. 1976. Genetic manipulation of neuroanatomical traits. In L. Petrinovich and J. L. McGaugh, eds. Knowing, thinking and believing. Plenum Publishing Corp., New York.

Rodgers, D. A. 1967. Behavior genetics and overparticularization: an historical perspective. In J. N. Spuhler, ed. Genetic diversity and human behavior. Aldine Publishing Co., Chicago.

Rodgers, D. A., and G. E. McClearn. 1962. Mouse strain differences in preference for various concentrations of alcohol. Q. J. Stud. Alc. **23**:26-33.

Rodgers, D. A., and G. E. McClearn. 1964. Sucrose versus ethanol appetite in inbred strains of mice. Q. J. Stud. Alc. **25**:26-35.

Rodgers, D. A., G. E. McClearn, E. L. Bennett, and M. Herbert. 1963. Alcohol preference as a function of its caloric utility in mice. J. Comp. Physiol. Psychol. **56**:666-672.

Roe, A. 1944. The adult adjustment of children of alcoholic parentage raised in foster homes. Q. J. Stud. Alc. **5**:378-393.

Roff, M. 1950. Intra-family resemblance in personality characteristics. J. Psychol. **30**:199-227.

Rogers, L. J. 1974. Persistence and search influenced by natural levels of androgens in young and adult chickens. Physiol. Behav. **12**:197-204.

Rogers, P. V., and Richter, C. P. 1948. Anatomical comparison between the adrenal glands of wild Norway, wild Alexandrine and domestic Norway rats. Endocrinology **42**:46-55.

Röll, A., and J. L. Entres. 1936. Zum Problem der Erbprognosebestimmung. Z. Gesamte Neurol. Psychiatr. **156**:169-202.

Rosanoff, A. J., L. M. Handy, and I. R. Plesset. 1935. The etiology of manic-depressive syndromes with special reference to their occurrence in twins. Am. J. Psychiatry **91**:725-762.

Rosanoff, A. J., L. M. Handy, and I. R. Plesset. 1937. The etiology of mental deficiency with special reference to its occurrence in twins. Psychol. Monogr. **48**(216).

Rosanoff, A. J., L. M. Handy, and I. A. Rosanoff. 1934. Criminality and delinquency in twins. J. Crim. Law Criminol. **24**:923-934.

Rose, A., and P. A. Parsons. 1970. Behavioural studies in different strains of mice and the problem of heterosis. Genetica **41**:65-87.

Rose, R. M., J. W. Holaday, and I. S. Bernstein. 1971. Plasma testosterone, dominance rank and aggressive behavior in male rhesus monkeys. Nature **231**:366-368.

Rosenberg, C. M. 1967. Familial aspects of obsessional neurosis. Br. J. Psychiatr. **113**:405-413.

Rosenthal, D. 1959. Some factors associated with concordance and discordance with respect to schizophrenia in monozygotic twins. J. Nerv. Ment. Dis. **129**:1-10.

Rosenthal, D. 1961. Sex distribution and the severity of illness among samples of schizophrenic twins. J. Psychiatr. Res. **1**:26-36.

Rosenthal, D. 1962. Problems of sampling and diagnosis in the major twin studies of schizophrenia. J. Psychiatr. Res. **1**:116-134.

Rosenthal, D. 1963. The Genain quadruplets. Basic Books, Inc., Publishers, New York.

Rosenthal, D. 1966. The offspring of schizophrenic couples. J. Psychiatr. Res. **4**:169-188.

Rosenthal, D. 1970. Genetic theory and abnormal behavior. McGraw-Hill Book Co., Inc., New York.

Rosenthal, D. 1972. Three adoption studies of heredity in the schizophrenic disorders. Int. J. Ment. Health **1**:63-75.

Rosenthal, D., and J. Van Dyke. 1970. The use of monozygotic twins discordant as to schizophrenia in the search for an inherited characterological defect. Acta Psychiatr. Scand. [Suppl.]**219**:183-189.

Rosenthal, D., P. H. Wender, S. S. Kety, F. Schulsinger, J. Welner, and L. Östergaard. 1968. Schizophrenics' offspring reared in adoptive homes. In D. Rosenthal and S. S. Kety, eds. The transmission of schizophrenia. Pergamon Press Ltd., Oxford.

Rosenthal, D., P. H. Wender, S. S. Kety, J. Welner, and F. Schulsinger. 1971. The adopted-away offspring of schizophrenics. Am. J. Psychiatry **128**:307-311.

Rothenbuhler, W. C. 1958. Genetics and breeding of the honey bee. Annu. Rev. Entomol. **3**:161-180.

Rothenbuhler, W. C. 1964a. Behaviour genetics of nest cleaning in honey bees. I. Responses of four inbred lines to disease-killed brood. Anim. Behav. **12**:578-583.

Rothenbuhler, W. C. 1964b. Behavior genetics of nest cleaning in honey bees. IV. Responses of F_1 and backcross generations to disease-killed brood. Am. Zool. **4**:111-123.

Rothenbuhler, W. C. 1967. Genetic and evolutionary considerations of social behavior of honey bees and some related insects. In J. Hirsch, ed. Behavior-genetic analysis. McGraw-Hill Book Co., New York.

Roubicek, C. G., and D. E. Ray. 1969. Genetic selection for adipsia and polydipsia in the rat. J. Hered. **60**:332-335.

Rowland, G. L., and P. J. Woods. 1961. Performance of the Tryon bright and dull strains under two conditions in a multiple T-maze. Can. J. Psychol. **15**:20-28.

Royce, J. R. 1955. A factorial study of emotionality in the dog. Psychol. Monogr. **69**(22):1-27.

Royce, J. R. 1957. Factor theory and genetics. Educ. Psychol. Meas. **17**:361-376.

Royce, J. R. 1973. The conceptual framework for a multi-factor theory of individuality. In J. R. Royce, ed. Multivariate analysis and psychological theory. Academic Press Inc. (London) Ltd., London.

Royce, J. R., A. Carran, and E. Howarth. 1970. Factor

analysis of emotionality in ten strains of inbred mice. Multivar. Behav. Res. **5:**19-48.

Royce, J. R., T. M. Holmes, and W. Poley. 1975. Behavior genetic analysis of mouse emotionality. III. The diallel analysis. Behav. Genet. **5:**351-372.

Royce, J. R., W. Poley, and L. T. Yeudall. 1973. Behavior-genetic analysis of mouse emotionality. I. Factor analysis. J. Comp. Physiol. Psychol. **83:**36-47.

Royce, J. R., L. T. Yeudall, and W. Poley. 1973. Diallel analysis of avoidance conditioning in inbred strains of mice. J. Comp. Physiol. Psychol. **76:**353-358.

Rucker, W. B. 1973. Is genetic separation of long- and short-term memory possible? American Philosophical Society Yearbook. American Philosophical Society, Philadelphia.

Rüdin, E. 1916. Zur Vererbung und Neuenstehung der Dementia praecox. Springer Verlag, Berlin.

Rüdin, E. 1923. Über Vererbung geistigen Störungen. Z. Gesamte Neurol. Psychiatr. **81:**459-496.

Rüdin, E. 1953. Ein Beitrag zur Frage der Zwangskrankheit insbesondere hereditären Beziehungen. Arch. Psychiatr. Nervenkr. **191:**14-54.

Rundquist, E. A. 1933. The inheritance of spontaneous activity in rats. J. Comp. Psychol. **16:**415-438.

Runner, M. N. 1954. Inheritance of susceptibility to congenital deformation-embryonic instability. J. Natl. Cancer Inst. **15:**637-649.

Rushton, W. A. H. 1966. Densitometry of pigments in rods and cones of normal and color defective subjects. Invest. Ophthalmol. **5:**233-241.

Russell, E. S. 1963. Problems and potentialities in the study of genic action in the mouse. In W. J. Burdette, ed. Methodology in mammalian genetics. Holden-Day, Inc., San Francisco.

Russell, W. L. 1941. Inbred and hybrid animals and their value in research. In G. B. Snell, ed. 1960. Biology of the laboratory mouse. Dover Publications, Inc., New York.

Rutter, M. 1966. Children of sick parents: an environmental and psychiatric study. Maudsley Monogr. No. 16.

Rutter, M., S. Korn, and H. G. Birch. 1963. Genetic and environmental factors in the development of primary reaction patterns. Br. J. Soc. Clin. Psychol. **2:**161-173.

Sahakian, W. S. 1970. Psychopathology today: experimentation, theory and research. F. E. Peacock Publishers, Inc., Itasca, Ill.

Sainsbury, P. 1968. Suicide and depression. In A. Coppen and A. Walk, eds. Recent developments in affective disorders. Br. J. Psychiatry Spec. Publ. No. 2.

St. John, R. D., and P. A. Corning. 1973. Maternal aggression in mice. Behav. Biol. **9:**635-639.

Sakai, T. 1967. Clinico-genetic study on obsessive-compulsive neurosis. In H. Mitsuda, ed. Clinical genetics in psychiatry. Igaku Shoin Ltd., Tokyo.

Saldanha, P. H., and J. Nacrur. 1963. Taste thresholds for phenylthiourea among Chileans. Am. J. Phys. Anthropol. **21:**113-119.

Sanders, J. 1934. Homosexuelle tweelingen. Ned. Tijdschr. Geneeskd. **78:**3346-3352.

Sants, H. J. 1964. Genealogical bewilderment in children with substitute parents. Br. J. Med. Psychol. **37:**133-141.

Sargent, F., and K. P. Weinman. 1966. Physiological individuality. Ann. N.Y. Acad. Sci. **134:**696-719.

Satinder, K. P. 1972. Behavior-genetic-dependent self-selection of alcohol in rats. J. Comp. Physiol. Psychol. **80:**422-434.

Satinder, K. P., and K. D. Hill. 1974. Effects of genotype and postnatal experience on activity, avoidance, shock threshold, and open-field behavior of rats. J. Comp. Physiol. Psychol. **86:**363-374.

Satinder, K. P., and W. R. Petryshyn. 1974. Interaction among genotype, unconditioned stimulus, d-amphetamine and one-way avoidance of rats. J. Comp. Physiol. Psychol. **86:**1059-1073.

Satinder, K. P., J. R. Royce, and L. T. Yeudall. 1970. Effects of electric shock, d-amphetamine sulfate and chlorpromazine on factors of emotionality in mice. J. Comp. Physiol. Psychol. **71:**443-447.

Saul, G. B., II, E. B. Garrity, K. Benirschke, and H. Valtin. 1968. Inherited hypothalamic diabetes insipidus in the Brattleboro strain of rats. J. Hered. **59:**113-117.

Sawin, P. B., and R. H. Curran. 1952. Genetic and physiological background of reproduction in the rabbit. I. The problem and its biological significance. J. Exp. Zool. **120:**165-201.

Sawin, P. B., and D. D. Crary. 1953. Genetic and physiological background of reproduction in the rabbit. II. Some racial differences in the pattern of maternal behavior. Behaviour **6:**128-145.

Scarr, S. 1969a. Environmental bias in twin studies. In M. Manosevitz, G. Lindzey, and D. Thiessen, eds. Behavioral genetics: method and research. Appleton-Century-Crofts, New York.

Scarr, S. 1969b. Social introversion-extraversion as a heritable response. Child Dev. **40:**823-832.

Scarr-Salapatek, S. 1971. Race, social class and IQ. Science **174:**1285-1295.

Schaffner, A., E. A. Lane, and G. W. Albee. 1967. Intellectual differences between suburban preschizophrenic children and their siblings. J. Consult. Psychol. **31:**326-327.

Schaie, K. W., V. E. Anderson, G. E. McClearn, and J. Money. 1975. Developmental human behavior genetics. D. C. Heath & Co., Lexington, Mass.

Scheinfeld, A. 1956. The new heredity and you. Chatto & Windus Ltd., London.

Scheinfeld, A. 1963. Biosocial effects on twinning incidences. II. The world situation. Proceedings of the Second International Conference of Human Genetics, Rome. Istituto Gregor Mendel, Rome.

Scheinfeld, A. 1967. Twins and supertwins. J. B. Lippincott Co., Philadelphia.

Scheinfeld, A., and J. Schacter. 1963. Biosocial effects on twinning incidences. I. Intergroup and generation differences, U.S. Proceedings of the Second International Conference of Human Genetics, Rome. Istituto Gregor Mendel, Rome.

Schlesinger, K., R. C. Elston, and W. Boggan. 1966. The genetics of sound induced seizures in inbred mice. Genetics 54:95-103.

Schlesinger, K., and B. J. Griek. 1970. The genetics and biochemistry of audiogenic seizures. In G. Lindzey and D. D. Thiessen, eds. Contributions to behavior-genetic analysis: the mouse as a prototype. Appleton-Century-Crofts, New York.

Schlesinger, K., R. Kakihana, and E. L. Bennett. 1966. Effects of tetraethylthiuramdisulfide (Antabuse) on metabolism and consumption of ethanol in mice. Psychosom. Med. 28:514-520.

Schlesinger, K., and S. K. Sharpless. 1975. Audiogenic seizures and acoustic priming. In B. E. Eleftheriou, ed. Psychopharmacogenetics. Plenum Publishing Corp., New York.

Schneider, C. W., S. K. Evans, M. B. Chenoweth, and F. L. Beman. 1973. Ethanol preference and behavioral tolerance in mice: biochemical and neurophysiological mechanisms. J. Comp. Physiol. Psychol. 82:466-474.

Schneirla, T. C. 1965. Aspects of stimulation and organization in approach withdrawal processes underlying vertebrate behavioral development. In D. S. Lehrman, R. A. Hinde, and E. Shaw, eds. Advances in the Study of Behavior. Academic Press, Inc., New York.

Schoenfeldt, L. F. 1969. Hereditary-environmental components of the project TALENT two-day test battery. Proceedings of the Sixteenth International Congress on Applied Psychology, London. Swets & Zeitlinger, Amsterdam.

Schoenheimer, R. 1942. The dynamic state of the body constituents. Harvard University Press, Cambridge, Mass.

Schooler, C. 1972. Birth order effects: not here, not now! Psychol. Bull. 78:161-175.

Schuckit, M. A., D. W. Goodwin, and G. Winokur. 1972. A study of alcoholism in half siblings. Am. J. Psychiatry 128:122-126.

Schull, W. J., H. Nagano, M. Yamomoto, and I. Komatsu. 1970. The effects of parental consanguinity and inbreeding in Hirado, Japan. I. Stillbirths and prereproductive mortality. Am. J. Hum. Genet. 22:239-262.

Schull, W. J., and J. V. Neel. 1965. The effects of inbreeding on Japanese children. Harper & Row, Publishers, New York.

Schuster, E., and E. M. Elderton. 1907. The inheritance of ability. Eugen. Lab. Mem. 1:1-42.

Schwartz, M., and J. Schwartz. 1976. Comment on "IQ's of identical twins reared apart." Behav. Genet. 6:367-368.

Schwesinger, G. C. 1933. Heredity and environment. MacMillan Book Co., New York.

Scott, J. P. 1942. Genetic differences in the social behavior of inbred strains of mice. J. Hered. 33:11-15.

Scott, J. P. 1943. Effects of single genes on the behavior of *Drosophila*. Am. Natur. 77:184-190.

Scott, J. P. 1949. Genetics as a tool in experimental psychological research. Am. Psychol. 4:526-530.

Scott, J. P. 1958. Animal behavior. University of Chicago Press, Chicago.

Scott, J. P. 1964. Genetics and the development of social behavior in dogs. Am. Zool. 4:161-168.

Scott, J. P. 1966. Agonistic behavior of mice and rats: a review. Am. Zool. 6:683-701.

Scott, J. P., and E. Fredericson. 1951. The causes of fighting in mice and rats. Physiol. Zool. 24:273-309.

Scott, J. P., and J. L. Fuller. 1963. Behavioral differences. In W. J. Burdette, ed. Methodology in mammalian genetics. Holden-Day, Inc., San Francisco.

Scott, J. P., and J. L. Fuller. 1965. Genetics and the social behavior of the dog. University of Chicago Press, Chicago.

Scott, W. A. 1958. Research definitions of mental health and mental illness. Psychol. Bull. 55:29-45.

Searle, L. V. 1949. The organization of hereditary maze-brightness and maze-dullness. Genet. Psychol. Monogr. 39:279-325.

Seemanová, E. 1971. A study of children of incestuous matings. Hum. Hered. 21:108-128.

Seiger, M. B. 1967. A computer simulation study of the influence of imprinting on population structure. Am. Natur. 101:47-57.

Selander, R. K. 1970. Behavior and genetic variation in natural populations. Am. Zool. 10:53-66.

Seligman, M. E. P. 1970. On the generality of the laws of learning. Psychol. Rev. 77:406-418.

Seligman, M. E. P., and J. L. Hager. 1972. Biological boundaries of learning. Appleton-Century-Crofts, New York.

Selmanoff, M. K., S. C. Maxson, and B. E. Ginsburg. 1976. Chromosomal determinants of intermale aggressive behavior in inbred mice. Behav. Genet. 6:53-69.

Sen Gupta, N. N. 1941. Heredity in mental traits. Macmillan & Co. Ltd., London.

Shaefer, V. H. 1959. Differences between strains of rats in avoidance conditioning without an explicit warning stimulus. J. Comp. Physiol. Psychol. 52:120-122.

Shapiro, A. P., J. Nicotero, J. Sopria, and E. T. Scheib. 1968. Analysis of the variability of blood pressure, pulse rate and catecholamine responsivity in identical and fraternal twins. Psychosom. Med. 30:506-520.

Shapiro, B. H., and A. S. Goldman. 1973. Feminine saccharin preference in the genetically androgen-insensitive male rat pseudohermaphrodite. Horm. Behav. 4:371-375.

Sharpe, R., and P. Johnsgard. 1967. Inheritance of behavioural characters in F_2 mallard and pintail ducks. Behaviour 27:259-272.

Sheppard, J. R., P. Albersheim, and G. E. McClearn. 1966. Enzyme activities and ethanol preference in mice. Biochem. Genet. 2:205-212.

Sheppard, J. R., P. Albersheim, and G. E. McClearn. 1970. Aldehyde dehydrogenase and ethanol preference in mice. J. Biol. Chem. 245:2876-2882.

Sheridan, C. L. 1965. Interocular transfer of brightness and pattern discriminations in normal and corpus callosum-sectioned rats. J. Comp. Physiol. Psychol. 59:292-294.

Shields, J. 1954. Personality differences and neurotic traits in normal twin school children. Eugen. Rev. 45:213-246.

Shields, J. 1962. Monozygotic twins brought up apart and brought up together. Oxford University Press, London.

Shields, J. 1968. Summary of the genetic evidence. In D. Rosenthal and S. S. Kety, eds. The transmission of schizophrenia. Pergamon Press Ltd., Oxford.

Shields, J., and I. I. Gottesman. 1972. Cross-national diagnosis of schizophrenia in twins. Arch. Gen. Psychiatry 27:725-730.

Shields, J., I. I. Gottesman, and E. Slater. 1967. Kallman's 1946 schizophrenic twin study in the light of new information. Acta Psychiatr. Scand. 43:385-396.

Shire, J. G. M., and S. G. Spickett. 1968. Genetic variation in adrenal structure: strain differences in quantitative characters. J. Endocrinol. 40:215-229.

Shuster, L. 1975. Genetic analysis of morphine effects: activity, analgesia and sensitization. In B. E. Eleftheriou, ed. Psychopharmacogenetics. Plenum Publishing Corp., New York.

Shuter, R. P. G. 1966. Hereditary and environmental factors in musical ability. Eugen. Rev. 58:149-156.

Sidman, R. L., J. B. Angevine, and P. E. Taber. 1971. Atlas of the mouse brain and spinal cord. Harvard University Press, Cambridge, Mass.

Sidman, R. L., M. C. Green, and S. H. Appel. 1965. Catalog of the neurological mutants of the mouse. Harvard University Press, Cambridge, Mass.

Siegel, P. B. 1965. Genetics of behavior: selection for mating ability in chickens. Genetics 52:1269-1277.

Siegel, P. B. 1972. Genetic analyses of male mating behavior in chickens. I. Artificial selection. Anim. Behav. 20:564-570.

Siegel, P. B., and E. L. Wisman. 1966. Selection for body weight at eight weeks of age. VI. Changes in appetite and food utilization. Poultry Sci. 45:1391-1397.

Siemens, H. W. 1924. Die Zwillingspathologie. Springer-Verlag, Berlin.

Silverman, W., F. Shapiro, and W. T. Heron. 1940. Brain weight and maze learning in rats. J. Comp. Psychol. 30:279-282.

Simmel, E. C., and B. E. Eleftheriou. 1977. Multivariate and behavior genetic analysis of avoidance of complex visual stimuli and activity in recombinant inbred strains of mice. Behav. Genet. 7:239-250.

Sims, V. M. 1931. The influence of blood relationship and common environment on measured intelligence. J. Educ. Psychol. 22:56-65.

Sines, J. O. 1959. Selective breeding for development of stomach lesions following stress in the rat. J. Comp. Physiol. Psychol. 52:615-617.

Sjögren, T. 1948. Genetic-statistical and psychiatric investigations of a West Swedish population. Acta Psychiatr. Neurol. Scand. [Suppl.]52.

Skeels, H. M. 1936. The mental development of children in foster homes. J. Genet. Psychol. 49:91-106.

Skeels, H. M. 1966. Adult status of children with contrasting early life experiences: a follow-up study. Child Dev. Monogr. 31(3): Serial No. 105.

Skinner, B. F. 1938. The behavior of organisms. Appleton-Century-Crofts, New York.

Skinner, B. F. 1940. A method of obtaining an arbitrary degree of hunger. J. Comp. Psychol. 30:139-145.

Skinner, B. F. 1957. Verbal behavior. Appleton-Century-Crofts, New York.

Skodak, M., and H. M. Skeels. 1949. A final follow-up study of one-hundred adopted children. J. Genet. Psychol. 75:85-125.

Slater, E. 1936. The inheritance of manic-depressive insanity: provisional report. Proc. R. Soc. Med. 29:981-990.

Slater, E. 1938. Zur Erbpathologie des manisch-depressiven Irreseins. Die Eltern und Kinder von Manisch-Depressiven. Z. Gesamte Neurol. Psychiatr. 163:1-47.

Slater, E. 1943. The neurotic constitution: a statistical study of two thousand neurotic soldiers. J. Neurol. Psychiatry 6:1-16.

Slater, E. 1953. Psychotic and neurotic illnesses in twins. Spec. Rep. Med. Res. Council No. 278. Her Majesty's Stationery Office, London.

Slater, E. 1958. The monogenic theory of schizophrenia. Acta Genet. Stat. Med. 8:50-56.

Slater, E. 1961. The thirty-fifth Maudsley lecture: "Hysteria 311." J. Ment. Sci. 107:359-381.

Slater, E. 1962. Birth order and maternal age of homosexuals. Lancet 1:69-71.

Slater, E. 1965. Diagnosis of "hysteria." Br. Med. J. 1:1395-1399.

Slater, E. 1966. Expectation of abnormality on paternal and maternal sides: a computational model. J. Med. Genet. 3:159-161.

Slater, E. 1968. A review of earlier evidence on genetic factors in schizophrenia. In D. Rosenthal and S. S. Kety, eds. The transmission of schizophrenia. Pergamon Press Ltd., Oxford.

Slater, E. 1972. The case for a major partially dominant gene. In A. R. Kaplan, ed. Genetic factors in "schizophrenia." Charles C Thomas, Publisher, Springfield, Ill.

Slater, E., and V. Cowie. 1971. The genetics of mental disorders. Oxford University Press, London.

Slater, E., and E. Glithero. 1965. A follow-up of patients diagnosed as suffering from "hysteria." J. Psychosom. Res. 9:9-13.

Slater, E., and J. Shields. 1969. Genetical aspects of anxiety. In M. H. Lader, ed. Br. J. Psychiatry Spec. Publ. No. 3.

Slater, E., and N. T. Tsuang. 1968. Abnormality on paternal and maternal sides: observations in schizophrenia and manic-depression. J. Med. Genet. 5:197-199.

Smart, J. L. 1970. Trial and error behaviour of inbred and F_1 hybrid mice. Anim. Behav. 18:445-453.

Smart, R. G. 1963. Alcoholism, birth order and family size. J. Abnorm. Soc. Psychol. 66:17-23.

Smith, A., and R. G. Record. 1955. Maternal age and birth rank in the aetiology of mongolism. Br. J. Prev. Med. **9**:51-55.

Smith, C. 1970. Heritability of liability and concordance in monozygous twins. Ann. Hum. Genet. **34**:85-91.

Smith, C. 1971. Recurrence risks for multifactorial inheritance. Am. J. Hum. Genet. **23**:578-588.

Smith, C. 1974. Concordance in twins: methods and interpretation. Am. J. Hum. Genet. **26**:454-466.

Smith, G. 1949. Psychological studies in twin differences. Stud. Psychol. Paedogog. Ser. Altera Invest. **3**. AB C. W. K. Gleerup Bokförlag, Lund, Sweden.

Smith, G. 1953. Twin differences with reference to the Müller-Lyer illusion. Lund Univ. Arsskrift N.F. Aud. 1, **50**:1-27.

Smith, R. T. 1965. A comparison of socioenvironmental factors in monozygotic and dizygotic twins: testing and assumption. In S. G. Vandenberg, ed. Methods and goals in human behavior genetics. Academic Press, Inc., New York.

Smith, S. M., and L. S. Penrose. 1955. Monozygotic and dizygotic twin diagnosis. Ann. Hum. Genet. **19**:273-289.

Snider, B. 1955. A comparative study of achievement test scores of fraternal and identical twins and siblings. Doctoral dissertation. Iowa State University, Ames, Iowa.

Snowdon, C. T., D. D. Bell, and N. D. Henderson. 1964. Relationship between heart rate and open field behavior. J. Comp. Physiol. Psychol. **58**:423-426.

Snyder, L. H. 1931. Inherited taste deficiency. Science **74**:151-152.

Snyder, L. H. 1932. The inheritance of taste deficiency in man. Ohio J. Sci. **32**:436-440.

Snyder, L. H., and P. R. David. 1953. Penetrance and expression. In A. Sorsby, ed. Clinical genetics. Butterworth & Co. (Publishers) Ltd., London.

Snygg, D. 1938. The relation between the intelligence of mothers and of their children living in foster homes. J. Genet. Psychol. **52**:401-406.

Sorensen, M. I., and H. D. Carter. 1940. Twin resemblances in community of free association responses. J. Psychol. **9**:237-246.

Sorsby, A. 1953. The eye. In A. Sorsby, ed. Clinical genetics. Butterworth & Co. (Publishers) Ltd., London.

Sorsby, A. 1970. Ophthalmic genetics, 2nd ed. Butterworth & Co. (Publishers) Ltd., London.

Southwick, C. H. 1968. Effect of maternal environment on aggressive behavior of inbred mice. Commun. Behav. Biol. A **1**:129-132.

Southwick, C. H., and L. H. Clark. 1968. Interstrain differences in aggressive behavior and exploratory activity of inbred mice. Commun. Behav. Biol. A **1**:49-59.

Spassky, B., and T. Dobzhansky. 1967. Response of various strains of *Drosophila pseudoobscura* and *D. persimilis* to light and gravity. Am. Natur. **101**:59-63.

Spearman, C. 1927. The abilities of man. Macmillan & Co. Ltd., London.

Sperry, R. W., and M. S. Gazzaniga. 1967. Language following surgical disconnection of the hemispheres. In F. L. Darley, ed. Brain mechanisms underlying speech and language. Grune & Stratton, Inc., New York.

Spiess, E. B., and B. Langer. 1964. Mating speed control by gene arrangements in *Drosophila pseudoobscura* homokaryotes. Proc. Natl. Acad. Sci. U.S.A. **51**:1015-1019.

Spiess, E. B., and H.-F. Yu. 1975. Relative mating activity of the sexes in homokaryotypes of *Drosophila persimilis* from a Redwoods population. Behav. Genet. **5**:203-216.

Spiess, L. D., and E. B. Spiess. 1969. Minority advantage in interpopulation matings of *Drosophila persimilis*. Am. Natur. **103**:155-172.

Spieth, H. T. 1951. Mating behavior and sexual isolation in the *Drosophila virilis* species group. Behaviour **3**:105-145.

Spieth, H. T. 1952. Mating behavior within the genus *Drosophila* (Diptera). Bull. Am. Mus. Nat. Hist. **99**:395-474.

Spong, P. 1962. Recognition and recall of retarded readers: a developmental study. Winifred Gembett Rep. Univ. Aukland, N.Z.

Spuhler, J. N. 1968. Sociocultural and biological inheritance in man. In D. Glass, ed. Genetics. Rockefeller University Press, New York.

Srb, A. M., and R. D. Owen. 1953. General genetics. W. H. Freeman & Co. Publishers, San Francisco.

Stabenau, J. R., W. Pollin, L. R. Mosher, C. Frohan, A. J. Friedhoff, and W. Turner. 1969. Study of monozygotic twins discordant for schizophrenia. Arch. Gen. Psychiatry **20**:145-158.

Stafford, R. E. 1965. Nonparametric analysis of twin data with the Mann-Whitney U-test. Research report No. 10, Louisville Twin Study, Child Development Unit, University of Louisville School of Medicine, Louisville, Ky.

Stalker, H. D. 1942. Sexual isolation studies in the species complex *Drosophila virilis*. Genetics **27**:238-257.

Stamm, J. S. 1954. Genetics of hoarding. I. Hoarding differences between homozygous strains of rats. J. Comp. Psychol. **47**:157-161.

Stamm, J. S. 1956. Genetics of hoarding. II. Hoarding behavior of hybrid and backcrossed strains of rats. J. Comp. Physiol. Psychol. **49**:349-352.

Stanton, H. M. 1922. Inheritance of specific musical capacities. Psychol. Monogr. **31**(140):157-204.

Starch, D. 1915. The inheritance of abilities in school studies. School Soc. **2**:608-610.

Starch, D. 1917. The similarity of brothers and sisters in mental traits. Psychol. Rev. **24**:235-238.

Stasik, J. H. 1970. Inheritance of T-maze learning in mice. J. Comp. Physiol. Psychol. **71**:251-257.

Stenstedt, A. 1952. A study in manic-depressive psychosis: clinical social and genetic investigations. Acta Psychiatr. Neurol. Scand. [Suppl.]**79**.

Stenstedt, A. 1966. Genetics of neurotic depression. Acta Psychiatr. Scand. **42**:398-409.

Stern, C. 1973. Principles of human genetics, 3rd ed. W. H. Freeman & Co. Publishers, San Francisco.

Stewart, J. 1969a. Biometrical genetics with one or two loci. I. The choice of a specific genetic model. Heredity **24**:211-224.

Stewart, J. 1969b. Biometrical genetics with one or two loci. II. The estimation of linkage. Heredity **24**:225-238.

Stewart, J., and R. C. Elston. 1973. Biometrical genetics with one or two loci: the inheritance of physiological characters in mice. Genetics **73**:675-693.

Stockard, C. R., O. D. Anderson, and W. T. James. 1941. Genetic and endocrinic basis for differences in form and behavior. Wistar Institute Press, Philadelphia.

Stocks, P., and M. N. Karn. 1933. A biometric investigation of twins and their brothers and sisters. Ann. Eugen. **5**:1-55.

Stockton, M. D., and G. Whitney. 1974. Effects of genotype, sugar and concentration on sugar preference of laboratory mice. J. Comp. Physiol. Psychol. **86**:62-68.

Stone, C. P. 1932. Wildness and savageness in rats. In K. S. Lashley, ed. Studies in the dynamics of behavior. University of Chicago Press, Chicago.

Strandskov, H. H. 1954. A twin study pertaining to the genetics of intelligence. Caryologia Suppl. Att. 9th Int. Cong. Genet. 811-813.

Strömgren, E. 1950. Statistical and genetical population studies within psychiatry. Congrès Internationale de Psychiatrie, vol. 6. Hermann, Paris.

Strong, E. K. 1943. Vocational interests in men and women. Stanford University Press, Stanford.

Stumpfl, F. 1936. Die Ursprünge des Verbrechens, dargestellt am Lebenslauf von Zwillingen. Georg Thieme, Leipzig.

Stumpfl, F. 1937. Untersuchungen an psychopathischen Zwillingen. Z. Gesamte Neurol. Psychiatr. **158**:480-482.

Sulston, J. E., and S. Brenner. 1974. The DNA of *Caenorhabditis elegans*. Genetics **77**:95-104.

Super, D. E. 1949. Appraising vocational fitness by means of psychological tests. Harper & Brothers, New York.

Surwillo, W. W. 1977. Interval histograms of period of electroencephalogram and the reaction time of twins. Behav. Genet. **7**:161-170.

Sutton, W. S. 1903. The chromosomes in heredity. Biol. Bull. **4**:231-251.

Sward, K., and M. B. Friedman. 1935. Jewish temperament. J. Appl. Psychol. **19**:70-84.

Symons, J. P., and R. L. Sprott. 1976. Genetic analysis of schedule-induced polydipsia. Physiol. Behav. **17**:837-839.

Szaz, T. S. 1960. The myth of mental illness. Am. Psychol. **15**:113-118.

Tallman, G. G. 1928. A comparative study of identical and non-identical twins with respect to intelligence resemblances. In Twenty-seventh Yearbook of the National Society for the Study of Education, Part I. Public School Publishing Co., Bloomington, Ill.

Tangherani, W., and L. Pardelli. 1958. The electroencephalogram in the neonatal period. II. Findings in twins. Lattante **29**:152-161.

Tarcsay, I. 1939. Testing of will-temperament in twins. Psychol. Stud. Univ. Budapest **3**:79-111.

Tardif, G. N., and M. R. Murnik. 1975. Frequency-dependent sexual selection among wild-type strains of *Drosophila melanogaster*. Behav. Genet. **5**:373-379.

Tellegen, A., and J. M. Horn. 1972. Primary aggressive motivation in three inbred strains of mice. J. Comp. Physiol. Psychol. **78**:297-304.

Terman, L. M., and M. H. Oden. 1959. The gifted group at midlife. Stanford University Press, Palo Alto, Calif.

Thiessen, D. D. 1965. The wabbler-lethal mouse: a study in development. Anim. Behav. **13**:87-100.

Thiessen, D. D. 1971. Reply to Wilcock on gene action and behavior. Psychol. Bull. **75**:103-105.

Thiessen, D. D. 1972. A move toward species-specific analyses in behavior genetics. Behav. Genet. **2**:115-126.

Thiessen, D. D., G. Lindzey, and K. Owen. 1970. Behavior and allelic variations in enzyme activity and coat color at the C locus of the mouse. Behav. Genet. **1**:257-268.

Thiessen, D. D., K. Owen, and M. Whitsett. 1970. Chromosomal mapping of behavioral activities. In G. Lindzey and D. D. Thiessen, eds. Contributions to behavior-genetic analysis: the mouse as a prototype. Appleton-Century-Crofts, New York.

Thiessen, D. D., and D. A. Rodgers. 1967. Behavior genetics as the study of mechanism-specific behavior. In J. N. Spuhler, ed. Genetic diversity and human behavior. Aldine Publishing Co., Chicago.

Thiessen, D. D., and P. Yahr. 1970. Central control of territorial marking in the Mongolian gerbil. Physiol. Behav. **5**:275-278.

Thoday, J. M. 1961. Location of polygenes. Nature **191**:368-370.

Thoday, J. M. 1967. New insights into continuous variation. In J. F. Crow and J. V. Neel, eds. Proceedings of the Third International Conference of Human Genetics. The Johns Hopkins University Press, Baltimore.

Thompson, J. S., and M. W. Thompson. 1967. Genetics in medicine. W. B. Saunders Co., Philadelphia.

Thompson, W. R. 1953a. Exploratory behavior as a function of hunger in "bright" and "dull" rats. J. Comp. Physiol. Psychol. **46**:323-326.

Thompson, W. R. 1953b. The inheritance of behavior: behavioral differences in fifteen mouse strains. Can. J. Psychol. **7**:145-155.

Thompson, W. R. 1954. The inheritance and development of intelligence. Proc. Assoc. Res. Nerv. Ment. Dis. **33**:209-231.

Thompson, W. R. 1957a. Influence of prenatal maternal

anxiety on emotionality in young rats. Science **125:** 698-699.

Thompson, W. R. 1957b. Traits, factors and genes. Eugen. Q. 4:8-16.

Thompson, W. R. 1967. Some problems in the genetic study of personality and intelligence. In J. Hirsch, ed. Behavior-genetic analysis. McGraw-Hill Book Co., New York.

Thompson, W. R. 1968. Genetics and social behavior. In D. Glass, ed. Genetics. Rockefeller University Press, New York.

Thompson, W. R., and D. Bindra. 1952. Motivational and emotional characteristics of "bright" and "dull" rats. Can. J. Psychol. 6:116-122.

Thompson, W. R., and J. L. Fuller. 1957. The inheritance of activity in the mouse. Am. Psychol. **12:**433. (Abstr.)

Thompson, W. R., and A. Kahn. 1955. Retroaction effects in the exploratory activity of "bright" and "dull" rats. Can. J. Psychol. 9:173-182.

Thompson, W. R., and G. J. S. Wilde. 1973. Behavior genetics. In B. Wolman, ed. Handbook of general psychology. Prentice-Hall, Inc., Englewood Cliffs, N.J.

Thorndike, R. L. 1951. Community variables as predictors of intelligence and academic achievement. J. Educ. Psychol. 42:321-338.

Thorne, F. C. 1940. Approach and withdrawal behavior in dogs. J. Genet. Psychol. 56:265-272.

Thorne, F. C. 1944. The inheritance of shyness in dogs. J. Genet. Psychol. 65:275-279.

Thuline, H. C. 1967. Inheritance of alcoholism. Lancet 1:274-275.

Thurstone, L. L. 1947. Multiple factor analysis. University of Chicago Press, Chicago.

Tienari, P. 1963. Psychiatric illness in identical twins. Acta Psychiatr. Scand. [Suppl]171:9-195.

Tienari, P. 1971. Schizophrenia and monozygotic twins. In K. A. Achté, ed. Psychiatria Fennica. Helsinki University Central Hospital, Helsinki.

Timon, V. M., and E. J. Eisen. 1970. Comparison of *ad libitum* and restricted feeding of mice selected and unselected for postweaning gain. I. Growth, food consumption and feed efficiency. Genetics 64:41-57.

Timon, V. M., E. J. Eisen, and J. M. Leatherwood. 1970. Comparisons of *ad libitum* and restricted feeding of mice selected and unselected for postweaning gain. II. Carcass composition and energetic efficiency. Genetics 65:145-155.

Tindell, D., and J. V. Craig. 1960. Genetic variation in social aggressiveness and competition effects between sire families in small flocks of chickens. Poultry Sci. 39:1318-1320.

Tobach, E., J. S. Bellin, and D. K. Das. 1974. Differences in bitter taste perception in three strains of rats. Behav. Genet. 4:405-410.

Tollman, J., and J. A. King. 1956. The effects of testosterone propionate on aggression in male and female C57BL/10 mice. Br. J. Anim. Behav. 4:147-149.

Tolman, E. C. 1924. The inheritance of maze learning in rats. J. Comp. Psychol. 4:1-18.

Tolman, E. C. 1932. Purposive behavior in men and animals. Appleton-Century-Crofts, New York.

Torrey, E. F. 1973. Is schizophrenia universal? An open question. Schizophrenia Bull. 7:53-59.

Trankell, A. 1955. Aspects of genetics in psychology. Am. J. Hum. Genet. 7:269-276.

Treiman, D. M., D. W. Fulker, and S. Levine. 1970. Interaction of genotype and environment as determinants of corticosteroid response to stress. Dev. Psychobiol. 3:131-140.

Trivers, R. L. 1971. The evolution of reciprocal altruism. Q. Rev. Biol. 46:35-57.

Tryon, R. C. 1929. The genetics of learning ability in rats. Preliminary report. Univ. Calif. Publ. Psychol. 4:71-89.

Tryon, R. C. 1930. Studies in individual differences in maze ability. I. The measurement of the reliability of individual differences. J. Comp. Psychol. 11:145-170.

Tryon, R. C. 1931a. Studies in individual differences in maze ability. II. The determination of individual differences by age, weight, sex, and pigmentation. J. Comp. Psychol. 12:1-22.

Tryon, R. C. 1931b. Studies in individual differences in maze ability. III. The community of function between two maze abilities. J. Comp. Psychol. 12:95-116.

Tryon, R. C. 1931c. Studies in individual differences in maze ability. IV. The constancy of individual differences: correlation between learning and relearning. J. Comp. Psychol. 12:303-348.

Tryon, R. C. 1931d. Studies in individual differences in maze ability. V. Luminosity and visual acuity as systematic causes of individual differences, and an hypothesis of maze ability. J. Comp. Psychol. 12:401-420.

Tryon, R. C. 1939. Studies in individual differences in maze ability. VI. Disproof of sensory components: experimental effects of stimulus variation. J. Comp. Psychol. 28:361-415.

Tryon, R. C. 1940a. Genetic differences in maze learning in rats. In Thirty-ninth Yearbook of the National Society for the Study of Education, Part I. Public School Publishing Co., Bloomington, Ill.

Tryon, R. C. 1940b. Studies in individual differences in maze ability. VII. The specific components of maze ability, and a general theory of psychological components. J. Comp. Psychol. 30:283-335.

Tryon, R. C. 1940c. Studies in individual differences in maze ability. VIII. Prediction validity of the psychological components of maze ability. J. Comp. Psychol. 30:535-582.

Tryon, R. C. 1941. Studies in individual differences in maze ability. X. Ratings and other measures of initial emotional responses of rats to novel inanimate objects. J. Comp. Psychol. 32:447-473.

Tryon, R. C., C. M. Tryon, and G. Kaznets. 1941. Studies in individual differences in maze ability. IX. Ratings of hiding, avoidance, escape and vocalization responses. J. Comp. Psychol. 32:407-435.

Tsuda, K. 1967. Clinico-genetic study of depersonalization neurosis. In H. Mitsuda, ed. Clinical genetics in psychiatry. Igaku Shoin Ltd., Tokyo.

Turpin, R., J. Lejeune, J. Lafourcade, P. Chigot, and C. Salmon. 1961. Presomption de monozygotisme en dèpit d'un dimorphisme sexuel: sujet masculin XY et sujet neutre Haplo-X. C. R. Acad. Sci. [D]252:2945-2946.

Tyler, A. 1969. "Masked" messenger RNA and the determination process in embryonic development. In H. J. Teas, ed. Genetics and developmental biology. University of Kentucky Press, Lexington.

Tyler, P. A., and G. E. McClearn. 1970. A quantitative genetic analysis of runway learning in mice. Behav. Genet. 1:57-70.

Urbach, P. 1974. Progress and degeneration in the "IQ debate." Br. J. Philos. Sci. 25:99-135.

Utsurikawa, N. 1917. Temperamental differences between outbred and inbred strains of the albino rat. J. Anim. Behav. 7:111-129.

Vale, J. R., D. Ray, and C. A. Vale. 1972. Interaction of genotype and exogenous neonatal androgen: agonistic behavior in female mice. Behav. Biol. 7:321-334.

Vale, J. R., D. Ray, and C. A. Vale. 1973. The interaction of genotype and exogenous androgen and estrogen: sex behavior in female mice. Dev. Psychobiol. 6:319-327.

Vale, J. R., and C. A. Vale. 1969. Individual differences and general laws in psychology. Am. Psychol. 24:1093-1108.

Vale, J. R., C. A. Vale, and J. P. Harley. 1971. Interaction of genotype and population number with regard to aggressive behavior, social grooming, and adrenal and gonadal weight in male mice. Commun. Behav. Biol. [A]6:209-221.

Valenstein, E. S., W. Riss, and W. C. Young. 1954. Sex drive in genetically heterogenous and highly inbred strains of male guinea pigs. J. Comp. Physiol. Psychol. 47:162-165.

Valenstein, E. S., W. Riss, and W. C. Young. 1955. Experiential and genetic factors in the organization of sexual behavior in male guinea pigs. J. Comp. Physiol. Psychol. 48:397-403.

van Abeelen, J. H. F. 1963a. Mouse mutants studied by means of ethological methods. I. Ethogram. Genetica 34:79-94.

van Abeelen, J. H. F. 1963b. Mouse mutants studied by means of ethological methods. II. Mutants and methods. Genetica 34:95-101.

van Abeelen, J. H. F. 1963c. Mouse mutants studied by means of ethological methods. III. Results with yellow, pink-eyed dilution, brown and jerker. Genetica 34:270-286.

van Abeelen, J. H. F. 1966. Effects of genotype on mouse behaviour. Anim. Behav. 14:218-225.

van Abeelen, J. H. F. 1970. Genetics of rearing behavior in mice. Behav. Genet. 1:71-76.

van Abeelen, J. H. F. 1974. Genotype and the cholinergic control of exploratory behaviour in mice. In J. H. F. van Abeelen, ed. The genetics of behaviour. North-Holland Publishing Co., Amsterdam.

van Abeelen, J. H. F., L. Gilissen, T. Hanssen, and A. Lenders. 1972. Effects of intrahippocampal injections with methylscopolamine and neostigmine upon exploratory behavior in two inbred mouse strains. Psychopharmacologia 24:470-475.

van Abeelen, J. H. F., A. J. M. Smits, and W. G. M. Raaijmakers. 1971. Central location of a genotype-dependent cholinergic mechanism controlling exploratory behavior in mice. Psychopharmacologia 19:324-328.

van Abeelen, J. H. F., and H. Strijbosch. 1969. Genotype-dependent effects of scopolamine and eserine on exploratory behavior in mice. Psychopharmacologia 16:81-88.

van Bemmelen, J. F. 1927. Heredity of mental faculties. Proc. R. Acad. Sci. Amsterdam 30:769-795.

Vandenberg, S. G. 1959. The primary mental abilities of Chinese students: a comparative study of the stability of a factor structure. Ann. N.Y. Acad. Sci. 79:257-304.

Vandenberg, S. G. 1962. The hereditary abilities study: hereditary components in a psychological test battery. Am. J. Hum. Genet. 14:220-237.

Vandenberg, S. G. 1966a. Contributions of twin research to psychology. Psychol. Bull. 66:327-352.

Vandenberg, S. G. 1966b. Hereditary factors in normal personality traits (as measured by inventories). Research report No. 19, Louisville Twin Study, Child Development Unit, University of Louisville School of Medicine, Louisville, Ky.

Vandenberg, S. G. 1968a. The nature and nurture of intelligence. In D. Glass, ed. Genetics. Rockefeller University Press, New York.

Vandenberg, S. G., ed. 1968b. Progress in human behavior genetics. The Johns Hopkins University Press, Baltimore.

Vandenberg, S. G. 1969. Contributions of twin research to psychology. In M. Manosevitz, G. Lindzey, and D. D. Thiessen. Behavioral genetics: method and research, New York: Appleton-Century-Crofts.

Vandenberg, S. G. 1972. Assortative mating, or who marries whom? Behav. Genet. 2:127-158.

Vandenberg, S. G., P. J. Clark, and I. Samuels. 1965. Psychophysiological reactions of twins: hereditary factors in galvanic skin resistance, heart beat and breathing rates. Eugen. Q. 12:7-10.

Vandenberg, S. G., and F. Falkner. 1965. Hereditary factors in human growth. Hum. Biol. 37:357-365.

Vandenberg, S. G., and L. Kelly. 1964. Hereditary components in vocational preferences. Acta Genet. Med. Gemellol. 13:266-277.

Vandenberg, S. G., R. E. Stafford, and A. M. Brown. 1968. The Louisville twin study. In S. G. Vandenberg, ed. Progress in human behavior genetics. The Johns Hopkins University Press, Baltimore.

van den Daele, L. D. 1971. Infant reactivity to redundant proprioceptive and auditory stimulation: a twin study. J. Psychol. 78:269-276.

Van Valen, L. 1972. Brain size and intelligence in man. Am. J. Phys. Anthropol. 40:417-424.

Vaughn, J. E., D. A. Matthews, R. P. Barber, C. C. Wimer, and R. E. Wimer. 1977. Genetically-associated variations in the development of hippocampal pyramidal neurons may produce differences in mossy fiber connectivity. J. Comp. Neurol. **173**:41-51.

Vernon, P. E. 1954. Intelligence tests in population studies. Eugen. Q. **1**:221-224.

Verplanck, W. S. 1955. Since learned behavior is innate, and vice versa, what now? Psychol. Rev. **62**: 139-144.

Verriest, G., ed. 1974, 1976. Colour vision deficiencies. II and III. S. Karger AG, Basel, Switzerland.

Vesell, E. S., J. G. Page, and G. T. Passanti. 1971. Genetic and environmental factors affecting ethanol metabolism in man. Clin. Pharmacol. Ther. **12**:192-201.

Vicari, E. M. 1929. Mode of inheritance of reaction time and degrees of learning in mice. J. Exp. Zool. **54**:31-88.

Vogel, F. 1957. Elektroencephalographische Untersuchungen an gesunden Zwillingen. Acta Genet. **7**: 334-337.

Vogel, F. 1970. The genetic basis of the human electroencephalogram (EEG). Humangenetik **10**:91-114.

von Bracken, H. 1936. Vererbungheit und Ordnung in Binnenleben von Zwillingspaaren. Z. Pädag. Psychol. **37**:65-81.

von Bracken, H. 1939. Wahrnehmungstaüschungen und scheinbare Nachbildgröze bei Zwillingen. Arch. Gesamte Psychol. **103**:203-230.

von Bracken, H. 1940. Untersuchungen an Zwillingen über die quantitätiven und qualitätiven des Merkmale des Schreibdrucks. Z. Ang. Psychol. **58**:367-384.

von Schilcher, F., and A. Manning. 1975. Courtship song and mating speed in hybrids between *Drosophila melanogaster* and *Drosophila simulans*. Behav. Genet. **5**:395-404.

von Tomasson, H. 1938. Further investigations on manic-depressive psychosis. Acta Psychiatr. Neurol. **13**:517-526.

Waaler, G. H. M. 1927. Über die Erblichkeitsverhaltnisse der verschiedenen Arten von angeborener Rotgrunblindheit. Z. Abstgs. Vererbslehre **45**:279-333.

Wahl, C. W. 1956. Some antecedent factors in the family histories of 568 male schizophrenics of the United States Navy. Am. J. Psychiatr. **113**:201-210.

Wahlsten, D. 1972. Phenotypic and genetic relations between initial response to electric shock and rate of avoidance learning in mice. Behav. Genet. **2**:211-240.

Wahlsten, D. 1974a. A developmental time scale for postnatal changes in brain and behavior of B6D2F2 mice. Brain Res. **72**:251-264.

Wahlsten, D. 1974b. Heritable aspects of anomalous myelinated fibre tracts in the forebrain of the laboratory mouse. Brain Res. **68**:1-18.

Wahlsten, D. 1975. Genetic variation in the development of mouse brain and behavior: evidence from the middle postnatal period. Dev. Psychobiol. **8**:371-380.

Wald, G. 1964. The receptors of human color vision. Science **145**:1007-1016.

Wald, G. 1966. Defective color vision and its inheritance. Proc. Natl. Acad. Sci. U.S.A. **55**:1347-1363.

Waller, J. H. 1971a. Achievement and social mobility: relationships among IQ score, education, and occupation in two generations. Soc. Biol. **18**:252-259.

Waller, J. H. 1971b. Differential reproduction: its relation to IQ test score, education, and occupation. Soc. Biol. **18**:122-136.

Walls, G. L. 1955. A branched-pathway schema for the color vision system and some of the evidence for it. Am. J. Ophthalmol. **39**:8-23.

Walls, G. L., and R. W. Matthews. 1952. New methods of studying color blindness and normal foveal color vision. Univ. Calif. Publ. Psychol. **7**:1-172.

Walsh, R. N., and R. A. Cummins. 1976. The open field test: a critical review. Psychol. Bull. **83**:482-504.

Walton, P. D. 1970. The genetics of phototaxis in *Drosophila melanogaster*. Can. J. Genet. Cytol. **12**: 283-287.

Ward, S. 1973. Chemotaxis by the nematode *Caenorhabditis elegans*: Identification of attractants and analysis of the response by the use of mutants. Proc. Natl. Acad. Sci. U.S.A. **70**:817-821.

Warren, J. M. 1953. Handedness in the Rhesus monkey. Science **118**:622-623.

Warren, J. M. 1973. Learning in vertebrates. In D. A. Dewsbury and D. A. Rethlingshafer, eds. Comparative psychology. McGraw-Hill Book Co., New York.

Warren, J. M., and A. Baron. 1956. The formation of learning sets by cats. J. Comp. Physiol. Psychol. **49**: 227-231.

Warriner, C. C., W. B. Lemmon, and T. S. Ray. 1963. Early experience as a variable in mate selection. Anim. Behav. **11**:221-224.

Watson, A., and R. Moss. 1971. Spacing as affected by territorial behavior, habitat and nutrition in red grouse *(Lagopus l. scoticus)*. In A. H. Esser, ed. Behavior and environment: the use of space by animals and men. Plenum Publishing Corp., New York.

Watson, J. D. 1970. Molecular biology of the gene, 2nd ed. W. A. Benjamin, Inc., New York.

Watson, J. D., and F. H. C. Crick. 1953. The structure of DNA. Cold Spring Harbor Symp. Quant. Biol. **18**:123-131.

Watt, N. F. 1972. Longitudinal changes in the social behavior of children hospitalized for schizophrenia as adults. J. Nerv. Ment. Dis. **155**:42-54.

Watts, C. A. H. 1966. Depressive disorder in the community. John Wright & Sons Ltd., Bristol, England.

Wechsler, D. 1971. Intelligence: definition, theory and the IQ. In R. Cancro, ed. Intelligence: genetic and environmental influences. Grune & Stratton, Inc., New York.

Wecker, S. C. 1963. The role of early experience in habitat selection by the prairie deer mouse, *Peromyscus maniculatus bairdii*. Ecol. Monogr. **33**:307-325.

Weinberg, I., and J. Lobstein. 1936. Beitrag zur Vererbung des manisch-depressiven Irreseins. Psychiatr. Neurol. Bull. **1**:339-372.

Weinberg, I., and J. Lobstein. 1943. Inheritance in schizophrenia. Acta Psychiatr. Neurol. **18**:93-140.

Weinberg, W. 1908. Über den Nachweis der Vererbung beim Menschen. Jahreshefte Verein vaterl. Naturk. Wurttemberg 64:368-382.

Weinberg, W. 1927. Mathematische Grundlagen der Probandmethode. Z. Induct. Abstam. Vererb. 48:179-228.

Weitze, M. 1940. Hereditary adiposity in mice and the cause of this anomaly. Store Nordeske Videnskabsboghandel, Copenhagen.

Wender, P. H., D. Rosenthal, and S. S. Kety. 1968. A psychiatric assessment of the adoptive parents of schizophrenics. In D. Rosenthal and S. S. Kety, eds. The transmission of schizophrenia. Pergamon Press Ltd., Oxford.

Wender, P. H., D. Rosenthal, S. Kety, F. Schulsinger, and J. Welner. 1973. Social class and psychopathology in adoptees: a natural experimental method for separating the roles of genetic and experiential factors. Arch. Gen. Psychiatry 28:318-325.

Wherry, R. J. 1941. Determination of the specific components of maze-ability for Tryon's bright and dull rats by means of factorial analysis. J. Comp. Psychol. 32:237-252.

Whitney, G. 1973. Vocalization of mice influenced by a single gene in a heterogeneous population. Behav. Genet. 3:57-64.

Whitney, G., G. E. McClearn, and J. C. DeFries. 1970. Heritability of alcohol preference in laboratory mice and rats. J. Hered. 61:165-169.

Wictorin, M. 1952. Bidrag till Räknefärdighetens Psykologi en Tvillingundersökning. Elanders, Gothenburg, Sweden.

Wilcock, J. 1969. Gene action and behavior: an evaluation of major gene pleiotropism. Psychol. Bull. 72:1-29.

Wilcock, J. 1971. Gene action and behavior: a clarification. Psychol. Bull. 75:106-108.

Wilcock, J., and P. L. Broadhurst. 1967. Strain differences in emotionality: open field and conditioned avoidance behaviour in the rat. J. Comp. Physiol. Psychol. 63:335-338.

Wilcock, J., and D. W. Fulker. 1973. Avoidance learning in rats: genetic evidence for two distinct behavioral processes in the shuttle box. J. Comp. Physiol. Psychol. 82:247-253.

Wilde, G. J. S. 1964. Inheritance of personality traits: an investigation into the hereditary determination of neurotic instability, extroversion, and other personality traits by means of a questionnaire administered to twins. Acta Psychol. 22:37-51.

Wilde, G. J. S. 1970. An experimental study of mutual behavior imitation and person perception in MZ and DZ twins. Acta Genet. Med. Gemellol. 19:273-279.

Willham, R. L., D. F. Cox, and G. G. Karas. 1963. Genetic variation in a measure of avoidance learning in swine. J. Comp. Physiol. Psychol. 56:294-297.

Willham, R. L., G. G. Karas, and D. C. Henderson. 1964. Partial acquisition and extinction of an avoidance response in two breeds of swine. J. Comp. Physiol. Psychol. 57:117-122.

Williams, C. D., S. Zerof, and R. Carr. 1962. Exploratory behavior of the crosses of three strains of rats. J. Comp. Physiol. Psychol. 55:121-122.

Williams, R. J. 1956. Biochemical individuality. John Wiley & Sons, Inc., New York.

Williams, R. J., L. J. Berry, and E. Beerstecher, Jr. 1949. Biochemical individuality. III. Genetotrophic factors in the etiology of alcoholism. Arch. Biochem. 23:275-290.

Williams, R. J., L. J. Berry, and E. Beerstecher, Jr. 1950. The concept of genetotrophic disease. Lancet 258:287-289.

Williams, R. J., R. B. Pelton, and L. L. Rogers. 1955. Dietary deficiencies in animals in relation to voluntary alcohol and sugar consumption. Q. J. Stud. Alc. 16:234-244.

Willingham, W. W. 1956. The organization of emotional behavior in mice. J. Comp. Physiol. Psychol. 49:345-348.

Willmer, E. N. 1946. Retinal structure and colour vision. Cambridge University Press, Cambridge, England.

Willott, J. F., and K. R. Henry. 1974. Auditory evoked potentials: developmental changes of threshold and amplitude following early acoustic trauma. J. Comp. Physiol. Psychol. 86:1-7.

Willoughby, R. R. 1927. Family similarities in mental test abilities. Genet. Psychol. Monogr. 11:234-277.

Wilson, E. O. 1975. Sociobiology. Harvard University Press, Cambridge, Mass.

Wilson, E. O., and W. H. Bossert. 1971. A primer of population biology. Sinauer, Stamford, Conn.

Wilson, P. T. 1934. A study of twins with special reference to heredity as a factor in determining differences in environment. Hum. Biol. 6:324-357.

Wilson, R. S. 1970. Blood typing and twin zygosity. Hum. Hered. 20:30-56.

Wilson, R. S. 1972. Twins: early mental development. Science 175:914-917.

Wilson, R. S. 1977. Mental development in twins. In A. Oliverio, ed. Genetics, environment and intelligence. Elsevier/North Holland Biomedical Press, Amsterdam.

Wilson, R. S., A. M. Brown, and A. P. Matheny, Jr. 1971. Emergence and persistence of behavioral differences in twins. Child Dev. 42:1381-1389.

Wimer, C. C., R. E. Wimer, and T. H. Roderick. 1971. Some behavioral differences associated with relative size of the hippocampus in the mouse. J. Comp. Physiol. Psychol. 76:57-65.

Wimer, R. E., L. Symington, H. Farmer, and P. Schwartzkroin. 1968. Differences in memory processes between inbred mouse strains C57BL/6J and DBA/2J. J. Comp. Physiol. Psychol. 65:126-131.

Wing, J. K., J. L. T. Birley, J. E. Cooper, P. Graham, and A. D. Isaacs. 1967. Reliability of a procedure for measuring and classifying "present psychiatric state." Br. J. Psychiatry 113:499-515.

Wingfield, A. H., and P. Sandiford. 1928. Twins and orphans. J. Educ. Psychol. 19:410-423.

Winokur, G. 1967. X-borne recessive genes in alcoholism. Lancet 1:466.

Winokur, G., and P. Clayton. 1967. Family history studies. I. Two types of affective disorders separated according to genetic and clinical factors. In I. J. Wortis, ed. Recent advances in biological psychiatry. Plenum Publishing Corp., New York.

Winokur, G., P. J. Clayton, and T. Reich. 1969. Manic-depressive illness. The C. V. Mosby Co., St. Louis.

Winston, H. D. 1963. Influence of genotype and infantile trauma on adult learning in the mouse. J. Comp. Physiol. Psychol. 56:630-635.

Winston, H. D. 1964. Heterosis and learning in the mouse. J. Comp. Physiol. Psychol. 57:279-283.

Winston, H. D., and G. Lindzey. 1964. Albinism and water escape performance in the mouse. Science 144:189-191.

Witelson, S. F. 1976. Sex and the single hemisphere: specialization of the right hemisphere for spatial processing. Science 193:425-427.

Witelson, S. F., and W. Pallie. 1973. Left hemisphere specialization for language in the newborn: neuroanatomical evidence of asymmetry. Brain 96:641-646.

Witkin, H. A., S. A. Mednick, F. Schulsinger, E. Bakkestrøm, K. O. Christiansen, D. R. Goodenough, K. Hirschhorn, C. Lundsteen, D. R. Owen, J. Philip, D. B. Rubin, and M. Stocking. 1976. Criminality in XYY and XXY men. Science 193:547-555.

Witt, G. M., and C. S. Hall. 1949. The genetics of audiogenic seizures in the house mouse. J. Comp. Physiol. Psychol. 42:58-63.

Woerner, P. I., and S. B. Guze. 1968. A family and marital study of hysteria. Br. J. Psychiatry 114:161-168.

Wolfer, J. A. 1963. Maze learning of bright and dull rats as a function of level of motivation. Diss. Abstr. 24:1718.

Wolfer, J. A., L. D. Reid, S. M. Gledhill, and P. B. Porter. 1964. Feeding and metabolic differences between Tryon bright and dull rats. J. Comp. Physiol. Psychol. 58:317-320.

Wolstenholme, G. E. W., and J. Knight, eds. 1970. Sensorineural hearing loss. Ciba Foundation Symposium. J. & A. Churchill, London.

Woo, T. L. 1928. Dextrality and sinistrality of hand and eye: second memoir. Biometrika 22A:79-148.

Woo, T. L., and K. Pearson. 1927. Dextrality and sinistrality of hand and eye. Biometrika 19:165-199.

Wood-Gush, D. G. M. 1958. Genetic and experimental factors affecting the libido of cockerels. Proc. R. Soc. Edinb. 27:6-7.

Wood-Gush, D. G. M. 1960. A study of sex-drive of two strains of cockerels through three generations. Anim. Behav. 8:43-53.

Woods, F. A. 1906. Mental and moral heredity in royalty. Holt, New York.

Woods, J. W. 1954. Some observations on adrenal cortical functions in wild and domesticated Norway rats. Doctoral dissertation. Johns Hopkins University, Baltimore.

Woolf, C. M. 1972. Genetic analysis of geotactic and phototactic behavior in selected strains of *Drosophila pseudoobscura*. Behav. Genet. 2:93-106.

Woolf, G. L. 1965a. Body composition and coat color correlation in different phenotypes of "viable yellow" mice. Science 147:1145-1147.

Woolf, G. L. 1965b. Hereditary obesity and hormone deficiencies in yellow dwarf mice. Am. J. Physiol. 209:632-636.

Woolf, G. L. 1971. Genetic modification of homeostatic regulation in the mouse. Am. Natur. 105:241-252.

Woolf, G. L., and H. C. Pitot. 1973. Influence of background genome on enzymatic characteristics of yellow $(A^y/-, A^{vy}/-)$ mice. Genetics 73:109-123.

Wright, S. 1920. The relative importance of heredity and environment in determining the piebald pattern of guinea pigs. Proc. Natl. Acad. Sci. U.S.A. 6:320-332.

Wright, S. 1931. Statistical methods in biology. J. Am. Stat. Assoc. [Suppl.] 26:155-163.

Wright, S. 1934a. The method of path coefficients. Ann. Math. Stat. 5:161-215.

Wright, S. 1934b. The results of crosses between inbred strains of guinea pigs differing in number of digits. Genetics 19:537-551.

Wright, S. 1952. The genetics of quantitative variability. In E. C. R. Reeve and C. H. Waddington, eds. Quantitative inheritance. Her Majesty's Stationery Office, London.

Wright, W. D. 1957. Diagnostic tests for colour vision. Ann. R. Coll. Surg. Engl. 20:177-191.

Wynne-Edwards, V. C. 1968. Population control and social selection in animals. In D. Glass, ed. Genetics. Rockefeller University Press, New York.

Yakovlev, P. I. 1964. Teleokinesis and handedness (an empirical generalization). In J. Wortis, ed. Recent advances in biological psychiatry. Plenum Publishing Corp., New York.

Yanai, J., and G. E. McClearn. 1972. Assortative mating in mice. I. Female mating preference. Behav. Genet. 2:173-184.

Yanai, J., and G. E. McClearn. 1973a. Assortative mating in mice. II. Strain differences in female mating preference, male preference, and the question of possible sexual selection. Behav. Genet. 3:65-74.

Yanai, J., and G. E. McClearn. 1973b. Assortative mating in mice. III. Genetic determination of female mating preference. Behav. Genet. 3:75-84.

Yanai, J., P. Y. Sze, and B. E. Ginsburg. 1975. Effects of aminergic drugs and glutamic acid on audiogenic seizures induced by early exposure to ethanol. Epilepsia 16:67-71.

Yeakel, E. H., and R. P. Rhoades. 1941. A comparison of the body and endocrine gland (adrenal, thyroid and pituitary) weights of emotional and nonemotional rats. Endocrinology 28:337-340.

Yen, T. T. T., L. Lowry, and J. Steinmetz. 1968. Obese locus in *Mus musculus*: a gene dosage effect. Biochem. Biophys. Res. Commun. 33:883-887.

Yerkes, A. W. 1916. Comparisons of the behavior of

stock and inbred albino rats. J. Anim. Behav. **6**:267-296.

Yerkes, R. M. 1913. The heredity of savageness and wildness in rats. J. Anim. Behav. **3**:286-296.

Yoon, C. H. 1955. Homeostasis associated with heterozygosity in the genetics of time of vaginal opening in the house mouse. Genetics **40**:297-309.

Yoshimasu, S. 1965. Criminal life-curves of monozygotic twin pairs. Acta Criminol. Med. Legal. Jpn. **31**:5-6.

Young, J. P. R., and G. W. Fenton. 1971. An investigation of the genetic aspects of the alpha attenuation response. Psychol. Med. **1**:365-371.

Yule, E. P. 1935. The resemblance of twins with regard to perseveration. J. Ment. Sci. **81**:489-501.

Zajonc, R. B., and G. B. Markus. 1975. Birth order and intellectual development. Psychol. Rev. **82**:74-88.

Zangwill, O. L. 1967. Speech and the minor hemisphere. Acta Neurol. Belg. **67**:1013-1020.

Zazzo, R. 1960. Les jumeaux, le couple et la personne. Presses Universitaires de France, Paris.

Zerbin-Rüdin, E. 1967. Endogene Psychosen. In P. E. Becker, ed. Humangenetik, ein kurzes Handbuch, vol. 2. Georg Thieme Verlag KG, Stuttgart, West Germany.

Zerbin-Rüdin, E. 1969. Zur Genetik der depressiven Erkrankungen. In H. Hippius and H. Selbach, eds. Das depressive Syndrom. Verlag Urban Schwarzenberg, Berlin.

Zigler, E. 1967. Familial mental retardation: a continuing dilemma. Science **155**:292-298.

Zucker, L. M., and T. F. Zucker. 1961. Fatty, a new mutation in the rat. J. Hered. **52**:275-278.

Zung, W. W. K., and W. P. Wilson. 1967. Sleep and dream patterns in twins: Markov analysis of a genetic trait. In J. Wortis, ed. Recent advances in biological psychiatry, vol. 9. Plenum Publishing Corp., New York.

Author index

Subject index